Exhumation of the North Atlantic Margin: Timing, Mechanisms and Implications for Petroleum Exploration

Special Publication reviewing procedures

The Society makes every effort to ensure that the scientific and production quality of its books matches that of its journals. Since 1997, all book proposals have been refereed by specialist reviewers as well as by the Society's Books Editorial Committee. If the referees identify weaknesses in the proposal, these must be addressed before the proposal is accepted.

Once the book is accepted, the Society has a team of Book Editors (listed above) who ensure that the volume editors follow strict guidelines on refereeing and quality control. We insist that individual papers can only be accepted after satisfactory review by two independent referees. The questions on the review forms are similar to those for *Journal of the Geological Society*. The referees' forms and comments must be available to the Society's Book Editors on request.

Although many of the books result from meetings, the editors are expected to commission papers that were not presented at the meeting to ensure that the book provides a balanced coverage of the subject. Being accepted for presentation at the meeting does not guarantee inclusion in the book.

Geological Society Special Publications are included in the ISI Science Citation Index, but they do not have an impact factor, the latter being applicable only to journals.

More information about submitting a proposal and producing a Special Publication can be found on the Society's web site: www.geolsoc.org.uk.

It is recommended that reference to all or part of this book should be made in one of the following ways:

DORÉ, A. G., CARTWRIGHT, J. A., STOKER, M. S., TURNER, J. P. & WHITE, N. (eds) 2002. *Exhumation of the North Atlantic Margin: Timing, Mechanisms and Implications for Petroleum Exploration*. Geological Society, London, Special Publications, **196**.

BLUNDELL, D. J. 2002. Cenzoic inversion and uplift of southern Britain. *In*: DORÉ, A. G., CARTWRIGHT, J. A., STOKER, M. S., TURNER, J. P. & WHITE, N. (eds) *Exhumation of the North Atlantic Margin: Timing, Mechanisms and Implications for Petroleum Exploration*. Geological Society, London, Special Publications, **196**, 85–101.

GEOLOGICAL SOCIETY SPECIAL PUBLICATION No. 196

Exhumation of the North Atlantic Margin: Timing, Mechanisms and Implications for Petroleum Exploration

EDITED BY

A. G. DORÉ
Statoil, UK

J. A. CARTWRIGHT
Cardiff University, UK

M. S. STOKER
British Geological Survey, Edinburgh, UK

J. P. TURNER
University of Birmingham, UK

and

N. WHITE
Bullard Laboratories, Cambridge, UK

2002
Published by
The Geological Society
London

THE GEOLOGICAL SOCIETY

The Geological Society of London (GSL) was founded in 1807. It is the oldest national geological society in the world and the largest in Europe. It was incorporated under Royal Charter in 1825 and is Registered Charity 210161.

The Society is the UK national learned and professional society for geology with a worldwide Fellowship (FGS) of 9000. The Society has the power to confer Chartered status on suitably qualified Fellows, and about 2000 of the Fellowship carry the title (CGeol). Chartered Geologists may also obtain the equivalent European title, European Geologist (EurGeol). One fifth of the Society's fellowship resides outside the UK. To find out more about the Society, log on to www.geolsoc.org.uk.

The Geological Society Publishing House (Bath, UK) produces the Society's international journals and books, and acts as European distributor for selected publications of the American Association of Petroleum Geologists (AAPG), the American Geological Institute (AGI), the Indonesian Petroleum Association (IPA), the Geological Society of America (GSA), the Society for Sedimentary Geology (SEPM) and the Geologists' Association (GA). Joint marketing agreements ensure that GSL Fellows may purchase these societies' publications at a discount. The Society's online bookshop (accessible from www.geolsoc.org.uk) offers secure book purchasing with your credit or debit card.

To find out about joining the Society and benefiting from substantial discounts on publications of GSL and other societies worldwide, consult www.geolsoc.org.uk, or contact the Fellowship Department at: The Geological Society, Burlington House, Piccadilly, London W1J 0BG: Tel. +44 (0)20 7434 9944; Fax +44 (0)20 7439 8975; Email: enquiries@geolsoc.org.uk.

For information about the Society's meetings, consult *Events* on www.geolsoc.org.uk. To find out more about the Society's Corporate Affiliates Scheme, write to enquiries@geolsoc.org.uk.

Published by The Geological Society from:
The Geological Society Publishing House
Unit 7, Brassmill Enterprise Centre
Brassmill Lane
Bath BA1 3JN, UK
(*Orders*: Tel. +44 (0)1225 445046
Fax +44 (0)1225 442836)
Online bookshop: http://bookshop.geolsoc.org.uk

British Library Cataloguing in Publication Data
A catalogue record for this book is available from the British Library.

ISBN 1-86239-112-2

Typeset by Alden Bookset, Somerset, UK
Printed by The Alden Press, Oxford, UK.

Distributors

USA
AAPG Bookstore
PO Box 979
Tulsa
OK 74101-0979
USA
Orders: Tel. + 1 918 584-2555
Fax +1 918 560-2652
E-mail *bookstore@aapg.org*

India
Affiliated East-West Press PVT Ltd
G-1/16 Ansari Road, Daryaganj,
New Delhi 110 002
India
Orders: Tel. +91 11 327-9113
Fax +91 11 326-0538
E-mail *affiliat@nda.vsnl.net.in*

Japan
Kanda Book Trading Co.
Cityhouse Tama 204
Tsurumaki 1-3-10
Tama-shi
Tokyo 206-0034
Japan
Orders: Tel. +81 (0)423 57-7650
Fax +81 (0)423 57-7651

Contents

Implications for petroleum exploration

Exhumation of the North Atlantic margin: introduction and background

A. G. DORÉ[1], J. A. CARTWRIGHT[2], M. S. STOKER[3], J. P. TURNER[4] & N. J. WHITE[5]

[1]*Statoil (UK) Ltd, 11a Regent Street, London SW1Y 4ST, UK (e-mail: agdo@statoil.com)*

[2]*Department of Earth Sciences, Cardiff University, PO Box 914, Cardiff CF10 3YE, UK*

[3]*British Geological Survey, Murchison House, West Mains Road, Edinburgh EH9 3LA, UK*

[4]*University of Birmingham, School of Earth Sciences, Edgbaston, Birmingham B15 2TT, UK*

[5]*Bullard Laboratories, Madingley Rise, Madingley Road, Cambridge CB3 0EZ, UK*

Since consolidation during the Caledonian and Variscan orogenies, NW Europe has undergone repeated episodes of exhumation (the exposure of formerly buried rocks) as a result of such factors as post-orogenic unroofing, rift-shoulder uplift, hotspot activity, compressive tectonics, eustatic sea-level change, glaciation and isostatic readjustment. Modern measurement techniques, such as apatite fission-track analysis, have helped to establish useful denudation chronologies for this entire time span. However, the main observational legacy of exhumation around the North Atlantic is preserved in the comparatively young (Mesozoic and Cenozoic) geological record of this region. This is clearly reflected by the unifying theme of this volume, which documents evidence for the widespread uplift and emergence of large sections of the North Atlantic margin in Cenozoic time.

All students of NW European geology are aware of the compelling palaeogeographical evidence for the transition at the end of the Cretaceous from shelf seas and low-relief landmasses to an area dominated by highlands and newly emergent landmasses, flanked by shelves dominated by rejuvenated clastic deposition. Similarly, it is also widely known that the highlands of Norway and Scotland do not represent the original Caledonian mountain range but must be instead a product of late emergence or uplift.

The Cenozoic uplift of Fennoscandia in particular has a long history of study. It is arguably one of the oldest debates in the history of systematic geology and featured prominently in Lyell's *Principles of Geology* (Lyell 1830–1875). All of this early work was, of course, based on onshore observations. By the late 19th century, it was realized that Norway was essentially a tableland, a plain that had been reduced to some basc level and subsequently uplifted (e.g. Beete-Jukes 1872). The ancient land surface, now considerably modified and incised by recent glacial and fluvial erosion, was termed the Paleic Surface by Reusch (1901) and subsequently described in detail by Gjessing (1967). Based solely on regional evidence, principally the Alpine-related uplift of large parts of central Europe, it was inferred that such a surface must have been formed in late Mesozoic or early Cenozoic time, and that uplift must have taken place at some later stage of Cenozoic time (see, e.g. Gregory 1913).

Overlapping with this work, similar planation surfaces and episodes of Cenozoic uplift were inferred in Scotland (see e.g. Godard 1962; George 1966; Hall 1991), and the Cenozoic emergence of southern Britain was obvious, based on widespread outcrops of Jurassic, Cretaceous and Eocene marine rocks.

Holtedahl (1953) made the critical observation that some of the highlands bordering the North Atlantic were probably complementary to areas of downwarp and deposition on the adjacent shelves. Although by no means obvious at the time, the hypothesis was quickly tested by the explosion in offshore hydrocarbon exploration, which confirmed that most of the surrounding shelves were characterized by Mesozoic–Cenozoic sedimentary basins. Consequently, an attempt could be made to match the supposed Cenozoic evolution of the land areas, including denudation in response to uplift, to the offshore sedimentary response. Furthermore, recognition of the importance of the Cenozoic evolution in the formation of the offshore hydrocarbon riches (e.g. Parker 1975) provided a commercial, as well as academic, motivation for continued research.

From: DORÉ, A.G., CARTWRIGHT, J.A., STOKER, M.S., TURNER, J.P. & WHITE, N. 2002. *Exhumation of the North Atlantic Margin: Timing, Mechanisms and Implications for Petroleum Exploration.* Geological Society, London, Special Publications, **196**, 1–12. 0305-8719/02/$15.00

Historical descriptions of the numerous strands of subsequent investigation have been given by Gabrielsen & Doré (1995), Stuevold & Eldholm (1996) and Japsen & Chalmers (2000). A summary of the key findings is as follows.

- As well as in mainland Norway and Britain, Cenozoic uplift and/or emergence took place in Spitsbergen (Harland 1969), Sweden and part of Denmark (e.g. Japsen & Chalmers 2000), Ireland (e.g. Naylor 1992), East Greenland (e.g. Johnson & Gallagher 2000) and West Greenland (e.g. Mathiesen 1998). In other words, it is a circum-North Atlantic phenomenon.
- Cenozoic exhumation also took place in basins peripheral to the landmasses. It was recognized early that the huge expanse of shelf forming the Barents Sea was exposed subaerially and eroded during late Cenozoic time (Nansen 1904; Harland 1969). To this have subsequently been added the Horda Platform, Stord Basin and Farsund Basin of offshore Norway (e.g. Ghazi 1992; Jensen & Schmidt 1993), the West Shetland Inner Moray Firth and East Irish Sea Basins in UK waters (described by, for example, Lewis *et al.* (1992), Parnell *et al.* (1999), Hillis *et al.* (1994) and Rowley & White (1998), respectively), the Slyne–Erris and North Celtic Sea Basins off the Irish coast (described by Scotchman & Thomas (1995) and Murdoch *et al.* (1995), respectively) and numerous others. Early Cenozoic uplift and subaerial exposure also took place along the volcanic highs marginal to the newly developing North Atlantic, areas now submerged to depths of a kilometre or more (e.g. Eldholm *et al.* 1989).
- From the numerous studies now carried out, both local and regional, it is clear that the circum-North Atlantic uplift and erosion was variable in magnitude, location and timing. It could thus perhaps be argued that the phenomenon represents a patchwork of effects deriving from many unrelated causes. Nevertheless, as integration between the various studies improves, it seems that at least two events had regional significance: (1) a Paleocene episode of widespread emergence in NW Europe coincident with North Atlantic opening and the initial effects of the Iceland Plume (e.g. White 1988; Brodie & White 1995); (2) a Neogene (mainly Plio-Pleistocene) episode with no obvious tectonic cause, emphasized by rapid glacial erosion and isostatic adjustment of landmasses and bordering shelves (e.g. Solheim *et al.* 1996; Japsen & Chalmers 2000), together with redeposition and a widespread change in the deep-water circulation pattern in the adjacent basins (Stratagem Partners 2002). The Paleocene event appears to have been particularly significant in the British Isles, whereas the late Neogene event seems to have extended from Scandinavia to the Atlantic margin of Britain and Ireland. Additionally, compressional uplift (inversion) associated with Alpine stress and/or ridge-push from Atlantic spreading is common throughout much of the area, although localized and very variable in effect (e.g. Murdoch *et al.* 1995; Doré & Lundin 1996).
- Critically for the petroleum industry, several North Atlantic basins containing commercially significant hydrocarbon resources were both uplifted and exhumed during Cenozoic time, with profound implications for the quantity and nature of the hydrocarbons discovered. These effects have been systematically studied in the Barents Sea (Nyland *et al.* 1992) and to some extent in the east Irish Sea (Cowan *et al.* 1999), but in terms of overall applicability to uplifted terranes are still underestimated.

Despite this rapid increase in the understanding of the exhumation of the North Atlantic borderlands, there are still many unknowns. The relative intensity of the various phases, and their variation in importance geographically, are still only understood in a very general sense. Although there is no shortage of postulated uplift mechanisms, there is still a scarcity of observational evidence and modelling studies to establish beyond reasonable doubt which are the primary causes of exhumation, and how these may vary from area to area. Tied to these problems is the larger-scale question of whether the circum-North Atlantic is unique or whether its behaviour is typical for passive margins.

There have been several attempts in recent years to bring together researchers to address these questions, and compilations have been published that are the direct antecedents of this book (see particularly Jensen *et al.* (1992), Solheim *et al.* (1996) and Chalmers & Cloetingh (2000)). These proceedings, however, have tended to focus on one particular geographical area or one particular exhumation phase. There is an acknowledged need to bring together disciplines that have traditionally remained apart (e.g. geomorphologists and offshore seismic interpreters; Paleogene and Neogene

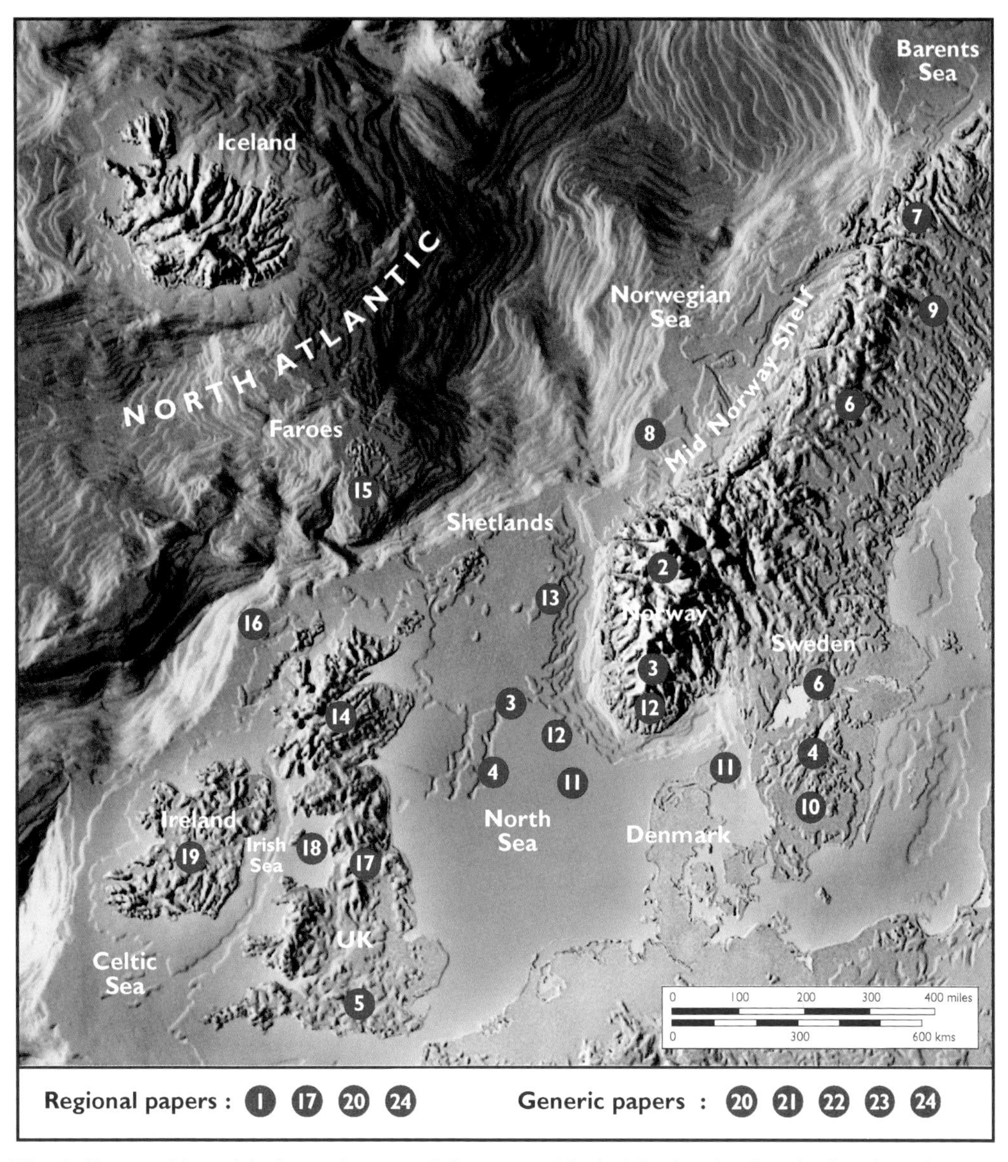

Fig. 1. Topographic and bathymetric map of the eastern North Atlantic, showing the location of studies represented in this volume. The papers are numbered as follows: 1, Jones *et al.*; 2, Rohrman *et al.*; 3, Nielsen *et al.*; 4, Graversen; 5, Blundell; 6, Lidmar-Bergström & Näslund; 7, Hendricks & Andriessen; 8, Evans *et al.*; 9, Stroeven *et al.*; 10, Cederbom; 11, Japsen *et al.*; 12, Huuse; 13, Faleide *et al.*; 14, Bishop & Hall; 15, Andersen *et al.*; 16, Stoker; 17, Green *et al.*; 18, Ware & Turner; 19, Allen *et al.*; 20, Doré *et al.*; 21, Price; 22, Parnell; 23, Cramer *et al.*; 24, Corcoran & Doré. For papers that attempt onshore–offshore event correlation, the area of interest is represented by the same number onshore and offshore.

workers; Scandinavian and British–Irish research schools) before an integrated story can emerge. By providing an interdisciplinary set of studies over a wide latitudinal range of the NW European margin (Fig. 1), this volume represents an initial step in this direction.

Exhumation and other terms: some definitions

Terms such as exhumation, erosion and denudation have a varied usage in the literature and are often used interchangeably. Numerous attempts have been made at deriving formal definitions for these terms, and the subtle differences between these definitions can render communication difficult. Furthermore, the way such terms are defined may depend on the concerns and research orientations of the users. So, for example, scientists primarily concerned with the lithospheric unloading of orogenic belts and core complexes (e.g. Ring *et al.* 1999) may emphasize different factors, and have different definitions, from those concerned with the behaviour of passive margins (e.g. those contributing to this volume).

We have not attempted to impose a rigid standardization on the papers in this volume, and have deliberately used the descriptive and less rigorously defined term 'exhumation' in the title. Nevertheless, it is worth while examining some alternative uses to determine whether any consensus can be drawn and as a background for the papers that follow.

Uplift

Uplift is a term in common usage and appeals to standardization are likely to fail. It is, however, useful to examine some ways in which the concept is formally used. England & Molnar (1990), Summerfield (1991) and Riis & Jensen (1992) all pointed to the existence of two types of uplift. These are *surface uplift*, referring to the upward movement of the Earth's surface (the land or sea bottom) with respect to a specific datum, usually mean sea level or the geoid, and *crustal uplift*, referring to upward movement of the rock column with respect to a similar datum. The relationship between surface uplift and crustal uplift depends upon the amount of denudation or deposition. If there is no denudation or deposition, then surface uplift and crustal uplift will be equal. If, as is likely, denudation occurs then crustal uplift will be greater than surface uplift, and if denudation is greater than crustal uplift surface elevation will be reduced. Surface uplift is mostly related to active tectonics, but crustal uplift can occur simply as the isostatic response to denudation. The term *net uplift* has been used by several of the 'Norwegian school' of workers to indicate the present elevation of a marker bed above its maximum burial depth (Nyland *et al.* 1992; Riis & Jensen 1992; Doré & Jensen 1996). It is a measure of the vertical distance a rock has been uplifted through a thermal frame of reference, and is thus particularly useful in studies of diagenesis and source rock maturation.

Erosion

Ring *et al.* (1999) defined erosion as 'the surficial removal of mass at a spatial point in the landscape by both mechanical and chemical processes'. Biological processes could also be added to this definition. Leeder (1999) provided a similar definition, but pointed out that erosion must be defined with reference to co-ordinates fixed beneath the surface, as the surface potential elevation may itself change because of tectonics. As indicated by Riis & Jensen (1992), erosion may be subaerial or submarine, thus also emphasizing the need for reference to fixed subsurface coordinates. *Erosion rates* may refer to local measured absolute values or to regional values integrated over a wider area. The difference between erosion and denudation is considered below.

Denudation

Denudation has been defined by Leeder (1999) as the loss of material from both surface *and subsurface* parts of a drainage basin or regional landscape by all types of weathering, physical and chemical. Leeder indicated that, like erosion, denudation should be defined with reference to fixed internal co-ordinates. So, according to Leeder, the important difference between erosion and denudation is that erosion is measured as an effect on the surface, whereas denudation can include subsurface processes such as chemical dissolution, and is not always accompanied by erosion. It should be noted that if the Leeder definition is used, it follows that denudation should be characterized as loss of mass rather than thickness. Ring *et al.* (1999) provided an important modifier to these definitions of denudation by including the loss of material by tectonic processes (e.g. thin-skinned extensional faulting in core complexes).

We advocate this consensual view of erosion and denudation, although loss of material by tectonic unloading does not play an important part in this volume. It is, however, worth pointing out that many other usages exist. For example, Summerfield (1991) from a geomorphological perspective, preferred to define erosion as *mechanical denudation* or the removal of solid particles, as opposed to *chemical denudation* (i.e. the removal of dissolved material). A very different emphasis was provided by Brown (1991) and Riis & Jensen (1992), arguing from a subsurface petroleum exploration perspective. They defined denudation as erosion (in their terms, decreasing thickness of overburden) seen in a thermal frame of reference and quantified by such measures as vitrinite reflectance and apatite fission-track trends (see also 'net uplift'). This definition is, however, specialized compared with most others in the literature. It is encumbered by the potential for thermal regimes to vary through time, thereby giving false values for denudation or erosion if a fixed thermal regime is assumed (see discussion given in this volume by Green *et al.*).

Exhumation

Exhumation is the most loosely defined of the terms used in this volume, but precisely because of its descriptive nature it is useful as an overall shorthand to describe the removal of material by any means from a basin or other terrane such that previously buried rocks are exposed. It can be characterized as a process or a history (thus, this definition from Ring *et al.* (1999): 'the unroofing history of a rock, as caused by tectonic and/or surficial processes'). It is seldom characterized as a measure, unlike erosion or denudation, which are often described in terms of loss of thickness mass or volume.

Inversion

Compressive reactivation with reversal of slip on formerly extensional fault systems is termed basin inversion and, ultimately, it leads to folding, thrusting and expulsion of the synrift fill (see, for example, Cooper & Williams (1989) and papers therein, and Buchanan & Buchanan (1995) and papers therein). Because inversion often causes the processes described above (uplift, erosion, denudation, exhumation) it is frequently used interchangeably with these terms, especially in NW Europe where compressive reactivation and general regional exhumation often coexist. This usage is misleading and should be avoided unless the underlying cause of the regional exhumation is known to be compression.

Towards an understanding of North Atlantic exhumation

The studies presented here are based on a variety of techniques that have been employed to address the main concerns of North Atlantic exhumation history, including timing, mechanisms and the sedimentary response of the continental margin. The 24 papers presented in this volume have accordingly been arranged in four sections to reflect the highly varied approach to this subject, and the commercial implications. Part 1 is mainly concerned with exhumation mechanisms; parts 2 and 3 present current research on the continental margin record of offshore Scandinavia and Britain, Ireland and the Faeroes, respectively; part 4 covers the implications of exhumation for hydrocarbon-bearing basins. Key aspects of this interdisciplinary approach are summarized below. The geographical spread of the papers is shown in Fig. 1.

Techniques

All of the main techniques for determining amount and timing of uplift, erosion and denudation are represented in this selection of studies. They are summarized in Table 1, which also indicates the advantages and disadvantages of each method and where it is discussed in this volume. The extreme ends of the spectrum are those studies that infer exhumation chronology from either exclusively onshore or offshore studies: the former utilizing modern techniques of geomorphological analysis (e.g. **Lidmar-Bergström & Näslund**, **Hall & Bishop**), the latter focusing on detailed seismic analysis of the offshore erosion products (e.g. **Evans *et al.***, **Faleide *et al.***, **Stoker**).

There is now full awareness of the need to integrate onshore and offshore studies to provide a more complete picture of uplift, erosion and sedimentary response. Available approaches include simple graphical reconstruction, that is, the projection of offshore stratigraphic breaks into onshore surfaces (e.g. **Graversen**, **Japsen *et al.***) or analysis of the mass balance between basins and hinterlands (**Jones *et al.***, **Andersen *et al.***). These methods, by their nature, provide only a coarse chronology and are most effective if combined with physico-chemical measurement techniques for reconstructing exhumation history.

The apatite fission-track technique is by far the main 'growth area' and is represented in this volume by several regional studies (**Hendricks & Andriessen**, **Cederbom**, **Green *et al.***, **Allen *et al.***) Fission-track work is proving highly effective in establishing the cooling history of a rock, but with both this and the vitrinite reflectance method (e.g. **Japsen *et al.***, **Green *et al.***) the potential exists for interpreting anomalously high palaeo-heat flows as higher burial depths in the past, and vice versa. As indicated by several workers, fission-track data are unreliable at low temperatures ($<60°C$) and can produce spurious results for geologically recent exhumation. Compaction studies (**Japsen *et al.***, **Ware & Turner**) benefit from plentiful

Table 1. *Summary of techniques for measuring amount and timing of uplift and erosion–denudation, with attractions, limitations and where they are represented in this volume*

Measurement technique	Attractions	Problems, limitations	Paper in this volume
Geomorphological analysis	Field-based, plentiful data, relates landform development to geological history	Difficult to obtain quantitative measures of erosion–denudation, and to constrain timing; most effective when used together with offshore analysis	6, 9, 10, 14
Graphical reconstruction	Simple technique to estimate eroded thickness, correlate surfaces and derive onshore–offshore relationships	Relies on long-distance extrapolation and assumption of thickness	4, 6, 11
Offshore sedimentary response	Provides indicator of denudation chronology of source areas; seismic data allow full sedimentary section to be observed	Erosion and redeposition of offshore successions complicate correlation with onshore denudation chronology; most effective when used together with onshore analysis	8, 12, 13, 16
Mass balance	Directly and quantitatively correlates denudation with offshore sedimentation	Loss of mass in solution; difficulty in assigning sediment to correct catchment area	1, 15
Vitrinite reflectance	Preserves record of higher temperatures–burial depths than fission track	Vitrinite absent from many sediments and all basement; vitrinite may be misidentified; difficult to separate changes in basal and transient heat flow from exhumation; reworked and oxidized vitrinite a problem; only a crude timing indicator	11, 17
Apatite fission track	Can establish detailed exhumation and burial chronology both onshore and offshore	Unreliable in establishing recent exhumation events (cooling below annealing temperature); difficult to separate changes in basal and transient heat flow from exhumation	7, 10, 17, 19
Cosmogenic nuclide	Used to obtain detail on geologically recent landform development not obtainable from fission track	Onshore only, mainly limited to chronology of present land surface	9
Compaction, sonic velocity	Plentiful source of well data (sonic logs); easier to distinguish changes in basal heat flow from exhumation compared with thermal methods	Unreliability of baseline compaction trends for a basin or a lithology; difficulty in identifying 'typical' lithologies for analysis; mechanical compaction retarded by overpressure, leading to erosion underestimates; only a crude timing indicator	11, 18

For the numbering of the papers see Fig. 1.

data and, given the right stratigraphical information and lithologies, can fill in some of these gaps. The more recent onshore denudation chronology is partially catered for by the rapidly developing cosmogenic nuclide technique (**Stroeven *et al.***), although this method is not capable of resolving large (kilometre-scale) amounts of denudation. The essential conclusion, clear from Table 1, is that the more techniques that can be brought to bear, the fuller the picture that is likely to emerge.

Mechanisms

Views on the nature and cause of the North Atlantic Cenozoic exhumation have tended to depend on where the studies have been carried out. Thus the older 'Scandinavian school', impressed by the evidence for rapid Neogene outbuilding and the concentric nature of the Cenozoic outcrops around the coasts, has tended to emphasize Neogene effects including glacio-isostasy. The 'British school', on the other hand, presented with evidence for high Paleocene sedimentation rates and contemporaneous volcanism around the islands, has mainly stressed Paleogene hotspot models to explain the uplift, although George (1966) is a notable exception to this in favouring a substantial amount of Neogene reshaping of the British and Irish 'Massifs'. This volume is an initial attempt at bringing these schools together, as a means of examining whether relationships exist and how the different mechanisms may vary in importance in time and space. A summary of the postulated mechanisms, together with their attractions and limitations, is provided in Table 2.

Part 1 of the book focuses primarily on mechanisms related to the Iceland Plume. **Jones *et al.*** provide mass balance evidence that the Iceland Plume caused significant exhumation in Early Cenozoic time, and equally impressive gravimetric evidence that the plume causes a significant deflection of the geoid over a very wide area today. It is widely believed that much of the Paleogene emergence can be attributed to underplating of the lithosphere from the plume (Brodie & White 1995), but there is still a requirement to explain the apparent absence of (or lack of evidence for) underplating in some areas, and the puzzling absence of igneous activity in classic uplifted areas such as Norway. Nevertheless, there is solid evidence from Bouguer gravity and tomography that even the most recent uplift of Norway occurred with mantle involvement, perhaps upwelling or asthenospheric flow without the direct impingement of a hotspot (**Rohrman *et al.***).

A variation on this theme by **Nielsen *et al.*** suggests that the impingement of the Iceland Plume head caused delamination of the lower lithosphere in areas with deep crustal roots such as mainland Norway, leading to isostatic uplift. A central tenet of the **Nielsen *et al.*** model is that, after initial Paleocene plume-related uplift, no further active tectonic mechanism is required to explain either the denudation chronology or the present-day elevation of Norway. Post-Paleocene pulses of erosion and deposition are explained solely in terms of eustatic base-level fall, glaciation and isostatic response to these events. This view is emphasized in a supporting paper on the offshore sedimentary response (**Huuse**) but rejected by **Japsen *et al.***, who firmly believe in the necessity for a Neogene tectonic uplift event. Detailed seismic analysis of the Cenozoic sequences in the northern North Sea by **Faleide *et al.*** also seems to support a tectonic cause for the late Neogene exhumation. Resolution of this fascinating debate requires considerably more modelling and research on detailed chronology and mass balance. **Graversen** provides a controversial view (again involving mantle upwelling) speculatively linking doming in southern Scandinavia with earlier (Mesozoic) doming in the central North Sea.

To complete the section on mechanisms, **Blundell** describes the role of Alpine stress in the Cenozoic inversion and uplift of southern Britain, supported in this paper by thermo-mechanical modelling of the lithosphere. **Blundell** remarks, however, that apparent late Neogene uplift in western parts of the area cannot be attributed to inversion. **Ware & Turner**, working in the East Irish Sea Basin, show how localization of inversion-related strain superimposed on a 'background' epeirogenic erosion signature caused rapid changes in the degree of exhumation within a single basin.

Continental margin record

The papers presented in parts 2 (Scandinavia) and 3 (the Faeroes, the British Isles and Ireland) illustrate the communication that is now occurring between the two regional research schools and the acknowledgement of a multi-phase Cenozoic denudation chronology for both areas. Potential now exists for a comparison between these areas to determine which events are regionally correlated, and which events are of local significance only. For example, the major geomorphological reviews of Scandinavia (**Lidmar-Bergströ m & Näslund**) and northern Scotland (**Hall & Bishop**) appear to detail a similar event chronology, although late Neogene

Table 2. *Summary of potential mechanisms for exhumation of the North Atlantic margin, with attractions, limitations and where they are represented in this volume*

General grouping and mechanism	Attractions	Problems limitations	Paper in this volume
Associated with mantle plume			
Dynamic uplift Underplating	Regionally applicable; supported by geomorphological, palaeogeographical, mass balance and regional gravity studies	Distribution of underplating not fully understood; absence of igneous activity in some uplifted areas, e.g. Fennoscandia	1, 3, 12, 13
Mantle upwelling without magmatism	Supported by Bouguer gravity and tomography, mainly southern Norway	Causal agency poorly understood	2, 4
Rayleigh–Taylor instability	Supported by Bouguer gravity, mainly Norway	New model requires more testing; connection between plume and delamination beneath cratons not clear	3
Compressive			
Inversion Intraplate stress	Inversion tectonics widespread on Atlantic margin; can explain rapid local variation in exhumation	Difficult to separate local from more regional background mechanisms; significant shortening occurs before 'classic' inversion geometries develop, hence difficult to quantify bulk strain; cannot explain uplift of cratons	5, 15, 18
Tectonically 'passive'			
Glacio-isostatic	Strong correlation in some areas between glaciation and offshore sediment flux	Emphasizes pre-existing topography, but does not explain it	12, 13, 16
Base-level change	Could eliminate need for enigmatic Neogene tectonic uplift mechanism	More studies correlating denudation, sediment flux and glacio-eustatic lowstands required to remove necessity for Neogene tectonic uplift	3, 12

For the numbering of the papers see Fig. 1.

doming is considered more prominent in the former, whereas the latter emphasizes the particular importance of Paleocene uplift.

The fission-track cooling chronologies established for Ireland (**Allen *et al.***), the UK (**Green *et al.***), Sweden (**Cederbom**) and northern Norway (**Hendricks & Andriessen**) also bear comparison. Allen *et al.*'s useful compilation of fission-track data shows that, although for any given time period Ireland was a patchwork of differing exhumation profiles, some general conclusions on intensity through time can be drawn. Cederbom's work on southern Sweden reaches the same conclusion. Following this logic, a synthesis of the latest fission-track work over all of NW Europe would provide valuable insights into the regionally significant exhumation signatures, especially if integrated with methods capable of resolving the most recent events such as (U–Th)/He thermochronology and perhaps even *in situ* and detrital cosmogenic nuclide techniques. The integration of these three techniques is the key aim of the recently initiated CRUST project (Constraining Regional Uplift, Sedimentation & Thermochronology), a collaborative project between the Universities of Glasgow, Edinburgh and Aberdeen, and the Scottish Universitites' Environmental Research Centre, East Kilbride.

As a final example, a similarly instructive view of regionally important events would be obtained by correlating the Cenozoic seismic sequence chronologies reported from the Mid-Norwegian shelf (**Evans *et al.***), North Sea (**Faleide *et al.***), Faeroes shelf (**Andersen *et al.***) and UK Atlantic margin (**Stoker**). To this end, a unified, regional Neogene (Miocene–Holocene) stratigraphy has recently been established for the entire Atlantic continental margin between the Lofoten Islands (off North Norway) and the Porcupine region (off southern Ireland), which includes adjacent deep-water basins such as the Norwegian Basin, Faeroe–Shetland Channel, Rockall Trough and Porcupine Basin (Stratagem Partners 2002). This study demonstrates that the entire NE Atlantic margin covered by this volume (Fig. 1) was affected by an early–'mid'-Pliocene event that resulted, most significantly, in the initiation of the Plio-Pleistocene prograding wedges.

Implications for petroleum exploration

Exhumation leads to cooling and lithostatic pressure decrease, and these factors add a degree of difficulty to petroleum exploration. The following effects are among the most important: (1) sealing horizons are removed and/or their effectiveness is severely reduced; (2) faults are often reactivated, causing them to become conduits for hydrocarbon leakage to the surface; (3) source rocks will be at a higher degree of maturation than expected from their present depth and will cease generation upon cooling; (4) potentially attractive reservoirs may, likewise, be overcompacted and downgraded; (5) pressure reduction during exhumation causes oil accumulations and formation water to exsolve gas, causing gas flushing and the spillage of oil accumulations; (6) regional tilting during uplift results in changes to trap configurations and fluid migration directions.

In spite of this, many NW European sedimentary basins that underwent severe late Mesozoic and Cenozoic exhumation have retained prospectivity. These include the Barents Sea off northern Norway and the East Irish Sea Basin, while in the Wessex basin of southern England, Europe's largest onshore oilfield at Wytch Farm is located in the footwall to a major compressionally reactivated fault zone.

In part 4 of the volume, five papers describe the significant changes to the hydrocarbon systems that occur in exhumed basins. **Doré *et al.*** catalogue these effects in terms of two of the standard procedures carried out in the oil industry, namely, petroleum resource evaluation and risk analysis. This paper documents both positive and negative implications, including the tendency towards gas-dominated systems, and references these to a selection of exhumed basins on the North Atlantic seaboard. **Price** considers the same phenomena in some of the well-documented uplifted basins in western North America. This paper is presented in abstract form only because of the untimely death of the author during the preparation of these proceedings, an event that robbed us of an important and entertaining contributor to the uplift debate.

The three final papers consider specific aspects of exhumed hydrocarbon systems. A review by **Parnell** examines the migration of fluids, both hydrocarbons and formation water, and the diagenetic changes that occur during uplift of a basin. **Cramer & Poelchau** draw attention to the potential for liberation of huge amounts of methane from formation water during pressure and temperature decrease, a process that has contributed significantly to the giant gas fields of Western Siberia, and by implication to many other gas accumulations worldwide. Complete or partial failure of seals as a result of brittleness, stress-induced fracturing and hydrofracturing is shown by **Corcoran & Doré** to be a frequent outcome of exhumation, leading to the prediction of underfilled traps and near-hydrostatic gradients. These inferences are again illustrated by case studies on the North Atlantic margin.

Concluding remarks

The North Atlantic margin described in this volume is, of course, part of a more global system of marginal uplifts bordering many oceanic basins. These include the almost classical area for geomorphologists of the southern African marginal escarpments, along with the marginal uplifted massifs of Eastern Australia and Antarctica. Although these other marginal uplifts were beyond the scope of this volume, many of the techniques developed and refined for the North Atlantic margins are applicable more generally, and may help to broaden and deepen the investigation of this global tectonic process.

Perhaps the key to really advancing our understanding of the processes involved in oceanic margin uplifts lies on a global scale, that is, to identify their spatial distribution within a high-resolution chronostratigraphic framework. There is thus a prime need for better resolution of the timing and magnitude of all these marginal uplifts, as part of any attempt to synthesize their distribution globally.

For the North Atlantic margins, significant progress has been made along these lines, as shown by many contributions in this volume. Additional effort is required to fully integrate the work of the 'Scandinavian' and 'British' schools, and also to forge closer links across disciplines. The sequence stratigraphers working on the offshore seismic record need greater awareness of the many contributions made by onshore geomorphologists and vice versa. Better resolution of climate variation through Cenozoic time is of paramount importance in constraining the long- and short-term rates of erosion and sediment transport. In this way, depositional systems analysis offshore can be linked more effectively to surface uplift onshore, and lags between uplift and sediment flux from the uplifted regions can be identified. This research problem is truly demanding and multidisciplinary, and although much is still to be done, much has already been achieved towards a better process understanding.

References

BEETE-JUKES, J. 1872. *The Student's Manual of Geology.* 3rd Edition. A. & C. Black, Edinburgh.

BRODIE, J. & WHITE, N.J. 1995. The link between sedimentary basin inversion and igneous underplating. *In*: BUCHANAN, J.G. & BUCHANAN, P.G. (eds) *Basin Inversion.* Geological Society, London, Special Publications, **88**, 21–38.

BROWN, R.W. 1991. Backstacking apatite fission-track 'stratigraphy': a method for resolving the erosional and isostatic rebound components of tectonic uplift histories. *Geology*, **19**, 74–77.

BUCHANAN, J.G. & BUCHANAN, P.G. (eds) 1995. *Basin Inversion.* Geological Society, London, Special Publications, **88**.

CHALMERS, J.A. & CLOETINGH, S. (eds) 2000. *Neogene Uplift and Tectonics around the North Atlantic. Global and Planetary Change, Special Issue*, **34** (3–4).

COOPER, M.A. & WILLIAMS, G.D. (eds) 1989. *Inversion Tectonics.* Geological Society, London, Special Publications, **44**.

COWAN, G., BURLEY, S.D., HOEY, A.N. & 5 OTHERS 1999. Oil and gas migration in the Sherwood Sandstone of the East Irish Sea Basin. *In*: Fleet, A.J. & Boldy, S.A.R. (eds) *Petroleum Geology of Northwest Europe: Proceedings of the 5th Conference.* Geological Society, London, 1383–1398.

DORÉ, A.G. & JENSEN, L.N. 1996. The impact of late Cenozoic uplift and erosion on hydrocarbon exploration: offshore Norway and some other uplifted basins. *Global and Planetary Change*, **12**, 415–436.

DORÉ, A.G. & LUNDIN, E.R. 1996. Cenozoic compressional structures on the NE Atlantic margin: nature, origin and potential significance for hydrocarbon exploration. *Petroleum Geoscience*, **2**, 299–311.

ELDHOLM, O., THEIDE, J. & TAYLOR, E. 1989. Evolution of the Vøring volcanic margin. *In*: ELDHOLM, O., THEIDE, J. & TAYLOR, E. (eds) *Proceedings of the Ocean Drilling Program, Scientific Results, 104.* Ocean Drilling Program, College Station, TX, 1033–1065.

ENGLAND, P. & MOLNAR, P. 1990. Surface uplift, uplift of rocks, and exhumation of rocks. *Geology*, **18**, 1173–1177.

GABRIELSEN, R.H. & DORÉ, A.G. 1995. History of tectonic models on the Norwegian continental shelf. *In*: HANSLIEN, S. (ed.) *Petroleum Exploration and Exploitation in Norway.* Norwegian Petroleum Society (NPF) Special Publication, **4**, 333–368.

GEORGE, T.N. 1966. Geomorphic evolution in Hebridean Scotland. *Scottish Journal of Geology*, **2**, 1–34.

GHAZI, S.A. 1992. Cenozoic uplift in the Stord Basin area and its consequences for exploration. *In*: JENSEN, L.N. & RIIS, F. (eds) *Post-Cretaceous Uplift and Sedimentation along the Western Fennoscandian Shield. Norsk Geologisk Tidsskrift.* vol **72**, 285–290.

GJESSING, J.P. 1967. Norway's paleic surface. *Norsk Geografisk Tidsskrift*, **21**, 69–132.

GODARD, A. 1962. Essais de corrélation entre l'altitudes des reliefs et les caractères pétrographique des roches dans les socles de l'Écosse du nord. *Comptes Rendus de l'Académie des Sciences*, **255**, 139–141.

GREGORY, J.W. 1913. *The Nature and Origin of Fjords*. John Murray, London.
HALL, A.M. 1991. Pre-Quaternary landscape evolution in the Scottish Highlands. *Transactions of the Royal Society of Edinburgh: Earth Sciences*, **82**, 1–26.
HARLAND, W.B. 1969. Mantle changes beneath the Barents Shelf. *Transactions of the New York Academy of Sciences, Series 2*, **31**, 25–41.
HILLIS, R.R., THOMSON, K. & UNDERHILL, J.R. 1994. Quantification of Tertiary erosion in the Inner Moray Firth using sonic velocity data from the Chalk and the Kimmeridge Clay. *Marine and Petroleum Geology*, **11**, 283–293.
HOLTEDAHL, O. 1953. On the oblique uplift of some northern lands. *Norsk Geografisk Tidsskrift*, **14**, 132–139.
JAPSEN, P. & CHALMERS, J.A. 2000. Neogene uplift and tectonics around the North Atlantic: overview. *In*: Chalmers, J.A. & Cloetingh, S. (eds) *Neogene Uplift and Tectonics around the North Atlantic. Global and Planetary Change*, **24** (3–4), 165–174.
JENSEN, L.N. & SCHMIDT, B.J. 1993. Neogene uplift and erosion offshore south Norway: magnitude and consequences for hydrocarbon exploration in the Farsund Basin. *In*: SPENCER, A.M. (ed.) *Generation, Accumulation and Production of Europe's Hydrocarbons, III*. Special Publication of the European Association of Petroleum Geoscientists, **3**, 79–88.
JENSEN, L.N., RIIS, F. & BOYD, R. (eds) 1992. Post-Cretaceous Uplift and Sedimentation along the Western Fennoscandian Shield. *Norsk Geologisk Tidsskrift, Symposium Issue*, **3**, 72.
JOHNSON, C. & GALLAGHER, K. 2000. A preliminary Mesozoic and Cenozoic denudation history of the North East Greenland onshore margin. *Global and Planetary Change*, **24**, 261–274.
LEEDER, M.R. 1999. *Sedimentology and Sedimentary Basins*. Blackwell Science, Oxford.
LYELL, C. 1830–1875. *Principles of Geology*, 12 editions. John Murray, London.
MATHIESEN, A. 1998. *Modelling of Uplift History from Maturity and Fission-track Data, Nuussauq, West Greenland*. Danmarks og Grønlands Geologiske Undersøgelse Rapport, **87**.
MURDOCH, L.M., MUSGROVE, F.W. & PERRY, J.S. 1995. Tertiary uplift and inversion history in the North Celtic Sea Basin and its influence on source rock maturity. *In*: CROKER, P.F. & SHANNON, P.M. (eds) *The Petroleum Geology of Ireland's Offshore Basins*. Geological Society, London, Special Publications, **93**, 297–319.
NANSEN, F. 1904. The bathymetrical features of the North Polar seas, with discussion of the continental shelves and previous oscillations of the shoreline. *In*: NANSEN, F. (ed.) *The Norwegian North Polar Expedition 1893–1896, Scientific Results*. Jacob Bydwad, Christiania (Oslo).
NAYLOR, D. 1992. The post-Variscan history of Ireland. *In*: PARNELL, J. (ed.) *Basins on the Atlantic Seaboard: Petroleum Geology, Sedimentology and Basin Evolution*. Geological Society, London, Special Publications, **62**, 255–275.
NYLAND, B., JENSEN, L.N., SKAGEN, J., SKARPNES, O. & VORREN, T. 1992. Tertiary uplift and erosion in the Barents Sea; magnitude, timing and consequences. *In*: LARSEN, R.M., BREKKE, H., LARSEN, B.T. & TALLERAAS, E. (eds) *Structural and Tectonic Modelling and its Application to Petroleum Geology*. Norwegian Petroleum Society (NPF) Special Publication, **1**, 153–162.
PARKER, J.R. 1975. Lower Tertiary sand development in the central North Sea. *In*: WOODLAND, A.W. (ed.) *Petroleum and the Continental Shelf of North-West Europe*. Applied Science, Barking, UK, 447–454.
PARNELL, J., CAREY, P.F., GREEN, P.F. & DUNCAN, W. 1999. Hydrocarbon migration history, west of Shetland: integrated fluid inclusion and fission track studies. *In*: FLEET, A.J. & BOLDY, S.A.R. (eds) *Petroleum Geology of Northwest Europe: Proceedings of the 5th Conference*. Geological Society, London, 613–626.
REUSCH, H. 1901. Nogle bidrag till forstaaelsen af hvorledes Norges dale og fjelde er blevne til. *Norges Geologiske Undersøgelse, Aarbog*, **14**, 96–102.
RIIS, F. & JENSEN, L.N. 1992. Introduction: measuring uplift and erosion—proposal for a terminology. *In*: Jensen, L.N. & Riis, F. (eds) *Post-Cretaceous Uplift and Sedimentation along the Western Fennoscandian Shield. Norsk Geologisk Tidsskrift*, **72**, 223–228.
RING, U., BRANDON, M.T., LISTER, G.S. & WILLETT, S.D. 1999. Exhumation processes. *In*: RING, U., BRANDON, M.T., LISTER, G.S. & WILLETT, S.D. (eds) *Exhumation Processes: Normal Faulting, Ductile Flow and Erosion*. Geological Society, London, Special Publications, **154**, 1–27.
ROWLEY, E. & WHITE, N. 1998. Inverse modelling of extension and denudation in the East Irish Sea and surrounding areas. *Earth and Planetary Science Letters*, **161**, 57–71.
SCOTCHMAN, I.C. & THOMAS, J.R.W. 1995. Tertiary uplift and inversion history in the North Celtic Sea Basin and its influence on source rock maturity. *In*: CROKER, P.F. & SHANNON, P.M. (eds) *The Petroleum Geology of Ireland's Offshore Basins*. Geological Society, London, Special Publications, **93**, 385–411.
SOLHEIM, A., RIIS, F., ELVERHOI, A., FALEIDE, J.I., JENSEN, L.N. & CLOETINGH, S. 1996. Impact of glaciations on basin evolution: data and models from the Norwegian margin and adjacent areas—introduction and summary. *Global and Planetary Change*, **12**, 1–9.
STEUVOLD, L.M. & ELDHOLM, O. 1996. Cenozoic uplift of Fennoscandia inferred from a study of the mid-Norwegian margin. *Global and Planetary Change*, **12** (1–4), 359–386.
STRATAGEM PARTNERS, 2002. The Neogene stratigraphy of the European margin between Lofoten and

Porcupine. Stoker, M.S. (compiler). A product of the EC-funded STRATAGEM project. Online. Available at http://www.stratagem-europe.org.

SUMMERFIELD, M.A. 1991. *Global Geomorphology*. Longman, Harlow, UK.

WHITE, R.S. 1988. A hot-spot model for Early Tertiary volcanism in the North Atlantic. *In*: MORTON, A.C. & PARSONS, L.M. (eds) *Early Tertiary Volcanism in the North Atlantic*. Geological Society, London, Special Publications, **39**, 241–252.

Present and past influence of the Iceland Plume on sedimentation

STEPHEN M. JONES[1], NICKY WHITE[1], BENJAMIN J. CLARKE[1,2], ELEANOR ROWLEY[1,3] & KERRY GALLAGHER[1,4]

[1]*Bullard Laboratories, Madingley Rise, Madingley Road, Cambridge CB3 0EZ, UK (e-mail: jones@esc.cam.ac.uk)*

[2]*Present address: Department of Geology and Geophysics, University of Edinburgh, King's Building, West Mains Road, Edinburgh EH9 3JW, UK*

[3]*Present address: Shell Eygpt, 6 Hassan El-Sherley Street, PO Box 2681 El Horreya, Heliopolis, Cairo, Eygpt*

[4]*TH Huxley School of Environment, Earth Science, and Engineering, Imperial College of Science, Technology and Medicine, South Kensington, London SW7 2AS, UK*

Abstract: The Cenozoic development of the North Atlantic province has been dramatically influenced by the behaviour of the Iceland Plume, whose striking dominance is manifest by long-wavelength free-air gravity anomalies and by oceanic bathymetric anomalies. Here, we use these anomalies to estimate the amplitude and wavelength of present-day dynamic uplift associated with this plume. Maximum dynamic support in the North Atlantic is 1.5–2 km at Iceland itself. Most of Greenland is currently experiencing dynamic support of 0.5–1 km, whereas the NW European shelf is generally supported by <0.5 km. The proto-Iceland Plume had an equally dramatic effect on the Early Cenozoic palaeogeography of the North Atlantic margins, as we illustrate with a study of plume-related uplift, denudation and sedimentation on the continental shelf encompassing Britain and Ireland. We infer that during Paleocene time a hot subvertical sheet of asthenosphere welled up beneath an axis running from the Faroes through the Irish Sea towards Lundy, generating a welt of magmatic underplating of the crust which is known to exist beneath this axis. Transient and permanent uplift associated with this magmatic injection caused regional denudation, and consequently large amounts of clastic sediment have been shed into surrounding basins during Cenozoic time. Mass balance calculations indicate agreement between the volume of denuded material and the volume of Cenozoic sediments deposited offshore in the northern North Sea Basin and the Rockall Trough. The volume of material denuded from Britain and Ireland is probably insufficient to account for the sediment in the Faroe–Shetland Basin and an excess of sediment has been supplied to the Porcupine Basin. We emphasize the value of combining observations from both oceanic and continental realms to elucidate the evolution of the Iceland Plume through space and time.

The free-air gravity field over the northern hemisphere is dominated by a long-wavelength high centred on Iceland, stretching from the Azores to Siberia and from Baffin Island to Denmark (Fig. 1). Both theoretical considerations and observations from the world's oceans suggest that long-wavelength (>1000 km) positive anomalies are generally associated with mantle upwelling and dynamic uplift (Sclater *et al.* 1975; McKenzie 1994). If these inferences also hold in the North Atlantic province, the areal extent of the gravity high suggests that the continental margins of NW Europe and eastern Greenland are dynamically supported at present. In the first part of this paper, we compare estimates of dynamic support derived from the bathymetry and gravity field of the North Atlantic oceanic realm to constrain the present magnitude of dynamic support of both oceanic crust and the adjacent continental margins.

Next, we consider the region that was affected by the plume in the past. Temporal and spatial variation of plume-related uplift has played an important role in the evolution of both margins, especially in controlling the generation and distribution of clastic sediments. Here, we consider the specific example of sediment mass balance around Britain and Ireland. Geochemical evidence has proved that this region experienced magmatic underplating of the crust during

From: DORÉ, A.G., CARTWRIGHT, J.A., STOKER, M.S., TURNER, J.P. & WHITE, N. 2002. *Exhumation of the North Atlantic Margin: Timing, Mechanisms and Implications for Petroleum Exploration*. Geological Society, London, Special Publications, **196**, 13–25. 0305-8719/02/$15.00

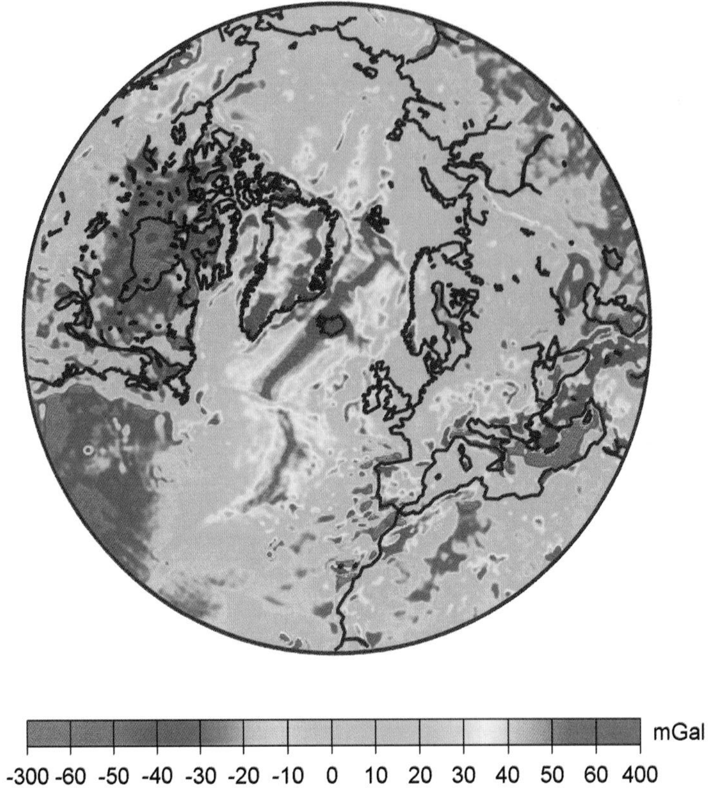

Fig. 1. Free-air gravity anomaly over part of the northern hemisphere, displayed using a Lambert azimuthal equal area projection centred on Iceland. The long-wavelength gravity high centred on Iceland, which extends from south of the Azores to Spitzbergen and from Baffin Island to Scandinavia, should be noted.

Paleocene time, leading to permanent uplift, which drove denudation (Thompson 1974; Brodie & White 1995). Mass balance provides a means of comparing and verifying complementary measures of Early Cenozoic denudation, estimated from the onshore record of denudation and the offshore record of sediment accumulation.

Two forms of surface uplift are associated with mantle plumes. Dynamic support always results when abnormally hot mantle is emplaced beneath the lithosphere. This support is transient and disappears when the thermal anomalies in the asthenosphere and lithosphere dissipate by convection and conduction. The North Atlantic region is dynamically supported today and it must also have been dynamically supported during Early Cenozoic time, as the volcanic record attests to abnormally hot mantle at that time. Permanent uplift may also occur if mantle thermal anomalies induce melting and the melt is injected into or just beneath the crust. Permanent uplift affected Britain and Ireland during Paleocene time, and the regions flanking the line of continental separation at the Paleocene–Eocene boundary. Comparison of the two themes we present in this paper therefore demonstrates the relative spatial extent and magnitude of transient and permanent uplift. The term epeirogenic uplift refers to uplift that could be permanent and/or transient. In general, epeirogenic uplift refers to uplift that is not generated by horizontal plate motions, and the term need not imply any particular mechanism. However, the present and past epeirogenic uplift of the North Atlantic region we discuss in this study is generated by the Iceland Plume.

Present-day dynamic support

The North Atlantic Ocean is anomalously shallow in the vicinity of Iceland. Anomalous topography culminates at Iceland itself, which rises to *c.* 2 km above sea-level, or *c.* 4.5 km above the average depth of the global mid-ocean ridge system. Two methods can be employed to investigate dynamic support of the region of oceanic crust round Iceland. The first method exploits the fact that the depth of oceanic crust away from the influence of mantle plumes varies with age in a well-understood manner, by comparing the present-day bathymetry around Iceland with this reference depth. The second method involves establishing a link between dynamic support and the long-wavelength free-air gravity anomaly. Analysis of gravity anomalies should always be treated with caution, as any gravity field can be explained by an infinite number of density distributions, each with different implications for dynamic support. In this section, we first establish that estimates of dynamic support derived from bathymetry and gravity are in general agreement over oceanic crust around Iceland. This result then gives us confidence in estimating dynamic support of the adjacent continental margins using gravity alone.

Estimates from bathymetry

Figure 2 is a plot of anomalous topography in the North Atlantic, calculated by subtracting the well-known age–depth model of Parsons & Sclater (1977) from the bathymetry. Unfortunately, the anomalous topography in Fig. 2 cannot be interpreted solely in terms of present-day dynamic support because it also includes a component of permanent topography caused by spatial variations in the thickness of oceanic crust. However, the magnitude of this permanent topography can be estimated given determinations of oceanic crustal thickness from wide-angle seismic experiments. In the oceans surrounding Iceland but away from the continental margins and the Greenland–Iceland–Faroes Ridge, anomalously hot mantle has generated crust 7–10 km thick (White 1997). Isostatic balancing shows that this variation of up to 3 km greater than the thickness of standard

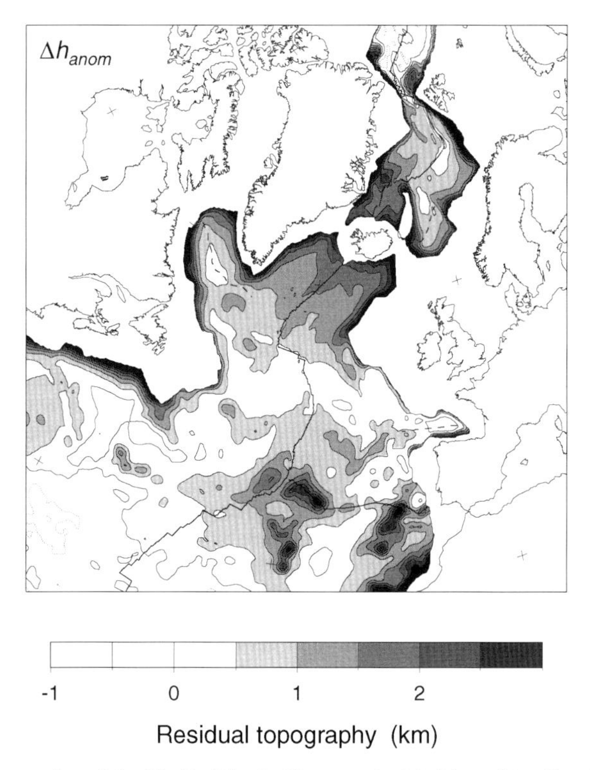

Fig. 2. Anomalous topography of the North Atlantic Ocean, calculated by subtracting the age–depth cooling relationship of Parsons & Sclater (1977) from the ETOPO5 bathymetry grid. The age of oceanic lithosphere was taken from Müller *et al.* (1997); the Greenland–Iceland–Faroes Ridge was excluded from the calculation because its age is not well known. Anomalous topography is corrected for sediment loading of oceanic basement using the method of Le Douaran & Parsons (1982) and the sediment thickness map of Laske & Masters (1997). It should be noted that the anomalous topography displayed here contains both a component of present-day dynamic support and a permanent component caused by crustal thickness variations.

oceanic crust (7 km, White *et al.* 1992) generates permanent topography of 0–0.5 km. This value of 0.5 km is smaller than the total variation in anomalous topography across these regions, suggesting that most of the anomalous topography is generated by present-day dynamic support. However, close to the continental margins and the Greenland–Iceland–Faroes Ridge, the thickness of oceanic crust is 10–15 km, varying over distances of <50 km (Barton & White 1997; Weir *et al.* 2001). In these regions, permanent topography probably accounts for most of the anomalous topography in Fig. 2.

To summarize the discussion above, anomalous topography of those regions away from the continental margins and the Greenland–Iceland–Faroes Ridge can be interpreted as an estimate of dynamic support with an error range of 0.5 km. However, short-wavelength variations in anomalous topography associated with continental margins and the Greenland–Iceland–Faroes Ridge are most likely to reflect permanent topography associated with crustal thickness variations. Bearing in mind these caveats, three important results concerning present-day dynamic support can be gleaned from Fig. 2. First, the amplitude of dynamic support close to Iceland itself is 1.5–2 km. Secondly, the amplitude of dynamic support is partially controlled by the active mid-ocean ridge system. This fact is clearly seen both to the north of Iceland, where the active Kolbeinsey Ridge is more anomalously elevated than the extinct Aegir Ridge, and to the south of Iceland, where a tongue of anomalous uplift extends along the Reykjanes Ridge. Thirdly, the continental margins of eastern Greenland and NW Europe are currently experiencing at least 0.5 km of dynamic support, implying that the adjacent continental shelves are also experiencing significant dynamic support at present.

Estimates from gravity

The free-air gravity field over a convecting layer results from two competing effects: density variations within the layer and surface deformation induced by the convective circulation. In the case of upwelling plumes, hot mantle has a lower density than the surrounding mantle, which acts to reduce the gravity. However, an upwelling plume deforms the Earth's surface upwards, which acts to increase the gravity. The sign of the total free-air gravity anomaly depends on the relative magnitude of these effects. At Rayleigh numbers appropriate to Earth, the effect of surface deformation outweighs the effect of the reduction in density, so upwelling regions in the mantle are characterized by positive free-air gravity anomalies at the surface.

The transfer function between the free-air gravity anomaly g and topography h is called the admittance and is given by

$$Z(k) = g(k)/h(k).$$

Admittance is a function of wavenumber k. At wavelengths shorter than *c.* 500 km, admittance is controlled by the mechanical properties of the lithosphere. Admittance decreases systematically with increasing wavelength (i.e. decreasing wavenumber), and the rate of decrease is dependent upon the effective elastic thickness of the lithosphere. McKenzie (1994) and McKenzie & Fairhead (1997) exploited this behaviour to estimate the effective elastic thickness of the lithosphere in the Pacific and Indian Oceans from the short-wavelength part of the gravity and topography fields. At wavelengths above *c.* 500 km the admittance calculated from the observed gravity and topography diverges from that calculated from theoretical models assuming an elastic plate. At these long wavelengths, the flexural strength of the lithosphere plays no part in supporting topography. Instead, topography is supported by stresses exerted on the base of the lithosphere by the convecting mantle and by long-wavelength variations in the density structure of the lithosphere, which are isostatically compensated. Thus, dynamic support produces long-wavelength anomalous topography Δh_{conv} that correlates with the long-wavelength free-air gravity anomaly. Isostatically compensated topography, such as the permanent anomalous topography that arises from crustal thickness variations in the North Atlantic, is not correlated with a measurable gravity anomaly.

The behaviour of the admittance function within the wavelength band 500–3000 km suggests a linear relationship between gravity and topography. If this simple relationship, observed in the Pacific and Indian Oceans, also holds in the North Atlantic then anomalous topography caused by dynamic convective support can be calculated from the long-wavelength free-air gravity field using

$$\Delta h_{\mathrm{conv}} = g/Z.$$

The value of Z to be used can be constrained by observations. It has long been recognized that $Z \approx 35\ \mathrm{mGal\,km^{-1}}$ is appropriate for Earth's oceans (Sclater *et al.* 1975). Another measure of the admittance to be used to calculate dynamic support can be obtained from numerical convection experiments. McKenzie (1994) summarized several numerical models of axisymmetric plumes, including the model of Watson & McKenzie (1991) that successfully matched the observed gravity, topography and melt production of the Hawaiian Plume. These convection models are characterized by admittances in the range 34.4 ± 2.2 mGal km^{-1}. Hence, observed and theoretical values for the admittance between topography and gravity of the oceans at long wavelengths are in good agreement.

The free-air gravity dataset used here is a compilation of point measurements over land, together with the satellite gravity dataset of Sandwell & Smith (1997) over the oceans, as described by McKenzie & Fairhead (1997). The short-wavelength part of the gravity field that is influenced by the flexural strength of the lithosphere was removed using a low-pass filter. The region of interest has dimensions comparable with the radius of Earth, so low-pass filtering was carried out using a spherical harmonic model of the gravity dataset. The long-wavelength gravity model was generated using spherical harmonic coefficients of degrees $0 < l = m \leqslant 53$ (equivalent to a low-pass filter of *c.* 750 km) with a suitable taper at the upper cut-off to prevent ringing.

Figure 3 shows the dynamic component of anomalous topography $\Delta h_{\rm conv}$ estimated from the free-air gravity field using two different values of Z. These estimates are simply a scaled version of the long-wavelength gravity field and resemble the original gravity field fairly closely (compare Figs 1 and 3). In the region surrounding Iceland, estimates of convective support from the gravity field agree reasonably well with independent estimates of convective support from topography (compare Figs 2 and 3). In particular, the dynamic support estimates from gravity reinforce the three observations derived from Fig. 2 and noted in the previous section. First, peak dynamic support is *c.* 1.8 km at Iceland. Secondly, the active spreading axis exerts an important control on the long-wavelength gravity field and on the magnitude of dynamic support. Dynamic support is centred on the Mid-Atlantic Ridge, and to the north of Iceland the active Kolbeinsey Ridge is experiencing greater support than the extinct Aegir Ridge. Thirdly, the continental margins are currently experiencing significant dynamic support of 0.5–1 km.

It is important to note that estimates of present-day dynamic support calculated directly from long-wavelength gravity anomalies are not always in agreement with estimates calculated from the bathymetry. Figure 3 suggests that

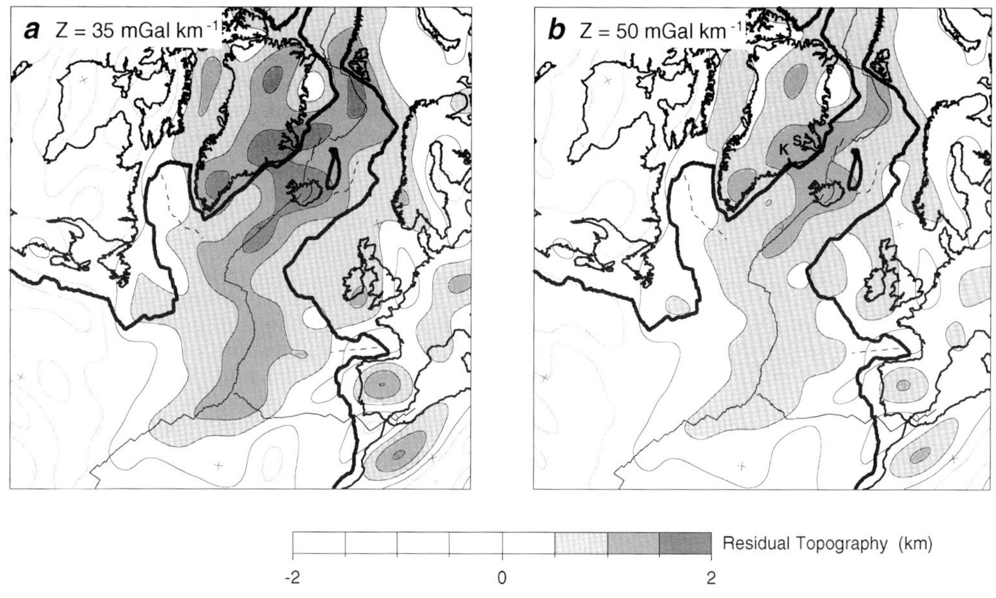

Fig. 3. Estimates of present-day dynamic support in the North Atlantic region, calculated by dividing the long-wavelength free-air gravity field by a constant admittance, as discussed in the text. Bold continuous lines indicate continent–ocean boundaries. (**a**) Dynamic support predicted using an admittance of $Z = 35$ mGal km^{-1} (appropriate for subaqueous regions); (**b**) dynamic support predicted using an admittance of $Z = 50$ mGal km^{-1} (equivalent value for subaerial regions). K, Kangerlussuaq; S, Scoresby Sund.

significant dynamic support occurs from Iceland to the Azores. However, Fig. 2 indicates no measurable dynamic support just south of the Charlie Gibbs fracture zone. Thus, the amplitude of dynamic support calculated from gravity data is unlikely to be correct in detail and should be treated with some caution. Nevertheless, the agreement between dynamic support estimates based on bathymetry and gravity in the vicinity of Iceland suggests that Fig. 3 provides a reasonable estimate of dynamic support of the Greenland–Iceland–Faroes Ridge and the adjacent continental shelves.

Free-air gravity anomalies may be used to estimate dynamic support of the continents in the same way as for the oceans. Whereas it is appropriate to employ an admittance of $Z = 35\,\mathrm{mGal\,km^{-1}}$ to estimate dynamic support of regions covered by water, an equivalent air-loaded value of $Z = 50\,\mathrm{mGal\,km^{-1}}$ should be used for subaerial regions. Dynamic support estimates calculated using both values of admittance are shown in Fig. 3. Greenland is currently experiencing dynamic support of 0.5–1 km. Dynamic support is greatest on the east coast, notably in the region between Kangerlussuaq and Scoresby Sund, adjacent to the Greenland–Iceland–Faroes Ridge. In contrast, the magnitude of dynamic support beneath the NW European shelf seldom exceeds 0.5 km. Dynamic support increases westwards from Scandinavia across the Norwegian continental shelf. The North Sea and southern England do not appear to be dynamically supported.

Cenozoic denudation of Britain and Ireland

In this section, we focus on one aspect of the problem of determining the Early Cenozoic shape of the Iceland Plume, namely the relationship between Cenozoic uplift, denudation and sedimentation on the continental shelf surrounding Britain and Ireland. Our long-term goal is to use mass balance calculations in conjunction with estimates of permanent uplift caused by magmatic underplating to determine the behaviour of the Iceland Plume throughout Cenozoic time.

We have chosen to carry out mass balance calculations for the British Isles for three important reasons. First, geochemical evidence collected from rocks of the British Cenozoic Igneous Province proves that the crust beneath much of Britain and Ireland has been thickened by a substantial amount of igneous material as we explain below, implying significant permanent uplift during Paleocene time. Secondly, a substantial body of information about the denudation of Britain and Ireland is available based on modelling of subsidence, vitrinite reflectance, apatite fission-track and sonic velocity datasets. Thirdly, the products of Cenozoic denudation have been carefully mapped in all of the surrounding sedimentary basins. Although detailed work has been carried out previously on both onshore denudation and offshore deposition, no attempts have yet been made to check these estimates by constructing a mass balance on a regional scale.

The evidence for magmatic underplating of the crust beneath Britain and Ireland is not widely recognized, despite the fact that the consequences of such an igneous addition, in terms of uplift, denudation and sedimentation, can explain many features of the surface geology and thus have obvious implications for the hydrocarbon industry. For decades it has been recognized that the composition of flood basalts from the Hebrides can be explained only by fractional crystallization of up to 70% of their original liquid mass (Thompson 1974). The crystallized residuum must therefore remain at depth. From a surface processes point of view the argument ends here, because if such material is added anywhere within the upper half of the lithosphere its density will be less than that of the asthenosphere it displaces, and isostatic balancing shows that permanent uplift will result. The depth at which crystallization occurred, and at which the crystallized residuum remains, can be established using geobarometry techniques, which determine the pressure at which the major-element compositions of both flood basalts and the phenocrysts they contain are in equilibrium. Pressures of *c.* 1 GPa are estimated, equivalent to a depth of around 30 km, i.e. the depth of the Moho (Thompson 1974; Brodie & White 1995). This result is not surprising, as the fact that basaltic melt is less dense than the mantle lithosphere but roughly the same density as the lower crust means that the melt should rise through the mantle and pond at the base of the crust, giving rise to the concept of magmatic underplating. Wide-angle seismic experiments across North Atlantic volcanic continental margins have imaged high-velocity bodies at the base of the continental crust, which are usually interpreted as pods of igneous material underplated beneath the crust at the time of continental separation (Barton & White 1997). Preliminary results from modelling of a wide-angle seismic line spanning the Irish Sea suggest that a high-velocity zone exists near the Moho, which probably represents the igneous material that we expect to see underplated beneath the onshore part of the

British Cenozoic Igneous Province (S. Al-Kindi, pers. comm.). However, as this discussion implies, seismic evidence alone can never directly reveal the age or nature of these high-velocity layers, so it is always necessary to interpret seismic evidence in conjunction with geochemical evidence. To conclude, it should be noted that crustal magmatic underplating is not a peculiar feature of the North Atlantic but is common to all continental flood basalt provinces (Cox 1993).

Mass balance calculations

The mass balance calculation has three parts: the extent and amount of denudation of the sediment source area or catchment (Fig. 4); the mass of sediment accumulated through time in the sedimentary basins immediately adjacent to the British Isles (Fig. 5); and the mass of material lost from the system by solution and by escape to the deep ocean.

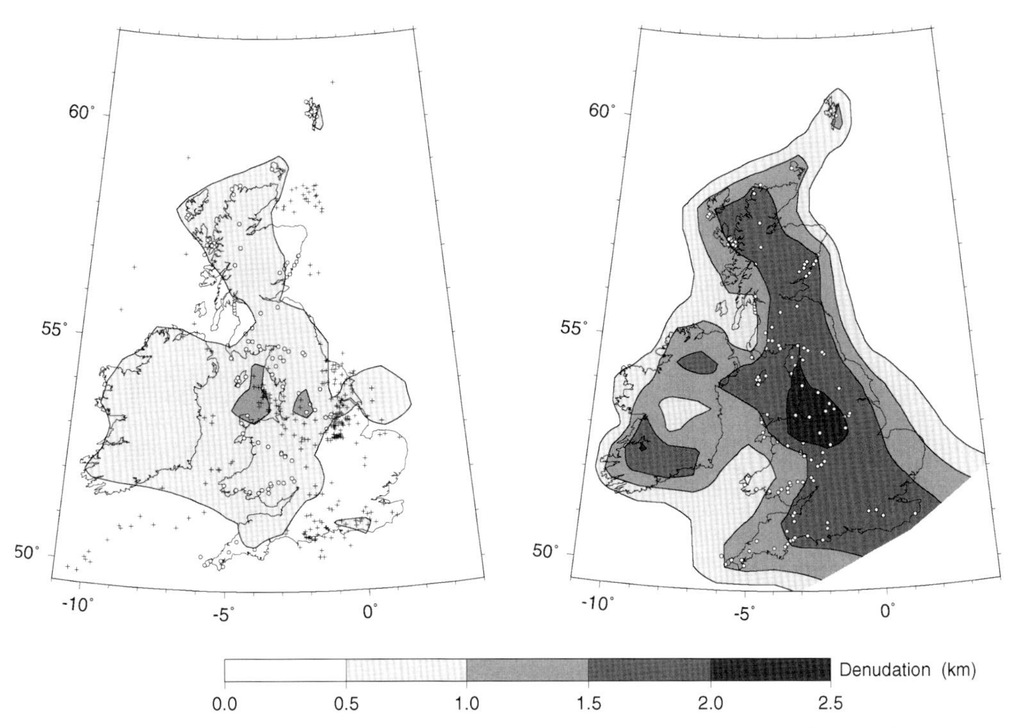

Fig. 4. Estimates of Cenozoic denudation of Britain and Ireland based on modelling subsidence histories and apatite fission-track length distributions. +, locations of well sections in extensional sedimentary basins used for subsidence modelling. The amount of missing post-rift basin fill was predicted by fitting a theoretical subsidence curve to the remnant synrift stratigraphy, assuming the standard lithospheric stretching model. This technique has been described fully by Rowley & White (1998), and results have been tabulated by Hall (1995) and Rowley (1998). O, locations of apatite samples. Apatite fission-track length distributions were modelled to find Mesozoic–Cenozoic thermal histories by K. Gallagher using the Laslett *et al.* (1987) annealing model for Durango apatite and the method described by Gallagher (1995); the results were reported by Rowley (1998). The amount of Cenozoic cooling was converted to a range of denudation estimates by Monte Carlo modelling using the range of geothermal gradients observed in the North Sea today (Rowley 1998). Data covering Ireland are from Allen *et al.* (2002). (**a**) Minimum denudation estimate found by contouring results derived from subsidence analysis, the lower bounds of the Rowley (1998) denudation estimates derived from modelling apatite fission-track length distributions, and the minimum denudation estimate for Ireland of Allen *et al.* (2002, fig. 10b). (**b**) Maximum denudation estimate found by contouring the modes of the Rowley (1998) denudation estimates and the maximum denudation estimate for Ireland of Allen *et al.* (2002, fig. 10a). These contour plots were generated by taking the mean of all estimates within blocks of dimension 1° longitude × 30′ latitude (*c.* 50 km × 50 km) and gridding the resulting values using the continuous curvature spline method of Smith & Wessel (1990). It should be noted that a further set of Cenozoic denudation estimates based on modelling of vitrinite reflectance profiles from a subset of the wells used for subsidence analysis yields a denudation estimate that lies between the minimum and maximum estimates illustrated here (Rowley 1998).

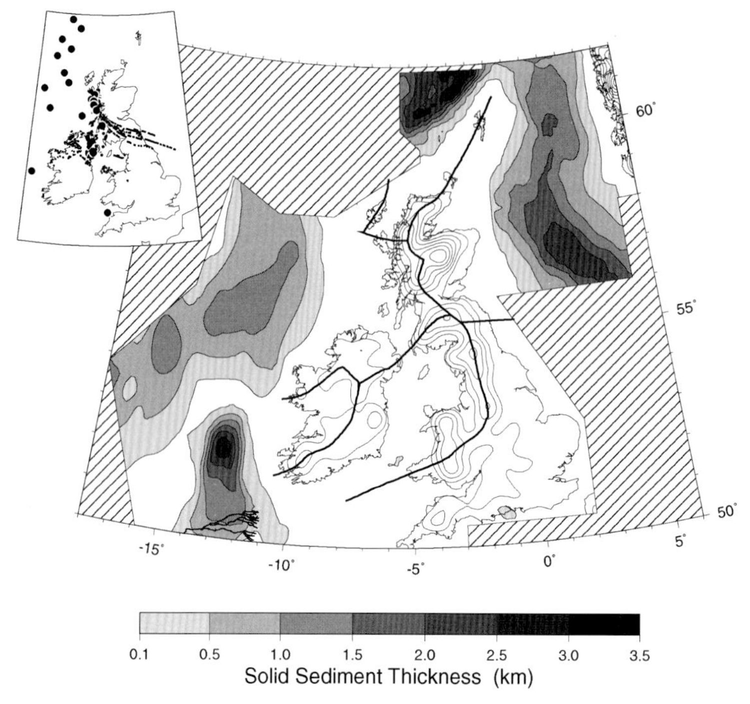

Fig. 5. Map of the NW European continental shelf showing the Cenozoic solid sediment thickness. This isopach map was constructed from a database of 2D and 3D seismic reflection surveys calibrated with well-log information. Solid thickness was calculated from the observed thickness by subtracting the pore-space volume predicted by the standard exponential relationship between fractional porosity ϕ and depth z given by $\phi = \phi_0 \exp(-z/\lambda)$, where $\phi_0 = 0.6$ is the depositional porosity and $\lambda = 2$ km is the compaction length scale. The present-day canyon system at the southern end of the Porcupine Basin is shown. Contours on land show topography smoothed using a filter of width 50 km and used to define the drainage catchments supplying each offshore sediment sink. Segments of catchment boundaries that are offshore at present were positioned using present-day bedload transport patterns. Inset shows central igneous complexes, flood basalts and dykes of the North Atlantic Igneous Province, mostly emplaced during Paleocene time.

The best-constrained element in the mass balance problem is the mass of Cenozoic sediment in the offshore basins. Here, the volume of solid sediment is used in place of the mass of sediment. The solid volume of the offshore sediment pile is calculated by removing the volume accounted for by porosity, predicted by the standard exponential porosity model. The principal source of error in calculating the solid volume is the depth conversion procedure. However, the error introduced by this uncertainty can be realistically quantified by using a range of velocity–depth functions, and the resulting error estimates are illustrated in Fig. 6. Parameterization of the porosity–depth relationship used in the compaction calculation is relatively well determined for the basins surrounding Britain and Ireland. For example, using parameters for the end-member lithologies of shale and sand determined by Sclater & Christie (1980) for the North Sea yields a variation in solid volume that is only 14% of the error range associated with the depth conversion calculation in that region. Figure 5 shows the solid sediment thickness accumulated around Britain and Ireland during Cenozoic time. During this time the main sediment sinks were the Porcupine Basin, the Rockall Trough, the Faroe–Shetland Basin and the North Sea Basin. The solid volume of Cenozoic deposits in southern England and offshore southern Ireland is negligible.

The total Cenozoic denudation is shown in Fig. 4, based on three independent lines of evidence. The first line of evidence depends on our knowledge of extensional sedimentary basins. Many extensional basins around Britain and Ireland contain a fault-controlled syn-rift stratigraphy but the anticipated post-rift stratigraphy is partially or entirely absent (Brodie & White 1995). The amount of missing post-rift stratigraphy can be predicted, based on our knowledge of the kinematics of extensional

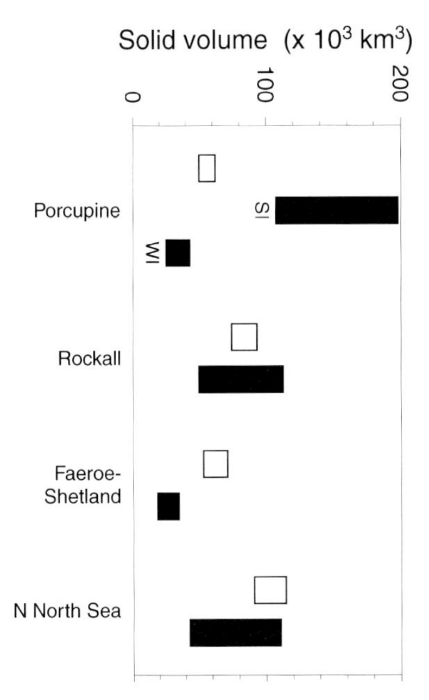

Fig. 6. Summary of mass balance calculations for NW European continental shelf. Open bars represent the amount of solid sediment accumulated offshore, calculated by integrating Fig. 5 over each offshore depocentre. Error ranges reflect uncertainties in depth conversion. The solid sediment volume for the North Sea plotted here has been halved to account for the fact that Scandinavia has also supplied sediment to the North Sea. Filled bars represent volume of rock denuded from onshore catchments, calculated by integrating the minimum and maximum estimates given in Fig. 4 over the regions marked in Fig. 5. The Porcupine sediment sink is supplied by two catchments: SI, southern Ireland and the Irish Sea; WI, western Ireland. The other sediment sinks have one catchment each.

basins (Rowley & White 1998). The other two methods of estimating denudation rely on thermal indicators within the sediment pile that retain a memory of their burial history. Reflectance of vitrinite increases with rising temperature by means of a non-reversible reaction, so vitrinite retains a memory of the highest temperature it has experienced. Denudation estimates can be obtained by comparing vitrinite reflectance profiles with a global reflectance dataset from non-inverted basins (Rowley 1998). Analysis of fission tracks in apatite also provides temperature histories that can be interpreted in terms of denudation through time. The most prominent feature of Fig. 4 is the peak in denudation centred on NW England. This region suffered denudation of 1–2.5 km during Cenozoic time. A peak in denudation centred on NW England has also been suggested by previous studies based on apatite fission-track data alone (Lewis *et al.* 1992). The magnitude of Cenozoic denudation generally decreases towards the present offshore areas. Another important message from Fig. 4 is that the error range on the denudation estimate for any particular region is *c.* 1 km.

It is difficult to quantify loss of mass from the system. Loss by solution depends strongly on the rock type being eroded. It would be difficult to include the amount of loss by solution in the mass balance calculations because of the uncertainty over the original lithologies being eroded and the variety of lithologies within each catchment. Here we apply no corrections to account for loss of mass from the system. When loss of mass from the system is neglected, the solid volume of sediment measured in offshore basins provides a lower bound on the denudation. If the solid sediment volume measured offshore is found to be greater than the onshore estimate of denudation then an additional, unidentified source of sediment to the offshore basin would be implied. On the other hand, if the offshore solid sediment volume is found to be significantly less than the onshore estimate of denudation then the difference between the two estimates may provide an estimate of the amount of material lost from the system.

Mass balance results

Figure 6 summarizes the result of balancing the volume of solid sediment accumulated offshore with direct estimates of denudation made onshore for the Porcupine, Rockall, Faroe–Shetland and northern North Sea systems. The drainage catchment matched with each offshore sediment sink is shown in Fig. 5. The boundaries of these catchments are based on the present-day topography and bedload transport patterns. Palaeotopographic reconstructions of Britain and Ireland during earliest Eocene time, at the time of maximum dynamic support by the Iceland Plume (see discussion later), suggest that the north–south drainage divide running through Scotland and England has remained relatively static since Paleocene time (Jones 2000). The positions of the west–east-oriented drainage divides are more likely to have altered position during Cenozoic time, and we consider the effect of such migration in the following discussion.

In the Rockall Trough and northern North Sea systems, the volume of Cenozoic sediment accumulated offshore is the same as the denudation measured onshore, within error.

However, in the Faroe–Shetland system, the volume of sediment accumulated offshore is greater than the denudation measured onshore by at least $2 \times 10^4 km^3$. This discrepancy implies that the Faroe–Shetland Basin has another source of Cenozoic sediment in addition to the catchment in the region of NW Scotland. The additional sediment source may be a catchment in the region of the Faroes or Greenland. Alternatively, the boundaries of the drainage catchment shown in Fig. 5 covering NW Scotland may be incorrect. The southern boundary of that catchment is not well constrained and it is possible that sediment was also sourced from western Scotland and channelled between the Outer Hebrides and the Scottish mainland towards the Faroe–Shetland Basin. A third possibility is that additional sediment was transported into the Faroe–Shetland Basin by currents running along the basin axis. Strong bottom currents flowing southwestwards along the Faroe–Shetland Trough were initiated in Oligocene time and persist to the present day (Davies *et al.* 2001).

Lack of a significant difference between the two estimates of denudation in the Rockall and northern North Sea systems suggests that little mass has been lost from those systems. However, the total sediment supplied to the Porcupine Basin by both the catchment covering western Ireland and the catchment covering southern Ireland and the Irish Sea is greater than the material accumulated offshore by at least $4 \times 10^4 km^3$. Thus, a significant mass of sediment has been lost from this system. The magnitude of this discrepancy shown in Fig. 6 is a lower bound because no account has been taken of sediment supplied to the western side of the Porcupine Basin by the Porcupine Bank and Ridge. Three effects have probably contributed to the loss of mass from the Porcupine system. First, Carboniferous limestone crops out over a large area of the Irish catchments, and erosion of this limestone produces negligible clastic sediment. Secondly, it is likely that most of the sediment supplied to the southern end of the Porcupine Basin has escaped directly to the deep-sea sediment fan observed on the North Atlantic oceanic abyssal plain via the canyon system in the SE of the Porcupine Basin (Fig. 5). This canyon system has been active since Early Oligocene time (Jones 2000). Thirdly, a variety of evidence from both the continental shelf surrounding Britain and Ireland and adjacent oceanic crust suggests that the head of the Iceland Plume expanded rapidly during earliest Eocene time, causing dynamic uplift centred NW of Britain and Ireland (see discussion below). The resulting southeasterly tilt of Britain and Ireland probably caused most of the material eroded from southern Ireland and the Irish Sea during Eocene time to be shed southwards into the Bay of Biscay, rather than westwards into the Porcupine Basin.

Cenozoic permanent and transient uplift

In this section, we discuss evidence for temporal and spatial variation in the Iceland Plume throughout Cenozoic time.

Paleocene permanent uplift

The history of sediment flux into the basins surrounding Britain and Ireland during Paleocene time adds detail to the denudation history established using the mass balance. The rate of sediment flux into an offshore basin is related to the size and the rate of denudation of the corresponding drainage catchment. The lag time between an increase in denudation rate and the corresponding increase in sediment flux offshore is likely to be <100 ka (Reading 1991; Burgess & Hovius 1998). Therefore, sediment flux histories in the basins surrounding Britain and Ireland can be directly related to uplift of their sediment source regions. Calculation of the volumes of Paleocene and Eocene sediment sequences in the northern North Sea and Faroe–Shetland Basins has shown that sediment flux into these basins grew through Paleocene time to a maximum at 59–58 Ma and then decreased into Eocene time (Reynolds 1994; Clarke 2002). Thus we may infer that epeirogenic uplift of Scotland and northern England was initiated in Early Paleocene time and the rate of uplift peaked during mid-Late Paleocene time. In contrast, sediment flux into the Porcupine Basin remained low throughout Paleocene time, suggesting that uplift rates were lower away from the region of Scotland and northern England at this time (Jones 2000).

The denudation history of Scotland inferred from the history of sediment flux around Scotland correlates with the history of onshore igneous activity, which also peaked at 59–58 Ma (White & Lovell 1997). As we have seen, the sediment flux data suggest that Paleocene epeirogenic uplift was confined to the region close to the present-day surface expression of onshore igneous activity. We therefore suggest that the Paleocene igneous activity, epeirogenic uplift and denudation were initiated when a subvertical sheet of unusually hot asthenosphere was injected beneath the present Faroes–Irish Sea–Lundy axis. The surface expression of onshore igneous activity is concentrated in western Scotland and Ulster, but this igneous activity is offset from the locus of maximum denudation, which is centred on northern England and the Irish Sea. If the denudation shown in Fig. 4 is driven mainly by permanent uplift resulting from magmatic underplating, the greatest amount of melt must have been added to the crust beneath this region. Preliminary modelling of a wide-angle seismic line crossing Ireland, the Irish Sea and northern England has imaged a high-velocity pod near the Moho, which is thickest beneath the centre of the profile and thins towards western Ireland and towards the North Sea (S. Al-Kindi, pers. comm.). This pod may represent the layer of magmatic underplating beneath the crust that is predicted by both petrological and sedimentological evidence.

Paleocene–Eocene dynamic support

The sedimentary records in all basins surrounding the denuded area of Britain and Ireland indicate a major regression–transgression cycle, with maximum regression corresponding to the upper Flett Formation and its lateral equivalents, which were deposited during earliest Eocene time (e.g. Milton *et al.* 1990; Ebdon *et al.* 1995). It is now generally believed that this regression–transgression cycle is related to a phase of transient dynamic uplift that peaked in earliest Eocene time (Nadin *et al.* 1997; Jones *et al.* 2001). The magnitude of dynamic support can be quantified within extensional sedimentary basins, where we can isolate epeirogenic vertical motions from tectonic vertical motions, which are ultimately caused by horizontal plate motions. Given the synrift subsidence history of an extensional basin, the post-rift subsidence history can be calculated using the well-established lithospheric stretching model. Post-rift marker horizons with sedimentologically well-constrained water depths are then compared with the anticipated post-rift subsidence curve to reveal the magnitude of dynamic support through time. The most important of these marker horizons around Britain and Ireland are the earliest Eocene delta-top coals of the upper Flett Formation and its lateral equivalents. The results of this subsidence analysis show that peak dynamic support was *c.* 0.5 km in the Porcupine and Faroe–Shetland Basins, and dynamic support decreased in a southeasterly direction to zero across southern England (Nadin *et al.* 1997; Jones 2000; Jones *et al.* 2001).

Peak dynamic support at the Paleocene–Eocene boundary was coeval with voluminous intrusive and extrusive igneous activity offshore NW of Britain and Ireland that was associated with break-up of Europe and Greenland above unusually hot asthenosphere. White (1997) collated oceanic crustal thickness measurements and showed that the hot asthenosphere of the Iceland Plume head extended at least 1000 km along the continent–ocean boundaries to the north and south of the present Greenland–Iceland–Faroes Ridge. Dynamic support estimates from subsidence analyses around Britain and Ireland suggest that the plume head extended a similar distance inboard of the NW European continental margin.

We might expect that as maximum transient dynamic support of the basins surrounding Britain and Ireland occurred during earliest Eocene time, the acme of denudation and offshore sediment flux should also have occurred at this time. However, as we discussed above, maximum sediment flux into the Faroe–Shetland and North Sea Basins actually occurred 3–4 Ma previously, during Late Paleocene time. Only in the Porcupine Basin, west of Ireland, was the peak sedimentation rate coeval with peak dynamic support in earliest Eocene time (Jones 2000). This discrepancy in timing of several million years between the two different measures of epeirogenic uplift is an important observation that has yet to be explained. One possible explanation is that there were two separate phases of mantle plume activity, the first an upwelling hot vertical sheet that led to magmatic underplating of the crust beneath the Faroes–Irish Sea–Lundy axis, and the second the growth of a mushroom-shaped plume head beneath a region over 1000 km in radius.

Oligocene–Recent dynamic support

Variations in crustal thickness and structure around Iceland record changes in the size of the thermal head of the Iceland Plume, measured at the Mid-Atlantic Ridge, through Cenozoic time. As we have already discussed, the radius of the plume head was >1000 km at the time of continental separation between Europe and Greenland in earliest Eocene time. However, the distribution of normal-thickness oceanic crust south of Iceland suggests that during Late Eocene times, the part of the plume head beneath the Mid-Atlantic Ridge extended <300 km from the present centre of Iceland (White 1997). The portion of the plume head beneath the ridge then increased to its present radius of *c.* 1000 km between Oligocene time and the present (Figs 2 and 3). Studies of plate motion with respect to the hotspot reference frame suggest that the centre of the Iceland Plume lay beneath Greenland during mid-Cenozoic time and has effectively moved eastwards towards Europe through time (Lawver & Müller 1994). We suggest that this relative motion between the Mid-Atlantic Ridge and a plume of relatively constant mass flux may account for the apparent increase in the size of the head of the Iceland Plume between Oligocene time and the present. This history of motion also agrees with our observation that Greenland is currently experiencing greater dynamic support than NW Europe (Fig. 3). The reason is that a greater volume of hot plume-head material accumulates beneath the plate that the plume stem is moving away from than accumulates beneath the plate that the plume stem is approaching (Ribe & Delattre 1998). As Rohrman & van der Beek (1996) have suggested, relative movement of the Iceland Plume towards Europe may also provide an explanation for the Miocene–Recent epeirogenic uplift and consequent increase in denudation that is known to have affected Scandinavia.

Conclusions

The Cenozoic evolution of the North Atlantic province has been dominated by interaction between the Iceland Plume convective system and sea-floor spreading between Europe and Greenland on a hierarchy of spatial and temporal scales. The continental sedimentary record seems to be influenced by the relative importance of dynamic support, which has varied through through time, and permanent uplift, which was driven by magmatic underplating of the crust. Further progress in understanding this relationship will depend upon measuring the distribution of Paleogene magmatic underplating and upon an improved quantitative understanding of Cenozoic denudation.

Our conclusions concerning the present shape and size of the Iceland Plume and concerning Cenozoic mass balance are:

(1) maximum dynamic support in the North Atlantic is 1.5–2 km at Iceland itself.

(2) The magnitude of dynamic support in the North Atlantic is influenced by the location of the active spreading centres.

(3) The continent–ocean boundary of NW Europe is currently experiencing dynamic uplift of *c.* 0.5 km, which decreases to zero across Scandinavia and the North Sea. The continent–ocean boundary of eastern Greenland is currently experiencing dynamic uplift of *c.* 1 km.

(4) Cenozoic mass balance validates onshore denudation estimates by showing they are compatible with the solid volume of sediment accumulated offshore.

S.M.J. was supported by an NERC studentship and B.J.C. was supported by a BP studentship. We are indebted to B. Mitchener and J. Perry of BP for providing data and support. A. Carter and T. Hurford provided the fission-track data that were modelled to produce Fig. 4, and P. Allen allowed us to use results reported elsewhere in this volume in the same figure. We thank H. Walford for help in producing Fig. 3, and M. Shaw-Champion for help in producing Fig. 5. A.G. Doré, J.-I. Faleide, B. Lovell and M. Rohrman provided helpful reviews. This paper is Department of Earth Sciences Contribution ES.6677.

References

ALLEN, P.A., CUNNINGHAM, M.J.M., BENNETT, S.D. & 6 OTHERS 2002. The post-Variscan thermal and denudational history of Ireland. *In*: DORÉ, A.G., CARTWRIGHT, J., STOKER, M.S., TURNER, J.P. & WHITE, N. (eds) *Exhumation of the North Atlantic Margin: Timing, Mechanisms and Implications for Petroleum Exploration*. Geological Society, London, Special Publications, **196**, 371–399.

BARTON, A.J. & WHITE, R.S. 1997. Crustal structure of Edoras Bank continental margin and mantle thermal anomalies beneath the North Atlantic. *Journal of Geophysical Research*, **102**, 3109–3129.

BRODIE, J. & WHITE, N. 1995. The link between sedimentary basin inversion and igneous underplating. *In*: BUCHANAN, J.G. & BUCHANAN, P.G. (eds) *Basin Inversion*. Geological Society, London, Special Publications, **88**, 21–38.

BURGESS, P.M. & HOVIUS, N. 1998. Rates of delta progradation during highstands: consequences for timing of deposition in deep-marine systems. *Journal of the Geological Society, London*, **155**, 217–222.

CLARKE, B.J. 2002. *Early Cenozoic denudation of the British Isles: a quantitative stratigraphic approach*. PhD dissertation, University of Cambridge.

COX, K.G. 1993. Continental magmatic underplating. *Philosophical Transactions of the Royal Society of London, Series A*, **342**, 155–166.

DAVIES, R., CARTWRIGHT, J., PIKE, J. & LINE, C. 2001. Early Oligocene initiation of North Atlantic Deep Water formation. *Nature*, **410**, 917–920.

EBDON, C.C., GRANGER, P.J., JOHNSON, H.D. & EVANS, A.M. 1995. prospectivity. *The Tectonics, Sedimentation and Palaeoceanography of the North Atlantic Region*. *In*: SCRUTTON, R.A., STOKER, M.S., SHIMMIELD, G.B. & TUDHOPE, A.W (eds) Geological Society, London, Special Publications, **90**, 51–69.

GALLAGHER, K. 1995. Evolving temperature histories from apatite fission track data. *Earth and Planetary Science Letters*, **136**, 421–435.

HALL, B.D. 1995. *Early Tertiary Tectonics and igneous activity adjacent to the N.W. European continental margin*. PhD dissertation, University of Cambridge.

JONES, S.M. 2000. *Influence of the Iceland Plume on Cenozoic sedimentation patterns*. Ph.D. dissertation, University of Cambridge.

JONES, S.M., WHITE, N. & LOVELL, B. 2001. Cenozoic and Cretaceous transient uplift in the Porcupine Basin and its relationship to a mantle plume. *In*: SHANNON, P.M., HAUGHTON, P.D.W. & CORCORAN, D. (eds) *The Petroleum Exploration of Ireland's Offshore Basins*. Geological Society, London, Special Publications, **188**, 345–360.

LASKE, G. & MASTERS, A. 1997. A global digital map of sediment thickness. *EOS Transactions, American Geophysical Union*, **78**, F483.

LASLETT, G.M., GREEN, P.F., DUDDY, I.R. & GLEADOW, A.J.W. 1987. Thermal annealing of fission tracks in apatite 2. A quantitative analysis. *Chemical Geology (Isotope Geosciences Section)*, **65**, 1–13.

LAWVER, L.A. & MÜLLER, R.D. 1994. Iceland hotspot track. *Geology*, **22**, 311–314.

LEWIS, C.L.E., GREEN, P.F., CARTER, A. & HURFORD, A.J. 1992. Elevated K/T palaeotemperatures throughout Northwest England: three kilometres of Tertiary erosion? *Earth and Planetary Science Letters*, **112**, 131–145.

LE DOUARAN, S. & PARSONS, B. 1982. A note on the correction of ocean floor depths for sediment loading. *Journal of Geophysical Research*, **87**, 4715–4722.

MCKENZIE, D. 1994. The relationship between topography and gravity on Earth and Venus. *Icarus*, **112**, 55–88.

MCKENZIE, D. & FAIRHEAD, D. 1997. Estimates of the effective elastic thickness of the continental lithosphere from Bouguer and free air gravity anomalies. *Journal of Geophysical Research*, **102**, 27523–27552.

MILTON, N.J., BERTRAM, G.T. & VANN, I.R. 1990. Early palaeogenetectonics and sedimentation in the Central North Sea. *In*: HARDMAN, R.F.P. & BROOKS, J. (eds) *Tectonic Events Responsible for Britain's Oil and Gas Reserves*. Geological Society, London, Special Publications, **55**, 339–351.

MÜLLER, R.D., ROEST, W.R., ROYET, J.-Y., GAHAGAN, L.M. & SCLATER, J.G. 1997. Digital isochrons of the world's ocean floor. *Journal of Geophysical Research*, **102**, 3211–3214.

NADIN, P., KUSZNIR, N. & CHEADLE, M. 1997. Early Tertiary plume uplift of the North Sea and Faroe–Shetland Basins. *Earth and Planetary Science Letters*, **148**, 109–127.

PARSONS, B. & SCLATER, J.G. 1977. An analysis of the variation of ocean floor bathymetry and heat flow with age. *Journal of Geophysical Research*, **82**, 803–827.

READING, H.G. 1991. The classification of deep-sea depositional systems by sediment calibre and feeder system. *Journal of the Geological Society, London*, **148**, 427–430.

REYNOLDS, R. 1994. Quantitative analysis of submarine fans in the Tertiary of the North Sea Basin. *Marine Petroleum Geology*, **11**, 202–207.

RIBE, N.M. & DELATTRE, W.L. 1998. The dynamics of plume–ridge interaction—III. The effects of ridge migration. *Geophysical Journal International*, **133**, 511–518.

ROHRMAN, M. & VAN DER Beek, P. 1996. Cenozoic postrift domal uplift of North Atlantic margins: an asthenospheric diapirism model. *Geology*, **10**, 901–904.

ROWLEY, E.J. 1998. *Quantifying Cenozoic exhumation across the British Isles*. PhD dissertation, University of Cambridge.

ROWLEY, E. & WHITE, N. 1998. Inverse modelling of extension and denudation in the East Irish Sea and surrounding areas. *Earth and Planetary Science Letters*, **161**, 57–71.

SANDWELL, D.T. & SMITH, W.H.F. 1997. Marine gravity anomaly from Geosat and ERS-1 satellite altimetry. *Journal of Geophysical Research*, **102**, 10039–10054.

SCLATER, J.G. & CHRISTIE, P.A.F. 1980. Continental stretching: an explanation of the post-mid-Cretaceous subsidence of the Central North Sea basin. *Journal of Geophysical Research*, **85**, 3711–3739.

SCLATER, J.G., LAWVER, L.A. & PARSONS, B. 1975. Comparison of long wavelength residual elevation and free-air gravity anomalies in the North Atlantic and possible implications for the thickness of the lithospheric plate. *Journal of Geophysical Research*, **80**, 1031–1052.

SMITH, W.H.F. & WESSEL, P. 1990. Gridding with continuous curvature splines in tension. *Geophysics*, **55**, 293–305.

THOMPSON, R.N. 1974. Primary basalts and magma genesis, I: Skye, Northwest Scotland. *Contributions to Mineralogy and Petrology*, **45**, 317–341.

WATSON, S. & MCKENZIE, D. 1991. Melt generation by plumes: a study of Hawaiian volcanism. *Journal of Petrology*, **32**, 501–537.

WEIR, N.R.W., WHITE, R.S., BRANDSDÓTTIR, B., EINARSSON, P., SHIMAMURA, H., SHIOBARA, H. & RISE FIELDWORK TEAM 2001. Crustal structure of the northern Reykjanes Ridge and Reykjanes Peninsula, south-west Iceland. *Journal of Geophysical Research*, **106**, 6347–6368.

WHITE, N. & LOVELL, B. 1997. Measuring the pulse of a plume with the sedimentary record. *Nature*, **387**, 888–891.

WHITE, R.S. 1997. Rift–plume interaction in the North Atlantic. *Philosophical Transactions of the Royal Society of London, Series A*, **355**, 319–339.

WHITE, R.S., MCKENZIE, D. & O'NIONS, R.K. 1992. Oceanic crustal thickness from seismic measurements and rare earth element inversions. *Journal of Geophysical Research*, **97**, 19683–19715.

Timing and mechanisms of North Atlantic Cenozoic uplift: evidence for mantle upwelling

MAX ROHRMAN[1,2], PETER A. VAN DER BEEK[3],
ROB D. VAN DER HILST[4] & PAUL REEMST[5]

[1]*Landmark Graphics, Stavanger, Norway*

[2]*Present address: Shell UK Exploration & Production, 1 Altens Farm Rd, Nigg, Aberdeen AB12 3FY, UK (e-mail: max.rohrman@expro.shell.co.uk)*

[3]*Laboratoire de Géodynamique des Chaînes Alpines, Université Joseph Fourier, Grenoble, France*

[4]*Department of Earth, Atmospheric and Planetary Sciences, MIT, Cambridge, MA 02139, USA*

[5]*Geologica AS, Stavanger, Norway (present address: NAM, Assen, The Netherlands)*

Abstract: Postrift domal uplift patterns are a distinct feature of northern North Atlantic margins. On the basis of apatite fission-track data, offshore seismic stratigraphy, geomorphology, gravity and seismic tomography, we argue that southern Norway is characterized by predominantly Neogene domal uplift. The uplift is tectonically driven and estimated at around 1.5 km. Low flexural rigidity (*c.* 10^{22} N m) and corresponding equivalent elastic thickness T_e (*c.* 15 km) values for the southern Norwegian lithosphere indicate that the lithosphere is relatively weak. Additionally, high temperature estimates derived from low-velocity mantle P- and S-wave seismic tomography below the dome suggest a thermal anomaly at depth. Therefore, the observed topography is most plausibly explained by mantle upwelling. This would supercede other previously proposed primary mechanisms such as eustasy, isostatic readjustment to glacial erosion, magmatic underplating and intraplate compression. Currently available data suggest similar processes for other uplifted regions such as Spitsbergen, northern Norway, the British Isles and parts of East Greenland.

During recent decades it has become apparent that the North Atlantic margins experienced substantial vertical movements in Cenozoic time (e.g. White & Lovell 1997; Doré *et al.* 1999; Japsen & Chalmers 2000) characterized by the coupled emergence of rift margins and anomalous subsidence of the nearby basins (e.g. Cloetingh *et al.* 1990). However, the mechanism as well as its temporal and spatial resolution has been a matter of debate. Recently, some progress has been made with respect to timing, and two phases that have influenced the whole North Atlantic region have been identified (e.g. Riis 1996; Martinsen *et al.* 1999). The first is a Paleogene phase primarily associated with Eocene rifting and contemporaneous volcanism generated by the Iceland mantle plume (e.g. Doré *et al.* 1999). There is considerable evidence that this event affected all regions around the North Atlantic (Britain, Norway and Greenland), although its magnitude differed from region to region (e.g. Green *et al.* 1993; Riis 1996; Dam *et al.* 1998; Doré *et al.* 1999). The second is a Neogene phase (Rohrman & van der Beek 1996; Japsen & Chalmers 2000), which has a more enigmatic cause. This event is characterized by substantial uplift of Britain, parts of Norway, Spitsbergen, the Faeroes and Greenland (Japsen & Chalmers 2000) and by associated subsidence of nearby basins (e.g. North Sea, Møre Basin, offshore Greenland, Sørvestnaget Basin, Rockall Trough, Porcupine Basin). Because of the association of uplift and subsidence, the mechanism driving these processes is probably tectonic (e.g. Doré *et al.* 1999; Japsen & Chalmers 2000). Various mechanisms have been proposed including intraplate compression (Cloetingh *et al.* 1990; Doré *et al.* 1999), mantle phase changes (Riis & Fjeldskaar 1992), magmatic underplating (Cox 1993; Brodie & White 1995) and small-scale asthenospheric convection (Vågnes & Amundsen 1993; Rohrman & van der Beek 1996; Stuevold & Eldholm 1996), among others.

In this paper, we briefly review proposed mechanisms for uplift around the North Atlantic.

From: DORÉ, A.G., CARTWRIGHT, J.A., STOKER, M.S., TURNER, J.P. & WHITE, N. 2002. *Exhumation of the North Atlantic Margin: Timing, Mechanisms and Implications for Petroleum Exploration*. Geological Society, London, Special Publications, **196**, 27–43. 0305-8719/02/$15.00

Subsequently, we focus on southern Norway, and review the evidence for the timing and magnitude of Neogene uplift and denudation. We calculate tectonic uplift from the elevation of plateau-surface remnants and an estimate of flexural rigidity from the coherence between Bouguer gravity and topography. These results are then used to differentiate between mechanisms proposed. Additionally, we present seismic tomographic images of P-wave velocities in the upper mantle around the North Atlantic and correlate these with other data, to arrive at the most plausible mechanism operating during Neogene time.

Mechanisms proposed

Early studies were mainly focused on Norway and primarily based on geomorphological observations (e.g. Holtedahl 1953; Torske 1972). The rugged mountains of southern Norway, spectacular fjords and preservation of plateau surfaces at high altitude (Gjessing 1967; Peulvast 1985; Riis 1996) strongly suggest a recent upwarp of the southern Norwegian landmass. Intensive hydrocarbon exploration of offshore Norway has provided further evidence from observed structural basinward dip of pre-Neogene strata and build-up of large clastic wedges (Jordt *et al.* 1995). Palaeogeographical reconstructions have added additional evidence for a late emergence (Doré 1992a, 1992b). Apatite fission-track (AFT) data constrain denudation to be a mainly Neogene event, starting at *c.* 30 Ma. Denudation occurred in a dome-like pattern with the amplitude of denudation decreasing radially outward from a maximum of 2.0 ± 0.5 km at the centre (Rohrman *et al.* 1995). The timing of the onset of denudation is consistent with offshore stratigraphic evidence (Rundberg & Smalley 1989; Jordt *et al.* 1995).

Any mechanism proposed, to be successful, has to explain this timing of events (van der Beek & Rohrman 1997). Some workers favour a Paleogene onset of uplift for southern Norway, but most available evidence strongly suggests a primarily Neogene event (Rundberg & Smalley 1989; Jensen & Schmidt 1993; Jordt *et al.* 1995; Rohrman *et al.* 1995; Riis 1996; Martinsen *et al.* 1999).

Early interpretations suggested that uplift of southern Norway and other margins around the North Atlantic was associated with Paleocene–Eocene break-up and plume activity (e.g. Torske 1972; Cox 1993). However, timing of domal uplift at *c.* 30 Ma, i.e. 25–30 Ma after onset of Eocene volcanism and rifting, precludes significant synrift uplift. Moreover, dynamic plume-generated uplift is transient, and should reverse into subsidence after break-up. Permanent uplift can be generated by magmatic underplating, but in the case of southern Norway there is no sign of onshore Cenozoic intrusions. Eocene magmatic activity took place 300–400 km offshore.

Riis & Fjeldskaar (1992) proposed that most of the Norwegian uplift was caused by Pliocene–Pleistocene isostatic readjustment to glacial erosion. However, their study showed that additional tectonic uplift is required to explain the present-day elevation of morphological surfaces. They explained this by mantle phase transitions as a result of erosional unloading, but the dynamics of mantle phase changes driving this tectonic uplift component are at present poorly understood. Non-tectonic mechanisms that have been suggested (Eyles 1996) focus on climatic deterioration and sea-level changes. However, the kilometre-scale magnitude of uplift is not readily explained by eustasy.

Others have proposed that Neogene uplift of western Fennoscandia is a result of thermal instability caused by large horizontal temperature gradients between the Fennoscandian mainland and the Norwegian–Greenland Sea (Theilen & Meissner 1979; Peulvast 1985; Vågnes & Amundsen 1993). This would set up secondary convection in the sub-lithospheric mantle, causing the hot asthenosphere to rise and subsequently generate tectonic uplift. This could be a viable model, and will be discussed in the following sections.

Finally, various workers have drawn attention to the synchronous timing of margin uplift and anomalous basin subsidence around Norway and the whole North Atlantic, suggesting that coupled uplift and subsidence is flexural in nature and induced by intraplate stress fluctuations (Cloetingh *et al.* 1990). However, flexure-induced uplift is not consistent with the observed correlation between topography and Bouguer gravity anomalies (up to −80 mgal) below the uplifted regions (Fig. 1). Although the gravity is largely isostatically compensated (Balling 1980), there is a mass deficit below the regions of highest elevation that cannot be explained by topography and differences in crustal thickness.

Timing of denudation on the eastern Atlantic margin

There are two principal types of constraint on the timing of Norwegian denudation: the structural

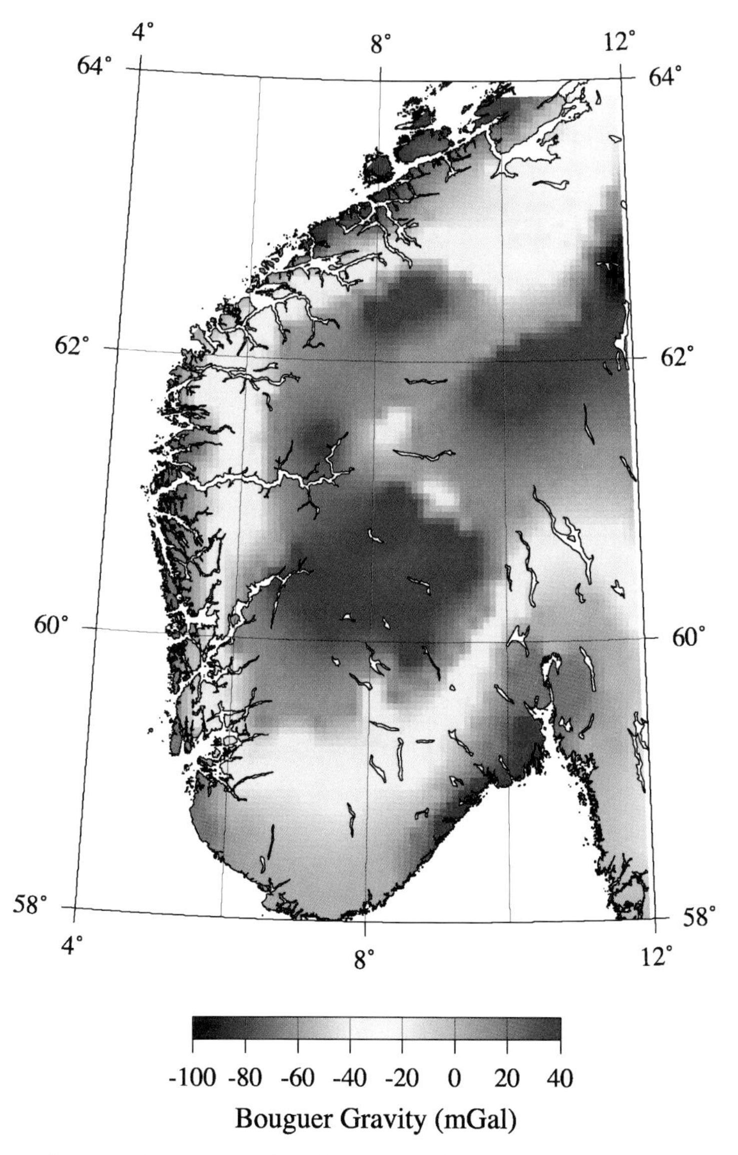

Fig. 1. Bouguer gravity anomaly map for southern Norway, data from Sveriges Geologiske Undersökelse (1985).

and stratigraphic relationship of Cenozoic strata offshore (Jordt *et al.* 1995) and planation surfaces onshore, and the AFT analysis of the Norwegian basement (Rohrman *et al.* 1995). AFT data from southern Norway define a structural dome with the youngest ages increasing radially from *c.* 100 Ma at sea level in the inner fjords, to *c.* 170 Ma at the top of the Jotunheimen peaks and around 200 Ma at elevations less than 500 m near the shorelines (Fig. 2). Mean track length distributions are more variable, but the younger ages generally correspond to low mean track lengths (*c.* 11.6 μm) and a lack of long (recent) tracks (Fig. 2). The latter indicate fast cooling from temperatures at the lower end of the partial annealing zone (*c.* 60–70 °C) to surface temperatures. Although our youngest AFT samples yield mixed ages, it is possible to extract thermal history information by using state-of-the-art modelling techniques. The youngest AFT samples (AFT age *c.* 100 Ma, mean track lengths 11.6 μm) suggest a predominantly Neogene onset of denudation (Rohrman *et al.* 1995).

Northern Norway shows similar AFT basement ages although they tend to be somewhat older (around 160 Ma) with distributions showing longer track lengths (Hendriks & Andriessen 2002). Meanwhile, Neogene clastic wedges offshore from Lofoten indicate more recent erosion (Mokhtari & Pegrum 1992). A recent AFT study of Spitsbergen by Blythe & Kleinspehn (1998) placed significant denudation around 36 Ma, which was mainly attributed by those workers to rift flank uplift, followed by Pliocene–Holocene glacial erosion. The data pattern yields strongly varying AFT ages (*c.* 27–56 Ma), sometimes within the same locality. This pattern suggests that the samples experienced other processes than simple uplift and erosion, possibly indicating migration of hot fluids. This interpretation is supported by the resetting of zircon fission-track ages and vitrinite reflection data, in the proximity of unreset samples at similar elevations. A possible explanation for these patterns might be volcanic and hydrothermal activity during Miocene and Quaternary time (e.g. Vågnes & Amundsen 1993). Similar fission-track patterns have been observed in the Permian Oslo Rift, where igneous activity is well documented (e.g. Rohrman *et al.* 1994). However, at sea level there seems to be a slight trend from west to east along the Longyearbyen fjord (Central Spitsbergen), which suggests a decrease in AFT ages from around 50 Ma near the basement on the west coast to 38 Ma in the Eocene deltaic strata of the Central Basin. This concurs with earlier studies (Nøttvedt *et al.* 1992), suggesting Neogene domal uplift based on geological observations, vitrinite reflectance and seismic velocity data. Evidently, more AFT samples of east Spitsbergen are needed to verify or refute this assumption.

Much of the British Isles lacks the high rugged topography of southern Norway. Elevations are more subdued and do not exceed 1 km in large parts of England, Wales and Ireland. Only the Scottish highlands reach higher elevations, up to 1343 m. Published AFT data from Scotland

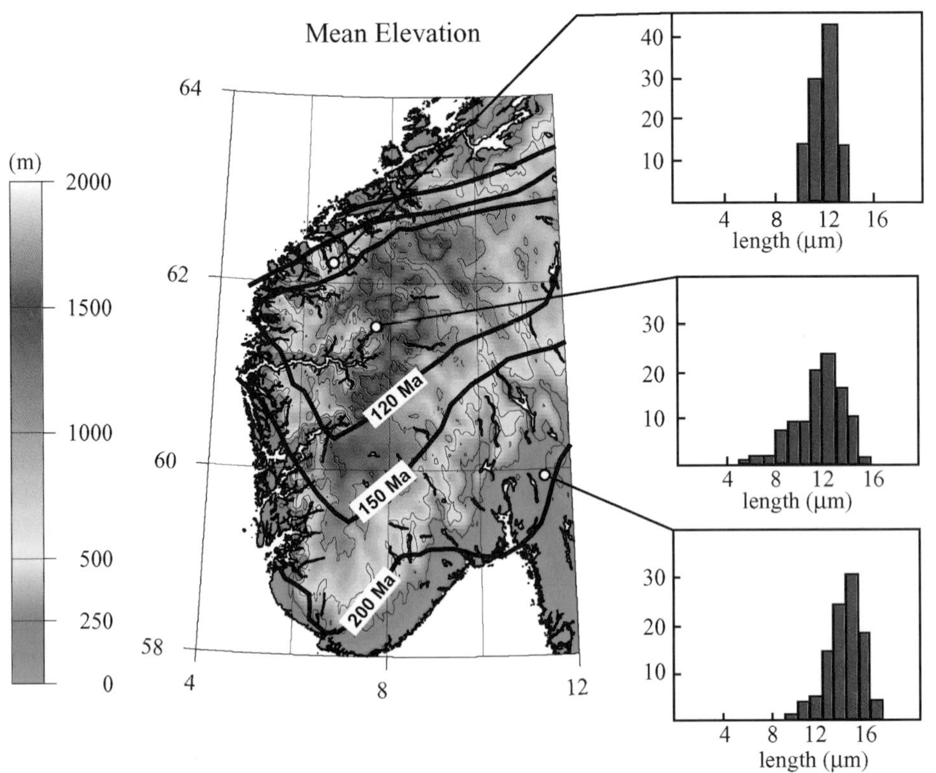

Fig. 2. Mean elevation and AFT isochrons of southern Norway drawn at sea level, defining a structural dome concordant with topography, from Rohrman *et al.* (1995). Typical track length distributions for selected samples are shown on the right.

(Lewis *et al.* 1992) suggest predominantly Triassic–Jurassic ages. Only near the Eocene intrusions are younger ages found. However, data are too sparse to infer any details for this region. AFT data from northern England and the Irish Sea area show AFT ages between 45 Ma and >400 Ma (Green *et al.* 1993, 1997, 2001), suggesting significant exhumation around 60 Ma. This uplift is possibly centred in the East Irish Sea and decreasing to the SE, concordant with the Mesozoic sedimentary outcrop pattern of southeastern England. Recently, Japsen (1997) suggested evidence for a Neogene denudation phase based on AFT data and compaction studies from eastern England and the western North Sea. He estimated that Paleogene and Neogene denudation were equal in magnitude (around 1 km each). However, the spatial pattern of both Neogene and Paleogene denudation for the entire United Kingdom requires further study.

Moreover, the area of highest denudation (East Irish Sea) is at present below sea level, in contrast to the dome-shaped topography of the other regions (southern Norway, northern Norway and Spitsbergen).

Quantification of Neogene uplift and denudation in southern Norway

Regional changes in surface elevation are the combined result of tectonic uplift, erosion and the isostatic response to erosion. To use the present-day elevation of an uplifted region to constrain tectonic processes, the components of elevation change that result from erosion and isostatic rebound must be quantified (England & Molnar 1990; Gilchrist *et al.* 1994). The tectonic uplift (u_T) is related to the present-day elevation (H_0) and the amount of denudation (ΔE) by

$$u_T = H_0 - H_i + \Delta E - I \quad (1)$$

where H_i is the initial elevation and I is isostatic rebound:

$$I = \left(\frac{\Delta E \rho_c}{\rho_a}\right)\left(1 + \frac{D k_E^4}{(\rho_a - \rho_c)\mathbf{g}}\right)^{-1} \quad (2)$$

where ρ_c is the density of the eroded crustal section, ρ_a is sublithospheric mantle density, k_E is the spatial wavenumber of erosional unloading, $\mathbf{g}$ is acceleration of gravity and D is flexural rigidity (van der Beek *et al.* 1994; see Table 1). As $D \rightarrow 0$, equation (2) simplifies to $I = \Delta E \rho_c / \rho_a$, the local isostatic solution. As $D \rightarrow \infty$ (for an infinitely strong lithosphere), $I \rightarrow 0$. A regional analysis of tectonic uplift therefore requires an estimate of H_i and D, as well as the ability to map out spatial variations in ΔE (Abbott *et al.* 1997; Small & Anderson 1998).

Whereas the AFT thermochronological data discussed above give us a high temporal resolution to decipher the denudation history of southern Norway, the spatial resolution of our data is rather coarse. We therefore use the elevation of preserved remnants of a plateau surface (the 'Palaeic surface') to spatially constrain the amounts of denudation, isostatic rebound, and tectonic uplift in southern Norway. The correlation of plateau remnants and their use in reconstructing landscape development is a relatively hazardous undertaking because of the general lack of temporal constraints (Brown *et al.* 1999; Summerfield 1999). In Norway, as elsewhere, there is no consensus on the age of the surface remnants, which have been varyingly interpreted as being of Jurassic to Paleogene age, nor on their correlation (Gjessing 1967; Torske

Table 1. *Parameter values employed*

Symbol	Description	Value
ρ_c	Crustal density	$2600\ kg\ m^{-3}$
ρ_a	Asthenospheric density	$3250\ kg\ m^{-3}$
$\mathbf{g}$	Gravitational acceleration	$9.8\ m\ s^{-2}$
D	Flexural rigidity of the lithosphere	8.9×10^{21} to $7.2 \times 10^{22}\ N\ m$
E	Young's modulus	$10^{11}\ N\ m^{-2}$
ν	Poisson ratio	0.25
f	Ratio of surface to base loading of lithosphere	1
k	Thermal conductivity	$0.0006\ cal\ (cm\ s\ °C)^{-1}$
α	Thermal diffusivity	$0.01\ cm^2\ s^{-1}$
L	Lithospheric thickness after thinning	80–110 km

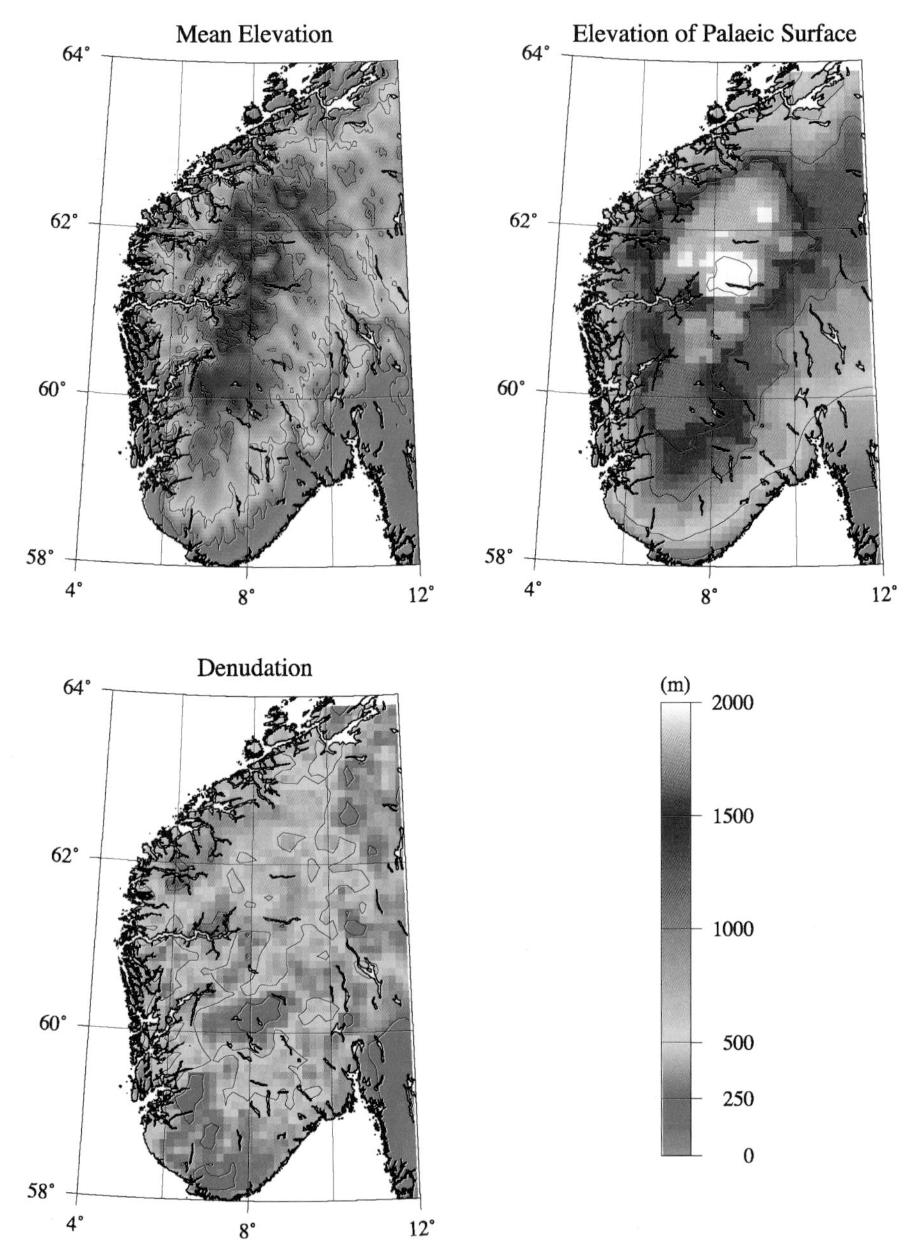

Fig. 3. Maps of mean elevation of southern Norway (from ETOPO-5 global topography data), elevation of the Palaeic surface (after Riis & Fjeldskaar 1992) and amount of erosion, calculated by subtracting mean elevation from the elevation of the Palaeic surface.

1972; Peulvast 1985; Doré 1992a; Riis & Fjeldskaar 1992; Riis 1996).

We follow the most recent correlation of Riis & Fjeldskaar (1992) and Riis (1996), who suggested the Palaeic surface to be of Paleogene age because (1) our fission-track data suggest that samples from close to the Palaeic surface in the Hardangervidda area reached surface temperatures in Paleogene times (Rohrman *et al.* 1995) and (2) the elevation of the Palaeic surface in the Jotunheimen area is consistent with the amount of Neogene denudation recorded by

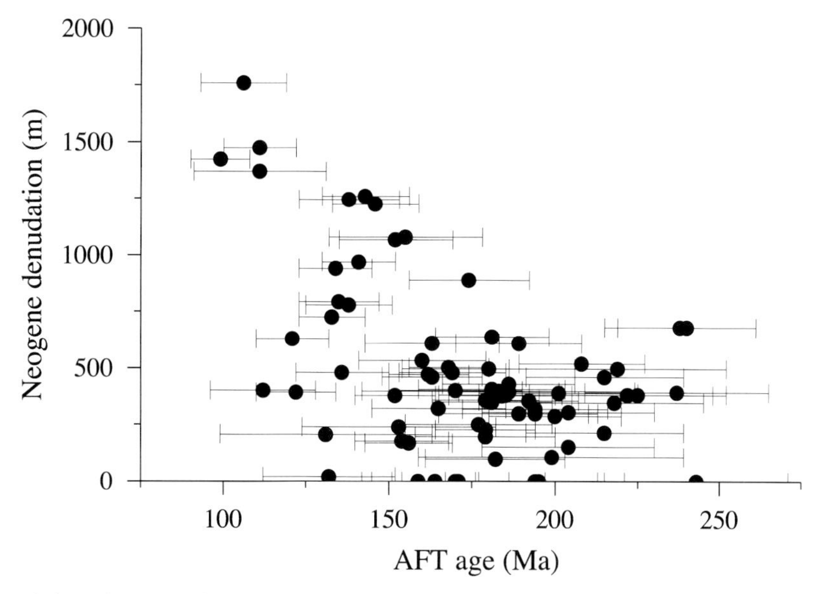

Fig. 4. Correlation of amount of Neogene denudation, calculated by subtracting present-day topography from elevation of Palaeic surface (see Fig. 3) with apatite fission-track ages (from Rohrman *et al.* 1994, 1995). Neogene denudation for fission-track samples is corrected for sample elevation with respect to mean elevation of the 5′ resolution topographic grid.

fission-track thermochronology of samples from close to sea level (2.0 ± 0.5 km; van der Beek 1995).

Figure 3 shows the present-day elevation of southern Norway as well as the elevation of the Palaeic surface. The latter was digitized from the map of Riis & Fjeldskaar (1992) and interpolated using a continuous curvature algorithm (Smith & Wessel 1990). The amount of denudation since the formation of the surface (i.e. since the end of Paleogene time) can be calculated by subtracting the present-day elevation H_0 from the elevation of the Palaeic surface. The result shows a mean amount of denudation of c. 400 m, with maxima of ≥1000 m in the inner fjords and along the northwestern coast. The pattern of denudation that is calculated in this manner is largely consistent with the pattern deduced from fission-track thermochronology. Figure 4 shows the correlation between the amount of denudation calculated using this approach and the AFT ages from Rohrman *et al.* (1995). Although the fission-track ages are mixed ages, with minima around 100 Ma, a clear trend emerges with samples with the youngest fission-track ages encountered in the regions of maximum Neogene denudation.

Late Paleocene and early Eocene marine diatoms are encountered on planation surfaces in Sweden and Finland (Fenner 1988), which may be correlated with the Palaeic surface in Norway. On the basis of this observation, Riis (1996) suggested that the surface was at or near sea level in early Cenozoic time. This assertion is consistent with the offshore sedimentation data (Jordt *et al.* 1995), which suggest very low sediment input from Fennoscandia before late Oligocene time. We therefore assume that the Palaeic surface can be used as a marker not only of Neogene denudation, but also of uplift.

If we suppose $H_i = 0$ in equation (1), then the present-day elevation of the Palaeic surface represents $H_0 + \Delta E = u_T + I$. In the case of local isostasy ($D = 0$), the amount of isostatic rebound reduces to $I = \Delta E \rho_c / \rho_a$ (*c.* $0.8\Delta E$ for the values in Table 1) and can be calculated directly from the pattern of denudation (Fig. 3). The tectonic uplift u_T then equals the present-day elevation of the Palaeic surface minus the isostatic uplift I. Figure 5 shows calculated amounts of Neogene tectonic uplift and isostatic rebound for the local isostatic case. Tectonic uplift in this case has the same domal pattern as the mean elevation and the elevation of the Palaeic surface, reaching maximum values of *c.* 1500 m for the high mountain areas (where Neogene denudation is negligible), whereas isostatic rebound reaches a maximum of *c.* 750 m in the inner fjords and on the NW coast. However, because the isostatic response to denudation is generally regional instead of local, the pattern and the amount of tectonic uplift will

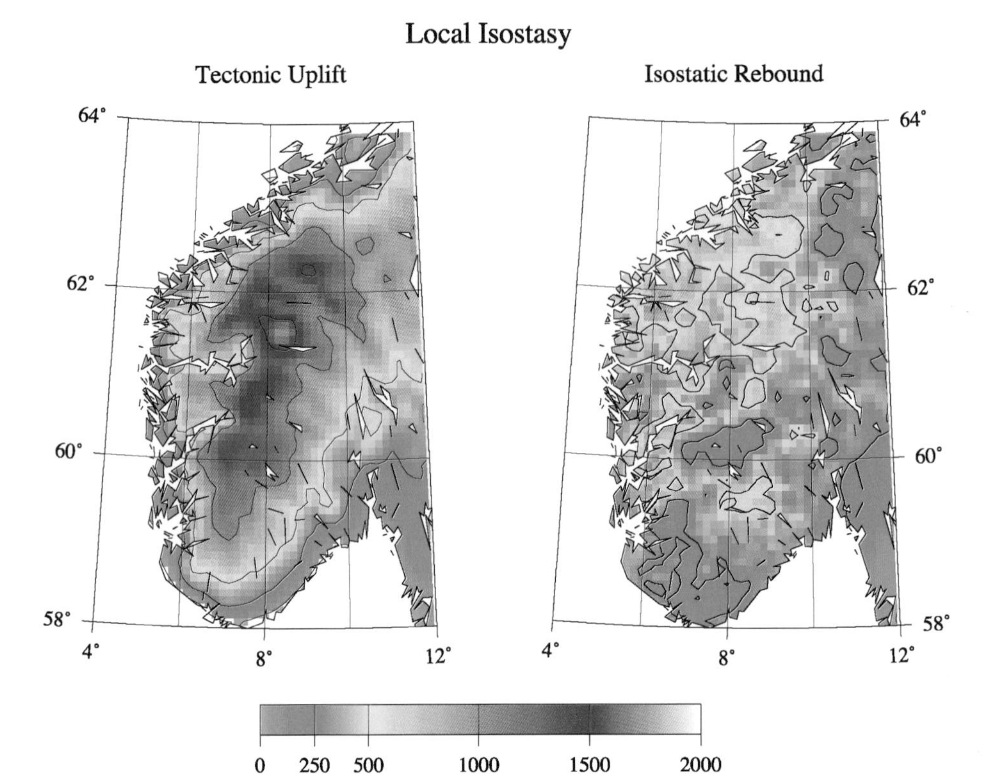

Fig. 5. Calculated tectonic uplift and isostatic rebound for a model of local isostatic response to erosion. In this case isostatic rebound equals 0.8 times the amount of erosion; tectonic uplift was calculated by subtracting this amount from the elevation of the Palaeic surface (see text).

also be a function of the flexural rigidity D of the southern Norwegian lithosphere.

Estimating flexural rigidity

To make a realistic assessment of the amount of tectonic uplift, the flexural rigidity, or corresponding equivalent elastic thickness T_e, must be estimated. Flexural rigidity may be readily estimated by an analysis of the coherence of topography and Bouguer gravity anomalies. Gravity and topography will be coherent at long wavelengths but not at shorter ones, the wavelength at which coherence breaks down being dependent on D. The observed coherence γ_0^2 is defined as (Forsyth 1985)

$$\gamma_0^2 = \frac{C^2_{(\bar{k})}}{E_{0(\bar{k})}E_{1(\bar{k})}} \tag{3}$$

where $E_{0(k)}$ is the average power of topography for a discrete wavenumber, $E_{1(k)}$ is the average power of gravity for that wavenumber, $C_{(k)}$ is the cross-spectral power of gravity and topography and the overbar indicates averaging over a wavenumber band. For a mechanically anisotropic lithosphere a similar concept can be used (Simons *et al.* 2000), but in this analysis we assume isotropy. The theoretical coherence for a plate loaded both at its surface and base is given by (Forsyth 1985)

$$\gamma^2 = \frac{\langle H_T W_T \rangle \langle H_B W_B \rangle}{\langle H_T^2 + H_B^2 \rangle \langle W_T^2 + W_B^2 \rangle} \tag{4}$$

where H_T is the amplitude of surface deflection as a result of surface loading, W_T is the amplitude of deflection of the base of the plate (i.e. the Moho) as a result of surface loading, H_B the amplitude of surface deflection resulting from loading at the base of the plate and W_B the amplitude of Moho deflection resulting from base loading. H_T, H_B, W_T and W_B depend on the wavenumber k and flexural rigidity D

via

$$H_{B(k)} = f\rho_c H_{T(k)}/\psi\Delta\rho$$

$$W_{T(k)} = -\rho_c H_{T(k)}/\psi\Delta\rho \quad (5)$$

$$W_{B(k)} = -\rho_c H_{B(k)}\phi/\Delta\rho$$

with f the ratio of surface to base loading; $\Delta\rho = (\rho_a - \rho_c)$; $\phi = 1 + (Dk^4/\rho_c \mathbf{g})$ and $\psi = 1+ (Dk^4/\Delta\rho\mathbf{g})$. The flexural rigidity D is related to the equivalent elastic thickness T_e by

$$D = \frac{ET_e^3}{12(1-\nu^2)} \quad (6)$$

where E is Young's modulus and ν is the Poisson ratio (see Table 1).

The gravity data we used for this analysis are from the Sveriges Geologiske Undersökelse 1985; Fig. 1) and topography from the ETOPO-5 database. Gravity and topography data were projected onto a 500 km × 720 km UTM grid. Data were clipped in the deep offshore areas and tapered towards the mean value along the sides of the grid, before being transformed into the wavenumber domain using a fast Fourier transform algorithm (see van der Beek 1995).

Results of the coherence analysis are shown in Fig. 6. Although a best-fitting f and T_e can be estimated independently from the data using a least-squares criterion (Forsyth 1985), we feel that this may put too much emphasis on a best-fitting number for the present data quality. Visual inspection indicates a best-fit T_e between 10 and 20 km ($D = 8.9 \times 10^{21}$ to 7.1×10^{22} N m), assuming $f = 1$. This value is at the low end of estimates of flexural rigidity for Fennoscandia from glacial rebound studies, which constrain T_e to be ≤ 50 km (Fjeldskaar & Cathles 1991). It is, however, consistent with regional T_e estimates from the post-glacial tilt of palaeoshorelines (Fjeldskaar 1997), which suggested $T_e \leq 20$ km, as well as with an independent coherence study by Poudjom Djomani *et al.* (1999), who found $8 \leq T_e \leq 18$ km for southern Norway.

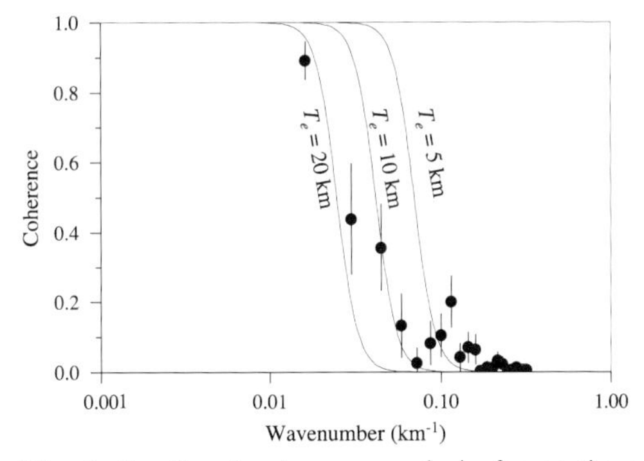

Fig. 6. Results of coherence analysis for southern Norway. Dots indicate observed coherence for average wavenumber bands, bars denote 1σ errors. Continuous lines are the various equivalent elastic thicknesses T_e.

We have calculated the isostatic response to denudation adopting constant T_e values of 10 and 20 km, using equation (2). The predicted isostatic rebound has a much smoother pattern than the local isostatic case, reaching a maximum of *c.* 700 m near the NW coastline (Fig. 7). The magnitude of isostatic rebound is, however, not much smaller than for the local isostatic case, as the inferred flexural rigidity is not very large. Resulting tectonic uplift patterns are also shown in Fig. 7; the models incorporating flexural rigidity show a similar dome-shaped uplift pattern as the local isostatic case. Uplift is centred on the regions of highest present-day elevation, reaching a maximum of *c.* 1500 m. This surprisingly high value is relatively insensitive to the adopted flexural rigidity because, as D increases, the amount of isostatic rebound decreases but also spreads more toward the regions of highest present-day elevation and contributes to the uplift of these regions. The critical assumption in this analysis is that the Palaeic surface was at sea level before Neogene uplift; if it was at some initial elevation H_i the inferred Neogene uplift will be overestimated by the same amount (see equation (1)).

Southern Norway: mechanisms of uplift

The relatively low flexural rigidity of the lithosphere (*c.* 10^{22} N m) and high tectonic uplift strongly suggest an endogenous cause for the uplift. This leaves us with essentially two possible solutions: intraplate compression and dynamic mantle upwelling. Intraplate processes require positive gravity anomalies below the dome regions, in contrast to the observations. Furthermore, intraplate compression cannot generate the amount of uplift required, as it is an order of magnitude too low (van der Beek 1995).

Mantle upwelling has been proposed by various workers for the North Atlantic margins (e.g. Vågnes & Amundsen 1993; Stuevold & Eldholm 1996; Rohrman & van der Beek 1996). The upwelling model seems plausible for Spitsbergen, where high heat flow (*c.* 130 mW m^{-2}) prevails and mantle xenoliths are present in Quaternary alkali basalts. This model seems less obvious for southern Norway, where low surface heat flow (*c.* 40 mW m^{-2})

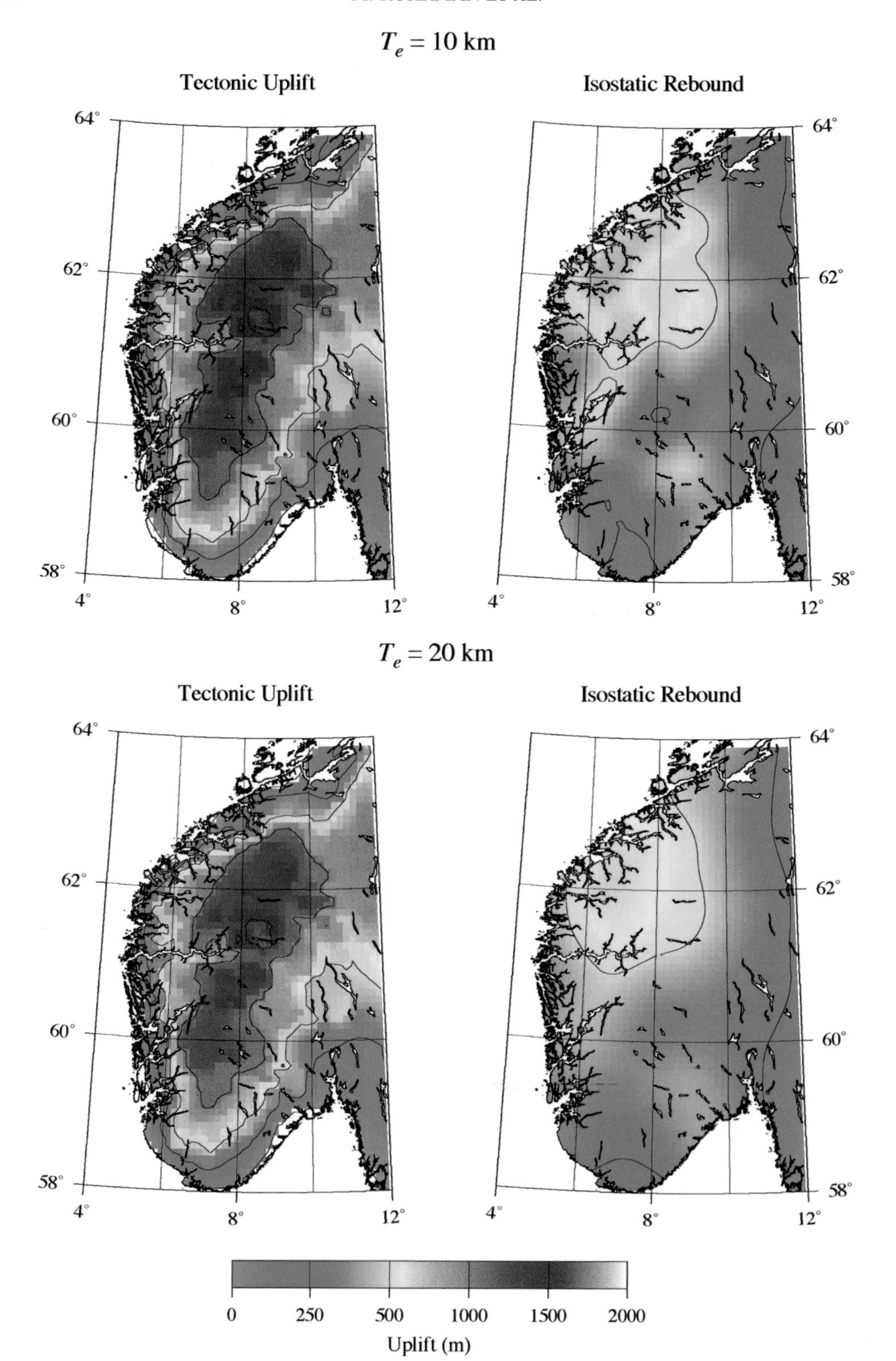

Fig. 7. Calculated tectonic uplift and isostatic rebound for a model of regional isostatic response to denudation and different flexural rigidities, corresponding to $T_e = 10$ km ($D = 8.9 \times 10^{21}$ Nm; top) and $T_e = 20$ km ($D = 7.1 \times 10^{22}$ N m; bottom). (See text for discussion.)

(Cermak 1979; Balling 1995) and lack of volcanism are not readily consistent with mantle upwelling. However, a recent study by Goes *et al.* (2000) proposed high sub-Moho temperatures in excess of 1000 °C based on P- and S-wave tomography below southern Norway (Fig. 8). These temperatures are similar to those in regions such as the Massif Central (Sobolev *et al.* 1996), where a mantle plume seems evident. If these temperature estimates are correct, the discrepancy between surface heat flow and elevated temperatures at depth can be explained by a time lag between equilibration of high lithospheric heat flow and surface heat flow (McGuire & Bohannon 1989). Assuming conductive heat loss, the change in surface heat flow Δq_o is related to a change in temperature at the base of the thinned lithosphere ΔT_l, the thickness of the lithosphere after thinning L, and the diffusion time t, according to

$$\Delta q_o = \frac{k\Delta T_l}{L}\left\{1 + 2\sum_{n=1}^{\infty}\cos(n\pi)\exp\left[\frac{-n^2\pi^2\alpha t}{L^2}\right]\right\} \quad (7)$$

where k is thermal conductivity and α the thermal diffusivity (Table 1). For L between 110 and 80 km and ΔT_l from 400 °C to 500 °C, it takes around 60 Ma to observe a *c.* 10 mW m^{-2} heat-flow change for southern Norway.

Upwelling material is most probably associated with advection, therefore we can use a 10 mW m^{-2} heat-flow change as a conservative estimate. As no regional heat-flow anomalies are observed in the southern Norwegian data, we assume that any dynamic mantle upwelling must be younger than 60 Ma. Along with the evidence of reduced P and S waves below southern Norway (Bannister *et al.* 1991; see also below), this strongly suggests that the southern Norwegian structural dome was generated by active mantle upwelling or diapirism in Neogene time.

Seismic tomography of the North Atlantic

One of the major advances in our understanding of the Earth has been the advent and subsequent development of seismic tomography, a class of imaging that now provides increasingly detailed 3D maps of seismic velocity variations that can be related to thermal and chemical variations in the mantle. Although it has proven relatively easy to image downwellings (i.e. subducting slabs), it seems much more difficult to image mantle upwellings (e.g. Grand *et al.* 1997). This is because upwellings are likely to occur in aseismic regions and are not as well sampled by seismic data, especially in the shallow mantle. Furthermore, use of first arrivals of seismic waves causes a natural bias toward fast anomalies, because annealing of wavefronts creates a tendency to underestimate slow anomaly amplitudes. Another important issue is that resolution depends on data coverage, which is uneven owing to the sparse distribution of sources (i.e. earthquakes) and receivers (i.e. seismological stations) and the 3D geometry of the ray paths along which the seismic waves propagate. Mantle structure has remained unresolved beneath large regions of the North Atlantic because of absence of recording stations and low levels of seismic activity.

With this in mind, we present P-wave velocity maps of the North Atlantic at various depths (Fig. 9), which provide a snapshot of the mantle

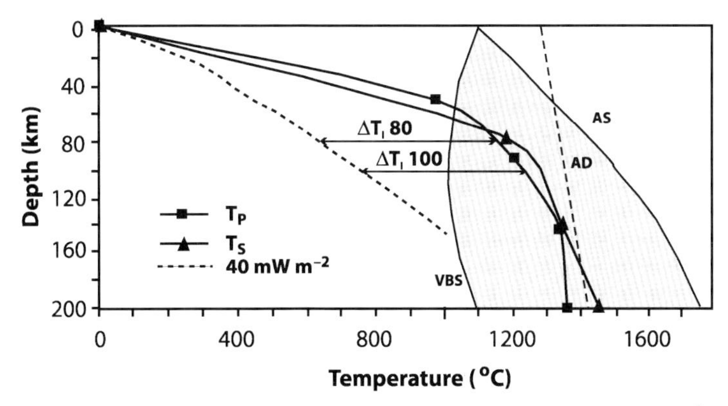

Fig. 8. Temperature–depth plot of three geotherms below southern Norway. The 40 mW m^{-2} geotherm (broken line) is based on surface heat-flow data. T_P and T_S are geotherms derived from inversion of P-wave and S-wave seismic velocities, respectively (Goes *et al.* 2000; Goes pers. comm.). The relative convergence of T_P and T_S should be noted. VBS, volatile-bearing peridotite solidus; AS, anhydrous peridotite solidus; AD, adiabat.

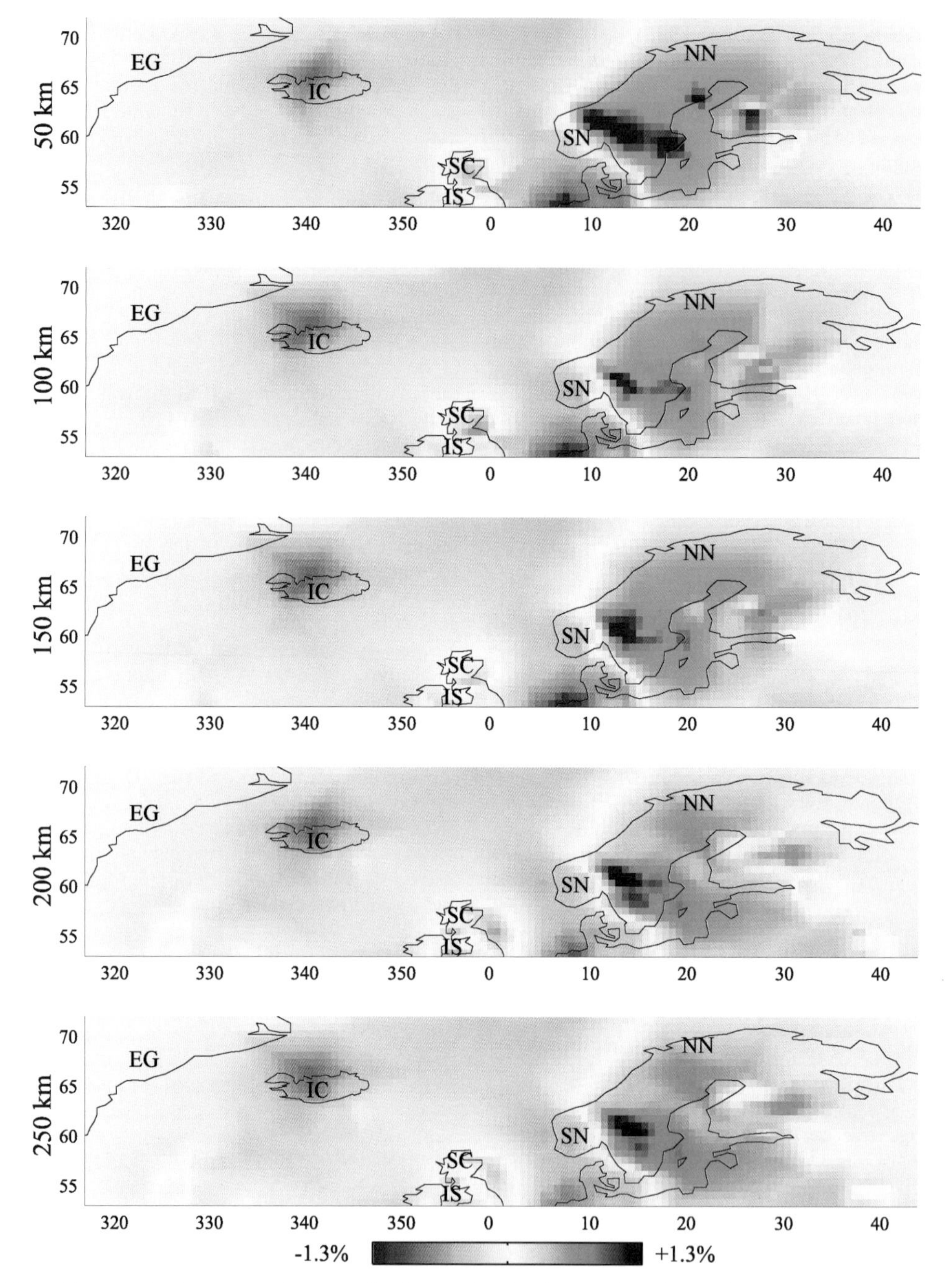

Fig. 9. P-wave seismic tomography maps for the North Atlantic at various depths. For the upper mantle low velocities (red) correspond to high temperatures, whereas high velocities (blue) correspond to low temperatures. EG, East Greenland; IC, Iceland; SN, southern Norway; NN, northern Norway; IS, Irish Sea; SC, Scotland.

as it is today. The image (Fig. 9) depicts P-wave velocity maps through the global model of Kárason & van der Hilst (2000) for the North Atlantic region. Figure 10 shows a resolution test for the velocity map at 150 km, using a chequerboard model. Resolution is rather poor below East Greenland (EG) and most of the North Atlantic Ocean. A better

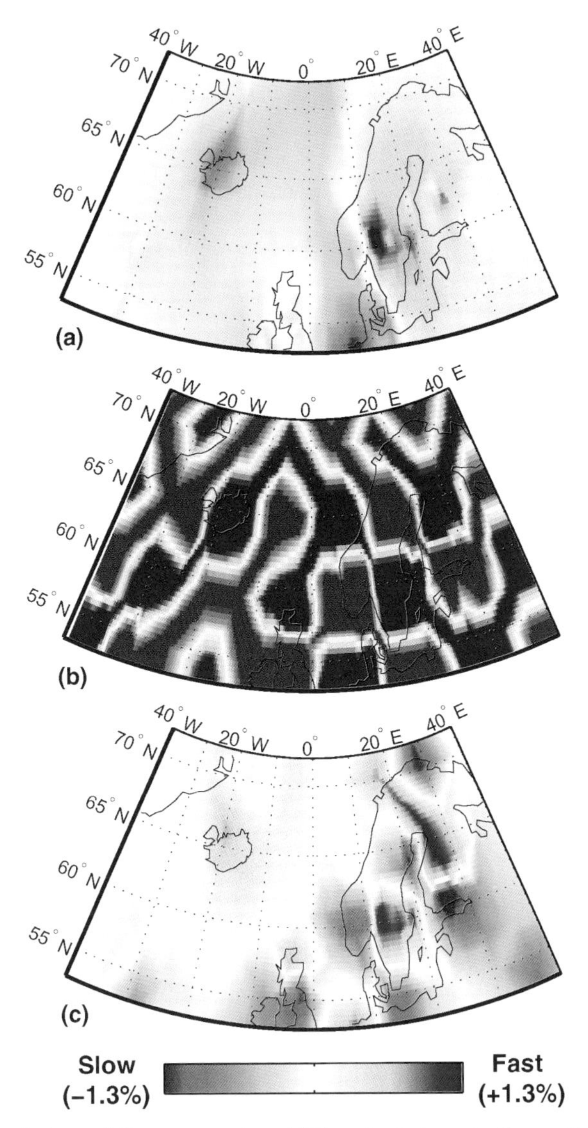

Fig. 10. (**a**) Lateral variations of P-wave speed at 150 km depth beneath the northern Atlantic. (**b**) Input 'chequerboard' model for resolution test. (**c**) Result of 'chequerboard' resolution test; as expected, our ability to image structure in the upper mantle beneath the oceanic regions is very poor owing to absence of earthquake sources and seismological stations, but beneath southern Norway the lateral resolution is reasonable. Vertical resolution is poor because of the small incidence angles of the seismic waves used for imaging in this region.

defined low-velocity anomaly is detected below Iceland (IC). This anomaly extends from near the surface to *c*. 400 km, but a deeper structure could have been overlooked owing to insufficient data coverage (e.g. Keller *et al.* 2000). Below southern Norway (SN) a low-velocity anomaly is observed that extends from >50 km depth to >250 km. The vertical resolution is poor in this part of the model, but the slow anomaly agrees well with previous P- and S-wave tomographic studies (Bannister *et al.* 1991; Zielhuis & Nolet 1994a, 1994b; Marquering & Snieder 1996; Bijwaard *et al.* 1998). Although this is not likely to be well resolved, the model suggests that a thin high-velocity layer overlies a low-velocity anomaly below southern Norway, which is consistent with our estimates of T_e and suggests a thinned and relatively weak lithosphere. Farther eastward we encounter the Baltic shield, evident as a thick (>250 km) high-velocity area. The change from a low-velocity zone in the upper mantle of southern Norway (SN) to the high velocities of the Baltic shield is dramatic and suggests some fundamental process that has so far been neglected in our current view of plate tectonics. We remark that similar features have been observed in southeastern Australia (Zielhuis & van der Hilst 1996). Here, a low-velocity upper-mantle anomaly is depicted below the mountains (up to 2230 m) of Eastern Australia, whereas the low-altitude (<200 m) shield towards the west shows higher velocities.

Regional tomographic studies (Bannister *et al.* 1991) indicate a low-velocity anomaly below northern Norway (NN), but in our global model resolution is too low to infer any details for this region. Similar resolution problems are observed below the British Isles, where faint low-velocity anomalies are present below the Irish Sea (IS) at *c.* 200 km depth. Low velocities below the Irish Sea were also observed in previous models (Marquering & Snieder 1996; Bijwaard *et al.* 1998). Scotland (SC) also suffers from the aforementioned resolution problems, but generally suggests higher velocities.

Discussion

Our results from integrating geological, AFT, geomorphological, Bouguer gravity and seismic tomography data strongly suggest that mantle upwelling is present below southern Norway. Furthermore, it seems evident that this upwelling was most active in Neogene time and directly responsible for generating most of the present-day topography. Whether other areas experienced similar upwellings is still uncertain, mainly because of lack of sufficient data. However, there are strong indications that similar mechanisms generated the domal topography of Spitsbergen, northern Norway and the East Greenland coast (Rohrman & van der Beek 1996). There are also some structural differences among the domes. Whereas Spitsbergen records Neogene volcanism, southern Norway is completely devoid of any Cenozoic magmatic activity. This is probably simply constrained by the amount of active lithosphere thinning by the upwelling. Spitsbergen's lithospheric thickness is estimated to be *c.* 50 km (Vågnes & Amundsen 1993), whereas the south Norwegian lithosphere is much thicker, probably *c.* 80–100 km. The Norwegian dome thus possibly represents an early stage of upwelling. Additionally, the P- and S-wave temperatures of Goes *et al.* (2000) for Europe suggest that divergence of T_P (P-wave temperature) and T_S (S-wave temperature) at depths <100 km could indicate the presence of melt, as shown for the Massif Central area. Southern Norway shows relative convergence between T_P and T_S (Fig. 8), possibly suggesting absence of melt and therefore no volcanism.

Apart from explaining observed uplift patterns, the mantle upwellings also offer a solution for offshore anomalous basin subsidence, where asthenospheric material flows from below the basins (North Sea, Møre Basin) toward the upwelling (southern Norway). Furthermore, in this interpretation the timing of upwelling becomes less of an issue, as the available evidence suggests that boundary conditions are generated by rifting over an anomalously hot asthenosphere. Subsequent uplift pulses might be inferred by changes in upwelling flux (Rohrman & van der Beek 1996; White & Lovell 1997).

Some concerns remain; mantle upwellings seem to follow a distinct sequence of events, starting with surface uplift, followed by basaltic volcanism and finally thermal subsidence. Alternatively, our results might suggest that not all upwellings cause surface volcanism, as upwelling temperature and velocity could be too low. Another question is the relationship between plate movement and upwelling. Traditional theory suggests that upwellings remain stationary with respect to each other, but our results suggest that this might not always be the case.

We thank Shell UK Exploration & Production for sponsoring the color figures in this article and Saski Goes for providing the temperature data below southern Norway

References

ABBOTT, L.D., SILVER, E.A., ANDERSON, R.S. & 7 OTHERS 1997. Measurement of tectonic surface uplift rate in a young collisional orogen. *Nature*, **385**, 501–507.

BALLING, N. 1980. The land uplift in Fennoscandia, gravity field anomalies and isostasy. *In*: MÖRNER, N.A. (ed.) *Earth, Rheology, Isostasy and Eustasy*. Wiley, Chichester, 297–321.

BALLING, N. 1995. Heat flow and thermal structure across the Baltic shield and northern Tornquist zone. *Tectonophysics*, **245**, 13–50.

BANNISTER, S.C., RUUD, B.O. & HUSEBYE, E.S. 1991. Tomographic estimates of sub-Moho seismic velocities in Fennoscandia and structural implications. *Tectonophysics*, **189**, 37–53.

BIJWAARD, H., SPAKMAN, W. & ENGDAHL, E.R. 1998. Closing the gap between regional and global travel time tomography. *Journal of Geophysical Research*, **103**, 30055–30078.

BLYTHE, A.E. & KLEINSPEHN, K.L. 1998. Tectonically versus climatically driven Cenozoic exhumation of the Eurasian plate margin, Svalbard: fission track analyses. *Tectonics*, **17**, 621–639.

BRODIE, J. & WHITE, N. 1995. The link between sedimentary basin inversion and igneous underplating. *In*: BUCHANAN, J.G. & BUCHANAN, P.G. (eds) *Basin Inversion*. Geological Society, London, Special Publications, **88**, 21–38.

BROWN, R.W., GALLAGHER, K., GLEADOW, A.J.W. & SUMMERFIELD, M.A. 1999. Morphotectonic evolution of the South Atlantic margins of Africa and South America. *In*: SUMMERFIELD, M.A. (ed.) *Global Tectonics and Geomorphology*. Wiley, Chichester, 257–283.

CERMAK, V. 1979. Heat flow map of Europe. *In*: CERMAK, V. & RYBACH, L. (eds) *Terrestrial Heatflow in Europe*. Springer, Berlin, 3–40.

CLOETINGH, S., GRADSTEIN, F.M., KOOI, H., GRANT, A.C. & KAMINSKI, M. 1990. Plate reorganization; a cause of rapid late Neogene subsidence and sedimentation around the North Atlantic. *Journal of the Geological Society, London*, **147**, 495–506.

COX, K.G. 1993. Continental magmatic underplating. *Philosophical Transactions of the Royal Society of London, Series A*, **342**, 155–166.

DAM, G., LARSEN, M. & SØRENSEN, J.C. 1998. Sedimentary response to mantle plumes: implications from Paleocene onshore successions West and East Greenland. *Geology*, **26**, 207–210.

DORÉ, A.G. 1992*a*. The Base Tertiary surface of southern Norway and the northern North Sea. *Norsk Geologisk Tidsskrift*, **72**, 259–265.

DORÉ, A.G. 1992*b*. The structural foundation and evolution of Mesozoic seaways between Europe and the Arctic. *Paleogeography, Paleoclimatology, Paleoecology*, **87**, 441–492.

DORÉ, A.G., LUNDIN, E.R., JENSEN, L.N., BIRKELAND, O., ELIASSEN, P.E. & FICHLER, C. 1999. Principal tectonic events in the evolution of the northwest European margin. *In*: FLEET, A.J. & BOLDY, S.A.R. (eds) *Petroleum Geology of Northwest Europe: Proceedings of the 5th Conference*. Geological Society, London, 41–61.

ENGLAND, P. & MOLNAR, P. 1990. Surface uplift, uplift of rocks, and exhumation of rocks. *Geology*, **18**, 1173–1177.

EYLES, N. 1996. Passive margin uplift around the North Atlantic and its role in Northern Hemisphere late Cenozoic glaciation. *Geology*, **24**, 103–106.

FENNER, J. 1988. Occurrences of pre-Quaternary diatoms in Scandinavia reconsidered. *Meyniana*, **40**, 133–141.

FJELDSKAAR, W. 1997. Flexural rigidity of Fennoscandia inferred from the post-glacial uplift. *Tectonics*, **16**, 596–608.

FJELDSKAAR, W. & CATHLES, L. 1991. Rheology of mantle and lithosphere inferred from post-glacial uplift in Fennoscandia. *In*: SABADINI, R., LAMBECK, K. & BOSSI, E. (eds) *Glacial Isostasy, Sea-Level and Mantle Rheology*. Kluwer, Dordrecht, 1–19.

FORSYTH, D.W. 1985. Subsurface loading and estimates of the flexural rigidity of continental lithosphere. *Journal of Geophysical Research*, **90**, 12623–12632.

GILCHRIST, A.R., SUMMERFIELD, M.A. & COCKBURN, H.A.P. 1994. Landscape dissection, isostatic uplift, and the morphologic development of orogens. *Geology*, **22**, 963–966.

GJESSING, J. 1967. Norway's Paleic surface. *Norsk Geografisk Tidsskrift*, **21**, 69–132.

GOES, S., GOVERS, R. & VACHER, P. 2000. Shallow mantle temperatures under Europe from P and S wave tomography. *Journal of Geophysical Research*, **105**, 11153–11169.

GRAND, S.P., VAN DER HILST, R.D. & WIDIYANTORO, S. 1997. Global seismic tomography: a snapshot of convection in the Earth. *GSA Today*, **7**, 1–7.

GREEN, P.F., DUDDY, I.R. & BRAY, R.J. 1997. Variation in thermal history styles around the Irish Sea and adjacent areas: implications for hydrocarbon occurrence and tectonic evolution. *In*: MEADOWS, N.S., TRUEBLOOD, S.P., HARDMAN, M. & COWAN, G. (eds) *Petroleum Geology of the Irish Sea and Adjacent Areas*. Geological Society, London, Special Publications, **124**, 73–93.

GREEN, P.F., DUDDY, I.R., BRAY, R.J. & LEWIS, C.L.E. 1993. Elevated paleotemperatures prior to early Tertiary cooling throughout the UK region. *In*: PARKER, J.R. (ed.) *Petroleum Geology of NW Europe: Proceedings of the 4th Conference*. Geological Society, London, 1067–1074.

GREEN, P.F., THOMSON, K. & HUDSON, J.D. 2001. Recognition of tectonic events in undeformed regions: contrasting results from the Midland Platform and east Midlands Shelf, Central England. *Journal of the Geological Society, London*, **158**, 59–73.

HENDRIKS, B.W.H. & ANDRIESSEN, P.A.M. 2002. Pattern and timing of the post-Caledonian exhumation of northern Scandinavia constrained by apatite fission-track thermochronology. *In*: DORÉ, A.G., CARTWRIGHT, J.A., STOKER, M.S., TURNER, J.P. & WHITE, N. (eds) *Exhumation of the North Atlantic Margin: Timing, Mechanism and Implications for Petroleum Exploration*. Geological Society, London, Special Publications, **196**, 117–137.

HOLTEDAHL, H. 1953. On the oblique uplift of some Northern lands. *Norsk Geografisk Tidsskrift*, **14**, 132–139.

JAPSEN, P. 1997. Regional Neogene exhumation of Britain and the western North Sea. *Journal of the Geological Society, London*, **154**, 239–247.

JAPSEN, P. & CHALMERS, J.A. 2000. Neogene uplift and tectonics around the North Atlantic: overview. *Global and Planetary Change*, **24**, 165–173.

JENSEN, L.N. & SCHMIDT, B.J. 1993. Neogene uplift and erosion in the northeastern North Sea; magnitude and consequences for hydrocarbon exploration in the Farsund Basin. *In*: SPENCER, A.M. (ed.) *Generation, Accumulation and Productions of Europe's Hydrocarbons*. European Association of Petroleum Geologists Special Publication, **3**, 79–88.

JORDT, H., FALEIDE, J.I., BJØRLYKKE, K. & IBRAHIM, M.T. 1995. Cenozoic sequence stratigraphy in the central and northern North Sea Basin: tectonic development, sediment distribution and provenance areas. *Marine and Petroleum Geology*, **12**, 845–880.

KÁRASON, H. & VAN DER HILST, R.D. 2000. Constraints on mantle convection from seismic tomography. *In*: RICHARDS, M.R., GORDON, R. & VAN DER HILST, R.D. (eds) *The History and Dynamics of Global Plate Motion*. Geophysical Monograph, American Geophysical Union, **121**, 277–288.

KELLER, W.R., ANDERSON, D.L. & CLAYTON, R.W. 2000. Resolution of tomographic models of the mantle beneath Iceland. *Geophysical Research Letters*, **27**, 3993–3996.

LEWIS, C.L.E., CARTER, A. & HURFORD, A.J. 1992. Low temperature effects of the Skye Tertiary intrusions on Mesozoic sediments in the Sea of Hebrides basin. *In*: PARNELL, J. (ed.) *Basins on the North Atlantic Seaboard: Petroleum Geology, Sedimentology, and Basin Evolution*. Geological Society, London, Special Publications, **62**, 175–188.

MARQUERING, H. & SNIEDER, R. 1996. Shear wave velocity structure beneath Europe, the northeastern Atlantic and western Asia from waveform inversions including surface-wave mode coupling. *Geophysical Journal International*, **124**, 283–304.

MARTINSEN, O.J., BØEN, F., CHARNOCK, M.A., MANGERUD, G. & NØTTVEDT, A. 1999. Cenozoic development of the Norwegian margin 60–64°N: sequences and sedimentary response to variable basin physiography and tectonic setting. *In*: FLEET, A.J. & BOLDY, S.A.R. (eds) *Petroleum Geology of Northwest Europe: Proceedings of the 5th Conference*. Geological Society, London, 293–304.

MCGUIRE, A.V. & BOHANNON, R.G. 1989. Timing of mantle upwelling: evidence for a passive origin for the Red Sea rift. *Journal of Geophysical Research*, **94**, 1677–1682.

MOKHTARI, M. & PEGRUM, R.M. 1992. Structure and evolution of the Lofoten continental margin, offshore Norway. *Norsk Geologisk Tidsskrift*, **72**, 339–355.

NØTTVEDT, A., LIVBJERG, F., MIDBØE, P.S. & RASMUSSEN, E. 1992. Hydrocarbon potential of the Central Spitsbergen basin. *In*: VORREN, T.O., BERGSAKER, E., DAHL-STAMNES, Ø.A., HOLTER, E., JOHANSEN, B., LIE, E. & LUND, T.B. (eds) *Arctic Geology and Petroleum Potential*. Norwegian Petroleum Society Special Publication, **2**, 333–361.

PEULVAST, J.-P. 1985. Post-orogenic morphotectonic evolution of the Scandinavian Caledonides during the Mesozoic and Cenozoic. *In*: GEE, D.G. & STURT, B.A. (eds) *The Caledonide Orogen—Scandinavia and Related Areas*. Wiley, Chichester, 979–995.

POUDJOM DJOMANI, Y.H., FAIRHEAD, J.D. & GRIFFIN, W.L. 1999. The flexural rigidity of Fennoscandia: reflection of the tectonothermal age of the lithospheric mantle. *Earth and Planetary Science Letters*, **174**, 139–154.

RIIS, F. 1996. Quantification of Cenozoic vertical movements of Scandinavia by correlation of morphological surfaces with offshore data. *Global and Planetary Change*, **12**, 331–358.

RIIS, F. & FJELDSKAAR, W. 1992. On the magnitude of the Late Tertiary and Quaternary erosion and its significance for the uplift of Scandinavia and the Barents Sea. *In*: LARSEN, R.M., BREKKE, H., LARSEN, B.T. & TALLERAAS, E. (eds) *Structural and Tectonic Modelling and its Application to Petroleum Geology*. Norwegian Petroleum Society Special Publication, **1**, 163–185.

ROHRMAN, M. & VAN DER BEEK, P. 1996. Cenozoic postrift domal uplift of North Atlantic margins: an asthenospheric diapirism model. *Geology*, **24**, 901–904.

ROHRMAN, M., VAN DER BEEK, P. & ANDRIESSEN, P.A.M. 1994. Syn-rift thermal structure and post-rift evolution of the Oslo Rift (SE Norway): new constraints from fission track thermochronology. *Earth and Planetary Science Letters*, **127**, 39–54.

ROHRMAN, M., VAN DER BEEK, P.A., ANDRIESSEN, P.A.M. & CLOETINGH, S. 1995. Meso-Cenozoic morphotectonic evolution of Southern Norway: Neogene domal uplift inferred from apatite fission track thermochronology. *Tectonics*, **14**, 704–718.

RUNDBERG, Y. & SMALLEY, P.C. 1989. High-resolution dating of Cenozoic sediments from the northern North Sea using $^{87}Sr/^{86}Sr$ stratigraphy. *AAPG Bulletin*, **73**, 298–308.

SIMONS, F.J., ZUBER, M.T. & KORENAGA, J. 2000. Isostatic response of the Australian lithosphere: estimation of effective elastic thickness and anisotropy using multitaper spectral analysis. *Journal of Geophysical Research*, **105**, 19163–19184.

SMALL, E. & ANDERSON, R.S. 1998. Pleistocene relief production in Laramide mountain ranges, western Unites States. *Geology*, **26**, 123–126.

SMITH, W.H.F. & WESSEL, P. 1990. Gridding with continuous curvature splines in tension. *Geophysics*, **55**, 293–305.

SOBOLEV, S.V., ZEYEN, H., STOLL, G., WERLING, F., ALTHERR, R. & FUCHS, K. 1996. Upper mantle temperature from teleseismic tomography of French Massif Central including effects of composition, mineral reactions, anharmonicity, anelasticity and partial melt. *Earth and Planetary Science Letters*, **139**, 147–163.

STUEVOLD, L.M. & ELDHOLM, O. 1996. Cenozoic uplift of Fennoscandia inferred from a study of the mid-Norwegian margin. *Global and Planetary Change*, **12**, 359–386.

SUMMERFIELD, M.A. 1999. Geomorphology and global tectonics: introduction. *In*: SUMMERFIELD, M.A. (ed.) *Geomorphology and Global Tectonics*. Wiley, Chichester, 3–11.

Sveriges Geologiske Undersökelse, 1985. Scandinavian Caledonides gravity anomaly map. *In*: GEE, D.G. & STURT, B.A. (eds) *The Caledonide Orogen—Scandinavia and Related Areas*. Wiley, Chichester.

THEILEN, F. & MEISSNER, R. 1979. A comparison of crustal and upper mantle features in Fennoscandia and the Rhenish Shield, two areas of recent uplift. *Tectonophysics*, **61**, 227–242.

TORSKE, T. 1972. Tertiary oblique uplift of western Fennoscandia; crustal warping in connection with rifting and break-up of the Laurasian continent. *Norges Geologiske Undersøkelse*, **273**, 43–48.

VÅGNES, E. & AMUNDSEN, H.E.F. 1993. Late Cenozoic uplift and volcanism on Spitsbergen: caused by mantle convection? *Geology*, **21**, 251–254.

VAN DER BEEK, P.A. 1995. *Tectonic evolution of continental rifts: inferences from numerical modelling and fission track thermochronology*. PhD thesis, Vrije Universiteit Amsterdam.

VAN DER BEEK, P.A. & ROHRMAN, M. 1997. Passive margin uplift around the North Atlantic region and its role in Northern Hemisphere late Cenozoic glaciation: Comment. *Geology*, **25**, 282.

VAN DER BEEK, P.A., CLOETINGH, S. & ANDRIESSEN, P.A.M. 1994. Mechanisms of extensional basin formation and vertical motions at rift flanks: constraints from tectonic modelling and fission-track thermochronology. *Earth and Planetary Science Letters*, **121**, 417–433.

WHITE, N. & LOVELL, B. 1997. Measuring the pulse of a plume with the sedimentary record. *Nature*, **387**, 888–891.

ZIELHUIS, A. & NOLET, G. 1994*a*. Deep seismic expression of an ancient plate boundary in Europe. *Science*, **265**, 79–81.

ZIELHUIS, A. & NOLET, G. 1994*b*. Shear-wave velocity variations in the upper mantle beneath central Europe. *Geophysical Journal International*, **117**, 695–715.

ZIELHUIS, A. & VAN DER HILST, R.D. 1996. Mantle structure beneath the eastern Australian region from partitioned waveform inversion. *Geophysical Journal International*, **127**, 1–16.

Paleocene initiation of Cenozoic uplift in Norway

S. B. NIELSEN[1], G. E. PAULSEN[1], D. L. HANSEN[1], L. GEMMER[1], O. R. CLAUSEN[1], B. H. JACOBSEN[1], N. BALLING[1], M. HUUSE[2] & K. GALLAGHER[3]

[1]*Department of Earth Sciences, Aarhus University, Finlandsgade 8, DK-8200 Aarhus N, Denmark (e-mail: sbn@geo.aau.dk)*

[2]*Department of Geology & Petroleum Geology, University of Aberdeen, King's College, Meston Building, Aberdeen AB24 3UE, UK*

[3]*Imperial College of Science, Technology & Medicine, South Kensington, London SW7 2AS, UK*

Abstract: The timing of Cenozoic surface uplift in NW Europe relies on the assumption that the sedimentary response in basins is synchronous with tectonic processes in the source areas. However, many of the phenomena commonly used to infer recent uplift may as well be a consequence of climate change and sea-level fall. The timing of surface uplift therefore remains unconstrained from the sedimentary record alone, and it becomes necessary to consider the constraints imposed by physically and geologically plausible tectonic mechanisms, which have a causal relation to an initiating agent. The gradual reversal of the regional stress field following the break-up produced minor perturbations to the thermal subsidence on the Norwegian Shelf and in the North Sea. Pulses of increased compression cannot be the cause of Cenozoic land surface uplift and accelerated Neogene basin subsidence.Virtually deformation-free regional vertical movements could have been caused by changes in the density column of the lithosphere and asthenosphere following the emplacement of the Iceland plume. A transient uplift component was produced as the plume displaced denser asthenosphere at the base of the lithosphere. This component decayed as the plume material cooled. Permanent uplift as a result of igneous underplating occurred in areas of a thin lithosphere (some Palaeozoic and Mesozoic basins) or for lithosphere under extension at the time of plume emplacement (the ocean–continent boundary). In areas of a thicker lithosphere (East Greenland, Scotland and Norway) plume emplacement may have triggered a Rayleigh–Taylor instability, causing partial lithospheric delamination and associated transient surface uplift at a decreasing rate throughout Cenozoic time. A possible uplift history for the adjacent land areas hence reads: initial transient surface uplift around the break-up time at 53 Ma caused by plume emplacement, and permanent tectonic uplift caused by lithospheric delamination and associated lithospheric heating. The permanent tectonic uplift increased through Cenozoic time at a decreasing rate. Denudation acted on this evolving topography and reduced the average surface elevation, but significantly increased the elevation of the summit envelope. The marked variations in the sedimentary response in the basins were caused by climatic variations and the generally falling eustatic level. This scenario bridges the gap between the ideas of Paleocene–Eocene uplift versus repeated Cenozoic tectonic activity: the tectonic uplift history was initiated by the emplacement of the Iceland plume, but continued throughout Cenozoic time as a consequence of early plume emplacement, with climatic and eustatic control on denudation. The mechanism is consistent with topography, heat flow, crustal structure, and the Bouguer gravity of Norway, and may be applicable also to East Greenland.

Geomorphological evidence and the age and structure of sediments point to the occurrence of large-scale Cenozoic vertical movements in NW Europe (Riis & Fjeldskaar 1992; Riis 1996; Stuevold & Eldholm 1996). The Norwegian continental shelf, the Viking Graben and the central North Sea all experienced subsidence, and apparently at an accelerating rate (Cloetingh *et al.* 1990, 1992), throughout Cenozoic time. Contemporaneously, adjacent continental areas and their inner shelves such as the British Isles (Japsen & Chalmers 2000), northern and southern Norway (Riis 1996; Lidmar-Bergström *et al.* 2000), southern Sweden (Lidmar-Bergström 1991), as well as the Barents Sea and the margins of the North Sea Basin (Doré 1992; Jensen &

From: DORÉ, A.G., CARTWRIGHT, J.A., STOKER, M.S., TURNER, J.P. & WHITE, N. 2002. *Exhumation of the North Atlantic Margin: Timing, Mechanisms and Implications for Petroleum Exploration*. Geological Society, London, Special Publications, **196**, 45–65. 0305-8719/02/$15.00

Schmidt 1992; Japsen 1998) experienced uplift and denudation. A similar pattern of vertical movements is observed at the western margin of the North Atlantic (Larsen 1990; Japsen & Chalmers 2000; Johnson & Gallagher 2000).

Continued Cenozoic basin subsidence can generally be understood in terms of a continuation of Palaeozoic–Mesozoic mainly rift-initiated basin subsidence (McKenzie 1978). However, much controversy exists about the causal relationships regarding the simultaneous Cenozoic uplift and denudation of the adjacent land areas. Was there an initial Paleocene–Eocene surface uplift related to the magmatically dominated opening of the North Atlantic, followed by a climatically controlled denudation response (Riis & Fjeldskaar 1992), or have a number of tectonic pulses, particularly in Paleocene and (mainly) in Neogene times (Riis 1996; Stuevold & Eldholm 1996), been active throughout Cenozoic time?

The role of climate and eustasy

The arguments for the timing of Cenozoic tectonics are predicated on the assumption that the sedimentary response in basins reflects contemporaneous tectonic processes in the source areas. However, denudation depends strongly on climate (which on a regional level controls precipitation and vegetation cover), on the erosional base level defined by the eustatic sea level, and the connectivity of the drainage basin to the regional base level. The question is then to what extent climate changes may have influenced the denudation process in the North Atlantic region.

Among the climate variations, obviously the Quaternary glaciations have had a major denudational impact in NW Europe and in Greenland. Other climate variations were more subtle; however, it is well documented that post Late Eocene time was characterized by a significant global climate cooling and a falling eustatic level of about 200–250 m (e.g. Haq *et al.* 1987; Miller *et al.* 1998). These changes can be correlated with a world-wide change in sedimentation style from carbonate-dominated environments to siliciclastic-dominated systems with high rates of sedimentation. More so, the mid-Miocene climate and eustatic changes signalled a further world-wide acceleration in the siliciclastic flux (Donnelly 1982; Bartek *et al.* 1991). The widespread distribution of these contemporaneous phenomena points to a global-scale cause, such as climate change and sea-level fall. Cenozoic climate changes in the North Sea region have been addressed in moderate detail by

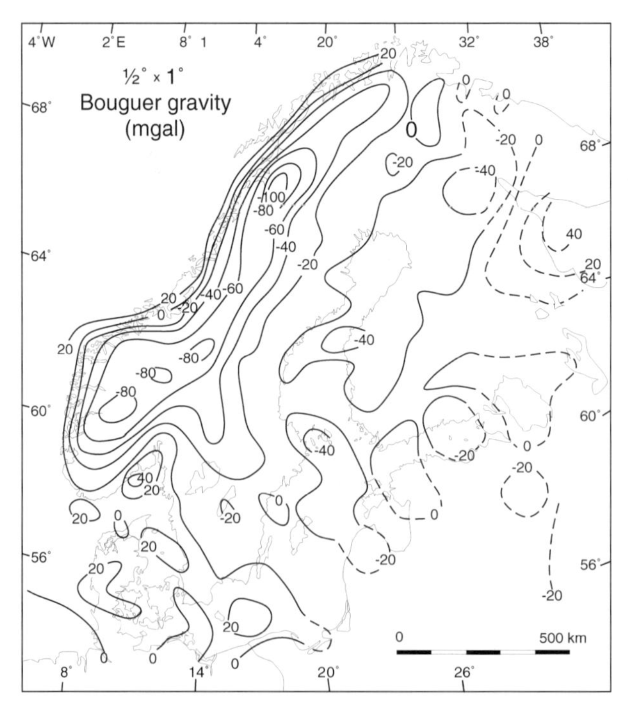

Fig. 1. Averaged Fennoscandian Bouguer gravity field. Redrawn after Balling (1984).

Buchardt (1978), who derived a $\delta^{18}O$ record based on macrofossils. His results are in general agreement with those derived from calcareous nannoplankton studies (Miller *et al.* 1998) and show the onset of Cenozoic cooling in late Eocene–Oligocene times (see Huuse 2002b).

Apart from blurring the causal relation between tectonics in the source area and sediment stratigraphy, the dependence of denudation on climate can effect a coupling between climatic deterioration and the uplift of mountain ranges. Molnar & England (1990) argued that the transition to a cooler and more erosive climate enhances relief in mountain ranges and raises mountain crests in isostatic response to the removal of load. This is in contrast to Whipple *et al.* (1999), who claimed that erosional processes in general reduce relief and that the isostatic peak uplift is negligible. However, if the Norwegian Caledonides were indeed peneplained at the onset of the North Atlantic opening it is clear that significant relief has been produced in the mean time (Fig. 2, below). We therefore believe that the arguments of Molnar & England (1990) must also be considered in the North Atlantic region. A further discussion has been given by Huuse (2002b).

Tectonics

Irrespective of the denudational response in basins, primary surface uplift must be caused by physically plausible tectonic mechanisms, which lend themselves to a quantitative description and hypothesis testing.

During mid–late Cenozoic time, mid-ocean ridge compressional stress produced large offshore domes on the Norwegian margin (e.g. the Helland Hansen and Ormen Lange Domes), and has been proposed to be an agent in Neogene surface uplift of Norway by either (1) the buckling of an elastic plate (e.g. Cloetingh *et al.* 1992; Doré 1992) or (2) thickening of the continental lithosphere by shortening.

(1) Elastic buckling would for the wavelength (200 km) and amplitude (1 km) require forces far beyond the compressional strength of continental lithosphere. Buckling of an elastic plate is therefore not a possible mechanism. Lambeck

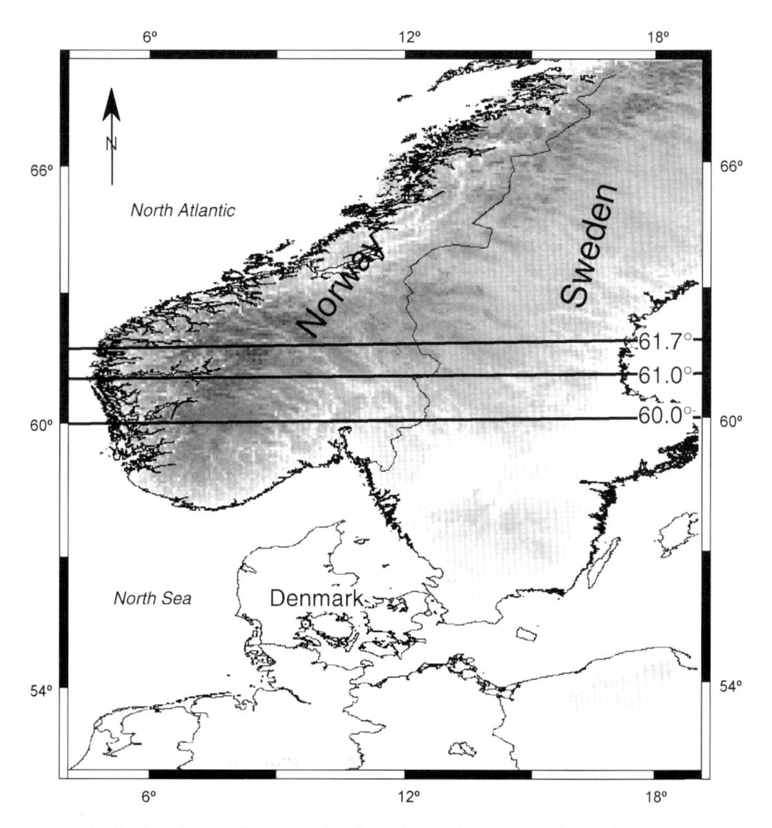

Fig. 2. Topographic relief of southern Norway with locations of topographic profiles.

(1983) argued that a viscoelastic plate with erosion and deposition produces long-wavelength large-amplitude deflections that resemble buckling for stresses that are much less than the elastic buckling stresses. However, this model fails the gravity and structural tests in Norway, as ridges and highlands would become associated with Bouguer gravity maxima because of an elevated Moho, whereas basins would become associated with Bouguer gravity minima because of a depressed Moho. As pointed out by Balling (1984), the Bouguer gravity of the Norwegian highlands (Fig. 1) shows significant minima (deeper than − 80 mgal) with an axis following closely the axis of the topography.

(2) The uplift of Scotland, the Norwegian Caledonides and East Greenland is virtually deformation free and cannot be caused by shortening of the continental lithosphere, as the following simple argument demonstrates (Brodie & White 1994). The excess crustal thickness, x, necessary to support topography of height h is given by $x = h\rho_m/\Delta\rho$, where ρ_m is the density of the upper mantle and $\Delta\rho$ is the density contrast between the upper mantle and the crustal root. For $\rho_m = 3300\,\mathrm{kg\,m^{-3}}$ and $\Delta\rho = 400\,\mathrm{kg\,m^{-3}}$, x becomes 8.25 km per km of topography h. For an initial crustal thickness of 35 km, the horizontal shortening needed to produce 1 km of topography amounts to 24%, and the final depth to Moho becomes 44.25 km, which is similar to the present-day Moho depth under the Norwegian Caledonides. However, 24% of horizontal shortening would have produced significant compression structures, which are not observed.

The reversal of the regional stress field that the Norwegian Shelf and the North Sea must have experienced does have consequences for the subsidence pattern, although not to the extent previously assumed. Below we discuss this in more detail.

Vertical movements on a regional scale occur as a consequence of changes in the density column of the lithosphere and asthenosphere and the associated isostatic adjustments, or as so-called dynamic topography supported by vertical stresses at the base of the lithosphere, which are generated by flow in the mantle. The detection of dynamic topography, particularly over the oceans, remains controversial (Lithgow-Bertelloni & Silver 1998). In this paper we limit our scope to

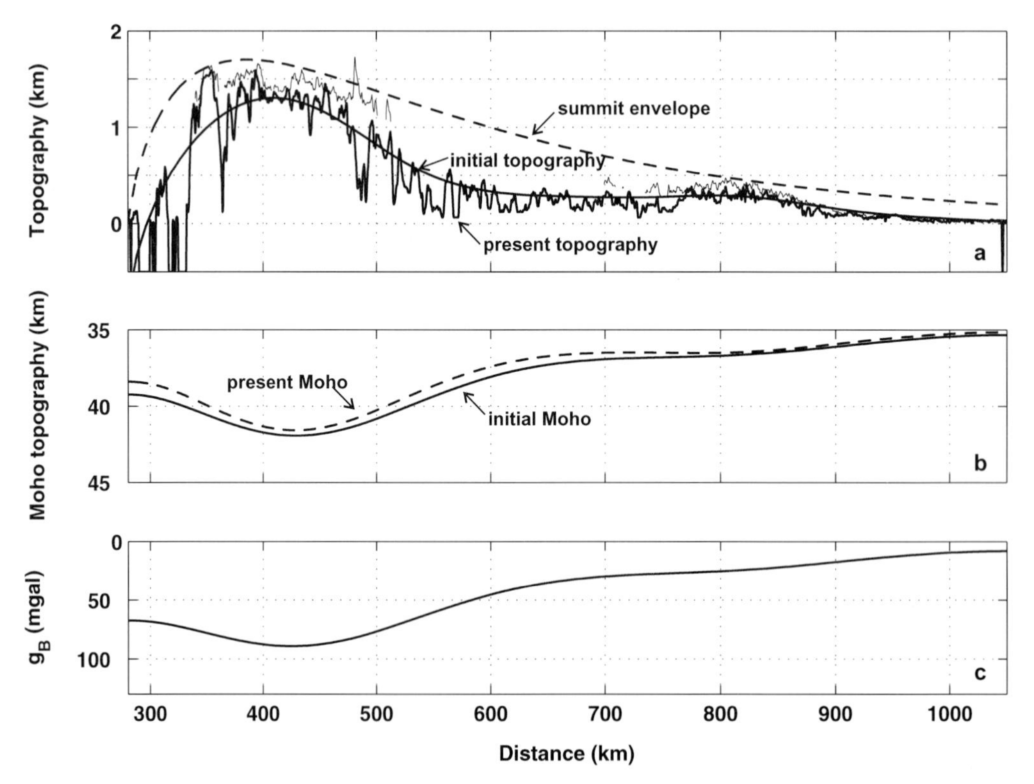

Fig. 3. (**a**) Present-day topography (60°N), summit envelope (dashed line), and initial surface before erosion; (**b**) initial and present (dashed) Moho position; (**c**) calculated Bouguer gravity.

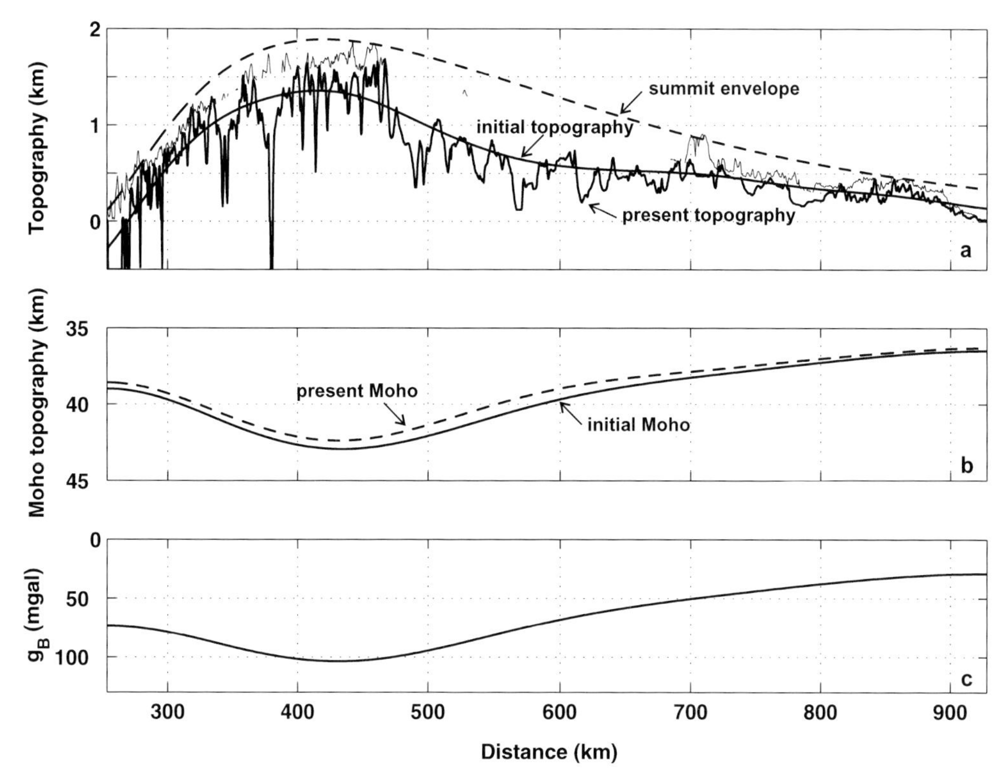

Fig. 4. (**a**) Present-day topography (61°N), summit envelope (dashed line), and initial surface before erosion; (**b**) initial and present (dashed) Moho position; (**c**) calculated Bouguer gravity.

isostatically supported vertical movements. Virtually deformation-free, isostatically supported and positive vertical movements may be caused by (1) magmatic underplating at crustal levels (McKenzie 1984), (2) lithospheric delamination (Bird 1979; Houseman *et al.* 1981), or (3) emplacement of a hot plume at the base of the lithosphere (Skogseid *et al.* 2000). In view of the magmatic-dominated opening of the North Atlantic these lithospheric–asthenospheric mechanisms therefore must be considered to be the candidates to produce the Cenozoic long-wavelength land surface uplift.

In the following section we produce a brief catalogue of mechanisms of passive and active surface uplift and present new results for later reference in the discussion.

Mechanisms of surface uplift

Passive mechanisms

Isostatic adjustments caused by erosion of existing topography. Erosion of existing topography results in lowering of the average topography but, as proposed by Molnar & England (1990), Riis & Fjeldskaar (1992) and Gilchrest *et al.* (1994), the summit envelope height may increase because of the isostatic uplift following the removal of mass between the peaks or interfluves. In this section this mechanism is discussed within the framework of unloading of an elastic lithosphere by erosion so as to assess the amount of tectonic uplift required to produce the topography of southern Norway. The approach is similar to that of Riis & Fjeldskaar (1992) except that, for a given amount of erosion, we determine the initial topography required to reproduce the present topography. Furthermore, we calculate the associated Bouguer gravity anomaly. The timing of the erosion is not addressed by this procedure.

Three profiles at 60, 61 and 61.7°N (Fig. 2) were extracted from the 1 km × 1 km digital topography database GLOBE (GLOBE Task Team *et al.* 1999). The summit envelope of each profile was constructed by passing a smooth curve through the summits contained in a belt of 40 km width surrounding each profile. The amount eroded is the difference between the

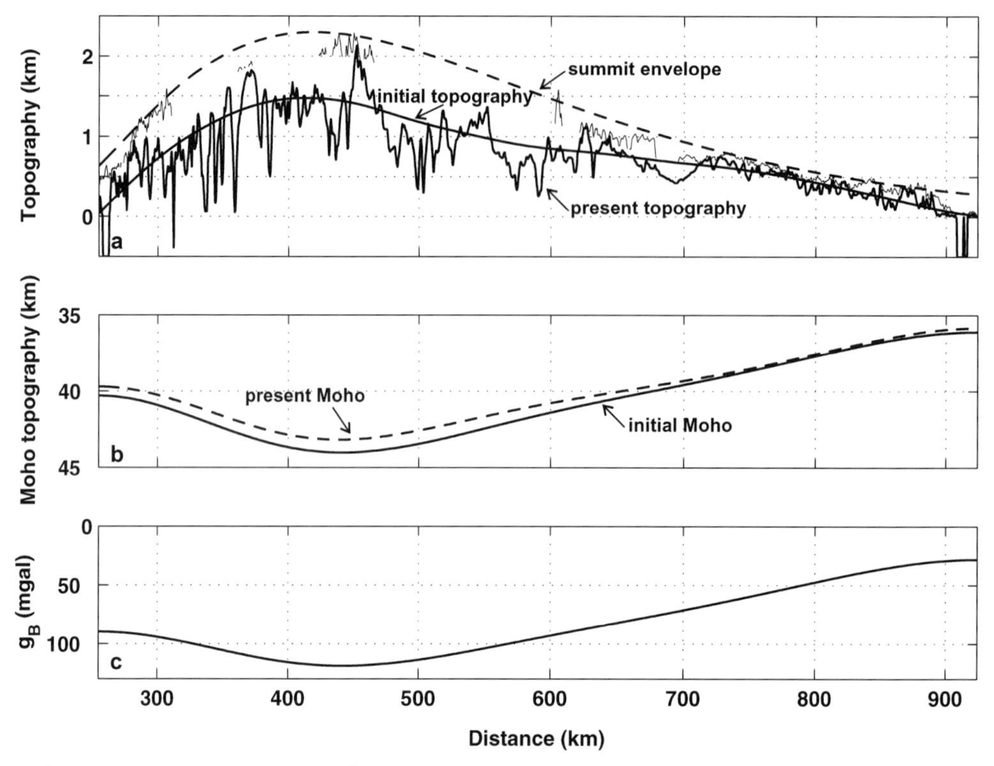

Fig. 5. (**a**) Present-day topography (61.7°N), summit envelope (dashed line), and initial surface before erosion; (**b**) initial and present (dashed) Moho position; (**c**) calculated Bouguer gravity.

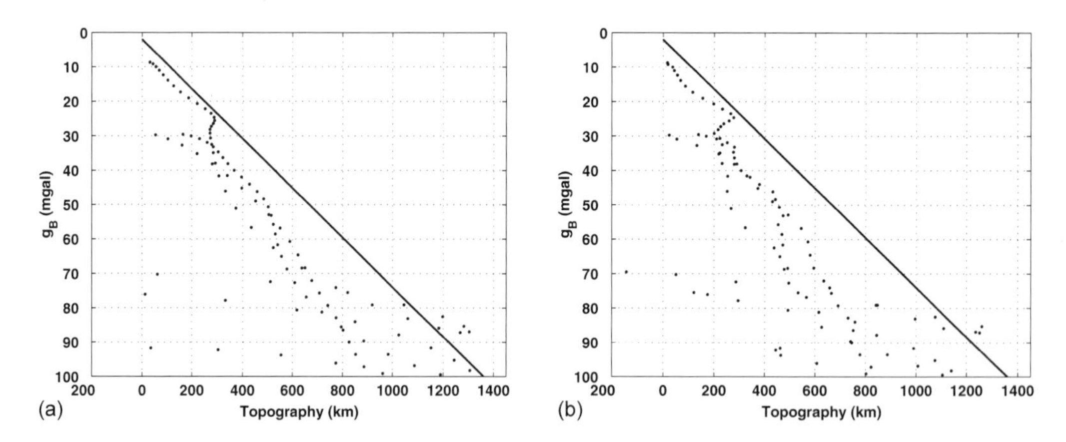

Fig. 6. The correlation between Bouguer gravity and average topographic height for the three profiles: (**a**) before erosion; (**b**) after erosion. Continuous line is the correlation derived by Balling (1984).

summit envelope and the present topography. Next the pre-erosion Moho depth and topography is determined so that after erosion of the specified amount and isostatic compensation, the model topography agrees with the present topography. It is assumed that isostatic compensation takes place at the crust–mantle interface with a density contrast of 400 kg m^{-3}, and that the upper mantle reacts by flexure to loading with an elastic thickness of 15 km. The average crustal density is 2800 kg m^{-3}.

Figures 3–5 show the results of applying this procedure. It appears that the process of erosion has raised the summits a maximum of *c.* 800 m. The amount is directly visible as the difference between initial and present Moho depths. The maximum initial topography required to reproduce the present topography is located close to the axis of maximum present-day topography and is in the range 1.3–1.5 km. To the west and east the initial topography gradually approaches zero. The initial topography is identical to the tectonic uplift that must have been produced by the mechanism responsible for the Cenozoic uplift of the Norwegian Caledonides, save for the possible existence of topography before the action of the mechanism. For a prior topography of *c.* 300 m, corresponding approximately to the topography that would emerge after the post-mid-Cretaceous sea-level fall, if topography then had been peneplained to sea level, the maximum necessary amount of tectonic uplift is in the range 1–1.2 km. If positive topography is allowed for during the time of maximum sea level the necessary tectonic uplift becomes correspondingly less.

Figure 6 shows the correlation between topography and the Bouguer gravity anomaly before (a) and after (b) erosion. The correlation is negative because topography is compensated by a crustal root, a fact that is not modified by a relief-producing denudational process. The modelled correlation is in agreement with the correlation determined by Balling (1984) for the western Fennoscandian Shield (Norway). The model Moho depths beneath southern Norway are qualitatively in agreement with the depths derived by Kink *et al.* (1993), but exhibit some deviations to the east in the Fennoscandian Shield, probably because of breakdown of the assumption of a laterally homogeneous crust.

Isostatic and erosional response of basin margins to sea-level fall. It is generally acknowledged that the eustatic level decreased from mid-Cretaceous time to the present day by 200–250 m (Haq *et al.* 1987; Miller *et al.* 1998), although smaller amplitudes have been suggested (Kominz *et al.* 1998). This eustatic fall exposed the margins of sedimentary basins, which may have been established under a former higher eustatic level, and caused erosional unloading and passive isostatic uplift (i.e. erosional rebound).

The subsidence diagram of Fig. 7 demonstrates this mechanism in action on a basin margin. The initial eustatic level at 100 Ma is 250 m, which is also the water depth at that time. Sedimentation has filled the basin to the brim at the onset (60 Ma) of a gradual eustatic fall of 250 m. A maximum of 750 m of sediment is accommodated by the initial 250 m of accommodation space because of sediment loading and compaction of the already existing sediment. As the eustatic fall exposes the sediment surface it is eroded. At the present day a total of 500 m of sediment has been eroded, and the sediment at the surface and at any depth below therefore shows an overburial (*sensu* Japsen 1988) of 500 m.

Figure 8a–c shows the results of combining the above principle with thermal subsidence, lateral sediment transport, and a zone of tectonic inversion in a 1300 km profile. The profile could simulate an east–west North Sea profile at 56°N. It represents the post-mid-Cretaceous filling from the sides of an intra-cratonic basin

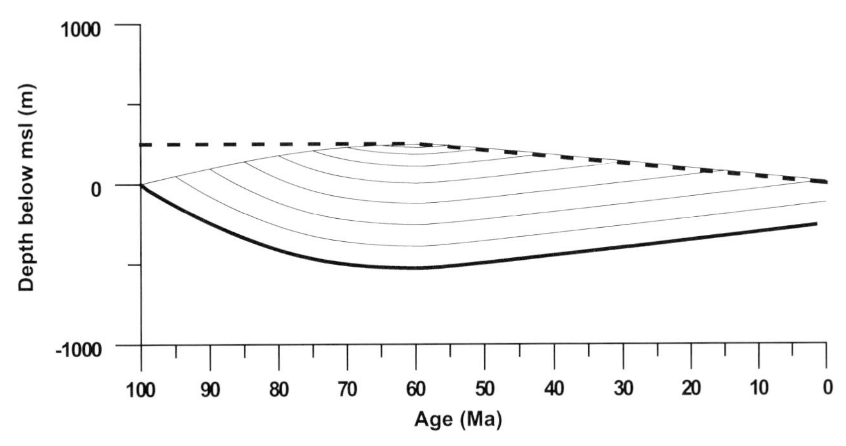

Fig. 7. Subsidence diagram showing the effect of falling sea level (dashed line) on the erosion of a basin margin.

dominated by thermal subsidence in the central parts, falling eustatic level, and localized tectonic inversion (e.g. the Sorgenfrei–Tornquist Zone) at the eastern margin.

The subsidence diagram (Fig. 8a) shows quiet sedimentation and subsidence until the clinoforms migrating from the basin margins reach the basin centre. For later reference it is noted that

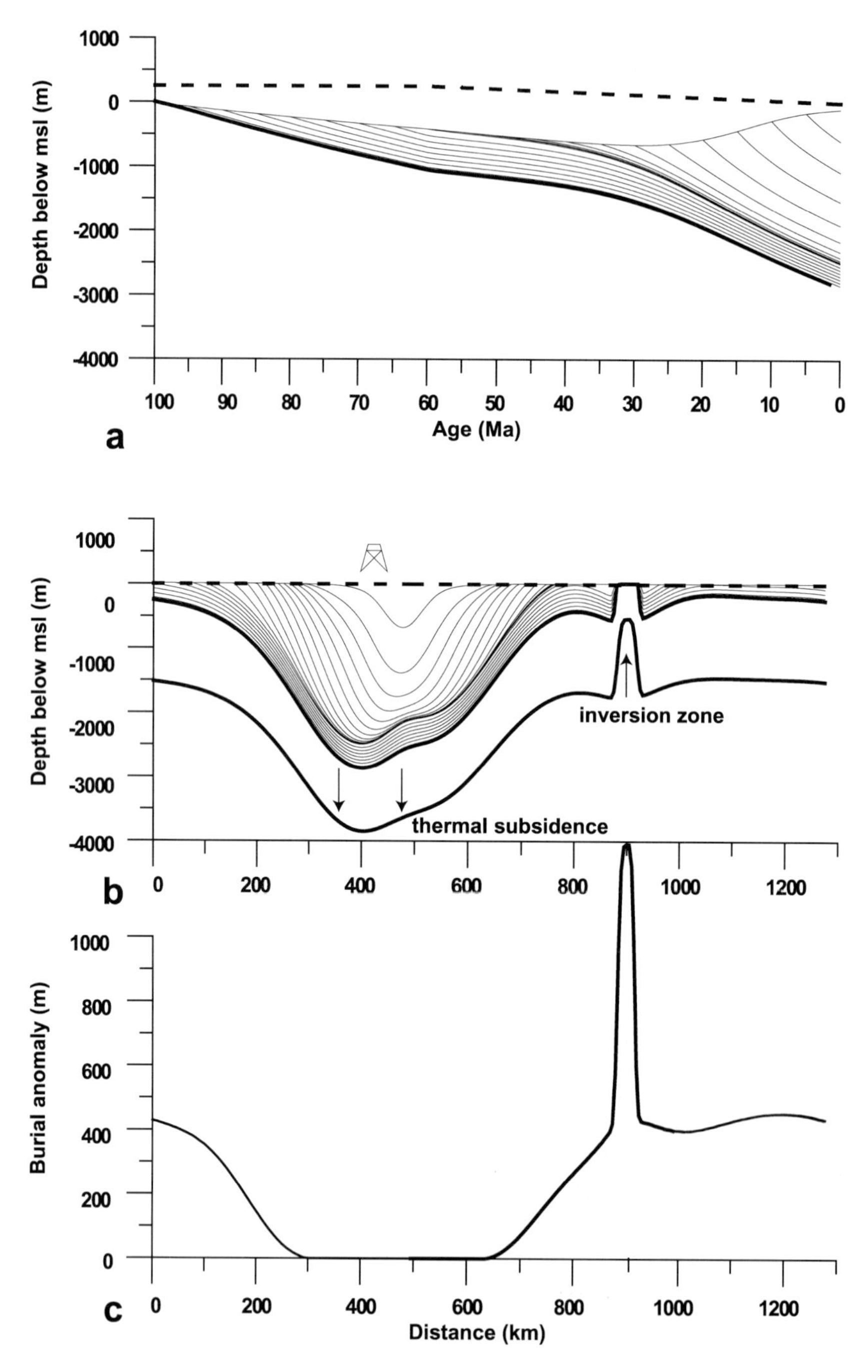

Fig. 8. Subsidence diagram (**a**), sediment structure at 0 Ma (**b**), and overburial (**c**) for a 1300 km profile representing an intra-cratonic basin with thermal subsidence in the centre, falling eustatic level, and sediment infilling from the sides. Also shown is a zone of structural inversion.

passive infilling of existing accommodation space, quiet thermal subsidence and sediment loading lead to an apparently accelerating Neogene subsidence. The present-day structure of the sediments (Fig. 8b) shows truncation of the clinoforms towards the margins as well as the inversion structure and its marginal troughs. The present-day burial anomaly along the profile (Fig. 8c) varies from 0 m in the basin centre, where thermal subsidence and sediment loading and compaction outpaced the reduction in accommodation space produced by the falling sea level and sedimentation, to more than 1 km in the inversion zone. The background value of overburial on the basin margins is caused by the mechanism of Fig. 7.

The above principles, including temporal variations in sediment source areas reflecting the clockwise rotation of source areas throughout Cenozoic time, have been built into a large-scale post-mid-Cretaceous North Sea model. By trial-and-error runs, in which the initial thermal anomaly and the sediment feeding was varied, this model was brought into good agreement with the present-day thickness distribution of the Upper Cretaceous and Cenozoic units (Gemmer *et al.* 2002; Nielsen 2002). Here we show two of the principal predictions of this model, which can be compared with observations: chalk overburial (Fig. 9) and the pre-Quaternary geological map (Fig. 10).

The regional pattern of chalk overburial in the North Sea Basin was quantified by Japsen 1998 by comparing measured compressional velocities of chalk with a reference depth–velocity model. He found that the overburial increases towards the eastern and western North Sea margins in a symmetrical pattern around a central zone aligned along the Central Graben area, in which chalk is at its maximum depth at present day. The overburial prediction of Fig. 6 reproduces the general pattern of overburial inferred by Japsen (1998). However, the predicted amplitudes are smaller than the amplitudes of Japsen (1998), except in the inversion zones, where very large values of overburial can occur also in the model.

The predicted pre-Quaternary map of Fig. 10 shows marked similarities with the pre-Quaternary onshore Denmark (Sorgenfrei & Berthelsen 1954) and with the compilation of Japsen (1998). The areal extent of the Oligocene outcrop of onshore Denmark, however, is too large as a consequence of erosion to sea level in the model of all topography in the basin. The curvature of the Cenozoic layer boundaries adjacent to the Sorgenfrei–Tornquist Zone agrees with observations and is caused by the filling of the flexural foredeep and the later erosional truncation.

The conclusion of this section is that much of the known post-mid-Cretaceous sedimentary structure of the North Sea Basin can be reproduced by a model that includes passive

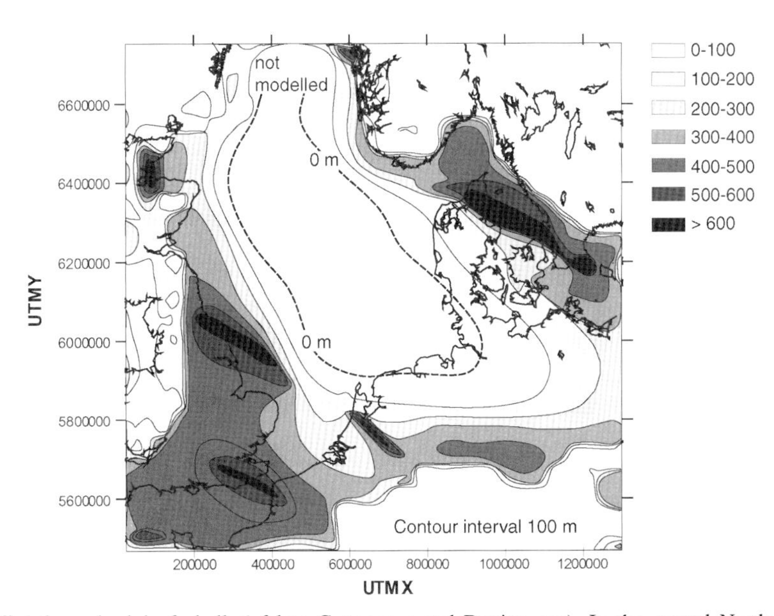

Fig. 9. Predicted overburial of chalk (of late Cretaceous and Danian age). In the central North Sea thermal subsidence and compaction of previous sediments keep pace with the reduction of accommodation space caused by falling sea level and sedimentation. Four zones of structural inversion have been included.

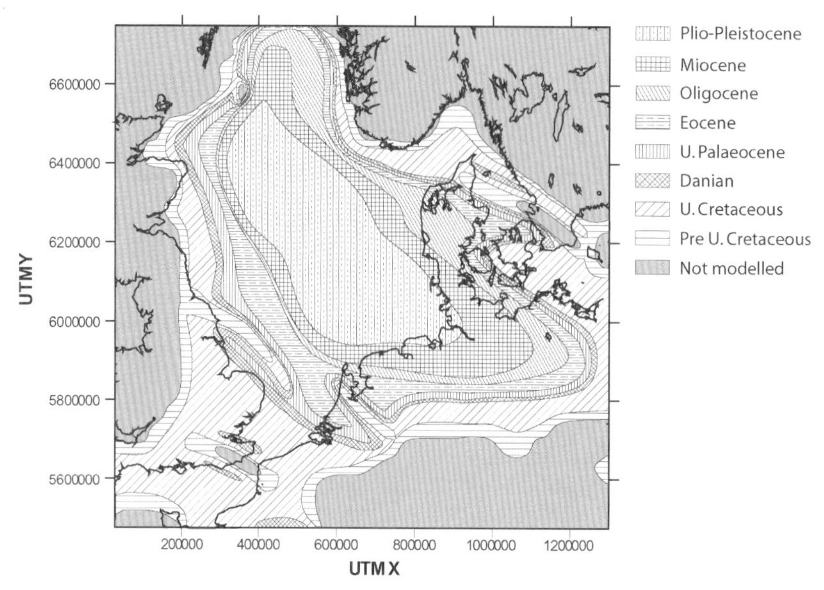

Fig. 10. Predicted pre-Quaternary geological map.

isostatic effects in response to thermal cooling, sea-level fall, erosion, and changing sediment source areas. The question that remains is whether the discrepancy of chalk burial amplitudes towards the margins is a consequence of an unknown mechanism of active surface uplift, which is not part of this numerical model, or can be explained by the reference chalk velocity model (Huuse 2002a, 2002b).

Active mechanisms

Changing in-plane stress. It is well known that fluctuating in-plane stress causes fluctuations of relative water depth on passive margins and in intra-cratonic basins because of the induced vertical deflections of the lithosphere (Cloetingh *et al.* 1985, 1990, 1992; Karner 1989). However, it is perhaps less well established whether an increase in compression causes a depression or an elevation of the basin floor. The principal reason for this controversy is that the thin elastic plate models applied in the majority of studies require preloading in order to respond by flexure to horizontal stress changes. This preloading produces an initial deflection, the amplitude of which is enhanced in compression or reduced in extension, as the product of the in-plane force and initial plate curvature acts in the same way as a vertical load. The result of a thin plate compression or extension experiment is hence given *a priori* by the modeller's preference about initial loading.

An indication of what might happen when the tectonic stress on a continental margin changes from extension to compression was presented by Braun & Beaumont (1989), who simulated passive margin formation by a numerical thermomechanical model. They found that a sudden relief of the extensional tectonic stress causes basin floor rise and rift flank subsidence as a new balance between in-plane stress and buoyancy forces is established.

Here we pursue this a little further by simulating the formation of a generic passive margin by lithospheric extension in Mesozoic time, and reversal of the stress field at the time of opening of the North Atlantic (Paulsen *et al.* 2001). We consider the inherited strain history, state of stress, and sediment and thermal loading of the margin at the time of stress reversal. The results also apply to rifts and intra-continental rift basins such as the Viking Graben and the central North Sea.

The numerical model is a dynamical large-strain Lagrangian finite element model with an elasto-viscoplastic rheology. Plasticity is modelled using non-associated plastic flow with a Drücker–Prager yield criterion. The viscoelastic deformation is modelled by a non-linear Maxwell rheology with viscosity depending on temperature, material type, and deviatoric stress. In general, the type of deformation occurring at a point in the model depends on temperature, deviatoric stress, confining pressure and material type. Figure 11 shows the initial undeformed

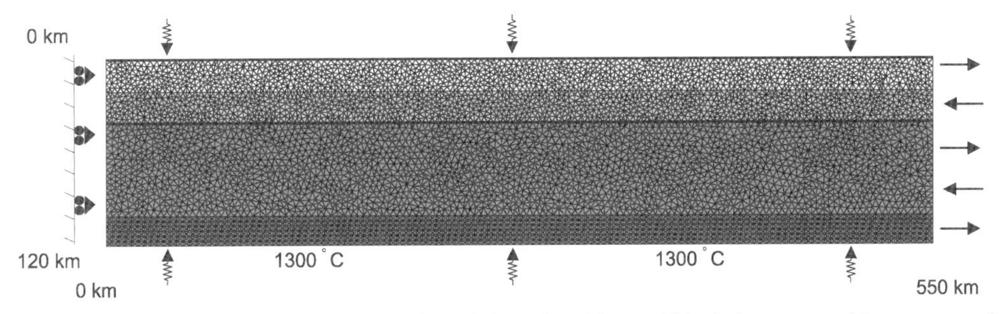

Fig. 11. Finite element model of passive margin–rift formation. The model includes upper and lower crust and mantle lithosphere. The boundary conditions are 0 °C at the surface and 1300 °C at a fixed depth. There is no flow of heat across the vertical boundaries. Loading by sedimentation and erosion occur at the surface. The bottom boundary condition is a hydrostatic pressure. A kinematic boundary condition causes extension or compression of the profile. Model parameters follow Paulsen *et al.* (2001).

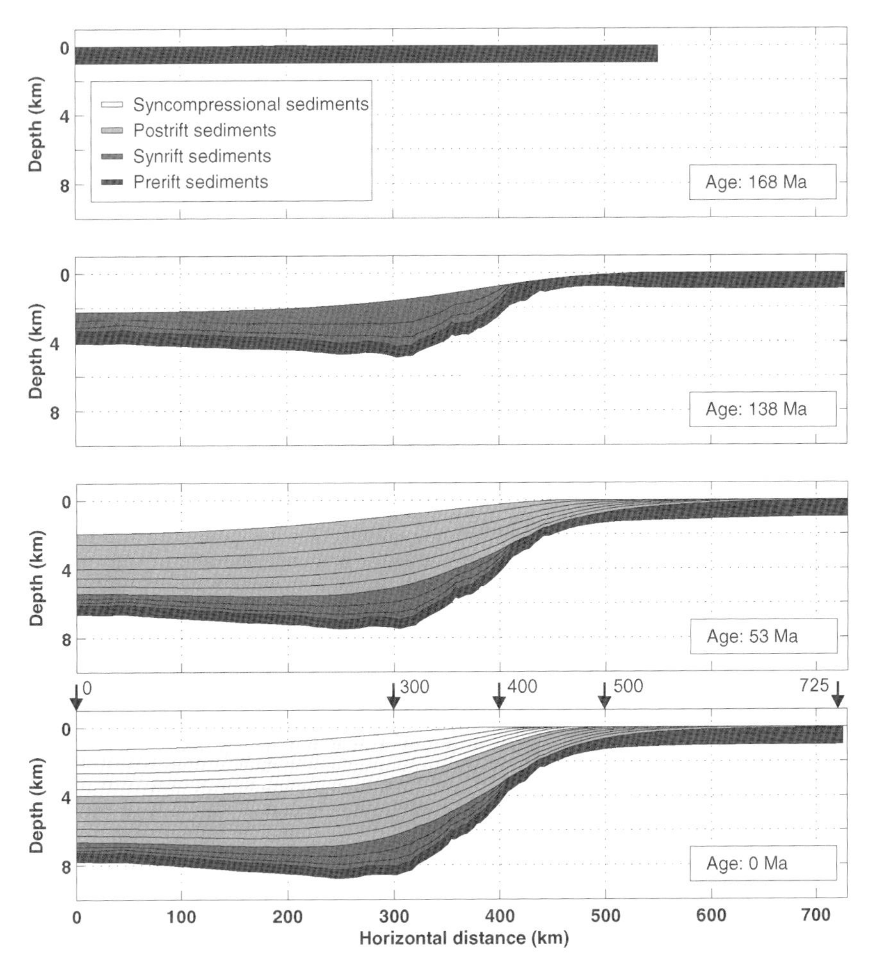

Fig. 12. Cross-sections of margin evolution with positions of tracked basement points indicated by arrows at locations 0, 50, 90 and 160 km: 0 Ma, initial sediment structure; 15 Ma, end of rifting; 105 Ma, onset of compression; 158 Ma, final sediment structure. The deposition of syncompressional sediments has been retarded between 20 and 120 km.

finite element grid and the thermal and mechanical boundary conditions. Localization of margin formation is achieved by assuming a weaker (wet) rheology in the region that is to become extended. Thermal weakening would yield similar results, but the exact mode of weakening is not the subject here.

The margin is created by extension for 32 Ma by moving the right model boundary at a constant velocity to the right (Fig. 12). A smaller rate of extension would result in more synrift and less postrift thermal subsidence, and vice versa. However, the duration of extension is not important to the principal results. After extension, thermal subsidence follows for a period of 85 Ma. Subsequently, the stress field is reversed by moving the right boundary to the left. This is equivalent to applying compressive stresses associated with the opening of the North Atlantic.

Figure 13a shows the force history applied to the profile. The force is obtained as the depth-averaged square root of the second invariant of the deviatoric stress at the right model boundary multiplied by the lithosphere thickness. This force measure is always positive, although the sign of the principal horizontal tectonic stress changes. The initial pulse of force produces the initial extension. After cessation of straining the force relaxes by viscous dissipation of deviatoric stresses in the ductile parts of the lithosphere. The gentle reversal yields a slow increase in the force. At 12 Ma a short duration pulse simulates the effects of reorganization of North Atlantic spreading. The total compressive force ends at *c.* 1.7×10^{12} N m^{-1}, which can be compared with an estimated magnitude of *c.* 3.9×10^{12} N m^{-1} of the ridge-push force in an oceanic lithosphere of 100 Ma age (Turcotte & Schubert 1982).

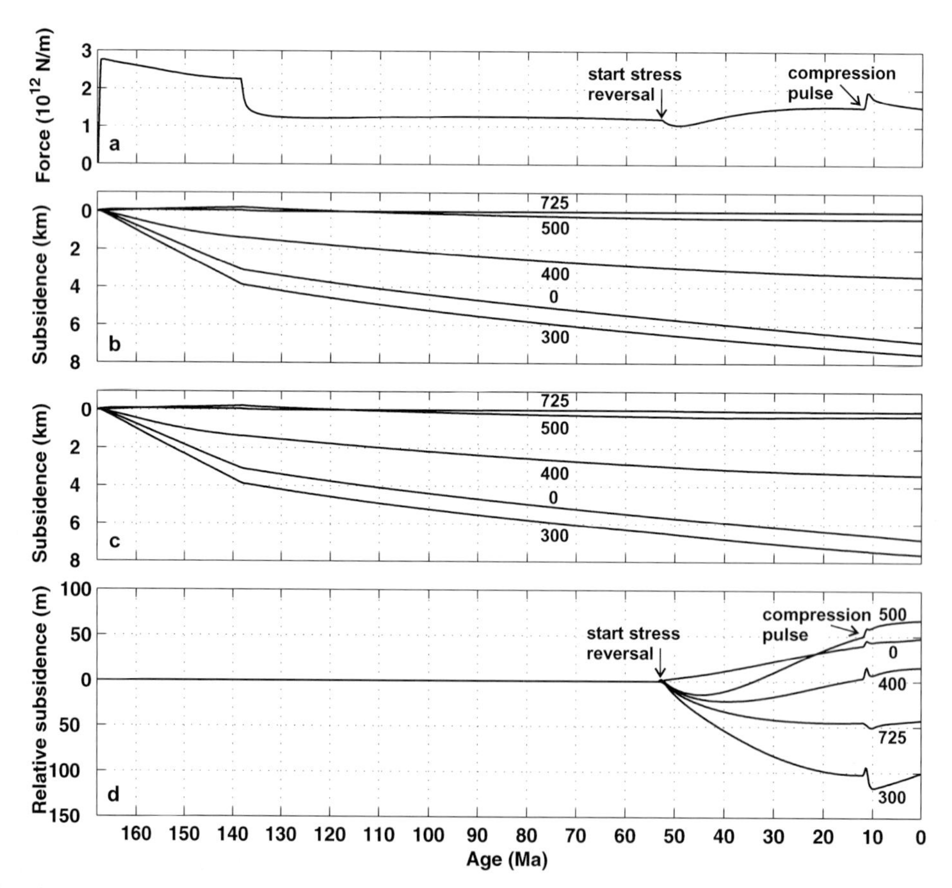

Fig. 13. (**a**) The force history applied to the profile; (**b**) subsidence history without compression; (**c**) subsidence history with compression; (**d**) difference between (**c**) and (**b**).

Figure 13b and c shows the vertical movements of fixed basement points along the profile in the cases without (Fig. 13b) and with stress reversal (Fig. 13c). The location of the basement points and the evolution of cross-sections are shown in Fig. 12 at age 0 Ma.

The initial margin formation results in basin subsidence as tracked by basin points 0 km, 300 km and 400 km. Point 500 km shows the formation of the rift flank during rifting. Point 725 km registers the far-field vertical movements.

Following extension there is thermal subsidence on the passive margin, enhanced by sediment loading. From comparison of Fig. 13b and c it becomes apparent that this general picture continues through the gradual stress reversal at 53 Ma. The small subsidence perturbations produced by the stress reversal are enhanced in Fig. 13d, which shows the difference between the subsidence curves of Fig. 13b and c. The subsidence rates of points 300 km and 400 km increase slightly in the initial phase of stress reversal because of the small elastic volume reduction associated with stress reversal and because of the continuous adjustments to a new equilibrium between in-plane force, buoyancy forces and evolving stresses. For continuing compression the subsidence of the basin points becomes retarded as compared with the case of pure thermal subsidence. The rift flank point (500 km) subsides at an increasing rate in the initial phase of compression, and later becomes slightly uplifted.

The response to the compressional pulse at 12 Ma depends on the position along the profile. Generally, there is a small elevation of the basin floor, which, however, in the case of point 300 km is followed by a minor increase in subsidence. A larger compressional pulse would produce larger-amplitude deflections of the same shape.

Our model applies to a passive margin with laterally smoothly changing rheological properties. In reality, the strain following stress reversal will be accommodated by the reactivation of major basement faults, resulting in localized shortening and the development of inversion structures or domes such as the Helland Hansen Dome and Ormen Lange on the Norwegian continental margin. Marginal troughs with local subsidence may develop in connection with such inversion zones (Nielsen & Hansen 2000).

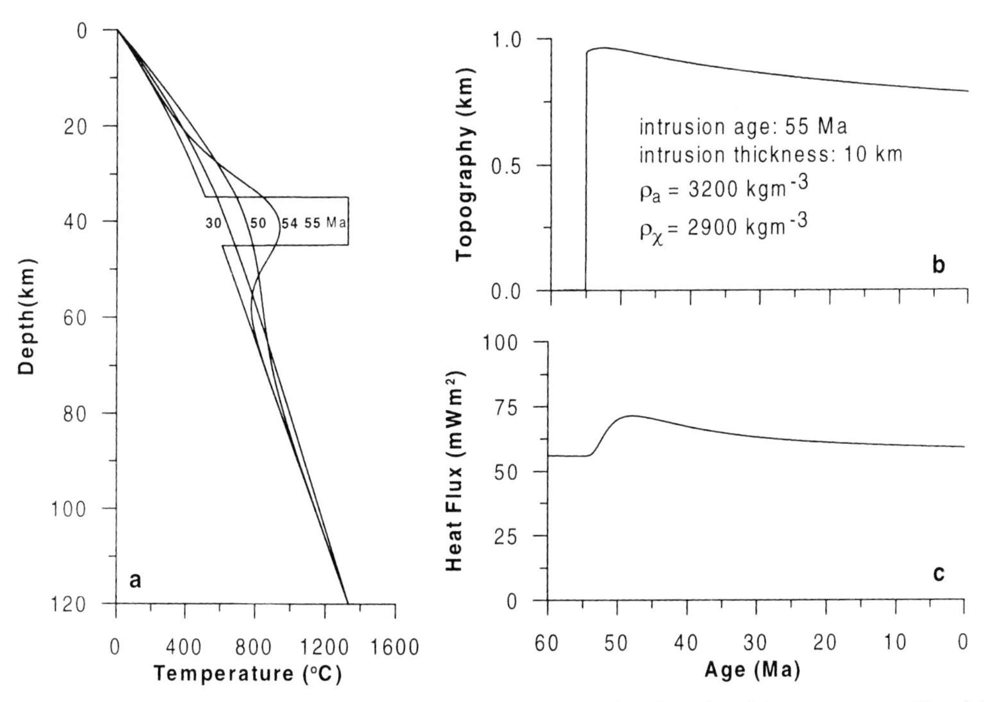

Fig. 14. Thermal and topographic response of lithospheric column to 10 km intrusion: (**a**) temperature profiles; (**b**) topography; (**c**) surface heat flux.

Magmatic underplating. It is well known that magmatic underplating provides a mechanism of immediate epeirogenic uplift (McKenzie 1984; Brodie & White 1994). Figure 14 shows the thermal and topographic response of a 1D lithospheric column to the introduction of a magma at the base of the crust. The intrusion causes immediate surface uplift followed by a minor thermal subsidence as the intrusion cools. The uplift, h, as a result of underplating of thickness χ is given by $h = (1 - \rho_\chi/\rho_a)\chi$, where ρ_χ is the density of the underplating material and ρ_a is the density of the asthenosphere. Erosion of the surface is not considered. The heat flow, initially at 50 mW m^{-2}, increases in less than 10 Ma to 62 mW m^{-2}. At 55 Ma after the intrusion the surface heat flow is essentially back to the initial value.

The underplating mechanism produces uplift in direct response to the underplating history. It is followed by minor thermal subsidence. Depending on the density contrast between the underplating magma and the mantle, *c.* 10 km of underplating is required to produce an initial uplift of 1 km.

Lithospheric delamination. Delamination of the lithosphere, or base lithosphere erosion, means loss in some way of a fraction of the base of the mantle lithosphere. This mechanism has been applied to explaining the sudden uplifts of the Colorado Plateau (Bird 1979) and the Himalayas (Houseman *et al.* 1981). In these cases delamination happens as a consequence of thickening and destabilizing of the mantle lithosphere in collision. In the North Atlantic case delamination must be related somehow to the interaction of the Iceland plume with the lithosphere. We suggest that a gravitational (Rayleigh–Taylor) instability (Conrad & Molnar 1997; Houseman & Molnar 1997) triggered by the arrival of the low-viscosity and low-density Iceland plume may have been responsible for convective removal of a fraction of the mantle lithosphere. For later reference we present here a 1D model of the thermal and topographic consequences of this process.

It is assumed that the lithosphere remains at the reduced thickness after delamination. Any recovery of the lithosphere thickness reduces the topographic response of the partial delamination mechanism. For a full recovery of lithosphere thickness the topography returns to its initial position, unless perhaps the delaminated lithosphere was chemically buoyant and distinct from the remaining lithosphere. The present scenario hence assumes that the lithosphere is

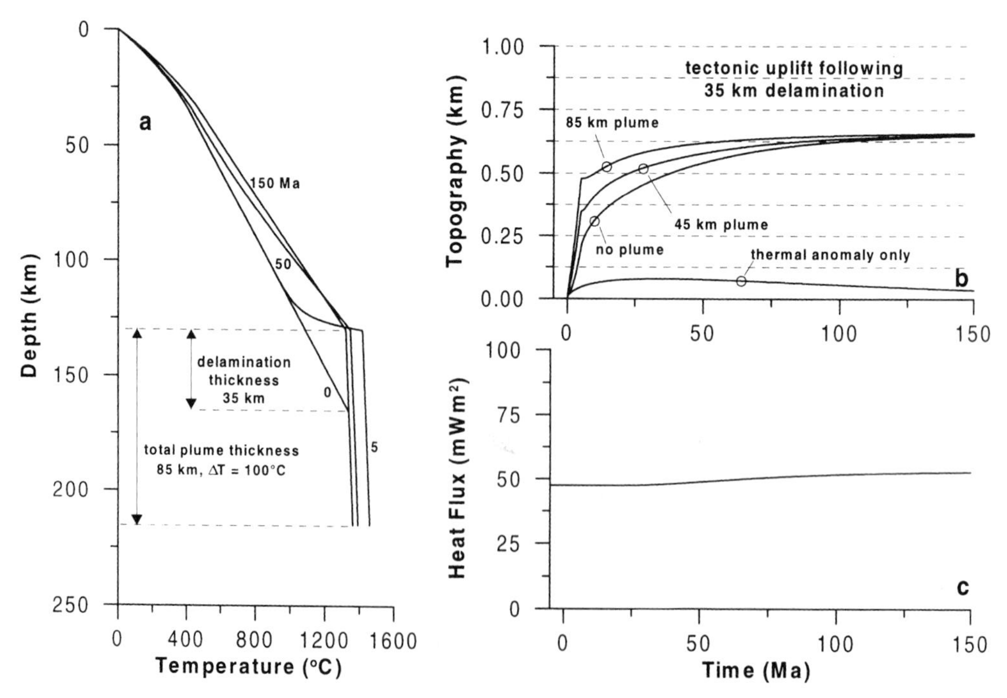

Fig. 15. Thermal and topographic response of lithosphere suffering 35 km of delamination: (**a**) temperature profiles (85 km plume); (**b**) topography in various cases; (**c**) surface heat flux (85 km plume).

delaminated after destabilization by plume emplacement to a new and thinner equilibrium thickness, which may be true if the lithosphere is initially thicker than the present equilibrium thickness.

Figure 15 shows the thermal and topographic response to delamination over 5 Ma of a 1D lithospheric column. Airy isostasy is assumed, and the initial topography is zero. At 0 Ma the plume arrives at the base of the lithosphere, initially 165 km thick. Over a time span of 5 Ma the plume grows in thickness to 45 or 85 km simultaneously with erosion of 35 km of the base of the lithosphere. Other growth times could have been chosen but do not significantly influence the principal results. The plume has an excess temperature of 100 °C, which decays after plume emplacement with a tentative time constant of 40 Ma. Complete modelling of the decay of the excess plume temperature involves modelling of whole-mantle convection, which is beyond the scope of this paper. The cases of delamination without plume thickening, but in the presence of the basal temperature anomaly, and the effect of the basal temperature anomaly alone are shown for comparison.

The lithospheric temperature profiles (Fig. 15a) are for the case of 85 km of plume. The tectonic uplift histories (Fig. 15b) show a phase of rapid initial uplift as plume material replaces denser asthenosphere and 35 km of the lower lithosphere, followed by increasing tectonic uplift caused by heating of the remaining lithosphere to a new thermal equilibrium situation. The magnitude of the initial uplift depends strongly on the plume thickness and the amount of lithospheric delamination. However, the plume contribution decays over time as the plume material cools. The resulting permanent tectonic uplift depends only on the amount of lithospheric delamination. The uplift rate following plume emplacement and delamination depends on the amount of delamination and the initial plume thickness. For given delamination thickness the rate increases with decreasing plume thickness. The surface heat flux (Fig. 15c) shows a minor slow increase. Unlike lithospheric stretching, which also thins the lithosphere, delamination causes no rapid squeezing of the isotherms and therefore shows a much quieter thermal response. For later reference it is noted that the initial, possibly rapid, disturbance of the lithosphere–asthenosphere systems causes tectonic surface uplift to evolve with a time constant of *c.* 60 Ma. A Paleocene–Eocene delamination hence produces increasing cumulative tectonic uplift at a decreasing rate throughout Cenozoic time.

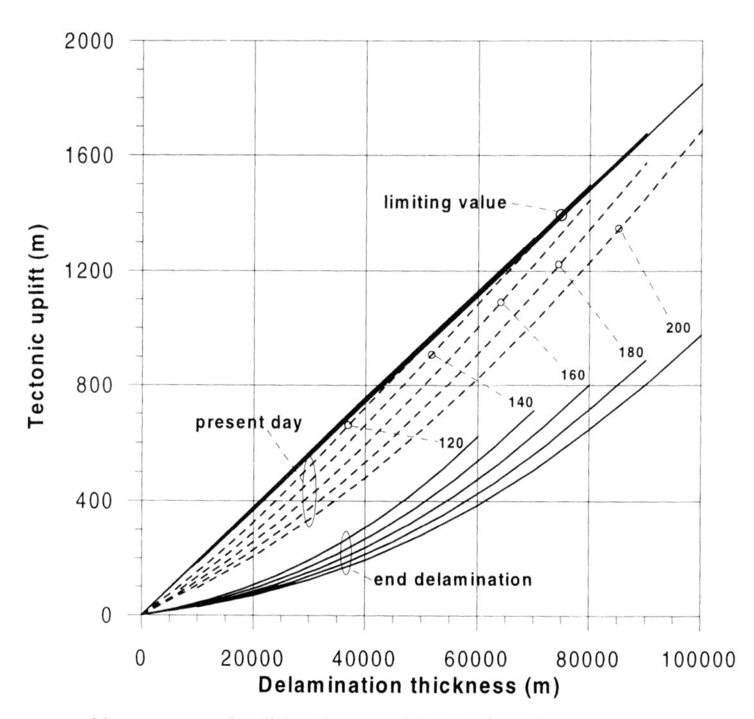

Fig. 16. The topographic response of a lithosphere to delamination. Curve parameter is initial lithosphere thickness.

Figure 16 summarizes the ability of the delamination mechanism to produce permanent tectonic uplift. For simplicity, the transient components caused by plume emplacement and the excess plume temperature are left out; they disappear anyway. The uplift at the end of delamination represents, therefore, the pure initial delamination uplift for the given lithosphere thickness. The dependence of initial uplift on lithosphere thickness is caused by the dependence of the lithospheric geotherm on lithospheric thickness. For a thick lithosphere the delaminated material has an average temperature that is closer to the asthenospheric temperature than it is in the case of a thin lithosphere. Replacing it with asthenosphere therefore yields less excess buoyancy. The present-day values are the tectonic surface uplift after thermal relaxation for 50 Ma following delamination. At that time, a thin lithosphere is almost in thermal equilibrium, whereas a thick lithosphere still has not adjusted to the new thermal equilibrium situation. The limiting value of surface uplift after infinite relaxation time depends linearly on the delamination thickness.

Discussion

Following Doré *et al.* (1999), the North Atlantic continental break-up at 53 Ma was the culmination of a *c.* 350 Ma period of Permo-Triassic, (mainly late) Jurassic, Early Cretaceous, 'mid'-Cretaceous, and latest Cretaceous–Early Eocene rifting in the area of the Caledonian orogen. After these events the topography of NW Europe was low and separated by shallow seas overlying deep Palaeozoic and Mesozoic sedimentary basins.

The final stage of latest Cretaceous–Early Eocene rifting led to break-up and became associated with intense magmatic activity as the Iceland plume arrived at the base of the North Atlantic lithosphere a few million years before break-up (Skogseid *et al.* 2000). The plume front may have extended more than 1500 km from the plume centre, encircling the North Atlantic Volcanic Province, including the British Tertiary Volcanic Province, the Labrador Sea volcanic provinces, and the Greenland, Scottish and Norwegian Caledonides (Skogseid *et al.* 2000).

In the British Isles, where ages of volcanic rocks in the British Tertiary Volcanic Province and large offshore thicknesses of Paleocene and Eocene sequences bear direct witness to the timing of the uplift, there is a good correlation between plume emplacement and clastic sediment influx, which has been used to infer surface uplift variations (White & Lovell 1997). However, regarding the surface uplift of the Norwegian (and Scottish) Caledonides, the direct role of the plume as an initiating agent at the time of emplacement has not been widely considered. On the Norwegian continental shelf and in large parts of the North Sea the direct association between sediment patterns and tectonic activity in the source areas yields minor Paleocene–Eocene tectonic activity with the major phase starting in Oligocene time (see Huuse 2002b) and accelerating through Neogene times. This leaves a time gap of *c.* 30 Ma between plume emplacement and the major phase of surface uplift, and produces major problems explaining the most recent uplift mechanism: why did the plume, which was most vigorous and hot during the short time of emplacement, after which it started to cool, take 30 Ma to produce significant surface uplift?

Sediment structure and fission tracks

We believe that the direct association of intensity of sedimentary response with contemporaneous tectonic activity in the source areas is problematic. Molnar & England (1990) argued that the phenomena commonly used to infer recent uplift may as well be a consequence of climate change. On the basis of examples from the Alps, Pyrenees and the Rocky Mountains of Colorado, Wyoming and Utah, they argued that the existence of geomorphologically young landscapes with deeply incised valleys and a jagged relief, which usually is taken as an indication of recent tectonic uplift, is a consequence of a change in the process of erosion, for example caused by climate change, and in the extreme by glaciers. The denudation process produces apparent uplift by raising the summit envelope by passive isostatic compensation. The North Sea area did feel the global climate change from late Eocene into Oligocene times and did experience deep glacial erosion and a general sea-level fall (Riis & Fjeldskaar 1992; Huuse 2002b). Therefore, in our opinion, the timing of an uplift history based on sediment ages must be tempered with the influence of climate and erosional base level on denudation rates.

The influence of climate and eustasy is not known in great detail, but the cooling history associated with denudation may be inferred directly from apatite fission tracks (AFTs). This technique therefore has been extensively applied in Norway (Rohrman *et al.* 1995) to infer recent cooling and has contributed to the concept of accelerating Neogene uplift of the southern and northern Norwegian domes. However, published models for annealing of fission tracks show a marked lack of sensitivity to temperatures below 60 °C, although it is clear that some annealing does occur at lower temperatures over geological time scales. When modelling real data, this can often lead to the spurious inference of recent cooling from around 60 °C (Gallagher & Brown 1999). The uncertainty inherent in constraining denudation (and any associated uplift) becomes most acute in recently glaciated areas, or where only 1 km or less of material may have been removed. This is a problem in many parts of southern Scandinavia. This recently recognized issue has not been addressed in AFT studies of the area and means that the inferred temperature histories do not properly represent the true cooling history produced by denudation. The recently revived (U–Th)/He technique, applied to apatite, which has a lower temperature sensitivity than fission tracks (e.g. House *et al.* 1998; Wolf *et al.* 1998), may provide the key timing information in this region.

Mechanism of surface uplift

In the absence of direct constraints from sediment ages and cooling histories, the discussion of the timing of active Cenozoic surface uplift around the North Atlantic and the simultaneous basin subsidence becomes most meaningful when the requirement of a physically plausible mechanism is considered.

Compressional stress is not a plausible agent. Our model example shows that the gradual change from extensional to compressional stress, which the peri-North Atlantic region must have experienced following the break-up, has had only minor influence on the subsidence of the basins and uplift of the land areas. For physically reasonable stress levels, stress reversal causes a small enhancement of the rate of thermal subsidence, and a sudden compressional pulse caused by, for example, spreading reorganization causes an elevation of the basin floor. We therefore think that the accelerating late Neogene tectonic subsidence derived by Cloetingh *et al.* (1990, 1992) for the North Atlantic margins and the North Sea area is not related to compressional pulses from spreading reorganization. The space needed to accommodate the Neogene sediments is much too large to be explained by compressional forces. Rather, we suggest that the derived tectonic subsidence may be a consequence of the use of too shallow palaeo-water depths in the tectonic subsidence calculation. Palaeo-water depths are notoriously difficult to assess, yet have significant influence on the results of tectonic subsidence calculations. In our opinion, the apparently accelerating late Neogene subsidence is simply therefore passive filling of already existing accommodation space enhanced by loading-induced subsidence (see Fig. 8a).

We now turn our attention to the role of the Iceland plume in explaining the Cenozoic surface uplift. We adopt the plume emplacement scenario of Skogseid *et al.* (2000): a flattened plume head with an excess temperature of 100 °C (density contrast of *c.* -11 kg m^{-3}) arrives a few million years before break-up time at the base of the North Atlantic lithosphere, which shows marked thickness variations following the Palaeozoic–Mesozoic rifting period.

As described by Skogseid *et al.* (2000), areas of relatively thin lithosphere became associated with the thickest plume column and therefore experienced the largest initial isostatic uplift. These are also the areas that are prone to decompression melting as the plume material rises higher, with the possibility of magmatic activity in the form of flood basalts, magmatic underplating of the crust, and intrusions. The Norwegian continental shelf and the northern North Sea, with a dense data coverage, provide suitable areas for quantification of the sudden isostatic uplift inflicted by plume emplacement.

In the neighbouring continental areas of thicker lithosphere such as western Fennoscandia, Scotland and East Greenland, plume emplacement caused a transient isostatic uplift component as hot plume material displaced denser asthenosphere. Furthermore, if partial delamination of the lithosphere took place, permanent surface uplift would have resulted. It is therefore important to consider the physical reason for delamination, and we propose that the Rayleigh–Taylor instability is a candidate for explaining the process of partial delamination.

Rohrman & van der Beek (1996) applied this principle to explaining the peri-North Atlantic domal uplift pattern. Their variant of the Rayleigh–Taylor instability focuses on the generation of diapirs in the buoyant lower layer, which rise into the rigid lid at selected locations and produce domal uplift. This application utilizes localized upwelling and is analogous to the generation of salt diapirs or thunderstorms. The timing of their scenario requires that plume emplacement beneath the Norwegian and Scottish Caledonides took place around 40–30 Ma, which is in contrast to the early plume emplacement scenario of Skogseid *et al.* (2000).

Our application of the Rayleigh–Taylor instability to the lithosphere–asthenosphere boundary follows the original concept.

Conrad & Molnar (1997) and Houseman & Molnar (1997) have shown in a linearized analysis how disturbances at the lithosphere–asthenosphere boundary with a wavelength between 100 and 200 km can become unstable, resulting in the base of the lithosphere being swept into drips or blobs of descending material. This ultimately results in thinning of the mantle lithosphere as the drop detaches and sinks through the low-viscosity asthenosphere, carrying mantle lithosphere away. The result is localized downwelling, which feeds on lithosphere material from the sides. There are no localized asthenospheric diapirs, but a rather broad asthenosphere upwelling to fill the space vacated by the lost lithosphere. The possible growth rate of the instability ranges from 10^{-14} s^{-1} to higher, corresponding to evolution times of 3 Ma or longer. Depending on the timing of the process it is possible that decompression melting may take place in the upwelling asthenosphere. It may therefore be theoretically possible that magmatic underplating also occurred under the adjacent land areas, explaining part of the uplift. Whether this has happened may be tested by the teleseismic receiver function experiments that are in progress in southern Norway.

The above-quoted analysis was based on continental lithosphere destabilized by compressional thickening. However, we suggest that the arrival of a hot, low-viscosity and buoyant plume at the base of the North Atlantic lithosphere could have triggered the Rayleigh–Taylor instability in locations of relatively deep lithospheric roots or in areas of a steep gradient of the lithosphere–asthenosphere boundary. The northern and southern Norwegian domes, the Scottish Caledonides and East Greenland, in this model, would have undergone particularly severe delamination, perhaps because the lithosphere under these areas after the Palaeozoic–Mesozoic rifting period was left relatively thick and with a steep gradient towards the basin areas. At present, the lithosphere is known to be thick beneath the Fennoscandian Shield to the east. Therefore, a transition to the normal thickness lithosphere on the Norwegian margin must occur somewhere under western Fennoscandia.

Assuming that delamination is possible, we now combine the possible tectonic uplift history inflicted by the Iceland plume on the adjacent land areas with the possibility of increasing rates of denudation induced by the changing Cenozoic climate and the falling erosional base level.

Figure 17 shows the topographic response to 35 km of lithospheric delamination and 45 km of plume emplacement around the time of North Atlantic break-up. The initial topography is set to 200 m, approximately corresponding to the topography that would emerge if the surface of the lithosphere was at sea level during the Cretaceous culmination of global sea level. It is assumed that the rate of denudation increased through Cenozoic time, simulating the effects of climate cooling. Changes of denudation rate in this simplified model are located at the Eocene–Oligocene transition, at mid-Miocene times and at the onset of glaciations.

The height of the average surface increases rapidly during the phase of plume emplacement and delamination, and continues to increase as the remaining lithosphere is heated. As denudation starts to remove material and the rate of lithospheric heating decreases, the rate of surface uplift decreases and eventually starts to decline with the progression of denudation.

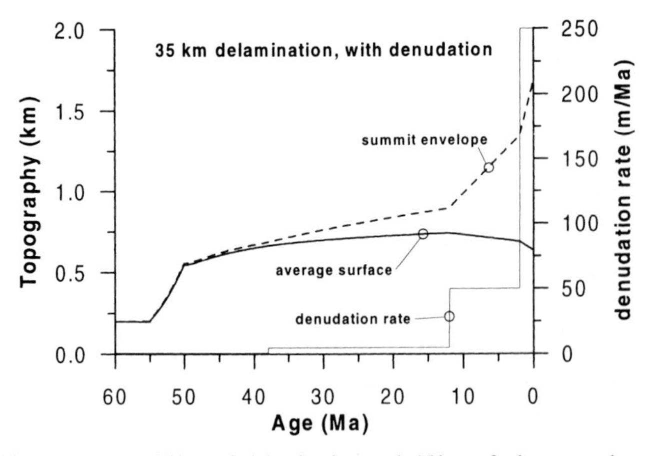

Fig. 17. Topographic response to 35 km of delamination and 45 km of plume emplacement, with varying denudation rate included.

The difference between the summit height envelope (Molnar & England 1990) and the average surface represents the average thickness of material removed by denudation. The rather limited reduction in average surface height is caused by the large isostatic compensation of removed mass and the continuous tectonic surface uplift through Cenozoic time (Fig. 15b). In reality, erosion depth shows large lateral variations from almost uneroded peaks to deeply incised valleys with bottoms far below the average surface height. The summit height envelope therefore shows the elevation at which uneroded isolated peaks could, in principle, occur, if indeed they exist.

The summit height envelope (Fig. 17) increases throughout Cenozoic time with a rate that primarily reflects the changes in the average denudation rate. In the case of Fig. 17 it reaches a maximum value of *c.* 1700 m for an average surface of *c.* 700 m. Other examples with different ratios of summit envelope and average surface height can be produced to fit local features of the Norwegian, Scottish, or East Greenland topography by varying the initial topography and lithosphere thickness, delamination thickness, duration of the delamination process, and denudation rates.

Gravity

The relatively deep and narrow Bouguer gravity anomaly (Fig. 1) indicates that topography in western Scandinavia is compensated mainly at Moho depth. In southern Norway in particular, there is a good correlation between average surface topography and Moho depth. In northern Norway, topography compensation by lateral density variations of the crust seems to play a more significant role. An outstanding question, not addressed in detail in this paper, is how this gravity anomaly is related to the possible mechanism of land surface uplift.

The emergence of the close correlation between topography and Bouguer gravity (Balling 1984) is most easily understood if magmatic underplating at basal crustal levels supports the topography. However, for the amount of underplating needed the absence of dykes and volcanic rocks of the relevant age at the surface is curious.

The net gravitational effect of partial delamination of the lithosphere is a significant negative Bouguer anomaly. Superimposing this anomaly on the gravitational field inherited from the Mesozoic rifting phase produces a negative Bouguer field correlated with topography. However, because of the depth of origin, the anomaly becomes long wavelength, unlike the relatively narrow anomaly observed. A preliminary forward calculation of the gravity response shows that the width of the gravity anomaly associated with delamination is significantly reduced when the initially domal topography is eroded to increasing depth with increasing distance from the centre of the uplift. This denudation pattern brings the Moho closer to the surface by isostatic compensation and adds a positive gravity contribution, which narrows the originally wide gravity anomaly so that it becomes consistent with observations. The model requires of the order of 1.0–2.0 km of Cenozoic denudation in the coastal areas of Norway, in concordance with the results of Riis & Fjeldskaar (1992) and Riis (1996).

Heat flow

The thermal effects associated with magmatic underplating (Fig. 14c) and delamination (Fig. 15c) are insignificant at the present day. We therefore should expect the surface heat flux in Norway and on the Norwegian passive margin to be close to the values of a standard passive margin. This is supported by the evidence presented by Balling (1993), who found heat flux in Norway to be 50–60 mW m^{-2}, similar to the flux in the rest of Scandinavia. The well-defined heat flux–oceanic age relationship found by Sundvor *et al.* (2000) only supports the picture of a generally normal passive margin thermal regime.

Conclusions

In this paper we have outlined a scenario for the Cenozoic surface uplift history of western Scandinavia, which is directly linked to plume emplacement and is consistent with topography, heat flow, crustal structure, and the Bouguer gravity of western Scandinavia. The scenario may also be applicable to East Greenland.

The fundamental surface uplift mechanism is furnished by partial lithospheric delamination by the Rayleigh–Taylor instability, which was triggered by the arrival of the Iceland plume at the base of the North Atlantic lithosphere. Delamination preferentially occurs in areas of deep lithospheric roots or at steep slopes of the lithosphere–asthenosphere boundary. The delamination mechanism allows for a delay of surface uplift according the finite time scale for evolution of the instability. Transient surface uplift continued after cessation of delamination because of lithospheric heating to a warmer equilibrium state.

This scenario calls for the occurrence of (1) significant climate and erosional base-level control on denudation, for the denudation response in the basins to become consistent with the observed sediment pattern, and (2) differential Cenozoic denudation, with denudation depths increasing away from the central highland areas towards the west coast of Norway, for the Bouguer gravity field associated with lithospheric delamination to become consistent with observations.

Continuing $\delta^{18}O$ profiling in the Paleocene–Oligocene interval in the Danish sector, receiver function experiments and joint apatite fission-track and (U–Th)/He analysis in southern Norway, and numerical modelling of the geodynamic processes and Bouguer gravity are expected to contribute to resolving these remaining questions.

References

BALLING, N. 1984. Gravity and isostasy in the Baltic Shield. *In*: GALSON, D.A. & MUELLER, S. (eds) *First EGT Workshop: The Northern Segment*. European Science Foundation, Strasbourg, 53–68.

BALLING, N. 1993. Heat flow and thermal structure of the lithosphere across the Baltic Shield and northern Tornquist Zone. *Tectonophysics*, **244**, 13–50.

BARTEK, L.R., VAIL, P.R., ANDERSON, J.B., EMMET, P.A. & WU, S. 1991. Effect of Cenozoic ice sheet fluctuations in Antarctica on the stratigraphic signature of the Neogene. *Journal of Geophysical Research*, **96**, 6753–6778.

BIRD, P. 1979. Continental delamination of the Colorado Plateau. *Journal of Geophysical Research*, 7561–7571.

BRAUN, J. & BEAUMONT, C. 1989. A physical explanation of the relationship between flank uplifts and the breakup unconformity at rifted continental margins. *Geology*, **17**, 760–764.

BRODIE, J. & WHITE, N. 1994. Sedimentary basin inversion caused by igneous underplating: Northwest European continental shelf. *Geology*, **22**, 147–150.

BUCHARDT, B. 1978. Oxygen isotope palaeotemperatures from the Tertiary period in the North Sea area. *Nature*, **275**, 121–123.

CLOETINGH, S., GRADSTEIN, F.M., KOOI, H., GRANT, A.C. & KAMINSKI, M. 1990. Plate reorganization: a cause of rapid late Neogene subsidence and sedimentation around the North Atlantic? *Journal of the Geological Society, London*, **147**, 495–506.

CLOETINGH, S., MCQUEEN, H. & LAMBECK, K. 1985. On a tectonic mechanism for regional sea level variations. *Earth and Planetary Science Letters*, **75**, 157–166.

CLOETINGH, S., REEMST, P., KOOI, H. & FANAVOLL, S. 1992. Intraplate stresses and the post-Cretaceous uplift and subsidence in northern Atlantic basins. *Norsk Geologisk Tidsskrift*, **72**, 229–235.

CONRAD, P.C. & MOLNAR, P. 1997. The growth of Rayleigh–Taylor instabilities in the lithosphere for various rheological and density structures. *Geophysical Journal International*, **129**, 95–112.

DONNELLY, T. 1982. Worldwide continental denudation and climatic deterioration during the late Tertiary: evidence from deep-sea sediments. *Geology*, **10**, 451–454.

DORÉ, A.G. 1992. The Base Tertiary Surface of southern Norway and the northern North Sea. *Norsk Geologisk Tidsskrift*, **72**, 259–265.

DORÉ, A.G., LUNDIN, E.R., JENSEN, L.N., BIRKELAND, Ø., ELIASSEN, P.E. & FICHLER, C. 1999. Principal tectonic events in the evolution of the northwest European Atlantic margin. *In*: FLEET, A.J. & BOLDY, S.A.R. (eds) *Petroleum Geology of Northwest Europe: Proceedings of the 5th Conference*. Geological Society, London, 41–61.

GALLAGHER, K. & BROWN, R. 1999. Denudation and uplift at passive margins: the record on the Atlantic margin of southern Africa. *Philosophical Transactions of the Royal Society of London*, **357**, 835–859.

GEMMER, L., NIELSEN, S.B. & LYKKE-ANDERSEN, H. 2002. Differential vertical movements in the eastern North Sea area from 3D thermo-mechanical finite element modelling. *Bulletin of the Geological Society of Denmark*, **49**, in press.

GILCHREST, A.R., SUMMERFIELD, M.A. & COCKBURN, H.A.P. 1994. Landscape dissection, isostatic uplift and the morphologic development of orogens. *Geology*, **22**, 963–966.

GLOBE TASK TEAM *et al.* (eds) 1999. *The Global Land One-kilometer Base Elevation (GLOBE) Digital Elevation Model, Version 1.0*. National Oceanic and Atmospheric Administration, National Geophysical Data Center, Boulder, CO. Digital database available at http://www.ngdc.noaa.gov/seg/topo/globe.shtml and as CD-ROMs.

HAQ, B.U., HARDENBOL, J. & VAIL, P. 1987. Chronology of fluctuating sea levels since the Triassic. *Science*, **235**, 1156–1167.

HOUSE, M.A., WERNICKE, B.P. & FARLEY, K.A. 1998. Dating topographic uplift of the Sierra Nevada, California, using apatite (U–Th)/He ages. *Nature*, **396**, 66–69.

HOUSEMAN, G.A. & MOLNAR, P. 1997. Gravitational (Rayleigh–Taylor) instability of a layer with non-linear viscosity and convective thinning of continental lithosphere. *Geophysical Journal International*, **128**, 125–150.

HOUSEMAN, G., MCKENZIE, D. & MOLNAR, P. 1981. Convective instability of a thickened boundary layer and its relevance for the thermal evolution of continental convergent belts. *Journal of Geophysical Research*, **86**, 6115–6132.

HUUSE, M. 2002*a*. Late Cenozoic palaeogeography of the eastern North Sea Basin: climatic vs. tectonic forcing of basin margin uplift and deltaic progradation. *Bulletin of the Geological Society of Denmark*, **49**, in press.

HUUSE, M. 2002*b*. Cenozoic uplift and denudation of southern Norway: insights from the North Sea Basin. *In*: DORÉ, A.G., CARTWRIGHT, J.A.,

STOKER, M.S., TURNER, J.P. & WHITE, N. (eds) *Exhumation of the North Atlantic Margin: Timing, Mechanisms and Implications for Petroleum Exploration*. Geological Society, London, Special Publications, **196**, 209–233.

JAPSEN, P. 1998. Regional velocity–depth anomalies, North Sea chalk: a record of overpressure and Neogene uplift and erosion. *AAPG Bulletin*, **82**, 2031–2074.

JAPSEN, P. & CHALMERS, J.A. 2000. Neogene uplift and tectonics around the North Atlantic: overview. *Global & Planetary Change*, **24**, 165–173.

JENSEN, L.N. & SCHMIDT, B. 1992. Late Tertiary uplift and erosion in the Skagerrak area: magnitude and consequences. *Norsk Geologisk Tidsskrift*, **72**, 275–279.

JOHNSON, C. & GALLAGHER, K. 2000. A preliminary Mesozoic and Cenozoic denudation history of the North East Greenland Margin. *Global and Planetary Change*, **24**, 261–274.

KARNER, G.D. 1989. Effects of lithospheric in-plane stress on sedimentary basin stratigraphy. *Tectonics*, **5**, 573–588.

KINK, J.J., HUSEBYE, E.S. & LARSSON, F.R. 1993. The Moho depth distribution in Fennoscandia and the regional tectonic evolution from Archean to Permian times. *Precambrian Research*, **64**, 23–51.

KOMINZ, M.A., MILLER, K.G. & BROWNING, J.V. 1998. Long-term and short-term global Cenozoic sea-level estimates. *Geology*, **26**, 311–314.

LAMBECK, K. 1983. Structure and evolution of the intracratonic basins of central Australia. *Geophysical Journal of the Royal Astronomical Society*, **74**, 843–886.

LARSEN, H.C. 1990. The East Greenland Shelf. *In*: GRANTZ, A., JOHNSON, L. & SWEENEY, J.F. (eds) *The Arctic Ocean Region*. Geological Society of America, Boulder, CO, 185–210.

LIDMAR-BERGSTRÖM, K. 1991. Phanerozoic tectonics in southern Sweden. *Zeitschrift für Geomorphologie, Neue Folge*, **82**, 1–16.

LIDMAR-BERGSTRÖM, K., OLLIER, C.D. & SULEBAK, J.R. 2000. Landforms and uplift history of southern Norway. *Global and Planetary Change*, **24**, 211–231.

LITHGOW-BERTELLONI, C. & SILVER, P.G. 1998. Dynamic topography, plate driving forces and the African superswell. *Nature*, **395**, 269–272.

MCKENZIE, D. 1978. Some remarks on the development of sedimentary basins. *Earth and Planetary Science Letters*, **40**, 25–32.

MCKENZIE, D. 1984. A possible mechanism for epeirogenic uplift. *Nature*, **307**, 616–618.

MILLER, K.G., MOUNTAIN, G.S. & BROWNING, J.V. 1998. & 5 OTHERS Cenozoic global sea level, sequences, and the New Jersey Transect: results from coastal plain and continental slope drilling. *Reviews of Geophysics*, **36**, 569–601.

MOLNAR, P. & ENGLAND, Ph. 1990. Late Cenozoic uplift of mountain ranges and global climate change: chicken or egg? *Nature*, **346**, 29–34.

NIELSEN, S.B. 2002. A post-mid Cretaceous North Sea model. *Bulletin of the Geological Society of Denmark*, **49**, in press.

NIELSEN, S.B. & HANSEN, D.L. 2000. The formation and evolution of inversion structures and marginal troughs. *Geology*, **28**, 875–878.

PAULSEN, G.E., HANSEN, D.L. & NIELSEN, S.B. 2001. Thermo-mechanical modelling of subsidence effects induced by changing in-plane stresses at passive continental margins and in intra-cratonic basins. (Abstract volume, EUG XI.). *Terra Nova*, 710.

RIIS, F. 1996. Quantification of Cenozoic vertical movements of Scandinavia by correlation of morphological surfaces with offshore data. *Global and Planetary change*, **12**, 331–357.

RIIS, F. & FJELDSKAAR, W. 1992. On the magnitude of the Late Tertiary and Quaternary erosion and its significance for the uplift of Scandinavia and the Barents Sea. *In*: LARSEN, R.M., BREKKE, H., LARSEN, B.T. & TALLERAAS, E. (eds) *Structural and Tectonic Modelling and its Application to Petroleum Geology*. Norwegian Petroleum Society Special Publication, **1**, 163–185.

ROHRMAN, M. & VAN DER BEEK, P. 1996. Cenozoic postrift domal uplift of North Atlantic margins: an asthenospheric diapirism model. *Geology*, **24**, 901–904.

ROHRMAN, M., VAN DER BEEK, P., ANDRIESSEN, P. & CLOETINGH, S. 1995. Meso-Cenozoic morphotectonic evolution of southern Norway: Neogene domal uplift inferred from apatite fission track thermochronology. *Tectonics*, **14**, 704–718.

SKOGSEID, J., PLANKE, S., FALEIDE, J.I., PEDERSEN, T., ELDHOLM, O., NEVERDAL, F., *et al.* 2000. NE Atlantic continental rifting and volcanic margin formation. *In*: NØTTVEDT, A. (ed.) *Dynamics of the Norwegian Margin*. Geological Society, London, Special Publications, **167**, 295–326.

SORGENFREI, T. & BERTHELSEN, O. 1954. *Geology and water well boring*. Geological Survey of Denmark, III. Series, no. 31.

STUEVOLD, L.M. & ELDHOLM, O. 1996. Cenozoic uplift of Fennoscandia inferred from a study of the mid-Norwegian margin. *Global and Planetary Change*, **12**, 359–386.

SUNDVOR, E., ELDHOLM, O., GLADCZENKO, T.P., PLANKE, S., *et al.* 2000. Norwegian–Greenland Sea thermal field. *In*: NØTTVEDT, A. (ed.) *Dynamics of the Norwegian Margin*. Geological Society, London, Special Publication, **167**, 397–410.

TURCOTTE, D.L. & SCHUBERT, G. 1982. *Geodynamics. Applications of Continuum Physics to Geological Problems*. Wiley, New York.

WHIPPLE, K.X., KIRBY, E. & BROCKLEHURST, S.H. 1999. Geomorphic limits to climate-induced increases in topographic relief. *Nature*, **401**, 39–43.

WHITE, N. & LOVELL, B. 1997. Measuring the pulse of a plume with the sedimentary record. *Nature*, **387**, 888–891.

WOLF, R.W., FARLEY, K.A. & KASS, D.A. 1998. Sensitivity analysis of the apatite (U–Th)/He chronometer. *Chemical Geology*, 105–114.

A structural transect between the central North Sea Dome and the South Swedish Dome: Middle Jurassic–Quaternary uplift–subsidence reversal and exhumation across the eastern North Sea Basin

OLE GRAVERSEN

Geological Institute, University of Copenhagen, Øster Voldgade 10, DK-1350 Copenhagen K, Denmark (e-mail: oleg@geo.geol.ku.dk)

Abstract: The Jurassic–Cenozoic structural evolution of the eastern North Sea Basin is influenced by the central North Sea Dome, the Danish Megablock, the Tornquist Zone and the South Swedish Dome. The central North Sea Dome is a composite dome comprising the Triple Junction Dome, the Central Graben Dome and the Friesland Dome. The Danish Megablock, newly recognized here, is a first-order tectonic element between the Central Graben and the Tornquist Zone. In Jurassic–Cretaceous time it was tilted towards the east during uplift of the Central Graben Dome, whereas the movement was reversed during the Cenozoic post-rift subsidence. Contemporaneous with the westward tilting of the Danish Megablock, the South Swedish Dome was uplifted to the east. The uplift–subsidence reversal across the eastern North Sea Basin links the collapse of the Central Graben Dome and the tilt reversal of the Danish Megablock with the uplift of the South Swedish Dome. The uplift followed by subsidence probably involved mass flow in the asthenosphere to account for the observed balance between post-rift subsidence and marginal uplift. The model explains the uplift of both the South Swedish Dome and southern England as the result of Cenozoic post-rift subsidence of the Mesozoic Central Graben Dome.

Uplift and erosion can only be inferred from the rock record, as material is removed during exhumation. As a result, studies of sediment distribution in adjoining basins and geomorphological studies of uplifted areas give an indication of the balance between uplift and subsidence (e.g. Riis 1996). Structural analysis may help to identify crustal movement and processes. Petrographic and geophysical methods (vitrinite reflectance, analysis of density and sonic velocity trends) allow estimates of maximum burial to be made and fission-track analysis points to the subsidence–uplift history (see Japsen & Chalmers 2000, and references therein).

A primary objective of this paper is to analyse the Cenozoic uplift and erosion of the eastern North Sea Basin and adjoining parts of the Baltic Shield in southern Sweden. This requires analysis of Mesozoic rifting in the central North Sea, which was the precursor to the Cenozoic post-rift sag basin. The paper presents a structural analysis of first-order tectonic elements, namely the central North Sea Dome (Ziegler 1990; Underhill & Partington 1993), the South Swedish Dome (Lidmar-Bergström 1988, 1991, 1993), and the Danish Megablock, which is established here as a new tectonic element (Fig. 1). The backbone of the paper is a geological cross-section that combines a cross-section based on a regional seismic line (RTD-81-22) across the central and eastern North Sea Basin with a cross-section across southern Sweden (Figs 1 and 2). From the cross-section and regional maps it is possible to demonstrate basin evolution from initial Jurassic rifting, accompanied by volcanicity and uplift of the central North Sea Dome, to Cenozoic basin subsidence and Neogene uplift of the South Swedish Dome.

Upper-crustal configuration of the eastern North Sea Basin

The present structure of the eastern North Sea Basin is outlined by the depth-structure of the uppermost pre-Zechstein surface and the configuration of overlying sedimentary basins (Ziegler 1990; Vejbæk & Britze 1994). Major tectonic elements are the Central Graben, the Danish Megablock, the Tornquist Zone and the

From: DORÉ, A.G., CARTWRIGHT, J.A., STOKER, M.S., TURNER, J.P. & WHITE, N. 2002. *Exhumation of the North Atlantic Margin: Timing, Mechanisms and Implications for Petroleum Exploration*. Geological Society, London, Special Publications, **196**, 67–83. 0305-8719/02/$15.00

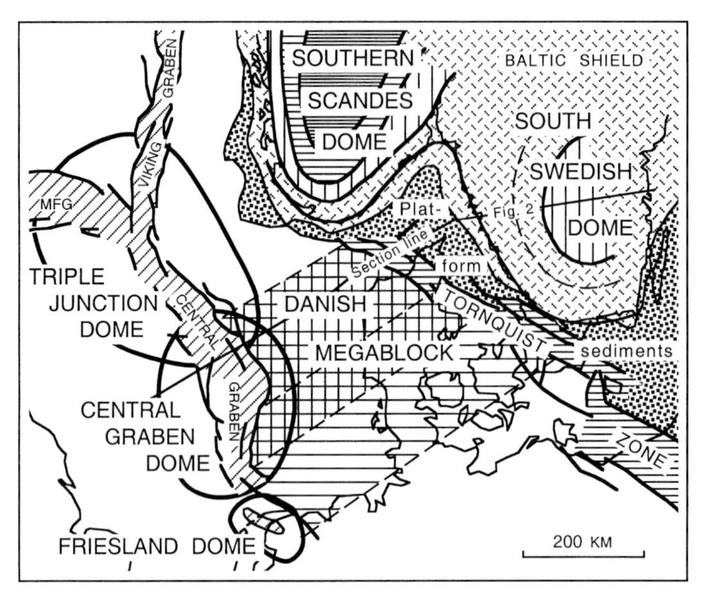

Fig. 1. Jurassic–Quaternary tectonic elements of the eastern North Sea Basin and adjoining border of the Fennoscandian High. Line indicates position of cross-section in Fig. 2. The Triple Junction Dome, the Central Graben Dome and the Friesland Dome are dome centres within the composite central North Sea Dome. MFG, Moray Firth Graben.

South Swedish Dome (Baltic Shield) rimmed by the Skagerrak–Kattegat Platform (Figs 1 and 2). The crystalline basement of the South Swedish Dome extends westwards into the Tornquist Zone and the Danish Megablock, where it subcrops the Mesozoic and Palaeozoic sediments (EUGENO-S Working Group 1988; Michelsen & Nielsen 1993; Vejbæk & Britze 1994). The Danish Megablock is thus separated from the South Swedish Dome by the fault-bounded Tornquist Zone.

The post-Palaeozoic sedimentary cover of the Danish Megablock can be divided into a Mesozoic series thinning towards the west and an overlying Cenozoic interval that thins towards the east (Figs 2 and 3). The opposed thinning directions relate to Jurassic–Cretaceous domal uplift along the Central Graben and eastward tilting of the Danish Megablock. This movement was reversed in Cenozoic time, when a westward tilting brought the uppermost pre-Zechstein surface back down to subhorizontal level. The uplift followed by subsidence demonstrates the impact of the central North Sea Dome (Ziegler 1990) on the Danish Megablock that occupied the eastern part of the composite dome. The central North Sea Dome, the Danish Megablock and the South Swedish Dome are thus major tectonic constituents in the evolution of the central and eastern North Sea Basin since Jurassic time.

The central North Sea Dome

The central North Sea Dome is a Jurassic arch or composite palaeo-dome situated in the central North Sea. The subaerial part of the dome can be identified by a regional early Middle Jurassic erosional unconformity (the 'Mid-Cimmerian' unconformity) (Ziegler 1990; Underhill & Partington 1993). The dome is cut by the Central Graben, the south Viking Graben and the Moray Firth Basin, meeting in a triple junction. The uplift of the Jurassic dome was initiated during the Early–Middle Jurassic transition (late Toarcian–early Aalenian time), and the largest area uplifted above the erosional base was apparently reached during Aalenian–early Bajocian time (Ziegler 1990; Underhill & Partington 1993) persisting until late Jurassic time. The centre of the dome was situated above the graben triple junction and the Bajocian–Bathonian volcanic centre to the north (Fig. 4). Whereas the Triple Junction Dome was (partly) flooded during Late Jurassic time (Underhill & Partington 1993), Palaeozoic sediments and basement subcrop the base Cretaceous unconformity along the margins of the southern Central Graben and the Friesland High (Fig. 4). Subcrop patterns on the pre-Cretaceous geological map can be used to identify two Late Jurassic dome centres to the south: one along the Danish Central Graben rift axis, the Central Graben

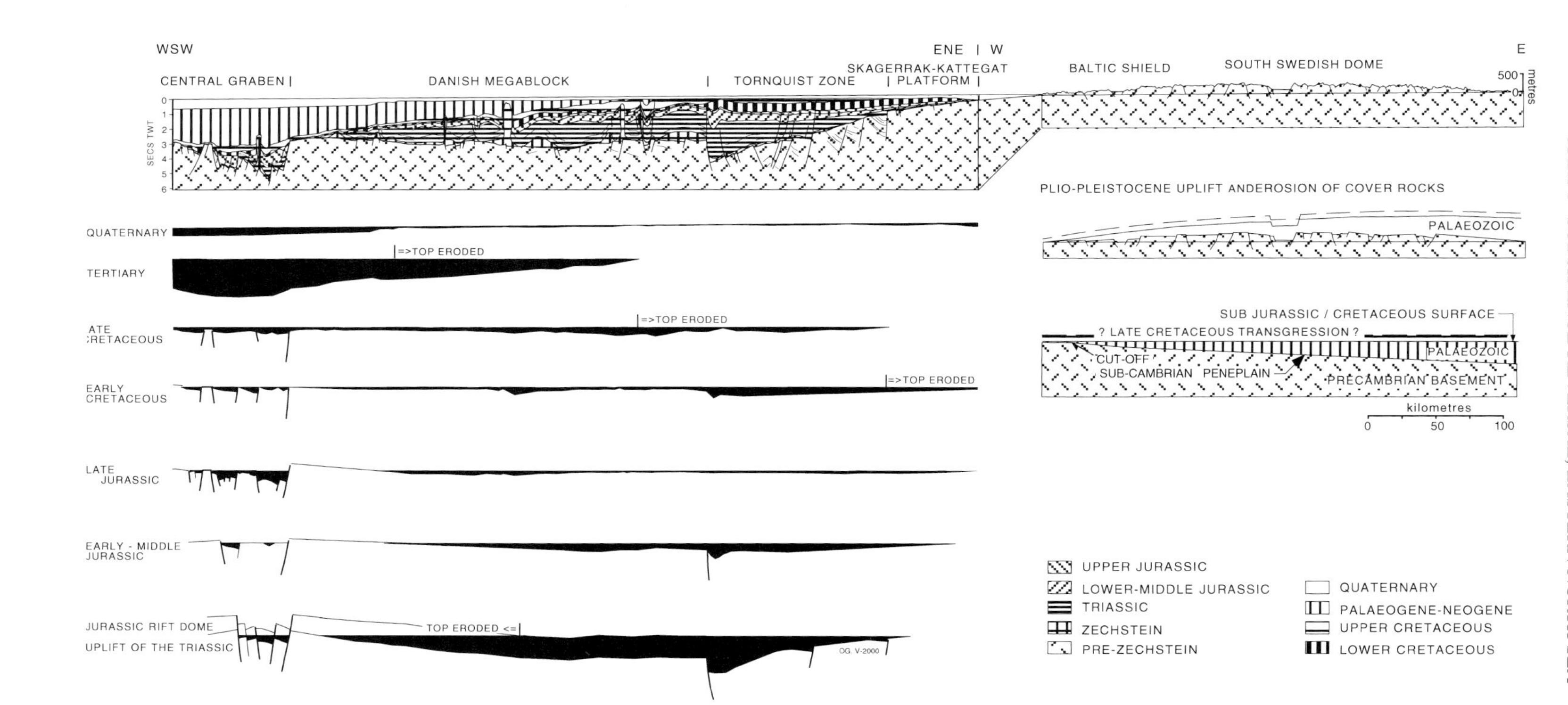

Fig. 2. Geosection across the eastern North Sea Basin and the South Swedish Dome on the Baltic Shield. The offshore interval is based on seismic line RTD-81-22 adopted from Vejbæk (1997); the onshore South Swedish Dome is adopted from Lidmar-Bergström (1988). Position of section line is indicated in Fig. 1.

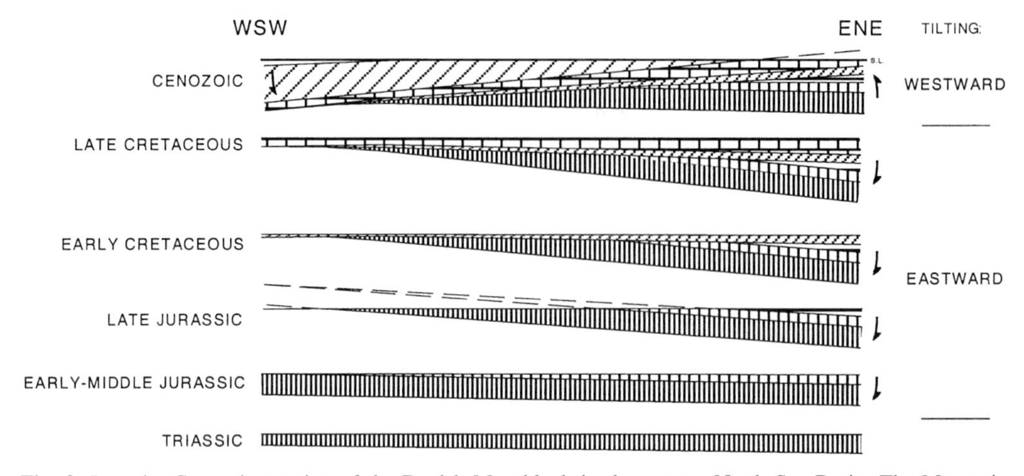

Fig. 3. Jurassic–Cenozoic rotation of the Danish Megablock in the eastern North Sea Basin. The Mesozoic movement was dominated by greater subsidence in the east with uplift and early erosion in the west. At the Mesozoic–Cenozoic transition the movement was reversed; the base Mesozoic surface subsided in the west, whereas uplift and erosion dominated in the east.

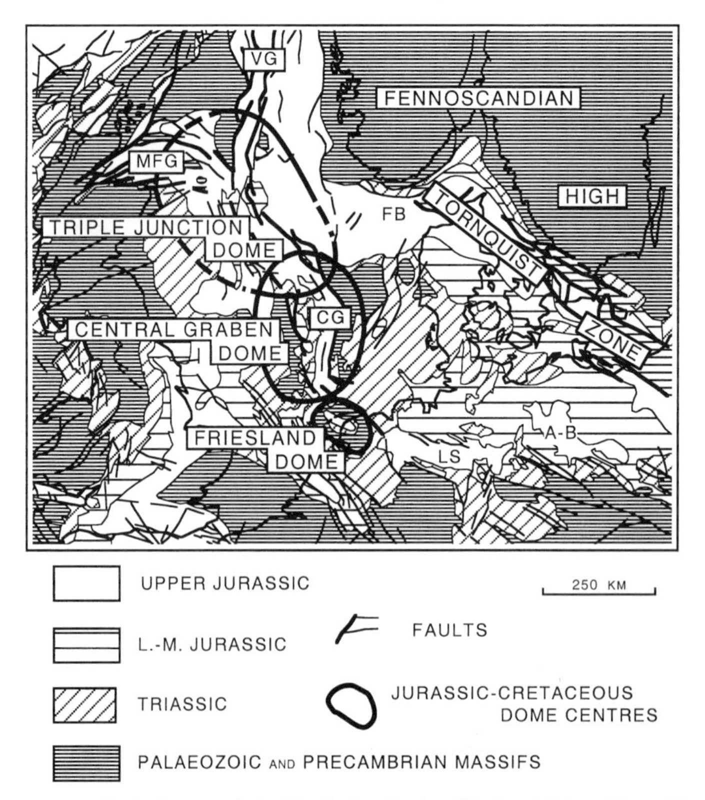

Fig. 4. Pre-Cretaceous geological map of the North Sea Basin (Ziegler 1990) with position of the Middle Jurassic–Cretaceous dome centres of the central North Sea Dome. (See text for further explanation.) A-B, Altmark–Brandenburg Basin; CG, Central Graben; FB, Farsund Basin; LS, Lower Saxony Basin; MFG, Moray Firth Graben; VG, Viking Graben.

Dome, and one along the axial depression of the Friesland High, the Friesland Dome (Fig. 4).

On the basis of the above observations it is concluded that the central North Sea Dome was a composite dome formed by the Triple Junction Dome to the north, the Central Graben Dome, and the Friesland Dome to the south (Fig. 4). The long axis of the oval-shaped domes changes with the orientation of the Central Graben axis. The Triple Junction Dome and the Friesland Dome trend NW–SE, and are connected by the Central Graben Dome, which shows an overall north–south trend with the Danish Megablock on its eastern flank. The rise of the composite dome was initiated to the north in the triple junction area (early Middle Jurassic time). The zone of uplift then gradually moved southward to the Central Graben Dome (early Late Jurassic time) and the Friesland Dome (late Late Jurassic time) whereas the Triple Junction Dome to the north suffered deflation.

The Danish Megablock

The Danish Megablock is a crustal block extending from the Central Graben in the west to the Tornquist Zone in the east (Figs 1 and 2). To the north it is bounded by the Farsund Basin (Fig. 4). The southern boundary is less distinct and its position is suggested towards the Lower Saxony and Altmark–Brandenburg basins.

During early Palaeozoic time, the area that became the Danish Megablock formed an integral part of the Baltica Platform (EUGENO-S Working Group 1988; Berthelsen *et al.* 1992). Decoupling of the Danish Megablock from the Baltica Platform–Baltic Shield took place in Triassic time. In Middle Jurassic–Cretaceous time, the Danish Megablock formed the eastern block of the Central Graben Dome as part of the central North Sea dome complex (Fig. 1).

The Jurassic–Cenozoic evolution of the Danish Megablock documents the evolution to the east of the rift axis of the Central Graben Dome. The main trends of the uplift or subsidence can be evaluated from regional erosional unconformities and isopach thicknesses. Analysis of the Danish Megablock has also included the pre-Cretaceous geological map (Fig. 4), as well as the Berriasian palaeotectonic map and isopach maps of Triassic, Lower Jurassic, and Lower and Upper Cretaceous rocks (Ziegler 1990). The restored intervals (by simple horizon flattening) illustrate the uplift and erosion of the Triassic deposits along the Central Graben rift, and show that the Jurassic intervals only partly covered the Central Graben Dome (Fig. 2). The Central Graben Dome was gradually flooded during Early Cretaceous time and remained submerged during Late Cretaceous time (Ziegler 1990). The westward thinning of the Cretaceous deposits (Fig. 2) indicates continued eastward tilting of the Danish Megablock during Cretaceous time. A schematic model of the rotation of the Danish Megablock is illustrated in Fig. 3.

In Cenozoic time the subsidence of the Danish Megablock shifted to the west along the Central Graben post-rift thermal sag basin (Fig. 2). The onset of the Cenozoic period thus marks a tilt reversal of the Danish Megablock towards the west, in contrast to the eastward tilting in Mesozoic time (Fig. 3). Progressively older pre-Quaternary sediments are encountered towards the NE until the Precambrian crystalline basement of the Baltic Shield is reached (Jensen & Michelsen 1992; Japsen 1998). This suggests that the regional uplift and erosion along the eastern margin of the North Sea Basin may have corresponded in part to the subsidence of the central North Sea.

The South Swedish Dome

The South Swedish Dome is a broad, dome-like structure identified by the topography of the present-day Precambrian surface in southern Sweden (Lidmar-Bergström 1988, 1991, 1993, 1999). The highest elevation of *c.* 400 m above sea level is encountered around the southern Vättern Graben. Lidmar-Bergström (1988, 1991, 1993) has distinguished three groups of palaeosurfaces: a sub-Cambrian peneplain, a pre-Cretaceous surface of supposed Permo-Triassic age that has 'sub-Mesozoic hilly relief', and the Paleogene South Småland Peneplain (Fig. 5a). Segments of the sub-Cambrian peneplain and the South Småland Peneplain radiate out from the centre of the dome and are cut out by the more steeply dipping pre-Cretaceous surface encountered along the margins (Fig. 5a and d). An exception to this general pattern is formed by an oval-shaped, block-faulted depression situated to the SE of the Vättern Graben. The structure is interpreted as a crestal collapse structure (Fig. 5a). Lower Palaeozoic cover rocks are found in a continuous belt to the east and as outliers in the central and northern part of the dome. The steeper inclination of the pre-Cretaceous surface relative to the sub-Cambrian peneplain indicates that uplift of the South Swedish Dome postdates the development of the pre-Cretaceous surface. It is believed that the Late Cretaceous transgression may have resulted in a cover of chalk being deposited over southern

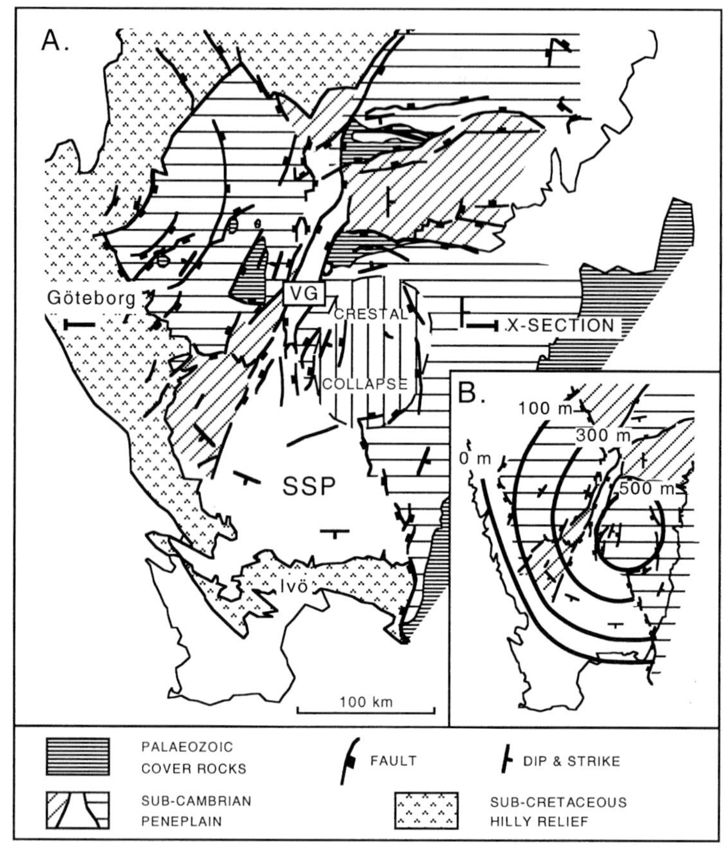

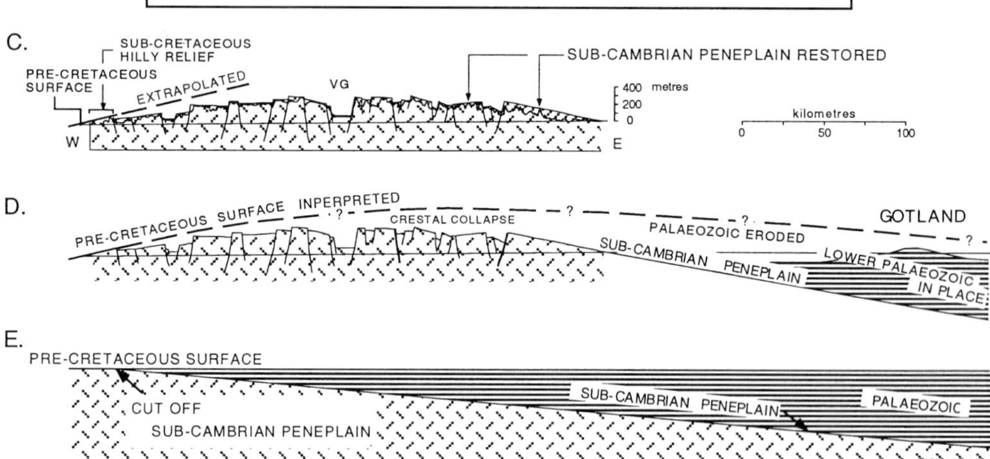

Fig. 5. The South Swedish Dome. (**a**) Morphological and structural elements based on Lidmar-Bergström (1994). Position of cross-section shown in (**c**) is indicated. SSP, South Småland Peneplain; VG, Vättern Graben. (**b**) Structure of the pre-Cretaceous surface (reconstruction). Elevation of structure contours above present sea level. (**c**) East–west cross-section of the South Swedish Dome. Reconstruction of the block-faulted sub-Cambrian peneplain cut out by the pre-Cretaceous surface to the west. (Interpretation based on data from Lidmar-Bergström 1988). (**d**) Interpretation of the pre-Cretaceous surface before exhumation of the sub-Cambrian peneplain. (**e**) Reconstruction of the pre-Cretaceous surface and the sub-Cambrian peneplain before uplift of the South Swedish Dome.

Sweden before the uplift of the dome (Norling 1994).

The sub-Cretaceous hilly relief along the dome margin is eroded down into the basement below the sub-Cambrian peneplain and has a kaolinitic surface caused by deep Mesozoic weathering. By contrast, the sub-Cambrian peneplain in the centre of the dome does not have these features. This indicates that the basement in the centre of the dome must have been covered by Palaeozoic sediments with a thickness exceeding the maximum weathering depth of 200 m during Mesozoic time (Lidmar-Bergström 1993, 1995). Extrapolation of the more steeply inclined pre-Cretaceous surface across the dome, in combination with the weathering depth and the thickness of the Lower Palaeozoic cover to the east, suggests that the pre-Cretaceous surface in the centre of the dome may have been at least 100–200 m above the sub-Cambrian peneplain before the post-Cretaceous erosion (Fig. 5d). This value is in agreement with the altitude of the Palaeozoic outliers in the western part of the dome, which rise up to 200 m above the sub-Cambrian peneplain. The uplift of the pre-Cretaceous surface at the centre of the South Swedish Dome may thus have been of the order of 500–600 m above present sea level judged from present-day elevation and estimated erosion. A reconstruction of the uppermost pre-Cretaceous surface of the South Swedish Dome before erosion is illustrated in both map view and cross-section (Fig. 5b and d). A restoration of the pre-Cretaceous surface before doming is shown in Fig. 5e.

The occurrence of gravelly saprolites in the Precambrian crystalline basement is confined to the crestal collapse structure in the eastern part of the South Swedish Dome (Lidmar-Bergström 1997; Lidmar-Bergström *et al.* 1997). The development of these saprolites is ascribed to weathering in a cool to cold climate in Plio-Pleistocene time (Lidmar-Bergström *et al.* 1997, 1999). This illustrates late uplift and fracturing of the dome accompanied by renewed weathering, part of the ?20–3 Ma denudation episode of Lidmar-Bergström (1996), resulting in exhumation of the sub-Cambrian peneplain and the sub-Cretaceous hilly relief. The proposed timing is in accordance with fission-track analysis, which indicates a late uplift and exhumation during the last 25 Ma (Cederbom *et al.* 2000).

In the marine platform area in the Hanö Bay to the south of the dome, the thickness of the

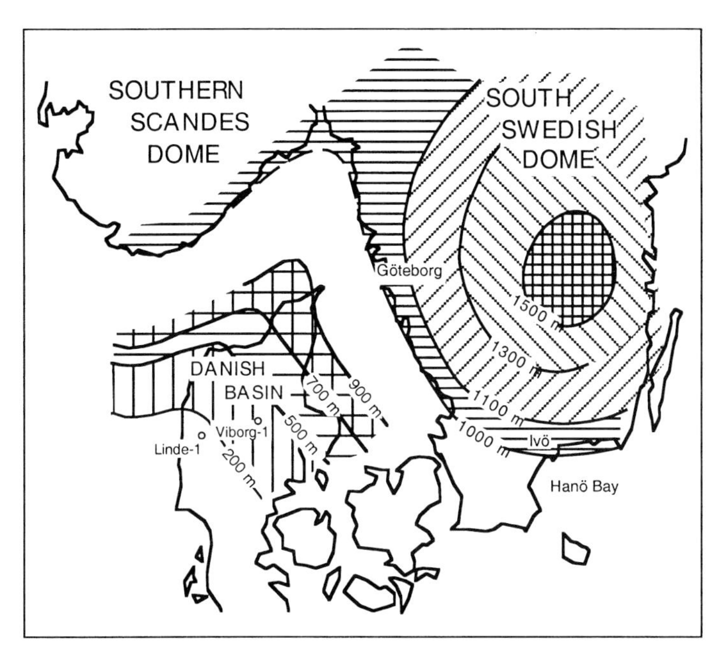

Fig. 6. Neogene uplift of the eastern North Sea Basin and the South Swedish Dome. Uplift in the Danish Basin is based on sonic velocities and vitrinite reflection (Jensen & Michelsen 1992; Japsen 1998; Graversen 1999). Uplift of the South Swedish Dome is based on uplift of the pre-Cretaceous surface modelled in this paper (Fig. 5b) added to *c.* 1 km uplift established at the eastern border of the Danish Basin and AFTA from Ivö (Cederbom 2002). (See text for discussion and references.)

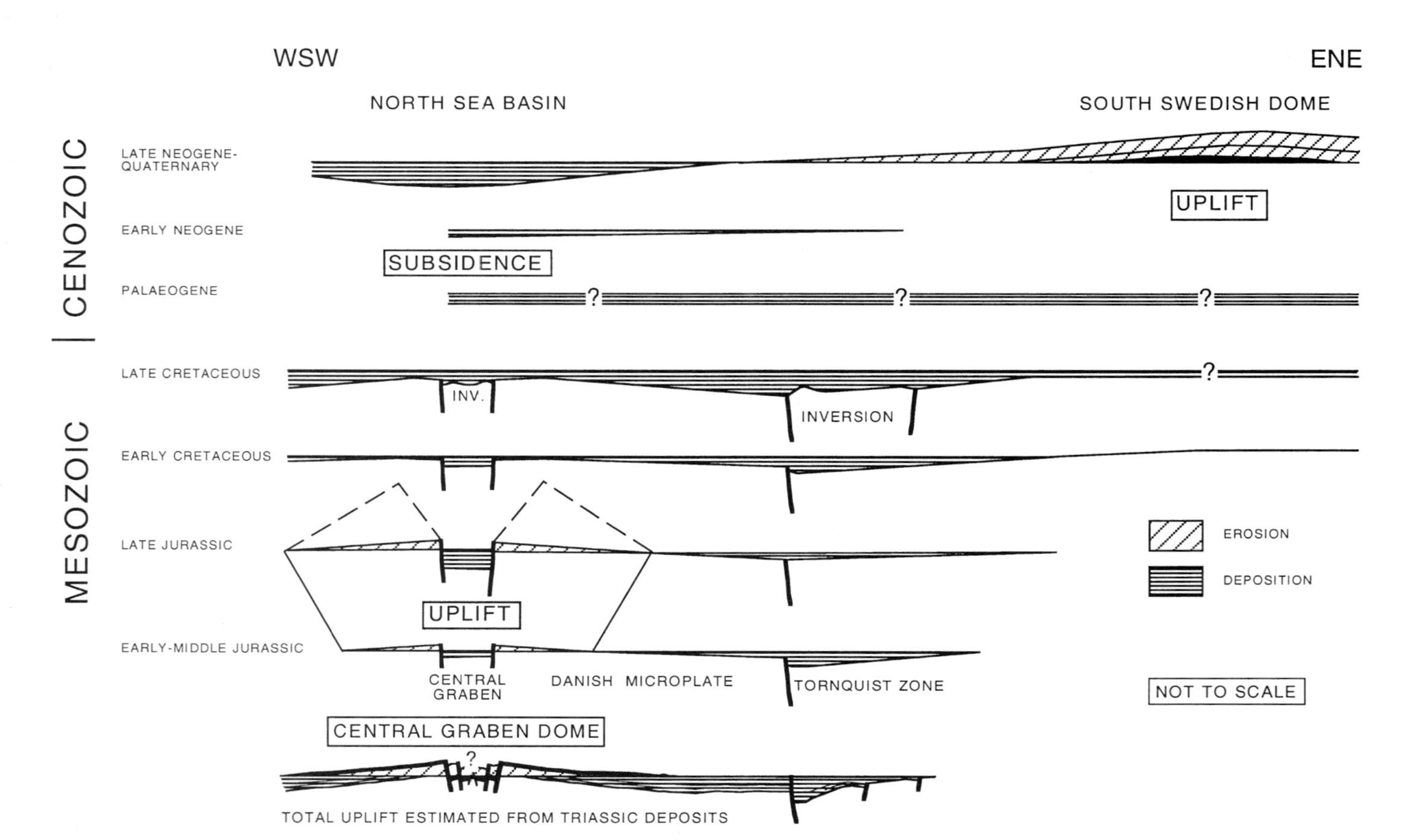

Fig. 7. Uplift–subsidence reversal across the eastern North Sea Basin and the South Swedish Dome. The Central Graben Dome was uplifted and eroded during Jurassic time. The dome was raised above sea level to a maximum in Late Jurassic time and was then gradually flooded during Early Cretaceous time. The Cenozoic post-rift subsidence was centred above the Central Graben whereas the South Swedish Dome was uplifted and eroded to the east.

Cretaceous sediments ranges up to *c.* 650 m (Norling & Bergström 1987) (Fig. 6). This should be regarded as a maximum value, as supposed Paleogene (Paleocene?) sediments are encountered above (Kumpas 1980; Norling & Bergström 1987; Sviridov *et al.* 1995). Also, erratics of Paleocene–Eocene marine sediments have been found in eastern Scania (Norling & Bergström 1987), and a marine diatomite of Eocene age has been reported from Finland (Fenner 1988), suggesting that Paleocene–Eocene sediments were also deposited in southern Sweden. An estimate of the thickness may be obtained from the Viborg-1 and Linde-1 wells in the Danish Basin to the west, where 200 m and 250 m, respectively, of Paleocene–Eocene sediments have been drilled (Dinesen *et al.* 1977; Heilmann-Clausen 1995) (Fig. 6). The Cretaceous and Paleogene sediments add up to a total thickness of up to 900 m that may have been deposited above the pre-Cretaceous surface. An evaluation of the uplift should also take pre-compaction thicknesses and water depth into account and this may add another 100–200 m. A Cenozoic uplift up to about 1 km may therefore be added to the pre-Cretaceous surface of the South Swedish Dome. This value should be regarded as an absolute maximum, and a realistic figure may be considerably less. Riis (1996) estimated Plio-Pleistocene erosion in southwestern Sweden of the order of 500–1000 m.

Analysis of reflectance data from the Jurassic Fjerritslev Formation (Jensen & Michelsen 1992) and sonic velocities from the Fjerritslev Formation and the Cretaceous Chalk Group from the Danish Basin and the Skagerrak–Kattegat Platform indicates an uplift of the order of 1 km or more (Japsen 1992, 1998; Jensen & Michelsen 1992; Michelsen & Nielsen 1993; Japsen & Bidstrup 1999). Apatite fission-track analysis (AFTA) indicates that Cretaceous–Paleogene sediments amounting to *c.* 650 m (Göteborg) and *c.* 1 km (Ivö) have been eroded from the western and southern border of the South Swedish Dome (Cederbom 2002).

On the basis of the structural analysis, the uplift in the centre of the dome should be expected to exceed the border uplift by 500–600 m. Although there are some differences between the values obtained from the various methods, the present morphological dome in southern Sweden rises to nearly 400 m and this should be regarded as a minimum uplift of the dome centre above the marginal areas. The uplift of the dome centre in relation the present sea level is estimated to be 500–600 m (Fig. 5b), and the total uplift, when adding the missing section along the border of the dome, may add another 1000 m when the sonic and fission-track analyses are taken into account. Figure 6 combines estimates of the uplift (missing section) from sonic velocity analysis and AFTA with the reconstructed structure of the pre-Cretaceous surface of the South Swedish Dome (Fig. 5b).

Jurassic–Cenozoic uplift–subsidence reversal across the eastern North Sea Basin

The structural evolution of the eastern North Sea Basin, i.e. the Danish sector and southern Sweden, is dominated by the Middle Jurassic–Cretaceous uplift of the Central Graben Dome followed by Cenozoic subsidence in the west and uplift of the South Swedish Dome in the east. This is summarized in Fig. 7.

Mesozoic structure and evolution of the Central Graben Dome

In Early–Middle Jurassic time, rifting and subsidence was concentrated along the Tornquist Zone in the east, whereas only minor subsidence and deposition took place in the Danish Central Graben (Fig. 7). The emergent Central Graben Dome expanded in Late Jurassic time and the main depocentre shifted to the Central Graben whereas subsidence diminished in the Tornquist Zone. Renewed differential subsidence took place in the Tornquist Zone in Early Cretaceous time; the Central Graben Dome was gradually flooded, and subsidence in the Central Graben slowed down. Through Late Cretaceous time, subsidence and crestal collapse along with structural inversion of the graben sediments characterized the Central Graben Dome. Eastward tilting of the Danish Megablock was still active and structural inversion was also encountered to the east in the Tornquist Zone. In Jurassic–Cretaceous time, the intensity of basin subsidence shifted several times across the Danish Megablock between the Central Graben and the Tornquist Zone. This may indicate an active function of the Danish Megablock during tilting that was possibly related to an episodic rise of the Central Graben Dome. In Mesozoic time the Danish Megablock and the Baltic Shield were decoupled along the Tornquist fault zone. During the Cenozoic subsidence, however, the Tornquist Zone in Skagerrak and the northern part of Kattegat was no longer active, and the Danish Megablock and the Baltic Shield may have acted together as a single mega-unit.

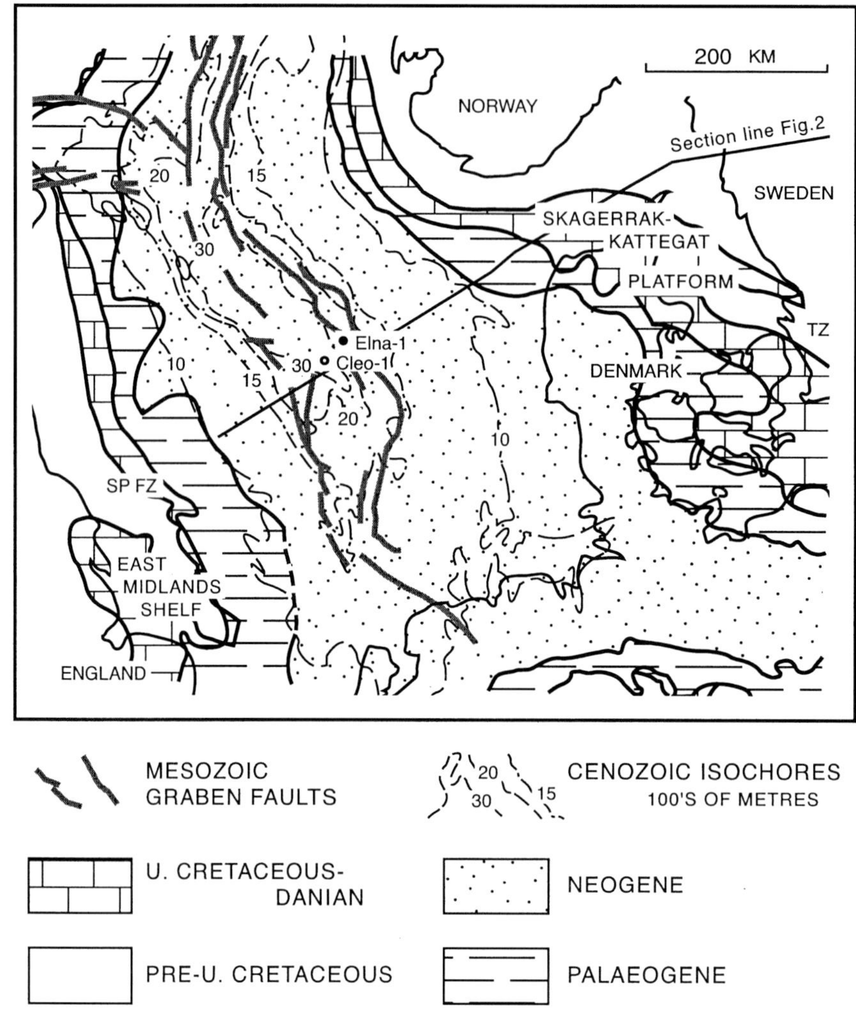

Fig. 8. Pre-Quaternary geology of the central and southern North Sea Basin and marginal areas. There is a conspicuous symmetry along the Cenozoic basin axis situated above the Mesozoic Central and Viking grabens. Compiled from Ziegler (1990) and Japsen (1998). SP FZ, Sole Pit fault zone; TZ, Tornquist Zone.

Cenozoic structure and evolution of the North Sea Basin

The Cenozoic post-rift sediments form a gentle syncline with the basin axis situated above the aborted Mesozoic rifts (Fig. 8). The geological map of the Upper Cretaceous–Neogene sequences exhibits a symmetrical pattern centred around the Neogene interval along the basin axis. Pre-Mesozoic basement crops out along the basin margins in Norway and Sweden, and in the British Isles. The basin margins were uplifted and eroded whereas the basin centre continued to subside (Japsen 1998).

In the central North Sea, Middle Miocene to Recent sediments represent about half of the Cenozoic sediment infill (Nielsen *et al.* 1986; Michelsen *et al.* 1998), and strong tectonic subsidence is observed in Mid–Late Miocene and Quaternary times (Kooi *et al.* 1991; Clausen *et al.* 1999). Accumulation in the basin centre (Fig. 9) is based on lithological thicknesses in wells (Nielsen & Japsen 1991) and isopach maps (Nielsen *et al.* 1986) along the modelled seismic profile (RTD-81-22, Fig. 2); the general trend demonstrates increasing accumulation and sedimentation rates through Cenozoic time with a culmination in Pliocene–Quaternary time. The Cenozoic

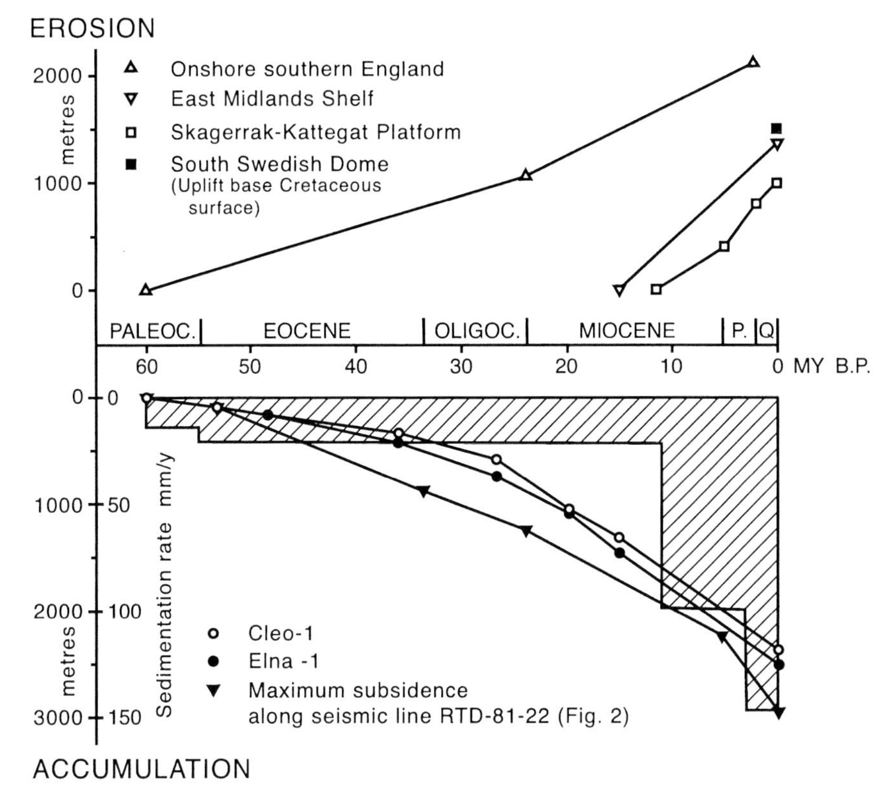

Fig. 9. Cenozoic accumulation, erosion and average sedimentation rates in the central North Sea Basin, the Skagerrak–Kattegat Platform, the East Midlands Shelf and southern England (see Fig. 8). Hatched area indicates average Tertiary sedimentation rates in the Danish and Norwegian Central Graben. Erosion in southern England assumes gradual Paleogene uplift. Compiled from Nielsen *et al.* (1986), Nielsen & Japsen 1991, Japsen (1997) and Japsen & Bidstrup (1999).

erosion (missing overburden) of the basin flanks before Quaternary time has been modelled by Japsen (1997, 1998, 2000) and Japsen & Bidstrup (1999). Erosion increased towards the basin flanks, where it exceeds 1 km, whereas maximum burial of the Chalk Group still occurs in the centre of the basin. In the South Swedish Dome and onshore areas in southern England, the total uplift amounts to 1.5–2 km (Figs 6 and 9) (Japsen 1997).

The timing of erosion has been modelled by Japsen (1997) and Japsen & Bidstrup (1999). At the SW border of the basin, maximum burial occurred on the East Midlands Shelf in mid-Miocene time (Fig. 9) (Japsen 1997). In the Danish area, the erosion was of Miocene–Pliocene age, and in the Skagerrak–Kattegat Platform on the NE border of the basin, the erosion is possibly not much earlier than late Miocene time (Japsen & Bidstrup 1999) (Fig. 9). In the South Swedish Dome, the late Neogene–Quaternary gravelly saprolites are weathering products of the crystalline basement formed during late uplift and exhumation in a cold climate (Lidmar-Bergström *et al.* 1997). In southern England tilting and uplift in Plio-Pleistocene time is recorded (Watts *et al.* 2000; Westaway *et al.* 2002). Late Neogene–Quaternary subsidence in the central North Sea is thus contemporaneous with, and counterbalanced by, uplift of the South Swedish Dome and southern England (Fig. 9).

Later uplift and erosion has removed direct evidence for the uplift of the Baltic Shield and evolution of the South Swedish Dome. Mapping of the Cenozoic sequences in the eastern North Sea Basin indicates that sediments from the Baltic Shield were routed into the basin from the north and NE in Paleogene time (Danielsen *et al.* 1997; Michelsen *et al.* 1998; Clausen *et al.* 2000). In Neogene time, the main transport route was from the NE in Miocene time and from the east in Pliocene time (Sørensen *et al.* 1997; Clausen *et al.* 1999). The clockwise rotation of the infill directions during Paleogene–Neogene time is believed to indicate that Cenozoic uplift of the Baltic Shield was initiated in southern Norway during Paleogene time and then changed to the South Swedish Dome in Neogene time (Riis 1996; Japsen 2000).

Lithosphere models of the Jurassic–Cenozoic North Sea Basin

The Mesozoic central North Sea domes were rift dominated, accompanied by volcanic activity and underlain by a thin crust (Ziegler 1990). In contrast, the southern Scandinavian Cenozoic domes are non-volcanic, developed along the SW rifted margin of the uplifted Precambrian–Palaeozoic crystalline basement where crustal thickness reaches 50 km (Thybo 1997, 2000).

During Mesozoic time, the Central Graben Dome and the Danish Megablock were decoupled from the Baltic Shield–Platform along the Tornquist Zone (Fig. 7). After the Late Cretaceous–Danian inversion differential subsidence–uplift in the northern Tornquist Zone stopped, and the Cenozoic subsidence of the Central Graben Dome was accompanied by uplift of the South Swedish Dome. This may indicate a deep-seated subcrustal reorganization and a possible mutual relationship between subsidence and uplift, as both the Central Graben Dome and the South Swedish Dome are interpreted as crustal structures involving the mantle lithosphere and/or the asthenosphere.

The Jurassic–Cretaceous uplift of the Central Graben Dome and the Cenozoic subsidence both amount to *c.* 2 km along the analysed section (Fig. 2). This value is similar to the proposed 1.5 km uplift of the South Swedish Dome. It is therefore suggested that there may be a link between the subsidence of the Central Graben Dome and the uplift of the South Swedish Dome to the ENE.

Models of synrift and post-rift basin evolution

Existing models of synrift and post-rift basin evolution are based on the McKenzie (1978) concept on sedimentary basin formation. The structure and stratigraphy of the post-rift basin depends on the rheological properties of the lithosphere. Watts *et al.* (1982) modelled the shape and synthetic stratigraphy of post-rift basins developed on either an elastic or a viscoelastic plate overlying a weak substratum. In the elastic model, the post-rift basin develops as a transgressive basin, with a so-called steershead geometry in cross-sectional view. By contrast, in the case of a viscoelastic plate, the post-rift basin develops as a regressive basin (Watts *et al.* 1982). Although the modelled maximum subsidence is of the same order in both settings, the viscoelastic model results in a post-rift subsidence *c.* 20% greater than the elastic model. The effective elastic thickness of the lithosphere determines whether the sediments are locally supported, i.e. short-wavelength, Airy-type isostatic compensation of a viscoelastic plate, or supported by longer-distance, lateral strength of the plate, i.e. long-wavelength, elastic compensation (Watts *et al.* 1982; Barton & Wood 1984).

Cloetingh *et al.* (1985) and Cloetingh (1986) introduced intraplate stresses as a tectonic mechanism in continental basin development. The model superimposes fluctuating lateral stresses on existing basins situated on an elastic lithosphere as a mechanism to explain observed vertical plate movements. The intraplate stress model may be viewed as a late short-term tectonic overprint on the long-term McKenzie (1978) concept of basin evolution and lithospheric flexure modelled by Watts *et al.* (1982) (Kooi & Cloetingh 1989; Cloetingh *et al.* 1990; Cloetingh & Kooi 1992).

Previous models of Mesozoic–Cenozoic evolution of the North Sea Basin

Previous models of synrift and post-rift evolution of the North Sea area were based on the McKenzie (1978) concept of lithosphere extension applied to the Mesozoic Central Graben rift basin, superseded by the Cenozoic post-rift thermal sag basin (Sclater & Christie 1980; Watts *et al.* 1982; Barton & Wood 1984; Cloetingh *et al.* 1985). Both Watts *et al.* (1982) and, for example, Cloetingh *et al.* (1990) have modelled the post-rift North Sea Basin to have developed on an elastic plate exhibiting a steershead geometry for the post-rift sediments. Watts *et al.* (1982) estimated the effective elastic thickness as the depth to the 450°C isotherm; in the case of the North Sea Basin this is of the order of 30–35 km. On the basis of gravity modelling, however, Barton & Wood (1984) arrived at the conclusion that the effective elastic thickness of the lithosphere is only of the order of 5 km, and that the sedimentary load was compensated by local, Airy-type isostasy, i.e. a viscoelastic plate.

Intraplate stresses have also been proposed to be involved in post-rift basin subsidence and uplift of the margins. Cloetingh *et al.* (1990, 1992), Ziegler (1990), Kooi *et al.* (1991), Cloetingh & Kooi (1992) and Doré (1992) described the gentle warping of the Base Tertiary surface during Cenozoic subsidence of the North Sea Basin along with uplift of southern Norway and the British Isles as initiated by sediment loading coupled with uplift and erosion of the basin margins; the vertical deflection was interpreted to have been amplified by intraplate compression.

The geometry of the Cenozoic post-rift sediments may give the impression that it possess a steershead geometry (Fig. 2). However, the thick interval above the Central Graben is due to a combination of post-rift compaction of the synrift sediments in the underlying graben, and the thick Plio-Pleistocene section in the centre of the North Sea Basin (Japsen 1998); the Tertiary "steershorn" extending to the ENE (Fig. 2) is mainly related to erosion during uplift of the South Swedish Dome. Furthermore, in the case of a steershead geometry, the synthetic stratigraphy should exhibit an onlapping, transgressive basin, whereas observations outline an overall regressive basin-fill prograding from the basin margins (Sørensen *et al.* 1997; Michelsen *et al.* 1998; Clausen *et al.* 1999). The Cenozoic subsidence rate in the centre of the southern North Sea Basin increased throughout Neogene time (Fig. 9) (Barton & Wood 1984; Kooi *et al.* 1991). This observation contrasts with most models of post-rift thermal subsidence that imply a decreasing subsidence with age.

New model: Cenozoic subsidence linked with uplift of marginal domes

The collapse of the Central Graben Dome caused reversal of the tilt direction of the Danish Megablock. Whereas the Mesozoic period witnessed an increasing tilt towards the ENE, the tilt was reversed in Cenozoic time: the uppermost pre-Zechstein–lowermost Triassic level along the analysed section is now almost back to subhorizontal (Figs 2 and 3). During Mesozoic tilting, the Danish Megablock was disconnected with the Baltic Shield along the Tornquist Zone (Fig. 7). In Cenozoic time, however, the discontinued faulting along the Tornquist Zone indicates that the megablock and the shield acted together as a single unit.

Both pre-Quaternary geology and the amount of erosion outline a symmetrical arrangement across the southern North Sea area (Fig. 8). The geometry illustrates Cenozoic subsidence along the basin axis and uplift of the margins. Subsidence and deposition in the centre (2 km) was counterbalanced by uplift and erosion in the South Swedish Dome and southern England (1.5–2 km) (Fig. 9). Deposition along the basin axis increased over time, and the curve of deposition is mirrored by the curve of uplift. The Skagerrak–Kattegat Platform and the East Midlands Shelf along the eastern and western margins of the North Sea Basin both underwent uplift and/or erosion during Miocene–Quaternary time; the gradients and magnitudes of uplift correspond to deposition in the basin centre (Fig. 9). The observation that deposition was counterbalanced by (pene)-contemporaneous uplift and/or erosion is interpreted to indicate that Cenozoic post-rift subsidence was linked with marginal uplift.

In a case where post-rift thermal subsidence has induced a flexural bulge in marginal areas, uplift would equal only a small fraction (6–7%) of the subsidence (Turcotte & Schubert 1982). If this was the situation, uplift of the southern North Sea margins should only amount to *c.* 150 m. In addition, Cenozoic post-rift subsidence increased (Fig. 9), whereas thermal modelling invokes decreasing subsidence over time (McKenzie 1978; Watts *et al.* 1982). These observations indicate that thermal subsidence alone is not able to account for the observed balance between subsidence and uplift (Fig. 9). It is suggested instead that the subsidence of the Central Graben Dome, accompanied by the backward tilt of the Danish Megablock, was associated with a reorganized flow pattern in the asthenosphere below the subsiding North Sea Basin (Fig. 10). In this case, mass movements directed towards the marginal domes may account for the observed uplift of the South Swedish Dome (Fig. 10a). It is difficult to evaluate the strength of the lithosphere during the collapse of the Central Graben Dome, as vertical movements of the lithosphere may be in interaction with, and supported by, the asthenosphere.

The South Swedish Dome and the Southern Scandes Dome in Norway are situated along individual segments of the rifted margin on the SW corner of the emergent Precambrian–Palaeozoic crust (Fig. 1). Uplift of southern Norway probably also has a deep-seated origin in the upper mantle and the asthenosphere (Rohrman & van der Beek 1996). The suggested interrelationship modelled for the South Swedish Dome and the Central Graben Dome may also be proposed for the southern Norway uplift and the Triple Junction Dome. Initial uplift in southern Norway took place in Paleogene time whereas the uplift of the South Swedish Dome peaked in Neogene–

Quaternary time (Riis 1996; Japsen 2000). The difference in timing of initial Cenozoic uplift may be related to the migration of early uplift and subsidence of the Triple Junction Dome relative to the Central Graben Dome to the SE.

Discussion and conclusions

The proposed model of linked subsidence and uplift shows that a better understanding of Cenozoic uplift in the North Sea Basin may be obtained when a link is made to the Mesozoic evolution. Analysis of the central North Sea Dome has demonstrated the composite character and differential uplift and subsidence of the dome. Rifting and uplift was initiated in Middle Jurassic time in the north in the Triple Junction Dome. This dome was flooded during Late Jurassic time as rifting and uplift moved south to the Central Graben Dome (in early Late Jurassic time) and then the Friesland Dome (in late Late Jurassic time). The Central Graben Dome was flooded during Early Cretaceous time.

The large-scale evolution of the Jurassic–Cretaceous and Cenozoic sedimentary basins in the eastern North Sea shows that the Mesozoic depocentres were situated in the Central Graben and along the Tornquist Zone whereas the Cenozoic depocentres were concentrated in the central North Sea above the aborted Mesozoic rifts (Figs 7 and 8). The changed subsidence pattern around the Mesozoic–Cenozoic transition indicates an uplift–subsidence reversal. The change of the overall basin morphology was governed by the first-order tectonic elements, i.e. the Central Graben Dome and the Danish Megablock, the Tornquist Zone, and the South Swedish Dome. These structures are several hundred kilometres across and involve vertical movements of between 1 and 2 km. The magnitude of the structures and the movement suggest causal mechanisms in the mantle involving the entire overlying crust. Jurassic doming was characterized by thinning of the crust and the development of rift basins with volcanic activity (Ziegler 1990), but by contrast the Cenozoic domes developed on a thick Caledonian and Precambrian crystalline crust, and there are no signs of volcanic activity and only limited faulting is observed (i.e. the crestal collapse structure of the South Swedish Dome). The Cenozoic domes all have a marginal position along the western faulted margin of the Fennoscandian High. The domes appear to have been stationary throughout their existence, and this suggests that the crustal structure was the locating factor. If the domes had developed above mantle plumes ascending from the deep mantle, they would have a more random distribution.

The subdivision of the central North Sea Dome into individual dome centres connects the Danish Megablock with the Central Graben Dome (Fig. 1). The tilt reversal of the Danish Megablock at the Mesozoic–Cenozoic transition thus suggests a mutual linkage between the subsidence of the Central Graben Dome and the uplift of southern Sweden (Fig. 7). The accelerated subsidence of the central North Sea Basin during late Neogene–Quaternary time (Fig. 9) (Nielsen *et al.* 1986; Ziegler 1990) is thus contemporaneous with the uplift of the South Swedish Dome and the exhumation of the sub-Cambrian peneplain established by Lidmar-Bergström (1995). Fission-track thermochronology supports a model of accelerated uplift during Neogene time (Cederbom *et al.* 2000; Cederbom 2002).

The Cenozoic post-rift subsidence in the central North Sea Basin does not follow the McKenzie (1978) model. Instead of the modelled decreasing subsidence over time, the observations outline increasing subsidence (Fig. 9). The Mesozoic–Cenozoic evolution of the North Sea area involved uplift or subsidence of the order of 2 km or more of large crustal plates; for example, the Danish Megablock covering the eastern half of the southern North Sea. The model proposed in the present paper attributes uplift of the South Swedish Dome to subsidence of the eastern North Sea Basin during collapse of the Central Graben Dome. The lithosphere carrying the Danish Megablock and adjoining Baltic Shield were linked in a common movement that involved a reorganized flow pattern in the asthenosphere (Fig. 10).

The uplift followed by subsidence probably involved mass flow in the asthenosphere to account for the observed balance between post-rift subsidence and marginal uplift. In this scenario the impact of thermal subsidence may be suppressed and not easy to model.

The regressive character of the Cenozoic basin (Michelsen *et al.* 1998) and limited occurrence of major faults (Ziegler 1990) suggests that the Cenozoic package was deposited on a viscoelastic plate (Watts *et al.* 1982). This conclusion is in accordance with the modelling of Barton & Wood (1984), who suggested an effective elastic thickness of the order of 5 km. The Cenozoic basin shows increasing subsidence rates over time (Fig. 9), and high subsidence rates in late

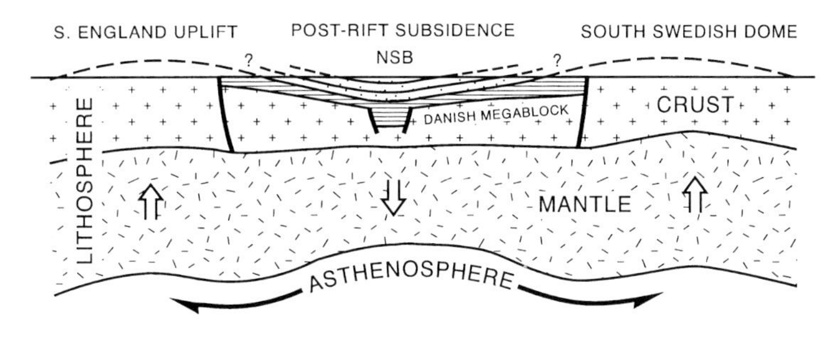

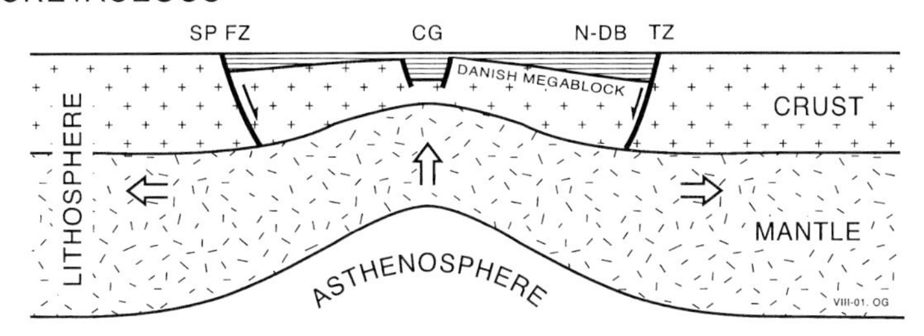

Fig. 10. Lithosphere model of crustal uplift–subsidence reversal across the southern North Sea Basin. CG, Central Graben; N-DB, Norwegian–Danish Basin; NSB, North Sea Basin; SP FZ, Sole Pit fault zone; TZ, Tornquist Zone. (**a**) During the Cenozoic post-rift subsidence, faulting along the Tornquist and Sole Pit fault zones was discontinued. The subsidence was accompanied by reverse tilting of the flanks, whereas southern Sweden and southern England were raised along the borders. It is suggested that Cenozoic subsidence of the lithosphere resulted in a revised flow pattern in the asthenosphere directed towards southern Sweden and southern England. (**b**) Mesozoic rifting and uplift of the Central Graben Dome was accompanied by crustal thinning and asthenosphere rising below the rifted crust. The uplift of the Central Graben Dome was decoupled from the adjoining areas along the Tornquist and Sole Pit fault zones.

Neogene–Quaternary time may be viewed as a natural part of the observed trend as an alternative to intraplate deformation.

The bimodal symmetry of the North Sea Basin along the central Mesozoic rifts and Cenozoic basin axis implies that the proposed model of the Cenozoic uplift in the east along the border of the Fennoscandian High may be mirrored in the west in the Cenozoic evolution of Britain (Figs 8 and 10). Whereas the uplift along the eastern margin of the North Sea Basin may be relatively simple, the uplift in Britain may interfere with deformations of the fault-bounded shelf basins to the south and west, as well as with the effects of the opening of the Atlantic. However, Neogene–Quaternary uplift of the South Swedish Dome corresponds to uplift in southern England (Fig. 9), and Paleogene uplift of Scotland based on geomorphological evidence (Hall 1991) may counterbalance the uplift in southern Norway.

I thank K. Lidmar-Bergström (University of Stockholm), who introduced me to the South Swedish Dome in May 1999 during a Nordic research excursion through southern Sweden; I also thank T. Doré, R. Gatliff, E. Håkansson and an anonymous reviewer for constructive reviews and language corrections of an early draft of the manuscript.

References

Barton, P. & Wood, R. 1984. Tectonic evolution of the North Sea basin: crustal stretching and subsidence. *Geophysical Journal of the Royal Astronomical Society*, **79**, 987–1022.

Berthelsen, A., Burollet, P., Dal Piaz, G.V., Franke, W. & Trümpy, R. 1992. Tectonics, Atlas Map 1. *In*: Blundell, D., Freeman, R. & Mueller, S. (eds) *A Continent Revealed: The European Geotraverse*. Cambridge University Press, Cambridge.

Cederbom, C. 2002. The thermotectonic development of southern Sweden during Mesozoic and Cenozoic time. *In*: Doré, A.G., Cartwright, J.A., Stoker, M.S., Turner, J.P. & White, N. (eds) *Exhumation of the North Atlantic Margin: Timing, Mechanisms and Implications for Petroleum Exploration.* Geological Society, London, Special Publications, **196**, 169–182.

CEDERBOM, C., LARSON, S.Å., TULLBORG, E.-L. & STIBERG, J.-P. 2000. Fission track thermochronology applied to Phanerozoic thermotectonic events in central and southern Sweden. *Tectonophysics*, **316**, 153–167.

CLAUSEN, O.R., GREGERSEN, U., MICHELSEN, O. & SØRENSEN, J.C. 1999. Factors controlling the Cenozoic sequence development in the eastern parts of the North Sea. *Journal of the Geological Society, London*, **156**, 809–816.

CLAUSEN, O.R., NIELSEN, O.B., HUUSE, M. & MICHELSEN, O. 2000. Geological indications for a Paleogene onset of the 'Neogene uplift' in the eastern North Sea area. *Global and Planetary Change*, **24**, 175–187.

CLOETINGH, S. 1986. Intraplate stresses: a new tectonic mechanism for the fluctuations of relative sea level. *Geology*, **14**, 617–620.

CLOETINGH, S. & KOOI, H. 1992. Intraplate stresses and dynamical aspects of rifted basins. *Tectonophysics*, **215**, 167–185.

CLOETINGH, S., GRADSTEIN, F.M., KOI, H., GRANT, A.C. & KAMINSKI, M. 1990. Plate reorganization: a cause of rapid late Neogene subsidence and sedimentation around the North Atlantic. *Journal of the Geological Society, London*, **147**, 495–506.

CLOETINGH, S., MCQUEEN, H. & LAMBECK, K. 1985. On a tectonic mechanism for regional sealevel variations. *Earth and Planetary Science Letters*, **75**, 157–166.

CLOETINGH, S., REEMST, P., KOOI, H. & FANAVOLL, S. 1992. Intraplate stresses and the post-Cretaceous uplift and subsidence in northern Atlantic basins. *Norsk Geologisk Tidsskrift*, **72**, 229–235.

DANIELSEN, M., MICHELSEN, O. & CLAUSEN, O.R. 1997. Oligocene sequence stratigraphy and basin development in the Danish North Sea sector based on log interpretations. *Marine and Petroleum Geology*, **14**, 931–950.

DINESEN, A., MICHELSEN, O. & LIEBERKIND, K. 1977. *A Survey of the Paleocene and Eocene Deposits of Jylland and Fyn*. Geological Survey of Denmark, Series B, **1**.

DORÉ, A.G. 1992. The base Tertiary surface of southern Norway and the northern North Sea. *Norsk Geologisk Tidsskrift*, **72**, 259–265.

EUGENO-S WORKING GROUP, 1988. Crustal structure and tectonic evolution of the transition between the Baltic Shield and the North German Caledonides (the EUGENO-S Project). *Tectonophysics*, **150**, 253–348.

FENNER, J. 1988. Occurrences of pre-Quaternary diatoms in Scandinavia. *Meyniana*, **40**, 133–141.

GRAVERSEN, O. 1999. Plio-Pleistocæn hævning af Det danske Bassin. *Dansk Geologisk Tidsskrift*, **1**, 25–26.

HALL, A.M. 1991. Pre-Quaternary landscape evolution in the Scottish Highlands. *Transactions of the Royal Society of Edinburgh: Earth Sciences*, **82**, 1–26.

HEILMANN-CLAUSEN, C. 1995. Palæogene aflejringer over danskekalken. *In*: NIELSEN, O.B. (ed.) *Aarhus geokompendier 1: Danmarks geologi fra Kridt til i dag*. Geologisk Institut, Aarhus, 71–114.

JAPSEN, P. 1992. Landhævningerne I Sen Kridt og Tertiær I det nordlige Danmark. *Geological Society of Denmark,Årsskrift*, **1990–91**, 169–182.

JAPSEN, P. 1997. Regional Neogene exhumation of Britain and the western North Sea. *Journal of the Geological Society, London*, **154**, 239–247.

JAPSEN, P. 1998. Regional velocity–depth anomalies, North Sea chalk: a record of overpressure and Neogene uplift and erosion. *AAPG Bulletin*, **82**, 2031–2074.

JAPSEN, P. 2000. Fra Kridthav til Vesterhav, Nordsøbassinets udvikling vurderet ud fra seismiske hastigheder. *Geologisk Tidsskrift*, **2**, 1–36.

JAPSEN, P. & BIDSTRUP, T. 1999. Quantification of late Cenozoic erosion in Denmark based on sonic data and basin modelling. *Bulletin of the Geological Society of Denmark*, **46**, 79–99.

JAPSEN, P. & CHALMERS, J.A. 2000. Neogene uplift and tectonics around the North Atlantic: overview. *Global and Planetary Change*, **24**, 165–173.

JENSEN, L.N. & MICHELSEN, O. 1992. Tertiær hævning og erosion i Skagerrak, Nordjylland og Kattegat. *Geological Society of Denmark,Årsskrift*, **1990-91**, 159–168.

KOOI, H. & CLOETINGH, S. 1989. Intraplate stresses and the tectono-stratigraphic evolution of the central North Sea. *In*: TANKARD, A.J. & BALKWILL, H.F. (eds) *Extensional Tectonics and Stratigraphy of the North Atlantic Margins*. American Association of Petroleum Geologists Memoir, **46**, 541–558.

KOOI, H., HETTEMA, M. & CLOETINGH, S. 1991. Lithospheric dynamics and the rapid Pliocene–Quaternary subsidence phase in the southern North Sea basin. *Tectonophysics*, **192**, 245–259.

KUMPAS, M.G. 1980. Seismic stratigraphy and tectonics in Hanö Bay, southern Baltic. *Stockholm Contributions in Geology*, **34**, 35–168.

LIDMAR-BERGSTRÖM, K. 1988. Denudation surfaces of a shield area in south Sweden. *Geografiska Annaler*, **70A** (4), 337–350.

LIDMAR-BERGSTRÖM, K. 1991. Phanerozoic tectonics in southern Sweden. *Zeitschrift für Geomorphologie, Neuer Folge*, **82**, 1–16.

LIDMAR-BERGSTRÖM, K. 1993. Denudation surfaces and tectonics in the southernmost part the Baltic Shield. *Precambrian Research*, **64**, 337–345.

LIDMAR-BERGSTRÖM, K. 1994. Berggrundens utformer. *In*: FRÉDEN, C. (ed.) *Berg och Jord.* Sveriges Nationalatlas Förlag, Stockholm, 44–54.

LIDMAR-BERGSTRÖM, K. 1995. Relief and saprolites through time on the Baltic Shield. *Geomorphology*, **12**, 45–61.

LIDMAR-BERGSTRÖM, K. 1996. Long term morphotectonic evolution in Sweden. *Geomorphology*, **16**, 33–59.

LIDMAR-BERGSTRÖM, K. 1997. A long-term perspective on glacial erosion. *Earth Surface Processes and Landforms*, **22**, 297–306.

LIDMAR-BERGSTRÖM, K. 1999. Uplift histories revealed by landforms of the Scandinavian domes. *In*: SMITH, B.J., WHALLEY, W.B. & WARKE, P.A. (eds) *Uplift, Erosion and Stability: Perspectives on Longterm Landscape Development*. Geological Society, London, Special Publications, **162**, 85–91.

LIDMAR-BERGSTRÖM, K., OLSSON, S. & OLVMO, M. 1997. Palaeosurfaces and associated saprolites in southern Sweden. *In*: WIDDOWSON, M. (ed.) *Palaeosurfaces: Recognition, Reconstruction and Palaeoenvironmental Interpretation*. Geological Society, London, Special Publications, **120**, 95–124.

LIDMAR-BERGSTRÖM, K., OLSSON, S. & ROALDSET, E. 1999. Relief features and palaeoweathering remnants in formerly glaciated Scandinavian basement areas. *In*: MEDARD, T. & REGINE, S.C. (eds) *Palaeoweathering, palaeosurfaces and related continental deposits*. International Association of Sedimentologists, Special Publications, **27**, 275–301.

MCKENZIE, D.P. 1978. Some remarks on the development of sedimentary basins. *Earth and Planetary Science Letters*, **40**, 25–32.

MICHELSEN, O. & NIELSEN, L.H. 1993. Structural development of the Fennoscandian Border Zone, offshore Denmark. *Marine and Petroleum Geology*, **10**, 124–134.

MICHELSEN, O., THOMSEN, E., DANIELSEN, M., HEILMANN-CLAUSEN, C., JORDT, H. & LAURSEN, G.V. 1998. Cenozoic sequence stratigraphy in the eastern North Sea. *In*: DE GRACIANSKY, P.-C., HARDENBOL, J., JACQUIN, T. & VAIL, P.R. (eds) *Mesozoic and Cenozoic Sequence Stratigraphy of European Basins*. Society of Economic Paleontologists and Mineralogists, Special Publications, **60**, 91–118.

NIELSEN, L. H. & JAPSEN, P. 1991. *Deep Wells in Denmark 1935–1990*. Geological Survey of Denmark, Series A, **31**.

NIELSEN, O.B., SØRENSEN, S., THIEDE, J. & SKARBØ, O. 1986. Cenozoic differential subsidence of North Sea. *AAPG Bulletin*, **70**, 276–298.

NORLING, E. 1994. Kontinentalsockelns berggrund. *In*: FRÉDEN, C. (ed.) *Berg och Jord*. Sveriges Nationalatlas Förlag, Stockholm, 38–40.

NORLING, E. & BERGSTRÖM, J. 1987. Mesozoic and Cenozoic tectonic evolution of Scania, southern Sweden. *Tectonophysics*, **137**, 7–19.

RIIS, F. 1996. Quantification of Cenozoic vertical movements of Scandinavia by correlation of morphological surfaces with offshore data. *Global and Planetary Change*, **12**, 331–357.

ROHRMAN, M. & VAN DER BEEK, P. 1996. Cenozoic postrift domal uplift of North Atlantic margins: an asthenospheric diapirism model. *Geology*, **24**, 901–904.

SCLATER, J.G. & CHRISTIE, P.A.F. 1980. Continental stretching: an explanation of the post-mid-Cretaceous subsidence of the central North Sea basin. *Journal of Geophysical Research*, **85**, 3711–3739.

SØRENSEN, J.C., GREGERSEN, U., BREINER, M. & MICHELSEN, O. 1997. High-frequency sequence stratigraphy of Upper Cenozoic deposits in the central and southeastern North Sea areas. *Marine and Petroleum Geology*, **14**, 99–123.

SVIRIDOV, N.I., FRANDSEN, J.V., LARSEN, T.H., FRIIS-CHRISTENSEN, V., MADSEN, K.E. & LYKKE-ANDERSEN, H. 1995. The geology of Bornholm Basin. *Aarhus Geoscience*, **5**, 15–35.

THYBO, H. 1997. Geophysical characteristics of the Tornquist Fan area, northwest Trans-European Suture Zone: indication of late Carboniferous to early Permian dextral transtension. *Geological Magazine*, **134**, 597–606.

THYBO, H. 2000. Crustal structure and tectonic evolution of the Tornquist Fan region as revealed by geophysical methods. *Bulletin of the Geological Society of Denmark*, **46**, 145–160.

TURCOTTE, D.L. & SCHUBERT, G. 1982. *Geodynamics. Applications of Continuum Physics to Geological Problems*. Wiley, New York.

UNDERHILL, J.R. & PARTINGTON, M.A. 1993. Jurassic thermal doming and deflation in the North Sea: implications of the sequence stratigraphic evidence. *In*: PARKER, J.R. (ed.) *Petroleum Geology of Northwest Europe: Proceedings of the 4th Conference*. Geological Society, London, 337–345.

VEJBÆK, O.V. 1997. Dybe strukturer i danske sedimentære bassiner. *Geological Society of Denmark, Geologisk Tidsskrift*, **4**, 1–31.

VEJBÆK, O.V. & BRITZE, P. (eds) 1994. *Top pre-Zechstein (twoway traveltime and depth), geological map of Denmark*, 1:750 000. Geological Survey of Denmark, Map Series, **45**.

WATTS, A.B., KARNER, G.D. & STECKLER, M.S. 1982. Lithospheric flexure and the evolution of sedimentary basins. *Philosophical Transactions of the Royal Society of London, Series A*, **305**, 249–281.

WATTS, A.B., MCKERROW, W.S. & FIELDING, E. 2000. Lithospheric flexure, uplift, and landscape evolution in south-central England. *Journal of the Geological Society, London*, **157**, 1169–1177.

WESTAWAY, R., MADDY, D. & BRIDGLAND, D. 2002. Flow in the lower continental crust as a mechanism for the Quaternary uplift of south-east England: constraints from the Thames terrace record. *Quaternary Science Reviews*, **21**, 559–603.

ZIEGLER, P.A. 1990. *Geological Atlas of Western and Central Europe*. Shell International, The Hague.

Cenozoic inversion and uplift of southern Britain

DEREK J. BLUNDELL

Department of Geology, Royal Holloway, University of London, Egham TW20 0EX, UK (e-mail: d.blundell@gl.rhul.ac.uk)

Abstract: Whereas significant exhumation of northern Britain took place during Paleocene time, probably as a consequence of uplift caused by a mantle plume, Paleogene basin inversion and uplift in southern Britain appears to be a consequence of Alpine tectonism. Recent publications demonstrate that inversion of Mesozoic basins across southern Britain was accompanied by subsidence of flanking basins in areas that had previously remained stable. Structures observed on seismic sections across the Weald Basin in SE England reveal that inversion occurred locally by north-directed reverse movements on pre-existing normal faults that cut down at a low angle deep into the basement. The overall effect of inversion of the Weald Basin, however, is a bulk deformation that produced a domal uplift, flanked by subsidence of the London and Hampshire–Dieppe basins. A 2D finite element thermo-mechanical model of continental lithosphere containing a region of reduced strength in the crust simulates Jurassic–Early Cretaceous extension to form the Weald Basin, followed by compression during the Tertiary to produce its inversion and the flanking basins. The timing of tectonic events across southern Britain correlates with times when Alpine stresses were transmitted into the foreland to the north sufficiently well to link them. Through most of Tertiary time, the landscape of southern England was of relatively low elevation and low-energy surface processes. However, late Neogene uplift, generally greater in the west, appears to have been part of a larger-scale uplift of land areas with hard rock at surface, which has no obvious tectonic explanation.

In simple terms, the surface geology outcrop pattern of the British Isles shows the youngest rocks in the SE and the oldest in the north and west. The level of erosion is such that the Phanerozoic succession is fully exposed at surface, along with Precambrian units dating back to Early Proterozoic time. Moreover, the extent of Tertiary sediments exposed onshore is limited to southern England and that of Neogene sediments is restricted to eastern East Anglia. In similar fashion, topographic elevation and relief are generally greater in the north and west, and least in the south and east. This range of surface geology and topography contrasts with other regions of NW Europe with similar, 30 km crustal thickness, such as the Netherlands, northern Germany, Denmark and Poland.

Brodie & White (1994) explained the denudation of northern Britain as a consequence of 8 km of magmatic underplating related to a mantle plume beneath NW Scotland, recognized at surface as the Tertiary Volcanic Province. White & Lovell (1997) correlated the associated uplift and erosion from a succession of volcanic pulses with the deposition of large quantities of clastic sediment as submarine fans in neighbouring basins (North Sea, West Shetlands) during the period 62–54 Ma. Their estimates of up to 3 km of exhumation, based on vitrinite reflectance and fission-track data, were independently confirmed by a careful study of denudation of the Irish Sea Basin by Rowley & White (1998). In this, they estimated the amount of denudation by calculating the stretching and thermal subsidence history of the basin to determine how much post-rift sediment thickness is missing. The mismatch between the present-day and the predicted depth to basement represents the amount of denudation. They obtained values for the East Irish Sea Basin ranging between 0.4 and 2.6 km, with values up to 1.5 km and 1.7 km for the neighbouring West Lancashire and Cheshire basins, respectively. Doubts remain about the extent of uplift modelling based on fission-track and vitrinite reflectance data, which depends on the value used for the geothermal gradient at the time of uplift, and could be halved to around 1.5 km. None the less, there is a strong case for explaining the exhumation of northern Britain as a result of an Early Tertiary plume, which moved away around 54 Ma when North Atlantic seafloor spreading was initiated, leaving the area above sea level to the present day. However, the influence of the plume did not extend across

From: DORÉ, A.G., CARTWRIGHT, J.A., STOKER, M.S., TURNER, J.P. & WHITE, N. 2002. *Exhumation of the North Atlantic Margin: Timing, Mechanisms and Implications for Petroleum Exploration*. Geological Society, London, Special Publications, **196**, 85–101. 0305-8719/02/$15.00

southern Britain, so that other explanations are needed to account for the basin inversion and uplift that is observed. These are generally ascribed (e.g. Chadwick 1993) to the deformational response to lithospheric stresses generated during the Alpine Orogeny that were transmitted laterally northwards. The purpose of this paper is to examine the nature and timing of basin inversion across southern Britain, and the Weald in particular, to assess whether they can be correlated with Alpine events and why the deformation is apparently more pronounced across southern Britain than elsewhere in NW Europe. To understand the Cenozoic inversion of sedimentary basins across southern Britain it is necessary first to appreciate their earlier evolution as extensional basins.

Evolution of sedimentary basins through the Mesozoic era

The Celtic Sea, Bristol Channel, Central English Channel, Wessex and Weald basins (see Fig. 1 for locations) were built upon a Variscan structural framework (Fig. 2). The northward movement of the Armorican Block towards the Midland Craton and London Platform resulted in compression of the intervening crust. This deformed dominantly by north-vergent, east–west-trending thrusts and folds that developed above a mid-crustal décollement (BIRPS & ECORS 1986; Chadwick 1986; Brooks *et al.* 1988) and by pure shear in the lower crust. Lateral ramps and strike-slip escape structures created NE–SW-trending faults to the west and NW–SE-trending faults to the east (Fig. 2). In the collapse of the Variscan Orogeny, sedimentary basins were initiated during Permian time, filled with terrestrial red-bed sequences in the hanging walls of former thrusts, reactivated as low-angle normal faults. These were usually bounded by steeper, short-cut normal faults, mainly downthrown to the south. Extensional basins developed through Triassic time and, in the Wessex and Central English Channel basins, were compartmentalized by NW–SE-trending faults inherited from Variscan structures. Further extension and subsidence during Jurassic–Early Cretaceous time resulted in the deposition of shallow marine sequences that include limestones, sandstones and organic-rich claystones. At the same time, changing stress conditions resulted in the eastward migration of depocentres within sub-basins of the Wessex and Central English Channel basins and the development of new en echelon fault sets and transtensional movements of the earlier normal east–west faults. Continuing subsidence during Early Cretaceous time resulted in the deposition locally of fluvial–lacustrine deposits. The structural and stratigraphic evolution of the Wessex Basin has been described by a number of workers, notably Chadwick (1986), Lake & Karner (1987), Butler (1998) and Hawkes *et al.* (1998), and has been well summarized by Underhill & Stonely (1998),

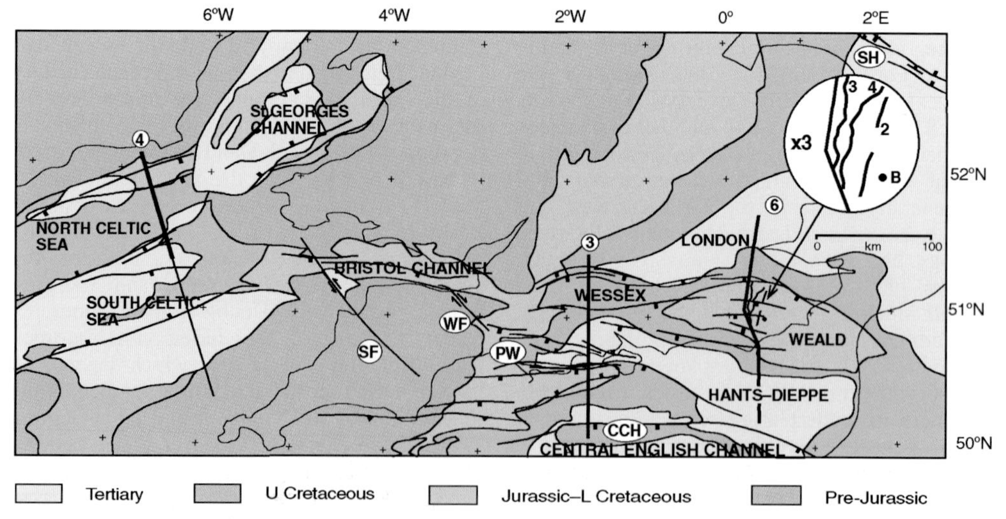

Fig. 1. Simplified geological map of southern England showing main faults and sedimentary basins. SF, Sticklepath Fault; SH, South Hewitt Fault; WF, Watchet–Cothelstone Fault; PW, Purbeck–Wight Disturbance; CCH, Central Channel High. 3, line of section shown in Fig. 3; 4, SWT4 profile, bold section shown in Fig. 4; 6, line of BGS section shown in Fig. 6. Inset: seismic lines C78-02, C78-03 and C78-04; B, Brightling well.

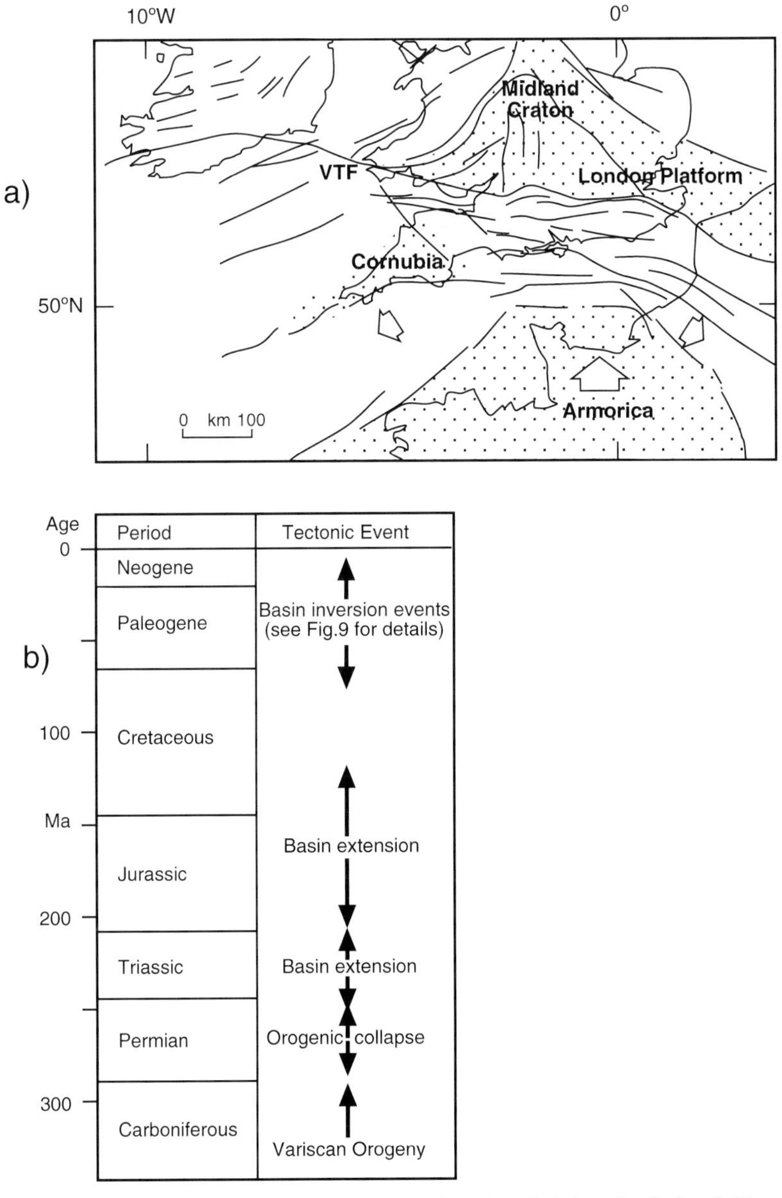

Fig. 2. (**a**) Map showing the Variscan structural framework of southern Britain (after Lefort & Max 1992). VTF, Variscan Thrust Front. (**b**) Summary of post-Variscan tectonic events affecting southern Britain.

who regarded the Permian–Lower Cretaceous succession as a single megasequence. Its upper bound is marked by a regional unconformity around Aptian time (124 Ma), which truncates successively older strata towards the west, cutting down to the Permo-Trias deposits. The Permian–Early Cretaceous structural and stratigraphic evolution of the North and South Celtic Sea basins has been described by Petrie *et al.* (1989), that of the Bristol Channel Basin by Brooks *et al.* (1988) and that of the Central English Channel Basin by Hamblin *et al.* (1992). These basins all have similar histories. Throughout this period the neighbouring stable blocks, such as the London Platform, Cornubian Massif and Armorican Block, had remained as positive regions with no history of subsidence. During Late Cretaceous time, a broadly westward progressing marine transgression led to the deposition of sandstone and claystone followed by the Chalk Group, which extended across the whole region and well beyond, including the former land areas. This is well illustrated in the palaeogeographical atlas of Ziegler (1990). Deposition of Chalk is known to have continued through to Maastrichtian time in the southern North Sea and the English Channel (Rawson 1992) but in most areas the youngest portion has been removed by erosion. A regional disconformity represents a period of uplift and erosion between 74 and 60 Ma, when the earliest recorded Tertiary sediments were deposited.

Cenozoic inversion

Extensive erosion of Chalk across southern England led to the development of a sub-Paleogene surface, in some areas covered by thin lateritic soil deposits (Green 1985). During Paleogene time, basin inversion is evident in various ways. In some areas it is concentrated on particular faults, whereas elsewhere it is characterized as a broad domal uplift (Chadwick 1993), with peripheral basins developed across areas that had remained as stable positive blocks during Permian–Late Cretaceous times.

Fault inversion

The most spectacular fault inversion is the Purbeck–Wight Disturbance, seen along the Dorset coast to the Isle of Wight. Underhill & Paterson (1998) described the structures in detail and demonstrated the effect of inversion upon the east–west-trending segmented fault system, linked by relay ramps, extending between Abbotsbury in the west and Whitecliff Bay, Isle of Wight, in the east (Fig. 1). Using information from seismic sections, wells and surface geology, they showed that the fault system acted as a normal growth fault, downthrown to the south, from Permian to Early Cretaceous time, but was inverted by reverse movement during the Tertiary. Although the fault system remains in net extension for the Permian–Lower Cretaceous succession, inverse movement is greatest where formerly extension was greatest. Deformation of incompetent hanging-wall strata against Chalk in the footwall, which acted as a semi-rigid buttress, implies that inversion resulted from northward-directed stress. Movement of hanging-wall sediments relative to footwall Chalk and footwall deformation of the Chalk can be used to estimate the amount of inversion, uplift and lateral shortening. For example, from a seismic section across the Abbotsbury–Ridgeway Fault (Butler 1998, Fig. 8) north of Weymouth, at least 350 m Tertiary displacement can be estimated. On the Isle of Wight, uplift of at least 1500 m has been recorded (Butler 1998, Fig. 13). In addition to localized uplift on inverted faults, there is a broad upwarp of the Wessex Basin of between 250 m (Chadwick 1993) and 500 m (Law 1998). Some, at least, of the variation in structural style can be attributed to the presence, or otherwise, of salt (Harvey & Stewart 1998).

Inversion in the Central English Channel Basin, to the south, is concentrated on the Central Channel High (Fig. 1), where Law (1998) has estimated over 1000 m of uplift from sonic velocity measurements in boreholes, supported by fission-track and vitrinite reflectance data. The structural evolution of the Central Channel High was investigated by Beeley & Norton (1998) using two north–south high-resolution seismic sections across it, with well ties for stratigraphic correlation and depth conversion. Sequential balanced section restoration of the two true-scale depth sections resulted in estimates of 1100–1500 m uplift of the hanging-wall sediments over the Central Channel High relative to the footwall to the south during Tertiary inversion. The fault geometry requires a northerly dip of 55° decreasing to 30° at about 12 km depth. From this, and the deep structure of the Wessex Basin identified byChadwick (1986, 1993), Beeley and Norton proposed that the inversion developed upon a linked set of crustal-scale faults that initiated as part of the Variscan thrust system, evolved during basin extension as low-angle normal faults and reactivated again during Tertiary time to produce the fault-controlled inversion. Their schematic cross-section is reproduced in Fig. 3. The implication from this is a north–south shortening by *c.* 4 km of some 200 km of crust.

There has been considerable debate about the timing of inversion. The sub-Paleogene regional unconformity, with differential erosion of the Chalk, indicates uplift across southern England, but as the Upper Paleocene–Eocene succession within the Hampshire–Dieppe Basin is itself involved in the fault inversion, for example on the Isle of Wight, at least part of the inversion must be younger. Chadwick (1993) argued strongly in support of a Miocene age for the main inversion episode, ‘corresponding to the main Alpine deformation events as continental collision occurred between Africa and Europe’. A succession of Tertiary sediments of 650 m thickness in the Hampshire–Dieppe Basin is well exposed in Alum Bay at the western end and in Whitecliff Bay at the eastern end of the Isle of Wight. Resting unconformably upon an eroded

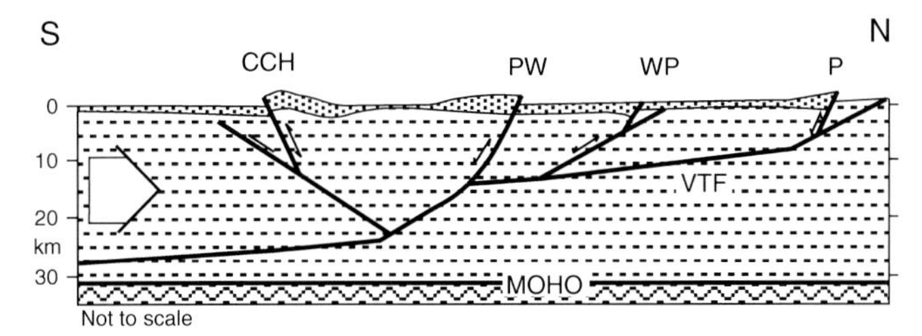

Fig. 3. Schematic crustal section by Beeley & Norton (1998) (location shown in Fig. 1), showing basin inversion on linked crustal-scale faults. CCH, Central Channel High; PW, Purbeck–Wight Disturbance; WP, Wardour–Portdown structure, P, Vale of Pewsey Fault; VTF, Variscan Thrust Front.

Chalk surface are red mottled soils (Reading Beds, *c.* 58 Ma) of Late Paleocene age. These are followed by an Eocene succession comprising, in turn, the marine London Clay, shallow marine and lagoonal sands and clays with occasional coals of the Bracklesham Group, and marine clays and sands of the Barton Group. These are followed by the Solent Group of nearshore and freshwater silts and clays, and a freshwater limestone. The youngest of these is of Early Oligocene age, *c.* 30 Ma. As all these sediments were deposited at or near sea level, sedimentation rates must have kept pace with subsidence, to fill the accommodation space within the Hampshire–Dieppe Basin.

A careful study by Gale *et al.* (1999) of the Eocene succession exposed in Whitecliff Bay established that clasts derived from successively older horizons occurred in successively younger strata, as the hanging-wall sediments of the Sandown Pericline, just a few hundred metres to the south (the eastern segment of the inverted fault system described by Underhill & Paterson (1998)), were being uplifted and eroded. Gale *et al.* (1999) found that uplift occurred in two main phases. Uplift of 200–300 m occurred between 47.5 and 44 Ma, with rates at times of 0.1 mm a^{-1}. A further 200–300 m uplift and erosion occurred during the period 42–36 Ma, probably in short pulses of less than 1 Ma duration. The presence of derived fossils and clasts within the uppermost beds in the succession suggests that uplift persisted into Early Oligocene time. The total uplift during the Eocene fault inversion of some 500 m is a substantial part of the displacement recorded on faults within the Wessex Basin. However, the Chalk and overlying Reading Beds and London Clay are vertical in both Alum Bay and Whitecliff Bay, the Bracklesham Group are steeply dipping to the north, but the dip reduces to a low angle of around 5° through the Barton and Solent Groups, some 800 m from the Chalk outcrop. Clearly, this deformation, which has up-ended the Paleogene sequence, is younger than the 30 Ma sediments involved.

Linked to the east–west-trending fault system of the Purbeck–Wight Disturbance via the Lytton Cheyney Fault is the NW–SE-trending Watchet–Cothelstone Fault (Fig. 1). Reviewing evidence of fault movements, Miliorizos & Ruffell (1998) concluded that it originated as a Variscan structure and has acted at various times subsequently as a strike-slip fault in both dextral and sinistral senses. East–west horsetail splays at either end suggest that the most recent movements have been sinistral, but their timing is constrained only to post-Jurassic time. Sixty-five kilometres to the west, the NW–SE-trending Sticklepath Fault (Fig. 1) appears to have had a similar history of strike-slip movements. Segmented along strike, relays have acted as releasing bends to produce pull-apart basins, the Lundy, Petrockstow and Bovey basins, which are filled with terrestrial sediments of Eocene–mid-Oligocene age. These include fluvial and lacustrine gravels, sands and clays with thin lignite bands. Although the Bovey Basin is at least 1000 m deep, and the oldest strata have not been sampled, the nature of the sediments implies that the tectonically driven subsidence was matched by the rate of sedimentation so that accommodation space was filled. Thus the sediments date the fault movement. Holloway & Chadwick (1986) argued that this movement was sinistral, with 6 km horizontal displacement. However, comparison with a scaled analogue model (Dooley & McClay 1997) suggests that the sinistral displacement was no more than 2 km. The ages of sediments in the smaller Petrockstow and Lundy basins are similar to those in the Bovey Basin and imply that movements of the Sticklepath Fault continued to mid-Oligocene time, around 28 Ma. This may also have been the case for the Watchet–Cothelstone Fault.

Inversion by bulk deformation and uplift

Although the near-surface mechanism of basin inversion may be through inverse displacements on individual faults, the bulk effect is the uplift of a basin to a domal structure. This is observed in the North Celtic Sea Basin and the Weald Basin. The SWAT 4 deep seismic section (BIRPS & ECORS 1986), Fig. 4 illustrates the overall effect of inversion of the North Celtic Sea Basin. The basin rests on the hanging wall of a low-angle, south-dipping, planar fault that can be traced from surface to 15 km depth, where it merges with the top of a highly reflective zone of subhorizontal reflection segments that extends down to the Moho. This fault can be correlated at surface in southern Ireland with the Variscan Thrust Front, but it has clearly been reactivated in extension as a low-angle normal fault to allow space for the formation of the North Celtic Sea Basin as a half-graben. The basin is filled with a thick Permo-Triassic terrestrial red-bed succession and a Jurassic–Cretaceous marine succession similar to that in the Wessex Basin (Petrie *et al.* 1989). Later compressive reactivation of the major fault has resulted in the inversion of faults within the North Celtic Sea Basin, producing localized anticlines and positive flower structures, together with a broad domal uplift of 1–2 km (Roberts 1989). Roberts argued that although some inversion may have occurred during Paleocene time, the major part is of Oligocene age. Erosion has resulted in the outcrop of Jurassic strata at the sea bed. Flanking the inverted North Celtic Sea Basin are sag basins filled with Paleogene sediments, one of which is shown in Fig. 1. It appears from their association that their subsidence accompanied the uplift of the North Celtic Sea Basin. A similar configuration is evident with the domal inversion of the Weald Basin and the subsidence of the flanking London and Hampshire–Dieppe basins, which are now examined in more detail.

Inversion of the Weald Basin

The Weald Basin initiated as an easterly prolongation of the Wessex Basin at the beginning of Jurassic time (208 Ma). It developed as an extensional basin, subsiding by means of normal growth faults of mainly east–west trend, the most active of which were close to its northern margin against the London Platform, which had acted as the undeformed foreland to the Variscan orogen. Thus the Weald Basin developed during Jurassic time mainly as an asymmetric basin with strong down-to-south normal faults along its northern margin, founded upon reactivated Variscan thrusts, which acted as low-angle extensional detachments. An element of transtensional fault movements during this

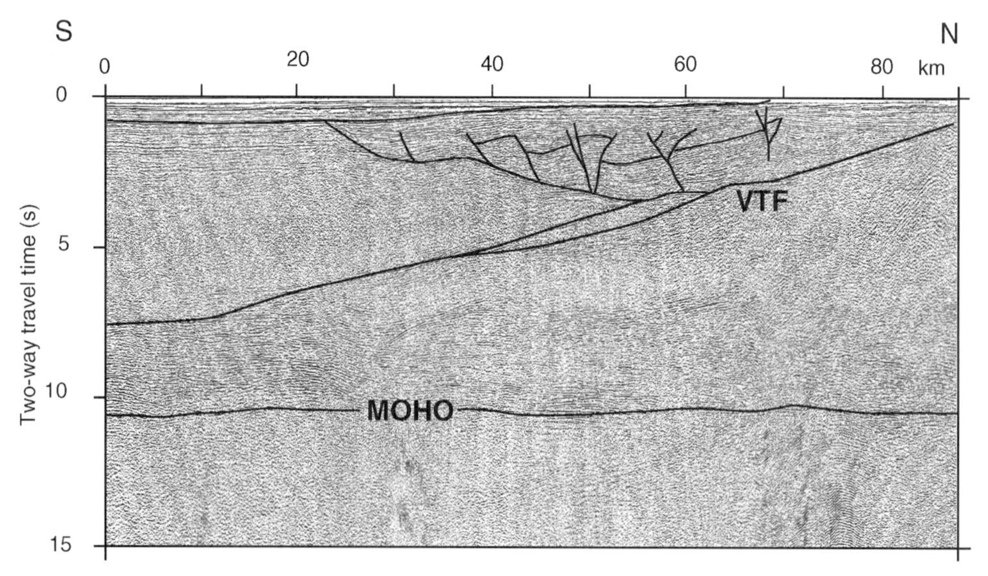

Fig. 4. Deep seismic profile SWAT 4 (BIRPS & ECORS 1986) (location shown in Fig. 1), interpreted to show the Variscan Thrust Front (VTF) reactivated as a low-angle normal fault to provide space for the Mesozoic North Celtic Sea Basin in its hanging wall, then inverted as indicated by structures within the basin. The broad uplift of the basin and the flanking Tertiary basin to the south should be noted.

period is marked by the en echelon geometry of the normal faults in plan view and the presence of associated WNW–ESE-trending faults (Fig. 1). Within the basin, lower Lias marine claystones and limestones rest directly upon a Devonian–Carboniferous basement. The Jurassic succession, although similar to that in the Wessex Basin, is dominated by carbonate rocks laid down in a shallow marine low-energy, wave-dominated environment, indicative of clear sea and a limited supply of clastic material from nearby land, particularly the London Platform to the north. Deposition through Early Cretaceous time of marine sandstones, mudstones and sabkha-type evaporites testified to continuing subsidence, at a rate greater than in the Wessex Basin, and deposition of clastic sediments in a nearshore environment. Sedimentation thus kept pace with subsidence. A mid-Cretaceous (124 Ma) unconformity marking the termination of this megasequence (compare the Wessex Basin) represents a relatively small time gap and sedimentation resumed, in continuity with the Wessex Basin, with marine sandstones and claystones succeeded by the Chalk. Although the youngest surviving Chalk is of late Campanian age (83 Ma), it is likely that deposition continued through Maastrichtian time to around 74 Ma, yielding a total thickness of Chalk of about 500 m. During the following 16 Ma, regional uplift and erosion led to the removal of a substantial portion of the Chalk, ranging, through differential erosion, between 100 and 350 m. A sub-Paleogene surface developed, upon which the commencement of subsidence of the London Basin to the north of the Weald, and the Hampshire–Dieppe Basin to the south, is recorded in the deposition of terrestrial, intertidal and shallow marine sediments of late Paleocene age (58 Ma). These were succeeded by the London Clay, representing marine conditions with no evidence of shoreline facies, which may have inundated much of southern England between 54 and 50 Ma. Interbedded shallow marine sands and clays were deposited in succession through Eocene time, the youngest preserved being of late Eocene age (*c.* 30 Ma). The maximum thickness of Tertiary deposits in the London Basin is 300 m and in the Hampshire Basin it is 650 m. Evidence that subsidence of the London Basin accompanied uplift of the Weald is less convincing than on the Isle of Wight. However, Gale *et al.* (1999) pointed out that chert clasts identified as being derived from the Lower Cretaceous Hythe Beds (*c.* 140 Ma) found in Lutetian age pebble beds in the London Basin imply that the Weald was being uplifted at the same time as the first phase of inversion of the Isle of Wight (47.5–44 Ma).

Inversion of the Weald can be observed on oil industry seismic sections now available for research use through the UK Onshore Geophysical Library. Figure 5 presents three north–south true-scale depth converted sections across the Weald showing the principal Jurassic horizons interpreted from the seismic data: line locations are shown in Fig. 1. Faults have been migrated within the vertical plane of section. Line 78-02 was shot with explosives, giving greater depth penetration than the vibroseis source used for the other lines. On this line, reflections from low-angle faults can be traced to 8 km depth, well into the crystalline basement. Similar south-dipping faults observed on the adjacent lines cannot be traced as far down. The low dip angles of these faults suggest that they originated as thrusts (probably Variscan), which then acted in extension during Jurassic time as the Weald Basin subsided. They were later inverted during Tertiary time, as is evident on the fault-related uplift structures within the Jurassic–Lower Cretaceous succession. The sense of movement of inversion is to the north. The Brightling well, offset 4 km east of line 78-02, provides stratigraphic correlation with seismic reflectors and, in addition, records a duplication by reverse faulting of the Lower Lias succession. Assuming the fault dips at 45° to the south, consistent with faults observed on the seismic section, implies a 350 m northward displacement as a result of Tertiary inversion, comparable with that on the Abbotsbury–Ridgeway Fault. Further examples of fault inversion in the Weald were presented by Butler & Pullen (1990), who also found NW–SE-trending faults that had undergone significant transpression. Although there is no direct evidence on the timing of inversion, Butler and Pullen reported the presence of ferroan calcite precipitated on the crests of Tertiary uplift structures that have Sr isotope ratios corresponding to a Late Oligocene–Early Miocene age (*c.* 24 Ma). However, such evidence is highly suspect.

Inverse model of the Weald

Although the seismic data indicate that the mechanism of inversion is based on reactivation of low-angle, south-dipping faults that cut deep into the crust, the overall effect of the inversion of the Weald is a broad domal uplift of some 1500 m, together with the downwarping of the flanking London and Hampshire–Dieppe basins, seen in Fig. 1. Figure 6a is a simplified version of

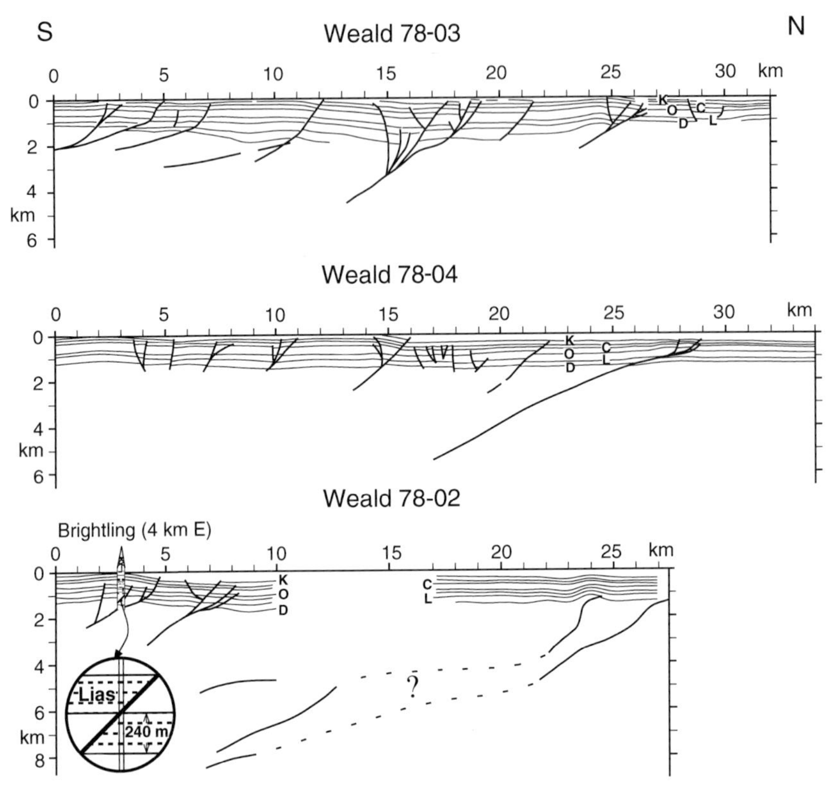

Fig. 5. True-depth interpretations of three north–south seismic sections across the Weald (location shown in Fig. 1). Formation tops: K, Kimmeridge; C, Corallian; O, Oxfordian; L, Lias; D, Devonian.

the cross-section produced by the British Geological Survey (BGS) with their 1:250 000 series solid geology maps, sheets 50N00 and 51N00(British Geological Survey 1988, 1989). The line of section is located in Fig. 1. In Fig. 6a, individual faults shown on the BGS section have been removed and the cross-section has been smoothed accordingly to focus on the effect of distributed bulk pure shear deformation, as a basis for preparing an inverse model. Critically for the modelling, the deep, crustal-scale faults provide a zone of low strength in the crust, laterally confined below the Weald Basin, with relatively strong crust beneath the flanking basins. A 10:1 vertical exaggeration emphasizes the domal uplift. Isochrons have been added to provide time markers that monitor the basin evolution. Using a knowledge of Tertiary stratigraphy and landscape evolution based on a reconstructed cross-section of the Weald by Jones (1999, Fig. 3), the cross-section has been reconstructed back to mid-Eocene time (40 Ma, Fig. 6b), at which time the flanking basins had subsided to a maximum extent. Because stratigraphic evidence indicates shallow seas, the reconstruction assumes the water depth is negligible. No account is taken of the land surface topography across the central Weald, where fault inversion would have created localized uplift. Reconstruction back to late Paleocene time (60 Ma, Fig. 6c) when the flanking basins began to subside, together with the broad domal uplift of the Weald, was accomplished by stripping off the younger strata and adjusting the geometry of the underlying strata by vertical movement. No account was taken of lateral compression from basin inversion. The upper surface of the cross-section represents the sub-Paleogene surface. The cross-section at latest Cretaceous time (68 Ma, Fig. 6d) represents the time when deposition of the Chalk had been completed and indicates a uniform thickness of Chalk across the section. The extent of uplift and erosion of Chalk between 68 and 60 Ma (Fig. 6c) is based on the estimates of Jones (1999). Using the work of Butler & Pullen (1990) for the Jurassic–Cretaceous succession, the reconstruction continued to the mid-Cretaceous unconformity at 124 Ma (Fig. 6e) and back to the initiation of the Weald Basin at the beginning

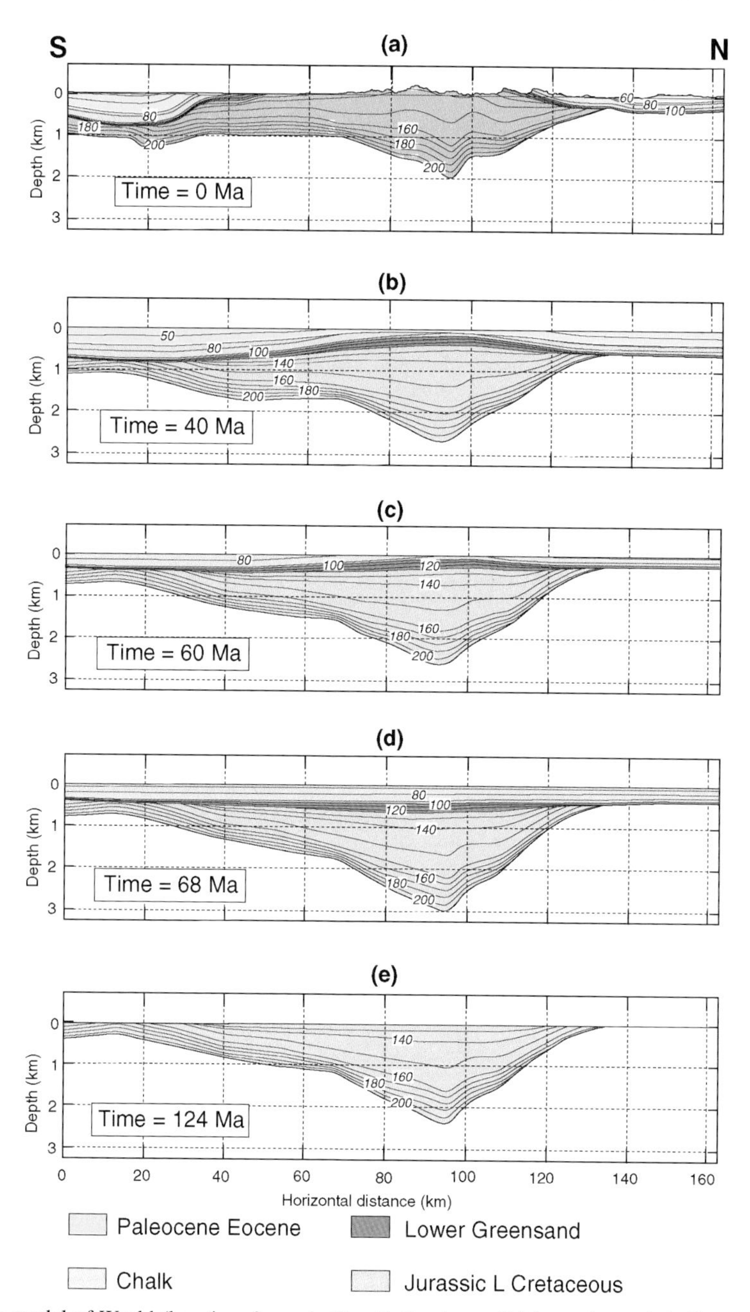

Fig. 6. Inverse model of Weald (location shown in Fig. 1). Isochrons (Ma) are shown in italics.

of Jurassic time (208 Ma). In this reconstruction, account was taken of the effects of compaction and decompaction of the Jurassic–Lower Cretaceous succession during subsidence and subsequent uplift, but this was regarded as unwarranted for the Upper Cretaceous and Tertiary deposits. No account was taken of any deeper crustal thermo-tectonic effects, as the purpose of the inverse model was to reconstruct the basin geometry in cross-section, which could act as a template for forward modelling.

Forward model

The purpose of forward modelling was to learn what caused the subsidence of the London and Hampshire–Dieppe basins at the same time as the Weald Basin was being inverted and uplifted. A 2D thermo-mechanical forward modelling scheme developed by Nielsen & Hansen (2000), based on the model of Braun & Beaumont (1987), provides a physical basis on which to simulate the subsidence of the Weald Basin and its subsequent inversion. Details of the mathematical expressions used for the model can be found in these publications. Figure 7 shows schematically how the model works. The upper-crust, lower-crust and upper-mantle lithosphere are defined by their physical properties, such as density, Young's modulus, Poisson's ratio, compressional strength, creep parameters, specific heat, heat production and thermal conductivity, appropriate for quartz-rich, feldspar-rich and olivine-rich mineral rheologies, respectively, as shown in Table 1. Their behaviour is dictated by the prevailing temperature, pressure and strain rate at each point in the model as it evolves, according to functions expressing the response of the material to stress in a viscoelastic or plastic manner. A constant heat flux of 35 mW m^{-2} is applied to the base of the model upward from the underlying mantle, which results in a surface heat flow of 60–70 mW m^{-2}, depending on the thickness of the crust. Buoyancy forces operate throughout to maintain isostatic equilibrium. The model consists of a finite element mesh that is closely spaced in the upper crust, intermediate in the lower crust and more open in the upper mantle. Time-variable conditions are imposed on the model, including heat flow upward from the mantle, surface rates of erosion and sediment deposition, and kinematic boundary conditions. These last conditions impose extension or compression from one end, whereas the other end remains pinned. Eustatic sea-level changes (Haq *et al.* 1987) are also incorporated into the model. In the model for the Weald, a central region of 100 km width is present in which the compressional strength of the upper crust is reduced to 40% of the values on either side, given in Table 1, and the creep parameter B of the lower crust is reduced to 85%, asymmetrically northward, of the values on either side (Fig. 7). Whereas the reduction of compressional strength ensures plastic yielding at lower stress levels, the reduction in B leads to accelerated creep as a result of lower viscosities. The model was first subjected to lateral extension at 0.1 mm a^{-1} between 210 and 170 Ma, rising to 0.5 mm a^{-1} between 165 and 155 Ma before reducing to zero by 145 Ma. This simulated the Jurassic–Early Cretaceous extension required to create the Weald Basin, amounting to 13 km. Thermal relaxation followed for 85 Ma until compression began at 60 Ma, rising to a rate of 0.25 mm a^{-1} between 50 and 20 Ma, reducing to 0.1 mm a^{-1} by 10 Ma, and continuing to the present, a total of 11.5 km. Because the model ignores fault movements, the amounts of extension and compression are overestimated by a factor of two so that the values given above should be halved to

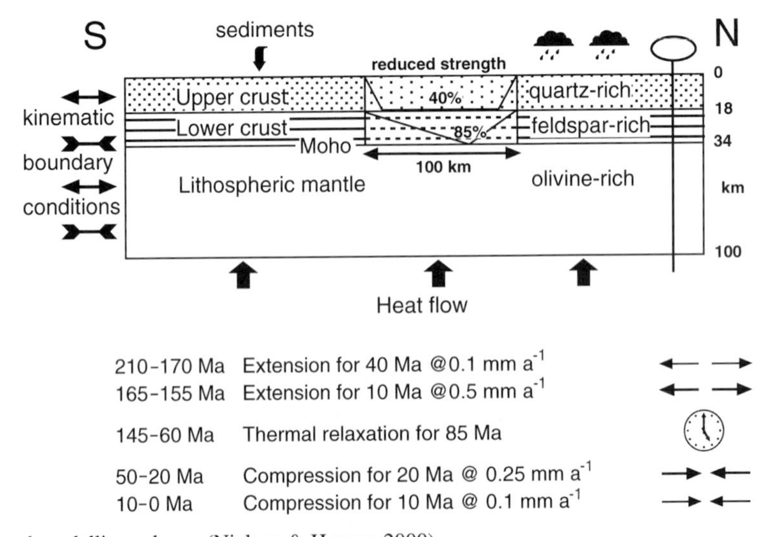

Fig. 7. Forward modelling scheme (Nielsen & Hansen 2000).

Table 1. *Parameters for the thermo-mechanical model of the Weald*

Parameter	Sediments	Upper crust	Lower crust	Mantle
Surface porosity ϕ_0	0.60			
Density (kg m^{-3})	2700	2700	2900	3300
Young's modulus (Pa)	10^{11}	10^{11}	10^{11}	10^{11}
Poisson's ratio	0.25	0.25	0.25	0.25
Creep parameter n	3.10	3.10	3.20	4.48
Creep parameter B (MPa $s^{1/n}$)	208	208	12.28	0.2628
Creep activation energy (kJ mol^{-1})	135	135	239	498
Compressional strength (MPa)	17.5	26.2	26.2	52.4
Thermal conductivity (W m^{-1} K^{-1})	2	3	2.3	4
Specific heat (J kg^{-1} K^{-1})	1000	850	900	1000
Heat production rate (μW m^{-3})	1.3	1.3	0.3	0.01

Sediment porosity ϕ varies with depth z (m) according to $\phi = \phi_0 \exp(-z/2000)$. Creep activation energy and parameters n and B relate to expressions for viscous deformation (Braun & Beaumont 1987).

be realistic. The progress of the model through this deformational history was monitored at 1 Ma intervals. Sedimentation rates of up to 0.1 mm a^{-1} ensured that sedimentation kept pace with subsidence. Accommodation space was largely filled and water depths were small. Similarly, erosion rates were such that erosion kept pace with uplift to keep surface elevation and relief low. The maximum erosion rate was 0.18 mm a^{-1} during the rapid fall of sea level in Tertiary time. Figure 8 presents the forward model at 124, 68, 60, 40 and 0 Ma for direct comparison with the inverse model, Fig. 6.

Because the model is an oversimplification of reality there is a limit to the extent to which it should attempt to replicate the details of the inverse model. However, the main features of the model offer new insights into the history of subsidence and inversion of the Weald and the reasons why the latter was accompanied by the formation of the London and Hampshire–Dieppe basins. The critical factor is the low strength of the crust across the Weald, sandwiched between the high strength of the London Platform to the north and the relatively high strength of the crust to the south, in the eastern part of the English Channel and northern France. This strength pattern was undoubtedly inherited from the Variscan Orogeny and the area of weak crust was exploited during Mesozoic time to control the locations of extensional basins across southern Britain. The model shows how the high strength of the crust in the areas flanking the Weald requires the development of compressional basins as a necessary accompaniment to the updoming of the Weald during inversion. Extension of the upper crust resulted in a 2 km rise of the Moho beneath the Weald Basin. Subsequent inversion thickened the crust beneath the basin so that the end result is a deepening of the Moho to 35 km depth below the zone of weakness. The Moho beneath the flanks remained at 34 km throughout. This model contrasts with one proposed by Cloetingh *et al.* (1990) in which lateral compression imposed on a rifted basin in lithosphere that has vertically variable but laterally uniform rheology results in a deepening of the basin and uplift of the flanks, the opposite of what is observed in the Weald.

The model simulates the subsidence history and geometry of the Weald Basin during Jurassic–Early Cretaceous time, which is indicated by the isochrons shown at 124 Ma in Fig. 8e. It also simulates the unconformity at 124 Ma as a response to the cessation of extension and the dominance of thermal relaxation. The model then simulates erosion of the Chalk between 68 and 60 Ma, but fails to include the commencement of uplift of the Weald at this time. Between 60 and 40 Ma, uplift of the Weald is accompanied by subsidence of the London and Hampshire–Dieppe basins. Between 40 Ma and the present, the Weald is further uplifted and eroded down to uppermost Jurassic level, the Weald dome becomes more pronounced and the flanking basins are further compressed and partially eroded. The latter is in part due to the eustatic fall in sea level of around 100 m since Miocene time (Haq *et al.* 1987).

Correlation of events in southern Britain with Alpine tectonics

A very extensive literature chronicles the tectonic evolution of the Alps as a consequence of the convergence of Africa with Europe over the past 120 Ma (Dewey *et al.* 1989; Pfiffner 1992).

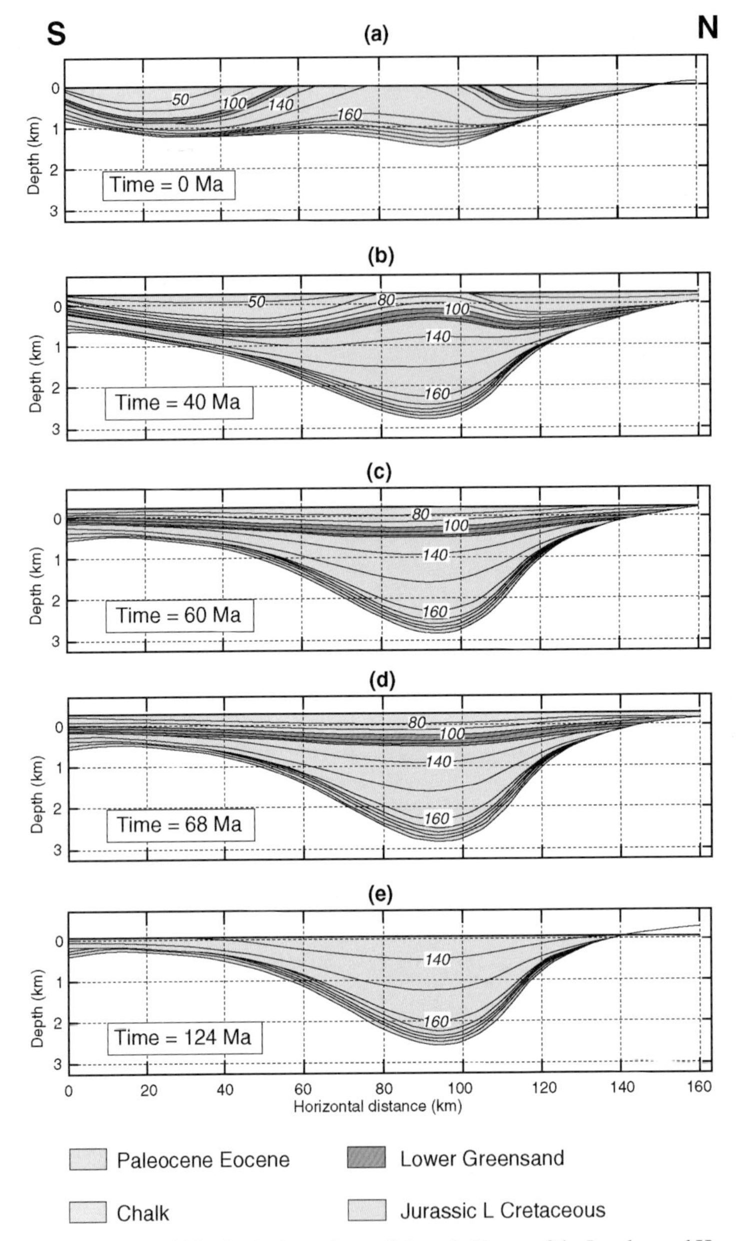

Fig. 8. Forward model of the Weald Basin, its inversion and the subsidence of the London and Hampshire–Dieppe basins produced by Hansen, based on the modelling scheme shown in Fig. 7. This may be compared with the inverse model, Fig. 6. Isochrons (Ma) are shown in italics.

Ziegler *et al.* (1995) examined in detail the effects of the Alpine collisional history upon the foreland to the north. They pointed out that 'forces related to collisional plate interaction appear to be responsible for the most important intra-plate compressional deformations' and concluded that 'the bulk of Late Cretaceous and younger intra-plate compressional deformations observed in western and central Europe developed in response to stresses that developed as a consequence of collisional coupling of the Alpine and Pyrenean orogens with their forelands'. The transmission of stresses into the foreland thus took place whenever they could not be accommodated within the orogen or the foreland basin. The evidence for the timing of such events can be found within the foreland basin, in particular the North Alpine Foreland Basin, with regard to north-directed stresses that could have affected southern Britain. Figure 9 summarizes the correlation, based on work by Ford *et al.* (1999) and Ford (pers. comm.) on the stratigraphic and structural evolution of the North Alpine Foreland Basin. She has recognized four stages of progressive deformation:

(1) a Mid–Late Eocene stage (46–36 Ma), when the convergence of the Apulian Plate with Europe, at a rate of $15\,\mathrm{mm\,a^{-1}}$, resulted in the migration of the flexural basin and the front of a low-angle external orogenic wedge, allowing the northward propagation of stress into the foreland;

(2) an Early–Late Oligocene stage (33–23 Ma), when the migration of the flexural basin and wedge front slowed significantly; no growth structures developed and shortening was accommodated within the thickness of the orogenic wedge; no stress would have propagated into the foreland;

(3) an Early–Mid-Miocene stage (16–11 Ma), when the system remained in much the same state as in the previous stage, with the wedge front stationary, so that little or no stress would have propagated into the foreland;

(4) a Late Miocene–Pliocene stage (11–3 Ma): around 11 Ma the outer orogenic wedge effectively collapsed and compressional deformation concentrated on the Jura fold belt, which detached on high-level Triassic evaporites, to accommodate some 30 km of NW shortening; beneath the Jura décollement, stress within the lithosphere is likely to have propagated into the foreland.

In addition to Alpine tectonics, extension and subsidence within the European Rift System, involving the Lower Rhenish Basin, the Rhine, Rhône and Limogne Graben, occurred mainly in Oligocene time through to Miocene time (Meier & Eisbacher 1991). As the graben are oriented north–south, rifting is inferred to have resulted from east–west tension.

As shown in Fig. 9, inversion events in southern Britain can be correlated in time with events in the Alps to explain: (1) the uplift and erosion of the Chalk between 68 and 60 Ma as due to stress generated by convergence between Africa and Europe; (2) the uplift of the Weald and subsidence of the London Basin and the uplift of the Sandown Pericline and subsidence of the Hampshire–Dieppe Basin in a succession of

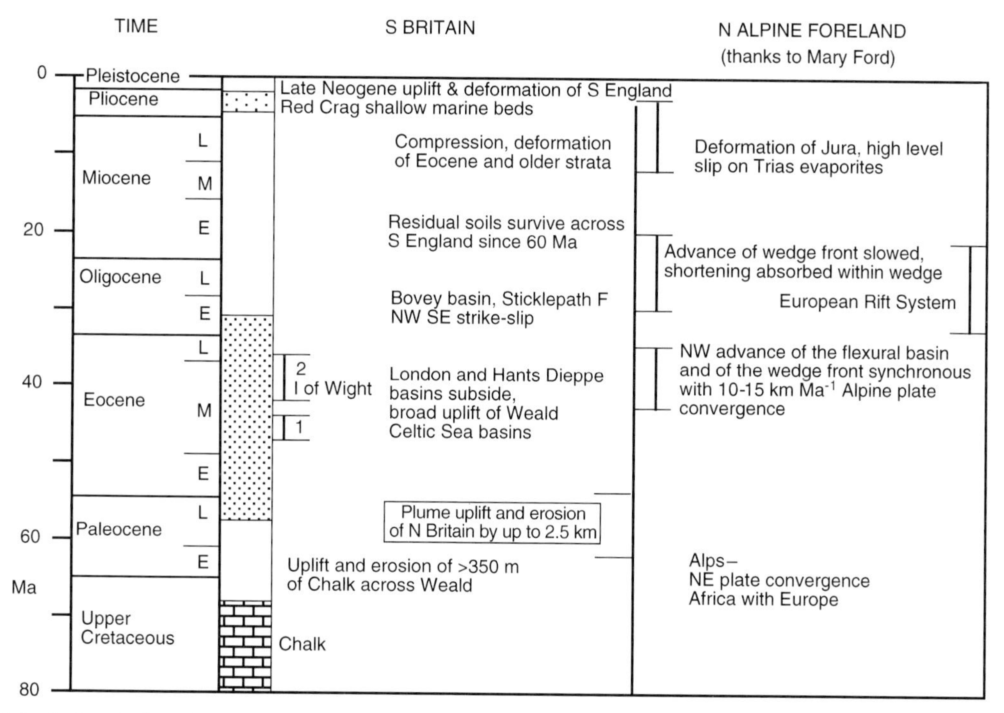

Fig. 9. Correlation of Tertiary tectonic events between southern Britain and the Alpine Foreland. Isle of Wight uplift phases 1 and 2 identified by Gale *et al.* (1999).

short pulses of deformation as a consequence of Alpine stresses that could not be accommodated within the North Alpine Foreland Basin, stage 1; (3) further uplift of the Weald, generation of thrusts extending into the London Basin capped by thrust anticlines (e.g. Windsor), inversion on faults in the Wessex Basin and tightening of folds, folding of Eocene strata (e.g. Isle of Wight), associated with the compression of the Jura fold belt and northward propagation of stress, stage 4.

The timing of strike-slip fault movements on the Sticklepath Fault (and probably the Watchet–Cothelstone Fault) could correlate with the generation of the European Rift System and its associated stress field. Alternatively, both they and the inversion of the Celtic Sea basins may be associated with changes in spreading rates in the North Atlantic. North of the Variscan Thrust Front, the St Georges Channel Basin is not inverted but is instead the site of a Tertiary basin flanking the North Celtic Sea Basin. Further north, Roberts *et al.* (1999) identified major compression events of Eocene to Oligicene age along the North Atlantic margin between the West Shetlands and Vøring that correspond to changes in spreading rates and plate boundaries in the Norwegian–Greenland Sea.

Inversion events in the southern North Sea, separated from southern Britain by the London Platform, appear to have acted independently. For example, Badley *et al.* (1989) interpreted the stratigraphic and structural evidence depicted on seismic sections to demonstrate that a Late Cretaceous (88 Ma) compressional event inverted the NW–SE-trending South Hewitt Fault (Fig. 1), resulting in typical harpoon structures, and a second compressional event in Miocene time superimposed a further inversion.

Tertiary landscape evolution across southern Britain

In his review of Tertiary landscape evolution, Jones (1999) drew attention to the development of a sub-Paleogene surface (68–65 Ma) upon which there developed a residual soil cover indicative of tropical conditions. This is preserved across the Haldon Hills, near Exeter, and plateau gravels are preserved in Dorset and across Salisbury Plain (Green 1985). Jones agrees with Green that the preservation of these soils and the lack of terrigenous sediment offshore means that, throughout Paleogene time, southern England formed a land area of low relief and low elevation (Green, pers. comm.), except in localized areas of tectonic uplift. In Neogene time, this low-lying ground was flooded from the east by a marine transgression at 2.6 Ma in which the Red Crag was deposited, succeeded by the Norwich Crag. These deposits have been subsequently deformed into basins and swells so that the Red Crag is now at −50 m Ordnance datum (OD) in East Anglia but over +100 m OD at the western end of its outcrop, 110 km to the west, near Bishop's Stortford. This is just part of more extensive evidence of Late Pliocene–Pleistocene uplift, tilting and deformation of southern England, which has placed the land surface of the Chalk Downs of SE England at 200 m OD, rising to surface elevations of 450 m OD in the west of England. Uplift of Thames terraces (Maddy 1997) is consistent with this general rise, which from their dating indicates an uplift rate of 0.07 mm a^{-1} during the past 2 Ma in the middle Thames Valley area.

Walsh *et al.* (1999) reviewed the evidence for a set of planation surfaces at elevations between 50 and 150 m OD developed across western Britain and Ireland, and linked them with a number of early Neogene sediment outliers and saprolite bodies. From this, they identified a sub-Neogene surface draped above the present topography of western Britain and Ireland that they regard as a land surface with subdued topography of vertical relief measured in tens of metres. This is supported by a map of Neogene and Quaternary uplift of northern England prepared by Fraser *et al.* (1990). The picture thus emerges of a British landscape of low elevation and subdued topography through Neogene time until about 2.6 Ma when a broad-scale uplift occurred, culminating across the western side of England and Scotland. The extent of this uplift has been quantified by Clayton & Shamoon (1999), from an analysis of the topography of Britain based on a 1 km square grid, as a regional uplift of up to 300–400 m across Wales, the Pennines, Lake District and Southern Uplands of Scotland, and up to 500–600 m across the Scottish Highlands. In particular, they calculated that the uplift of mountain peaks was amplified by the isostatic response to denudational offloading, created by the deep dissection of topography related to the most resistant rocks. Their calculation of denudational isostatic uplift accounts for approximately half the actual mean elevation, leaving the remaining uplift to be explained. The late Neogene uplift and deformation of southern England appears to have been part of a larger-scale uplift that affected all the land area of Britain. Clearly, there are exceptions, highlighted by Battiau-Quency (1999), such as the 500–800 m contrast in relief between Anglesey, where elevations are close to sea level, and the mountains of Snowdonia, which are separated by a major fault along the Menai Straits. Differential uplift can be related to bounding faults that are seismically active, indicative of modern differential movements.

The mountains of Norway resulted mainly from late Plio-Pleistocene uplift and are deeply incised, with denudation accompanied by high sedimentation rates in offshore basins, where deposits are up to 1500 m thick (Riis & Fjeldskaar, 1992). Before 2.5 Ma, the land surface was one of low relief. Riis and Fjeldskaar calculated the amount of isostatic uplift to be expected from denudational offloading, using flexural rigidity with an effective elastic thickness of 22 km. They also allowed for the effect of phase changes at depth resulting from pressure changes owing to loading and offloading at surface. Their calculated uplift surface accounts for less than 70% of the observed elevation.

In northern France, a general uplift is observed from the Armorican block to the Rhine Graben (Guillocheau *et al.* 2000). Differential uplift is responsible for large-scale relief development of the Armorican massif during Quaternary time (Bonnet *et al.* 2000). Lagarde *et al.* (2000) have demonstrated how small-scale deformation structures confirm that differential uplift is controlled by fault movements.

The land areas of Britain, northern France and Norway have risen during the past 2.5 Ma to create the present-day uplands. For a long time before that the landscape was generally of low relief and low elevation. The uplands are all areas where hard rocks existed at surface before the uplift 2.5 Ma ago. Climate change to glacial conditions increased geomorphological energy, erosion and dissection of the landscape in hard rock areas. Isostatic rebound from denudational offloading enhanced topographic range and raised the summit level, but can account for only 50–70% of the observed uplift. Uplift is differential and fault controlled. Present seismicity indicates that a number of the faults involved are currently active as normal faults. The cause of the uplift is unclear. It is difficult to explain simply as continental margin uplift originating from body forces at the continent–ocean crust boundary, as this formed west of Britain 30 Ma ago. Nor can it be attributed easily to plate margin stresses from the Alps or Mid-Atlantic Ridge, transmitted laterally, which were much greater earlier during Tertiary time than in the past 2.5 Ma. Differential erosion between hard and soft rock, creating a difference in surface elevation of 400 m between upland and trough from an initial flat surface, produces a deviatoric stress from differential offloading of 10 MPa. This appears to be sufficient to induce fault movements (Hardebeck & Hauksson 2001) and, hence, differential uplift. This effect is one of strain relief, in addition to isostatic rebound, and is akin to rock bursts in mines, but on a larger scale. It is climatically induced, resulting from the much increased rates of erosion from glacial processes. But it remains to be seen whether this is sufficient to account for the hitherto unexplained difference between the observed surface uplift and the amount calculated as due to isostatic rebound.

The author is deeply indebted to D. L. Hansen for using the forward modelling program which he and S. Nielsen had developed to simulate the evolution of the Weald, London and Hampshire–Dieppe basins, and to S. Nielsen for inviting him to a workshop at Aarhus in June 1999, which provided the impetus for much of the work in this paper. He has enjoyed and benefited greatly from stimulating discussions with colleagues J. Rose, C. Green and K. McClay, and from helpful advice from M. Ford, D. Roberts, S. Egan and M. Butler. He is very grateful to the UK Onshore Geophysical Library for the provision of seismic sections across the Weald, and to the Leverhulme Trust for the award of an Emeritus Fellowship, which made this research possible.

References

BADLEY, M.E., PRICE, J.D. & BACKSHALL, L.C. 1989. Inversion, reactivated faults and related structures: seismic examples from the southern North Sea. *In*: COOPER, M.A. & WILLIAMS, G.D. (eds) *Inversion Tectonics*. Geological Society, London, Special Publications, **44**, 201–219.

BATTIAU-QUENCY, Y. 1999. Crustal anisotropy and differential uplift: their role in long-term landform development. *In*: SMITH, B.J., WHALLEY, W.B. & WARKE, P.A. (eds) *Uplift, Erosion and Stability: Perspectives on Long-term Landscape Development*. Geological Society, London, Special Publications, **162**, 65–74.

BEELEY, H.S. & NORTON, M.G. 1998. The structural development of the Central Channel High: constraints from section restoration. *In*: UNDERHILL, J.R. (ed.) *Development, Evolution and Petroleum Geology of the Wessex Basin*. Geological Society, London, Special Publications, **133**, 283–298.

BIRPS & ECORS, 1986. Deep seismic reflection profiling between England, France and Ireland. *Journal of the Geological Society, London*, **143**, 45–52.

BONNET, S., GUILLOCHEAU, F., BRUN, J.-P. & VAN DEN DRIESSCHE, J. 2000. Large-scale relief development related to Quaternary tectonic uplift of a Proterozoic–Palaeozoic basement: the Armorican Massif, NW France. *Journal of Geophysical Research*, **105**, 19273–19288.

BRAUN, J. & BEAUMONT, C. 1987. Styles of continental rifting: results from dynamical models of lithospheric extension. *In*: BEAUMONT, C. & TANKARD, A.J. (eds) *Sedimentary Basins and Basin Forming Mechanisms*. Canadian Society of Petroleum Geologists, Memoirs, **12**, 241–258.

BRITISH GEOLOGICAL SURVEY 1988. *Sheet 50N00, Dungeness–Boulogne*, 1:250 000 series of solid geology maps. BGS, Keyworth.

BRITISH GEOLOGICAL SURVEY 1989. *Sheet 51N00, Thames Estuary*, 1:250 000 series of solid geology maps. BGS, Keyworth.

BRODIE, J. & WHITE, N. 1994. Sedimentary basin inversion caused by igneous underplating, Northwest European continental shelf. *Geology*, **22**, 147–150.

BROOKS, M., TRAYNER, P.M.P. & TRIMBLE, T.J. 1988. Mesozoic reactivation of Variscan thrusting in the Bristol Channel area, UK. *Journal of the Geological Society, London*, **145**, 439–444.

BUTLER, M. 1998. The geological history of the Wessex Basin: a review of new information from oil exploration. *In*: UNDERHILL, J.R. (ed.) *Development, Evolution and Petroleum Geology of the Wessex Basin*. Geological Society, London, Special Publications, **133**, 67–86.

BUTLER, M. & PULLEN, C.P. 1990. Tertiary structures and hydrocarbon entrapment in the Weald Basin of southern England. *In*: HARDMAN, R.F.P. & BROOKS, J. (eds) *Tectonic Events Responsible for Britain's Oil and Gas Reserves*. Geological Society, London, Special Publications, **55**, 371–391.

CHADWICK, R.A. 1986. Extension tectonics in the Wessex Basin, southern England. *Journal of the Geological Society, London*, **143**, 465–488.

CHADWICK, R.A. 1993. Aspects of basin inversion in southern Britain. *Journal of the Geological Society, London*, **150**, 311–322.

CLAYTON, K. & SHAMOON, N. 1999. A new approach to the relief of Great Britain III. Derivation of the contribution of neotectonic movements and exceptional regional denudation to the present relief. *Geomorphology*, **27**, 173–189.

CLOETINGH, S., GRADSTEIN, F.M., KOOI, H., GRANT, A.C. & KAMINSKI, M. 1990. Plate reorganization: a cause of rapid late Neogene subsidence and sedimentation around the North Atlantic? *Journal of the Geological Society, London*, **147**, 495–506.

DEWEY, J.F., HELMAN, M.L., TURCO, E., HUTTON, D.H.W. & KNOTT, S.D. 1989. Kinematics of the western Mediterranean. *In*: COWARD, M.P., DIETRICH, D. & PARK, R.G. (eds) *Alpine Tectonics*. Geological Society, London, Special Publications, **45**, 265–283.

DOOLEY, T. & MCCLAY, K. 1997. Analog modeling of pull-apart basins. *AAPG Bulletin*, **81**, 1804–1826.

FORD, M., LICKORISH, W.H. & KUSZNIR, N.J. 1999. Tertiary foreland sedimentation in the Southern Subalpine Chains, SE France: a geodynamic appraisal. *Basin Research*, **11**, 315–336.

FRASER, A.J., NASH, D.F., STEELE, R.P. & EBDON, C.C. 1990. A regional assessment of the intra-Carboniferous play of Northern England. *In*: BROOKS, J. (ed.) *Classic Petroleum Provinces*. Geological Society, London, Special Publications, **50**, 417–440.

GALE, A.S., JEFFREY, P.A., HUGGETT, J.M. & CONNOLLY, P. 1999. Eocene inversion history of the Sandown Pericline, Isle of Wight, southern England. *Journal of the Geological Society, London*, **156**, 327–339.

GREEN, C.P. 1985. Pre-Quaternary weathering residues, sediments and landform development: examples from southern Britain. *In*: RICHARDS, K.S. & ARNETT, R.R., ELLIS, S. (eds) *Geomorphology and Soils*. George, Allen & Unwin, London, 58–77.

GUILLOCHEAU, F., ROBIN, C., ALLEMAND, P. & 16 OTHERS 2000. Meso-Cenozoic geodynamic evolution of the Paris Basin: 3D stratigraphic constraints. *Geodinamica Acta*, **13**, 189–246.

HAMBLIN, R.J.O., CROSBY, A., BALSON, P.S., JONES, S.M., CHADWICK, R.A., PENN, I.E. & ARTHUR, M.J. 1992. *United Kingdom Offshore Regional Report: the Geology of the English Channel*. HMSO, London.

HAQ, B.U., HARDENBOL, J. & VAIL, P.R. 1987. Chronology of fluctuating sea levels since the Triassic. *Science*, **235**, 1156–1167.

HARDEBECK, J.L. & HAUKSSON, E. 2001. Crustal stress field in southern California and its implications for fault mechanics. *Journal of Geophysical Research*, **106B**, 21 859–21 882.

HARVEY, M.J. & STEWART, S.A. 1998. Influence of salt on the structural evolution of the Channel Basin. *In*: UNDERHILL, J.R. (ed.) *Development, Evolution and Petroleum Geology of the Wessex Basin*. Geological Society, London, Special Publications, **133**, 241–266.

HAWKES, P.W., FRASER, A.J. & EINCHCOMB, C.C.G. 1998. The tectono-stratigraphic development and exploration history of the Weald and Wessex Basins, southern England. *In*: UNDERHILL, J.R. (ed.) *Development, Evolution and Petroleum Geology of the Wessex Basin*. Geological Society, London, Special Publications, **133**, 39–65.

HOLLOWAY, S. & CHADWICK, R.A. 1986. The Sticklepath–Lustleigh fault zone: Tertiary sinistral reactivation of a Variscan strike-slip fault. *Journal of the Geological Society, London*, **143**, 447–452.

JONES, D.K.C. 1999. On the uplift and denudation of the Weald. *In*: SMITH, B.J., WHALLEY, W.B. & WARKE, P.A. (eds) *Uplift, Erosion and Stability: Perspectives on Long-term Landscape Development*. Geological Society, London, Special Publications, **162**, 25–43.

LAGARDE, J.-L., BAIZE, S., AMORESE, D., DELCAILLAU, B., FONT, M. & VOLANT, P. 2000. Active tectonics, seismicity and geomorphology, with special reference to Normandy (France). *Journal of Quaternary Science*, **15**, 745–758.

LAKE, S.D. & KARNER, G.D. 1987. The structure and evolution of the Wessex Basin, southern England: an example of inversion tectonics. *Tectonophysics*, **137**, 347–378.

LAW, A. 1998. Regional uplift in the English Channel: quantification using sonic velocity logs. *In*:

UNDERHILL, J.R. (ed.) *Development, Evolution and Petroleum Geology of the Wessex Basin*. Geological Society, London, Special Publications, **133**, 187–197.

LEFORT, J.P. & MAX, M.D. 1992. Structure of the Variscan belt beneath the British and Armorican overstep sequences. *Geology*, **20**, 979–982.

MADDY, D. 1997. Uplift-driven valley incision and river terrace formation in southern England. *Journal of Quaternary Science*, **12**, 539–545.

MEIER, L. & EISBACHER, G.H. 1991. Crustal kinematics and deep structure of the Northern Rhine Graben. *Tectonics*, **10**, 621–630.

MILIORIZOS, M. & RUFFELL, A. 1998. Kinematics and geometry of the Watchet–Cothelstone–Hatch Fault System: implications for the structural history of the Wessex Basin and adjacent areas. *In*: UNDERHILL, J.R. (ed.) *Development, Evolution and Petroleum Geology of the Wessex Basin*. Geological Society, London, Special Publications, **133**, 311–330.

NIELSEN, S.B. & HANSEN, D.L. 2000. Physical explanation of the formation and evolution of inversion zones and marginal basins. *Geology*, **28**, 875–878.

PETRIE, S.H., BROWN, J.R., GRANGER, P.J. & LOVELL, J.P.B. 1989. Mesozoic history of the Celtic Sea Basins. *In*: TANKARD, A.J. & BALKWILL, H.R. (eds) *Extensional Tectonics and Stratigraphy of the North Atlantic Margins*. Memoirs, American Association of Petroleum Geologists, **46**, 433–444.

PFIFFNER, A. 1992. Alpine orogeny. *In*: BLUNDELL, D.J., FREEMAN, R. & MUELLER, S. (eds) *A Continent Revealed: the European Geotraverse*. Cambridge University Press, Cambridge, 180–190.

RAWSON, P.F. 1992. The Cretaceous. *In*: DUFF, P.McL.D. & SMITH, A.J. (eds) *Geology of England and Wales*. Geological Society, London, 355–388.

RIIS, F. & FJELDSKAAR, W. 1992. On the magnitude of the Late Tertiary and Quaternary erosion and its significance for the uplift of Scandinavia and the Barents Sea. *In*: LARSEN, R.M., BREKKE, H., LARSEN, B.T. & TALLERAAS, E. (eds) *Structural and Tectonic Modelling and its Application to Petroleum Geology*. Norwegian Petroleum Society (NPF) Special Publication, **1**, 163–185.

ROBERTS, D.G. 1989. Basin inversion in and around the British Isles. *In*: COOPER, M.A. & WILLIAMS, G.D. (eds) *Inversion Tectonics*. Geological Society, London, Special Publications, **44**, 131–150.

ROBERTS, D.G., THOMPSON, M., MITCHENER, B., HOSSACK, J., CARMICHAEL, S. & BJØRNSETH, H.-M. 1999. Palaeozoic to Tertiary rift and basin dynamics: mid-Norway to the Bay of Biscay—a new context for hydrocarbon prospectivity in the deep water frontier. *In*: FLEET, A.J. & BOLDY, S.A.R. (eds) *Petroleum Geology of Northwest Europe: Proceedings of the 5th Conference*. Geological Society, London, 7–40.

ROWLEY, E. & WHITE, N. 1998. Inverse modelling of extension and denudation in the Irish Sea and surrounding areas. *Earth and Planetary Science Letters*, **161**, 57–71.

UNDERHILL, J.R. & PATERSON, S. 1998. Genesis of tectonic inversion structures: seismic evidence for the development of key structures along the Purbeck–Isle of Wight disturbance. *Journal of the Geological Society, London*, **155**, 975–992.

UNDERHILL, J.R. & STONELY, R. 1998. Introduction to the development, evolution and petroleum geology of the Wessex Basin. *In*: UNDERHILL, J.R. (ed.) *The Development, Evolution and Petroleum Geology of the Wessex Basin*. Geological Society, London, Special Publications, **133**, 1–18.

WALSH, P., BOULTER, M. & MORAWIECKA, I. 1999. Chattian and Miocene elements in the modern landscape of western Britain and Ireland. *In*: SMITH, B.J., WHALLEY, W.B. & WARKE, P.A. (eds) *Uplift, Erosion and Stability: Perspectives on Long-term Landscape Development*. Geological Society, London, Special Publications, **162**, 45–63.

WHITE, N. & LOVELL, B. 1997. Measuring the pulse of a plume with the sedimentary record. *Nature*, **387**, 888–891.

ZIEGLER, P.A. 1990. *Geological Atlas of Western and Central Europe*. 2nd edition; Shell Internationale Petroleum Maatschappij, The Hague.

ZIEGLER, P.A., CLOETINGH, S. & VAN WEES, J.-D. 1995. Dynamics of intra-plate compressional deformation: the Alpine foreland and other examples. *Tectonophysics*, **252**, 7–59.

Landforms and uplift in Scandinavia

K. LIDMAR-BERGSTRÖM & J. O. NÄSLUND

Department of Physical Geography and Quaternary Geology, Stockholm University, SE-106 91 Stockholm, Sweden (e-mail: karna@natgeo.su.se)

Abstract: The relation between Scandinavian landforms and Cenozoic uplift events is examined by analysis of digital elevation data in a regional geological context as well as in a geomorphological process perspective. Re-exposed flat sub-Cambrian and sub-Mesozoic hilly relief aids in deciphering uplift and erosional events. The highly dissected mountains of the Northern Scandes (NS) rise maximally 1500 m above a slightly tilted lowest level continuing in the Muddus plains eastwards at 300–550 m above sea level (a.s.l.). This level is correlated with the lowest, slightly warped level of thc Palaeic relief at 1000–1300 m a.s.l. of the Southern Scandes (SS), over which mountains of similar height rise. This lowest surface is thought to be the end result of Paleogene erosion to the general base level. Northern Scandinavia with the NS and the Muddus plains acted as a block that was progressively tilted to the SE, whereas the Southern Scandes experienced continuous doming, with a major uplift event of about 1000 m in Neogene time causing deep valley incision in the uplifted plateau. The South Swedish Dome emerged from its Palaeozoic and Mesozoic cover in Neogene time and still retains well-preserved re-exposed palaeosurfaces.

Uplift along continental margins has lately become a topic of common interest for geologists, geochronologists and geomorphologists (Japsen & Chalmers 2000; Summerfield 2000). Scandinavia is located close to the Atlantic margin. Its large-scale relief is characterized by three domes, the Northern Scandes (NS) reaching about 2000 m above sea level (a.s.l.), the Southern Scandes (SS) reaching 2500 m in south Norway, and the South Swedish Dome (SSD) reaching 375 m in south Sweden (Fig. 1). The uplift of the Scandes has been discussed since the beginning of the century (Reusch 1901; Ahlmann 1919) and Neogene uplift of the SS has lately been supported by a fission-track study (Rohrman *et al.* 1995). It has been suggested on different grounds that the domes have different uplift histories with a main uplift in Paleogene time in the north and in Neogene time in the south (Riis 1996; Lidmar-Bergström 1999). In this paper we examine and compare the topography of the domes in more detail and correlate surfaces between the domes to reveal areas of Neogene uplift and subsidence.

Methods

The relief was examined by analysing height layer maps, slope maps and topographic profiles. All maps and profiles were constructed from a digital elevation model (DEM) of Scandinavia. For Sweden the elevation data have a true spatial resolution of 500 m × 500 m (National Land Survey of Sweden), whereas original elevation data for surrounding areas had a resolution of 1000 m × 1000 m (Statens Kartverk in Norway, and ETOPO5). The latter data were subsequently resampled to 500 m × 500 m. Both resolutions are suitable for the study of large-scale morphology.

The major relief features are described with the aid of a height layer map. Local topography is analysed from slope maps and evaluated in a regional geological context as well as in a geomorphological process perspective following recent advances in knowledge on the effect of deep weathering in the shaping of relief (Thomas 1994).

Three major palaeosurfaces are identified and correlated between the domes: the sub-Cambrian peneplain, a Mesozoic surface and a Tertiary surface. The profiles are located to elucidate the suggested correlations of palaeosurface between domes and surrounding terrain. Further, the different degrees of valley dissection of the domes are examined.

Analysis of maps

Major shape of the Northern and Southern Scandes

The Northern Scandes form an elongated dome, 1000 km long and 270 km or 165 km wide (at the

From: Doré, A.G., Cartwright, J.A, Stoker, M.S., Turner, J.P. & White, N. 2002. *Exhumation of the North Atlantic Margin: Timing, Mechanisms and Implications for Petroleum Exploration*. Geological Society, London, Special Publications, **196**, 103–116. 0305-8719/02/$15.00

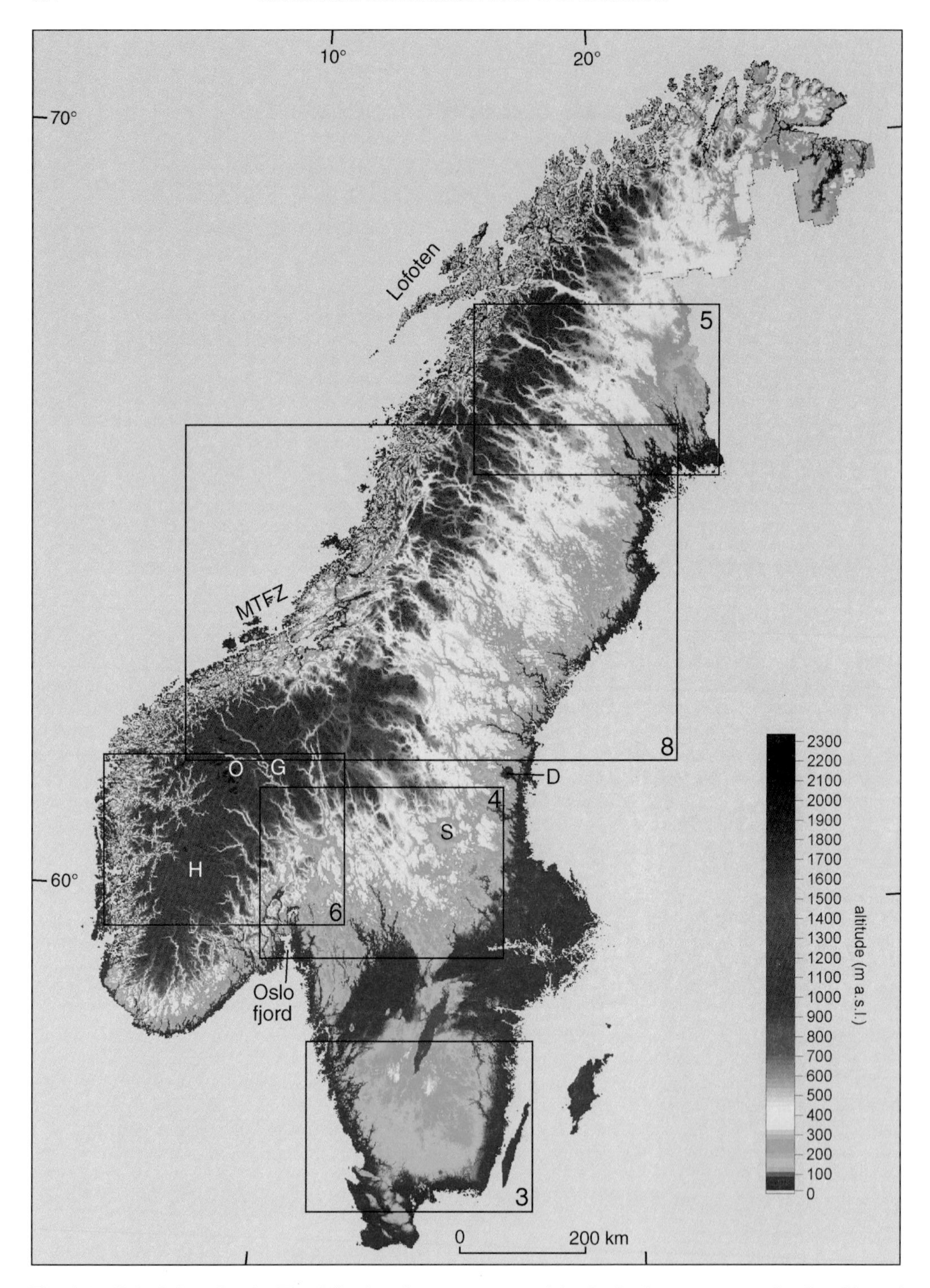

Fig. 1. Relief of Scandinavia. The following features, expressed in the landscape as a result of etching of geological structures, should be noted: the Dellen meteorite impact (D), the Siljan ring meteorite impact (S), and the plutonic rocks of the Oslo rift. The locations of Figs 3–6 and 8 are indicated by squares. MTFZ, Møre–Trøndelag Fault Zone; H, Hardangeroidda; O, Otta Valley; G, Gud brandsdalen valley.

600 m or 1000 m levels) (Fig. 1). The Southern Scandes form a more oval and somewhat bent dome, 680 km long and 400 or 265 km wide (at the 600 m or 1000 m levels). Thus the SS are about 100 km wider and 300 km shorter than the NS.

The NS are cut by valleys in a NW–SE direction, leaving intact interfluves above 1000 m a.s.l. with a maximum width of 20 km. The maximum length of the interfluves is 40 km. This is in contrast to the SS, which have an area with 430 km length from south to north above 1000 m a.s.l. This elevated area is cut only by two major valleys, the Gudbrandsdalen valley and its major tributary valley, the Otta valley (Fig. 1). Here, it is possible to walk on the so-called Palaeic relief for over 300 km without descending into a valley below the 1000 m level. In the east–west to

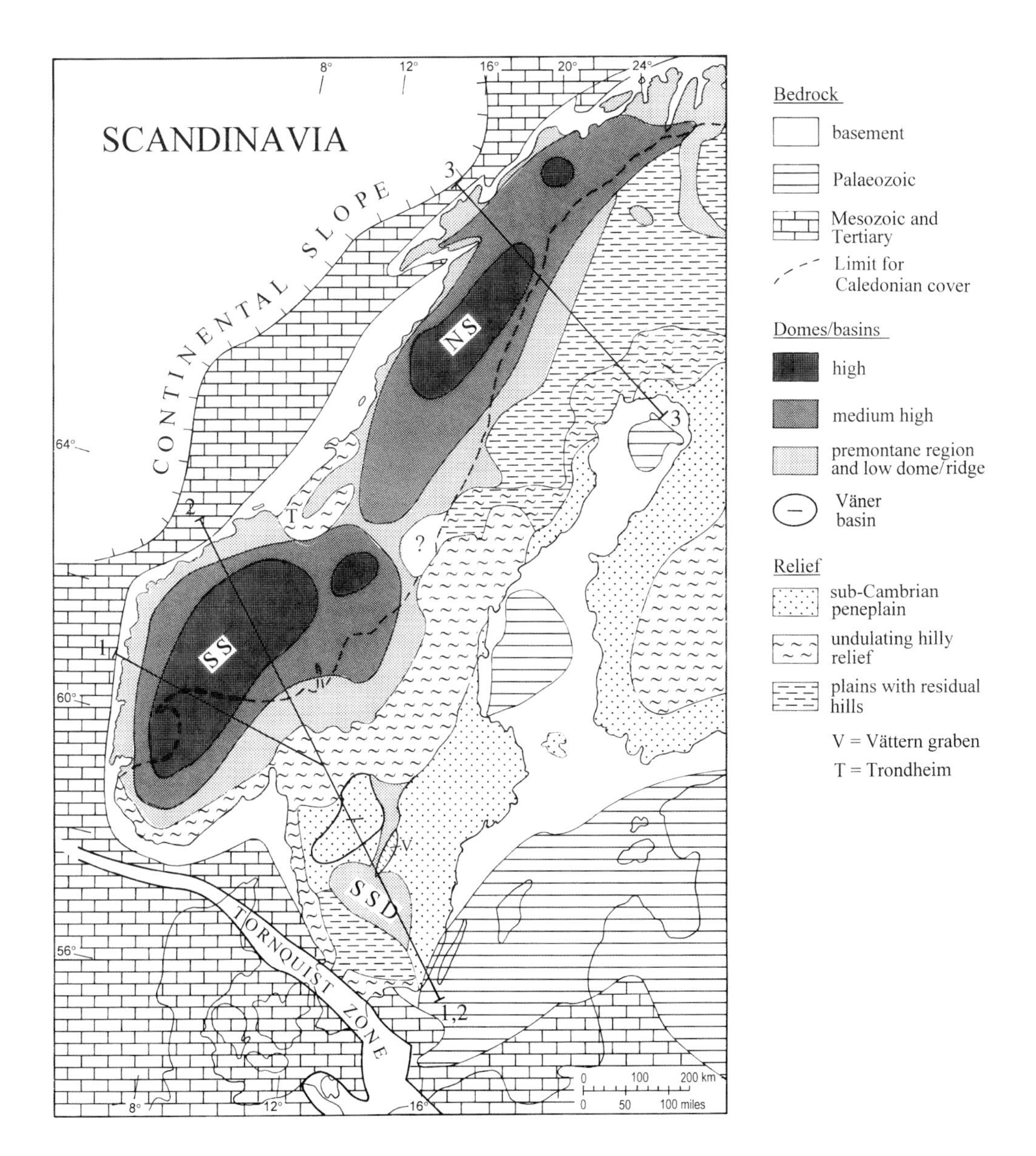

Fig. 2. Domes and generalized palaeosurfaces of Scandinavia. The Palaeic relief is located within the SS above about 1000 m a.s.l. (see Fig. 1). Location for profiles 1, 2 and 3 in Fig. 7 are indicated. Modified from Lidmar-Bergström (1999).

SE–NW directions, the interfluves are here unbroken for up to 130 km. The origin of the Palaeic relief is discussed below.

Thus the domes of the Northern Scandes and the Southern Scandes show characteristically different shapes in terms of width, length and valley incision. In addition, the main valleys in both areas are widened and deepened to varying degrees by glacial erosion.

Scandinavian palaeosurfaces formed by etching and planation

Four characteristic landscape types, formed by etching and planation, were studied in the maps: (1) sub-Cambrian peneplain; (2) undulating hilly relief (etch surfaces); (3) plains with residual hills (Muddus plains); (4) the Palaeic relief of southern Norway (Fig. 2).

Sub-Cambrian peneplain. Within the Precambrian part of Fennoscandia the bedrock was denuded to an almost level plain at the end of Proterozoic time (Högbom 1910). Vendian, Cambrian and Ordovician strata were successively deposited directly on the flat basement. The surface was to some extent overridden by Caledonian nappes in the west and buried below thick covers of sedimentary strata in the east during long periods of time (Koark *et al.* 1978; Zeck *et al.* 1988; Lidmar-Bergström 1995; Cederbom *et al.* 2000). This surface has been re-exposed and over large areas it has been totally obliterated by subsequent denudation. In other areas it is still well preserved, and is called the sub-Cambrian peneplain. It is encountered more or less intact in eastern and south–central Sweden from sea level to over 300 m a.s.l. (Lidmar-Bergström 1988, 1996) and also in contact with Cambrian cover rocks on Hardangervidda, south Norway, at about 1100–1350 m a.s.l. (Schipull 1974). The sub-Cambrian peneplain is met with along the eastern part of the NS, in the south at about 300 m a.s.l and in the north at over 1000 m a.s.l. (see

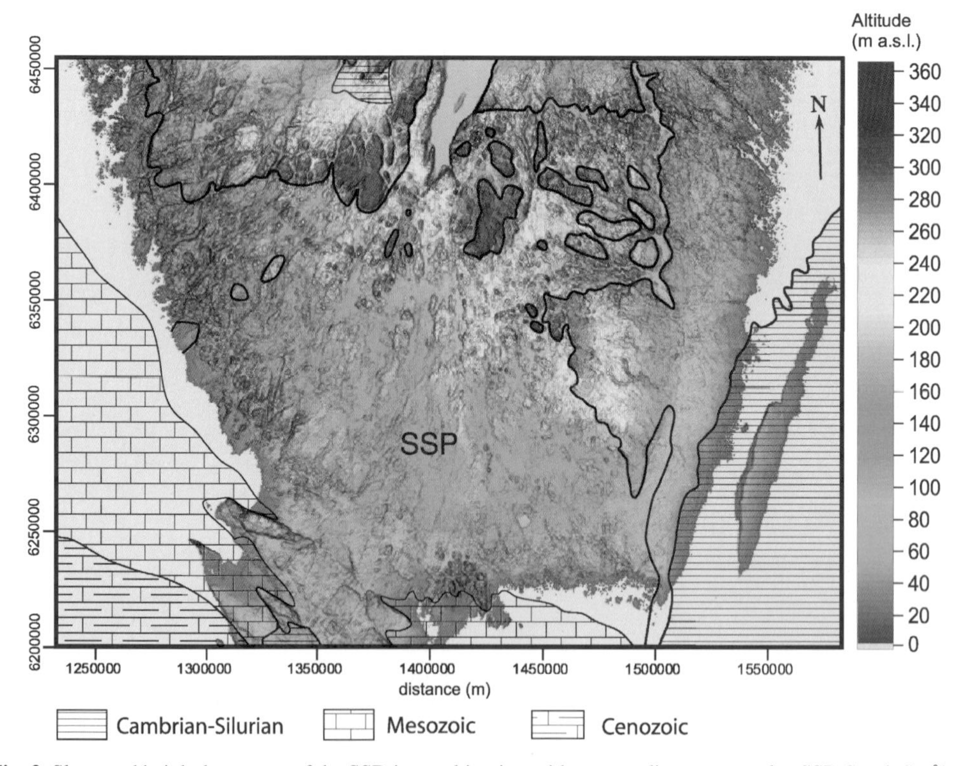

Fig. 3. Slope and height layer map of the SSD in combination with surrounding cover rocks. SSP, South Småland Peneplain. A generalized picture of the extent of the sub-Cambrian peneplain is shown by the black line. It is almost intact in the SE up to 300 m a.s.l. In the northern part of the map it reaches from below Cambrian cover rocks at 200 m a.s.l. to summits further south at 350 m a.s.l. Small areas in the west have low relief and Cambrian fissure fillings indicating long-lasting Cambrian cover. Exhumed sub-Cretaceous hilly relief extends from below Cretaceous cover rocks in the south and west. Coordinates are from the national grid of Sweden.

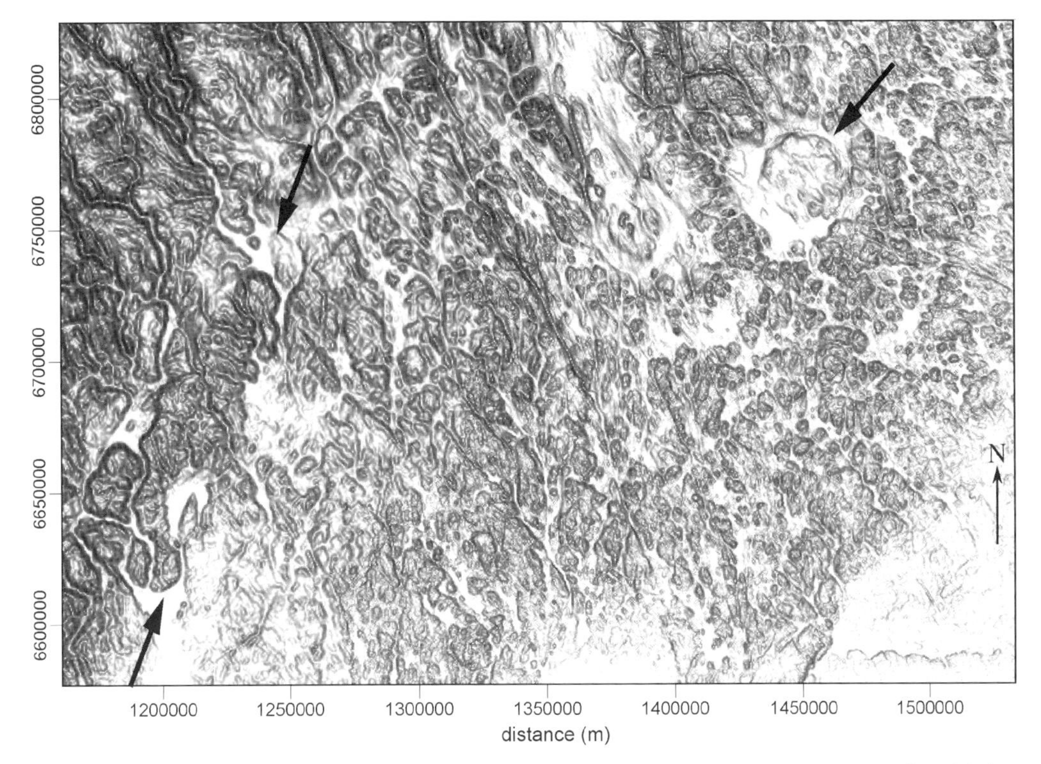

Fig. 4. Slope map of central Sweden and southeastern Norway. Features to be noted are the overall undulating, hilly relief of etch character, the Siljan Ring (meteorite impact), the volcanic rocks of the Oslo rift, and the exhumed sub-Cambrian peneplain with distinct faults in the southeastern corner.

Ljungner 1950). The present land surface in Precambrian rocks in Sweden coincides with or is as much as 600 m below this surface, which can be looked upon as the primary peneplain (Lidmar-Bergström 1995, 1996). Where the sub-Cambrian peneplain is well preserved the landscape shows an extremely flat topography without residual hills (Figs 3 and 4).

Undulating hilly relief. Along the eastern flank of the Southern Scandes the topography is hilly and the differences in bedrock composition and structure are well expressed in the relief (Figs 1 and 4), just as within the re-exposed sub-Cretaceous relief in southernmost Sweden (Fig. 3). The latter landscape was formed in Late Mesozoic time by deep weathering (etching) and subsequent stripping of the weathering mantle (Lidmar-Bergström 1989). East of the Southern Scandes the result of denudation of meteoric impacts such as the Late Palaeozoic (Bottomley *et al.* 1978;Åberg inWickmann 1988) Dellen structure (Fig. 1) and the Cretaceous (Deutsch *et al.* 1992) Siljan ring (Figs 1 and 4) is clearly seen in the topography. The resistant plutonic rocks in the Oslo field give rise to massive hills surrounded by low areas with Palaeozoic sedimentary rocks (Figs 1 and 4). The relief at the southern tip of Norway is often interpreted to have a rather well-preserved sub-Cambrian surface on interfluves between a few major joint-aligned valleys (e.g. Riis 1996). The peneplain is not intact and along the coast the relief is of undulating character (Rudberg 1960) and classified as an etch surface (Lidmar-Bergström *et al.* 2000). Along the west coast of south Norway the geological structures are well expressed in the relief, and this is mainly the case along the entire coast of western Norway. Most of the eastern flank of the Southern Scandes is located in Sweden and this part contains one of Sweden's ore provinces. These ores are called soft ores because of the deep weathering they have experienced (Vivallo & Broman 1993). Other clayey weathering residues are known from this area, as well as along the coast of Norway (Lidmar-Bergström *et al.* 1999). Deep weathering and subsequent stripping of saprolites are a major cause for the expression of geological structures in the relief (Thomas 1994). The undulating hilly relief on the eastern flank of the SS is thus interpreted to be of etch character, maybe of Mesozoic age (Reusch 1903; Lidmar-Bergström 1995; Cederbom *et al.* 2000). The structurally controlled relief along the west coast of Norway is also thought to have an etch origin, but with further incision along the etched structures after uplift. The Mesozoic age of the etching east of the SS is not confirmed but possible (Cederbom *et al.* 2000). Along the coast of SW Sweden the basement surface emerges from below Mesozoic strata, which date the etching here.

Grus saprolites and landforms. Besides the remnants of kaolinitic saprolites associated with the re-exposed Mesozoic undulating hilly relief, gravelly saprolites (grus) are of common occurrence within many parts of Fennoscandia (Lidmar-Bergström *et al.* 1999). They are interpreted to have developed mainly in Plio-Pleistocene time but may date back to Miocene time. Within south Sweden they are often associated with an etched landscape formed at the expense of the re-exposed sub-Cambrian peneplain (Lidmar-Bergstöm *et al.* 1997).

Plains with residual hills. In contrast to the dome flanks in the south, the relief east of the NS is characterized by plains with residual hills, the so-called Muddus plains, situated mainly at 300–550 m a.s.l. (Fig. 5) (Wråk 1908). The hills rise to 300 m above the plains. The plains are interpreted to have developed from undulating hilly relief (etch topography) by pedimentation processes during more arid periods of Tertiary time (Lidmar-Bergström 1995). The plains have acted as base levels for the valleys that penetrate the mountains in the west (Fig. 5). Individual plains are separated by low steps into a number of separate levels (Rudberg 1954; Lidmar-Bergström 1996). Similar plains (the South Småland Peneplain, SSP) with relatively few and low residual hills occur at the southwestern flank of the SSD above the exhumed sub-Cretaceous hilly relief (Fig. 3). The plains here cut off the re-exposed sub-Cretaceous etched relief and are thus of Tertiary age (Lidmar-Bergström 1982).

Palaeic surface (or relief) of southern Norway. Travellers in southern Norway can observe that most of the higher ground above about 1000 m a.s.l. is occupied by a high plateau

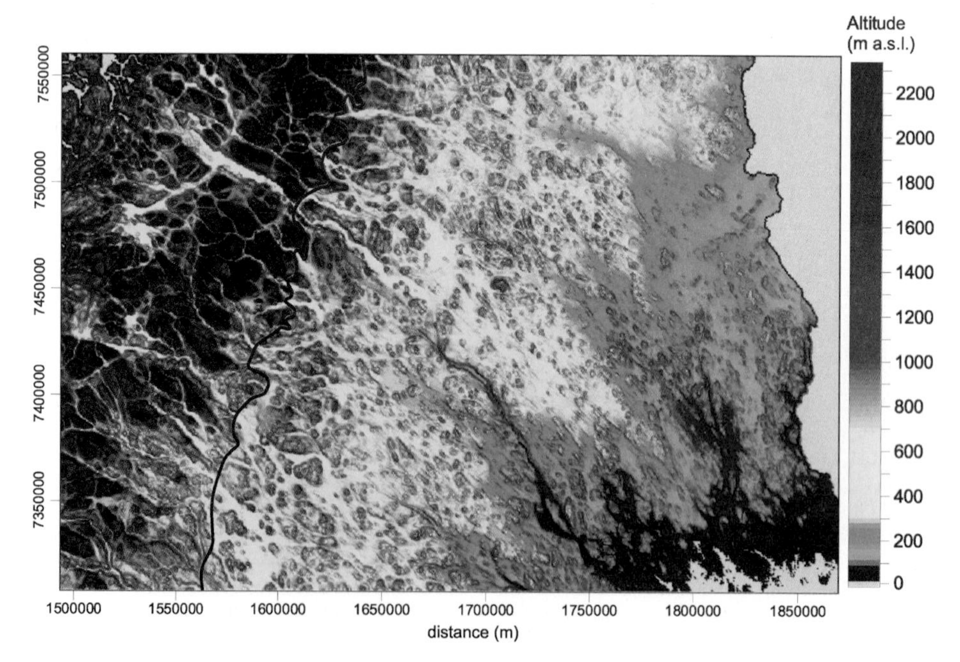

Fig. 5. Slope map of region in northern Sweden showing the Paleogene Muddus plains in yellow (300–700 m a.s.l.) in the east and their continuation as valleys into the mountains in the west. The line marks the border between Precambrian basement and Caledonian nappes (see Fig. 6).

(Fig. 6). From the west deep valleys cut far inland into this plateau, named the Palaeic surface by Reusch (1901). The difference between the high plateau and the deeply incised valleys was interpreted to reflect a late uplift (Reusch 1901; Ahlmann 1919; Peulvast 1978, 1985). The Palaeic surface is separated into different levels thought to be induced by renewed valley incision as a result of continued warping during several Mesozoic–Paleogene uplift events (Lidmar-Bergström *et al.* 2000). The Palaeic surface is composed of several surfaces separated by distinct steps and is better referred to as the Palaeic relief. The lowest level of the Palaeic relief, situated at about 1000–1200 m a.s.l., has a vast extent on the eastern side of the high dome in southern Norway. This level has acted as base level for the river systems penetrating the higher ground (Figs 1 and 6) and can be followed westwards along the major river systems. The 1000 m level occurs also along the western side of south Norway (Lidmar-Bergström *et al.* 2000). In detail, the Palaeic relief is characterized by slightly undulating plains with residual hills with surrounding pediments (Fig. 6). Deep weathering and pedimentation processes are thought to have been important in the formation of this relief (Gjessing 1967).

Zone of incised valleys

Around the plateau with Palaeic relief in the SS there is a zone with deeply incised valleys (Figs 1 and 6). The valleys are deeper on the western side, where they extend below sea level. A difference in relief of up to 800 m between valley bottoms of incised valleys and adjacent lowest shallow valley level of the Palaeic relief is common. The valleys are glacially deepened by more than 1000 m in the Sognefjord on the western, Atlantic side, and by *c.* 250 m on the eastern side, e.g. in the Mjösa area (Ahlmann 1919).

Comparison between the lowest level of the Palaeic relief and the Muddus plains

The lowest level of the Palaeic relief constitutes a plateau strikingly similar to the Muddus plains but at a higher level (Figs 5 and 6). They both have acted as base levels for river systems penetrating westwards. In detail, there are differences between the two, which can be explained by differences in bedrock composition. In the Palaeic relief depicted in Fig. 6, the

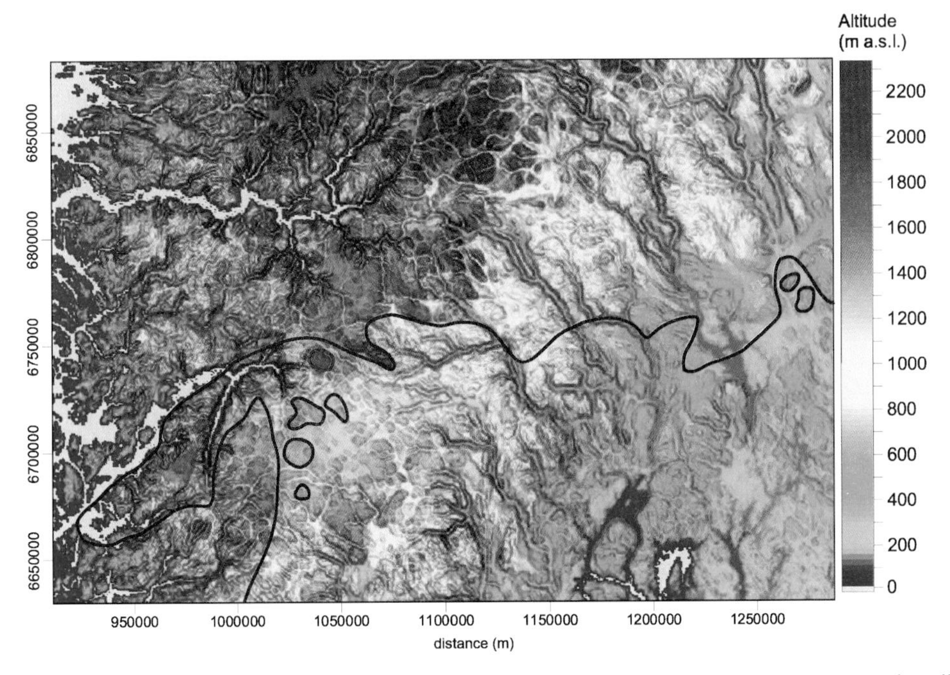

Fig. 6. Slope and height layer map showing the lowest level of the Palaeic relief of southern Norway in yellow (900–1200 m a.s.l.). The line marks the border between Precambrian basement and Caledonian nappes (see Fig. 5).

southern part has developed by erosion of the Caledonian nappes and now is mainly shaped in Precambrian rocks. The associated residual hills often include remnants of the Caledonian cover. Further to the NE (Fig. 6) the Palaeic relief has formed in highly variable Caledonian rocks, which have resulted in a somewhat different morphology with some resistant rocks giving rise to high summits. In contrast, the corresponding Muddus plains in northern Sweden (Fig. 5) were formed on Precambrian rocks, which had long since lost their Palaeozoic cover. In addition, the different appearance of the relief in Figs 5 and 6 is also to a minor extent the result of the use of elevation data with different resolutions in the construction of the maps. However, most of the difference in appearance is caused by the difference in geology.

It is likely that the Muddus plains of northern Sweden and the lowest level of the Palaeic relief of southern Norway developed at the same level and at the same time. The individual forms are the end result of etching, stripping and pedimentation in warm climates. This uniform land surface was subsequently deformed by differential uplift. This, including the timing of events, is further discussed below.

Analysis of profiles

The locations of the profiles (Fig. 2) are chosen to illustrate the correlation and deformation of the three palaeosurfaces, namely, the sub-Cambrian peneplain, the Mesozoic undulating hilly relief, and the Tertiary plains with residual hills (corresponding to the lowest level of the Palaeic relief). The vertical position of the palaeosurfaces is illustrated in the profiles and used for a discussion on uplift. Profiles 1 and 2 are used to discuss the relationship between the SSD and the SS, as well as the deformation of the sub-Cambrian peneplain, the inferred Mesozoic surfaces, and the lowest Palaeic level. Profile 3 is used for discussing the NS and its eastern flank.

Profile 1 (Figs 2 and 7)

The sub-Cambrian peneplain extends from below Cambrian cover rocks in the SE. The peneplain forms a low dome, constituting the SSD, which is interrupted by the Vättern Graben, filled with up to 1000 m of Late Proterozoic sedimentary rock (Axberg & Wadstein 1980), and the Hökensås Horst. Palaeozoic remnants occur on the slopes down to Lake Vänern but do not occur on the bottom of the lake. Further to the NW the peneplain is downwarped and forms the Väner Basin. From Lake Vänern the sub-Cambrian peneplain rises to the NW and can be followed in the summits for some distance. Profile 1 then follows a more westerly direction (Fig. 2). The sub-Cambrian peneplain is met with again below Palaeozoic strata in the Oslo rift and then encountered at about 1250–1300 m a.s.l. on the Hardangervidda. Here it is identified with the aid of Palaeozoic outliers. The peneplain rises towards the NW and thereafter it is downfaulted, where Caledonian rocks meet the Precambrian basement along a steeply dipping front at H in the profile (Fig. 7).

Northwest of Lake Vänern the sub-Cambrian peneplain is replaced by a hilly relief (an etch surface), which is tentatively interpreted as a Mesozoic surface (see above). This hilly relief is correlated with an inferred surface along the highest summits further to the west. The weathering-resistant plutonic rocks of the Oslo rift extend above this surface.

The present lowest level of the Palaeic relief at Hardangervidda partly coincides with but is mainly slightly below the sub-Cambrian peneplain. In the westernmost parts shallow valleys are incised in the lowest level, here at about 1100 m a.s.l., to slightly below 1000 m a.s.l. Structurally controlled deep valleys of the Hardangerfjord system penetrate the Palaeic relief in the NW. Deep valleys, Tinnsjö and Numedalen, are incised along the southeastern flank.

Profile 2 (Fig. 7)

Profile 2 is identical to profile 1 in the SE but then continues straight towards the NW (Fig. 2). In the area of hilly relief, NW of Lake Vänern, the sub-Cambrian peneplain has disappeared and is met with again below the Palaeozoic strata in the Oslo rift. Thereafter, it directly disappears below the Caledonian rocks and has no further influence on the present topography along the profile.

In this profile a correlation has also been made between the undulating hilly relief and the highest summits of the SS. The inferred Mesozoic surface has experienced doming and been uplifted about 2000 m in the NW. Subsequently, it has been successively dissected by major valleys, which has resulted in a stepped pattern of the present summit surfaces (Lidmar-Bergström *et al.* 2000), indicated by the three summit levels in the profile.

A slightly warped surface is seen at about 1000–1200 m a.s.l., dissected by deep valleys (Eikesdalen draining westwards; Gudbrandsdalen and other valleys draining southeastwards). This is the lowest level of the Palaeic relief,

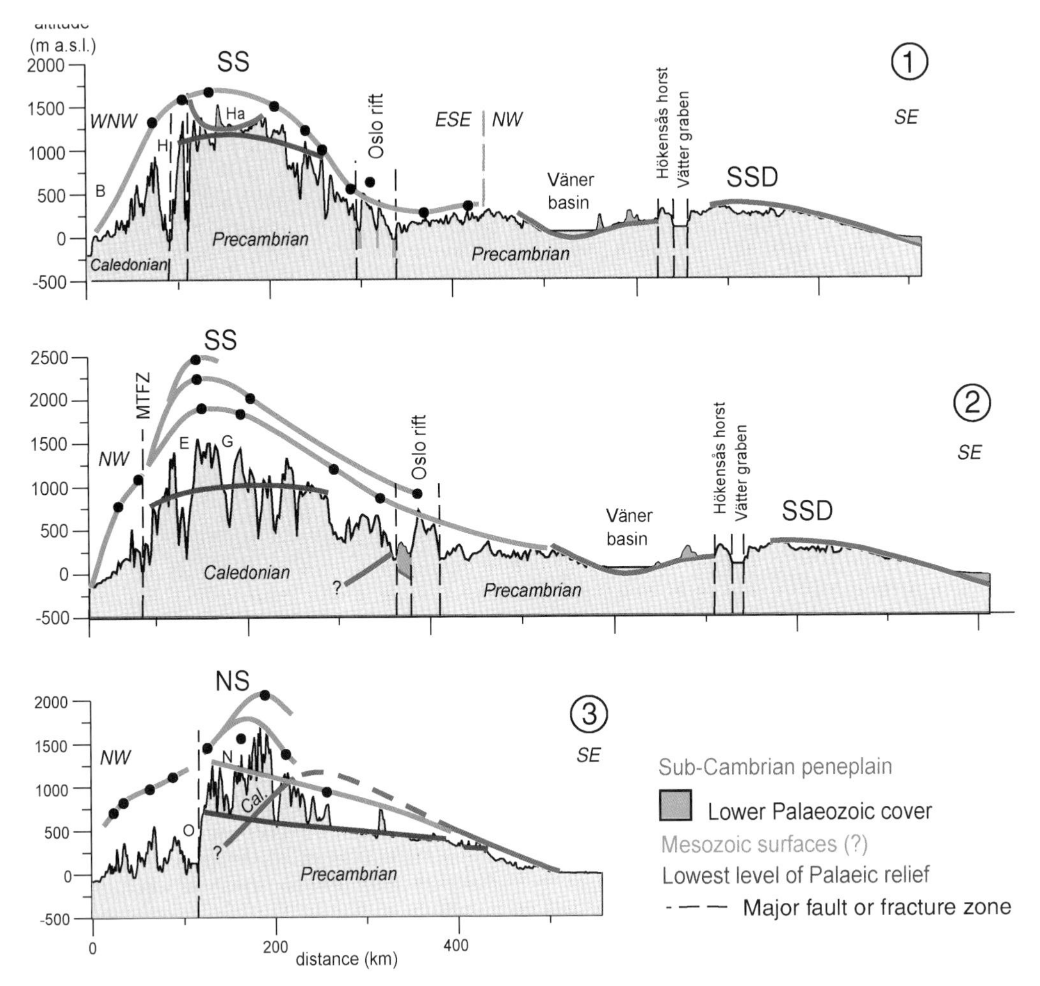

Fig. 7. Topographic profiles showing Scandinavian domes and palaeosurfaces. Location of profiles is shown in Fig. 2. Profile 1 crosses the southern part of the SS, Väner Basin and SSD. B, Bergen; H, Hardangerfjord; Ha, Hardangervidda. Profile 2 crosses the northern part of SS, Väner Basin and SSD. E, Eikesdalen; G, Gudbrandsdalen. Profile 3 crosses the NS and the Muddus plains in the east. O, Ofotfjorden; N, Norddalen; Cal, Caledonian.

comparable with the lowest level of Hardangervidda in profile 1. Above this lowest level, mountains rise to heights of 1500–2000 m a.s.l. The highest peaks of south Norway reach 1500 m above this lowest level, and are indicated by a point showing their vertical position in the profile. Before the profile reaches the Atlantic it crosses the Møre–Trøndelag Fault Zone (MTFZ).

Profile 3 (Fig. 7)

The exhumed sub-Cambrian peneplain is seen at the eastern coast and can be followed in some summits towards the west. The inferred sub-Cambrian peneplain is bent down, below the Caledonian rocks.

The eastern parts of the NS are here formed in Precambrian basement rock with the Caledonian nappes following westwards. The highest mountains are formed of rocks relatively resistant to deep weathering and their summit surfaces may date back to Mesozoic time (Lidmar-Bergström 1996). Mesozoic surfaces have tentatively been placed across the highest peaks along the profile. A lowest Mesozoic(?) level is tentatively shown as a surface inclined to the east along the whole profile (see Wråk 1908).

A few steps are incised below the sub-Cambrian surface close to the coast in the east and then follow westward the main plains with residual hills (the Muddus plains), which extend to the Scandes. Within the mountains this level can be followed as the base for valley incision. Mountains rise to 1500 m above this level.

Comparison and correlation of the Scandinavian domes from the profile data

Profile 3 (Fig. 7) shows the highly dissected mountains of the NS rising from a slightly tilted lowest level, continuing in the Muddus plains eastwards. In the map analysis (Fig. 5), the Muddus plains were correlated with the lowest level of the Palaeic relief in the SS (Fig. 6), and they are shown as the same surface in profiles 1 and 2 (Fig. 7). The correlation between the Muddus plains and the lowest level of the Palaeic relief is strongly supported by the fact that the vertical distance from these levels to the highest summits of the Scandes are 1500 m in both regions (Fig. 7, profiles 2 and 3). This suggests that the two domes were of the same height before the major uplift of the SS.

Discussion

Uplift of the Scandes

Riis (1996) suggested that the uplift along the western margin of Fennoscandia did not occur simultaneously in the north and the south. On the basis of offshore geology and identification of a Paleogene surface it was suggested that the main uplift in the north occurred in Paleogene time, whereas it took place in Neogene time in the south. Lidmar-Bergström (1999) agreed with this interpretation, as a result of analysis of the relation between the mountains and the two types of relief that occur along the eastern flanks of the NS and the SS.

The Northern Scandes and their eastern flank. The shape of the NS and the stepped morphology of the Muddus plains indicates tilting towards the east. The main relief within the mountains is the result of valley incision since the initiation of uplift, whereas the successively widened outer valleys formed in a tectonically relatively stable environment close to a general base level in the east with the Muddus plains as the end result. It is possible that the Precambrian surface, where not in contact with the sub-Cambrian peneplain, was exposed during Mesozoic time and that the Muddus plains ultimately were formed in Paleogene time, as indicated by finds of redeposited marine Eocene diatomaceans (Cleve-Euler 1941). Continuing apatite fission-track analysis (AFTA) in the area will shed more light on this. So far, these studies suggest a tilt of the area along the profile beginning in Cretaceous time and accelerating in Paleogene time (Hendriks & Andriessen 2001). A very high relative relief occurs west of the Ofotfjorden fault line, and, in detail, the relief is highly irregular. It is likely that this was originally an etch surface of Mesozoic (or maybe older) age, as Mesozoic (or older?) rocks are still preserved in a downfaulted basin on a kaolinitized basement surface (Sturt *et al.* 1979). Uplift has caused incison of the etched structures. Analysis of fission-track data indicates several phases of vertical movement in this region (Hendriks & Andriessen 2002).

The Southern Scandes. The surface forms of the SS are cut across both Caledonian and Precambrian basement. In the southern part of the dome the present surface is relatively close to the sub-Cambrian peneplain, whereas in the north the sub-Cambrian peneplain disappears below the Caledonian nappes. The Mesozoic surface inferred in the profiles indicates that the line through the highest summits represents a warped surface of this age. In the Oslo rift and southeastwards it becomes more likely that the present relief is part of a re-exposed sub-Cretaceous etch surface. Apatite fission-track modelling shows that this interpretation is viable (Cederbom *et al.* 2000). Profile 2 crosses the MTFZ. This is an old Caledonian structure that has subsequently been reactivated (Doré *et al.* 1999). It is possible that vertical movements along the MTFZ have affected the Mesozoic surface and contributed to a slight asymmetry of the SS. The lowest level of the Palaeic relief has a position at about 1000–1200 m a.s.l. compared with the tilted plain from which the NS rise at 300–700 m a.s.l. It is a slightly domed surface, which supports the idea of a continuation of the Mesozoic doming. The Neogene uplift of the SS can thus be estimated to be about 1000 m (Fig. 7 and Lidmar-Bergström *et al.* 2000). The valleys along the coast are mainly structurally controlled and it is suggested that the deep incision followed structures etched out during Mesozoic time.

A major hinge line across central Scandinavia? A line from the MTFZ towards the

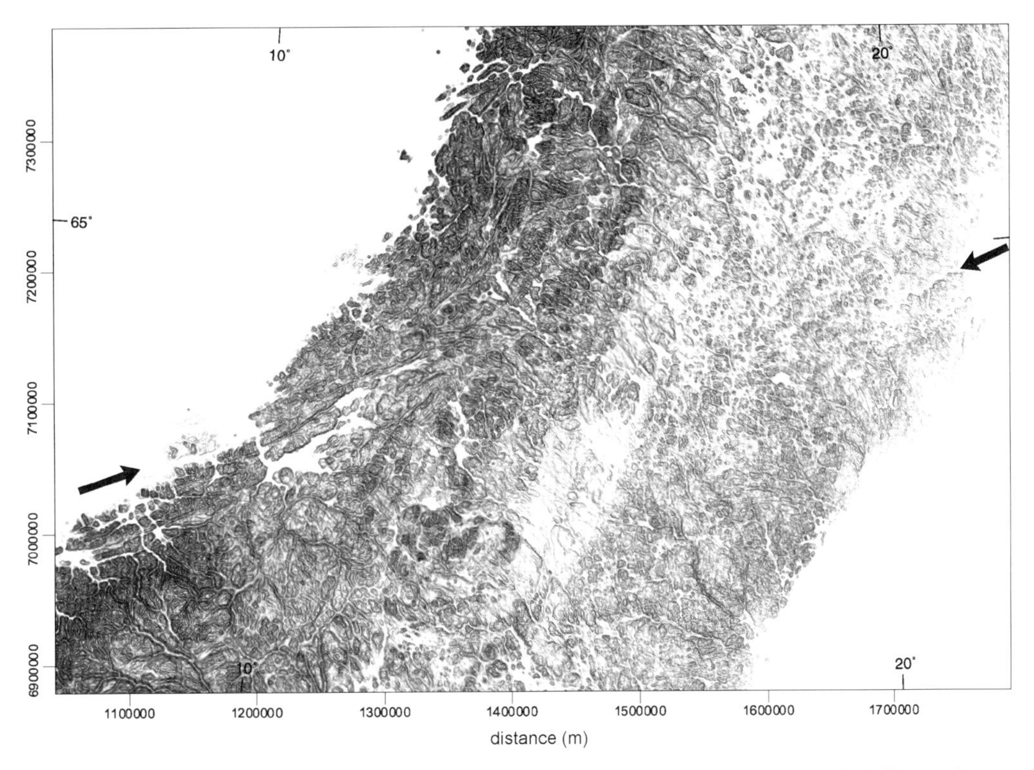

Fig. 8. Slope map of central Scandinavia. The Møre–Trøndelag Fault Zone (MTFZ) and the fault lines at the east coast of Sweden suggest a hinge line in a SW–NE direction. This line marks the border between the undulating hilly relief to the south and the Muddus plains to the north.

WSW–ENE-trending fault lines of the sub-Cambrian peneplain in north Sweden approximately coincides with the border between the plains with residual hills (the Muddus plains) and the undulating hilly relief (Fig. 8). In the region where this line crosses the mountain chain the Precambrian basement has a low position (300 m a.s.l.). This is in contrast to both in the NS and SS, where the basement reaches above 1000 m a.s.l. This line may approximately mark the border between areas with different uplift histories.

South Swedish Dome and Väner Basin

The SSD developed its present shape in Tertiary time, when the basement successively was exposed after erosion of Palaeozoic cover in the north and east and Mesozoic cover in the south and west (Lidmar-Bergström 1991, 1993). We suggest that in connection with the Neogene uplift of the SS, the Väner Basin was slightly depressed and the SSD uplifted. No Palaeozoic rocks occur on the bottom of Lake Vänern and the cover is therefore supposed to have been eroded before the downwarp. A Neogene rise of the SSD at the Oligocene–Miocene boundary with subsequent erosion of its cover is supported by a fission-track study of south Sweden (Cederbom 2002) and the observation of a large amount of erosion of Cretaceous and Paleogene strata from nearby Jylland and the Skagerrak–Kattegat Platform (Japsen & Bidstrup 1999).

Summary of uplift and relief development

The Precambrian shield of Scandinavia has been covered by Palaeozoic rocks. In some parts they were eroded during Mesozoic time and etched surfaces with undulating hilly relief were formed in the basement. Mesozoic denudation also caused relief development across Caledonian rocks and the Palaeozoic cover. Jurassic and (probably mainly) Cretaceous strata were then deposited over the area to an unknown extent. In some cases they were deposited directly on the etched basement and protected its Mesozoic relief for a long time. Large parts of the Caledonian areas experienced continuous relief development without any Mesozoic temporary cover. At the end of the Paleogene period, plains with residual hills had formed over the eastern parts of Scandinavia with valley systems penetrating mountainous areas in the west. The mountains reached about 1500 m above this lowest level of the Palaeic surface. In southern Norway the sub-Cambrian peneplain was successively re-exposed and the Precambrian basement here experienced further relief development.

The tectonic uplift differed between the NS and the SS. The NS mainly acted as a block, which experienced simple tilt, whereas the SS were characterized by doming. Neogene uplift caused a slight further tilt in the north and a major uplift with continued doming in the SS with re-exposure of sub-Mesozoic relief along their flanks. The deeply incised valleys of the SS were mainly formed in Neogene time as a consequence of uplift. Subsequent, Late Cenozoic ice sheet erosion has further widened and deepened these valleys.

Neogene uplift with a centre south of Lake Vättern caused the formation of the SSD. The buried sub-Cambrian peneplain was successively re-exposed. The re-exposure of this surface at the top of the dome caused reactivation of the weathering systems along the fracture systems, which were successively expressed in the topography as joint aligned valleys. Re-exposure of etched Mesozoic relief in the SW caused total stripping of the remains of the kaolinitic saprolites by the SW-flowing drainage system, down to *c.* 125 m a.s.l. Here the SSP gradually formed, governed by sea level as the base for its erosion. Its detailed forms probably developed in semiarid climates that promoted pedimentation (Lidmar-Bergström 1988). Neogene grus saprolites are still of common occurrence in the eastern part of the dome above this level and testify to continued etching.

The latest rise, probably in Pliocene time, caused the re-exposure of the sub-Cretaceous etch surfaces below 125 m a.s.l. in the south and west, and successive re-exposure of the extremely flat sub-Cambrian rock surfaces in the north and east.

Conclusions

(1) The main uplift of the Northern Scandes is older than the uplift of the Southern Scandes, as the NS are considerably more dissected by deep valleys. As a result of the Mesozoic Paleogene uplift and tilt of the NS, the Muddus plains formed and they acted as base levels for the development of the deep valleys of the mountain range. In Neogene time the area experienced additional minor uplift and tilting.

(2) The main uplift of the Southern Scandes took place in Neogene time, the uplift amounting to *c.* 1000 m. The Neogene uplift was a continuation of Mesozoic–Paleogene doming.

(3) The Muddus plains and their continuation in the valleys of the NS are tentatively correlated with the lowest level of the Palaeic relief of the SS, because of strong similarities in morphology.

(4) A hinge from the MTFZ to NE Sweden separates the NS with the Muddus plains from the SS and the undulating hilly relief on its eastern flank.

(5) In Neogene time the South Swedish Dome was elevated, with subsequent development of the South Småland Peneplain. At the same time the Väner Basin was downwarped.

(6) Genetically interpreted landforms are important datasets in morphotectonic analyses, complementary to studies of the sedimentary records and thermotectonic evolution of the bedrock.

The study was supported by a grant from the Swedish Natural Science Research Council. We also want to thank P. Japsen, S. A. Cloetingh and P. Andreissen for encouraging discussions and input. Elevation model data over Sweden courtesy of Swedish National Land Survey 2000. Excerpt from GSD-elevation database, case no. L2000/646. Elevation model data over Norway courtesy of Statens Kartverk, 3504 Hønefoss, Norway.

References

AHLMANN, H.W. 1919. Geomorphological studies in Norway. *Geografiska Annaler*, **1**, 1–20.

AXBERG, S. & WALDSTEIN, P. 1980. Distribution of the sedimentary bedrock in Lake Vättern, southern Sweden. *Stockholm Contributions in Geology*, **34** (2), 15–25.

BOTTOMLEY, R.J., YORK, D. & GRIEVE, R.A.F. 1978. $^{40}Ar-^{38}Ar$ ages of Scandinavian impact structures: I Mien and Siljan. *Contributions to Mineralogy and Petrology*, **68**, 79–84.

CEDERBOM, C. 2002. The thermotectonic development of southern Sweden during Mesozoic and Cenozoic time. *In*: DORÉ, A.G., CARTWRIGHT, J.A., STOKER, M.S., TURNER & J.P., WHITE, N. (eds) *Exhumation of the North Atlantic Margin: Timing, Mechanisms and Implications for Petroleum Exploration.* Geological Society, London, Special Publications, **196**, 169–182.

CEDERBOM, C., LARSSON, S.É., TULLBORG, E.-L. & STIBERG, J.-P. 2000. Fission track thermochronology applied to Phanerozoic thermotectonic events in central and southern Sweden. *Tectonophysics*, **316**, 153–167.

CLEVE-EULER, A. 1941. Alttertiäre Diatomeen und Silicoflagellaten im inneren Schwedens. *Palaeontographica, 92A*, 165–208.

DEUTSCH, A., BUHL, D. & LANGENHORST, F. 1992. On the significance of crater ages: new ages for Dellen (Sweden) and Araguainha (Brazil). *Tectonophysics*, **216**, 205–218.

DORÉ, A.G., LUNDIN, E.R., JENSEN, L.N., BIRKELAND, Ø., ELIASSEN, P.E. & FICHLER, C. 1999. Principal tectonic events in the evolution of the northwest European Atlantic margin. *In*: FLEET, A.J. & BOLDY, S.A.R. (eds) *Petroleum Geology of Northwest Europe: Proceedings of the 5th Conference*. Geological Society, London, 41–61.

GJESSING, J. 1967. Norway's paleic surface. *Norsk Geografisk Tidskrift*, **21**, 69–132.

HENDRIKS, B.W.H. & ANDRIESSEN, P.A.M. 2001. Pattern and timing of the part-Caledonian denudation of northern Scandinavia constrained by apatite fission-track thermochronology. *In*: DORÉ, A.G., CARTWRIGHT, J., STOKER, M.S., TURNER, J.P. & WHITE, N. (eds) *Exhumation of the North Atlantic Margin: Timing, Mechanisms and Implications for Petroleum Exploration*. Geological Society, London, Special Publications, **196**, 117–137.

HÖGBOM, A.G. 1910. Precambrian geology of Sweden. *Bulletin Geological Institute Upsala*, **10**, 1–80.

JAPSEN, P. & BIDSTRUP, T. 1999. Quantification of late Cenozoic erosion in Denmark based on sonic data and basin modelling. *Bulletin of the Geological Society of Denmark*, **46**, 79–99.

JAPSEN, P. & CHALMERS, J.A. 2000. Neogene uplift and tectonics around the North Atlantic: overview. *Global and Planetary Change*, **24**, 165–173.

KOARK, H.J., MÄRK, T.D., PAHL, M., PURTSCHELLER, F. & VARTANIAN, R. 1978. Fission-track dating of apatites in Swedish Precambrian apatite iron ores. *Bulletin, Geological Institute Uppsala, New Series*, **7**, 103–108.

LIDMAR-BERGSTRÖM, K. 1982. Pre-Quaternary Geomorphological Evolution in southern Fennoscandia. *Sveriges Geologiska Undersökning, Serie C*, 785.

LIDMAR-BERGSTRÖM, K. 1988. Denudation surfaces of a shield area in south Sweden. *Geografiska Annaler*, **70A** (4), 337–350.

LIDMAR-BERGSTRÖM, K. 1989. Exhumed Cretaceous landforms in south Sweden. *Zeitschrift für Geomorphologie, Neue Folge, Supplementband*, **72**, 21–40.

LIDMAR-BERGSTRÖM, K. 1991. Phanerozoic tectonics in southern Sweden. *Zeitschrift für Geomorphologie, Neue Folge, Supplementband*, **82**, 1–16.

LIDMAR-BERGSTRÖM, K. 1993. Denudation surfaces and tectonics in the southernmost part of the Baltic Shield. *Precambrian Research*, **64**, 337–345.

LIDMAR-BERGSTRÖM, K. 1995. Relief and saprolites through time on the Baltic Shield. *Geomorphology*, **12** (1), 45–61.

LIDMAR-BERGSTRÖM, K. 1996. Long term morphotectonic evolution in Sweden. *Geomorphology*, **16**, 33–59.

LIDMAR-BERGSTRÖM, K. 1999. Uplift histories revealed by landforms of the Scandinavian domes. *In*: SMITH, B.J., WHALLEY, W.B. & WARKE, P.A. (eds) *Uplift, Erosion and Stability: Perspectives on Long-term Landscape Development*. Geological Society, London, Special Publications, **162**, 85–91.

LIDMAR-BERGSTRÖM, K., OLLIER, C.D. & SULEBAK, J.R. 2000. Landforms and uplift history of southern Norway. *Global and Planetary Change*, **24**, 211–231.

LIDMAR-BERGSTRÖM, K., OLSSON, S. & OLVMO, M. 1997. Palaeosurfaces and related saprolites in southern Fennoscandia. WIDDOWSON, M. (ed.) *Palaeosurfaces: Recognition, Reconstruction and Palaeoenvironmental Interpretation*. Geological Society, London, Special Publications. *In*: vol **120**, 95–123.

LIDMAR-BERGSTRÖM, K., OLSSON, S. & ROALDSET, E. 1999. Relief features and palaeoweathering remnants in formerly glaciated Scandinavian baement areas. *In*: THIRY, M. & SIMON-COINÇON, R. (eds) *Palaeoweathering, Palaeosurfaces and Related Continental Deposits*. International Association of Sedimentologists, Special Publications, **27**, 275–301.

LJUNGNER, E. 1950. Urbergsytans form vid fjällranden. *Geologiska Föreningens i Stockholm Förhandlingar*, **72**, 269–300.

PEULVAST, J.P. 1978. Le bourrelet Scandinave et les Calédonides: un essai de reconstitution des modalités de la morphogenèse en Norvège. *Geographie Physique Quaterniaire*, **32**, 295–320.

PEULVAST, J.P. 1985. Postorogenic morphotectonic evolution of the Scandinavian Caledonides during the Mesozoic and Cenozoic. *In*: GEE, D.G. & STURT, B.A. (eds) *The Caledonide Orogen—Scandinavia and Related Areas*. Wiley, Chichester, 979–995.

REUSCH, H. 1901. Nogle bidrag till forstaaelsen af hvorledes Norges dale og fjelde er blevne til. *Norges Geologiske Undersøgelse, Aarbog (1900)*, **32**, 124–263.

REUSCH, H. 1903. Glommens bøjning ved Kongsvinger. *Norges Geografiske Selskab, Aarbog*, **14**, 96–102.

RIIS, F. 1996. Quantification of Cenozoic vertical movements of Scandinavia by correlation of morphological surfaces with offshore data. *Global and Planetary Change*, **12**, 331–357.

ROHRMAN, M., VAN DER BEEK, P., ANDRIESSEN, P. & CLOETINGH, S. 1995. Meso-Cenozoic morphotectonic evolution of southern Norway: Neogene domal uplift infered from apatite fission track thermochronology. *Tectonics*, **14**, 704–718.

RUDBERG, S. 1954. *Västerbottens berggrundsmorfologi*. Geographica, **25**.

RUDBERG, S. 1960. Geology and geomorphology. *In*: SØMME, A. (ed.) *A Geography of Norden*. J. W. Cappelens, Oslo, 27–40.

SCHIPULL, K. 1974. *Geomorphologische Studien in zentral Südnorwegen mit Beiträgen über Regelungs- und Stuerungssysteme in der Geo-*

morphologie. Hamburger Geographischer Studien, **31**.

STURT, B., DALLAND, A. & MITCHELL, J. 1979. The age of the Sub-Mid-Jurassic tropical weathering profile of Andøya, northern Norway, and the implications for the Late Palaeozoic palaeogeography in the North Atlantic region. *Geologische Rundschau*, **68**, 523–542.

SUMMERFIELD, M. 2000. *Geomorphology and Global Tectonics*. Wiley, Chichester.

THOMAS, M.F. 1994. *Geomorphology in the Tropics*. Wiley, Chichester.

VIVALLO, W. & BROMAN, C. 1993. Genesis of the earthy ores at Garpenberg, south central Sweden. *Geologiska Föreningens i Stockholm förhandlingar*, **115**, 209–214.

WICKMANN, F.E. 1988. Possible impact structures in Sweden. *In*: BODEN, A. & ERIKSSON, K.G. (eds) *Deep Drilling in Crystalline Bedrock*. Springer, Berlin, **1**, 299–327.

WRÍK, W. 1908. Bidrag till Skandinaviens reliefkronologi. *Ymer*, **28**, 141–191.

ZECK, H.P., ANDRIESSEN, P.A.M., HANSEN, K., JENSEN, P.K. & RASMUSSEN, B.L. 1988. Palaeozoic palaeo-cover of the southern part of the Fennoscandian shield—fission track constraints. *Tectonophysics*, **149**, 61–66.

Pattern and timing of the post-Caledonian denudation of northern Scandinavia constrained by apatite fission-track thermochronology

BART W. H. HENDRIKS & PAUL A. M. ANDRIESSEN

Department of Isotope Geochemistry, Faculty of Earth Sciences, Vrije Universiteit Amsterdam, De Boelelaan 1085, 1081 HV Amsterdam, The Netherlands (e-mail: henb@geo.vu.nl)

Abstract: Apatite fission-track thermochronology has been used to study the post-Caledonian denudation history of northern Scandinavia. Post-orogenic denudation progressively shifted from the interior of the continent towards the North Atlantic margin. The present-day area of maximum elevation in the Northern Scandes mountain range has experienced continuous denudation at least since Jurassic time. In Jurassic–Cretaceous time, the area north and east of this region experienced either no denudation at all or some denudation followed by a transient thermal event with a peak temperature in late Cretaceous time. Final denudation of the area to the east of the Northern Scandes probably started in late Cretaceous–Paleogene time and possibly accelerated in Neogene time. The denudation history of northern Scandinavia can be explained by scarp retreat of an uplifted rift flank. The pattern and timing of denudation of the Northern Scandes is different from that of the Southern Scandes, which experienced domal-style, late-stage postrift uplift in Neogene time. Geomorphological observations, offshore data from the Atlantic and Barents Sea margins, and scarce stratigraphical information from the mainland are in general agreement with the new thermochronological data.

Because of the almost complete absence of post-Caledonian sediments on the Scandinavian mainland, denudation ('uplift of rocks relative to the surface' according to the definition of Summerfield & Brown (1998)) of Scandinavia has been studied mainly by analysis of geomorphology (Lidmar-Bergström 1993, 1999) and its correlation with the offshore geology (Riis 1996). Low-temperature thermochronology has been applied successfully in southern Scandinavia (Lehtovaara 1976; Andriessen & Bos 1986; Zeck *et al.* 1988; Rohrman 1995; Hansen *et al.* 1996; Larson *et al.* 1999; Cederbom *et al.* 2000), but until now no comparable study has been undertaken in northern Scandinavia.

The geomorphology of Scandinavia can be characterized as a plateau with two large domes in the west and north and a smaller dome in southern Sweden (Fig. 1). The south Swedish dome reaches 377 m above sea level and is partly covered by Cambrian and Mesozoic cover rocks. The southern Norwegian dome, usually referred to as the Southern Scandes, reaches 2469 m above sea level on Jotunheimen. The more elongated northern Scandinavian dome, usually referred to as the Northern Scandes and the study area here, has a maximum elevation of 2113 m on Kebnekaise.

Riis (1996) and Lidmar-Bergström (1999) both concluded that the denudation history of the Southern Scandes is different from that of the Northern Scandes mountain range. According to both studies, the latest uplift phase in the north was earlier (in late Cretaceous to Paleogene time) than in the south (in Neogene time). This is in agreement with apatite fission-track data from Rohrman (1995) and from the present study. According to Rohrman (1995), the central Southern Scandes experienced Triassic–Jurassic erosion of 2.4 ± 1.1 km and a Neogene denudation of 2.0 ± 0.5 km, decreasing radially outward to less than 0.5 km near the coastline.

West of the Northern Scandes mountain range, the Lofoten and Vesterålen island groups (Fig. 2) are part of a horst and graben system that is structurally very complex and poorly understood. According to Riis (1996), final uplift of this area occurred in Neogene time, with a strong Plio-Pleistocene component. Inverse modelling of the small amount of apatite fission-track data we have from Lofoten and Vesterålen at present, indicates considerable Neogene denudation as well. A detailed study applying apatite fission-track analysis and (U–Th)/He thermochronometry to this region, aiming at unravelling the denudation history of the various structural

From: DORÉ, A.G., CARTWRIGHT, J.A, STOKER, M.S., TURNER, J.P. & WHITE, N. 2002. *Exhumation of the North Atlantic Margin: Timing, Mechanisms and Implications for Petroleum Exploration*. Geological Society, London, Special Publications, **196**, 117–137. 0305-8719/02/$15.00

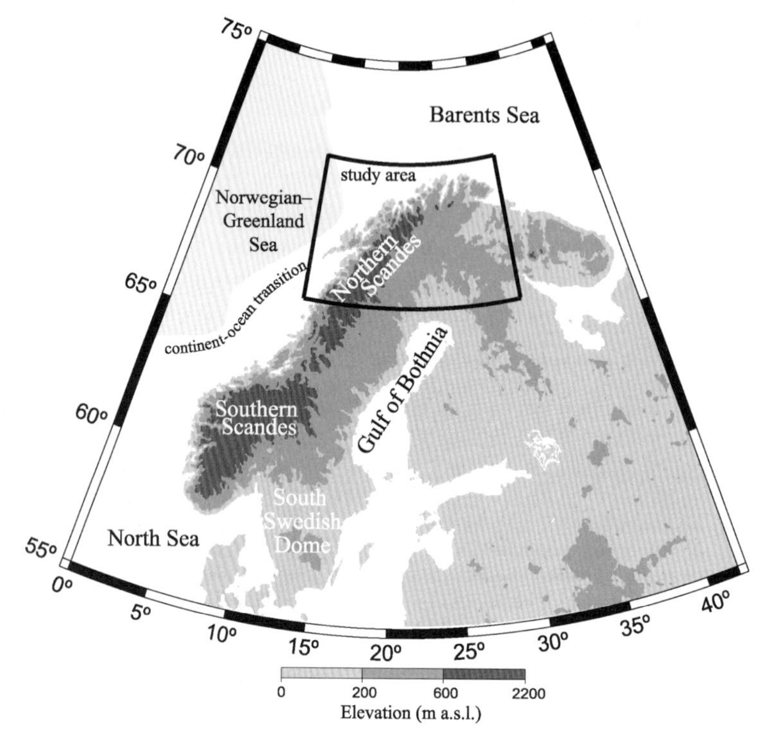

Fig. 1. Map of Scandinavia and surrounding areas, showing the location of the study area and indicating the Southern Scandes, Northern Scandes and South Swedish Dome.

blocks and dating fault movements, is in progress.

Geological setting

During the Caledonian orogeny the Fennoscandian shield was overthrust from the (north)west (Roberts & Gee 1985). At present, several windows in the Caledonian cover rocks expose parts of the Fennoscandian shield to the west of the Caledonian front on Lofoten, Vesterålen, Senja and to the SW and SE of Narvik (Geological Surveys of Finland, Norway and Sweden 1987).

In the immediate post-orogenic period, in Devonian time, fault-bounded molasse basins developed in the precursor Norwegian–Greenland rift system and continental molasse continued to accumulate in half-grabens bordering the remnant Caledonian mountains until mid-Permian time (Doré & Gage 1987). Crustal extension in the Norwegian–Greenland rift system accelerated during late Permian and Triassic time and resulted in low-relief doming of the flanking areas, particularly during late Triassic time (Ziegler 1987). Jurassic regional uplift on the northern margins of Laurentia and Baltica and mid-Jurassic reactivation of the margins and other source areas within the Atlantic Rift domain shed thick coarse clastic deposits onto surrounding areas (Doré 1991). Major rifting affected the entire Northern Atlantic domain at the Jurassic–Cretaceous transition (Faleide *et al.* 1993). In late Cretaceous time, the Barents Sea domain was tectonically decoupled along the Senja–Hornsund fault system from the Atlantic Rift domain where downwarping continued (Doré 1991). Following break-up in the Norwegian–Greenland Sea at the Paleocene–Eocene transition (Srivastava & Tapscott 1986), the western Barents Sea margin developed as a shear margin (Faleide *et al.* 1993).

According to Riis (1996), late Cretaceous–early Tertiary uplift of northern and western Fennoscandia caused deep erosion, and Paleogene uplift reached a maximum value of almost 1500 m in northern Scandinavia. Neogene tectonic uplift caused doming in southern Norway and was of the order of 1–1.5 km in the area of highest topography (Rohrman 1995). Lofoten and Vesterålen experienced two phases of Tertiary uplift (Stuevold & Eldholm 1996), with the Neogene uplift component being of the order of 1000 m (Riis 1996).

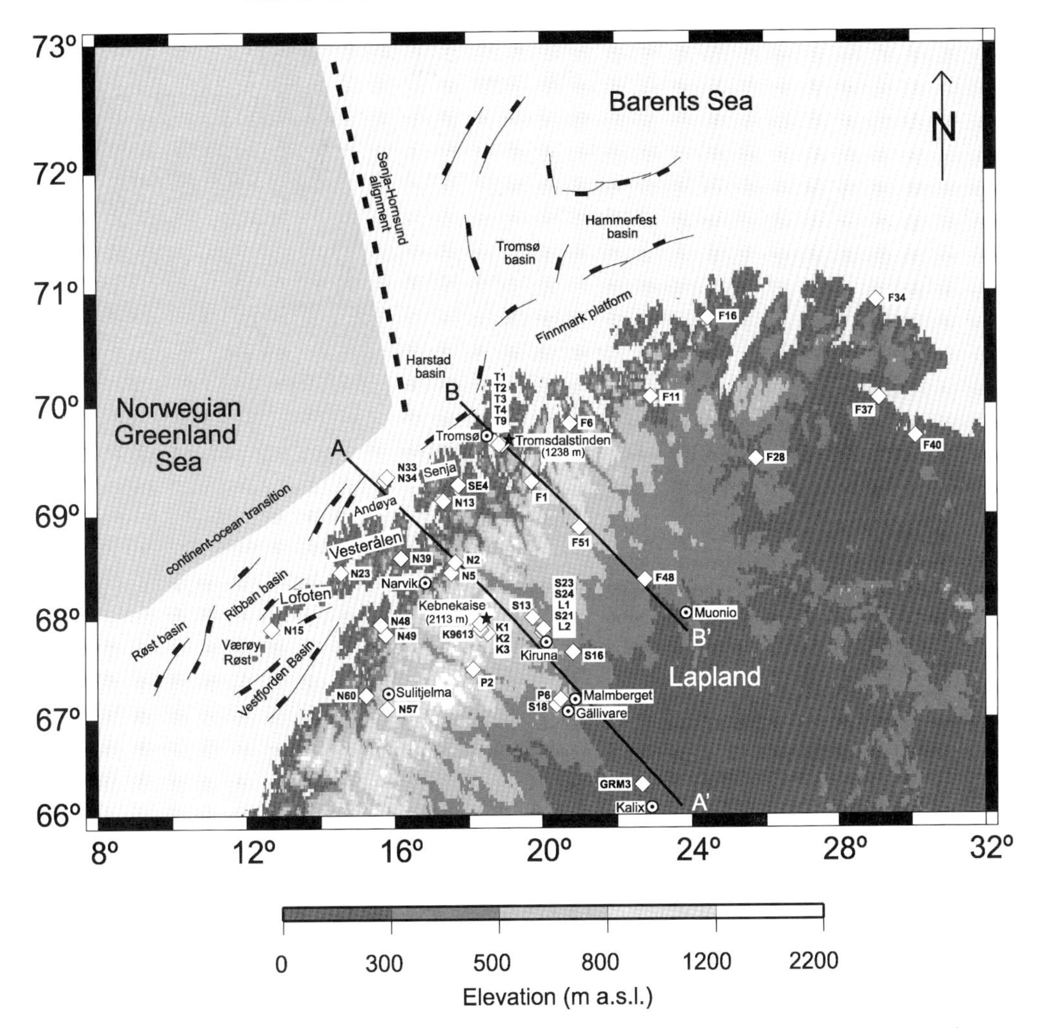

Fig. 2. Map of the study area showing topographic elevation, locations of samples and transects, and other features referred to in this paper. Locations of faults and basins after Faleide *et al.* (1993), Olesen *et al.* (1997) and Brekke (2000).

Plio-Pleistocene glacial erosion caused unloading and uplift of Fennoscandia and modified the older fluvial drainage system to a considerable extent, but did not totally obliterate the landforms that are typical of uplifted landmasses along passive continental margins (Lidmar-Bergström *et al.* 2000). Glacial erosion also affected Svalbard (Blythe & Kleinspehn 1998) and the western Barents Sea, which was subaerial in preglacial times. Tectonic uplift played an important role in this region before and possibly also during the glaciations (Dimakis *et al.* 1998). Postglacial crustal doming is estimated to have reached a maximum uplift value of 850 m in the centre of Fennoscandia (Gudmundsson 1999).

Apatite fission-track thermochronology

Annealing concept

Fission tracks are damage zones in the crystal lattice that are formed by the spontaneous fission of ^{238}U (Wagner 1968; Fleischer *et al.* 1975). The density of spontaneous fission tracks is proportional to the elapsed time and the uranium content. Except for rapidly cooled rocks, fission-track ages are systematically younger compared with other radiogenic age determinations. This is a result of the instability of fission tracks at high temperatures. Within a mineral-specific temperature range, called the partial annealing zone (PAZ), the tracks begin to anneal until they are completely erased at the upper temperature boundary. For apatite, the most commonly used mineral, the PAZ is usually defined as a temperature range between *c.* 60 °C and *c.* 130 °C (Gleadow & Duddy 1981; Naeser 1981; Wagner & Van den Haute 1992), but even at room temperature low-rate annealing takes place (Donelick *et al.* 1990). The temperature range of the PAZ will be different for apatites with different chemical compositions (e.g. Carlson *et al.* 1999). For extreme compositions the upper boundary of the PAZ, meaning total annealing, may be as low as 90 °C or as high as 200 °C (Ketcham *et al.* 1999). Fission tracks formed in the PAZ, and tracks that have been heated into the PAZ after formation at lower temperatures, have reduced lengths, compared with their initial length l_0 of between 15.8 and 16.6 μm. Track length reduction within the PAZ results in a mean length shorter than l_0 and a negatively skewed distribution towards the smaller tracks. The annealing characteristics of fission tracks thus allow the reconstruction of the thermal history of a sample from the track length distribution (Gleadow *et al.* 1986a, 1986b). Reheating of a sample by a transient thermal event can be inferred, but it is difficult, on the basis of fission-track data alone, to differentiate this from a purely cooling history. This is because the fission-track system is essentially one that records cooling, and even in a thermal history that has included some reheating, most of the record will come from the cooling segments (Gleadow & Brown 2000).

Analytical procedure

After mineral separation, apatite grains were mounted in epoxy, polished and etched in 7% HNO_3 at 20 °C to reveal spontaneous fission tracks. Etch time varied between 25 and 40 s, depending on the etching rate of the apatite. For all samples the external detector method was applied using micas that were heated for 48 h at 600 °C to erase any existing tracks. All mounts were irradiated together with dosimeter glass CN5 in the LFR facility of the ECN at Petten, the Netherlands. Micas covering the apatites were subsequently etched in 48% HF at 20 °C for 12 min. Micas covering the dosimeter glass were etched in 48% HF at 20 °C for 30 min. The counting of fission tracks in the apatites and external detectors and the measurement of confined fission-track lengths were performed at 1000 × magnification using 100 × dry objective with a numerical aperture of 0.90. Only apatites with sharp polishing scratches were used for counting and measuring.

Age and track length statistics

The fission-track ages were calculated using the ζ age calibration method (Hurford & Green 1983). A ζ value of 355 ± 15 (1σ) was obtained from Mt. Dromedary, Fish Canyon and Durango apatite standards using CN5 dosimeter glass. All samples displayed $P(\chi^2) > 5\%$, indicating that the data are consistent with a single population of ages in each sample (Galbraith 1981). Therefore all fission-track ages in this paper are reported as pooled ages, with a 1σ error. Initial fission-track length was determined on confined induced tracks of Fish Canyon apatite, giving a mean value of 16.3 μm with a standard error of 0.1 μm and a 1σ value of 0.49 μm.

Inverse modelling of fission-track data

Inverse modelling of the age and length measurement results was carried out using the AFTSOLVE 1.2.1a program (Ketcham *et al.* 2000). Compared with AFTSOLVE, many other fission-track modelling programs (Corrigan 1991; Lutz & Omar 1991; Gallagher 1995) tend to generate simpler histories because they implicitly or explicitly favour paths with fairly few degrees of freedom (Ketcham *et al.* 2000). AFTSOLVE yields a much wider array of possible modelling results. The annealing model of Crowley *et al.* (1991) for F-apatites and the annealing model of Laslett *et al.* (1987), which is based on laboratory studies of Durango apatite, have been used for inverse modelling. Compared with the annealing model of Crowley *et al.* (1991), the annealing model of Laslett *et al.* (1987) is less sensitive to low-temperature annealing and overemphasizes the importance of recent cooling (van der Beek 1995). The annealing model of Laslett *et al.* (1987) therefore often reveals thermal histories with a rapid late cooling event, which in many cases is a modelling artefact. However, the part of the thermal history below *c.* 60 °C is only loosely constrained by any annealing model and one should always be very critical about any interpretations based on this part of the modelled thermal history. At high temperatures the annealing model of Crowley *et al.* (1991) is much more retentive than the annealing model of Laslett *et al.* (1987). Extrapolation of the annealing model of Crowley *et al.* (1991) to geological time scales predicts F-apatite to be more resistant to annealing than the more Cl-rich Durango apatite, in contrast to geological observations (Gallagher *et al.* 1998). The Laslett *et al.* (1987) annealing model is the most widely used annealing model, but the Crowley *et al.* (1991) model for F-apatites has been used in many fission-track studies in Scandinavia (Rohrman 1995; Larson *et al.* 1999; Cederbom *et al.* 2000). To compare our results with studies using either the Laslett *et al.* (1987) or the Crowley *et al.* (1991) annealing model, all data have been modelled with both annealing models. In addition, it is possible to test the sensitivity of the thermal histories obtained from inverse modelling to the annealing model applied.

Modelling strategies for AFTSOLVE have been outlined by Ketcham *et al.* (2000). Initial track length was fixed at 16.3 μm. No other constraints were used than a present-day surface temperature of 0–10 °C. For samples S21 and P6, which were collected in the LKAB Kiruna and Malmberget mines at 765 m and 815 m below the surface, respectively, a present-day temperature of 10–25 °C was used. The outer envelope of the thermal histories presented in this paper represents thermal histories that give a value larger than 0.05 for both the age goodness-of-fit and the Kolmogorov–Smirnov test for the length distribution. This type of history cannot be ruled out by the data (Ketcham *et al.* 2000). The inner envelope represents thermal histories with a value larger than 0.5 for both the age goodness-of-fit and the Kolmogorov–Smirnov test for the length distribution. This type of history is supported by the data. Also, the best-fitting thermal history is depicted for all models presented in this paper. During the first inverse modelling runs for each sample, with the Crowley *et al.* (1991) annealing model as well as with the Laslett *et al.* (1987) annealing model, AFTSOLVE was restricted to produce purely cooling histories alone. If modelling with this restriction yielded only thermal histories that cannot be ruled out by the data, or very few thermal histories that are supported by the data, many different time intervals of reheating were tested for each sample. Special attention was given to make sure that the boundaries of the time interval did not force the model result in a certain direction, which could exclude possible thermal histories.

Sampling

Most samples are concentrated in two transects through the Northern Scandes mountain range. Transect A–A′ (Fig. 2) runs from Andenes on the northernmost tip of the Vesterålen islands towards the Gulf of Bothnia. This transect includes two vertical profiles. One is on Kebnekaise, ranging from 575 to 1530 m above sea level. The four samples in this vertical profile are all within 8 km in the horizontal direction. No post-Caledonian or neotectonic fault activity has been reported for this area. The second vertical profile in transect A–A′ is in the LKAB mine in Kiruna. The three lowermost samples in this profile, from −344 m to 196 m above sea level, are almost perfectly vertically aligned inside the mine itself. The top two samples, at 502 and 712 m above sea level, are about 3 km to the NE of the mine on a nearby hill, Luossavaara. Neotectonic activity has been reported for the Kiruna area (Dehls *et al.* 2000), but again no other post-Caledonian fault activity has been reported for this area to our knowledge. Another subsurface sample in this transect was collected inside the LKAB mine in Malmberget, at 815 m below the surface, at 191 m below sea level. Transect B–B′ (Fig. 2) runs between Tromsø, Norway, and Muonio, Finland. This transect includes a vertical profile on Tromsdalstinden, SE of Tromsø, from sea level to 1238 m elevation. This vertical profile includes five samples, and except for the sample at sea level, which is about 8 km to the west of the summit of Tromsdalstinden, all samples are no more than 3 km from each other in the horizontal direction. Two SW–NE-trending faults of unknown age lie within 5–10 km to the SE and NW of Tromsdalstinden (Zwaan *et al.* 1998). The one to the NW of Tromsdalstinden exhibits neotectonic activity (Dehls *et al.* 2000). However, we do not know of any post-Caledonian faults that would cut and might disturb the vertical profile. Block rotation between the two faults, however, may have caused rotation of the vertical profile through time.

In addition to the samples of the two transects, many surface samples were collected across the study area. Some of these samples were taken inside fjords, which can be up to 2 km deep. The fjords are glacially overdeepened valleys and the original fluvial incision may be much older than the Plio-Pleistocene glaciations (Lidmar-Bergström *et al.* 2000). Because of the perturbation effect of eroding topography on isotherms in the crust (Stüwe *et al.* 1994), the results from samples that were collected inside fjords cannot be immediately interpreted in the same way as those from 'normal' surface samples.

Several samples were collected especially to investigate the effects of a pronounced negative gravity anomaly centred around Sulitjelma, in the SW of the study area (Olesen *et al.* 1997). In contrast, the region encompassing Røst, Værøy and Lofoten is characterized by a strong positive gravity anomaly. Samples have been collected from this region, but results are not available yet.

Table 1. *Fission-track results*

Sample name	Elevation (m a.s.l.)	Latitude (N)	Longitude (E)	Pooled FT age ± SE (Ma)	$P(\chi^2)$ (%)	ρ_s (N_s) (10^6 tracks cm^{-2})	ρ_i (N_i) (10^6 tracks cm^{-2})	ρ_d (N_d) (10^6 tracks cm^{-2})	Number of grains	Mean track length (μm)	SE MTL (μm)	SD MTL (μm)	Number of lengths	Rock type
Transect A–A′														
N33	10	69.31	16.08	128 ± 14	23	0.368 (396)	0.491 (528)	0.957 (12033)	30	13.1	0.2	1.8	86	Gneiss
N34	360	69.28	16.01	142 ± 14	83	0.767 (553)	0.924 (666)	0.957 (12033)	30	11.6	0.2	2.2	175	Gneiss
N39	5	68.56	16.44	180 ± 21	28	0.330 (180)	0.308 (168)	0.957 (12033)	25	13.3	0.2	1.1	51	Granite
N2	380	68.52	17.89	134 ± 16	76	0.393 (240)	0.501 (306)	0.957 (12033)	22	–	–	–	–	Schist
N5	12	68.42	17.78	124 ± 12	9	0.802 (615)	1.104 (846)	0.957 (12033)	28	12.3	0.2	2.0	101	Granite
K9613	1530	67.93	18.54	220 ± 25	30	0.339 (395)	0.271 (316)	0.994 (12054)	30	12.9	0.2	1.6	100	Granite
K1	1070	67.88	18.54	205 ± 24	36	0.133 (349)	0.110 (288)	0.951 (11870)	36	13.7	0.2	1.2	58	Schist
K2	815	67.87	18.57	157 ± 21	91	0.155 (182)	0.167 (196)	0.951 (11870)	22	12.9	0.3	1.3	23	Schist
K3	575	67.84	18.75	113 ± 14	30	0.263 (207)	0.396 (312)	0.951 (11870)	21	12.8	0.3	1.9	36	Gneiss
S13	450	67.99	19.93	243 ± 23	32	1.937 (1375)	1.352 (960)	0.957 (12033)	55	13.0	0.2	1.7	100	Dolerite
S23	712	67.88	20.23	303 ± 36	26	0.773 (404)	0.427 (223)	0.951 (11870)	22	13.5	0.3	1.4	20	Volcanite
S24	502	67.87	20.21	268 ± 29	94	0.346 (508)	0.227 (333)	0.994 (12054)	52	12.9	0.3	1.7	23	Volcanite
L1	196	67.84	20.18	225 ± 22	80	1.019 (784)	0.763 (587)	0.951 (11870)	26	13.0	0.3	1.0	17	Iron ore
S21	−29	67.84	20.18	251 ± 30	100	0.270 (359)	0.176 (234)	0.927 (14206)	37	12.8	0.2	1.8	127	Iron ore
L2	−344	67.84	20.18	212 ± 24	45	0.503 (428)	0.400 (340)	0.951 (11870)	23	–	–	–	–	Iron ore
S16	418	67.65	21.00	322 ± 36	100	0.627 (536)	0.317 (271)	0.927 (14206)	28	13.0	0.2	2.0	109	Gneiss
P6	−191	67.18	20.67	264 ± 25	17	1.570 (1449)	1.045 (965)	0.994 (12054)	31	13.4	0.1	1.2	139	Iron ore
S18	430	67.13	20.57	318 ± 29	8	3.196 (2220)	1.695 (1177)	0.957 (12033)	33	13.3	0.1	1.3	100	Granodiorite
GRM3	40	66.31	22.82	268 ± 23	79	0.847 (508)	0.510 (306)	0.927 (12904)	20	12.6	0.2	1.6	100	Granite
Transect B–B′														
T1	1238	69.61	19.15	230 ± 26	84	0.216 (442)	0.159 (324)	0.951 (11870)	46	13.7	0.2	1.0	28	Gneiss
T2	753	69.61	19.11	241 ± 26	12	0.139 (522)	0.097 (364)	0.951 (11870)	51	13.7	0.2	1.1	40	Gneiss
T3	530	69.60	19.11	189 ± 19	81	0.492 (783)	0.460 (732)	0.994 (12054)	53	13.3	0.1	1.0	101	Gneiss
T4	262	69.61	19.07	203 ± 20	75	0.714 (864)	0.622 (753)	0.994 (12054)	40	13.2	0.1	1.2	100	Gneiss
T9	5	69.63	18.95	216 ± 22	96	0.273 (655)	0.213 (511)	0.951 (11870)	47	13.4	0.1	1.3	183	Gabbro
F1	15	69.26	19.92	170 ± 15	60	2.509 (3235)	2.492 (3213)	0.951 (11870)	49	12.8	0.1	1.3	204	Schist
F51	465	68.84	21.17	247 ± 22	53	0.667 (2408)	0.454 (1639)	0.951 (11870)	54	13.2	0.1	1.3	204	Gneiss
F48	300	68.35	22.90	365 ± 34	24	1.416 (1956)	0.647 (894)	0.951 (11870)	27	14.0	0.1	1.1	150	Granite
Other														
F6	1	69.79	20.94	214 ± 24	50	0.478 (384)	0.393 (316)	0.994 (12054)	32	13.8	0.1	0.8	18	Granitoid
F11	40	70.03	23.07	268 ± 36	98	0.351 (229)	0.230 (150)	0.994 (12054)	34	–	–	–	–	Volcanite
F16	5	70.71	24.59	205 ± 19	61	0.573 (1617)	0.472 (1332)	0.951 (11870)	34	12.9	0.1	1.4	107	Granite
F28	140	69.47	25.85	297 ± 26	15	1.111 (3415)	0.628 (1930)	0.951 (11870)	53	14.3	0.1	1.2	200	Gneiss
F34	10	70.86	29.11	269 ± 27	33	1.171 (774)	0.766 (506)	0.994 (12054)	34	13.0	0.1	1.1	100	Phyllite
F37	15	70.01	29.17	306 ± 28	84	1.044 (2418)	0.597 (1382)	0.994 (12054)	41	13.0	0.1	1.1	100	Gneiss

Table 1 – *continued*

Sample name	Elevation (m a.s.l.)	Latitude (N)	Longitude (E)	Pooled FT age ± SE (Ma)	$P(\chi^2)$ (%)	ρ_s (N_s) (10^6 tracks cm^{-2})	ρ_i (N_i) (10^6 tracks cm^{-2})	ρ_d (N_d) (10^6 tracks cm^{-2})	Number of grains	Mean track length (μm)	SE MTL (μm)	SD MTL (μm)	Number of lengths	Rock type
F40	40	69.67	30.16	237 ± 22	84	2.484 (1253)	1.719 (867)	0.927 (14206)	26	13.0	0.1	1.3	100	Granite
N13	10	69.09	17.58	168 ± 21	61	0.142 (245)	0.145 (249)	0.957 (12033)	42	12.5	0.2	1.5	53	Schist
N15	15	67.88	12.99	129 ± 15	72	0.385 (255)	0.509 (337)	0.957 (12033)	22	13.6	0.2	1.6	71	Anorthosite
N23	10	68.42	14.84	149 ± 15	43	0.234 (582)	0.267 (666)	0.957 (12033)	50	13.5	0.1	1.4	101	Mangerite
N48	5	67.92	15.89	89 ± 10	25	0.195 (338)	0.374 (648)	0.957 (12033)	50	12.5	0.2	1.7	104	Gneiss
N49	430	67.83	16.02	90 ± 8	66	0.621 (1070)	1.184 (2041)	0.957 (12033)	31	11.8	0.2	1.7	50	Granite
N57	380	67.10	16.05	114 ± 12	14	0.976 (467)	1.472 (704)	0.957 (12033)	28	–	–	–	–	Phyllite
N60	20	67.23	15.51	144 ± 17	100	0.367 (286)	0.437 (340)	0.957 (12033)	25	13.0	0.2	1.6	101	Gneiss
P2	375	67.48	18.35	243 ± 24	97	0.899 (328)	0.597 (218)	0.927 (7825)	22	13.3	0.2	1.6	100	Quartzite
SE4	2	69.24	17.98	173 ± 17	41	1.043 (731)	0.995 (697)	0.927 (14206)	30	12.0	0.1	1.5	152	Schist

m a.s.l., metres above sea level. Coordinates in decimal degrees. $P(\chi^2)$ is the chi-squared probability of the single-grain ages in per cent. ρ_s, ρ_i and ρ_d are the density of spontaneous tracks and induced tracks for the sample and for the induced tracks of the dosimeter glass, respectively. The number of actual counted fission tracks is shown in parentheses. MTL, mean track length.

Results

Apatite fission-track analysis has been performed on 43 samples. Despite the generally low uranium concentration in the samples, reflected by low ρ_i values (Table 1), counting statistics for all samples are at an acceptable level. All samples displayed $P(\chi^2) > 5\%$, indicating that the data are consistent with a single population of ages in each sample (Galbraith 1981). The standard error of the pooled age was generally lower than 10% and never more than 13%, and for all samples more than 20 grains could be counted. The low uranium concentration did, however, make it difficult to obtain enough length measurements of horizontal confined tracks for several samples. One hundred length measurements are considered necessary to obtain the statistically reliable length distribution needed for confidence in the inverse modelling of the fission-track data (Grist & Ravenhurst 1992). Six samples contained only between 50 and 100 confined tracks (N33, N39, K1, N13, N15 and N49) but nevertheless inverse modelling was also undertaken for these samples. The results are very similar to those from samples nearby that contained 100 or more confined tracks. Therefore these models are included in this paper, but should be considered supporting evidence for the nearby samples and conclusions should not be based on these samples individually. For samples that contained fewer than 50 confined tracks, only the mean track length can be used because inverse modelling would result in large time–temperature fields rather than time–temperature paths. Samples that do not have any track length information in Table 1 did not contain enough apatite to prepare a mount for the length measurements.

The oldest fission-track ages (>250 Ma) are from samples most distant from the Atlantic and southwestern Barents Sea margins. The fission-track ages in transect A–A′ range from 113 ± 14 to 322 ± 36 Ma (Fig. 3a). In transect B–B′ (Fig. 3b) the fission-track ages range from 170 ± 15 to 365 ± 34 Ma. The along-transect variation of the fission-track ages in transect A–A′ and transect B–B′ clearly indicates a decrease of the fission-track ages from SE to NW. Closer to both margins the fission-track ages decrease, but more so towards the Atlantic margin than towards the SW Barents Sea margin. As a result of complicated thermal histories, reflected by complex confined track length distributions, there is no obvious pattern for the mean track length in the study area. The highest mean track lengths were obtained from samples F28 and F48 (14.3 and 14.0 μm, respectively)

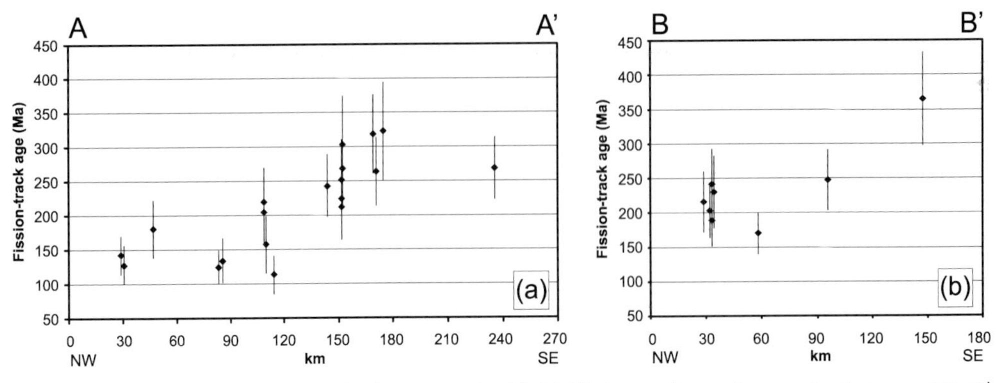

Fig. 3. (**a**) Fission-track ages for samples in transect A–A′. (**b**) Fission-track ages for samples in transect B–B′.

and these two samples are also among the oldest samples in the study area.

Figure 4 plots fission-track ages v. elevation for the vertical profiles of Tromsdalstinden (Fig. 4a), Kebnekaise (Fig. 4b) and Kiruna (Fig. 4c). All fission-track ages in Fig. 4 are displayed with 2σ errors. There is no obvious trend for the fission-track ages v. elevation in the vertical profile of Tromsdalstinden (Fig. 4a). Near-invariant apatite fission-track ages over elevation ranges of 1–2 km have been interpreted as the result of very high erosion rates, but can also result from cooling of the footwall during normal faulting (Gallagher *et al.* 1998). Because the age of most brittle faults in the area of Tromsdalstinden is unknown (Zwaan *et al.* 1998), it is difficult to determine what mechanism is responsible for the near-invariant fission-track ages of the Tromsdalstinden vertical profile. The fission-track age v. elevation plots for the vertical profiles of Kebnekaise (Fig. 4b) and Kiruna (Fig. 4c) show an increase of the apatite fission-track ages with higher elevation. Unfortunately, many of the samples from the vertical profiles of Tromsdalstinden, Kebnekaise and Kiruna yielded only small amounts of apatite that generally also were of poor quality and had a low concentration of uranium (reflected by low values for ρ_i in Table 1). Therefore the track length information from these samples is very limited, which severely reduces the amount of information that can be extracted from the vertical profiles, such as an estimate of the (palaeo)geothermal gradient.

Interpretation of fission-track data and inverse modelling

Thermal histories obtained from inverse modelling are non-unique and the uncertainties

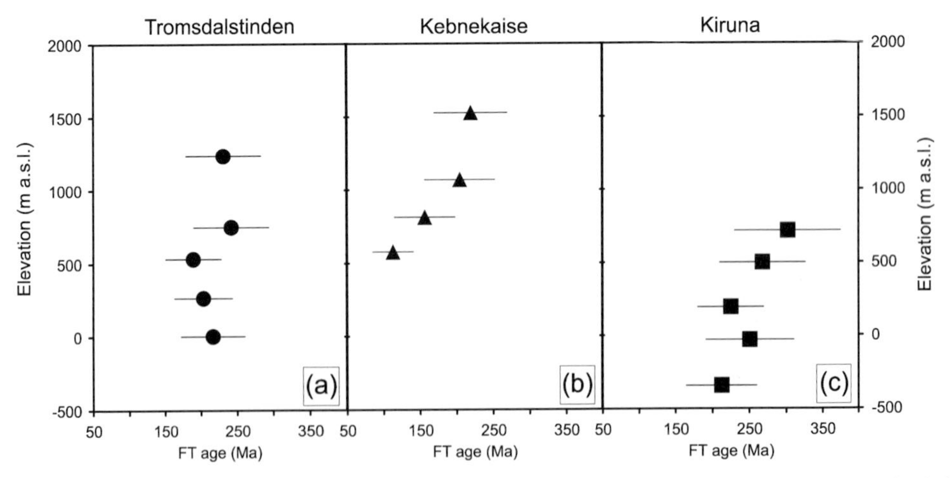

Fig. 4. Fission-track ages (2σ error) v. elevation for the vertical profiles of (**a**) Tromsdalstinden, (**b**) Kebnekaise and (**c**) Kiruna.

in the solutions for individual samples are large. However, the combination of modelled thermal histories of samples from different locations and different elevations makes it possible to reconstruct a regional denudation history with confidence. Because the annealing models used for inverse modelling of apatite fission-track data, in this study the Laslett *et al.* (1987) annealing model and the Crowley *et al.* (1991) annealing model, do not accurately mimic the behaviour of fission tracks at temperatures below 60 °C, this part of the modelled thermal history is not considered in the interpretation.

Post-orogenic cooling

Inverse modelling of samples from transects A–A′ and B–B′ indicates post-orogenic cooling progressively shifting towards the west (Figs 5 and 6). Thermal histories from the Laslett *et al.* (1987) annealing model (Figs 5b and 6b) tend toward slightly lower temperatures and indicate cooling to commence somewhat earlier than in the thermal histories from the Crowley *et al.* (1991) annealing model (Figs 5a and 6a). Taking into account, from both annealing models, the thermal histories that are supported by the data, it is clear that the easternmost samples in both transects show cooling from Devonian time onwards. The westernmost samples did not cool to temperatures within the PAZ before Triassic–early Jurassic time. This Triassic–early Jurassic timing probably is related to doming of the flanking areas of the Norwegian–Greenland rift system (Ziegler 1987). The high values for the mean track length of samples F28 and F48 (14.3 and 14.0 μm, respectively) reflect that they cooled rapidly through the PAZ in Carboniferous and Devonian time, respectively, and thereafter have not been reheated into the PAZ. This indicates rapid post-orogenic downwearing of the Caledonides, followed by a long period of relative tectonic stability of the interior of the continent.

Jurassic–Cretaceous denudation of the Northern Scandes

Samples from high elevations in the Northern Scandes mountain range (K9613, K1) recorded continuous cooling in Jurassic and Cretaceous times (Fig. 5). The Laslett *et al.* (1987) annealing model (Fig. 5b) again tends towards lower temperatures than the Crowley *et al.* (1991) annealing model (Fig. 5a), but both annealing models indicate a very similar purely cooling trend. Although sample K1 is from a lower level in the vertical profile on Kebnekaise than sample K9613, the thermal history solutions presented in Fig. 5 show that the temperature of K1 has been higher than that of sample K9613 for most of its thermal history. This does not make much sense, and shows that the modelled thermal histories of samples from which only such a small number of track length measurements could be obtained (58 confined track lengths for K1), have to be treated with caution. Besides samples K1 and K9613, samples N5, N48, N49 and N60 also recorded cooling in Cretaceous time (Figs 5 and 7). They probably experienced Jurassic cooling too, but except for sample N60, in that period they were still at temperatures too high to be constrained by apatite fission-track thermochronology. When samples N48, N49 and N60 were modelled with the Crowley *et al.* (1991) annealing model and only cooling was allowed, few thermal histories that are supported by the data were found for each of these samples. Without this restriction, the model indicated rising temperatures for late Cretaceous–Paleogene time for these samples (Fig. 7a). With the Laslett *et al.* (1987) annealing model, however, many purely cooling histories supported by the data were found (Fig. 7b). Although the annealing models do not agree on the late Cretaceous–Paleogene temperature history, both predict cooling during most of Cretaceous time for samples N48, N49 and N60.

For many of the samples to the east of the Northern Scandes, inverse modelling runs with cooling only, using the Crowley *et al.* (1991) annealing model as well as the Laslett *et al.* (1987) annealing model, generally resulted only in thermal histories that cannot be ruled out by the data. Very few thermal histories that are supported by the data were obtained in this way and for most of these samples none at all. Therefore reheating was allowed during later runs. However, it is important to keep in mind that the fission-track system is essentially one that records cooling, and that it is difficult to differentiate a thermal history that has included some reheating from a purely cooling history (Gleadow & Brown 2000).

Thermal histories from the Crowley *et al.* (1991) annealing model indicate that samples to the east of the Northern Scandes (GRM3, S16, P6, S21 and S13 in Fig. 5a; F51 in Fig. 6a; P2 in Fig. 7a), may have experienced heating during Jurassic–Cretaceous time. Also, most of these samples probably have experienced temperatures lower than 60 °C in this period. For samples GRM3 and S13, the Laslett *et al.* (1987) model also shows Jurassic–Cretaceous reheating, with a late Cretaceous peak temperature somewhat higher than 60 °C (Fig. 5b). This model also

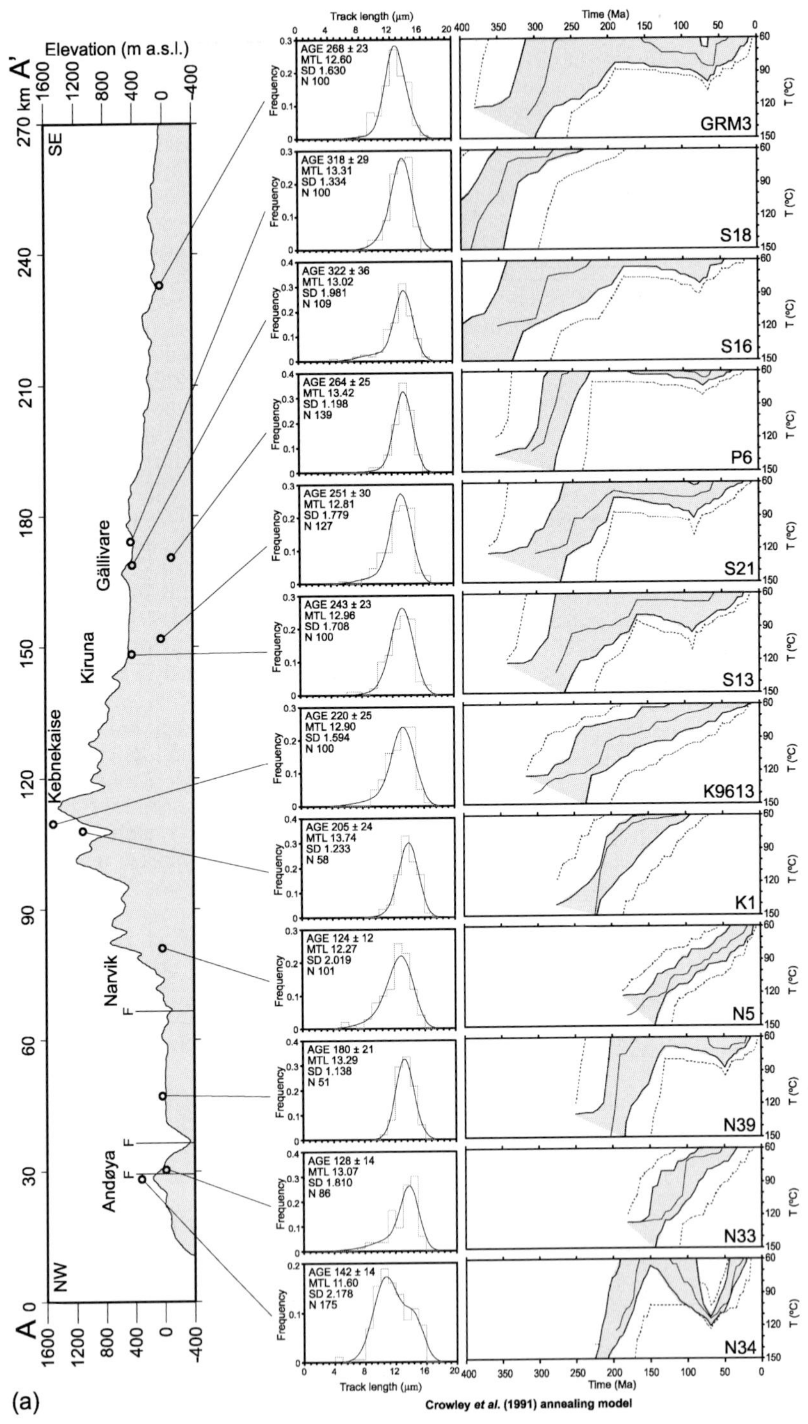

Fig. 5. Transect A–A′. Results of inverse modelling of fission-track data with (**a**) the Crowley *et al.* (1991) annealing model and (**b**) the Laslett *et al.* (1987) annealing model, together with resulting fit for track length distributions. Dashed line in diagram of thermal histories defines envelope for thermal histories with acceptable fit; grey field shows envelope for thermal histories with good fit; continuous line inside grey field indicates best-fitting thermal history. Diagram for track length distribution shows histogram of track length measurements (dotted line) and resulting fit from the best-fitting thermal history (continuous line). AGE, pooled fission-track age with 1σ error; MTL, mean track length (in μm); SD, standard deviation of mean track length (in μm); N, number of track length measurements; F, fault. Frequency: relative frequency of track lengths in 1 μm bins. Locations of samples have been projected onto transect line.

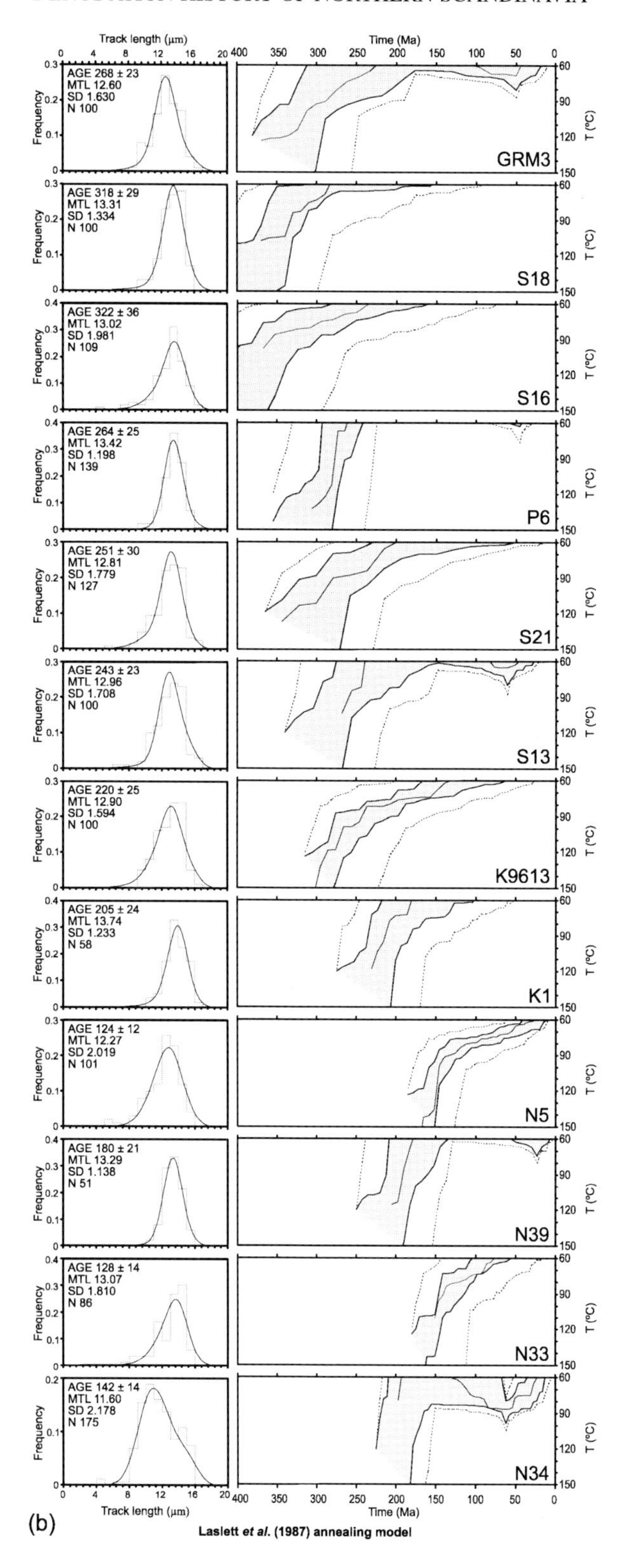

Fig. 5. *continued*

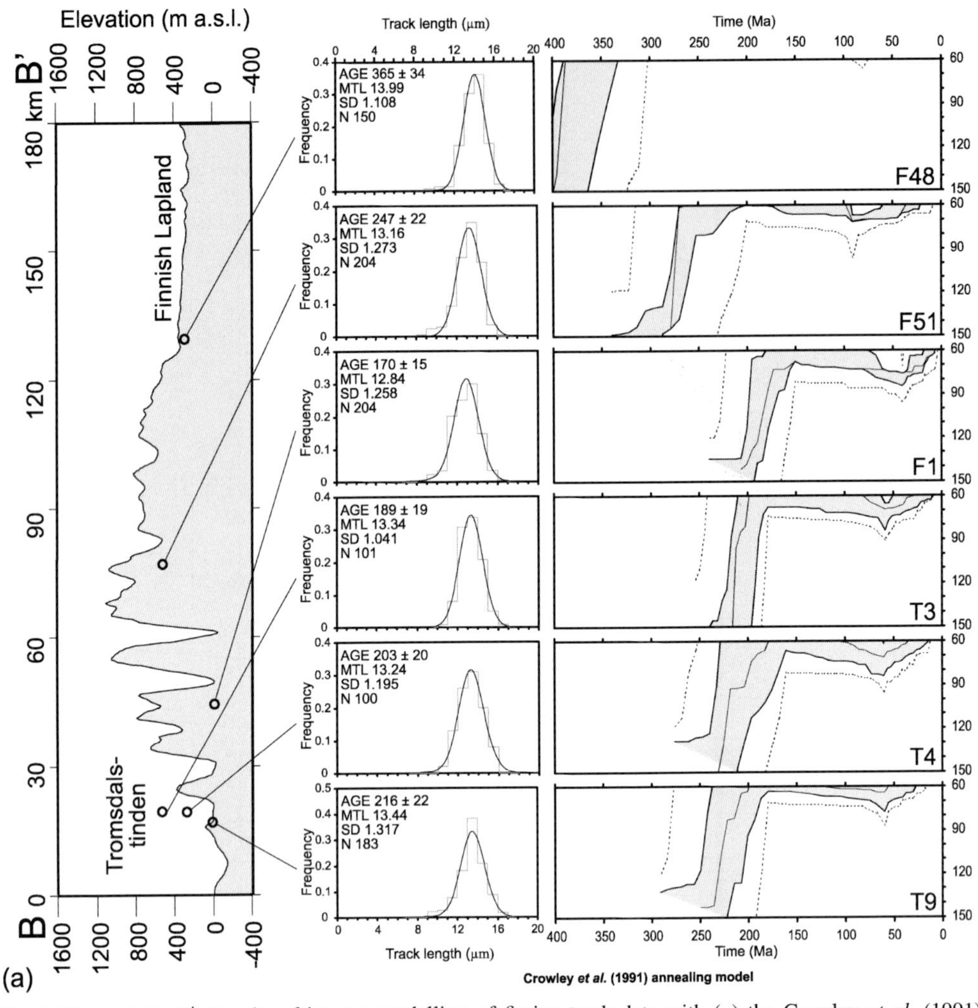

Fig. 6. Transect B–B′. Results of inverse modelling of fission-track data with (**a**) the Crowley *et al.* (1991) annealing model and (**b**) the Laslett *et al.* (1987) annealing model, together with resulting fit for track length distributions. (For other details, see caption of Fig. 5.)

indicates that all other samples to the east of the Northern Scandes were below 60 °C during the last 150 Ma. Thus together, the annealing models indicate that, during Jurassic–Cretaceous time, most of the samples to the east of the Northern Scandes experienced temperatures of around 60 °C or less. Samples GRM3 and S13 probably were still inside the PAZ in late Cretaceous time. This means they may have experienced temperatures below 60 °C during Jurassic–Cretaceous time and a late Cretaceous thermal event with a peak temperature somewhat higher than 60 °C, or that they experienced a more or less stable temperature also somewhat higher than 60 °C during all of Jurassic–Cretaceous time.

Also for samples T3, T4, T9 and F1 (Fig. 6) and for samples F16 and F37 (Fig. 7), the Crowley *et al.* (1991) and the Laslett *et al.* (1987) annealing models indicate that during Jurassic–Cretaceous time they were at temperatures of around 60 °C or less. For these samples, again, it was almost impossible, with both annealing models, to obtain purely cooling thermal histories that are supported by the data. For samples F34 and F40, in the northeastern corner of the study area, it was possible to obtain thermal histories that are supported by the data with the Crowley *et al.* (1991) annealing model when only cooling was allowed (Fig. 7a). But for these samples it was necessary to allow for

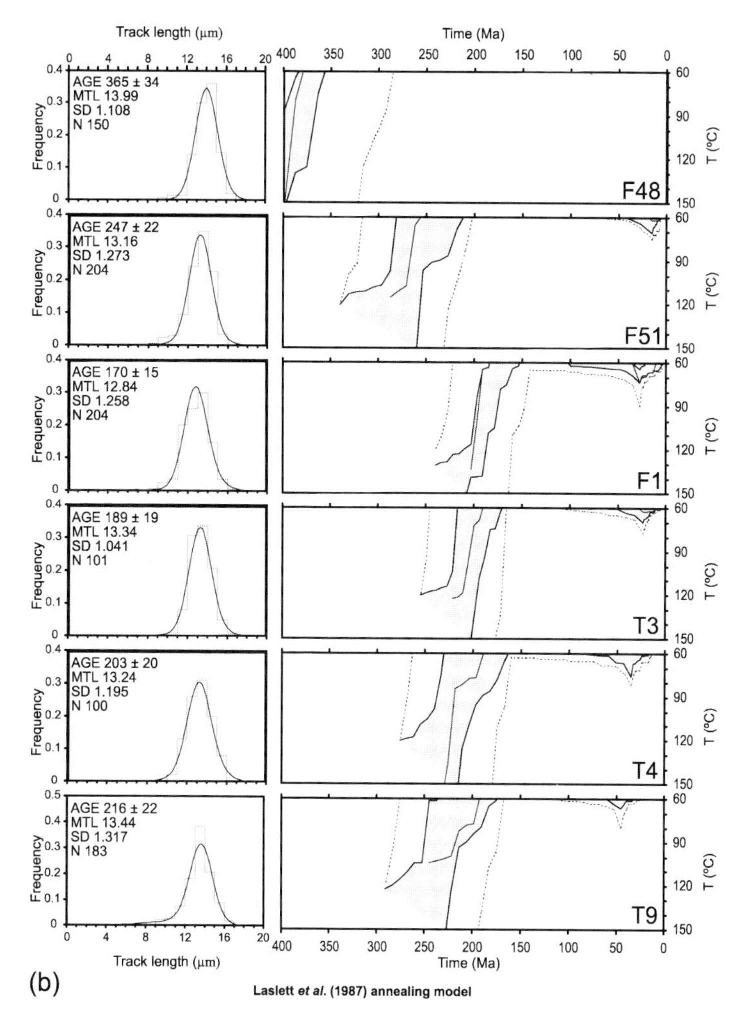

Fig. 6. *continued*

reheating to obtain thermal histories that are supported by the data with the Laslett *et al.* (1987) annealing model (Fig. 7b). However, both annealing models indicate that samples F34 and F40 experienced a temperature of around 60 °C or less, with maybe some cooling or reheating, during Jurassic–Cretaceous time. For samples SE4 and N13, also, it was difficult to obtain thermal histories that are supported by the data when only cooling was allowed. Modelled thermal histories of samples SE4 and N13 (Fig. 7), to the SW of Tromsø, are similar to those of samples from the vertical profile on Tromsdalstinden (Fig. 6). The main difference is that samples SE4 and N13 generally have a temperature somewhat higher than that of samples T3, T4 and T9. For samples SE4 and N13 both annealing models indicate temperatures within the PAZ in late Cretaceous–Paleogene time. This means that samples SE4 and N13 experienced either temperatures below 60 °C during Jurassic–Cretaceous time and a late Cretaceous–Paleogene thermal event with a peak temperature inside the PAZ, or a more or less stable temperature also within the PAZ during all of Jurassic–Cretaceous time.

Samples from high elevations in the Northern Scandes mountain range (K1 and K9613) and samples from the Atlantic margin (N5, N48, N49 and N60), recorded Jurassic–Cretaceous cooling. This could be the result of lowering of the geothermal gradient. Unfortunately, we cannot obtain any information on the geothermal gradient directly from our fission-track data.

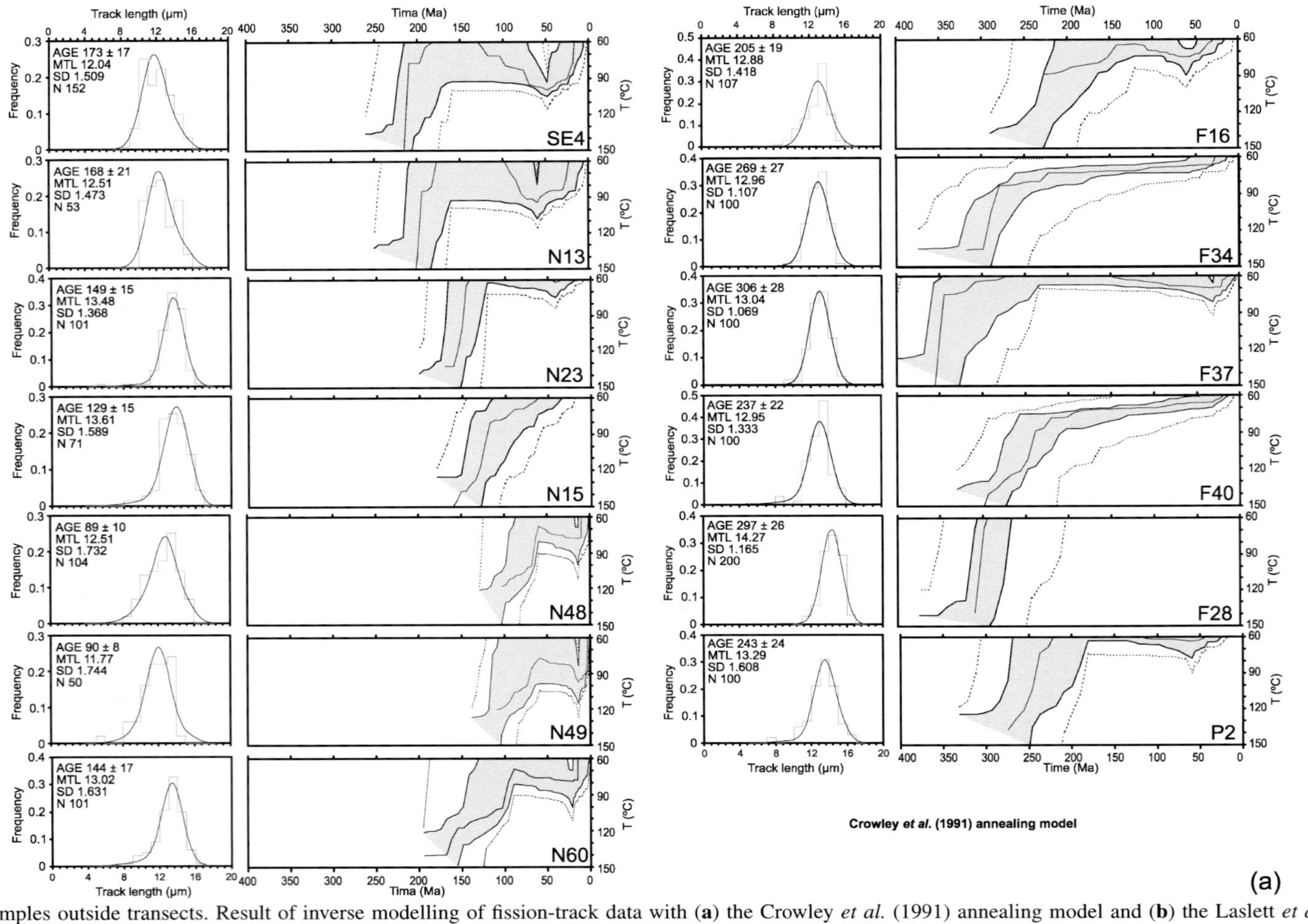

Fig. 7. Samples outside transects. Result of inverse modelling of fission-track data with (**a**) the Crowley *et al.* (1991) annealing model and (**b**) the Laslett *et al.* (1987) annealing model, together with resulting fit for track length distributions. (For other details, see caption of Fig. 5.)

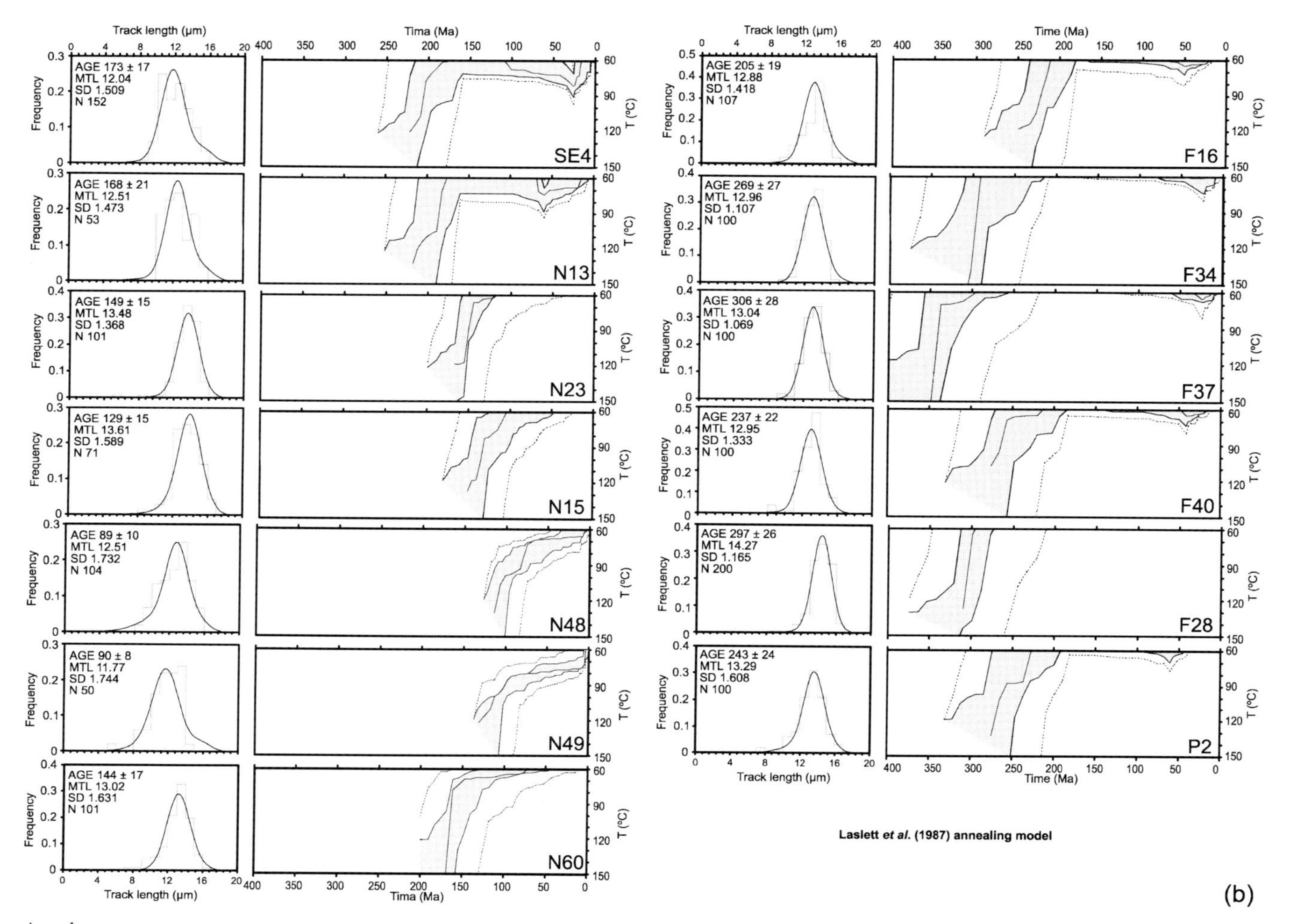

Fig. 7. *continued*

There is extension and crustal thinning offshore (Brekke 2000), but there is no indication of extension in the area where these samples were collected, and therefore one would not expect a significant lowering of the geotherm to occur. This means it is unlikely that lowering of the geotherm is the cause of the observed cooling of samples K1, K9613, N5, N48, N49 and N60. Generally, subsurface thermal effects of extension are largely restricted to the region undergoing extension, crustal thinning and subsidence (Gallagher *et al.* 1998). Away from this region, the low-temperature thermal history of rocks is primarily controlled by denudation (Gallagher & Brown 1997; Summerfield & Brown 1998). If cooling is primarily the result of denudation, it follows that the present-day area of maximum elevation and part of the Atlantic margin experienced Jurassic–Cretaceous denudation. Most of the samples from the rest of the mainland were around or below 60 °C during this period, and, at least in late Cretaceous time, some probably experienced temperatures within the PAZ. This means that in Jurassic–Cretaceous time the rest of the mainland did not experience denudation, or that following some denudation, it was affected by a transient thermal event that reached its peak temperature in late Cretaceous time. A possible transient thermal event could be the deposition of a sedimentary cover. This cover subsequently must have been removed, because there are no Jurassic–Cretaceous sediments on the northern Scandinavian mainland at present. A small amount of Jurassic–Cretaceous sediment is preserved, however, on Andøya, in the north of the Vesterålen island group (Dalland 1980).

The amount of denudation of the present-day area of maximum elevation can be calculated from the thermal history of sample K9613. This sample stayed within the PAZ for almost the entire Jurassic–Cretaceous period, and the thermal histories from the Crowley *et al.* (1991) annealing model and the Laslett *et al.* (1987) annealing model are very similar for this sample. For samples N5, N48, N49 and N60 the differences between the thermal histories from the Crowley *et al.* (1991) annealing model and the Laslett *et al.* (1987) annealing model are significant. For sample K1 only 58 contained track lengths could be measured. Therefore the amount of denudation has not been calculated from these five samples. According to the best-fitting thermal history from the Laslett *et al.* (1987) annealing model, sample K9613 experienced 20–25 °C of Jurassic–Cretaceous cooling (Fig. 5b). The best-fitting thermal history from the Crowley *et al.* (1991) annealing model indicates *c.* 30 °C of Jurassic–Cretaceous cooling (Fig. 5a). To calculate the amount of denudation, the (palaeo)geothermal gradient has to be known. Because we cannot calculate the (palaeo)geotherm from our vertical profiles, we used the Mesozoic geotherm of 28 ± 8 °C that Rohrman (1995) calculated from fission-track data from a vertical profile on Jotunheimen, southern Norway. Using this geotherm to calculate the amount of denudation of the present-day area of maximum elevation in our study area, we arrive at an estimate of 0.6–1.3 km with the Laslett *et al.* (1987) annealing model, and 0.8–1.5 km with the Crowley *et al.* (1991) annealing model. Taking into account the uncertainties from both annealing models, the Jurassic–Cretaceous denudation of the present-day area of maximum elevation is estimated to be 1 ± 0.5 km.

Tertiary denudation

To obtain detailed information about the post-Cretaceous thermal history of our samples from inverse modelling, the time-step size for the Cenozoic period was always kept small compared with that for the older part of the thermal history. This was done by increasing the number of nodal points for this time interval. In this way, there are fewer restrictions in terms of where in time a late cooling event can occur (Ketcham *et al.* 2000).

Inverse modelling with the Crowley *et al.* (1991) annealing model indicates that by the time of the Cretaceous–Tertiary transition, all samples to the east of the Northern Scandes were cooling inside the PAZ or were already below 60 °C (GRM3, S18, S16, P6, S21 and S13 in Fig. 5a; F48 and F51 in Fig. 6a; P2 in Fig. 7a). The Laslett *et al.* (1987) model indicates that only samples S13 and GRM3 were still inside the PAZ at that time, and that they were cooling (Fig. 5b). Because the Laslett *et al.* (1987) annealing model is renowned for producing anomalous late cooling events (van der Beek 1995), many scenarios for the Cenozoic thermal history of each sample have been tested. It was found that because, according to this model, only samples S13 and GRM3 were still inside the PAZ at the end of Cretaceous time, this mattered for the interpretation of these two samples only. When the time-step size was decreased only for Neogene time and not for Paleogene time, the Laslett *et al.* (1987) annealing model found many thermal histories that are supported by the data that indicate a Neogene onset for the final cooling of samples S13 and GRM3. But when the time-step size was decreased for all of Cenozoic time, the model no longer indicated a Neogene onset for the final cooling of the two samples, but tended more towards a Paleogene onset. Clearly, with the Laslett *et al.* (1987) annealing model, scenarios with a Paleogene or a Neogene onset of the final cooling phase both produce thermal histories that are supported by the data. However, one would certainly expect that the Laslett *et al.* (1987) annealing model would indicate a Neogene onset of the final cooling phase for a sample that really experienced Neogene cooling, even when the time-step size was decreased for all of Cenozoic time. This is not the case, and therefore it is considered likely that the final cooling of samples S13 and GRM3 started before Neogene time. This then means that denudation of the area to the east of the Northern Scandes probably resumed in late Cretaceous–Paleogene time. This leaves open the possibility of a Neogene accelerated cooling.

For most of the samples close to the Atlantic margin (N5 in Fig. 5; N48, N49, N13 and SE4 in Fig. 7) a Neogene onset for the final cooling phase is predicted by both annealing models, although the Crowley *et al.* (1991) model also allows for an earlier cooling of samples N13 and SE4. For sample N60 (Fig. 7a) the Crowley *et al.* (1991) model also predicts a Neogene onset of cooling, whereas the Laslett *et al.* (1987) model indicates a sub-PAZ temperature for this sample already in Paleogene time. For samples close to the Barents Sea margin (F1, T3, T4 and T9 in Fig. 6; F16, F34, F37 and F40 in Fig. 7) the Laslett *et al.* (1987) model indicates that these samples were below 60 °C already in Paleogene time, or that their final cooling out of the PAZ started in Neogene time. The Crowley *et al.* (1991) annealing model predicts a Neogene onset of cooling for samples F1 (Fig. 6a) and F34, F37 and F40 (Fig. 7a) as well. For samples T3, T4 and T9 (Fig. 6a) the Crowley *et al.* (1991) annealing model seems to indicate a Paleogene onset of cooling, but a Neogene onset of cooling is certainly possible too, within the limits set by thermal histories that are supported by the data.

Therefore for samples on the Atlantic margin as well as on the Barents Sea margin, the Laslett *et al.* (1987) annealing model predicts a Neogene onset of the final cooling phase, or indicates that they were already below 60 °C. The Crowley *et al.* (1991) annealing model indicates a Neogene phase of final cooling for the Atlantic margin and for the part of the Barents Sea margin in the northeastern corner of the study area as well. Although it seems to indicate a Paleogene onset of cooling for the part of the mainland adjacent to where the Atlantic and Barents Sea margins come together, it certainly does not exclude the possibility of a Neogene onset of the final cooling phase for this area. Assuming once again that cooling is the result of denudation (Summerfield & Brown 1998), it follows that the Atlantic and the Barents Sea margins probably experienced significant Neogene denudation. In the part of the mainland adjacent to where the Atlantic and Barents Sea margins meet, denudation may have started earlier.

It should be noted that almost all samples from the margins are taken inside fjords, which may be incised as deep as 2 km. Fjords are probably not simply glacial features, but remnants of an older fluvial drainage system. Consecutive glaciations preferentially exploited the pre-existing valleys (Lidmar-Bergström *et al.*2000). It is not immediately clear what the coastal samples actually record: regional denudation, or rapid fluvial incision and glacial erosion that could locally disturb the thermal structure of the crust (Stüwe *et al.* 1994).

Lofoten and Vesterålen

Samples from the Lofoten and Vesterålen islands have thermal histories very different from those of most other samples (Figs 5 and 7). Structurally this area is very complex, with many undated faults and probably many more brittle structures than are currently known. These islands therefore are now being studied in a separate project, which aims at resolving the timing of faulting and denudation of the various structural blocks in this area with fission-track analysis and (U–Th)/He thermochronometry. However, the few available fission-track data (samples N15, N23, N33, N34 and N39) suggest that this area experienced considerable Neogene denudation, supporting the observations of Riis (1996).

Pattern of denudation

Samples in the northeastern corner of the study area, close to the southern margin of the Barents Sea, have fission-track ages in the range of 200–300 Ma. These ages correspond to those of samples from Lapland, east of the Northern Scandes range, rather than to ages from samples close to the Atlantic margin, which are of the order of 90–150 Ma. The inverse models and the pattern of fission-track ages (Fig. 3) indicate that the greatest Mesozoic and Cenozoic cooling was experienced by the part of the study area closest to the Atlantic margin. It is evident that the denudation of the Northern Scandes mountain range did not produce a domal pattern similar to the one recognized in the topography and apatite fission-track data of the Southern Scandes (Rohrman 1995). Instead, the Atlantic margin has experienced much more Mesozoic and Cenozoic denudation than the present-day area of maximum elevation, which in turn was more affected than the interior of the continent.

Correlation with offshore geology

Because provenance is unclear for most of the sediments in basins surrounding the study area, it is difficult to study the relationship between the evolution of the onshore part of the study area and the adjacent offshore area. The complex structural evolution of the offshore areas also obscures the direct connection between on- and offshore. A further complicating factor is the limited time-resolution of the inverse models from apatite fission-track data, compared with the detailed stratigraphical information that is available for most basins. However, despite all this, there is a clear link in space and time between the off- and onshore geological evolution. Rapid Triassic–Jurassic denudation of samples in the Tromsø region (T3, T4, T9 and F1) was contemporaneous with deposition of Triassic–Jurassic sequences (Faleide *et al.* 1993; Reemst 1995) in the nearby Tromsø and Hammerfest basins (Fig. 2). The Cretaceous sequences in these basins covered at least part of the Finnmark platform as well (Faleide *et al.* 1993). Cretaceous deposits are now missing from the Finnmark platform and the mainland, and have been tilted away from the mainland in the basins further north and west. The inferred Cenozoic denudation of samples from the Tromsø region then implies a hinge zone close to the present-day coastline. The Vestfjorden, Ribban, Røst and Harstad basins are located offshore the Atlantic margin of the study area (Fig. 2). All four basins contain thick Cretaceous deposits that are underlain by thinner Jurassic sediments (Faleide *et al.* 1993; Olesen *et al.* 1997; Brekke 2000). Late Jurassic–early Cretaceous sediments are also preserved on Andøya, Vesterålen (Dalland 1980). These sediments thus were deposited at a time of inferred strong denudation of the present-day area of maximum elevation and the Atlantic margin, which indicates they may have been derived from this region of Jurassic–Cretaceous denudation. Except for the Harstad basin, the basins offshore the Atlantic margin of the study area contain relatively small amounts of Tertiary sediments compared with the thick Cretaceous sequences. Thin Paleocene sequences are present in the Vestfjorden, Ribban and Røst basins, which are unconformably overlain by Plio-Pleistocene sediments (Brekke 2000). This reflects the important role Neogene denudation and glacial erosion played in this region by removing earlier Tertiary deposits, but makes it very difficult to study the correlation with the Tertiary geological evolution onshore.

Discussion

Rohrman & van der Beek (1996) explained the domal style of late-stage postrift uplift of the Southern Scandes with an asthenospheric diapir model wherein a hot 'Icelandic' asthenosphere layer meets cold cratonic lithosphere. This mechanism would explain more or less simultaneous domal-style uplift of the Southern Scandes, Northern Scandes and also Svalbard. However, a comparison of the denudation history reconstructed on the basis of apatite fission-track data for the Southern Scandes (Rohrman 1995) range with that of the Northern Scandes (this study) makes clear that there are important differences between the two in both the pattern and timing of denudation. Inverse modelling of apatite fission-track data from the Southern Scandes suggests two distinct phases of denudation (Rohrman 1995). Triassic–Jurassic denudation of *c.* 1.3–3.5 km, probably as a consequence of base-level lowering and rift flank uplift, was followed by much slower denudation rates during Cretaceous–Paleogene time. Rapid denudation caused by tectonic uplift started from *c.* 30 Ma onward and produced the domal pattern recognized in the topography and apatite fission-track ages. Therefore, there are similarities in timing of denudation for the Southern and Northern Scandes until late Cretaceous time, but certainly the pattern of denudation and probably also the timing of denudation in Cenozoic time are different for the two ranges.

Although denudation is not necessarily connected to any tectonically driven surface uplift (Summerfield & Brown 1998), it is tempting to explain the results of the fission-track data from the study area by passive margin uplift followed by scarp retreat. This type of model predicts the maximum amount of denudation near the margin and moderate to low amounts of denudation in the interior; this characteristic produces a strong gradient in apatite fission-track age, with the oldest ages occurring in the interior and decreasing towards the coast (Gallagher & Brown 1997; Gallagher *et al.* 1998). Rift shoulders are thus often characterized by a marked large-scale asymmetry (Beaumont *et al.* 2000). The study area clearly displays all these characteristics. Another important prediction of a scarp retreat model is that the timing of maximum denudation decreases inland from the coast toward the final position of the escarpment

(Gallagher & Brown 1999). Nowhere on the Atlantic margin in the study area is the sampling density high enough to convincingly prove that the margin displays this characteristic. Sample N57 is from *c*. 25 km further from the margin, and also from 360 m higher than sample N60. Indeed, the fission-track age of sample N57 is 30 Ma younger than that of sample N60. But of course these are only two samples, and this is also the only place on the margin where we can test this prediction of the scarp retreat model. Passive margins are also generally associated with thick synrift sediments and a postrift unconformity (Braun & Beaumont 1989). Although the Norwegian shelf is a very complex system of basins and highs and this complexity makes it difficult to recognize some of these characteristics readily, it clearly displays these features (Brekke & Riis 1987; Reemst 1995; Brekke 2000). It appears therefore that rift flank uplift resulting from passive margin development followed by scarp retreat can explain the reconstructed denudation history of the Northern Scandes range. This certainly is not the case for the peculiar denudation history of the Southern Scandes (Rohrman 1995).

The atypical thermotectonic development of the Southern Scandes has implications for the accuracy of the estimate of the amount of Jurassic–Cretaceous denudation of the Northern Scandes. The Mesozoic geotherm of $28 \pm 8\,^{\circ}C\,km^{-1}$ (Rohrman 1995) from the Southern Scandes region has been used in this calculation and seems to be high compared with the present-day geotherm of less than $20\,^{\circ}C\,km^{-1}$ on the Scandinavian mainland (Balling 1990). The use of too high a value for the geotherm will lead to underestimation of the amount of denudation. The estimate given in this paper for the amount of Jurassic–Cretaceous denudation of the Northern Scandes could therefore be too low.

Summary and conclusions

Post-Caledonian cooling and denudation in northern Scandinavia progressively shifted from the interior of the continent towards the North Atlantic margin. In the Northern Scandes mountain range, the present-day area of maximum elevation has experienced continuous cooling and denudation at least since Jurassic time. The combined Jurassic–Cretaceous denudation of this region was 1 ± 0.5 km. Except for most of the North Atlantic margin, the northern Scandinavian mainland either was not affected by denudation in Jurassic–Cretaceous time, or experienced some denudation followed by a transient thermal event that reached its peak temperature in late Cretaceous time. This could indicate the occurrence of a Jurassic–Cretaceous sedimentary cover on the mainland. It subsequently must have been removed, because there are no Jurassic–Cretaceous sediments preserved on the northern Scandinavian mainland. Final denudation of the eastern flank of the Northern Scandes had probably already started in late Cretaceous–Paleogene time and possibly accelerated in Neogene time. The fission-track record of the Atlantic and Barents Sea margins indicates a Neogene onset for the final phase of cooling and denudation, although denudation may already have affected the area where the two margins meet before that time. The combined Mesozoic–Cenozoic denudation was strongest on the Atlantic margin. The interior of the continent experienced the least Mesozoic–Cenozoic denudation. The pattern and timing of denudation of the Northern Scandes mountain range is different from that of the Southern Scandes range, which experienced domal-style late-stage postrift uplift in Neogene time. The reconstructed denudation history of northern Scandinavia can be explained by scarp retreat of an uplifted rift flank.

This research is funded by Norsk Hydro, Statoil and the Norwegian Petroleum Directorate. It is part of the NSG and ISES research schools and has been performed in the Department of Isotope Geochemistry of the Vrije Universiteit Amsterdam. We thank P. Green and P. Bishop for comments and critical review of the manuscript. We acknowledge the ECN at Petten, The Netherlands, for irradiating the fission-track samples. We are grateful to LKAB Kiruna and LKAB Malmberget for providing deep samples from their mines, and to O. Svenningsen for providing us with samples from Kebnekaise. This paper is Publication 20010501 of the Netherlands Research School of Sedimentary Geology.

References

ANDRIESSEN, P.A.M. & BOS, A. 1986. Post-Caledonian thermal evolution and crustal uplift in the Eidfjord area, western Norway. *Norsk Geologisk Tidsskrift*, **66**, 243–250.

BALLING, N. 1990. Heatflow and lithospheric temperature along the northern segment of the European geotraverse. *In*: FREEMAN, R. & MUELLER, S. (eds) *Proceedings of the 6th EGT Workshop*. European Science Foundation, Strasbourg, 405–416.

BEAUMONT, C., KOOI, H. & WILLETT, S. 2000. Coupled tectonic–surface process models with applications to rifted margins and collisional orogens. *In*: SUMMERFIELD, M.A. (ed.) *Geomorphology and Global Tectonics*. Wiley, Chichester, 29–55.

BLYTHE, A.E. & KLEINSPEHN, K.L. 1998. Tectonically versus climatically driven Cenozoic exhumation of the Eurasian plate margin, Svalbard: fission track analysis. *Tectonics*, **17** (4), 621–639.

BRAUN, J. & BEAUMONT, C. 1989. A physical explanation of the relation between flank uplifts and the breakup unconformity at rifted continental margins. *Geology*, **17**, 760–764.

BREKKE, H. 2000. The tectonic evolution of the Norwegian Sea continental margin with emphasis on the Vøring and Møre Basins. *In*: NØTTVEDT, A. (ed.) *Dynamics of the Norwegian Margin*. Geological Society, London, Special Publications, **167**, 327–378.

BREKKE, H. & RIIS, F. 1987. Tectonics and basin evolution of the Norwegian shelf between 62°N and 72°N. *Norsk Geologisk Tidsskrift*, **67**, 295–321.

CARLSON, W.D., DONELICK, R.A. & KETCHAM, R.A. 1999. Variability of apatite fission-track annealing kinetics: I. Experimental results. *American Mineralogist*, **84**, 1213–1223.

CEDERBOM, C., LARSON, S.É., TULLBORG, E.-L. & STIBERG, J.-P. 2000. Fission track thermochronology applied to Phanerozoic thermotectonic events in central and southern Sweden. *Tectonophysics*, **316**, 153–167.

CORRIGAN, J.D. 1991. Inversion of apatite fission track data for thermal history information. *Journal of Geophysical Research*, **96**, 10347–10360.

CROWLEY, K.D., CAMERON, M. & SCHAEFFER, R.L. 1991. Experimental studies of annealing of etched fission tracks in fluorapatite. *Geochimica et Cosmochimica Acta*, **55**, 1449–1465.

DALLAND, A. 1980. Mesozoic sedimentary succession at Andøy, Northern Norway, and relation to structural development of the North Atlantic area. *In*: KERR, J.W., FERGUSSON, A.J. & MACHAN, L.C. (eds) *Geology of the North Atlantic Borderlands*. Canadian Society of Petroleum Geologists, Memoirs, **7**, 563–584.

DEHLS, J.F., OLESEN, O., BUNGUM, H., HICKS, E.C., LINDHOLM, C.D. & RIIS, F. 2000. *Neotectonic Map: Norway and Adjacent Areas*. Geological Survey of Norway, Trondheim.

DIMAKIS, P., BRAATHEN, B.I., FALEIDE, J.I., ELVERHØI, A. & GUDLAUGSSON, S.T. 1998. Cenozoic erosion and the preglacial uplift of the Svalbard–Barents Sea region. *Tectonophysics*, **300**, 311–327.

DONELICK, R.A., RODEN, M.K., MOOERS, J.D., CARPENTER, B.S. & MILLER, D.S. 1990. Etchable track length reduction of induced fission tracks in apatite at room temperature (~23 °C): crystallographic orientation effects and 'initial' mean lengths. *Nuclear Tracks and Radiation Measurements*, **17**, 261–265.

DORÉ, A.G. 1991. The structural foundation and evolution of Mesozoic seaways between Europe and the Arctic. *Palaeogeography, Palaeoclimatology, Palaeoecology*, **87**, 441–492.

DORÉ, A.G. & GAGE, M.S. 1987. Crustal alignments and sedimentary domains in the evolution of the North Sea, North-east Atlantic Margin and Barents Shelf. *In*: BROOKS, J. & GLENNIE, K. (eds) *Petroleum Geology of North West Europe*. Graham & Trotman, London, 1131–1148.

FALEIDE, J.I., VÍGNES, E. & GUDLAUGSSON, S.T. 1993. Late Mesozoic–Cenozoic evolution of the southwestern Barents Sea in a regional rift–shear tectonic setting. *Marine and Petroleum Geology*, **10**, 186–214.

FLEISCHER, R.L., PRICE, P.B. & WALKER, R.M. 1975. *Nuclear Tracks in Solids: Principles and Applications*. University of California Press, Berkeley, CA.

GALBRAITH, R.F. 1981. On statistical models for fission track counts. *Mathematical Geology*, **13** (6), 471–478.

GALLAGHER, K. 1995. Evolving temperature histories from apatite fission-track data. *Earth and Planetary Science Letters*, **136**, 421–435.

GALLAGHER, K. & BROWN, R. 1997. The onshore record of passive margin evolution. *Journal of the Geological Society, London*, **154**, 451–457.

GALLAGHER, K. & BROWN, R. 1999. Denudation and uplift at passive margins: the record on the Atlantic Margin of southern Africa. *Philosophical Transactions of the Royal Society of London, Part A*, **357**, 835–859.

GALLAGHER, K., BROWN, R. & JOHNSON, C. 1998. Fission track analysis and its applications to geological problems. *Annual Review of Earth and Planetary Sciences*, **26**, 519–572.

GEOLOGICAL SURVEYS OF FINLAND, NORWAY AND SWEDEN. 1987. *Geological Map of Northern Fennoscandia*, 1:1 000 000. Geological Surveys of Finland, Norway and Sweden, Helsinki.

GLEADOW, A.J.W. & DUDDY, I.R. 1981. A natural long-term track annealing experiment for apatite. *Nuclear Tracks*, **5**, 169–174.

GLEADOW, A.J.W. & BROWN, R.W. 2000. Fission-track thermochronology and the long-term denudational response to tectonics. *In*: SUMMERFIELD, M.A. (ed.) *Geomorphology and Global Tectonics*. Wiley, Chichester, 57–75.

GLEADOW, A.J.W., DUDDY, I.R., GREEN, P.F. & HEGARTY, K.A. 1986*a*. Fission track lengths in the apatite annealing zone and the interpretation of mixed ages. *Earth and Planetary Science Letters*, **78**, 245–254.

GLEADOW, A.J.W., DUDDY, I.R., GREEN, P.F. & LOVERING, J.F. 1986*b*. Confined fission track lengths in apatite: a diagnostic tool for thermal history analysis. *Contributions to Mineralogy and Petrology*, **94**, 405–415.

GRIST, A.M. & RAVENHURST, C.E. 1992. A step-by-step laboratory guide to fission track thermochronology at Dalhousie University. *In*: ZENTILLI, M. & REYNOLDS, P.H. (eds) *Low Temperature Thermochronology*. Short Course Handbook, Mineralogical Association of Canada, **20**, 189–201.

GUDMUNDSSON, A. 1999. Postglacial crustal doming, stresses and fracture formation with application to Norway. *Tectonophysics*, **307**, 407–419.

HANSEN, K., PEDERSEN, S., FOUGT, H. & STOCKMARR, P. 1996. Post-Sveconorwegian denudation and cooling history of Evje area, southern Setesdal,

Central South Norway. *Norges Geologiske Undersøkelse Bulletin*, **431**, 49–58.

HURFORD, A.J. & GREEN, P.F. 1983. The zeta age calibration of fission-track dating. *Isotope Geoscience*, **1**, 285–317.

KETCHAM, R.A., DONELICK, R.A. & CARLSON, W.D. 1999. Variability of apatite fission-track annealing kinetics: III. Extrapolation to geological time scales. *American Mineralogist*, **84**, 1235–1255.

KETCHAM, R.A., DONELICK, R.A. & DONELICK, M.B. 2000. AFTSolve: a program for multi-kinetic modeling of apatite fission-track data. *Geological Materials Research*, **2**, 1–32.

LARSON, S.É., TULLBORG, E.-L., CEDERBOM, C. & STIBERG, J.-P. 1999. Sveconorwegian and Caledonian foreland basins in the Baltic Shield revealed by fission-track thermochronology. *Terra Nova*, **11**, 210–215.

LASLETT, G.M., GREEN, P.F., DUDDY, I.R. & GLEADOW, A.J.W. 1987. Thermal annealing of fission tracks in apatite 2. A quantitative analysis. *Chemical Geology*, **65**, 1–13.

LEHTOVAARA, J. 1976. Apatite fission track dating of Finnish Precambrian intrusives. *Annales Academiae Scientiarum Fennicae*, **117**, 1–94.

LIDMAR-BERGSTRÖM, K. 1993. Denudation surfaces and tectonics in the southernmost part of the Baltic Shield. *Precambrian Research*, **64**, 337–345.

LIDMAR-BERGSTRÖM, K. 1999. Uplift histories revealed by landforms of the Scandinavian domes. *In*: SMITH, B.J., WHALLEY, W.B. & WARKE, P.A. (eds) *Uplift, Erosion and Stability: Perspectives on Long-term Landscape Development*. Geological Society, London, Special Publications, **162**, 85–91.

LIDMAR-BERGSTRÖM, K., OLLIER, C.D. & SULEBAK, J.R. 2000. Landforms and uplift history of southern Norway. *Global and Planetary Change*, **24**, 211–231.

LUTZ, T.M. & OMAR, G. 1991. An inverse method of modeling thermal histories from apatite fission-track data. *Earth and Planetary Science Letters*, **104**, 181–195.

NAESER, C.W. 1981. The fading of fission tracks in the geologic environment—data from deep drill holes. *Nuclear Tracks*, **5**, 248–250.

OLESEN, O., TORSVIK, T.H., TVETEN, E., ZWAAN, K.B., LØSETH, H. & HENNINGSEN, T. 1997. Basement structure of the continental margin in the Lofoten–Lopphavet area, northern Norway: constraints from potential field data, on-land structural mapping and palaeomagnetic data. *Norsk Geologisk Tidsskrift*, **77**, 15–30.

REEMST, P. 1995. *Tectonic modelling of rifted continental margins—basin evolution and tectono-magmatic development of the Norwegian and NW Australian margin*. PhD thesis, Vrije Universiteit Amsterdam.

RIIS, F. 1996. Quantification of Cenozoic vertical movements of Scandinavia by correlation of morphological surfaces with offshore data. *Global and Planetary Change*, **12**, 331–357.

ROBERTS, D. & GEE, D.G. 1985. An introduction to the structure of the Scandinavian Caledonides. *In*: GEE, D.G. & STURT, B.A. (eds) *The Caledonide Orogen—Scandinavia and Related Areas*. Wiley, Chichester, 55–68.

ROHRMAN, M. 1995. *Thermal evolution of the Fennoscandian region from fission track thermochronology—an integrated approach*. PhD thesis, Vrije Universiteit Amsterdam.

ROHRMAN, M. & VAN DER BEEK, P. 1996. Cenozoic postrift domal uplift of North Atlantic margins: an asthenospheric diapirism model. *Geology*, **24**, 901–904.

SRIVASTAVA, S.P. & TAPSCOTT, C.R. 1986. Plate kinematics of the North Atlantic. *In*: VOGT, P.R. & TUCHOLKE, B.E. (eds) *The Western North Atlantic Region, M*. Geological Society of America, Boulder, CO, 379–404.

STUEVOLD, L.M. & ELDHOLM, O. 1996. Cenozoic uplift of Fennoscandia inferred from a study of the mid-Norwegian margin. *Global and Planetary Change*, **12**, 359–386.

STÜWE, K., WHITE, L. & BROWN, R. 1994. The influence of eroding topography on steady-state isotherms. Application to fission track analysis. *Earth and Planetary Science Letters*, **124**, 63–74.

SUMMERFIELD, M.A. & BROWN, R.W. 1998. Geomorphic factors in the interpretation of fission-track data. *In*: VAN DEN HAUTE, P. & DE CORTE, F. (eds) *Advances in Fission-Track Geochronology*. Kluwer, Dordrecht, 269–284.

VAN DER BEEK, P.A. 1995. *Tectonic evolution of continental rifts—inferences from numerical modelling and fission track thermochronology*. PhD thesis, Vrije Universiteit Amsterdam.

WAGNER, G.A. 1968. Fission track dating of apatites. *Earth and Planetary Science Letters*, **4**, 114–145.

WAGNER, G.A. & VAN DEN HAUTE, P. 1992. *Fission-Track Dating*. Kluwer, Dordrecht.

ZECK, H.P., ANDRIESSEN, P.A.M., HANSEN, K., JENSEN, P.K. & RASMUSSEN, B.L. 1988. Paleozoic paleo-cover of the southern part of the Fennoscandian shield—fission track constraints. *Tectonophysics*, **149**, 61–66.

ZIEGLER, P. 1987. Evolution of the Arctic—North Atlantic Borderlands. *In*: BROOKS, J. & GLENNIE, K. (eds) *Petroleum Geology of North West Europe*. Graham & Trotman, London, 1201–1204.

ZWAAN, K.B., FARETH, E. and GROGAN, P.W. 1998. *Geologisk kart over Norge, berggrunnskart Tromsø, M 1:250 000*. Norwegian Geological Survey, Trondheim.

Along-slope variation in the late Neogene evolution of the mid-Norwegian margin in response to uplift and tectonism

D. EVANS[1], S. McGIVERON[2], Z. HARRISON[2], P. BRYN[3] & K. BERG[3]
[1]*British Geological Survey, Murchison House, West Mains Road, Edinburgh EH9 3LA, UK*
[2]*Svitzer Ltd, Morton Peto Road, Great Yarmouth NR31 0LT, UK*
[3]*Norsk Hydro ASA, PO Box 200, N-1321, Stabekk, Norway*

Abstract: As part of the Norwegian Deep Water Programme, a regional geological and geophysical interpretation of the mid-Norwegian margin resulted in the establishment of a late Paleogene to Holocene stratigraphic framework for the margin, and identification and mapping of a range of possible geohazards, including slides. At the Vøring margin in the north, there has been the build-out of a huge prograding wedge of sediment in Plio-Pleistocene times. The sediments of the wedge are assigned to the Naust Formation, which has been subdivided into eight units (A–H), and includes only a few palaeoslides. To the south of the Vøring margin lies the Storegga Slide Complex and the North Sea Fan, and the whole of this southern region shows evidence of several major palaeoslides. The sediments are also referred to the Naust Formation, which here are subdivided into nine units (O–W) that have been partly correlated with equivalent Naust units to the north. The oldest Naust unit in the south, Naust Unit W, is largely made up of slide deposits that provide evidence for the earliest identified large-scale slope instability in the Storegga Slide Complex. This instability was penecontemporaneous with the initiation of the prograding wedge to the north, and both features are postulated to be the result of uplift of the Norwegian mainland. In the case of the Storegga Slide Complex, which lies close to the main area of uplift, oversteepening of the margin, together with seismicity associated with the Jan Mayen Fracture Zone and Møre–Trønderlag Fault Complex, may have initiated sliding that has since occurred intermittently up to Holocene times, over a time interval when there has additionally been much glacio-isostatic movement.

In the south, the shelf off western Norway is narrow, but it widens considerable to the north in the Haltenbanken and beyond the Traenadjupet (Fig. 1). In the south the limit of the shelf north of the North Sea Fan is marked by the scarp, of 290 km length (Bugge *et al.* 1987), of the Storegga Slide, whereas in the north a more gentle slope leads down to the Vøring Basin, beyond which it rises slightly at the Vøring Plateau.

The mid-Norwegian margin became an important area for hydrocarbon exploration following the deep-water 15th Norwegian round of licensing in 1996. To gain a better understanding of the shallow geology and slope stability of the area, several companies jointly formed the Seabed Project, which was a component of the Norwegian Deep Water Programme (Bryn *et al.* 1998). This project was managed by Norsk Hydro, and one aspect of the project was geological and geophysical interpretation, which was carried out in three phases. The second and third phases were contracted to Svitzer Ltd between 1997 and 1999, with the British Geological Survey acting as consultants. An understanding of margin development and slope stability in the area is very important to the hydrocarbon industry as deep-water exploration continues and development of major new fields begins.

The summarized results of Phase II of the project were published by McNeill *et al.* (1998), who described the Cenozoic stratigraphic framework and discussed geohazards relevant to the oil industry. Bryn *et al.* (1998) concentrated on issues of slope stability. This assessment was updated during Phase III of the project, for which many additional data were available. The complete database included 10 000 km of new and reprocessed seismic and high-resolution seismic, mini-airgun, deep-tow boomer, 3D seismic data in licence blocks (Fig. 1), towed ocean bottom instrument (TOBI) sidescan sonar data, three Ocean Drilling Program (ODP) or Deep Sea Drilling Project (DSDP) sites, three exploration wells and four geotechnical boreholes.

From: DORÉ, A.G., CARTWRIGHT, J.A., STOKER, M.S., TURNER, J.P. & WHITE, N. 2002. *Exhumation of the North Atlantic Margin: Timing, Mechanisms and Implications for Petroleum Exploration*. Geological Society, London, Special Publications, **196**, 139–151. 0305-8719/02/$15.00

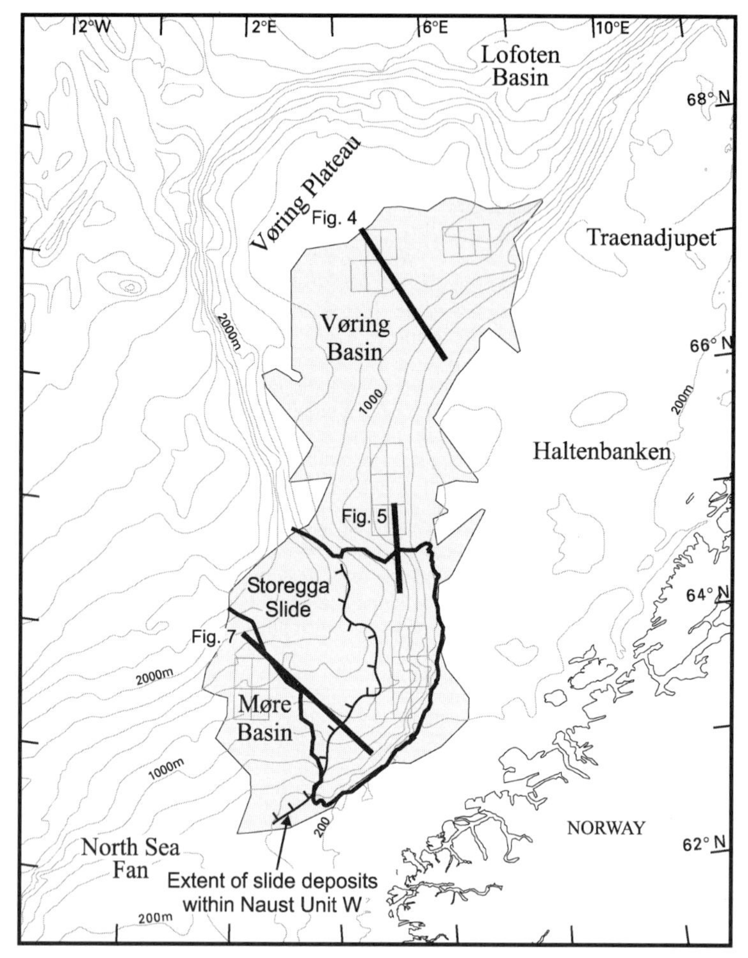

Fig. 1. The bathymetry (in metres) and location of the study area in relation to the mid-Norwegian shelf and the Storegga Slide. Also shown are the 15th Round licence blocks (see also Fig. 6), the location of seismic sections, and the landward limit of the slide deposits that form part of Naust Unit W.

This was a regional interpretation in which only the outermost portion of the shelf was included in the area of study. The area of study covers over 100 000 km^2, and extends along the continental slope from 62°15′N to 68°N. The westward extent of the area is approximately at the Vøring Escarpment in the north in water depth of about 1200–1300 m, whereas in the south it extends to depths of 2500 m within the Storegga Slide. It is important to note that in this paper the term 'Storegga Slide' refers only to the most recent movement on the slide as described by Bugge (1983) and Bugge *et al.* (1987, 1988), whereas the term 'Storegga Slide Complex' is used to describe an area of longer-term instability centred around the Storegga Slide. It must also be emphasized that this paper is based on the results of the Seabed Project up to 1999; many commercial data have been collected since that time, including 3D seismic data, but the newer data are not used here.

Although the project considered the whole post-Paleocene succession, this paper will focus on the Plio-Pleistocene history, with particular emphasis on the late Pliocene to early Pleistocene sedimentary response. This includes, in the northern part of the area, the build-out of a major prograding wedge, which is comparable with similar features observed on glaciated margins world-wide (e.g. Larter & Barker 1991; Clausen 1998; Kristofferson *et al.* 2000) and has commonly been described as a response to uplift of Scandinavia (Poole & Vorren 1993; Hendriksen & Vorren 1996; Riis 1996; Stuevold & Eldholm 1996). However, a marked along-slope contrast between the north and south of the study area will be described, and some possible reasons for these differences discussed.

Stratigraphic framework

The stratigraphic framework established for the area is based on seismic interpretation, age information from the project database, and published information. The framework illustrated in Fig. 2 is adapted from McNeill *et al.* (1998) following work in Phase III of the project. The formation names are those of Dalland *et al.* (1988), although the age-range of their Pliocene Naust Formation has been extended to include the Pleistocene deposits.

It can be seen that the Eocene to Oligocene Brygge Formation extends the length of the area. This formation was folded during late Eocene to Oligocene compression and inversion (Doré & Lundin 1996; Swiecicki *et al.* 1998), and Vågnes *et al.* (1998) have argued that inversion is continuing to the present. However, younger deposits do not have this continuity. Kai units B and C are thought on the basis of similarity of seismic character to have been originally continuous beneath the Storegga Slide, but Kai Unit A is at present found only to the north of Storegga. The shale-prone Brygge and Kai formations in the study area were both deposited in relatively deep-water basins, probably largely beyond the contemporary shelf (Gradstein & Backström 1996; Eidvin *et al.* 1998). It can be seen from Fig. 3 that these formations are commonly thin or absent on the shelf, although they typically have a combined thickness of over 1000 m in the Vøring and Møre basins.

Above the Kai Formation, there is a marked change in the style of sedimentation, for in late Pliocene times (Eidvin *et al.* 2000) the large prograding wedge of the Lower Naust and Naust formations built out to form the present-day shelf. This wedge has resulted in the advancement of the shelf by up to 100 km locally in the north (Hendriksen & Vorren 1996); the former extent in the south is unknown because of erosion at the Storegga Slide Complex. There is also a significantly different stratigraphic breakdown

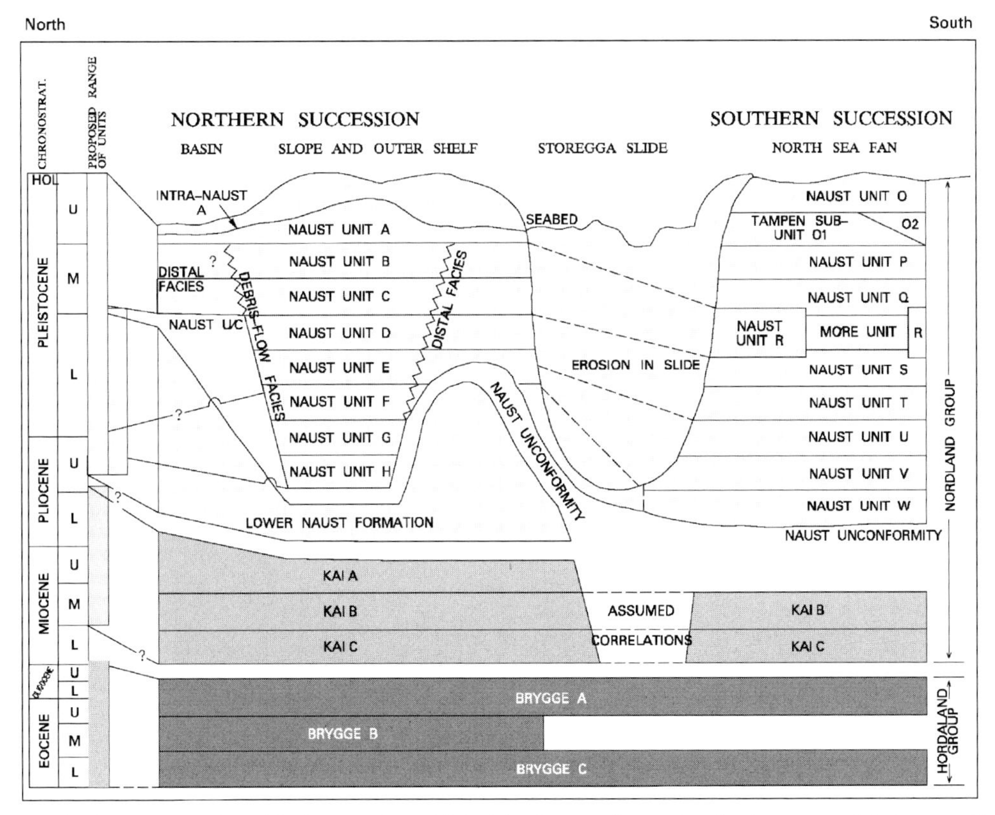

Fig. 2. The seismostratigraphic framework established for the area, illustrating the time range of the Brygge, Kai, Naust and Lower Naust formations. This is a modification of the stratigraphy proposed by McNeill *et al.* (1998).

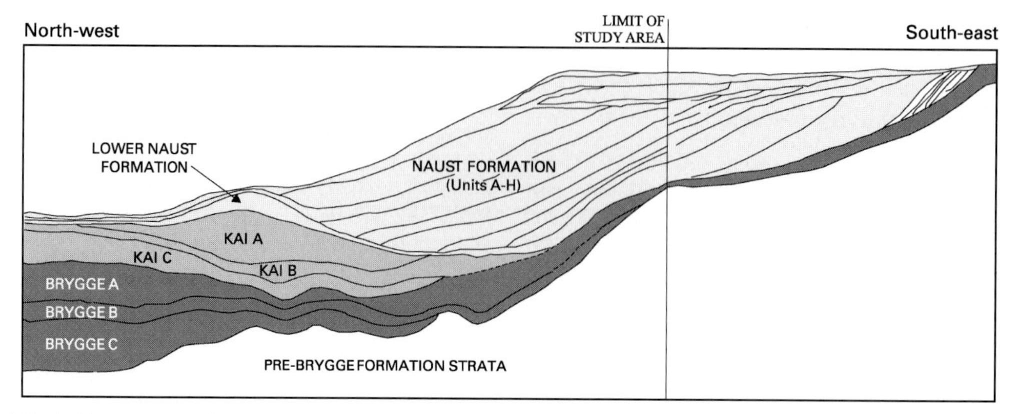

Fig. 3. Cross-section illustrating the general relationships of the Neogene stratigraphic units in the study area, with schematic extension across the shelf to the SE based on Rokoengen *et al.* (1995). The location of the portion within the study area is close to that of Fig. 4, with the schematic extension towards the Norwegian coast.

to the north and south of the northern flank of the Storegga Slide (Fig. 2), with only a limited seismic correlation established between the two successions.

North of the Storegga Slide Complex

In the north, the basal component above the Kai Formation is the informal Lower Naust formation, a unit that within the study area forms the thin, distal facies of the early build-out of the prograding wedge (Fig. 3). The formation is, however, locally thicker in mounded sections (Fig. 4), probably as a result of preferential fine-grained sedimentation under the influence of contour currents. The Lower Naust formation unconformably overlies older deposits, and is itself unconformably overlain by the down-lapping units of the Naust Formation on the slope, whereas in the basin it is overlain by the

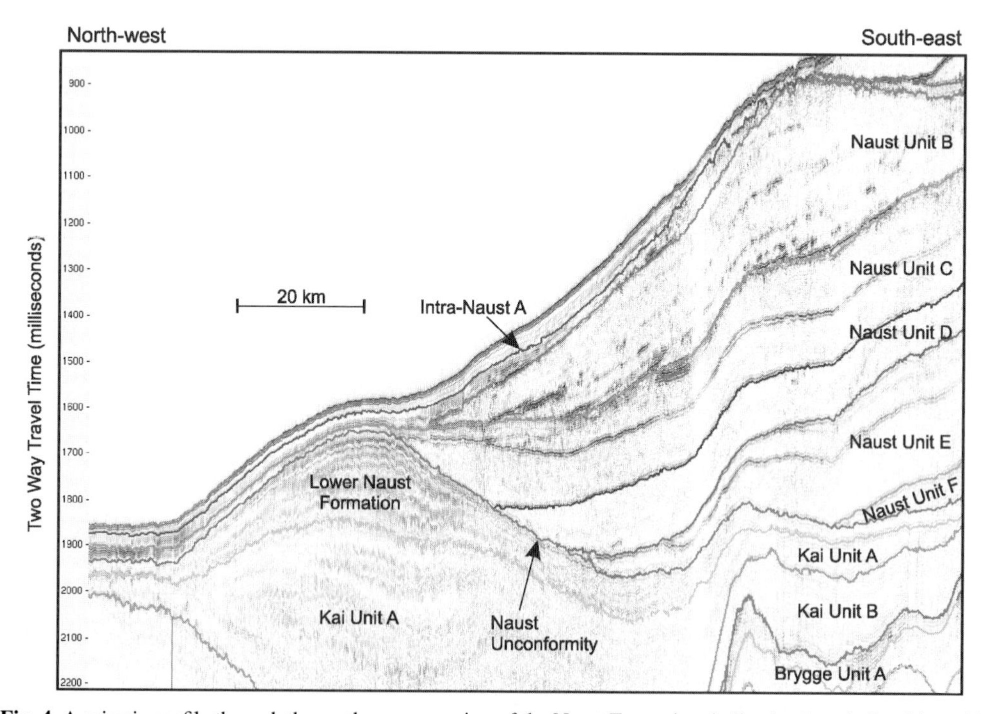

Fig. 4. A seismic profile through the northern succession of the Naust Formation, indicating its relationships with underlying formations. (For location, see Fig. 1.)

distal facies of the Naust Formation. The Naust Formation has been subdivided into units A–H, and the discrete, westerly downlapping packages of units B–H form the diachronous base to the formation. Their landward extents are truncated at a glacial unconformity at the base of Naust Unit A, although glacial erosion of the shelf probably pre-dates Naust Unit A (Haflidason *et al.* 1991). This glacial erosion surface was termed the URU (upper regional unconformity) by Hendriksen & Vorren (1996).

The Naust Formation in the north is up to 1500 m thick, and it has been estimated that its total volume in the north is 80 000 km^3 (Evans *et al.* 2000). A typical seismic section presented in Fig. 4 (see also McNeill *et al.* 1998; Evans *et al.* 2000) illustrates the downlapping character of the units, and shows that each unit is made up of one or more packages and is separated by well-defined reflectors. These packages may show bedding, but are generally acoustically structureless, although adjacent to the Storegga Slide Complex they are locally of an acoustically well-bedded facies (Fig. 5).

From the data available to the study, it is apparent that there is evidence for only a limited number of major palaeoslides within this sector, and no evidence of such erosion in the older part of the succession. The Traenadjupet and NE Nyk slides are found in the far north of the study area (Laberg *et al.* 1999), but these lie where the slope extends down to the Lofoten Basin to the north of the Vøring margin sector. A relatively small palaeoslide at the northern flank of the Storegga Slide has been described by Evans *et al.* (1996), and a part of Palaeoslide-2 has been mapped on the flank of the Storegga Slide Complex and in the far west of the study area. The only example on the Vøring slope during this project is the Traenabanken Slide to the north of the Helland Hansen licence area, a slide that took place after Naust Unit B times and commonly employed the top of Naust Unit C as a glide plane. The location of the headwall of this slide is evident in Fig. 6 from the removal of mid-Pleistocene sediments from the middle of the main northern depocentre.

South of the Storegga Slide Complex

The latest movements of the Storegga Slide as described by Bugge *et al.* 1987 were Slides I, II and III, and these formed the present topography of the region. Slide I was described as pre-dating the last glaciation, with Slides II and III

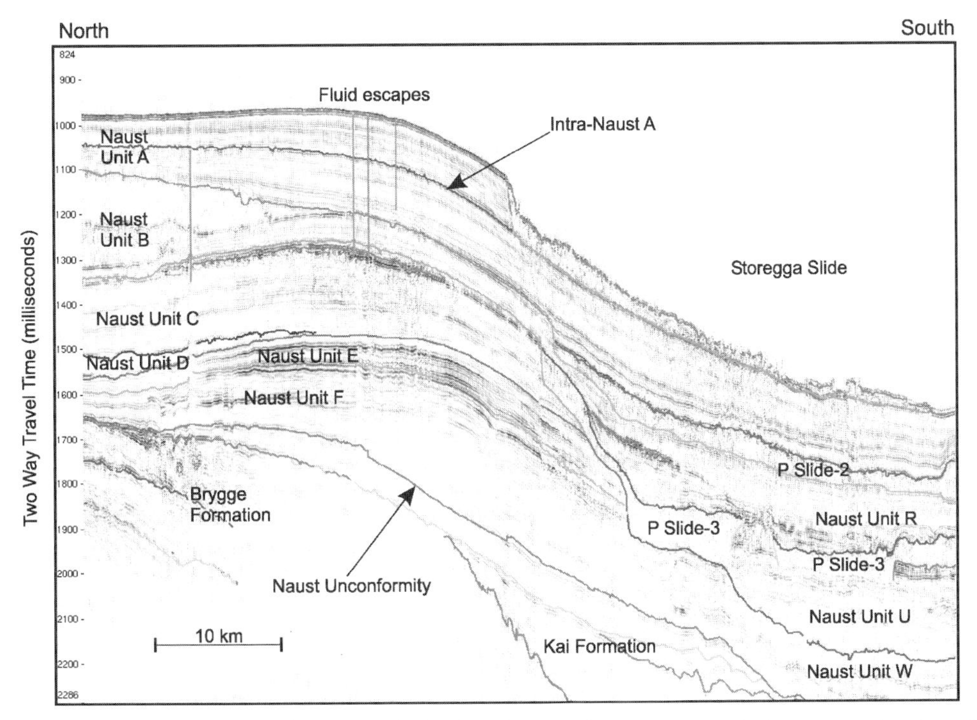

Fig. 5. A seismic section across the northern flank of the Storegga Slide illustrating the largely acoustically well-bedded nature of the Naust Formation immediately north of the sidewall. Palaeoslides can be seen at depth below the Storegga Slide. (For location, see Fig. 1.)

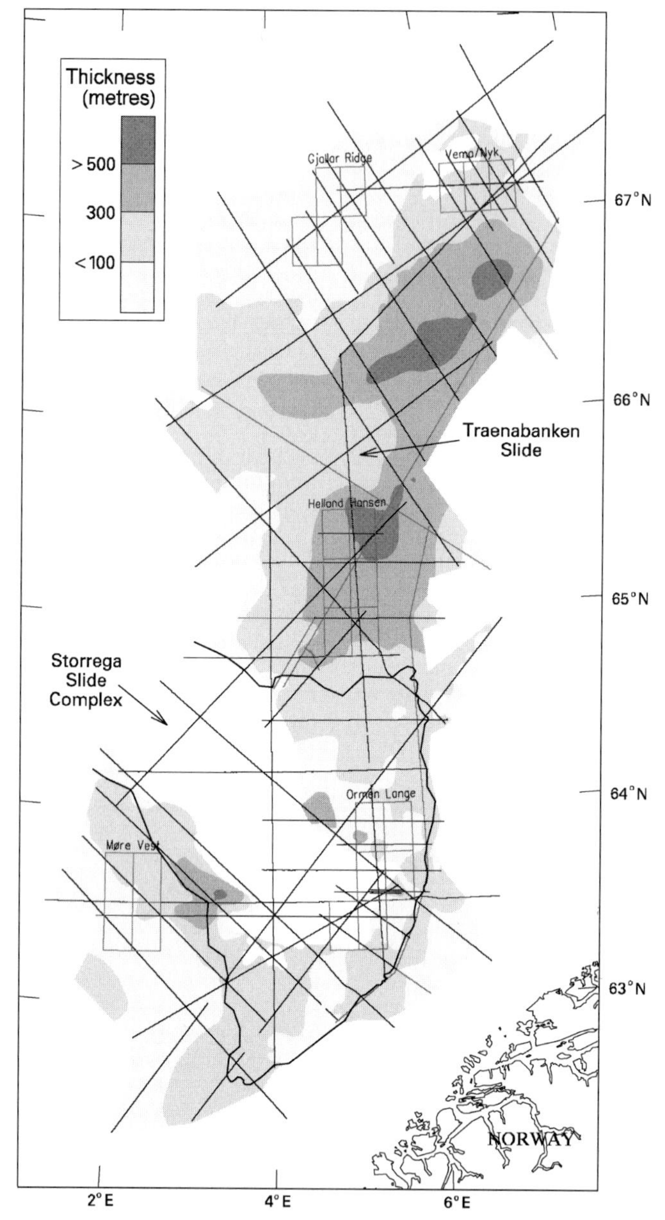

Fig. 6. Map showing the thickness of middle Pleistocene sediments (Naust units B and C) in the study area. The sediments have been wholly or partly removed from the Storegga Slide Complex region, and partially removed by the Traenabanken Slide. Also shown is the 2D seismic grid available to the study, and the 15th Round deep-water licence blocks.

occurring about 7000–7200 BP (Svendsen & Bondevik 1995; Bondevik *et al.* 1997). However, the deposits of Slide I are not blanketed by a well-defined sedimentary cover (Fig. 5) as would be expected if that slide pre-dated the last glaciation, and it is now considered that all three movements post-date the last glaciation and probably all occurred at around the same time. Haflidason *et al.* (2000) suggested that they took place between 6000 and 8000 BP. This suggestion is in accordance with an alternative interpretation originally proposed by Bugge (1983).

To the south of the northern flank of the Storegga Slide, Kai Unit A is absent and there is no direct equivalent of the Lower Naust formation. A correlation has been established between the top of Naust Unit W in the southern succession and the top of Naust Unit F in the north (Fig. 2), a stratigraphic level that from limited well evidence may approximate to the top of the Pliocene sequence, but is thought more likely to be of early Pleistocene age. Figure 5 shows the equivalence of the top of Naust Unit W (the oldest Naust unit of the southern succession in this region) with the top of Naust Unit F at the junction of the two successions beneath the northern sidewall of the Storegga Slide.

The character of Naust Unit W is in marked contrast to that of its coeval units (Naust units F, G and H) in the north. The base of the unit is in many places severely erosional; the example in Fig. 7 shows that one of the erosional scarps at its base exceeds 250 ms (*c*. 250 m) in height. Although the scale of erosion is small compared with the downcutting associated with the Holocene movements of the Storegga Slide, it none the less represents erosion on a large scale. Elsewhere the Unit W sediments have the slightly mounded and acoustically opaque internal character that is commonly associated with slide deposits, although not all profiles show evidence of a slide-related origin.

The extent of the sliding in Naust Unit W has been mapped and is shown in Fig. 1, whereas Fig. 7 shows the development of a headwall. The poorly defined trace of the headwall lies seaward of the present-day Storegga Slide scarp, and is seen to be of comparable length, indicating the magnitude of the erosion at that time, although the mass-movement may represent many separate events. It is considered that Naust Unit W was formed by more than one process, but that major slide erosion was a significant component. Importantly, the sliding associated with Naust Unit W is the oldest so far clearly identified in the study area, and has a close geographical association with the present-day bathymetry of the latest movements on the Storegga Slide as documented by Bugge *et al.* (1987, 1988). Bearing in mind the absence of Kai Unit A and the Lower Naust formation, it is possible that earlier instability may have caused the irregularities seen locally beneath the base of the Naust Formation in Fig. 7, but this remains unclear.

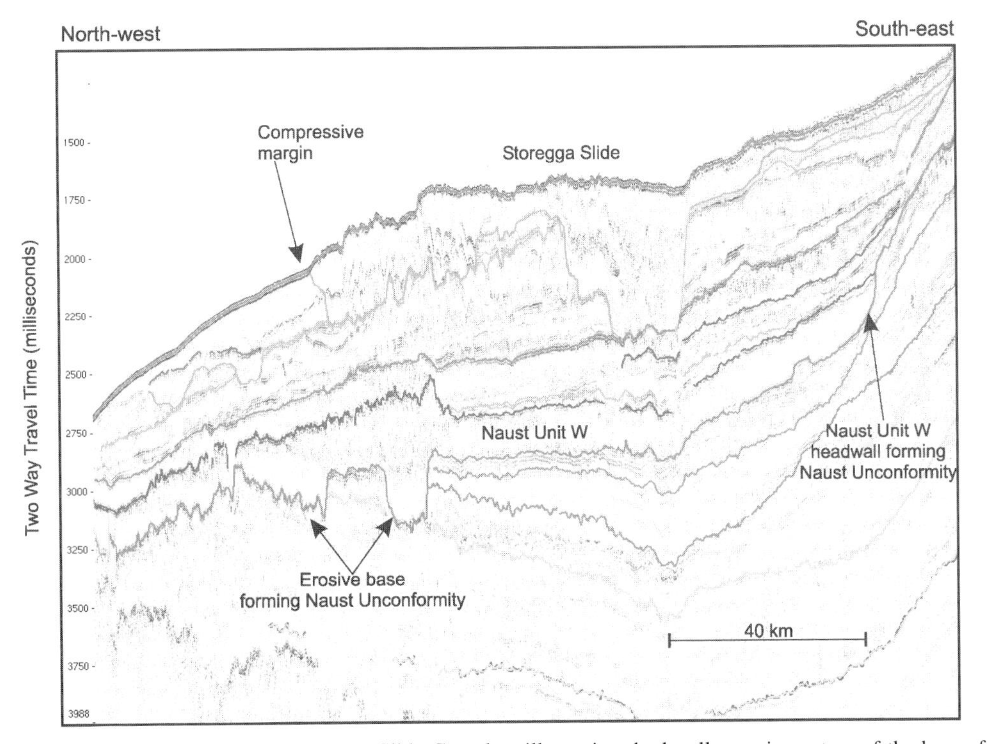

Fig. 7. Seismic profile from the Storegga Slide Complex, illustrating the locally erosive nature of the base of Naust Unit W, as well as deep erosion to the base of the sediments deposited during Holocene movement on the slide. (For location, see Fig. 1.)

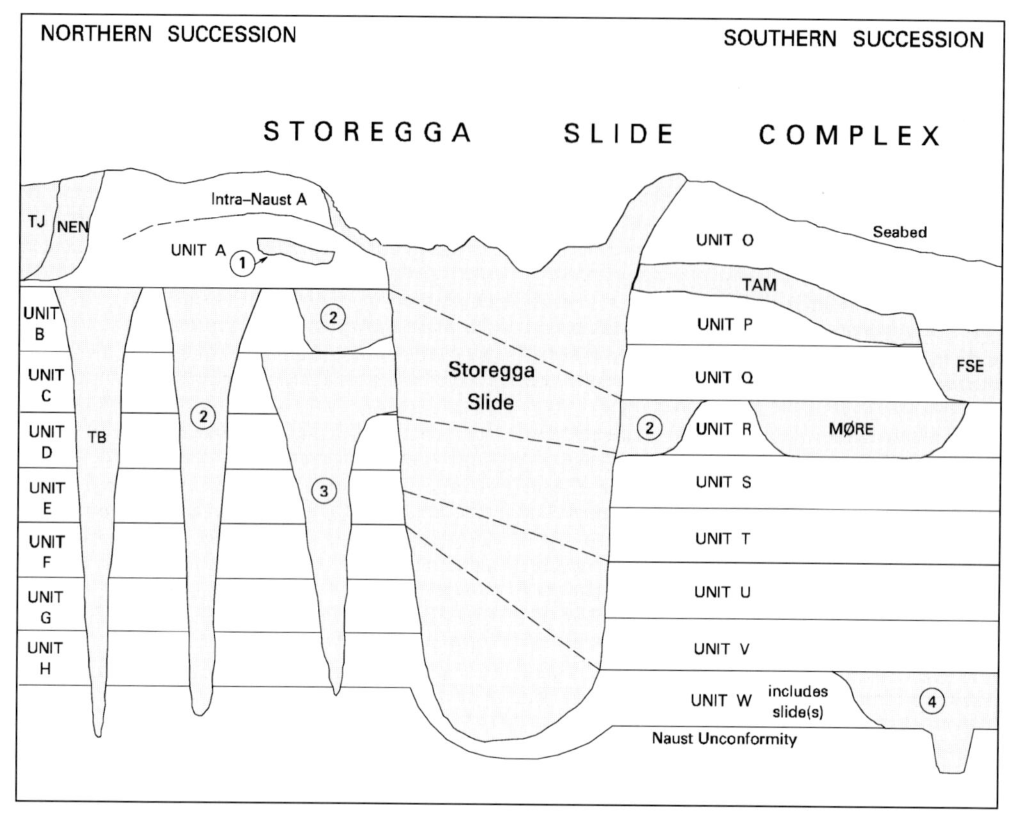

Fig. 8. The relationship of identified slides and palaeoslides to the stratigraphic subdivisions of the Naust Formation. TJ, Tranadjupet Slide; NEN, NE Nyk Slide; TB, Traenabanken Slide; TAM, Tampen Slide; FSE, Faeroe–Shetland Escarpment Slide; 1, 2, 3 and 4 are Palaeoslides-1, -2, -3 and -4. The Vigra and North Sea Fan Slide-1 were not specifically identified during this study, but were recognized by Evans *et al.* (1996) and King *et al.* (1996) within Naust units V–S.

The southern succession in the North Sea Fan region to the south of the Storegga Slide is of comparable thickness to the northern succession, but there are pronounced differences in its seismic character. The fan became increasingly important as a depocentre in mid- to late Pleistocene time as vast quantities of sediment were carried to the shelf break along the Norwegian Channel (Sejrup *et al.* 1996). The units show a variety of acoustic characteristics ranging from opaque to well bedded, but a key characteristic of the succession is that it includes evidence for a number of major translational slides. Evans *et al.* (1996) and King *et al.* (1996) have described North Sea Fan Slide-1, the Vigra Slide, the Møre Slide and the Tampen Slide on the fan.

The present study has identified the sliding associated with Naust Unit W, the Faeroe–Shetland Escarpment Slide that lies close to the eponymous escarpment, and Palaeoslides-1, -2, -3 and -4 (Fig. 8). In the present seismic correlation, Palaeoslide-2 is equivalent in age to the Traenabanken and Møre slides, and has been traced along the northern flank of the Storegga Slide as well as in the south (Fig. 6). Isolated thick remnants of mid-Pleistocene deposits seen in Fig. 6 adjacent to the area of complete removal testify to their probable former widespread presence before their removal by Palaeoslide-2 and contemporaneous events.

The precise extent of these slides is unknown, largely because significant parts of them may have been removed by later slide erosion, but some were undoubtedly very large, even compared with the latest Storegga Slide movements, which are the most recent major erosional events in this Møre Basin region. All these slides are located around the Storegga Slide as described byBugge *et al.* (1987), and it is clear that the term Storegga Slide does not adequately describe the long history of movements. The term Storegga Slide Complex is therefore proposed, with the term Storegga Slide used to define only the post-glacial movements and consequent bathymetric expression of those events.

Discussion

The above descriptions show that the history of Plio-Pleistocene margin development has been significantly different either side of a line that is approximately equivalent to the northern flank of the Storegga Slide. This line is coincident with the location of the Jan Mayen Lineament, which has controlled the tectonic development of the area since Cretaceous times (Brekke 2000). In particular, there was a marked difference during late Pliocene or early Pleistocene time, when the prograding wedge was initiated in the north during the deposition of the Lower Naust formation and its more proximal equivalents on the inner shelf, and Naust units F, G and H on the slope. However, a significant degree of progradation may also have occurred in the Storegga region, and may have been subsequently removed during Unit W or later erosion.

The initiation of the wedge represented a fundamental change in sedimentary architecture as the depocentre moved to the inner shelf from the earlier deep-water basin of the Kai and Brygge formations (Fig. 3). Workers such as Riis (1996) and Stuevold & Eldholm (1996) have related this change to a period of uplift of mainland Norway that led to increased erosion of the mountainous regions and an enhanced sediment supply to the shelf and the generation of accommodation space seaward of a hinge line. Climatic deterioration, with more effective fluvial erosion and the onset of upland glaciation, probably also played its part in the increased rate of erosion, and it is difficult to separate the two forces (Lidmar-Bergström *et al.* 2000). Any mechanism proposed for vertical movements in the mid-Norwegian area (e.g. Cloetingh *et al.* 1990; Riis & Feldskaar 1992) needs to be consistent with observations indicating late Neogene uplift in the broader North Atlantic region (Stoker 1995, 2002; Andersen *et al.* 2000; Chalmers 2000; Japsen & Chalmers 2000).

The Vøring margin would have been fed by sediments derived from central Norway, which, according to Riis (1996) was an area of lesser

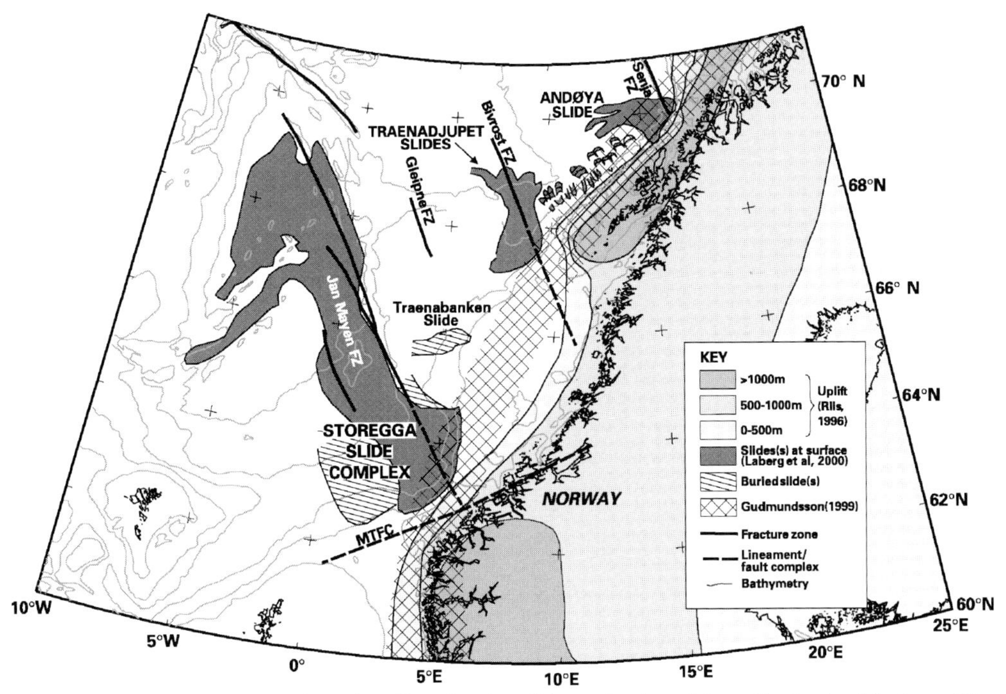

Fig. 9. Map showing the general relationships of major slides and palaeoslides with structural features, Plio-Pleistocene uplift (Riis 1996) and a zone of maximum post-glacial compressive stress (Gudmundsson 1999). Surface slides in the north are from Laberg *et al.* (2000). MTFC, Møre–Trøndelag Fault Complex (Gabrielsen *et al.* 1999), which is thought to be a particularly significant zone of seismicity.

uplift than southern and northern Norway at that time (Fig. 9). Although evidence of progradation in the south is lacking, it is clear that a vast quantity of sediment was delivered to the sea, and certainly led to the pronounced progradation and seaward advancement of the shelf break in the north.

At the same time as the wedge was being initiated, there was both sedimentation and major sliding in the Møre Basin to the south, although the Unit W sliding could have occurred as late as Unit F times. Given that the southern part of the area lies close to the region of maximum onshore uplift it is likely that there was an abundant sediment supply in late Pliocene to earliest Pleistocene times. Indeed, rapid sediment supply may have been a factor in triggering the slides that characterize Naust Unit W, as may have been the increase in slope that is likely to have been experienced at the continental margin as a result of uplift of the land and subsidence offshore. Present evidence points to late Pliocene uplift as being a factor in the initiation of the slides at that time, as well as in their continued occurrence throughout Quaternary time, but what was the trigger for the slides?

Earthquakes are a commonly quoted trigger for slides in this region (e.g.Bugge *et al.* 1987; King *et al.* 1996; Laberg *et al.* 1999). Figure 9 shows that the Storegga Slide Complex lies along the Jan Mayen Lineament (Blystad *et al.* 1995) and the Jan Mayen Fracture Zone, as has previously been noted by Bugge *et al.* (1987). Although this fracture zone is not a strong focus of modern earthquakes (Dehls *et al.* 2000), some seismic events have been recorded along it offshore, and are particularly common where the lineament meets the Norwegian coast (Bungum *et al.* 1991). Brekke (2000) considers it to have acted as a transfer zone during Cenozoic time, when it controlled the occurrence of compressional tectonics. Another factor is that the headwall (Fig. 9) is broadly coincident with the still seismically active Møre–Trøndelag Fault Complex (Gabrielsen *et al.* 1999). There is a clear relationship between the location of major structures or fracture zone and the position of the Storegga Slide Complex, and it is therefore likely that there has been late Pliocene or early Pleistocene tectonic movement in the vicinity of the slide complex that could have created significant seismic events. Although the Jan Mayen Fracture Zone may no longer be active, the lineament has continued to exert a control over the location of major slides during Quaternary time.

This indication is strengthened by a study of the distribution of similar features on the Vøring margin and farther north around Lofoten. Figure 9 shows that there are no crustal lineaments or fracture zones cutting the Vøring margin, which lies adjacent to a zone of lesser onshore uplift. This is the region with little evidence of palaeoslides; the only significant mapped feature south of Traenadjupet is the Traenabanken Slide, which interestingly lies on the projected line of the Gleipne Fracture Zone.

To the north of the Vøring margin off Lofoten, the zone of major uplift lies close to the shelf break, and this is a margin that displays ample evidence of downslope movement (Dowdeswell & Kenyon 1997; Taylor *et al.* 2000). In particular there are two major slides, the Traenadjupet and Andoya slides (Laberg *et al.* 1999, 2000), that respectively lie along the Bivrost Fracture Zone/Bivrost Lineament and the Senja Facture Zone (Blystad *et al.* 1995). The full history of these slides is not known, and although, as with the Storegga Slide Complex, their most recent movements were during Holocene time, older movements probably did occur but have yet to be fully documented. Nevertheless, they do contribute to a well-defined correlation between the location of major crustal structures and the distribution of large slides along the Norwegian margin.

In mid- to late Pleistocene times, glaciations became stronger and ice sheets extended onto the shelf with greater frequency (Williams *et al.* 1988; Mangerud *et al.* 1996; Valen *et al.* 1996; Vorren & Laberg 1996). During these times, movements along the Norwegian margin are likely to have been increasingly affected by isostatic movements related to the onset and removal of the ice cover from the shelf. These events increase the risk of seismicity, especially when superimposed on pre-existing crustal stresses related to plate tectonics (Talbot & Slunga 1988). Although they noted that observed stresses in Norway are consistent with uplift of Fennoscandia, Fejerskov & Lindholm (2000) considered that ridge-push associated with seafloor spreading is the primary cause of compressional stress. Figure 9 shows the present-day zone of maximum compressive stress around Norway as a result of post-glacial uplift as calculated by Gudmundsson (1999). This zone, which is likely to have been the locus of earthquakes that could trigger slides, covers the headwall of all three major Holocene slides. It has been estimated that seismic events as strong as M_w 7.9 may have occurred in Scandinavia during the last ice retreat (Muir Wood 1988), and given the occurrence of weak layers in the Plio-Pleistocene sedimentary column, slides could well have been generated by this mechanism.

Other mechanisms may also have generated slides, or facilitated their initiation; such factors include the presence of gas or gas hydrates (Bugge *et al.* 1987; Henriet & Mienert 1998; Bouriak *et al.* 2000), although these factors may be of more local importance than the regional view taken in this paper.

Conclusions

Late Neogene uplift of Norway had a pronounced influence on margin sedimentation, changing the pattern from one of slow deposition in deep-water basins to more rapid sedimentation on the inner shelf as a prograding wedge was initiated in response to increased sediment supply and the generation of accommodation space on the shelf.

Stratigraphic analysis shows that there has been a significant along-slope difference in the latest Cenozoic history of margin development between the Vøring and Møre margins. The former is characterized by the deposition of a vast prograding wedge with little evidence of major instability, whereas the latter has a long history of sliding, which has largely removed any evidence of wedge development, and may have received less sediment later in Quaternary time.

The oldest identified sliding in the Møre Basin was penecontemporaneous with the early stages of development of the prograding wedge in late Pliocene and earliest Pleistocene times, and both occurrences are considered to be related to uplift of the mainland at this time. Uplift was greatest in the south adjacent to the Møre margin, which has experienced a long history of instability. As isostatic uplift is continuing today, as suggested by Riis (1996), this may have implications for present-day slope stability.

Three large slides on the Norwegian margin lie at the junctions of oceanic fracture zones with the continental crust, or along crustal lineaments or major fault zones. They also lie adjacent to the zones of maximum onshore uplift. This suggests strong structural control on the location of slides on this margin, although there is little evidence of modern seismicity along the fracture zones, and it may be that the Møre–Trøndelag Fault Complex is a particularly significant structure.

The extent of structural control is emphasized by the observation that the largest slide area with the longest history of movement, the Storegga Slide Complex, lies at the conjunction of the largest oceanic fracture zone, the zone of maximum Plio-Pleistocene uplift, a major fault zone, and the zone of maximum post-glacial compressive stress.

The authors would like to thank the oil companies involved in the Seabed Project (BP Norge, Esso Norge, Mobil, Norske Conoco, Norsk Hydro, Shell and Statoil) for the opportunity to participate in the project and for their permission to publish the data included in this paper. We are also grateful to E. Gillespie for producing the diagrams. The paper benefited significantly both from the early comments of M.S. Stoker and D. Long as well as those of referees T. Eidvin and A.G. Doré. The contribution of D.E. is made with the permission of the Director of the British Geological Survey (NERC).

References

ANDERSEN, M.S., NIELSEN, T., SØRENSEN, A.B., BOLDREEL, L.O. & KUIJPERS, A. 2000. Cenozoic sediment distribution and tectonic movements in the Faroe region. *Global and Planetary Change*, **24**, 239–259.

BLYSTAD, P., BREKKE, H., FÆRSETH, R.B., LARSEN, B.T., SKOGSEID, J. & TØRUDBAKKEN, B. 1995. *Structural Elements of the Norwegian Continental Shelf, Part II: The Norwegian Sea Region.* Norwegian Petroleum Directorate Bulletin, No. 8.

BONDEVIK, S., SVENDSEN, J.I. & MANGERUD, J. 1997. Tsunami sedimentary facies deposited by the Storegga tsunami in shallow marine basins and coastal lakes, western Norway. *Sedimentology*, **44**, 1115–1131.

BOURIAK, S., VANNESTE, M. & SAOUTKINE, A. 2000. Inferred gas hydrates and clay diapirs near the Storegga Slide on the southern edge of the Vøring plateau, offshore Norway. *Marine Geology*, **163**, 125–148.

BREKKE, H., *et al.* 2000. The tectonic evolution of the Norwegian Sea Continental Margin with emphasis on the Vøring and Møre Basins. In: NØTTVEDT, A. (ed.) *Dynamics of the Norwegian Margin.* Geological Society, London, Special Publications, **167**, 327–378.

BRYN, P., ØSTMO, S.R., LIEN, R., BERG, K. & TJELTA, T.I. 1998. Slope stability in the deep water areas off mid Norway. *OTC paper 8640 presented at the 1998 Offshore Technology Conference.* Houston, TX, May 4–7.

BUGGE, T. 1983. *Submarine Slides on the Norwegian Continental Margin, with Special Emphasis on the Storegga area.* Institutt For Kontinentalsokkelundersøkelser Publications, 110.

BUGGE, T., BEFRING, S., BELDERSON, R.H. & 5 OTHERS 1987. A giant three-stage submarine slide off Norway. *Geo-Marine Letters*, **7**, 191–198.

BUGGE, T., BELDERSON, R.H. & KENYON, N.H. 1988. The Storegga Slide. *Philosophical Transactions of the Royal Society of London, Series A*, **325**, 357–388.

BUNGUM, H., ALSAKER, A., KVAMME, L.B. & HANSEN, R.A. 1991. Seismicity and seismotectonics of Norway and nearby continental shelf areas. *Journal of Geophysical Research*, **96**, 2249–2265.

CHALMERS, J.A. 2000. Offshore evidence for Neogene uplift in central West Greenland. *Global and Planetary Change*, **24**, 311–318.

CLAUSEN, L. 1998. The Southeast Greenland glaciated margin: 3D stratal architecture of shelf and deep sea. *In*: STOKER, M., EVANS, D. & CRAMP, A. (eds) *Geological Processes on Continental Margins: Sedimentation, Mass-wasting and Stability*. Geological Society, London, Special Publications, **129**: 173–203.

CLOETINGH, S., GRADSTEIN, F.M., KOOI, H., GRANT, A.C. & KAMINSKI, M. 1990. Plate reorganisation: a case for rapid late Neogene subsidence and sedimentation around the North Atlantic. *Journal of the Geological Society, London*, **147**, 495–506.

DALLAND, A., WORSLEY, D. & OFSTAD, K. 1988. A Lithostratigraphic Scheme for the Mesozoic and Cenozoic Succession Offshore Mid- and Northern Norway. *Norwegian Petroleum Directorate*, Bulletin 4.

DEHLS, J.F., OLESEN, O., BUNGUM, H., HICKS, E.C., LINDHOLM, C.D. & RIIS, F. 2000. Neotectonic Map: Norway and Adjacent Areas. Geological Survey of Norway, Trondheim.

DORÉ, A.G. & LUNDIN, E.R. 1996. Cenozoic compressional structures on the NE Atlantic margin: nature, origin and potential significance for hydrocarbon exploration. *Petroleum Geoscience*, **2**, 299–311.

DOWDESWELL, J.A. & KENYON, N.H. 1997. Long-range side-scan sonar (GLORIA) imagery of the eastern continental margin of the glaciated Polar North Atlantic. *In*: DAVIES, T.A., BELL, T., COOPER, A.K. & 5 OTHERS (eds) *Glaciated Continental Margins: an Atlas of Acoustic Images*. Chapman and Hall, London, 260–263.

EIDVIN, T., BREKKE, H., RIIS, F. & RENSHAW, D.K. 1998. Cenozoic stratigraphy of the Norwegian Sea continental shelf, 64 N–68 N. *Norsk Geologisk Tidsskrift*, **78**, 125–151.

EIDVIN, T., JANSEN, E., RUNDBERG, Y., BREKKE, H. & GROGAN, P. 2000. The Upper Cainozoic of the Norwegian continental shelf correlated with deep sea record of the Norwegian Sea and the North Atlantic. *Marine and Petroleum Geology*, **17**, 579–600.

EVANS, D., KING, E.L., KENYON, N.H., BRETT, C. & WALLIS, D.G. 1996. Evidence for long-term instability in the Storegga Slide region off western Norway. *Marine Geology*, **130**, 281–292.

EVANS, D., MCGIVERON, S., MCNEILL, A.E., HARRISON, Z., OSTMO, S.R. & WILD, J.B.L. 2000. Plio-Pleistocene deposits on the mid-Norway margin and their implications for late Cenozoic uplift of the Norwegian mainland. *Global and Planetary Change*, **24**, 233–237.

FEJERSKOV, M., LINDHOLM, C., *et al.* 2000. Crustal stress in and around Norway: an evaluation of stress-generating mechanisms. In: NØTTVEDT, A. (ed.) *Dynamics of the Norwegian Margin*. Geological Society, London, Special Publications, **167**, 451–467.

GABRIELSEN, R.H., ODINSEN, T. & GRUNNALEITE, I. 1999. Structuring of the Northern Viking Graben and the Møre Basin; the influence of basement structural grain, and the particular role of the Møre–Trøndelag Fault Complex. *Marine and Petroleum Geology*, **16**, 443–465.

GRADSTEIN, F. & BACKSTRÖM, S.A. 1996. Cenozoic biostratigraphy and palaeobathymetry, northern North Sea and Haltenbanken. *Norsk Geologisk Tidsskrift*, **76**, 3–32.

GUDMUNDSSON, A. 1999. Postglacial crustal doming, stresses and fracture formation with application to Norway. *Tectonophysics*, **307**, 407–419.

HAFLIDASON, H., AARSETH, I., HAUGEN, J.-E., SEJRUP, H.P., LØVLIE, R. & REITHER, E. 1991. Quaternary stratigraphy of the Draugen area, mid-Norwegian shelf. *Marine Geology*, **101**, 125–146.

HAFLIDASON, H., SEJRUP, H.P., BRYN, P. & MIENERT, J. 2000. Glide planes and slide frequency in the Storegga Slide Complex off mid-Norway. *Geoscience*, **2000**, 16.

HENDRIKSEN, S. & VORREN, T. 1996. Late Cenozoic sedimentation and uplift history on the mid-Norway continental shelf. *Global and Planetary Change*, **12**, 171–199.

HENRIET, J.P.; MIENERT, J. 1998. *Gas Hydrates: Relevance to World Margin Stability and Climatic Change*. Geological Society, London, Special Publications, **137**.

JAPSEN, P. & CHALMERS, J.A. 2000. Neogene uplift and tectonics around the North Atlantic: overview. *Global and Planetary Change*, **24**, 165–174.

KING, E.L., SEJRUP, H.P., HAFLIDASON, H., ELVERHØI, A. & AARSETH, I. 1996. Quaternary seismic stratigraphy of the North Sea Fan: glacially-fed gravity flow aprons, hemipelagic sediments, and large submarine slides. *Marine Geology*, **130**, 293–315.

KRISTOFFERSON, Y., WINTERHALTER, B. & SOLHEIM, A. 2000. Shelf progradation on a glaciated continental margin, Queen Maud Land, Antarctica. *Marine Geology*, **165**, 109–122.

LABERG, J.S., VORREN, T.O., DOWDESWELL, J.A., KENYON, N.H. & TAYLOR, J. 2000. The Andøya Slide and the Andøya Canyon, north-eastern Norwegian–Greenland Sea. *Marine Geology*, **162**, 259–275.

LABERG, J.S., VORREN, V.O., MIENERT, J., KENYON, N.H., EVANS, D., HENDRIKSEN, S. & DOWDESWELL, J.A. 1999. The Traenadjupet Slide area offshore Norway. *In*: MARTINSEN, O.J. & DREYER, T. (eds) *Sedimentary Environments Offshore Norway—Palaeozoic to Recent*. Norwegian Petroleum Society, Extended Abstracts. 223–226.

LARTER, R.D. & BARKER, P.F. 1991. Neogene interaction of tectonic and glacial processes at the Pacific margin of the Antarctic Peninsula. *In*: MACDONALD, D.I.M. (ed.) *Sedimentation, Tectonics and Eustasy: Sea-level Changes at Active Margins*, International Association of Sedimentologists, Special Publications, **12**, 165–186.

LIDMAR-BERGSTRÖM, K., OLLIER, C.D. & SULEBAK, J.R. 2000. Landforms and uplift history of southern Norway. *Global and Planetary Change*, **24**, 211–232.

MANGERUD, J., JANSEN, E. & LANDVIK, J.Y. 1996. Late Cenozoic history of the Scandinavian and Barents Sea ice sheets. *Global and Planetary Change*, **12**, 11–26.

MCNEILL, A.E., SALISBURY, R.S.K., ØSTMO, S.R., LIEN, R. & EVANS, D. 1998. A regional shallow stratigraphic framework off mid Norway and observations of deep water 'special features'. *OTC paper 8639 presented at the 1998 Offshore Technology Conference*, Houston, TX, May 4–7.

MUIR WOOD, R. 1988. Extraordinary deglaciation reverse faulting in northern Fennoscandia. *In*: GREGERSEN, S. & BASHAM, P.W. (eds) *Earthquakes at North Atlantic Passive Margins: Neotectonics and Postglacial Rebound. Proceedings of the NATO Advanced Research Workshop, Vordingbord, Denmark, 9–13 May 1988*. Kluwer, Dordrecht, 141–173.

POOLE, D.A.R. & VORREN, T.O. 1993. Miocene to Quaternary palaeoenviroments and uplift history on the mid-Norwegian shelf. *Marine Geology*, **115**, 173–205.

RIIS, F. 1996. Quantification of Cenozoic vertical movements of Scandinavia by correlation of morphological surfaces with offshore data. *Global and Planetary Change*, **12**, 331–357.

RIIS, F. & FELDSKAAR, W. 1992. On the magnitude of the late Tertiary and Quaternary erosion and its significance for the uplift of Scandinavia and the Barents Sea. *In*: LARSEN, R.M., BREKKE, H., LARSEN, B.T. & TELLERAAS, E. (eds) *Structural and Tectonic Modelling and its Application to Petroleum Geology*. Elsevier, Amsterdam, 163–185.

ROKOENGEN, K., RISE, L., BRYN, P., FRENGSTAD, B., GUSTAVSEN, B., HYGAARD, E. & SÆTTEM, J. 1995. Upper Cenozoic stratigraphy on the Mid Norwegian Continental Shelf. *Norsk Geologisk Tidsskrift*, **75**, 88–104.

SEJRUP, H.P., KING, E.L., AARSETH, I., HAFLIDASON, H. & ELVERHØI, A. 1996. Quaternary erosion and depositional processes: western Norwegian fjords, Norwegian Channel and North Sea Fan. *In*: DE BATIST, M. & JACOBS, P. (eds) *Geology of Siliciclastic Shelf Seas*. Geological Society, London, Special Publications, **117**, 187–202.

STOKER, M.S. 1995. The influence of glacigenic sedimentation on slope-apron development on the continental margin west of Britain. *In*: SCRUTTON, R.A., SHIMMIELD, G.B., STOKER, M.S. & TUDHOPE, A.W. (eds) *The Tectonics, Sedimentation and Palaeoceanography of the North Atlantic Region*. Geological Society, London, Special Publications, **90**, 159–177.

STOKER, M.S. 2002. Late Neogene development of the UK Atlantic Region. *In*: DORÉ, A.G., CARTWRIGHT, J.A., STOKER, M.S., TURNER, J.P. & WHITE, N. (eds) *Exhumation of the North Atlantic Margin: Timing, Mechanisms and Implications for Petroleum Exploration*. Geological Society, London, Special Publications, **196**, 313–329.

STUEVOLD, L.M. & ELDHOLM, O. 1996. Cenozoic uplift of Fennoscandia inferred from a study of the mid-Norwegian margin. *Global and Planetary Change*, **12**, 359–386.

SVENDSEN, J.I. & BONDEVIK, S. 1995. *Palaeotsunamis in the Norwegian and North Seas*. Report of the University of Bergen.

SWIECICKI, T., GIBBS, P.B., FARROW, G.E. & COWARD, M.P. 1998. A tectonostratigraphic framework for the mid-Norway region. *Marine and Petroleum Geology*, **15**, 245–276.

TALBOT, C.J. & SLUNGA, R. 1988. Patterns of active shear in Fennoscandia. *In*: GREGERSEN, S. & BASHAM, P.W. (eds) *Earthquakes at North Atlantic Passive Margins: Neotectonics and Postglacial Rebound. Proceedings of the NATO Advanced Research Workshop, Vordingbord, Denmark, 9–13 May 1988*. Kluwer, Dordrecht, 441–466.

TAYLOR, J., DOWDESWELL, J.A. & KENYON, N.H. 2000. Canyons and late Quaternary sedimentation on the North Norwegian margin. *Marine Geology*, **166**, 1–9.

VÅGNES, E., GABRIELSEN, R.H. & HAREMO, P. 1998. Late Cretaceous–Cenozoic intraplate contractional deformation at the Norwegian continental shelf: timing, magnitude and regional implications. *Tectonophysics*, **300**, 29–46.

VALEN, V., MANGERUD, J., LARSEN, E. & HUFTHAMMER, A.K. 1996. Sedimentology and stratigraphy in the cave Hamnsund helleren, western Norway. *Journal of Quaternary Science*, **11**, 185–201.

VORREN, T.O. & LABERG, J.S. 1996. Late glacial air temperature, oceanographic and ice sheet interactions in the southern Barents Sea region. *In*: ANDREWS, J.T., AUSTIN, W.E.N., BERGSTEN, H. & JENNINGS, A.E. (eds) *Late Quaternary Palaeoceanography of the North Atlantic Margins*. Geological Society, London, Special Publications, **111**, 303–321.

WILLIAMS, D.F., THUNELL, R.C., TAPPA, E., RIO, D. & RAFFI, I. 1988. Chronology of the Pleistocene oxygen isotope record: 0–1.88 m.y. *Palaeogeography, Palaeoclimatology, Palaeocology*, **64**, 221–240.

Reconstructing the erosion history of glaciated passive margins: applications of *in situ* produced cosmogenic nuclide techniques

ARJEN P. STROEVEN[1], DEREK FABEL[2], JON HARBOR[3], CLAS HÄTTESTRAND[1] & JOHAN KLEMAN[1]

[1]*Department of Physical Geography and Quaternary Geology, Stockholm University, S-106 91 Stockholm, Sweden (e-mail: arjen@geo.su.se)*

[2]*School of Earth Sciences, University of Melbourne, Parkville, Vic. 3052, Australia*

[3]*Department of Earth and Atmospheric Sciences, Purdue University, West Lafayette, IN 47907-1397, USA*

Abstract: Offshore sediment accumulations provide an intriguing record of the net sediment output resulting from geomorphological evolution of the circum-Atlantic continental margin since the commencement of Neogene glaciation. However, the onshore record of the timing, pattern and amount of bedrock erosion that produced these sediments is comparatively poorly constrained and understood, although there are good general models of glaciation history. The geomorphology of circum-Atlantic continental margin mountains, as assessed from remote sensing data and field observations, includes palimpsest landforms and landscapes that reflect a complex pattern of spatial and temporal variations in the impact of glacial, fluvial and periglacial processes. Perhaps most surprising is that, despite having been repeatedly overridden by large ice sheets, parts of the landscape appear to be relict, with nonglacial morphology. This has important implications both for glaciological conditions under ice sheets, and for sediment source areas and erosion rates. Conventional dating and analysis have provided an excellent way to begin unravelling the timing and pattern of erosion, landform development, and possible landform preservation under ice. However, testing hypotheses developed from current models, and addressing critical unresolved questions, requires additional approaches. The use of *in situ* cosmogenic nuclide production in bedrock is a new approach for investigating landscape evolution in mountainous areas. With careful interpretation of geomorphological settings, cosmogenic nuclides can be used to determine apparent surface exposure age and landscape preservation, and constrain erosion depths and duration of burial by ice. Here we provide a framework for the interpretation of cosmogenic nuclide concentrations in bedrock surfaces of landscapes affected by glacial, fluvial and periglacial processes, illustrated with examples from the northern Swedish mountains. This demonstrates potential uses of cosmogenic nuclide techniques, and provides a foundation for attempts to improve geomorphologically based reconstructions of relict landscapes, to reconstruct and analyse the dynamics of landscape change in glacial times, and to define the consequences of different process regimes in terms of erosion patterns, sediment transport, and the supply of sediments that are deposited offshore.

The evolution of the circum-Atlantic continental margin since the commencement of Neogene glaciation is reflected in large offshore sediment accumulations (e.g. Solheim *et al.* 1996). The thickness of offshore sediments, important quantities of which were generated by glacial processes, is one key ingredient considered in hydrocarbon exploration. However, the onshore record of the amount, timing and pattern of bedrock erosion that was the source of these offshore sediments is not well constrained, except at the most generalized level (e.g. Riis 1996). In particular, current techniques of estimating onshore exhumation patterns and rates do not resolve margin exhumation during late Cenozoic time (Hendriks & Andriessen 2002).

The broad-scale glaciation history of the circum-Atlantic continental margin is well understood in general terms (e.g. Shackleton *et al.* 1984; Jansen & Sjøholm 1991; Hölemann & Henrich 1994; Mangerud *et al.* 1996; Jansen *et al.* 2000). In Scandinavia, for example, ice sheets centred west of the mountain elevation

From: DORÉ, A.G., CARTWRIGHT, J.A., STOKER, M.S., TURNER, J.P. & WHITE, N. 2002. *Exhumation of the North Atlantic Margin: Timing, Mechanisms and Implications for Petroleum Exploration*. Geological Society, London, Special Publications, **196**, 153–168. 0305-8719/02/$15.00

axis were the dominant form of glaciation between 2.0 and 0.7 Ma ago (Ljungner 1949; Kleman 1992). An important threshold was passed at *c.* 0.9–0.7 Ma (e.g. Porter 1989; Raymo *et al.* 1989; Imbrie *et al.* 1992, 1993; Clark & Pollard 1998), after which ice sheets grew larger and became centred east of the mountain elevation axis (Kleman & Stroeven 1997). However, what is not well established is the timing and pattern of erosion and landscape development associated with this glacial chronology.

It has long been recognized that the typical geomorphology of circum-Atlantic continental margin mountains is a patchwork of glaciated terrain and relict upland surfaces (Gjessing 1967; Sugden 1968; Sugden & Watts 1977; Hall & Sugden 1987; Ballantyne 1994; Kleman & Stroeven 1997). For example, remnant surfaces were recognized in the Norwegian and Swedish mountains early in the twentieth century by Reusch (1901), Wråk (1908) and Ahlmann (1919). Glacial terrain is characterized by the presence of U-shaped valleys, cirques, horns and arrêtes, valley and mountain truncation, mountain asymmetry, and the ubiquitous presence of lakes (e.g. Sugden & John 1976). Relict (or remnant) upland surfaces, on the other hand, are characterized by winding V-shaped valleys, mountain symmetry, tors, weathering mantles and an absence of (water-filled) rock basins (Figs 1 and 2). A patchwork occurrence of glacially scoured and relict surfaces indicates that late Cenozoic subglacial erosion must have been areally variable (Sugden 1968, 1974). This patchwork pattern of glacial erosion and preservation could either result from restricted ice extents (glaciers limited to major valleys) or, where we know that large ice sheets covered the mountains, from complex patterns of ice sheet basal thermal regimes.

On the basis of observations of a patchwork of glacially scoured and relict surfaces, typical for glaciated passive margin mountains, the subglacial thermal regime of average Quaternary ice sheets (Porter 1989) was frozen on the uplands and melting in the main valleys, where outlet glaciers and ice-streams formed (Sugden 1968, 1974). Relict surfaces are best preserved at intermediate elevations, low enough not to have been covered by cirque glaciers, and apparently high enough not to have experienced melted-bed conditions and subglacial erosion during ice sheet overriding events (Kleman & Stroeven 1997). Hence, the morphological erosional impact of glaciers and ice sheets overriding and expanding through these circum-Atlantic continental margin mountains left a distinct landform pattern that is both areally and altitudinally variable.

The most prominent valleys of the preglacial landscape were presumably exploited by early ice sheets as primary routes of ice drainage, and hence became primary locations for subglacial bedrock erosion and valley deepening. Subsequent glaciations would then have preferentially exploited these deepened valleys, because they would have been more efficient for ice drainage with each successive glaciation cycle (Sugden & John 1976). The implication of this model is that intervening highlands were covered by relatively stagnant ice, ensuring that subglacial frozen conditions, and hence landscape preservation, prevailed. The most visible alteration of relict upland surfaces since the commencement of glaciation in the circum-Atlantic region has been the deepening of V-shaped valleys terminating in glacial troughs. The deepening of these V-shaped valleys presumably happened in interglacial times, as these rivers adjusted their profiles by vertical erosion towards the new base-level conditions produced by trough deepening during previous glaciation(s) (Ahlmann 1919; Rudberg 1992). The implication of this geomorphological history for interpreting circum-Atlantic continental margin sediments during Neogene glaciation is that offshore sediments were derived primarily from areally and altitudinally restricted source areas, rather than from equal erosion across much of the landscape.

This proposed model of the geomorphological history is based on field and remote sensing-based mapping and interpretation of geomorphological features (e.g. Kleman & Stroeven 1997). This interpretation leads to the following testable implications: (1) large-scale relict bedrock morphology should have surface exposure ages much older than adjacent glacially cut surfaces (potential age differences of 10^5–10^6 years); (2) bedrock erosion on glacially cut surfaces has been orders of magnitude higher than on relict surfaces (ranging from about 10 to 10^3 m); (3) many bedrock surfaces have undergone a complex history of multiple burials (underneath non-erosive ice) and re-exposures. These implications, and thus the larger model for the geomorphological history of the circum-Atlantic mountains, can now be tested using an approach based on measuring multiple *in situ* cosmogenic nuclides in exposed bedrock.

This paper serves to establish a framework for the interpretation of *in situ* cosmogenic nuclide concentrations in terms of the timing, patterns and magnitude of bedrock erosion of landscapes affected by glacial, fluvial and periglacial process systems. Formulating such a framework in advance minimizes the use of special pleading to interpret forthcoming data from glaciated passive margin mountains. The practical application of this framework to the geological and geomorphological 'traces' of the last glacial cycle establishes a set of attributes (typically, the timing, amount and patterns of erosion) that can also serve as a basis for interpreting geomorphological evidence from prior glacial cycles. First, we provide an overview of cosmogenic nuclide techniques and theoretical considerations. We then discuss the geomorphological aspects of landscape surface reconstruction on glaciated passive margins. Finally, we illustrate potential uses of cosmogenic nuclide techniques in testing erosion histories for glaciated passive margins.

Applications of *in situ* produced cosmogenic ^{10}Be, ^{26}Al and ^{21}Ne

The cosmogenic radionuclides ^{10}Be and ^{26}Al and the cosmogenic stable nuclide ^{21}Ne are produced in rocks near the ground surface by reactions with secondary and tertiary cosmic-ray neutrons and muons (Lal & Peters 1967). These nuclides are commonly used in studies of landscape evolution (reviewed by Nishiizumi *et al.* 1993; Bierman 1994; Cerling & Craig 1994; Fabel & Harbor 1999). This is because all three isotopes are produced within a few metres of the Earth's surface in quartz, a ubiquitous mineral in crustal rocks and sediments, which has a simple ^{16}O and ^{28}Si target chemistry and a tight crystal structure that minimizes diffusion and contamination.

Cosmogenic nuclide production rates vary with latitude and altitude because of the dissipation of cosmic radiation within the Earth's atmosphere and the dependence of the cosmic-ray flux on the strength of the Earth's magnetic field (Lal 1991; Robinson *et al.* 1995; Dunai 2000). Other variables influencing cosmogenic nuclide production rates are shielding of incoming cosmic rays by surrounding terrain, the geometry of the sample surface, and erosion and/or burial of the sample site (Lal 1987, 1991). Provided the geomorphological context of the sample is understood, appropriate corrections for each of these variables are available (Nishiizumi *et al.* 1989; Lal 1991; Dunne *et al.* 1999; Fabel & Harbor 1999).

Surface exposure dating and erosion rates

With prolonged exposure, cosmogenic nuclides accumulate within rock as a function of time and depth below the surface. The time elapsed since initial exposure of the rock surface can be calculated from cosmogenic nuclide concentrations in the rock, using known rates of production. If the surface undergoes erosion, depth profiles of nuclide concentrations, ratios of different nuclides in the rock, and nuclide concentrations in sediments can all provide measures of the erosion rate (Cerling & Craig 1994; Granger *et al.* 1996). The comparison of concentrations of stable and radioactive isotopes can even facilitate the unravelling of complex histories of exposure and burial such as occurred on glaciated passive margins.

Erosion of a rock surface leads to removal of accumulated cosmogenic nuclides and hence a

Fig. 1. Landscape typical for glacial erosion. Cirque glaciers have deepened and widened bedrock depressions in the Pårte Massif, Sarek National Park, northern Sweden, a mountain massif that was also a prominent part of the preglacial landscape. However, the glacier forefield, although riddled with lakes, has not been significantly eroded below its preglacial elevation.

Fig. 2. Landscape shaped by non-glacial processes. Upland surfaces, Tarrekaise Massif, northern Sweden, are thought to retain relict morphology. The fluvial valley, with interlocking spurs, is considered to be younger in age and cut in response to (local) base-level lowering by glacier erosion. These interlocking spurs also show that glacial erosion played no significant role in the evolution of valley pattern.

reduction in nuclide concentration (Fig. 3). Because production decreases roughly exponentially with depth below the surface, the accumulated cosmogenic radionuclide concentration in a mineral grain records the speed with which that grain has been uncovered; slower erosion rates imply longer exposure times near the surface, and thus higher concentrations (Lal 1991). Therefore, to calculate the exposure time from a measured cosmogenic radionuclide concentration in a sample requires that the erosion rate is known. If independent erosion rate evidence is not available, there is no unique solution to the exposure age calculation with a single radionuclide. However, a maximum steady-state erosion rate can be calculated. The cosmogenic radionuclide concentration can be used to calculate the steady-state erosion rate for the surface if the exposure time can be independently constrained, or vice versa (see Lal (1991) for equations).

The measurement of ^{10}Be and ^{26}Al concentrations in the same sample can provide a means of estimating both the exposure time and the steady-state erosion rate of the sample because the ratio of the ^{26}Al and ^{10}Be concentrations in an eroding horizon changes sensitively with the rate of erosion (Lal & Arnold 1985). The ^{26}Al/^{10}Be production rate ratio in quartz is 6.0 ± 0.3 (Nishiizumi *et al.* 1989), regardless of the absolute production rate. Because ^{26}Al decays more rapidly than ^{10}Be, the ^{26}Al/^{10}Be ratio decreases with increasing exposure time. For a continuously exposed sample with no erosion, the ^{26}Al/^{10}Be ratio will follow a smoothly varying trajectory reaching an end point where production and radioactive decay are balanced for both isotopes (Fig. 4; constant exposure curve). If the same sample is subject to steady-state erosion it is losing mass from the surface and the ^{26}Al/^{10}Be ratio should lie on the steady-erosion curve at a point determined by the erosion rate (Fig. 4). Provided the cosmic-ray intensity has remained constant the ^{26}Al/^{10}Be ratio will plot between these curves (steady-state erosion island) for any simple exposure history under conditions of steady-state erosion. It is therefore possible to determine the steady-state erosion rate and exposure time of a sample by measuring two cosmogenic radionuclides with different half-lives in the same sample.

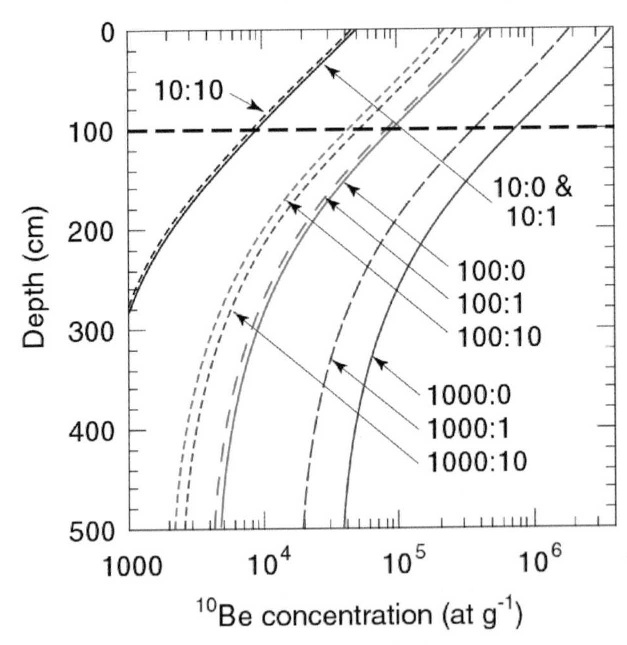

Fig. 3. ^{10}Be concentration v. depth for three different exposure times and three different steady-state erosion rates. The first number in the ratios is the exposure time (ka) and the second is the steady-state erosion rate (cm ka^{-1}). The curves include ^{10}Be production by muons and were calculated according to Granger & Smith (2000) for a ^{10}Be production rate of 5.1 atoms g^{-1} (SiO_2) yr^{-1} and a density of 2.7 g cm^{-3}. If a glacial erosion event removes 100 cm of bedrock (horizontal dashed line) from the surface after 10, 100 and 1000 ka exposure with zero steady-state erosion (continuous curves), the inherited ^{10}Be concentrations in the resulting 'new' surface are equivalent to apparent ^{10}Be ages of *c.* 2 ka, *c.* 18 ka and *c.* 155 ka, respectively.

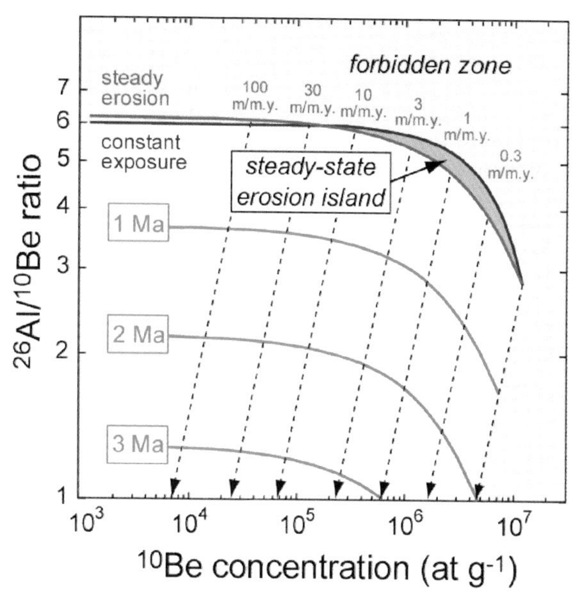

Fig. 4. $^{26}Al/^{10}Be$ ratio plotted against ^{10}Be concentration. ^{26}Al (half-life 705 ka) decay is relatively rapid with respect to ^{10}Be (half-life 1500 ka), forcing the $^{26}Al/^{10}Be$ ratio of a sample to decrease exponentially over time (Lal & Arnold 1985; Klein *et al.* 1986). If there is no erosion, $^{26}Al/^{10}Be$ ratios will fall somewhere along the blue line, yielding an apparent exposure age. If there is steady-state erosion (measured here in metres per million years), the ratio lies on the red steady-state erosion curve at a point determined by the steady-state erosion rate. The area between the two curves is called the steady-state erosion island (Lal 1991). If ratios plot below that island, then burial is inferred. Upon burial, ratios will fall in the direction of the dashed black arrows and burial time can be inferred when measured ratios are related to the ^{10}Be concentration in the sample (green lines).

Erosion depths from cosmogenic radionuclide inheritance

Passive continental margins in glaciated regions, such as those in Scandinavia, Scotland, Greenland, Canada and Antarctica, have apparently repeatedly experienced hundreds of metres of erosion underneath outlet glaciers whereas adjacent uplands remained unmodified (Sugden 1968, 1974; Hall & Sugden 1987; Glasser & Hall 1997; Kleman & Stroeven 1997). Hence, there is a remarkable spatial variation in subglacial erosion that may be reflected in cosmogenic radionuclide inheritance. Inheritance refers to the remnant cosmogenic radionuclide concentration from a prior exposure history. Thus far our discussion has largely dealt with cosmogenic nuclide accumulation in the absence of, or under conditions of steady-state erosion. Glacial erosion is a non-steady-state event. When ice overrides a rock surface it effectively shields that surface from cosmic rays and any further cosmogenic nuclide accumulation. If the overriding event is erosive, cosmogenic nuclides accumulated before the glaciation will be removed. Depending on the depth of glacial erosion, this cosmogenic nuclide removal may not be complete. In this case the surface retains some of the cosmogenic nuclides and hence an inheritance from the previous exposure event (Fig. 3). The depth of erosion required to remove the entire cosmogenic nuclide inventory of a previously exposed rock depends on the duration and erosion rate of the prior exposure history (Fig. 3) but is of the order of up to several metres. In cases where the timing of glaciations is independently constrained cosmogenic nuclide inheritance can be used to calculate the amount of rock removed by the glacial event (Briner & Swanson 1998; Fabel & Harbor 1999).

Complex exposure and shielding histories

Another challenge in dating landscape surfaces on glaciated passive continental margins is that the effects of burial by ice in a complex exposure and burial history for long-lived surfaces need to be addressed. Many bedrock outcrops on glaciated passive margin mountains underwent a complex history of burials (typical duration of *c.* 100 ka) and re-exposures (typical duration of *c.* 10 ka) (e.g. Kleman & Stroeven 1997). We can approach unravelling complex burial histories by comparing multiple cosmogenic radionuclides with different production rates and half-lives (Bierman *et al.* 1999; Fabel & Harbor 1999).

The cosmogenic nuclide concentration and $^{26}Al/^{10}Be$ ratio in quartz at the ground surface should plot somewhere within the steady-state erosion island (Fig. 4). If the surface is suddenly shielded from cosmic radiation by ice, then ^{10}Be and ^{26}Al production ceases. Radioactive decay then lowers the inherited $^{26}Al/^{10}Be$ ratio over time along the dashed black arrows (Fig. 4) because ^{26}Al decays faster than ^{10}Be (Lal 1991). Burial time (green curves in Fig. 4) can be inferred when measured ratios are related to the ^{10}Be concentration in the sample (Granger & Muzikar 2001). As in the case of obtaining steady-state erosion rates and exposure duration, burial dating requires steady-state conditions. A sample experiencing a non-steady-state erosion event after prolonged exposure at low steady-state erosion rates will also plot below the steady-state erosion island, but in this case burial cannot be inferred. This serves to illustrate the need for careful geomorphological interpretation when selecting sample sites.

Although surface exposure ages, erosion rates, and burial ages can be obtained for well-constrained situations, complex surface exposure histories are far more challenging. Interpretation of cosmogenic radionuclide concentrations is much more complicated if a surface has experienced multiple periods of exposure, erosion and burial, each of variable length (Bierman *et al.* 1999). Stable nuclides such as ^{21}Ne provide an important additional tool to assist in investigating complex exposure histories. Because ^{21}Ne is stable, there is no reduction in concentration during periods of burial by ice, in contrast to the radionuclides. Thus measurements of ^{21}Ne provide an indication of total exposure time, with loss owing to erosion, but not to burial. Measurement of ^{21}Ne permits studies of exposure histories beyond the limit of a few million years imposed by the ^{10}Be and ^{26}Al half-lives, and comparison with ^{10}Be and ^{26}Al concentrations should provide further insight into burial history. Thus, cosmogenic ^{21}Ne can provide additional glaciological information of pertinence to landscape history and landscape surface reconstruction (Niedermann *et al.* 1993).

Geomorphological principles for landscape surface reconstruction: examples from northern Sweden

The presence of relict surface remnants in the northern Swedish mountains provides a reference surface against which subsequent glacial erosion can be quantified. A comparison with Pliocene–Pleistocene offshore deposits requires that the amount, pattern and timing of this glacial erosion is better understood. As a first step we need to consider the principles for reconstructing the reference surface across areas that were dissected by glacial erosion.

On the largest landscape scale that we consider, that of individual mountain blocks, glacial valleys and the mountain foreland, reconstructing the elevation and shape of the reference surface involves considerations of (1) the configuration of the preglacial fluvial drainage system, (2) fluvial incision during interglacial periods, (3) glacial erosion of valleys by selective linear erosion, (4) glacial erosion by areal scouring, (5) localized subglacial modification of preglacial upland surfaces, and (6) lowering of preglacial upland surfaces by nonglacial processes. We will first consider the significance of each of these erosive phases for preglacial landscape modification, and then consider the application of *in situ* produced cosmogenic nuclide techniques to illuminate the magnitude, pattern and timing of these erosion phases.

Configuration of the preglacial fluvial drainage system

The preglacial evolution of the fluvial drainage pattern of the passive-margin mountain range in northern Scandinavia has been controlled primarily by uplift-induced base-level changes in the Baltic depression (Wråk 1908; Rudberg 1954; Lidmar-Bergström & Näslund 2002). The eastern mountain foreland is characterized by a stepped 'plains with residual hills' morphology generally at 300–400 and 400–550 m above sea level (e.g. Fredén 1994, pp. 50–51), referred to as the Muddus Plains (Wråk 1908). The Muddus Plains are the youngest preglacial fluvial surfaces of the region, the youngest of which was the last base level for the evolution of the preglacial fluvial drainage system in the mountain range and the one that must be used in any reconstruction of upstream mountain geomorphology.

Fluvial incision during interglacial periods

Fluvial incision of the preglacial mountain morphology continued after the Neogene onset of glaciation. For example, Kleman & Stroeven 1997, Figs 10 and 11) showed the distribution of what are presumably Plio-Pleistocene fluvial valleys in the northwestern Swedish mountains.

These valleys are all contained within relict surface remnants, are lowered relative to this surface by <300 m, and join prominent glacially modified U-shaped valleys (Fig. 5). Potentially, the size, depth and gradient of these fluvial valleys reflect the amount of time that was available for cutting them. Rivers probably cut these valleys during interglacial times and in

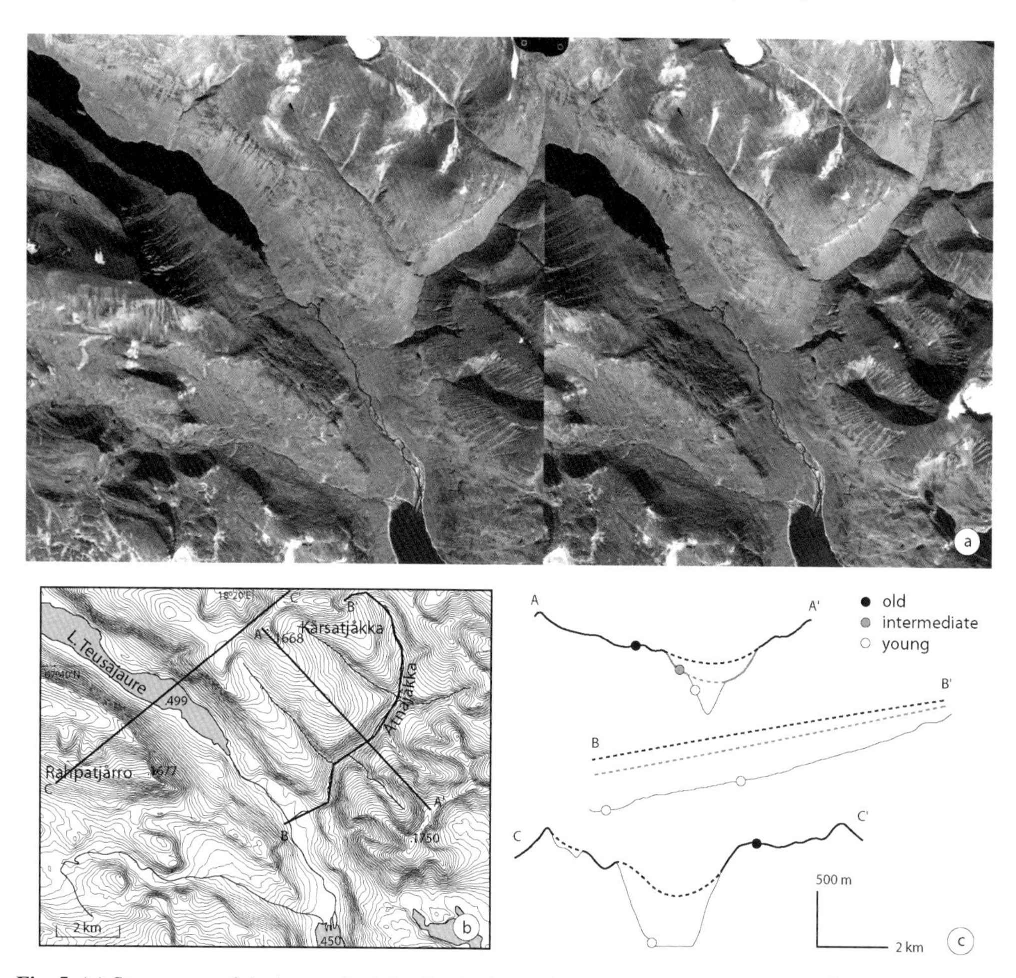

Fig. 5. (**a**) Stereogram of the largest fluvial valley in the northern Swedish mountain range, Ätnajåkka, joining a glacial valley, the Teusajaure valley. The V-shaped valley has been lowered by 300 m below the valley bench at its mouth (panel b). The river, of 9 km length, has a concave valley profile and is graded to a level that is 110 m above the Teusajaure valley floor (Kleman & Stroeven 1997). Above the 'old' valley bench (panel c) the preglacial surface extends towards the summits of the Kårsatjåkka mountain block. The Teusajaure valley, on the other hand, has been lowered by glacial erosion, initiating the accelerated downcutting of Ätnajåkka, leaving a typical valley floor (ice-moulded bedrock surfaces and elongated lakes occupying overdeepenings) and valley wall glacial morphology. The occurrence of preglacial surface remnants helps in reconstructing the preglacial morphology. Cosmogenic nuclide concentrations of bedrock surfaces in each of these three settings (preglacial, glacial and fluvial) would potentially yield widely different results. This is because, on the basis of the geomorphological interpretation of Kleman & Stroeven (1997), (1) preglacial surfaces were preserved underneath cold-based ice and would tend to yield high concentrations of isotopes and low surface erosion rates, (2) glacial surfaces underwent intense modification at some point during glaciation, resetting the cosmogenic isotope clock, and (3) fluvial surfaces have recorded maximum surface lowering during interglacial periods and would tend to yield the lowest cosmogenic isotope concentrations and highest surface erosion rates (Copyright: National Land Survey of Sweden 2002). (**b**) Topographical map with profiles (Courtesy of the National Land Survey, 2002. Excerpt from GSD-elevation data, case number L2000/646). (**c**) Topographical profiles illustrating potential cosmogenic nuclide sampling sites (dots, expected relative ages indicated) and possible preglacial valley reconstructions (dashed). For Profiles A–A′ and B–B′, two alternatives have been given, reflecting an uncertainty in the geomorphological interpretation of the extent of the preglacial surface.

response to base-level lowering as a result of glacial erosion (Ahlmann 1919; Rudberg 1992). Kleman & Stroeven (1997) hypothesized that the largest fluvial valley, Ätnajåkka (Fig. 5), is the oldest valley, which implies that the U-shaped valley which Ätnajåkka joins, the Teusajaure valley, should be one of the oldest glacial valleys in the mountain range. Extending this logic to conclude that other glacial valleys (some of which are much larger) must be younger in age,

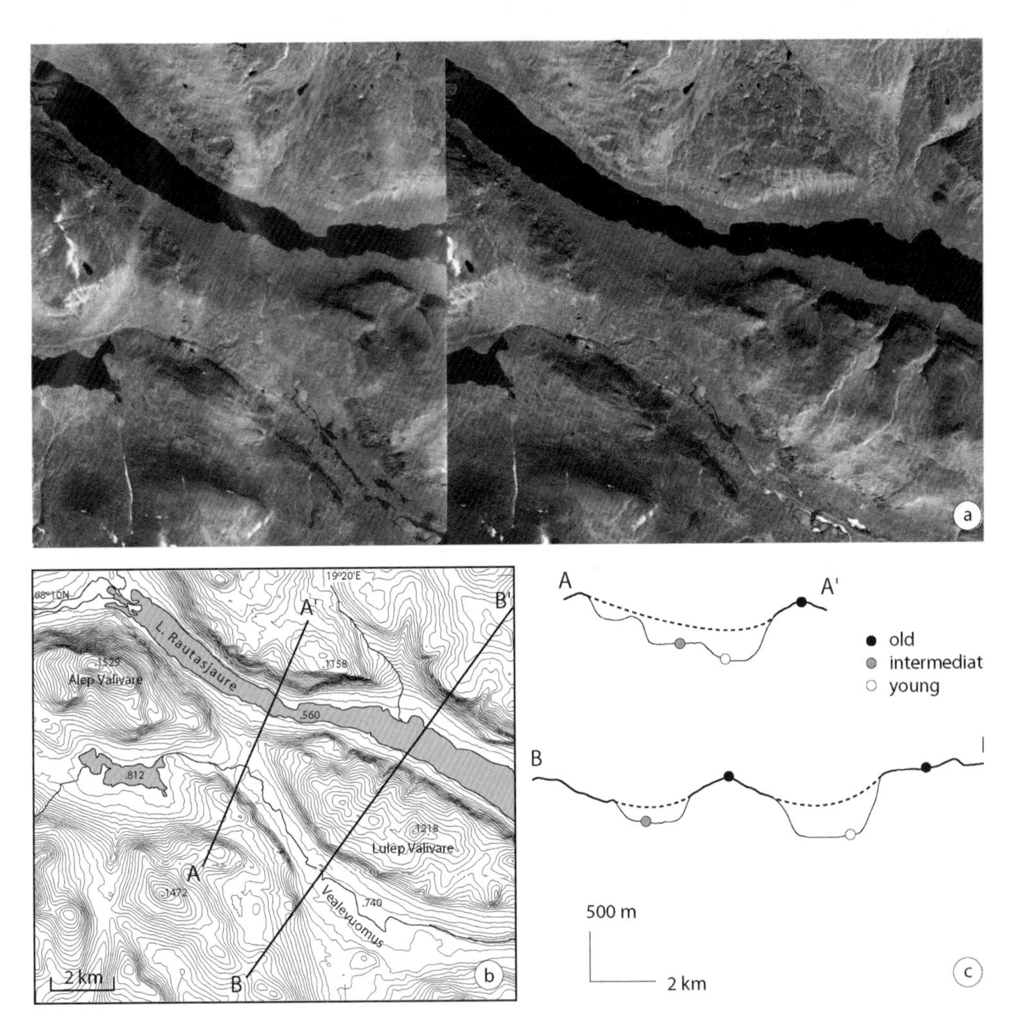

Fig. 6. (**a**) Stereogram of the junction between a small glacial valley, Vealevuomus, in the lower centre, and a much larger glacial valley, Rautas, in the upper centre (water-filled). The floor of Vealevuomus is hanging 180 m above the floor of Rautas valley (panel b). Hanging valleys are typical features of glaciated landscapes (Sugden & John 1976) and often a consequence of selective linear erosion. Although ice modified both these preglacial valleys, it did so more severely in Rautas valley, presumably because it was deeper, wider and better aligned for ice flow. Preglacial surface remnants occur on higher surfaces surrounding this junction and can aid in reconstructing the preglacial valley topography (panel c). Cosmogenic nuclide concentrations of bedrock surfaces in these valleys could potentially yield different results. This is because, given this difference in topography, Rautas valley may have been significantly modified by subglacial erosion, which would reset the cosmogenic isotope clock, whereas basal ice in Vealevuomus may have remained below the pressure melting point and inhibited erosion. Hence cosmogenic isotope concentrations on multiple isotopes could address differences in glacial erosion rates (Copyright: National Land Survey of Sweden 2002). (**b**) Topographical map with profiles (Courtesy of the National Land Survey, 2002. Excerpt from GSD-elevation data, case number L2000/646). (**c**) Topographical profiles illustrating potential cosmogenic nuclide sampling sites (dots, expected relative ages indicated) and possible preglacial valley reconstructions (dashed).

on the basis of the size of the fluvial valleys draining into them, represents one possible approach. However, these fluvial valleys may be the upper-reach remnants of fluvial valleys that were once much larger, but for which the lower reaches have been truncated as subsequent glaciers widened their valleys. A detailed geomorphological study of the valley benches may reveal the depth to which the preglacial valley was cut before glacially induced accelerated erosion commenced.

Glacial erosion by selective linear erosion

Glacial modification of landscape surfaces occurs in two distinctive patterns, by selective linear erosion of pre-existing depressions or areas less resistant to erosion, and by areal scouring of topographically less variable terrain (Sugden 1968; Sugden & John 1976; Harbor 1995). Selective linear erosion occurs when there are extreme spatial variations in rates of erosion under an ice sheet. Because pressure-melting conditions at the base of an ice sheet are necessary for both basal sliding and extensive glacial erosion, and these conditions are most common under thicker ice, pre-existing depressions and the deepest of preglacial valleys are lowered preferentially by glacial erosion. This enhances relief and has a positive-feedback effect on subglacial temperature and pressure-melting patterns (Oerlemans 1984; Mazo 1991). Thus pre-existing depressions and valleys aligned with the ice flow direction are preferentially eroded, and other areas are subject to much less or no erosion.

Two features typical of passive margin landscapes that can be explained in terms of selective linear erosion are sharp glacial surface–preglacial surface boundaries (Fig. 5) and hanging valleys (Fig. 6). The total relief in the northern Swedish mountains is >1 km but relief within existing upland remnants is *c.* 600 m (Kleman & Stroeven 1997). Hence, there was ample preglacial relief to establish strong subglacial temperature gradients favouring preferential deepening of pre-existing fluvial valleys or preglacial depressions. It is possible to constrain the original depth of the fluvial valleys or preglacial depressions based on (1) the inferred amount of glacial lowering from the depth of 'glacial-age' fluvial valleys joining them, (2) the shape and gradient of bordering relict slopes, and (3) fluvial valley gradients between rare preserved 'upland' preglacial-age valley floors and the youngest lowland Muddus Plain. On the basis of these lines of reasoning it is apparent that the larger glacial valleys have not been deepened by more than *c.* 400 m (Fig. 5b and c), although some valleys appear to have experienced more erosion locally (Kleman & Stroeven 1997).

Glacial erosion by areal scouring

On an ice-sheet scale, zones of areal scouring occur where the subglacial thermal effect of topographical convergence and divergence is of insufficient magnitude to counteract the overall subglacial melting regime. These conditions occur where topography is not pronounced compared with the thickness of the ice sheet, and where the heat of ice deformation suffices to initiate basal melting. These conditions were apparently present west of the elevation axis in the northern Swedish mountains, probably because ice accelerated towards the fjordal coast of Norway (Kleman & Stroeven 1997). West of this region of areal scour, ice flow crossed a threshold in basal topography and ice thickness, leading to conditions of selective linear erosion and the preservation of high-elevation perched relict surfaces along the Norwegian coast (Wråk 1908; Dahl 1966; Peulvast 1985).

Reconstructing the original elevation and shape of landscapes that subsequently underwent areal scouring is especially challenging because there are few, if any, undisturbed surfaces against which to compare modified surfaces. One approach is to view the areally scoured landscape as, at minimum, the result of stripping of material produced by preglacial chemical weathering. Hence, the minimum amount of erosion was approximately equal to the thickness of the weathering mantles that were removed (Glasser & Hall 1997). Limited observations in the northern Swedish mountains indicate a weathering mantle thickness of <10 m (e.g. Lundqvist 1985; Peulvast 1985; Hirvas *et al.* 1988; Olsen *et al.* 1996; Rea *et al.* 1996). In addition, glacial erosion may have extended well below the preglacial weathering base. Careful examination of overridden features, such as degraded cirque headwalls (Kleman & Stroeven 1997), may yield an additional estimate of the amount of solid bedrock removed. Finally, convergence must be sought between these former estimates of weathering mantle and bedrock erosion and an interpolated surface between the westernmost occurrences of preglacial remnants in Sweden and the easternmost outliers in Norway.

Localized subglacial modification of preglacial upland surfaces

Detailed studies of surface morphology (e.g. Kleman 1992; Clarhäll & Kleman 1999) indicate that subglacial reorganization of some areas within relict upland surfaces has occurred. Typically, this involves sporadic minor erosion, transport and deposition of the weathering mantle (lee-side scarps, stoss-side moraines), reshaping of a weathering mantle or till into till lineations (fluting), erosion of bedrock (striae), and deposition of till. Within relict areas evidence for subglacial reworking is generally confined to shallow depressions (Kleman *et al.* 1999). Boundaries between nonglacial and glacial surfaces are sometimes extremely sharp and have been interpreted as longitudinal and transverse subglacial sliding boundaries (Kleman & Borgström 1990; Kleman 1992). In the latter case, erosion occurred on the lee side of relict hills (creating a scarp) whereas zones of deposition (of the eroded weathering mantle) occurred on the stoss side of relict hills (Kleman *et al.* 1999). Currently, the timing of landscape modification can only be hypothesized from the association of glacial landforms (for example, the transverse lee-side scarps) with the inferred ice flow direction (and its age relative to other ice flow events).

Lowering of preglacial upland surfaces by nonglacial processes

The lowering of preglacial upland surfaces by nonglacial processes is probably mainly by periglacial activity as indicated by (1) the absence of linear features of fluvial origin, (2) the ubiquitous presence of gentle convex–concave slope profiles, (3) the presence of sporadic permafrost, especially at higher elevations, which promotes periglacial surface activity, (4) the ubiquitous presence of active and relict periglacial phenomena across uplands, and (5) the presence of tors on interfluves. The absence of fluvial erosive features on relict upland surfaces, such as dendritic river patterns with interlocking spurs, indicates that fluvial processes left no significant geomorphological imprint on the upland preglacial landscape until accelerated fluvial incision occurred during interglacial periods. However, in conjunction with the sporadic presence of permafrost (Lundqvist 1962;Sollid *et al.* 2000), transport of solutes in (melt)water through the active layer probably promoted widespread but low rates of long-term landscape lowering.

The most important processes of landscape denudation in these relict areas, it appears, are frost weathering and periglacial slope processes, such as solifluction. For example, stone stripes and other sorted phenomena resulting from creep processes on slopes are ubiquitous, as is mountain-top detritus on upland flats (e.g. Rea *et al.* 1996). The presence of preserved periglacial phenomena within upland relict landscapes (e.g. Clarhäll & Kleman 1999) as well as active forms in lowland locations indicates that landscape lowering by periglacial processes dominates and is probably of interstadial age. Tors are also conspicuous features of periglacial landscapes. Generally, it is considered that tors form as deep chemical weathering features (an uneven weathering surface), which are subsequently uncovered by stripping processes (e.g. Linton 1955). Such tors exhibit clear sheeting and rounded edges and are found in the northern Swedish lowlands, where they survived despite ice overriding (e.g. Hättestrand & Stroeven 2002; Stroeven *et al.* 2002). However, tors in the northern Swedish mountains exhibit an angular frost-shattered appearance, and this type of tor has previously been considered to have formed or to have been uncovered in a periglacial environment (Palmer & Radley 1961).

Application of *in situ* cosmogenic nuclide techniques to passive margin erosion history

Measurements of multiple cosmogenic nuclide concentrations of bedrock surfaces can potentially be used to reconstruct the timing and rates of landscape change of formerly glaciated passive continental margins. In particular, they can be used to (1) distinguish between surfaces of different age, (2) distinguish between surfaces that experienced different erosion histories, and (3) address complex exposure and shielding histories. Of these, calculations of long-term surface erosion rates are particularly useful in aiding reconstructions of the original surface morphology across areas of glacial erosion.

The most fundamental postulate that can be tested using cosmogenic isotope concentrations in bedrock is that ‘preglacial surfaces’ were preserved underneath ice sheets and thus can be used as a reference horizon against which to measure the magnitude and pattern of glacial erosion. If these surfaces were preserved underneath cold-based ice, then the cosmogenic nuclide implication is that the difference between the relative concentrations of stable ^{21}Ne and radioactive ^{10}Be (i.e. the ^{21}Ne$-^{10}$Be contrast) approaches the maximum possible given the glaciation history of the passive margin (Fig. 7) and that ^{26}Al/^{10}Be ratios are lower than expected for the other surfaces. If these surfaces are in fact

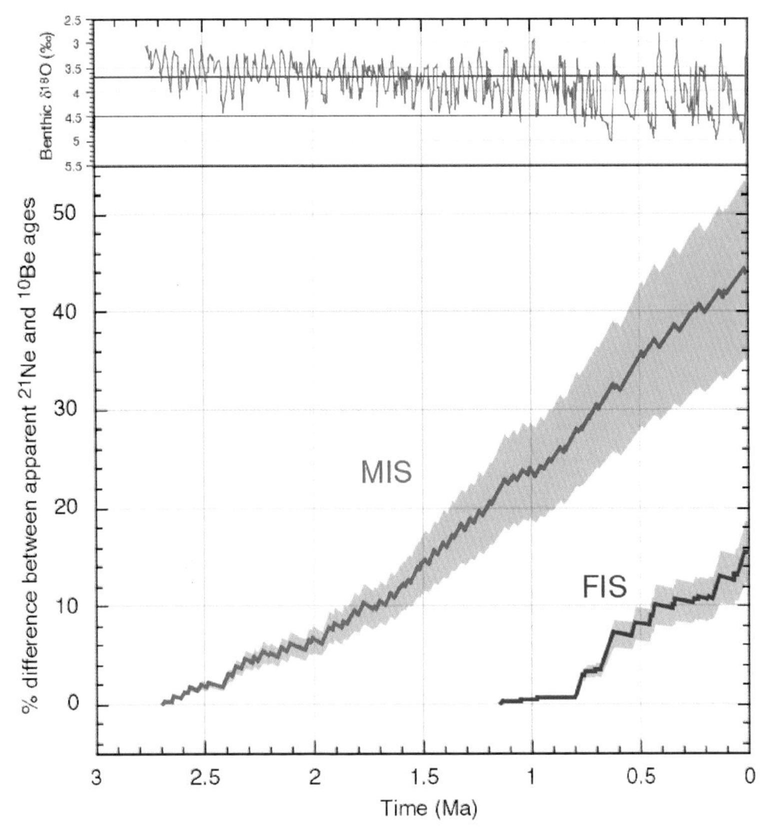

Fig. 7. Plot of the difference (%) between ^{21}Ne and ^{10}Be apparent exposure ages for a surface that has undergone a complex burial and exposure history imposed by the δ^{18}O record from DSDP Site 607 (top). The two curves are based on the assumption that Mountain Ice Sheets (MIS) covered the northern Swedish mountains when δ^{18}O was between 3.7 and 4.5‰, and Fennoscandian Ice Sheets (FIS) covered mountains and lowlands when δ^{18}O was >4.5‰. The grey shaded areas for both curves indicate 1σ errors if we assume that analytical precision for ^{10}Be and ^{21}Ne measurements are 5% and 20%, respectively.

not relict at all, and were formed either in postglacial times or by significant glacial erosion, then there should be no relative difference between these three isotopes; that is, they should give the same apparent exposure age.

There are three fundamentally different bedrock surfaces that could be contrasted by their cosmogenic nuclide concentrations (Fig. 5): relict surfaces (unmodified, slightly modified), glacial surfaces (selective linear erosion, areal scouring), and fluvial surfaces (preglacial age, glacial age). One would predict the following: (1) the unmodified relict surfaces would have the highest concentrations of isotopes, the lowest ^{26}Al/^{10}Be ratios, the highest ^{21}Ne–^{10}Be contrasts, and the lowest surface erosion rates; (2) glacial surfaces of selective linear erosion (U-shaped valleys) would yield deglaciation ages or older (depending on the amount of glacial erosion and, therefore, the inheritance signal), have ^{26}Al/^{10}Be ratios close to or equal to six, and small ^{21}Ne–^{10}Be contrasts; (3) fluvial surfaces, if actively formed during interglacial periods (Rudberg 1992; Kleman & Stroeven 1997), including the present interglacial, would have the lowest cosmogenic isotope concentrations, ^{26}Al/^{10}Be ratios equal to six, ^{21}Ne–^{10}Be contrasts of zero, and the highest current surface erosion rates. If the stability of upland surfaces is supported by the isotopic data, additional geomorphological interpretations of landscape change can be tested, and preglacial landscape reconstructions based on geomorphology can be strengthened.

In the case of hanging U-shaped valleys, cosmogenic nuclide approaches provide potential for differentiating between (1) glacial valleys of different age, where the hanging valley is left 'high and dry' preserved underneath erosionally ineffective ice, and (2) valleys that are contemporaneous but experienced different erosion rates as a result of differing subglacial conditions (the amount of time available for erosion and/or the rate of erosion). In the case of the Vealevuomus and Rautas valleys (Fig. 6), if both valleys underwent significant erosion during the same glacial event (i.e. (2) above), then bedrock samples from the two locations should yield similar exposure ages, ^{26}Al/^{10}Be ratios equal to six, and ^{21}Ne$-$^{10}Be contrasts of zero. However, if basal ice in Vealevuomus remained below the pressure melting point and inhibited erosion (i.e. (1) above), then bedrock samples in this valley should have an inheritance signal and yield older apparent exposure ages, lower ^{26}Al/^{10}Be ratios, and larger ^{21}Ne$-$^{10}Be contrasts than samples from the Rautas valley.

Finally, cosmogenic radionuclide measurements on bedrock samples can also contribute to resolving issues of glacial modification and nonglacial lowering of preglacial upland surfaces. For example, tors can be used to infer minimum rates of preglacial landscape lowering. Cosmogenic nuclide concentrations of multiple isotopes from tor summit flats should indicate the timing of tor exposure and long-term surface erosion rates. These long-term surface erosion rates on tors provide minimum long-term erosion rates for the surrounding summit flats. This is because the erosion rates of tors, assuming their formation or uncovering occurred in a periglacial environment, have to be lower than those of the surrounding summit flats, otherwise the tors would not exist. A cosmogenic radionuclide study of upland terrain in the Rocky Mountains, USA, yielded a long-term surface lowering rate of $10\,m\ Ma^{-1}$ for summit flats (Small *et al.* 1997). This value appears reasonable for northern Swedish conditions when compared with the height of individual tors above their surrounding (generally $<10\,m$) and the amount of erosion by fluvial and glacial processes on adjacent surfaces (an order of magnitude higher).

A framework for interpreting cosmogenic isotope concentrations in terms of patterns, rates and timing of landscape change, as is presented here, and developed before interpretation of actual data, gives maximum credibility to data interpretation and helps in efforts to resolve differences in regional datasets to arrive at a common interpretation. The practical application of this framework to the geological and geomorphological traces of the last glacial cycle, will help establish a set of attributes (typically, erosion rates and patterns) that serve as a working model for interpretations further back in time.

Concluding remarks: uneven sediment production on glaciated passive margins

The glacial chronologies of several passive margin mountains are well understood at a general level (e.g. Mangerud *et al.* (1996), Kleman *et al.* (1997) and Kleman & Stroeven (1997) for Scandinavia). A patchwork of glaciated terrain and preglacial relict upland surfaces in these mountains reflects (1) the total erosive impact of the late Cenozoic glaciers and ice sheets that covered the mountains, and (2) that this subglacial erosion must have been areally restricted. What is not well established is the timing and pattern of erosion and landscape development associated with this glacial chronology, an issue of critical importance when attempting a comparison of the onshore erosion histories and offshore sediment accumulations (e.g. Glasser & Hall 1997).

Conventional dating and analysis have provided an excellent way to begin unravelling the timing and pattern of erosion, landform development, and possible landform preservation under ice. However, attempts to utilize geomorphology and remote sensing techniques have often been frustrated by the limitations of conventional dating techniques. We describe a new approach for investigating landscape evolution in mountainous areas, the *in situ* production of cosmogenic nuclides in bedrock surfaces of landscapes affected by glacial, fluvial (preglacial or interglacial) and periglacial (interstadial) process systems (Cerling 1990; Nishiizumi *et al.* 1993; Bierman 1994; Cerling & Craig 1994; Small & Anderson 1995; Fabel & Harbor 1999). Cosmogenic nuclides produced in rocks near the ground surface by reactions with cosmic rays can be used to determine apparent surface exposure age and landscape preservation, and constrain erosion depths and duration of burial by ice.

This paper presents a coherent framework for interpreting measurements of *in situ* produced cosmogenic nuclides in bedrock in terms of timing, rates and patterns of landscape change on glaciated passive margins. The application of cosmogenic nuclide techniques to the reconstruction of preglacial surfaces provides new information to complement that obtained using traditional geomorphological approaches (Andersen & Nesje 1992; Nesje & Whillans 1994; Riis 1996; Glasser & Hall 1997). The cosmogenic nuclide technique has previously been used to investigate aspects of landscape change in glaciated regions such as the age of specific glacial events (e.g. Phillips *et al.* 1990, 2000; Brook *et al.* 1995; Stone *et al.* 1996; Ivy-Ochs *et al.* 1997; Bierman *et al.* 1999; Jackson *et al.* 1999; Schäfer *et al.* 1999), erosion rates over glacial surfaces (Nishiizumi *et al.* 1989; Brook *et al.* 1995; Briner & Swanson 1998), landscape denudation rates (Small *et al.* 1997; Summerfield *et al.* 1999), and the elevation of former ice sheet surface profiles (see Brook *et al.* 1996; Stone *et al.* 1998; Ackert *et al.* 1999; Kaplan *et al.* 2001). However, compared with these other studies, a strength of the structure presented here is that it is developed before, rather than as a response to, data interpretation. We now have a framework within which to interpret nuclide concentrations in terms of apparent exposure age and surface erosion rates, and which provides an overview of sampling strategies to determine the timing, rate and pattern of landscape change on glaciated passive margins. This is particularly important as complex burial–shielding and exposure histories have probably affected all samples where substantial preservation length has been inferred from geomorphological evidence.

Given the number of glacial cycles (*c*. 40) that have affected the landscape in the last 2.7 Ma, we cannot expect cosmogenic nuclide techniques to resolve individual events beyond the last glacial cycle. However, through cosmogenic dating of key landscape elements we can establish the pattern and selectivity of erosion by the last ice sheet, thereby establishing a model for the erosional functionality of ice sheets in rugged terrain. Once such a model is erected, it can be used to provide insight into process patterns and interrelationships across the glacial and interglacial cycles of the Quaternary period. This provides a solid foundation for attempts to improve geomorphologically based reconstructions of preglacial surfaces, reconstruct and analyse the dynamics of landscape change in glacial time, and define the consequences of different process regimes in terms of erosion patterns, sediment transport and the supply of sediments that are deposited offshore. The thickness of offshore sediments is one key ingredient considered in hydrocarbon exploration. However, despite the fact that important quantities of sediment were generated by glacial processes, current techniques of estimating onshore exhumation patterns and rates do not resolve margin exhumation during late Cenozoic time (e.g.Lidmar-Bergström & Näslund 2001). It is these offshore sediments, often used in generalized recontructions of passive margin development (e.g. Riis 1996) that, in the future, will provide the link between offshore patterns of passive margin sedimentation and onshore reconstructions of passive margin evolution. Furthermore, studies of the erosional history of upland landscapes will also aid in resolving climate debates where the source area of eroded sediments is of key importance, such as the mechanisms leading to, and the implications of Heinrich events for circum-Atlantic climate change and landform development.

This paper has benefited from a detailed formal review by P. Bishop, and from comments by K. Lidmar-Bergström (informal review) and A.G. Doré. Funding for part of the research reported here was provided by the Swedish Natural Science Research Council (NFR), Grants G-AA/GU 12034-300 and G-AA/GU 12034-301, and by the US National Science Foundation, Grant OPP 9818162. This manuscript was completed while J.H. was supported by the New Zealand–United States Educational Foundation as a Fulbright Senior Scholar.

References

ACKERT, R.P. JR, BARCLAY, D.J. JR, BORNS, H.W. JR, CALKIN, P.E. JR, KURZ, M.D. JR, FASTOOK, J.L. JR & STEIG, E.J. JR 1999. Measurements of past ice sheet elevations in interior West Antarctica. *Science*, **286**, 276–280.

AHLMANN, H.W. 1919. Geomorphological studies in Norway. *Geografiska Annaler*, **1**, 1–148.

ANDERSEN, B.G. & NESJE, A. 1992. Quantification of late Cenozoic glacial erosion in a fjord landscape. *Sveriges Geologiska Undersökning, Serie Ca*, **81**, 15–20.

BALLANTYNE, C.K. 1994. Scottish landform examples—10. *The tors of the Cairngorms. Scottish Geographical Magazine*, **110**, 54–59.

BIERMAN, P.R. 1994. Using *in situ* produced cosmogenic isotopes to estimate rates of landscape evolution: a review from the geomorphic perspective. *Journal of Geophysical Research*, **99**, 13885–13896.

BIERMAN, P.R., MARSELLA, K.A., PATTERSON, C., DAVIS, P.T. & CAFFEE, M. 1999. Mid-Pleistocene cosmogenic minimum-age limits for pre-Wisconsinan glacial surfaces in southwestern Minnesota and southern Baffin Island: a multiple nuclide approach. *Geomorphology*, **27**, 25–39.

BRINER, J.P. & SWANSON, T.W. 1998. Using inherited cosmogenic ^{36}Cl to constrain glacial erosion rates of the Cordilleran ice sheet. *Geology*, **26**, 3–6.

BROOK, E.J., BROWN, E.T., KURZ, M.D., ACKERT, R.P. JR, RAISBECK, G.M. JR & YIOU, F. JR 1995. Constraints on age, erosion, and uplift of Neogene glacial deposits in the Transantarctic Mountains determined from *in situ* cosmogenic ^{10}Be and ^{26}Al. *Geology*, **23**, 1063–1066.

BROOK, E.J., NESJE, A., LEHMAN, S.J., RAISBECK, G.M. & YIOU, F. 1996. Cosmogenic nuclide exposure ages along a vertical transect in western Norway: implications for the height of the Fennoscandian ice sheet. *Geology*, **24**, 207–210.

CERLING, T.E. 1990. Dating geomorphic surfaces using cosmogenic ^{3}He. *Quaternary Research*, **33**, 148–156.

CERLING, T.E. & CRAIG, H. 1994. Geomorphology and *in situ* cosmogenic isotopes. *Annual Review of Earth and Planetary Sciences*, **22**, 273–317.

CLARHÄLL, A. & KLEMAN, J. 1999. Distribution and glaciological implications of relict surfaces on the Ultevis plateau, northwestern Sweden. *Annals of Glaciology*, **28**, 202–208.

CLARK, P.U. & POLLARD, D. 1998. Origin of the middle Pleistocene transition by ice sheet erosion of regolith. *Paleoceanography*, **13**, 1–9.

DAHL, R. 1966. Block fields, weathering pits and tor-like forms in the Narvik mountains, Nordland, Norway. *Geografiska Annaler*, **48A**, 55–85.

DUNAI, T.J. 2000. Scaling factors for production rates of *in situ* produced cosmogenic nuclides: a critical reevaluation. *Earth and Planetary Science Letters*, **176**, 157–169.

DUNNE, J., ELMORE, D. & MUZIKAR, P. 1999. Scaling factors for the rates of production of cosmogenic nuclides for geometric shielding and attenuation at depth on sloped surfaces. *Geomorphology*, **27**, 3–11.

FABEL, D. & HARBOR, J. 1999. The use of *in-situ* produced cosmogenic radionuclides in glaciology and glacial geomorphology. *Annals of Glaciology*, **28**, 103–110.

FREDÉN, C. 1994. *Geology. National Atlas of Sweden*. SNA Publishing, Stockholm.

GJESSING, J. 1967. Norway's paleic surface. *Norsk Geografisk Tidsskrift*, **21**, 69–132.

GLASSER, N.F. & HALL, A.M. 1997. Calculating Quaternary glacial erosion rates in northeast Scotland. *Geomorphology*, **20**, 29–48.

GRANGER, D.E. & MUZIKAR, P.F. 2001. Dating sediment burial with *in situ*-produced cosmogenic nuclides: theory, techniques, and limitations. *Earth and Planetary Science Letters*, **188** (1–2), 269–281.

GRANGER, D.E. & SMITH, A.L. 2000. Dating buried sediments using radioactive decay and muogenic production of ^{26}Al and ^{10}Be. *Nuclear Instruments and Methods in Physics Research B: Beam Interactions with Materials and Atoms*, **172**, 822–826.

GRANGER, D.E., KIRCHNER, J.W. & FINKEL, R. 1996. Spatially averaged long-term erosion rates measured from *in situ*-produced cosmogenic nuclides in alluvial sediment. *Journal of Geology*, **104**, 249–257.

HALL, A.M. & SUGDEN, D.E. 1987. Limited modification of mid-latitude landscapes by ice-sheets: the case of north-east Scotland. *Earth Surface Processes and Landforms*, **12**, 531–542.

HARBOR, J.M. 1995. Development of glacial-valley cross sections under conditions of spatially variable resistance to erosion. *Geomorphology*, **14**, 99–107.

HÄTTESTRAND, C. & STROEVEN, A.P. 2002. A relict landscape in the centre of Fennoscandian glaciation: geomorphological evidence of minimal Quaternary glacial erosion. *Geomorphology*, **44**, 127–143.

HENDRIKS, B.W.H. & ANDRIESSEN, P.A.M. 2002. Pattern and timing of the post-Caledonian denudation of northern Scandinavia constrained by apatite fission-track thermochronology. *In*: DORÉ, A.G., CARTWRIGHT, J.A., STOKER, M.S., TURNER, J.P. & WHITE, N. (eds) *Exhumation of the North Atlantic Margin: Timing, Mechanisms and Implications for Petroleum Exploration*. Geological Society, London, Special Publications, **196**, 117–137.

HIRVAS, H., LAGERBÄCK, R., MÄKINEN, K., NENONEN, K., OLSEN, L., RODHE, L. & THORESEN, M. 1988. The Nord-Kalott Project: studies of Quaternary geology in northern Fennoscandia. *Boreas*, **17**, 431–437.

HÖLEMANN, J.A. & HENRICH, R. 1994. Allochthonous versus autochthonous organic matter in Cenozoic sediments of the Norwegian Sea: evidence for the onset of glaciations in the northern hemisphere. *Marine Geology*, **121**, 87–103.

IMBRIE, J., BOYLE, E.A., CLEMENS, S.C. & 15 OTHERS 1992. On the structure and origin of major glaciation cycles. 1. Linear responses to Milankovitch forcing. *Paleoceanography*, **7**, 701–738.

IMBRIE, J., BERGER, A., BOYLE, E.A. & 16 OTHERS 1993. On the structure and origin of major glaciation cycles. 2. The 100,000-year cycle. *Paleoceanography*, **8**, 699–735.

IVY-OCHS, S., SCHLÜCHTER, C., PRENTICE, M., KUBIK, P.W. & BEER, J. 1997. ^{10}Be and ^{26}Al exposure ages for the Sirius Group at Mount Fleming, Mount Feather and Table Mountain and the plateau surface at Table Mountain. *In*: RICCI, C.A. (ed.) *The Antarctic Region: Geological Evolution and Processes*. Terra Antartica Publication, Siena, 1153–1158.

JACKSON, L.E. JR, PHILLIPS, F.M. JR & LITTLE, E.C. JR 1999. Cosmogenic ^{36}Cl dating of the maximum limit of the Laurentide Ice Sheet in southwestern Alberta. *Canadian Journal of Earth Sciences*, **36**, 1347–1356.

JANSEN, E. & SJØHOLM, J. 1991. Reconstruction of glaciation over the past 6 Myr from ice-borne deposits in the Norwegian Sea. *Nature*, **349**, 600–603.

JANSEN, E., FRONVAL, T., RACK, F. & CHANNELL, J.E.T. 2000. Pliocene–Pleistocene ice rafting history and cyclicity in the Nordic Seas during the last 3.5 Myr. *Paleoceanography*, **15** (6), 709–721.

KAPLAN, M.R., MILLER, G.H. & STEIG, E.J. 2001. Low-gradient outlet glaciers (ice streams?) drained the Laurentide ice sheet. *Geology*, **29**, 343–346.

KLEIN, J., GIEGENGACK, R., MIDDLETON, R., SHARMA, P. & UNDERWOOD, J.R. 1986. Revealing histories of exposure using *in situ* produced ^{26}Al and ^{10}Be in Libyan desert glass. *Radiocarbon*, **28**, 547–555.

KLEMAN, J. 1992. The palimpsest glacial landscape in northwestern Sweden—Late Weichselian deglaciation landforms and traces of older west-centered ice sheets. *Geografiska Annaler*, **74A**, 305–325.

KLEMAN, J. & BORGSTRÖM, I. 1990. The boulder fields of Mt. Fulufjället, west–central Sweden—Late

Weichselian boulder blankets and interstadial periglacial phenomena. *Geografiska Annaler*, **72A**, 63–78.

KLEMAN, J. & STROEVEN, A.P. 1997. Preglacial surface remnants and Quaternary glacial regimes in northwestern Sweden. *Geomorphology*, **19**, 35–54.

KLEMAN, J., HÄTTESTRAND, C., BORGSTRÖM, I. & STROEVEN, A.P. 1997. Fennoscandian paleoglaciology reconstructed using a glacial geological inversion model. *Journal of Glaciology*, **43**, 283–299.

KLEMAN, J., HÄTTESTRAND, C. & CLARHÄLL, A. 1999. Zooming in on frozen-bed patches: scale dependent controls on Fennoscandian ice sheet basal thermal zonation. *Annals of Glaciology*, **28**, 189–194.

LAL, D. 1987. Cosmogenic nuclides produced *in situ* in terrestrial rocks. *Nuclear Instruments and Methods in Physics Research*, **B29**, 238–245.

LAL, D. 1991. Cosmic ray labeling of erosion surfaces: *in situ* nuclide production rates and erosion models. *Earth and Planetary Science Letters*, **104**, 424–439.

LAL, D. & ARNOLD, J.R. 1985. Tracing quartz through the environment. *Proceedings of the Indian Academy of Sciences (Earth and Planetary Sciences)*, **94**, 1–5.

LAL, D. & PETERS, B. 1967. Cosmic-ray produced radioactivity on the Earth. *Handbook of Physics*, **46**, 551–612.

LIDMAR-BERGSTRÖM, K. & NÄSLUND, J.O. 2002. Landforms and uplift in Scandinavia. *In*: DORÉ, A.G., CARTWRIGHT, J.A., STOKER, M.S., TURNER, J.P. & WHITE, N. (eds) *Exhumation of the North Atlantic Margin: Timing, Mechanisms and Implications for Petroleum Exploration*. Geological Society, London, Special Publications, **196**, 103–116.

LINTON, D.L. 1955. The problem of tors. *Geographical Journal*, **121**, 470–487.

LJUNGNER, E. 1949. The east–west balance of the Quaternary ice caps in Patagonia and Scandinavia. *Bulletin of the Geological Institute of Uppsala*, **33**, 11–96.

LUNDQVIST, J. 1962. *Patterned Ground and Related Frost Phenomena in Sweden*. Sveriges Geologiska Undersökning, **C583**.

LUNDQVIST, J. 1985. Deep-weathering in Sweden. *Fennia*, **163**, 287–292.

MANGERUD, J., JANSEN, E. & LANDVIK, J.Y. 1996. Late Cenozoic history of the Scandinavian and Barents Sea ice sheets. *Global and Planetary Change*, **12**, 11–26.

MAZO, V.L. 1991. Interactions of Ice Sheets: Instability and Self-organization. *International Association of Hydrological Sciences Publication*, **208**, 193–205.

NESJE, A. & WHILLANS, I.M. 1994. Erosion of Sognefjord, Norway. *Geomorphology*, **9**, 33–45.

NIEDERMANN, S., GRAF, T. & MARTI, K. 1993. Mass spectrometric identification of cosmic-ray-produced neon in terrestrial rocks with multiple neon components. *Earth and Planetary Science Letters*, **118**, 65–73.

NISHIIZUMI, K., WINTERER, E.L., KOHL, C.P., KLEIN, J., MIDDLETON, R., LAL, D. & ARNOLD, J.R. 1989. Cosmic ray production rates of ^{10}Be and ^{26}Al in quartz from glacially polished rocks. *Journal of Geophysical Research*, **94**, 17907–17915.

NISHIIZUMI, K., KOHL, C.P., ARNOLD, J.R. & 5 OTHERS 1993. Role of *in situ* cosmogenic nuclides ^{10}Be and ^{26}Al in the study of diverse geomorphic processes. *Earth Surface Processes and Landforms*, **18**, 407–425.

OERLEMANS, J. 1984. Numerical experiments on large-scale glacial erosion. *Zeitschrift für Gletscherkunde und Glazialgeologie*, **20**, 107–126.

OLSEN, L., MEJDAHL, V. & SELVIK, S.F. 1996. Middle and late Pleistocene stratigraphy, chronology and glacial history in Finnmark, North Norway. *Norsk Geologiske Undersøkelse Bulletin*, **429**, 1–111.

PALMER, J. & RADLEY, J. 1961. Gritstone tors of the English Pennines. *Zeitschrift für Geomorphologie, Neue Folge*, **5**, 37–52.

PEULVAST, J.-P. 1985. *In situ* weathered rocks on plateaus, slopes and strandflat areas of the Lofoten–Vesterålen, North Norway. *Fennia*, **163**, 333–340.

PHILLIPS, F.M., ZREDA, M.G., SMITH, S.S., ELMORE, D., KUBIK, P.W. & SHARMA, P. 1990. Cosmogenic chlorine-36 chronology for glacial deposits at Bloody Canyon, eastern Sierra Nevada. *Science*, **248**, 1529–1532.

PHILLIPS, W.M., SLOAN, V.F., SCHRODER, J.F. JR, SHARMA, P., CLARKE, M.L. & RENDELL, H.M. 2000. Asynchronous glaciation at Nanga Parbat, northwestern Himalaya mountains, Pakistan. *Geology*, **28**, 431–434.

PORTER, S.C. 1989. Some geological implications of average Quaternary glacial conditions. *Quaternary Research*, **32**, 245–261.

RAYMO, M.E., RUDDIMAN, W.F., BACKMAN, J., CLEMENT, B.M. & MARTINSON, D.G. 1989. Late Pliocene variation in northern hemisphere ice sheets and North Atlantic deep water circulation. *Paleoceanography*, **4**, 413–446.

REA, B.R., WHALLEY, W.B., RAINEY, M.M. & GORDON, J.E. 1996. Blockfields, old or new? Evidence and implications from some plateaus in northern Norway. *Geomorphology*, **15**, 109–121.

REUSCH, H. 1901. Nogle bidrag til forstaaelsen af hvorledes Norges dale og fjelde er blevne til. *Norges Geologiske Undersøkelse*, **32**, 124–217.

RIIS, F. 1996. Quantification of Cenozoic vertical movements of Scandinavia by correlation of morphological surfaces with offshore data. *Global and Planetary Change*, **12**, 331–357.

ROBINSON, C., RAISBECK, G.M., YIOU, F., LEHMAN, B. & LAJ, C. 1995. The relationship between ^{10}Be and geomagnetic field strength records in central North Atlantic sediments during the last 80 ka. *Earth and Planetary Science Letters*, **136**, 551–557.

RUDBERG, S. 1954. *Västerbottens bergggrundsmorfologi—ett försök till rekonstruktion av preglaciala erosionsgenerationer i Sverige*. Geographica, **25**.

RUDBERG, S. 1992. Multiple glaciation in Scandinavia—seen in gross morphology or not? *Geografiska Annaler*, **74A**, 231–243.

SCHÄFER, J.M., IVY-OCHS, S., WIELER, R., LEYA, I., BAUR, H., DENTON, G.H. & SCHLÜCHTER, C. 1999. Cosmogenic noble gas studies in the oldest landscape on Earth: surface exposure ages of the Dry Valleys, Antarctica. *Earth and Planetary Science Letters*, **167**, 215–226.

SHACKLETON, N.J., BACKMAN, J., ZIMMERMAN, H. & 14 OTHERS 1984. Oxygen isotope calibration of the onset of ice-rafting and history of glaciation in the North Atlantic region. *Nature*, **307**, 620–623.

SMALL, E.E. & ANDERSON, R.S. 1995. Geomorphologically driven late Cenozoic rock uplift in the Sierra Nevada, California. *Science*, **270**, 277–280.

SMALL, E.E., ANDERSON, R.S., REPKA, J.L. & FINKEL, R. 1997. Erosion rates of alpine bedrock summit surfaces deduced from *in situ* ^{10}Be and ^{26}Al. *Earth and Planetary Science Letters*, **150**, 413–425.

SOLHEIM, A., RIIS, F., ELVERHØI, A., FALEIDE, J.I., JENSEN, L.N. & CLOETINGH, S. 1996. Impact of glaciations on basin evolution: data and models from the Norwegian margin and adjacent areas—introduction and summary. *Global and Planetary Change*, **12**, 1–9.

SOLLID, J.L., HOLMLUND, P., ISAKSEN, K. & HARRIS, C. 2000. Deep permafrost boreholes in western Svalbard, northern Sweden and southern Norway. *Norsk Geografisk Tidsskrift*, **54** (4), 186–191.

STONE, J., LAMBECK, K., FIFIELD, L.K., EVANS, J.M. & CRESSWELL, R.G. 1996. A late glacial age for the main rock platform, western Scotland. *Geology*, **24**, 707–710.

STONE, J.O., BALLANTYNE, C.K. & FIFIELD, L.K. 1998. Exposure dating and validation of periglacial weathering limits, northwest Scotland. *Geology*, **26**, 587–590.

STROEVEN, A.P., FABEL, D., HÄTTESTRAND, C. & HARBOR, J. 2002. A relict landscape in the centre of Fennoscandian glaciation: cosmogenic radionuclide evidence of tors preserved through multiple glacial cycles. *Geomorphology*, **44**, 145–154.

SUGDEN, D.E. 1968. The selectivity of glacial erosion in the Cairngorm mountains, Scsotland. *Transactions of the Institute of British Geographers*, **45**, 79–92.

SUGDEN, D.E. 1974. Landscapes of glacial erosion in Greenland and their relationship to ice, topographic and bedrock conditions. *In*: BROWN, E.H. & WATERS, R.S. (eds) *Progress in Geomorphology*. Institute of British Geographers Special Publication, **7**, 177–195.

SUGDEN, D.E. & JOHN, B.S. 1976. *Glaciers and Landscape: a Geomorphological Approach*. Edward Arnold, London.

SUGDEN, D.E. & WATTS, S.H. 1977. Tors, felsenmeer, and glaciation in northern Cumberland Peninsula, Baffin Island. *Canadian Journal of Earth Sciences*, **14**, 2817–2823.

SUMMERFIELD, M.A., STUART, F.M., COCKBURN, H.A.P., SUGDEN, D.E., DENTON, G.H., DUNAI, T. & MARCHANT, D.R. 1999. Long-term rates of denudation in the Dry Valleys, Transantarctic Mountains, southern Victoria Land, Antarctica based on *in-situ*-produced cosmogenic ^{21}Ne. *Geomorphology*, **27**, 113–129.

WRÅK, W. 1908. Bidrag till Skandinaviens reliefkronologi. *Ymer*, **28**, 141–191.

The thermotectonic development of southern Sweden during Mesozoic and Cenozoic time

CHARLOTTE CEDERBOM

The CRUST Consortium, Department of Geology and Geophysics, Edinburgh University, Grant Institute, King's Buildings, Edinburgh EH9 3JW, UK (e-mail: cederbom@glg.ed.ac.uk)

Abstract: Late Carboniferous–Early Mesozoic exhumation of southern Sweden has previously been traced using apatite fission-track thermochronology. In addition, the morphotectonic development of the region has been studied using geomorphology. The aim of this study is to attain further knowledge of the Mesozoic and Cenozoic thermotectonic development of southern Sweden by integrating results from these methods.

Well-dated re-exposed palaeosurfaces and sedimentary records in the surrounding areas were used as constraints in the modelling of apatite fission-track data from the Precambrian basement. The obtained modelled thermal histories suggest that southern Sweden can be divided into three main tectonic areas associated with different cooling histories. In Triassic and Jurassic time, low to moderate exhumation in the central part was accompanied by more rapid exhumation in the SE and NW. Additionally, individual block movements may have occurred in the NW. It has also been possible to estimate the heating effect of renewed Cretaceous–Paleogene burial to 20–35 °C on the west and SE coasts.

Final Cenozoic unroofing of the basement is indicated by the modelled thermal histories. Areas around the southern tip of Lake Vättern together with the SE coast experienced the most pronounced exhumation compared with the surrounding parts.

Southern Sweden is characterized by a Precambrian basement, which emerges as a dome-shaped structure from below lower Palaeozoic cover rocks in the north and east and Mesozoic cover rocks in the south and west. Large-scale Phanerozoic events in the area can therefore be identified only by thermal and relief studies of the basement. Palaeozoic heating of central and southern Sweden as a result of the development of a Caledonian foreland basin has been established (Larson *et al.* 1999; Cederbom *et al.* 2000; Cederbom 2001). In addition, large-scale Palaeozoic to Early Mesozoic exhumation, perhaps accompanied by tectonism, has been demonstrated based on apatite fission-track (AFT) data from southern Sweden (Cederbom 2001). From another direction, Mesozoic and Tertiary morphotectonic events have been inferred based on studies of the relief and its relation to remnants of the cover rocks (Lidmar-Bergström 1994, 1996). In this study, results from AFT thermochronology are integrated with geomorphological data to attain further insight into the Mesozoic and Cenozoic thermotectonic development of southern Sweden.

Palaeosurfaces

The palaeosurfaces of southern Sweden (Fig. 1) have been analysed and roughly dated by means of their relative position, remnants of Palaeozoic and Mesozoic cover rocks, and saprolite occurrences (e.g. Lidmar-Bergström 1982, 1988, 1995). In addition, a model for the long-term morphotectonic evolution of southern Sweden has been suggested (Lidmar-Bergström 1994, 1996). The main types of palaeosurfaces, defined and interpreted by Lidmar-Bergström, are presented in Fig. 1 and summarized below (see Lidmar-Bergström 1996, Fig. 2).

In the north and east, Cambrian strata rest unconformably on Precambrian basement rocks. The relief on this basement surface is extremely flat. It is possible to trace this exhumed Sub-Cambrian Peneplain (SCP) from below lower Palaeozoic deposits in the Lake Vättern and Lake Vänern area and along the east coast, to the summits in the central part of southern Sweden (Fig. 1; Lidmar-Bergström 1988).

A hilly etch surface emerges from below Upper Cretaceous cover rocks in the SE and SW (Fig. 1). The hilly relief is associated with thick

From: DORÉ, A.G., CARTWRIGHT, J.A., STOKER, M.S., TURNER, J.P. & WHITE, N. 2002. *Exhumation of the North Atlantic Margin: Timing, Mechanisms and Implications for Petroleum Exploration*. Geological Society, London, Special Publications, **196**, 169–182. 0305-8719/02/$15.00

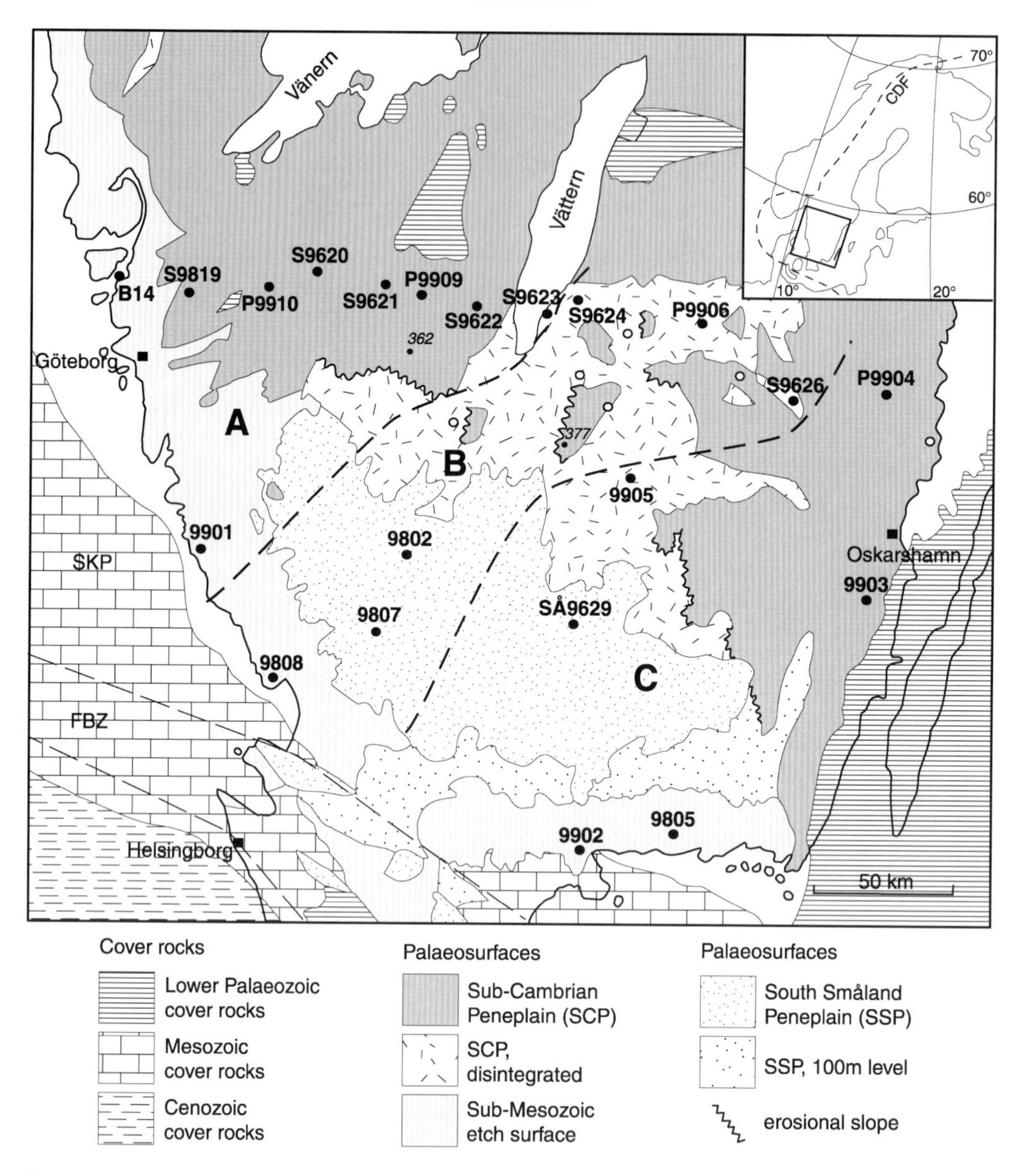

Fig. 1. Simplified map of palaeosurfaces and Phanerozoic sediments in southern Sweden (modified from Lidmar-Bergström 1996). Sample localities for apatite samples modelled in this study (●) are shown together with sample localities for additional apatite samples not used in this study (○) (Larson *et al.* 1999; Cederbom *et al.* 2000; Cederbom 2001). A, B and C indicate the three main tectonic areas discussed in the text (borders marked with bold dashed lines). The elevations for the two highest points in the area are given in metres above sea level (a.s.l.). CDF, Caledonian Deformation Front; FBZ, Fennoscandian Border Zone; SKP, Skagerrak–Kattegat Platform.

kaolinitic saprolite remnants, consistent with deep weathering during warm and humid conditions (Lidmar-Bergström 1995). This type of relief continues along the entire west coast and is interpreted as a re-exposed sub-Mesozoic palaeosurface (Lidmar-Bergström 1994).

A disintegrated part of the SCP is documented at *c.* 200 m elevation in the central part of southern Sweden (Fig. 1). Farther south, an almost horizontal plain with low relief is defined as the South Småland Peneplain (SSP). This surface truncates the sub-Mesozoic etch surface

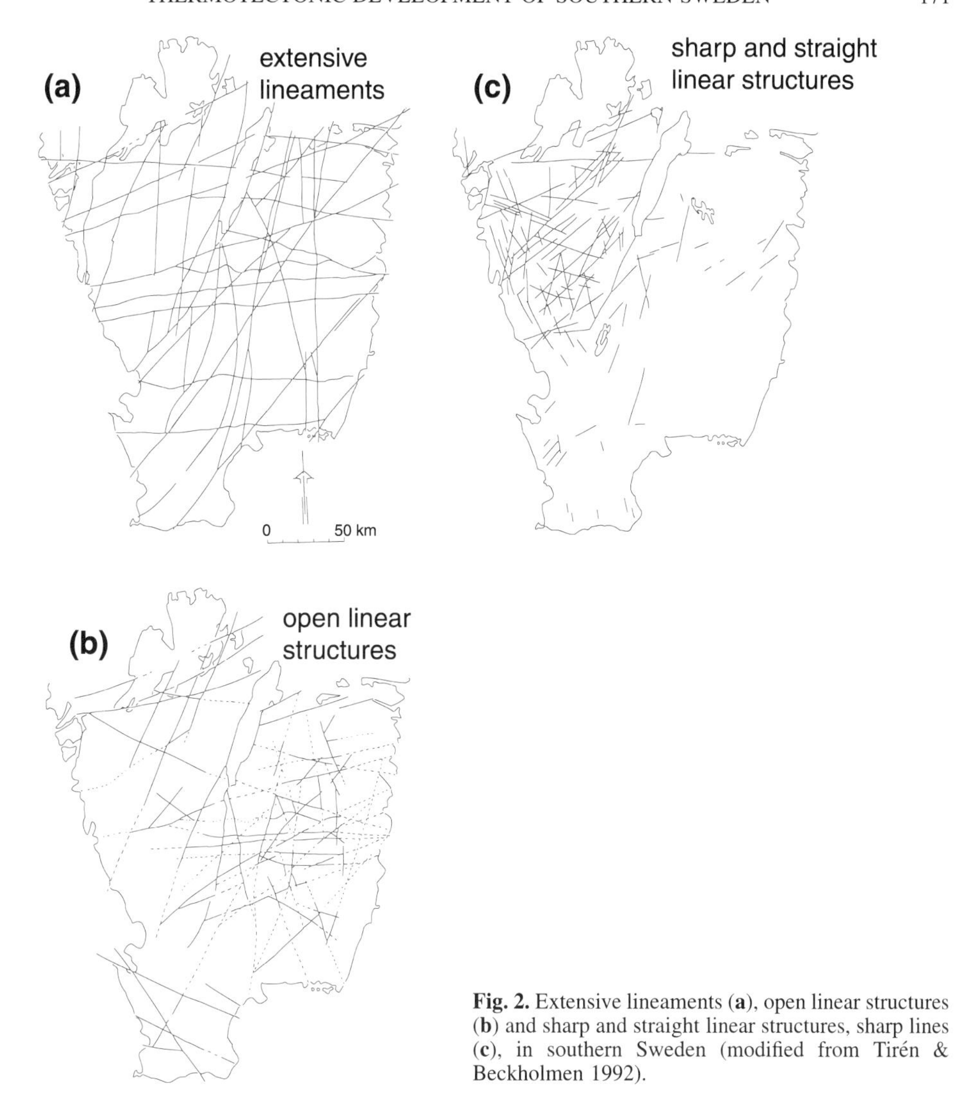

Fig. 2. Extensive lineaments (**a**), open linear structures (**b**) and sharp and straight linear structures, sharp lines (**c**), in southern Sweden (modified from Tirén & Beckholmen 1992).

(Lidmar-Bergström 1982) at about 100–125 m above sea level (a.s.l.) and is therefore interpreted as a younger surface. Frequent remnants of gravelly saprolites, consistent with weathering during cold and humid conditions, are associated with these surfaces (Lidmar-Bergström *et al.* 1997). The age of the gravelly saprolites is difficult to constrain, but they are certainly older than Late Weichselian time, and their position in areas protected from glacial erosion gives support for at least a Plio-Pleistocene age. Possibly, they had formed already in Miocene time (Lidmar-Bergström *et al.* 1997). The SSP and the disintegrated part of the SCP are both interpreted as Tertiary surfaces that were formed by stepwise exhumation and exposure of the basement during warm and dry conditions (Lidmar-Bergström 1991). Furthermore, the vast extent of the SSP indicates a considerable time for its formation.

Faults and lineaments

The regional occurrence and extent of Phanerozoic faults in southern Sweden is not well documented. Ahlin (1987) presented a study of Phanerozoic faults in the northwestern part of southern Sweden, based on field work and the reconstruction of the SCP. Frequent NE–SW-trending Phanerozoic faults with an offset of <100 m have been observed, and in Lake Vänern, an offset of 'some 100 metres' is recorded (Ahlin 1987). In addition, large-scale north–south-trending fracture zones appear along the eastern side of Lake Vättern (Persson & Wikman 1986). The lake is proposed to be a graben structure (e.g. Lind 1972; Persson *et al.* 1985), and it contains a *c.* 1000 m thickness of Late Proterozoic sediments (e.g. Vidal 1984). However, in most areas of southern Sweden only lineaments and fracture zones, which appear either on aeromagnetic maps and/or in the field, have been mapped. It is unknown whether some of them constitute faults or not. They are presented in a few 1:50 000 bedrock maps (covering *c.* 20% of the investigation area) and six 1:250 000 temporary general bedrock maps published by the Swedish Geological Survey, most of which are not digitized.

Phanerozoic faults and fractures have also been interpreted from the Relative Relief Map of Sweden (Lantmäteriverket 1986), contour maps and vertical topographic profiles (Elvhage & Lidmar-Bergström 1987; Lidmar-Bergström 1991, 1996). Relative relief maps are constructed by computerized hill shading, i.e. oblique illumination of a terrain model that is based on topographic maps (Elvhage & Lidmar-Bergström 1987). From the Relative Relief Map, the topographically well-expressed fracture lines, rather than the complete fracture pattern, were identified (Elvhage & Lidmar-Bergström 1987, p. 347). In comparison, on the basis of the same Relative Relief Map, Tirén & Beckholmen (1992) presented an analysis of the tectonic rock block pattern in southern Sweden together with a detailed analysis of the various fracture sets. To date, this is the only published compilation of linear structures for the whole of southern Sweden.

In Fig. 2, three types of structures are illustrated (extensive lineaments, open linear structures, sharp and straight linear structures), which were presented and interpreted by Tirén & Beckholmen (1992). The uncertainty of individual structures is high, but some general trends can be discerned from the lineament pattern. The plot of extensive lineaments (Fig. 2a) illustrates that a belt of *c.* 50 km width, including Lake Vättern, runs in a N20E direction, dividing southern Sweden into two halves (Tirén & Beckholmen 1992). Southwards the belt widens and deflects westwards, out into the sea. To the south, the orientation of lineaments is influenced by the NW–SE-oriented Fennoscandian Border Zone (Fig. 1). Open structures (Fig. 2b), which represent flexures, the termination of the SCP and wide, trough-like valleys, are scattered throughout southern Sweden. In contrast, sharp and straight linear structures (Fig. 2c) are recognized mainly in the northwestern part of southern Sweden. The surface nature of the sharp lineaments is not known, but they are probably the youngest structures detected from the Relative Relief Map (Tirén & Beckholmen 1992).

Modelling of AFT data

The dating method

Fission tracks are created by the spontaneous fission of ^{238}U and are formed continously through time. They become instantaneously annealed at high temperatures, and the number of tracks (i.e. the track density) is therefore proportional to the time that has passed since the grain cooled below a certain temperature and track retention started. For apatite, this temperature varies by several tens of degrees around 100 °C depending on the duration of heating and the mineral composition (Gleadow & Duddy 1981; Naeser 1981; Carlson *et al.* 1999).

Furthermore, it is possible to investigate the thermal history in apatite by studying the distribution of track lengths. Relatively slowly cooled basement samples are characterized by wide, often negatively skewed track length distributions with shorter mean track lengths and higher standard deviations compared with, for example, volcanic samples that have cooled very rapidly. Mixed track length distributions are formed when tracks from earlier events have survived partial reheating and a new population of tracks from the last period of cooling is added to the older generation of tracks (Gleadow *et al.* 1986a, 1986b).

Forward modelling of AFT data

Forward modelling of AFT data is the testing of potential thermal histories by comparing modelled and observed AFT results. The annealing model for apatite presented by Ketcham *et al.* (1999) and the AFTSolve modelling program constructed by Ketcham *et al.* (2000) were used in this study. In contrast to earlier published apatite annealing models (i.e. Laslett *et al.* 1987; Carlson 1990; Crowley *et al.* 1991; Laslett & Galbraith 1996), the Ketcham *et al.* (1999) model attempts to account for the variation in annealing behaviour that exists between different apatites as a result of compositional variations. When comparing the Ketcham *et al.* (1999) annealing model with the precursor models, it turned out to be more sensitive to both low-temperature and high-temperature annealing than its precursors. Generally, significantly lower Late Palaeozoic–Early Jurassic and Paleogene temperatures were obtained when using the Ketcham *et al.* (1999)

Table 1. *Modelling results*

Sample number	No. of grains	Observed FTA (Ma)	Modelled FTA (Ma)	GOF	No. of lengths	Observed MTL (μm)	Modelled MTL (μm)	K–S test
Area A								
B14	20	313 ± 25	322	0.72	50	13.6 ± 0.2	13.7 ± 0.2	0.58
S9819a	27	208 ± 10	209	0.92	100	12.7 ± 0.2	12.7 ± 0.2	0.31
S9819b	27	208 ± 10	214	0.55	100	12.7 ± 0.2	12.6 ± 0.2	0.30
P9910	20	161 ± 9	160	0.94	100	12.7 ± 0.2	12.7 ± 0.2	0.26
S9620	23	313 ± 17	313	1.0	100	13.1 ± 0.2	13.2 ± 0.2	0.50
S9621	20	220 ± 14	226	0.68	110	12.7 ± 0.2	12.6 ± 0.2	0.51
P9909a	20	149 ± 8	150	0.86	59	13.1 ± 0.2	13.1 ± 0.2	0.71
P9909b	20	149 ± 8	149	0.98	59	13.1 ± 0.2	13.1 ± 0.2	0.74
S9622	10	211 ± 16	218	0.65	100	12.4 ± 0.2	12.5 ± 0.2	0.64
S9623	20	188 ± 10	188	0.99	100	12.8 ± 0.2	12.8 ± 0.2	1.0
9901	20	175 ± 12	176	0.90	90	13.6 ± 0.1	13.6 ± 0.2	0.39
Area B								
S9624a	11	261 ± 17	264	0.86	92	13.0 ± 0.2	13.0 ± 0.2	0.89
S9624b	11	261 ± 17	262	0.97	92	13.0 ± 0.2	13.1 ± 0.2	0.97
P9906a	20	213 ± 14	215	0.87	100	12.4 ± 0.2	12.6 ± 0.3	0.11
P9906b	20	213 ± 14	212	0.91	100	12.4 ± 0.2	12.5 ± 0.2	0.70
P9906c	20	213 ± 14	216	0.81	100	12.4 ± 0.2	12.5 ± 0.2	0.81
S9626	20	282 ± 15	283	0.93	100	13.3 ± 0.2	13.4 ± 0.2	0.45
9802	21	231 ± 12	230	0.90	97	13.4 ± 0.1	13.4 ± 0.2	0.38
9807	20	240 ± 12	240	1.0	100	13.1 ± 0.2	13.1 ± 0.2	0.51
9808	20	231 ± 11	232	0.96	100	13.6 ± 0.1	13.5 ± 0.2	0.27
Area C								
P9904	20	196 ± 12	198	0.84	80	13.0 ± 0.2	13.0 ± 0.2	0.81
9905	14	188 ± 15	188	0.96	22	12.8 ± 0.2	12.9 ± 0.2	0.76
9903a	20	165 ± 10	165	0.99	52	12.8 ± 0.2	12.8 ± 0.2	0.77
9903b	20	165 ± 10	169	0.63	52	12.8 ± 0.2	13.1 ± 0.2	0.29
SÅ9629a	21	172 ± 9	174	0.86	166	13.5 ± 0.1	13.5 ± 0.2	0.60
SÅ9629b	21	172 ± 9	171	0.93	166	13.5 ± 0.1	13.4 ± 0.2	0.51
9902	20	160 ± 9	161	0.90	100	12.3 ± 0.2	12.4 ± 0.2	0.62
9805	23	189 ± 9	188	0.90	100	12.9 ± 0.1	12.9 ± 0.2	0.44

No. of grains, number of grains dated for each sample; FTA, apatite fission-track age; GOF, goodness of fit; no. of lengths, number of track length measurements in each track length distribution; MTL, mean track length; K–S test, Kolmogorov–Smirnov test. The dating results were originally published by Cederbom *et al.* (2000) (for SÅ9629) and by Cederbom (2001).

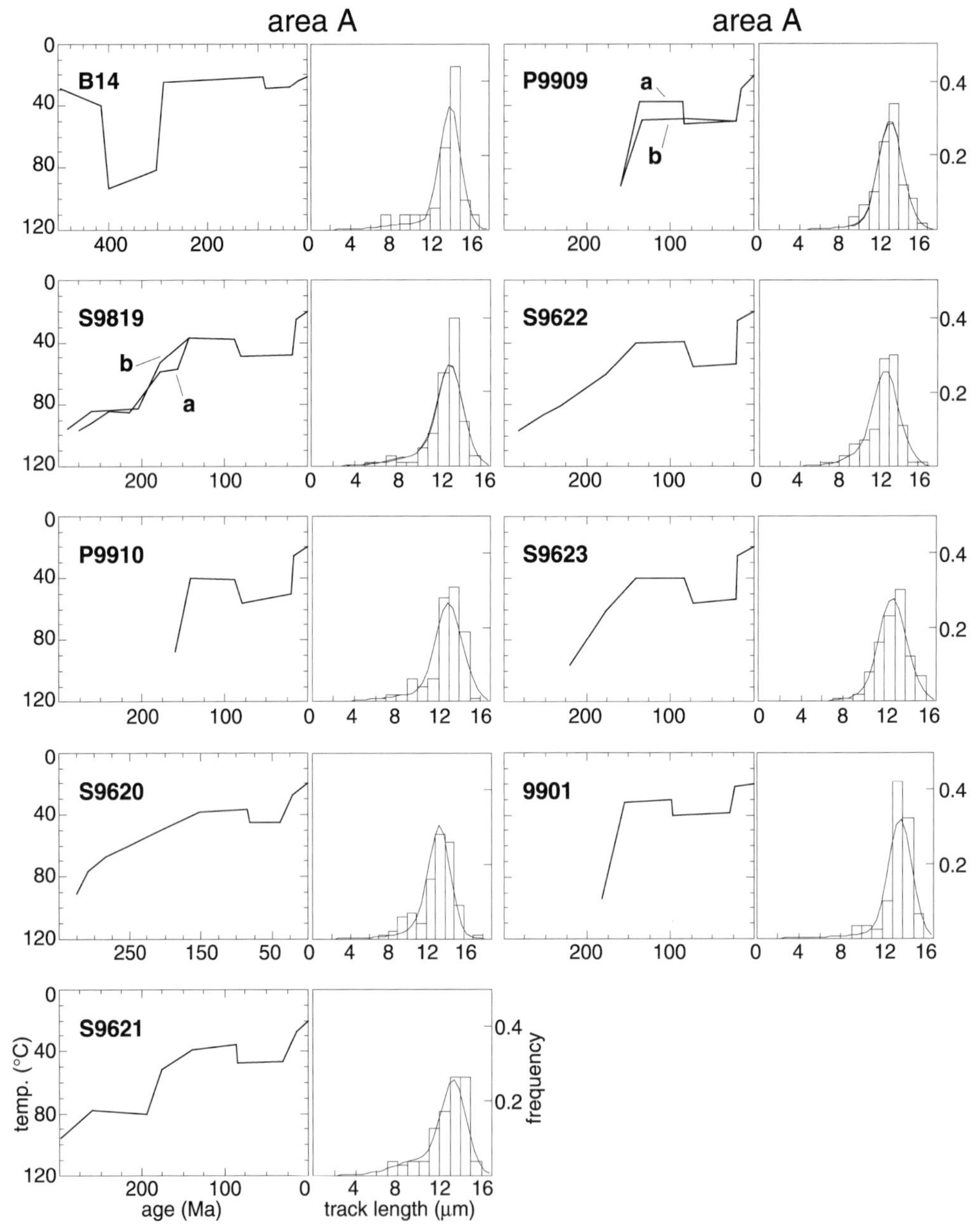

Fig. 3. Modelled thermal histories for apatite samples from southern Sweden (for sample localities, see Fig. 1). Only the 'discernible part' of the modelled Phanerozoic history is illustrated for each sample.

annealing model compared with when the Laslett *et al.* (1987) or the Crowley *et al.* (1991) annealing model was adopted. A detailed comparison between the annealing models was presented by Ketcham *et al.* (1999).

The dataset used in this study does not contain any kinetic parameters (i.e. track length orientations or etch pit widths), and the initial track length was therefore set to a fixed value of 16.2 μm. It is important to have in mind that the thermal histories presented in this study are not unique. In forward modelling, as well as in inverse modelling, somewhat different thermal histories that have not been tried may also fit the observed AFT results.

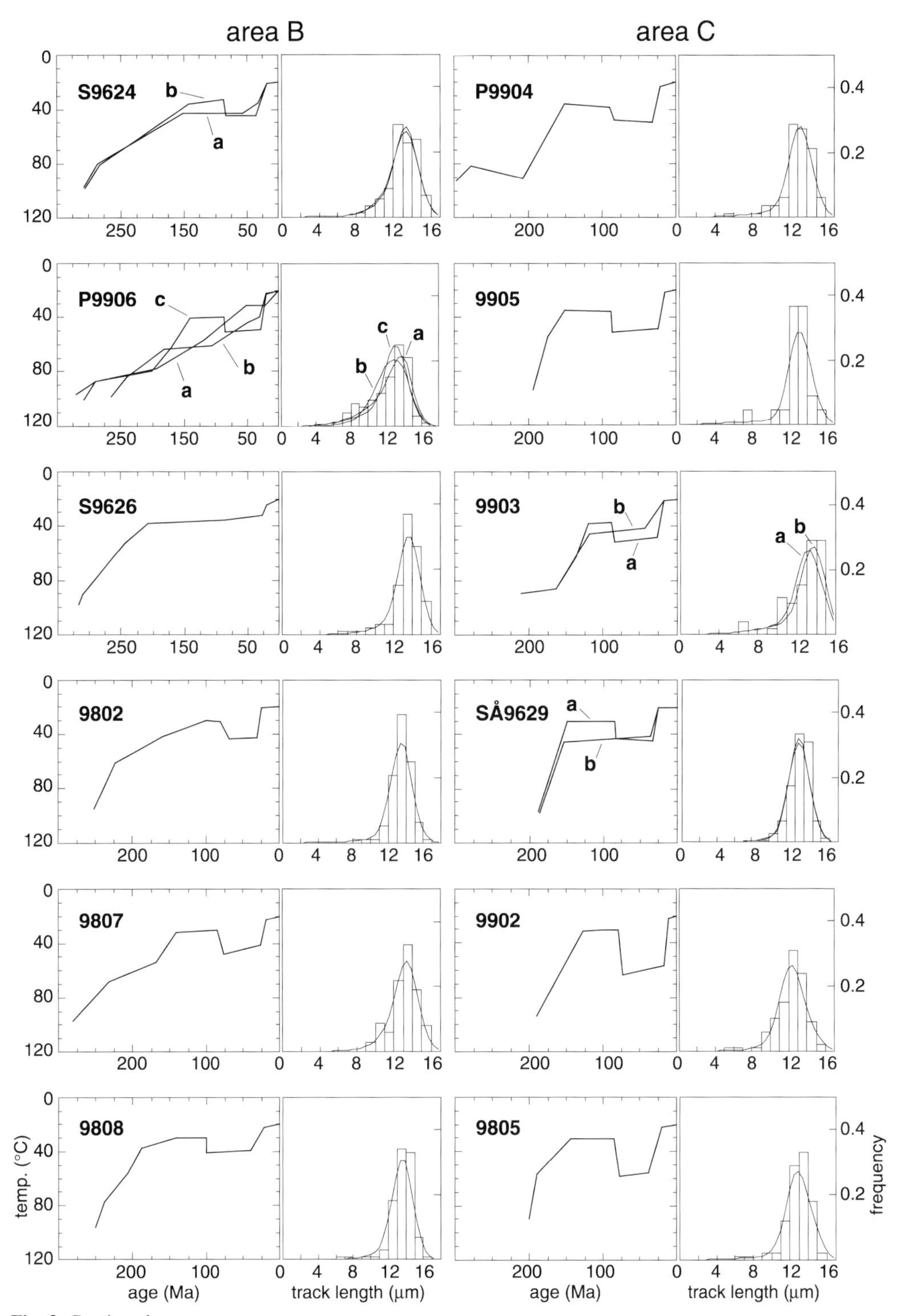

Fig. 3. Continued

In general, 20 grains should be dated to obtain a statistically reliable AFT age and, similarly, 100 track lengths should be measured. Modelled thermal histories for samples having fewer single grain age determinations and/or track length measurements (see Table 1) must be interpreted with caution.

Two statistical methods are used to gauge how well the data and the model results fit. The Kolmogorov–Smirnov test (K–S test) is used to compare the measured and the predicted track length distributions and the GOF is the 'goodness of fit' between the age data and the age predicted by the model. For both statistics, a probability value >0.5 implies that the modelled thermal history is 'supported' by the data and a probability value between 0.5 and 0.05 indicates that the modelled thermal history is 'not ruled out' by the data (Ketcham *et al.* 2000). Therefore, a value of 0.05 is regarded as the lower limit for acceptance, although a value >0.5 was aimed for when modelling. A detailed description of the statistical methods has been given by Ketcham *et al.* (2000).

Geological and geomorphological modelling constraints

In modelling, several constraints were set. Because all samples were collected at or close to the SCP, they are known to have experienced near-surface temperatures during Early Cambrian time. By the end of the Caledonian orogeny, southern and central Sweden were covered by thick foreland basin deposits (Larson *et al.* 1999; Cederbom *et al.* 2000; Cederbom 2001), resulting in total annealing of all AFT samples in southern Sweden. This heating was not set as a constraint when modelling, but was necessary to obtain a fit between observed and calculated AFT data. Additionally, heating to at least 90 °C was required.

Documented Cretaceous remnants, lying directly on basement on the west and SE coast of southern Sweden, in combination with a sub-Mesozoic relief, reveal that these areas experienced surface-level temperatures at *c.* 140 Ma. The Cretaceous palaeosurface temperature was estimated to have been *c.* 30 °C. Surface level conditions continued until Late Cretaceous sedimentation started at *c.* 85 Ma. In several cases, the final cooling event was assigned an age of at least 15 Ma, because this is the expected minimum time for the Tertiary etch surfaces to have developed (Lidmar-Bergström 1991). Samples from areas where the SCP is preserved probably experienced final exhumation in Neogene time. The recent (0 Ma) surface-level temperature was set to *c.* 20 °C. High surface-level temperatures (30 °C and 20 °C, respectively) were deliberately set to attain a minimum value of the modelled Cretaceous–Paleogene reheating.

The AFT dataset

Modelled thermal histories for 21 apatite samples were achieved using AFT data from Cederbom *et al.* (2000) and Cederbom (2001). The sampling localities are presented in Fig. 1. All samples were collected from the outcropping Precambrian basement (granitoids and gneisses), and they all experienced total annealing during Late Palaeozoic and/or Early Mesozoic time (Cederbom *et al.* 2000; Cederbom 2001). The AFT ages for the 21 apatite samples used in this study all passed the χ^2 test (Cederbom *et al.* 2000; Cederbom 2001), indicating that the grains in each sample belong to a single population.

There are minor differences in AFT results between the samples that may derive from kinetic variations among the apatites. The apatite composition has not been examined, so compositional variations cannot be excluded. Nevertheless, all samples passed the χ^2 test and no major difference in etch pit width was observed when analysing the grains.

There is no correlation between elevation and AFT age in southern Sweden. Instead, three areas characterized by different trends in the AFT results can be discerned (Cederbom 2001). The southeastern part of southern Sweden (area C) is characterized by AFT ages younger than 200 Ma. A SW–NE-trending belt (area B) includes samples with ages between 231 and 282 Ma, whereas area A in the northwestern part of southern Sweden represents samples with ages ranging from 149 to 315 Ma (see Fig. 2; Cederbom 2001). None of the samples have mixed track length distributions. To summarize, areas A, B and C are defined based on differences in the AFT results that are too large to be caused solely by compositional variations.

Modelling results

The modelling results for the 21 apatite samples are shown in Fig. 3 and Table 1. The Phanerozoic thermal history was modelled for all samples. However, only the discernible part of the thermal history is presented in Fig. 3. When modelling samples P9904 and 9905, for example, heating to at least 90 °C for the 200–300 Ma time interval is required for both samples (see Fig. 3). However, heating above 90 °C is not accepted for sample P9904 during this time interval, whereas heating above 90 °C makes no difference in the calculated AFT results of sample 9905.

There are minor differences between the modelled thermal histories that may derive from kinetic variations among the apatites. However, several significant observations can be made.

First, the modelling results can be separated into three groups supporting the existence of areas A, B and C in Fig. 1 (see Cederbom 2001). The samples from area C have relatively uniform cooling histories. They all cooled below c. 90 °C during Early Jurassic time and they are characterized by cooling rates of *c.* 15 °C per 10 Ma. Area B is characterized by samples with older AFT ages than the samples in area C. Likewise, modelled thermal histories for the samples in area B indicate much earlier cooling and much lower cooling rates. Cooling below 90 °C had occurred by Early Triassic time in the SW, and as early as Late Carboniferous time in the NE. With two exceptions the cooling rates decrease from *c.* 5 °C per 10 Ma in the SW to *c.* 3 °C per 10 Ma in the NE. Samples S9626 and 9808 at the margins of area B show slightly higher cooling rates, 7 °C per 10 Ma and 8 °C per 10 Ma, respectively. Area A includes samples showing a remarkable heterogeneity. The point of time when the samples cooled below 90 °C ranges from Late Carboniferous to Late Jurassic time. Additionally, the cooling rates vary between *c.* 40 °C per 10 Ma and *c.* 3 °C per 10 Ma.

Second, it is possible that all samples were reheated during Late Cretaceous time. However, such reheating can be established only for samples collected from the sub-Mesozoic palaeosurface, which are known to have experienced surface conditions before and/or during Cretaceous time. The modelling results for sample 9902 and 9805 indicate *c.* 35° reheating in SE Sweden, whereas heating of the order of *c.* 20 °C is indicated for the samples from the west coast (i.e. sample B14, 9901 and 9808) (Figs 1 and 3). Modelling of the remaining samples shows that Cretaceous reheating of southern Sweden as a whole is possible, but cannot be taken as certain. Alternative thermal histories are presented for a few samples (P9909, S9624, P9906, 9903, SÅ9629) in Fig. 3 and Table 1.

Third, Oligocene–Miocene final cooling is possible in all three areas according to the modelling. For occasional samples, Late Eocene final cooling is also allowed, although it is never required.

Discussion

The forward modelling of AFT results from southern Sweden confirms the conclusions of Cederbom (2001), i.e. that southern Sweden experienced differentiated cooling after the Caledonian orogeny and the related foreland basin formation. Furthermore, a pattern illustrating local variations in both onset of cooling and cooling rate appears and is discussed below.

The Palaeozoic and Mesozoic cooling was probably a result of erosion of cover rocks (unroofing), but changes in the geothermal gradient may have also had an influence. It is likely that the thermal conductivity and heat flow of the crystalline basement and sedimentary cover have differed, both regionally and through time during the Palaeozoic and Mesozoic eras. It can be noted that Triassic and Jurassic sediments deposited in shallow-water and fresh-water environments are recorded within the Fennoscandian Border Zone in southernmost Sweden (Norling & Bergström 1987; Guy-Ohlson & Norling 1988), supporting the idea of an Early Mesozoic cover in southern Sweden. The observed Cretaceous reheating was probably caused by sediment deposition, and the final cooling event is explained by Cenozoic final exhumation.

The onset of unroofing is not necessarily the same as the point in time when the samples cooled below 90 °C. If a higher temperature regime originally prevailed, exhumation may have started much earlier than when fission tracks started to accumulate. According to earlier published conodont alteration and organic maturation studies of Lower Palaeozoic remnants (Bergström 1980; Buchardt *et al.* 1997), however, a temperature regime above 90 °C is not likely for the area between Lake Vänern and Lake Vättern.

Triassic and Jurassic exhumation

After the Caledonian orogeny and the formation of a foreland basin, unroofing started. The basement, with its heavy pile of sedimentary strata, may not have acted as a single entity in southern Sweden. The first recorded unroofing is indicated in areas A and B. Within area A, Late Carboniferous, Early Jurassic and Late Jurassic first recorded exhumation timings are obtained. In area B the recorded exhumation started in Late Carboniferous to Early Triassic time, with a younging trend towards the SW. Early Jurassic first recorded exhumation is obtained in area C. A compilation of modelled cooling rates for three points in time is presented in Fig. 4 together with

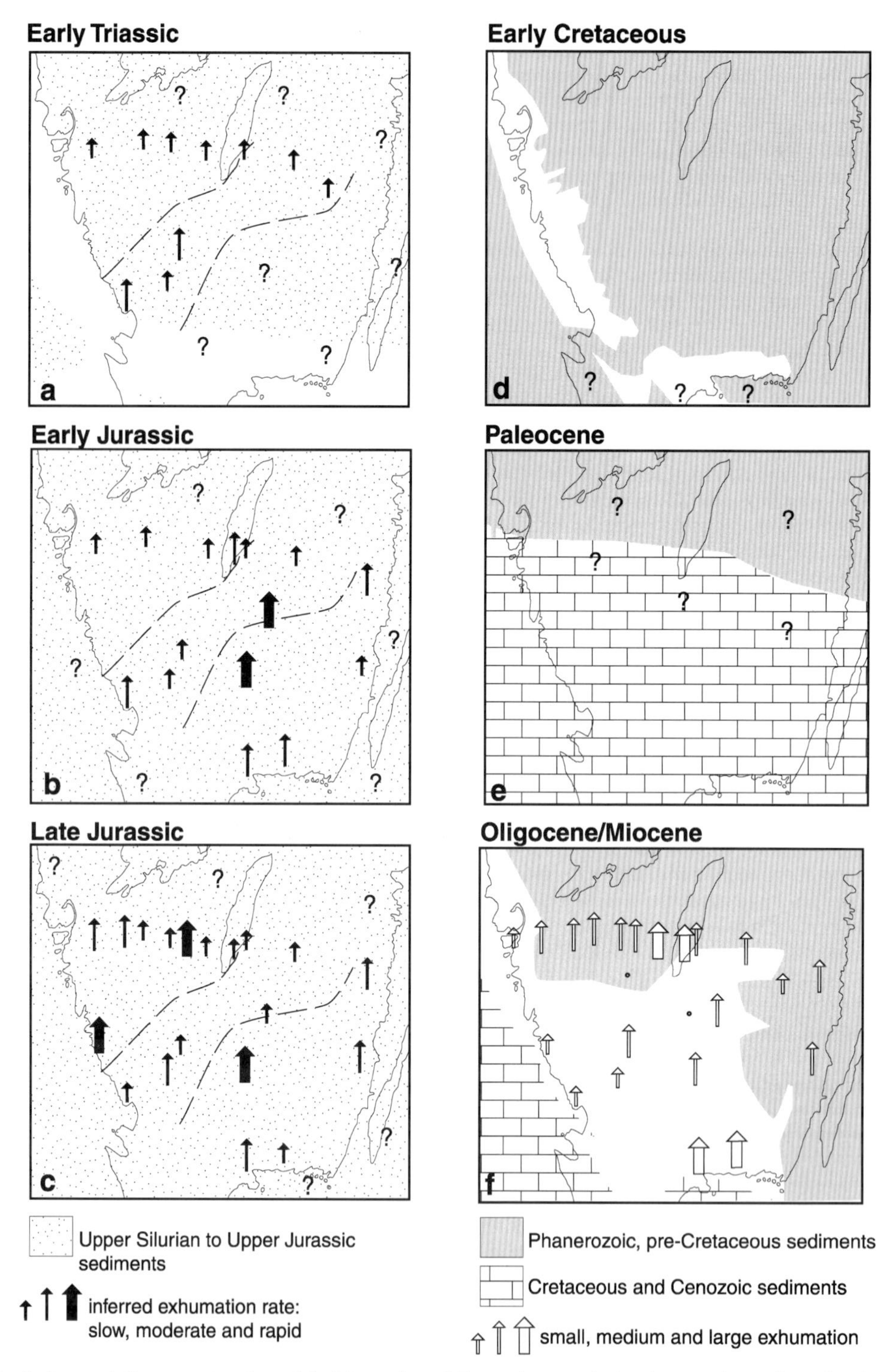

Fig. 4. A speculative reconstruction of the Mesozoic and Cenozoic tectonic development in southern Sweden. A heterogeneity in Triassic–Jurassic exhumation between the three main tectonic areas of southern Sweden, and between individual samples in the northwestern part of southern Sweden can be discerned. Oligocene–Miocene final exhumation was most pronounced around the southern tip of Lake Vättern and on the SE coast.

the probable extent of Upper Silurian to Upper Jurassic sediments (i.e. foreland basin related sediments). As mentioned above, minor differences between the modelled thermal histories may be due to compositional variations among the apatites. Nevertheless, Fig. 4 illustrates that there is a heterogeneity in Mesozoic exhumation among the three areas of southern Sweden.

In the beginning of Triassic time (Fig. 4a), it is clear that most of areas A and B were still covered by thick, foreland basin related deposits. Slow to moderate exhumation rates are recorded in areas A and B, with an increasing trend towards the SW. At the same time, the Skagerrak–Kattegat Platform (SKP) (Fig. 1) basement was exposed; seismic offshore records (e.g. Vejbæk 1997) reveal that the basement was exposed before Triassic sedimentation started. The AFT data for sample B14 also support surface temperature conditions. In contrast, the AFT results for sample 9808 reveal temperatures of 90 °C, indicating thick covering of the basement in the SW in Early Triassic time. To solve this contradiction, additional studies of both the sedimentary record and tectonics close to shore on the SKP are needed, together with denser AFT sampling.

It is probable, but not established, that area C and central Sweden were covered by sediments in Early Triassic time. In Early Jurassic time (Fig. 4b), however, it is evident that area C was covered by thick piles of sediment, as temperatures >90 °C are recorded by the AFT data and exhumation is recorded in the area. Cooling as a result of exhumation is also detected in all three areas for Late Jurassic time (Fig. 4c).

It is interesting to note that samples from the NE–SW-trending area B are consistent with slow to moderate cooling rates throughout Triassic and Jurassic time. Meanwhile, areas B and C behaved in a different manner, and large variations are recorded within area A. In area A, block movements may have caused uneven erosion and redeposition on downfaulted blocks. It is possible that area C acted as a temporary store for reworked sediments from the NW. The unroofing seems to have accelerated in earliest Jurassic time, when area C experienced rapid exhumation, followed by block uplifts in area A.

Before or during Early Cretaceous time, a NW–SE-trending axis of elevation developed, resulting in exposure of the west and SE coasts. The Palaeozoic cover had been removed in the Båstad and Kristianstad areas on the west coast and the SE coast when Early Cretaceous sedimentation started (Lidmar-Bergström 1982). In Fig. 4d, areas with exposed basement are illustrated together with the probable extent of Phanerozoic, pre-Cretaceous sediments by that time; erosion of the Upper Silurian to Late Jurassic sediments may have caused exposure of lower Palaeozoic sediments. However, most parts of the south Swedish basement surface were still protected from weathering (Lidmar-Bergström 1991).

The Cretaceous and Paleogene cover

Late Cretaceous and Paleogene deposits covered the southern part of the Fennoscandian Shield, but the extent of these sediments to the north is not established. The modelling results reveal that the west and SE coasts of southern Sweden were buried, and they illustrate that the central part of southern Sweden may also have been covered (Fig. 4e). An estimate of the thickness of these deposits on the west and SE coast can be made. The modelled thermal histories indicate that the Cretaceous–Paleogene cover was thicker on the SE coast (samples 9902, 9805) than on the west coast (9808, 9901), as *c.* 35 °C and *c.* 20 °C temperature increases are recorded for the SE and west coasts, respectively. An estimate of the thickness of these sediments involves an unconstrained estimate of the geothermal gradient. If the geothermal gradient during Cretaceous time was 30 °C km^{-1}, a sedimentary thickness of *c.* 650 m in the west and >1000 m in the SE is indicated. In comparison, sonic data and basin models for data from Danish wells support the idea that the Cretaceous–Paleogene cover on the SKP exceeded 1000 m thickness before Neogene uplift and erosion (Japsen & Bidstrup 1999).

Cenozoic exhumation

Where the SCP is well preserved it cannot have been re-exposed until Late Tertiary time, but it is uncertain when the basement was re-exposed within the highest parts to the south and SE of Lake Vättern. According to the modelled thermal histories, the basement, with or without an additional Cretaceous cover on top of the remnant Palaeozoic cover, did not reappear at the surface until at least Oligocene or Miocene time. Areas with a well-preserved SCP were still covered by sediments when the SSP was developed (Fig. 4f). In Fig. 4f, the relative amount of final exhumation indicated by the modelling results is also shown. This is a very speculative picture, as several alternative modelled histories match the AFT data. Large exhumation is indicated for the southern tip of Lake Vättern, where the highest altitudes are found today, and for the coast in the SE.

Cenozoic uplift and doming of southern Sweden has been supported by Lidmar-Bergström (1996, 1999) and Japsen & Bidstrup (1999). Cenozoic uplift of southern Sweden has also been discussed in a broader perspective, together with Cenozoic uplift and doming of the southern Scandes in southern Norway (e.g. Rohrman *et al.* 1995; Riis 1996; Lidmar-Bergström 1999). The modelled AFT data indicate that large-scale Cenozoic exhumation of southern Sweden has occurred, which is consistent with Cenozoic uplift. However, the speculative exhumation pattern illustrated in Fig. 4f does not support symmetrical doming of the basement in southern Sweden, as not only the central part of southern Sweden, but also the SE coast, experienced a large amount of late exhumation.

Large-scale Phanerozoic tectonism

The main differences in the AFT data and the modelled thermal histories for samples from areas A, B and C are too large to be explained solely by compositional variations in the apatites. Neither is long-term partial annealing a probable explanation for the differences, according to the observed track length distributions. If the basement in southern Sweden cooled as an entity during Late Palaeozoic to Late Jurassic time, large-scale Cretaceous and Cenozoic tectonic movements must have occurred. This seems unlikely considering the present-day topography. Alternatively, southern Sweden experienced differentiated exhumation accompanied by tectonism during Late Palaeozoic to Late Jurassic time. In addition, large differences in the AFT results and the modelled thermal histories recorded for samples within area A indicate individual block movements in the northwestern part of southern Sweden.

The proposed extent of the three main tectonic areas A, B and C is based on a coarse grid of apatite samples (see Fig. 1), and additional sampling is required before the existence of these areas can be verified. A comprehensive investigation of the occurrence and extent of Phanerozoic faults in southern Sweden has not been published. However, the study area where several Phanerozoic faults were observed by Ahlin (1987) is situated within area A. In addition, two fracture zones are mapped along the eastern side of Lake Vättern (Persson & Wikman 1986). One of them is situated between the sample points S9623 and S9624. Block movements have been observed along the eastern side of the lake (Persson & Wikman, 1986), but individual blocks have not been identified.

It is also interesting to compare the suggested extent of the three areas, which is based solely on AFT data, with the general trends in the lineament pattern presented by Tirén & Beckholmen (1992) (Fig. 2). The sharp lines, interpreted as the youngest features (Fig. 2c) are found in a restricted area south of Lake Vänern, corresponding more or less to area A. Furthermore, the NW–SE-trending belt of extensive lineaments that divides southern Sweden into two halves (Fig. 2a) may have a correspondence in the areal extent of area B.

The existence of three main tectonic units corresponding to areas A, B and C and individual block movements within at least northwestern Sweden is supported by the studies of faults, fracture zones and linear structures published to date. However, large-scale Phanerozoic tectonic block movements of the order of >100 m have previously been suggested only for Lake Vänern (Ahlin 1987).

Conclusions

On their own, published AFT data have revealed a general picture of the thermotectonic development in southern Sweden for Palaeozoic and Early Mesozoic time. However, by forward modelling the AFT data in combination with studies of palaeosurfaces and relief, further information on exhumation rates, sedimentary thicknesses and the Mesozoic to Cenozoic thermotectonic history has been derived.

Southern Sweden can be divided into three main tectonic areas, which were characterized by different onset of recorded unroofing and different exhumation rates during Late Palaeozoic to Late Jurassic time. Individual block movements in one of these areas, i.e. the northwestern part of southern Sweden, are suggested for this time interval.

There is a contradiction between the thermotectonic development in SW Sweden and on the SKP further west. Thick deposits covered the basement in SW Sweden during earliest Triassic time, whereas the basement on the SKP was exposed. To solve this contradiction, additional studies of offshore sedimentary records and nearshore tectonics, and denser AFT sampling are needed.

Southern Sweden experienced reburial during Late Cretaceous and Paleogene time. A temperature rise of *c.* 35 °C and *c.* 20 °C has been detected for the west coast and the SE coast, respectively. The temperature difference indicates that the sedimentary cover was thicker on the west coast than on the SE coast.

Final exhumation of southern Sweden during Cenozoic time is supported by the modelling results. This is consistent with previous studies of Cenozoic uplift of southern Sweden (Riis 1996; Lidmar-Bergström 1999; Japsen & Bidstrup 1999). However, the pattern of final exhumation obtained from the modelling of AFT data does not support a symmetrical doming of the basement in southern Sweden.

Finally, large-scale Phanerozoic tectonic movements in southern Sweden cannot be rejected based on fault and fracture zone studies published to date. Additionally, there are similarities between major features observed in the lineament pattern of southern Sweden and the suggested areal extent of the three main tectonic areas. Nevertheless, a more detailed and better constrained pattern of the tectonic development in southern Sweden demands denser apatite sampling and a regional study of the occurrence of Phanerozoic faults.

K. Lidmar-Bergström is gratefully acknowledged for fruitful criticism. The manuscript was improved by comments from K. Gallagher, A.G. Doré, H. Sinclair and an anonymous reviewer. This project has been financially supported by the Nuclear Waste and Management Co. (SKB) and the Royal Swedish Academy of Sciences.

References

AHLIN, S. 1987. Phanerozoic faults in the Västergötland basin area, SW Sweden. *Geologiska Föreningen i Stockholms Förhandlingar*, **109**, 221–227.

BERGSTRÖM, S. 1980. Conodonts as paleotemperature tools in Ordovician rocks of the Caledonides and adjacent areas in Scandinavia and the British Isles. *Geologiska Föreningen i Stockholms Förhandlingar*, **102**, 377–392.

BUCHARDT, B., NIELSEN, A.T. & SCHOVSBO, N.H. 1997. Alun Skiferen i Skandinavien. *Geologisk Tidsskrift*, **3**, 1–30.

CARLSON, W.D. 1990. Mechanisms and kinetics of apatite fission-track annealing. *American Mineralogist*, **75**, 1120–1139.

CARLSON, W.D., DONELICK, R.A. & KETCHAM, R.A. 1999. Variability of apatite fission-track annealing kinetics: I. Experimental results. *American Mineralogist*, **84**, 1213–1223.

CEDERBOM, C.E. 2001. Phanerozoic, pre-Cretaceous thermotectonic events in southern Sweden revealed by fission track thermochronology. *Earth and Planetary Science Letters*, **188**, 199–209.

CEDERBOM, C.E., LARSON, S.-Å., TULLBORG, E.-L. & STIBERG, J.-P. 2000. Fission track thermochronology applied to Phanerozoic thermotectonic events in central and southern Sweden. *Tectonophysics*, **316**, 153–167.

CROWLEY, K.D., CAMERON, M. & SCHAEFER, R.L. 1991. Experimental studies of annealing of etched fission tracks in fluorapatite. *Geochimica et Cosmochimica Acta*, **55**, 1449–1465.

ELVHAGE, C. & LIDMAR-BERGSTRÖM, K. 1987. Some working hypotheses on the geomorphology of Sweden in the light of a new relief map. *Geografiska Annaler*, **69A**, 343–358.

GLEADOW, A.J.W. & DUDDY, I.R. 1981. A natural long-term track annealing experiment for apatite. *Nuclear Tracks*, **5**, 169–174.

GLEADOW, A.J.W., DUDDY, I.R., GREEN, P.F. & HEGARTY, K.A. 1986*a*. Fission track lengths in the apatite annealing zone and the interpretation of mixed ages. *Earth and Planetary Science Letters*, **78**, 245–254.

GLEADOW, A.J.W., DUDDY, I.R., GREEN, P.F. & LOVERING, J.F. 1986*b*. Confined fission track lengths in apatite: a diagnostic tool for thermal history analysis. *Contributions to Mineralogy and Petrology*, **94**, 405–415.

GUY-OHLSON, D. & NORLING, E. 1988. *Upper Jurassic Litho- and Biostratigraphy of NW Scania, Sweden*. Sveriges Undersökning Serie Ca, **72**.

JAPSEN, P. & BIDSTRUP, T. 1999. Quantification of late Cenozoic erosion in Denmark based on sonic data and basin modelling. *Bulletin of the Geological Society of Denmark*, **46**, 79–99.

KETCHAM, R.A., DONELICK, R.A. & CARLSON, W.D. 1999. Variability of apatite fission-track annealing kinetics: III. Extrapolation to geological time scales. *American Mineralogist*, **84**, 1235–1255.

KETCHAM, R.A., DONELICK, R.A. & DONELICK, M.B. 2000. AFTSolve: a program for multi-kinetic modeling of apatite fission-track data. *Geological Materials Research*, **2** (1), 1–32.

LARSON, S.Å., TULLBORG, E.-L., CEDERBOM, C.E. & STIBERG, J.-P. 1999. Sveconorwegian and Caledonian foreland basins in the Baltic Shield revealed by fission-track thermochronology. *Terra Nova*, **11**, 210–215.

LASLETT, G.M. & GALBRAITH, R.F. 1996. Statistical modelling of thermal annealing of fission tracks in apatite. *Geochimica et Cosmochimica Acta*, **60**, 5117–5131.

LASLETT, G.M., GREEN, P.F., DUDDY, I.R. & GLEADOW, A.J.W. 1987. Thermal annealing of fission tracks in apatite 2. A quantitative analysis. *Chemical Geology*, **65**, 1–13.

LIDMAR-BERGSTRÖM, K. 1982. *Pre-Quaternary Geomorphological Evolution in Southern Fennoscandia*. Sveriges Undersökning Serie C, **785**.

LIDMAR-BERGSTRÖM, K. 1988. Denudation surfaces of a shield area in south Sweden. *Geografiska Annaler*, **70A** (4), 337–350.

LIDMAR-BERGSTRÖM, K. 1991. Phanerozoic tectonics in southern Sweden. *Zeitschrift für Geomorphologie, Neue Folge Supplement*, **82**, 1–16.

LIDMAR-BERGSTRÖM, K. 1994. Morphology of the bedrock surface. *In*: FREDÉN, C. (ed.) *Geology. National Atlas of Sweden*. SNA Publishing, Stockholm, 44–54.

LIDMAR-BERGSTRÖM, K. 1995. Relief and saprolites through time on the Baltic Shield. *Geomorphology*, **12**, 45–61.

LIDMAR-BERGSTRÖM, K. 1996. Long term morphotectonic evolution in Sweden. *Geomorphology*, **16**, 33–59.

LIDMAR-BERGSTRÖM, K. 1999. Uplift histories revealed by landforms of the Scandinavian domes. *In*: SMITH, B.J., WHALLEY, W.B. & WARKE, P.A. (eds) *Uplift, Erosion and Stability: Perspectives on Long-term Landscape Development*. Geological Society, London, Special Publications, **162**, 85–91.

LIDMAR-BERGSTRÖM, K., OLSSON, S. & OLVMO, M. 1997. Palaeosurfaces and associated saprolites in southern Sweden. *In*: WIDDOWSON, M. (ed.) *Palaeosurfaces: Recognition, Reconstruction and Palaeoenvironmental Interpretation*. Geological Society, London, Special Publications, **120**, 95–124.

LIND, G. 1972. The gravity and geology of the Vättern area, southern Sweden. *Geologiska Föreningen i Stockholms Förhandlingar*, **94**, 245–257.

LANTMÄTERIVERKET. 1986. *Sveriges Relief*. Lantmäteriverket, Gävle.

NAESER, C.W. 1981. The fading of fission tracks in the geologic environment—data from deep drill holes. *Nuclear Tracks*, **5**, 248–250.

NORLING, E. & BERGSTRÖM, J. 1987. Mesozoic and Cenozoic tectonic evolution of Scania, southern Sweden. *Tectonophysics*, **137**, 7–19.

PERSSON, L. & WIKMAN, H. 1986. *Provisoriska Översiktliga Berggrundskartan Jönköping, Map Sheet and Description*. Swedish Geological Survey **Af 39**.

PERSSON, L., BRUUN, A. & VIDAL, G. 1985. *Berggrundskartan Hjo SO, Map Sheet and Description*. Swedish Geological Survey **Af 134**.

RIIS, F. 1996. Quantification of Cenozoic vertical movements of Scandinavia by correlation of morphological surfaces with offshore data. *Global and Planetary Change*, **12**, 331–357.

ROHRMAN, M., VAN DER BEEK, P.A., ANDRIESSEN, P.A.M. & CLOETHING, S. 1995. Meso-Cenozoic morphotectonic evolution of southern Norway: Neogene domal uplift inferred from fission track thermochronology. *Tectonics*, **14**, 704–718.

TIRÉN, S.A. & BECKHOLMEN, M. 1992. Rock block map analysis of southern Sweden. *Geologiska Föreningen i Stockholms Förhandlingar*, **114**, 253–269.

VEJBÆK, O.V. 1997. *Dybe strukturer i sedimentære bassiner*. Geologisk Tidsskrift, **4**.

VIDAL, G. 1984. Lake Vättern. *Geologiska Föreningen i Stockholms Förhandlingar*, **106**, 397.

Neogene uplift and erosion of southern Scandinavia induced by the rise of the South Swedish Dome

PETER JAPSEN[1], TORBEN BIDSTRUP[1] & KARNA LIDMAR-BERGSTRÖM[2]

[1]*Geological Survey of Denmark and Greenland (GEUS), Øster Voldgade 10, DK-135 København K, Denmark (e-mail: pj@geus.dk)*

[2]*Department of Physical Geography and Quaternary Geology, Stockholm University, SE-10691 Stockholm, Sweden*

Abstract: Basin modelling and compaction studies based on sonic data from the Mesozoic succession in 68 Danish wells were used to estimate the amount of section missing due to late Cenozoic erosion. The missing section increases gradually towards the coasts of Norway and Sweden from zero in the North Sea to *c.* 500 m in most of the Danish Basin, but over a narrow zone it reaches *c.* 1000 m on the Skagerrak–Kattegat Platform in northernmost Denmark. The increasing amount of erosion matches the increase in the hiatus at the base of the Quaternary, where Neogene and older strata are truncated, and the Mesozoic succession is thus found to have been more deeply buried by c. 500 Paleocene–Miocene sediments in large parts of the area. These observations suggest that the onset of erosion occurred during the Neogene, and that the Skagerrak–Kattegat Platform was affected by tectonic movements prior to glacial erosion. In southern Sweden just east of the Kattegat, the exposed basement of the South Swedish Dome attains altitudes of almost 400 m. The formation of the Dome started in the Late Palaeozoic, but geomorphological investigations have led to the conclusion that a rise of the Dome occurred during the Cenozoic. We find that the pattern of late Cenozoic erosion in Denmark agrees with a Neogene uplift of the South Swedish Dome and of the Southern Scandes in Norway. This suggestion is consistent with major shifts in sediment transport directions during the late Cenozoic observed in the eastern North Sea, and with formation of a new erosion surface as well as re-exposure of sub-Cambrian and sub-Cretaceous surfaces in southern Sweden. The Neogene uplift and erosion of southern Scandinavia appears to have been initiated in two phases, an early phase of ?Miocene age and a better-constrained later phase that began in the Pliocene. Neogene uplift of the South Swedish Dome with adjoining areas in Denmark fits into a pattern of late Cenozoic vertical movements around the North Atlantic.

Recognition of the Neogene uplift and erosion of Denmark and Sweden is difficult because of its regional extent, and because the effects are overprinted by the erosion during the subsequent Quaternary glaciations (Fig. 1). Consequently, only few relevant observations were presented in the literature before the 1990s. Studies of the Miocene Vejle Fjord Formation led Larsen & Dinesen (1959) to conclude that considerable parts of Fennoscandia, including not only basement but also sedimentary formations, were eroded in Neogene time. Spjeldnæs (1975) found that uplift of the Fennoscandian Shield in late Oligocene–Miocene time resulted in a significant change in the sedimentary environment and in the drift of the coastline towards the SW.

The effect of late Cenozoic erosion in Denmark has been quantified by a number of workers, who estimated erosion to be from 1 to 2 km in several wells (Jensen & Schmidt 1992, 1993; Japsen 1993, 1998; Michelsen & Nielsen 1993). This paper reports the results of a study of the late Cenozoic erosion of cover rocks in Danish wells located outside the late Cenozoic depocentre in the central North Sea and outside the Bornholm area in the southern Baltic Sea (Fig. 2) (Japsen & Bidstrup 1999). Estimates of erosion have been based on basin modelling and sonic data from several stratigraphic units, resulting in maximum values of *c.* 1000 m, which are considerably lower than those reported in the above earlier studies.

All studies find erosion to increase from the eastern North Sea towards the coasts of Norway and Sweden. This increasing amount of erosion matches the increase in the hiatus at the base of the Quaternary succession, where Neogene and older strata are truncated. These observations suggest that the onset of erosion occurred during

From: Doré, A.G., Cartwright, J.A., Stoker, M.S., Turner, J.P. & White, N. 2002. *Exhumation of the North Atlantic Margin: Timing, Mechanisms and Implications for Petroleum Exploration*. Geological Society, London, Special Publications, **196**, 183–207. 0305-8719/02/$15.00

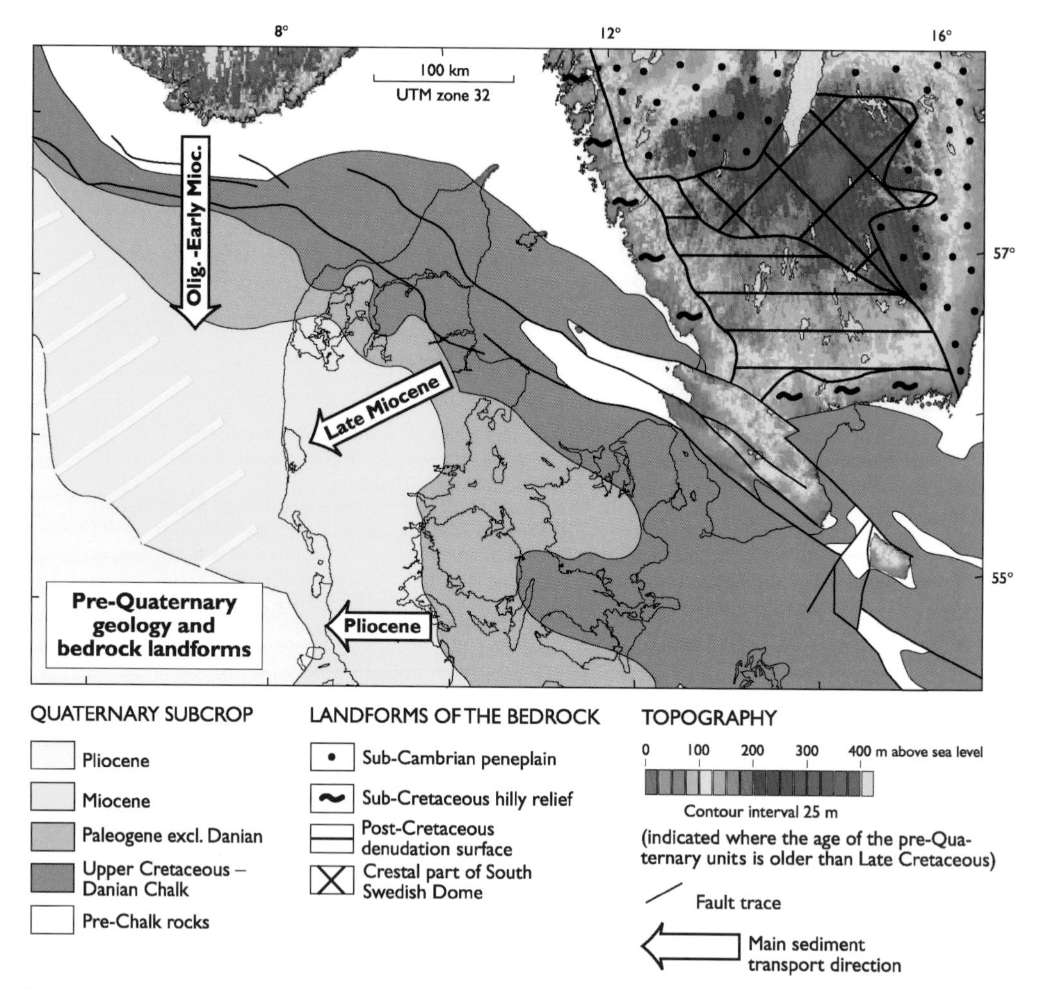

Fig. 1. Pre-Quaternary geology of southern Scandinavia and landforms of the bedrock across the South Swedish Dome. The hiatus at the base of the Quaternary (*c.* 2.4 Ma, Zagwijn 1989) and the change in sediment transport direction from Oligocene to Pliocene time agrees with Neogene uplift of the South Swedish Dome. Compare the increasing age of the Quaternary subcrop towards the exposed basement in Norway and Sweden with the corresponding deepening of the estimated erosion (Fig. 13). The mountains of the Southern Scandes constitute the main part of southern Norway. Modified after Fredén (1994), Vejbæk & Britze (1994), Lidmar-Bergström (1996), Japsen (1998) and Clausen *et al.* (1999).

Neogene time, and indicate that uplift and erosion have affected not only Norway and Denmark, but also southern Sweden (Japsen 1993). The geological record of south Scandinavia is thus of great importance to understanding the Neogene development of the whole of Scandinavia and the Atlantic margins as such. In this area it is easier than in most places to compare data from the Cenozoic sedimentary cover (partly onshore) with observations from exposed basement where pre-glacial landforms are well preserved. Distances are small, and data as well as geoscientific studies are abundant.

The present study concludes that the Mesozoic succession has been *c.* 500 m more deeply buried than today in most of the area, where the

---

Fig. 2. Location maps. (**a**) Place names and profiles ABCD (Fig. 15) and EF (Fig. 14). (**b**) Location of the 68 Danish and three Norwegian wells used in the study. (**c**) Structural elements. Basement highs indicated with grey. See also well location map of Nielsen & Japsen (1991).

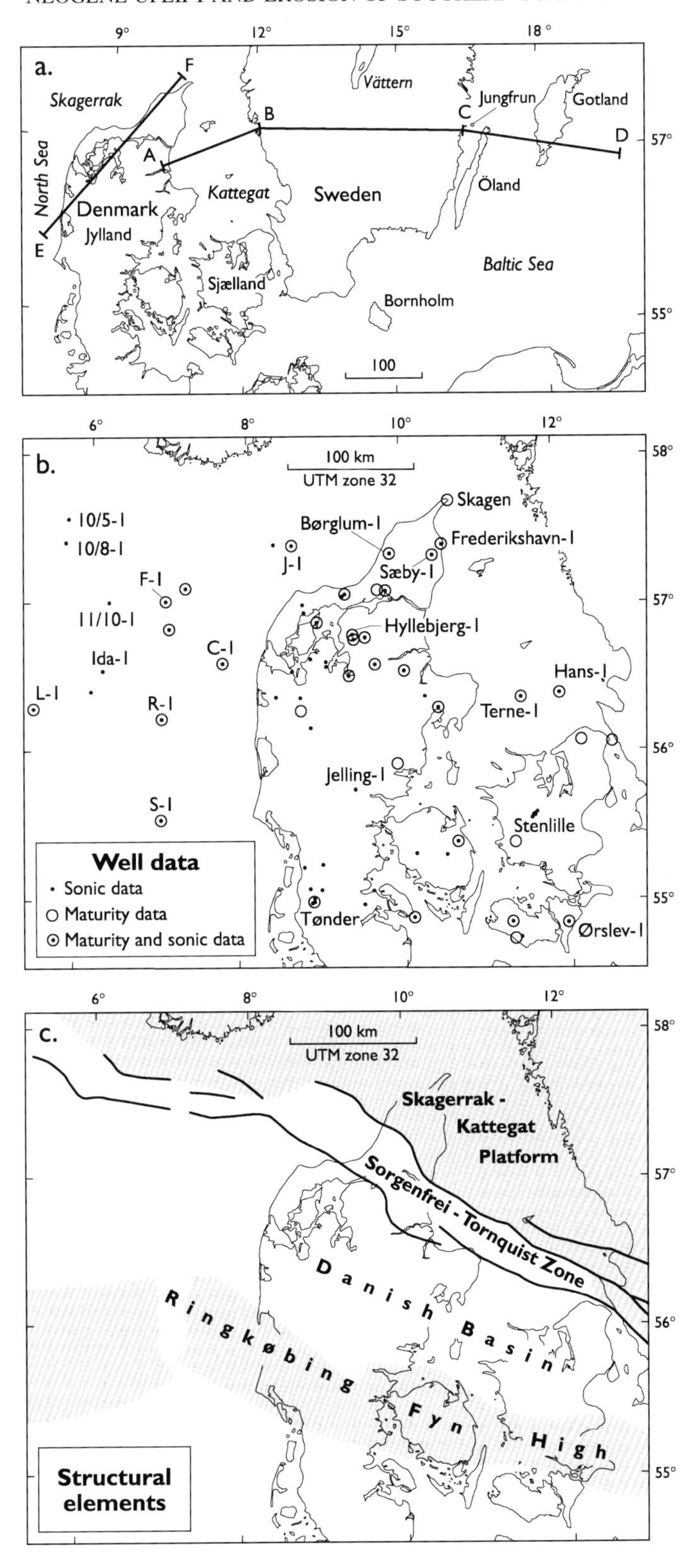
a.
Skagerrak
North Sea
Denmark
Jylland
Kattegat
Sjælland
Sweden
Vättern
Jungfrun
Gotland
Öland
Baltic Sea
Bornholm
100
b.
100 km
UTM zone 32
Skagen
Børglum-1
Frederikshavn-1
Sæby-1
Hyllebjerg-1
10/5-1
10/8-1
J-1
F-1
11/10-1
Ida-1
C-1
L-1
R-1
Hans-1
Terne-1
Jelling-1
S-1
Stenlille
Tønder
Ørslev-1
Well data
Sonic data
Maturity data
Maturity and sonic data
c.
100 km
UTM zone 32
Skagerrak - Kattegat Platform
Sorgenfrei - Tornquist Zone
Danish Basin
Ringkøbing Fyn High
Structural elements

succession generally is overlain by Paleogene strata, and that a section of mainly Paleocene–Miocene age must have been removed from this area during late Cenozoic time. We combine the model of late Cenozoic erosion in Denmark with observations from the exposed basement in Sweden, just east of the Kattegat. Here, geomorphological investigations have led to the conclusion that a rise of the South Swedish Dome (the uplands of Småland with altitudes up to 400 m) occurred during Cenozoic time (Lidmar-Bergström 1995, 1996). We therefore find that Neogene uplift of southern Scandinavia, centred around the South Swedish Dome, could explain both the geophysical and the geomorphological observations.

Erosion estimated from sonic data

Derivation of normal velocity–depth trends

A normal velocity–depth trend (velocity baseline), $V_N(z)$, describes in a functional form how the sonic velocity of a relatively homogeneous sedimentary formation saturated with brine increases with depth when porosity is reduced during normal compaction. The pressure of the formation is hydrostatic during normal compaction, and the formation is at maximum burial depth; i.e. the thickness of the overburden has not been reduced by erosion (e.g. Bulat & Stoker 1987; Hillis 1995, Japsen 1998, 2000). Simple boundary conditions for such trends are that the normal velocity at the surface equals the velocity of the sediment when it was first deposited, and that velocity at infinite depth approaches the matrix velocity of the rock whereas the velocity–depth gradient approaches zero.

The derivation of a normal velocity–depth trend involves three steps of generalization: (1) identification of a relatively homogeneous lithological unit; (2) selection of data points representing normal compaction; (3) assignment of a functional expression to the velocity–depth trend. Velocity baselines may thus be difficult to establish, and different trends have been assigned to identical units by different workers (compare Bulat & Stoker (1987) and Japsen (2000)).

(1) 'Relatively homogeneous' refers to those properties that are important for the macroscopic acoustic behaviour of the unit. However, we do not always know if data from a well represent the typical development of the unit or which mineralogical differences may be of importance for its acoustic behaviour, e.g. the clay content in sandstones or chalks.

(2) 'Normal compaction' may be a difficult condition to prove, because we do not always know if formation pressure is hydrostatic or if the formation has been buried deeper before erosion (Fig. 3). Late Cenozoic erosion along the margins of the North Sea Basin and over-

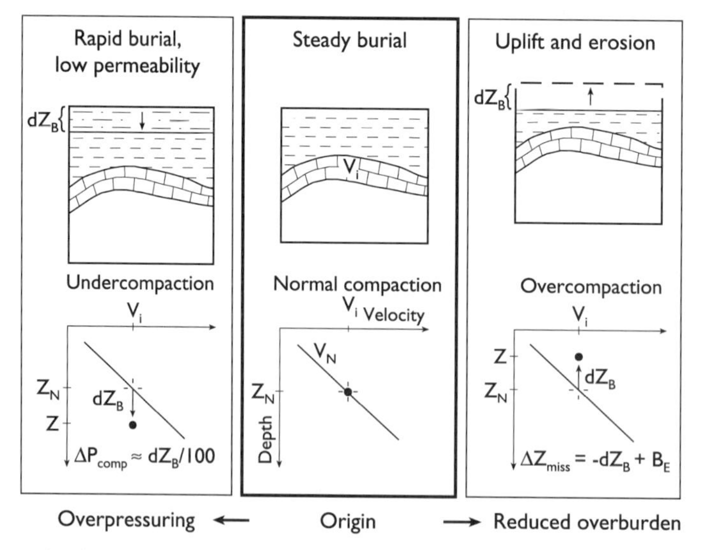

Fig. 3. Burial anomaly, $dZ_B(m)$, relative to a normal velocity–depth trend, V_N. Uplift and erosion reduce the overburden thickness and result in overcompaction expressed as anomalously high velocities relative to present-day depth (negative dZ_B). However, post-exhumational burial, B_E, will mask the magnitude of the missing section, Δz_{miss} (Eq. 2). Undercompaction as a result of rapid burial and low permeability causes overpressure, ΔP_{comp} (MPa), and low velocities relative to depth (positive dZ_B). Modified after Japsen (1998).

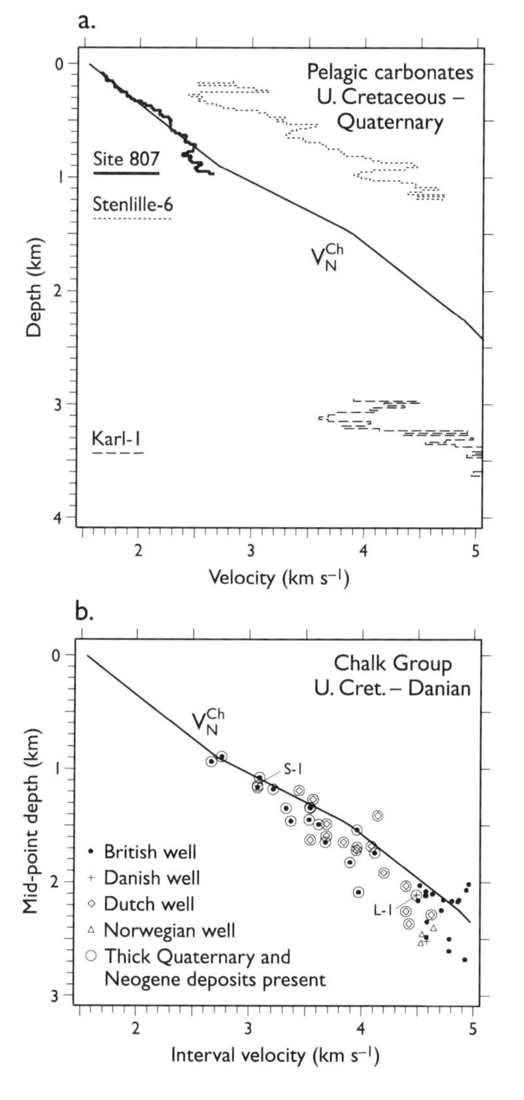

Fig. 4. Outline of the derivation of the normal velocity–depth trend for the North Sea Chalk (V_N^{Ch}, Eq. 3 in Appendix). The shallow part of the trend is constrained by sonic data from pelagic carbonate deposits of Recent age (**a**). The deeper part of the trend is defined by the upper bound for interval velocity data from wells where the Chalk is at maximum burial and at hydrostatic pressure (**b**). (**a**) Sonic logs from pelagic carbonate deposits of Eocene to Pleistocene age drilled in hole 807, Ocean Drilling Program (ODP) Leg 130 (Shipboard Scientific Party 1991), and the Chalk Group in the Danish Stenlille-6 (location shown in Fig. 2) and the Karl-1 wells (central North Sea; only the upper third of the log is shown). (**b**) Interval velocity v. mid-point depth for the Chalk Group for wells where thick Quaternary and Neogene sediments are present in areas with limited or no overpressure and for wells with maximum velocity for $z > 2000$ m (55 out of 845 wells in Chalk velocity database). In (**a**), the depth-shift should be noted between the three sonic logs that all represent pelagic carbonates of very uniform composition. The shift is suggested to be caused by overcompaction as a result of removal of overburden along the margin of the North Sea Basin during late Cenozoic time (Stenlille-6) and by undercompaction as a result of rapid burial in the central North Sea during late Cenozoic time (Karl-1; Chalk formation overpressure is 15 MPa in a nearby well). (Compare Fig. 3.) Modified after Japsen (1998, 2000).

pressuring as a result of rapid, late Cenozoic burial in the basin centre have resulted in a systematic variation of burial anomalies (see below). These anomalies are within ± 1 km for the Upper Cretaceous–Danian Chalk and lower Cenozoic sediments whereas the upper Cenozoic sediments are close to normal compaction (the anomalies for the Cenozoic sediments are calculated relative to a baseline for marine shale; see Eq. 4 in the Appendix) (Japsen 1998, 1999). Thus a regression line fitted to velocity–depth data may not represent a physical model of the subsurface if such anomalies are not taken into account.

(3) 'Assignment of a functional expression' to an observed velocity–depth trend is not straightforward because of the limited range of any dataset. An observed trend may, however, be extrapolated to range from the surface to infinite depth provided that the extrapolated trend complies with the above boundary conditions. Assignment of an arbitrary mathematical velocity–depth relation for a formation may lead to identification of an erroneous baseline, for example, the frequently applied formula for shale trends, $tt = 1/V = a\exp(-b/z)$, that predicts both velocity and velocity gradient to increase towards infinity with depth (tt is transit time (s m^{-1}), and a (s m^{-1}) and b (m) are parameters). Erosion may thus be underestimated if such a trend is applied to single data points at great depth.

Formulation of velocity baselines is thus not an arbitrary choice of mathematical functions and regression parameters, but should be considered as setting up a physical model for a given lithology. First, baselines should be established for formations that are relatively homogeneous with regard to macroscopic acoustic properties, e.g. chalk or marine shale dominated by smectite–illite. Second, baselines should reflect normal compaction, and burial anomalies relative to the trend should be in agreement with other estimates of erosion and overpressure. Consequently, a baseline for a

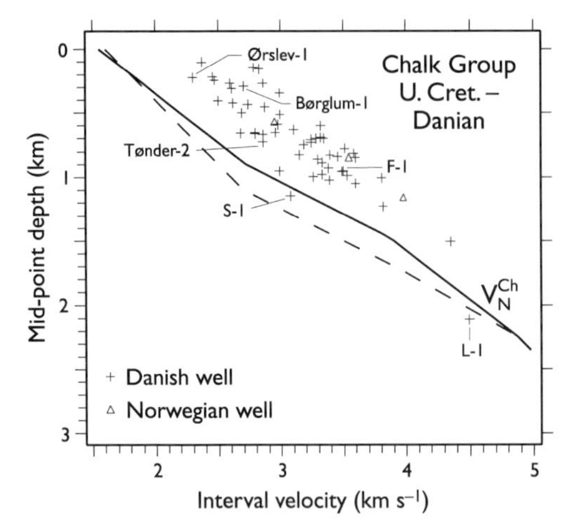

Fig. 5. Interval velocity v. mid-point depth for the Chalk Group in the wells studied, and the revised normal velocity–depth trend for the Chalk, V_N^{Ch} (Eq. 3 in Appendix) (Japsen 2000). Most data points reveal high velocities relative to the normal trend, and this is suggested generally to be due to overburden reduction. Estimates of erosion based on chalk sonic data correlate well with estimates based on basin modelling (Fig. 12a). The dashed line indicates the original baseline of Japsen (1998).

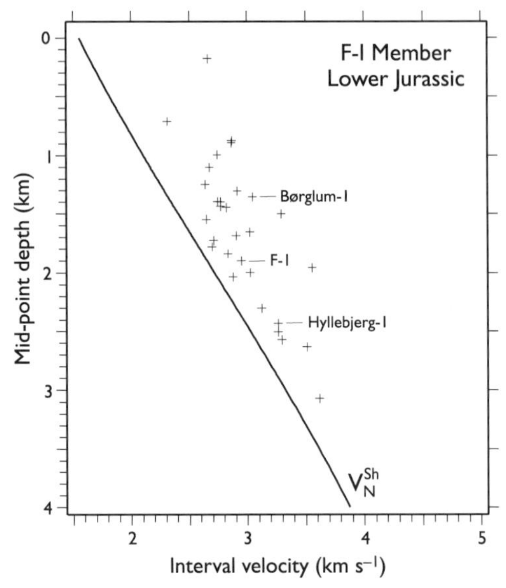

Fig. 6. Interval velocity v. mid-point depth for the Lower Jurassic F-I Member in the wells studied, and the normal velocity–depth trend for Lower Jurassic shale, V_N^{Sh}, suggested to be characteristic for marine shale dominated by smectite–illite (Eq. 4 in Appendix). All data points reveal high velocities relative to the normal trend, and this is suggested to be due to overburden reduction and to a high content of sand or kaolin in shale deposited close to the exposed basement of the Scandinavian Shield during earliest Jurassic time. Estimates of erosion based on sonic data for the Lower Jurassic shale are overestimated relative to estimates based on chalk data in NE Denmark (Fig. 12b).

given lithology should be constrained at the surface by the velocity of recent deposits of the sediment, at shallow depths by the lower bound for velocity–depth data for which the effect of overcompaction as a result of erosion is minimum, and at greater depths along the upper bound for data for which the effect of undercompaction as a result of overpressuring is minimum. Third, the mathematical formulation of baselines should be constrained by simple boundary conditions at the surface and at infinite depth.

The shape of baselines for uniform sedimentary formations reflects the fact that the compaction processes depend on the mineralogical composition of the formations (the derivation of normal velocity–depth trends for three uniform formations is outlined in the Appendix; see Fig. 4).

(1) The chalk baseline reveals a moderate velocity increase for depths less than 1 km, whereas the velocity gradient increases at greater depths until it is gradually reduced at depths below 1.5 km (Fig. 5; Eq. 3 in the Appendix). This variation is in agreement with the preservation of chalk porosities of *c*. 40% to depths of 1 km during normal compaction before the onset of calcite cementation and the consequent increase of velocity (Borre & Fabricius 1998; Japsen 1998).

(2) The moderate velocity gradient of the baseline for marine shale dominated by smectite–illite may also be related to mineralogical composition (Fig. 6; Eq. 4). Smectite–illite particles are separated by water molecules (Bailey 1980), and this interlayer water is adsorbed to the particles even during deep burial (van Olphen 1966). Japsen (1999, 2000) argued that the water adsorbed on the smectite–illite particles could lead to weak mechanical grain contacts, and thus to the low sonic velocity observed for the marine shale at depth.

(3) The baseline for the continental Bunter Shale (Lower Triassic) is found to be similar to the normal trend of the Bunter Sandstone (Fig. 7; Eq. 5 in the Appendix). Japsen (2000) suggested that this similarity could be related to the high kaolin content of the Bunter Shale. Kaolin has little adsorbed water and it builds up thick flakes that are up to a thousand times larger than smectite–illite particles that are separated by water molecules (Bailey 1980; Lindgreen, pers.

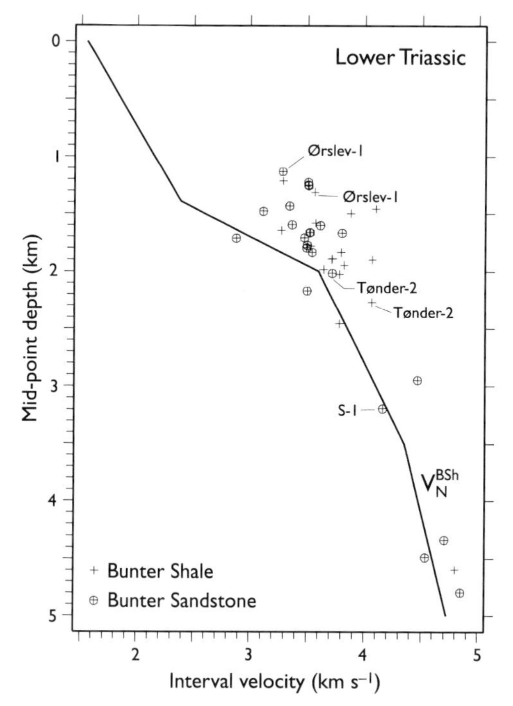

Fig. 7. Interval velocity v. mid-point depth for the Bunter Sandstone and Bunter Shale in the wells studied, and the Bunter Shale trend, V_N^{BSh}, which is suggested to be characteristic for lithologies dominated by quartz and/or kaolin (Eq. 5 in Appendix). Considerable lithological variations are likely within these formations, but the plot shows a number of data points close to the normal trend and others that plot above the trend generally as a result of overburden reduction. Estimates of erosion based on Bunter Shale and Sandstone data correlate well with estimates based on Chalk data (Fig. 12c). The clear distinction between the trend for the continental Bunter Shale and the trend for marine Lower Jurassic shale in Fig. 6 should be noted.

comm.). A shale dominated by well-packed flakes of kaolin could thus acquire rock physical properties similar to those of a consolidated sandstone. An alternative explanation could, however, be related to dominance of quartz in both the Bunter Sandstone and the Bunter Shale. The sharp increase of velocity around 2 km for these formations could thus be related to onset of quartz cementation, which has been reported to develop at depths below 2.5 km in, for example, the North Sea (Bjørlykke & Egeberg 1993). Further studies are required to clarify these issues.

Finally, it should be noted that the chalk baseline given by Eq. 3 is in agreement with that of Hillis (1995) for the depths relevant for estimating erosion, and that the trend for marine shale given by Eq. 4 is in agreement with the suggestions of Scherbaum (1982) and with that of Hansen (1996) for the upper *c.* 2 km of the shale trend. Suggested baselines for the Bunter Shale differ significantly, but the trend given by Eq. 5 agrees with the results of Marie (1975) and Bulat & Stoker (1987) for the velocity interval from which most data are available (see discussion by Japsen (2000)).

Burial anomaly and missing section

Velocity–depth studies have proven useful because they are based on easily accessible data with a wide areal coverage, and thus allow for setting up simple constraints on both physical and geological parameters. Over- and under-compaction relative to a baseline may be studied by computing burial anomalies, dZ_B (m), for a formation as the difference between present depths and depths corresponding to normal compaction for the measured velocity (z and z_N (m), respectively, V is sonic velocity (m s^{-1}); Fig. 3) (Japsen 1998) (Eq. 1):

$$dZ_B = z - z_N. \qquad (1)$$

A baseline may be given as a linear trend, $V = V_0 + k \cdot z$, where V_0 is velocity at the surface and k (m s^{-1} m^{-1}, or s^{-1}) is the velocity gradient (see Eq. 3). The burial anomaly thus becomes (Japsen 1993, 1998):

$$dZ_B = V_0/k + z_t - \Delta z(e^{k\Delta T/2} - 1)^{-1}$$

where Δz is layer thickness, ΔT (s) two-way traveltime thickness and z_t is depth to the top of the layer.

A baseline may also be formulated as a constrained, exponential transit time-depth trend, $tt = (tt_0 - tt_\infty)e^{-z/b_2} + tt_\infty$, where $tt = 1/V$ (s m^{-1}) is transit time, tt_0 and tt_∞ are transit time at the surface and at infinite depth, respectively and b_2 (m) is an exponential constant (see Eq. 4), In this case the burial anomaly may be approximated by the following expression if layer thickness and velocity gradient are moderate (Japsen 1999):

$$dZ_B = (tt_0 - tt_\infty) \cdot \ln \frac{tt - tt_\infty}{(tt - tt_\infty) \cdot e^{-z/b_2}}.$$

Low velocities relative to depth give positive burial anomalies, which may indicate under-compaction as a result of overpressure (see Japsen 1998). High velocities relative to depth give negative burial anomalies, which may be caused by a reduction in overburden thickness ('apparent uplift', Bulat & Stoker 1987; 'net uplift and erosion', Riis & Jensen 1992; 'apparent exhumation', Hillis 1995). It must, however, be noted that any post-exhumational burial, B_E (m), will mask the magnitude of the missing overburden section, Δz_{miss} (m), and we obtain

$$\Delta z_{miss} = -dZ_B + B_E \qquad (2)$$

where the minus indicates that erosion reduces depth (Hillis 1995; Japsen 1998).

Whether a burial anomaly is a measure of erosion or is caused by other factors (e.g. lithological changes) is subject to an integrated evaluation of the area in question. Apart from being in agreement with other estimates of erosion, the burial anomalies should also correspond geographically to the extent of a section missing from the stratigraphic record. Erosion may be underestimated if the compaction process is reversible when the load of the overburden is reduced. However, we find that previous estimates of erosion based on sonic data from the study area are exaggerated by up to 1000 m, mainly because of the above-mentioned problems with identifying valid baselines for homogeneous formations (see section 'Comparison with other studies'). Overestimation of erosion from sonic data thus seems to be a typical problem in such studies, and underestimation because of the reversibility of the compaction process seems to be only of minor importance.

Summing up, a large areal and stratigraphic data coverage is of crucial importance for evaluating the validity of estimates of erosion from sonic data. Moreover, both velocity baselines and estimates of erosion should be in agreement with stratigraphy in the area, formation overpressure in adjacent basins and estimates of erosion based on independent methods. It is thus an important test of the validity of the derived baselines that the known overpressure in the central North Sea may be predicted from burial anomalies relative to the trends for chalk and for marine shale (Japsen 1998, 1999).

Sonic data

Velocity–depth data from 60 Danish wells form part of the database for this study (Fig. 2). The data were presented by Nielsen & Japsen (1991), apart from the Ida-1 and Jelling-1 wells. Fifty-two of the wells have interval velocities for the Chalk Group, 31 for the F-I Member of the Lower Jurassic Fjerritslev Formation (Michelsen 1989), and 22 for the Bunter Sandstone or the Bunter Shale (Bertelsen 1980); 42 wells have data from the Chalk as well as from the pre-Chalk interval (Figs 5–7). Data from three Norwegian wells with Chalk velocity data are included to support the contouring.

Chalk burial anomalies were calculated relative to the revised normal velocity–depth trend for the Chalk developed by Japsen (1998, 2000) (Eq. 3). Anomalies for the pre-Chalk formations were calculated relative to baselines suggested by Japsen (2000). Burial anomalies for the Lower Jurassic F-I Member are calculated relative to the shale trend given by Eq. 4, and those for the Bunter Sandstone and the Bunter Shale relative to the Bunter Shale trend are given by Eq. 5. A single burial anomaly as a result of late Cenozoic erosion was estimated on the basis of the available sonic data for each well, and corrected for the Quaternary reburial to obtain an estimate of the missing section (Eq. 2) (see Japsen & Bistrup (1999) for details).

Erosion estimated from basin modelling

Model description and input data

The basin development of the study area has been modelled from 35 wells with a commercial, 1D forward modelling program (Yükler 1978; Iliffe & Dawson 1996). The program starts simulation of the geological development from the base of the sedimentary section and performs a calculation of parameters such as formation thickness, pressure, temperature and vitrinite reflectance as a function of time. From the geological input at the well location (thickness and ages of sediments, estimated magnitude and timing of erosion and estimated heat-flow history) the program calculates vitrinite reflectance, temperature and pressure as a function of time. The calculated values of these parameters for the present-day situation are then compared with data that can be divided into two groups: (1) data that constrain the thermal history: present-day temperature (BHT, bottom hole temperatures) and thermal maturity indicators (mainly vitrinite reflectance values in the study area); (2) data that constrain the compaction of the sediments: pressure and porosity (no overpressure is encountered within the study area).

If the measured and calculated values do not match, the input parameters are changed and the program is run again, until a satisfactory match is obtained. A general match to many data points and a laterally consistent heat-flow and erosion model was preferred because of the regional character of the study, the varying data quality and the sometimes conflicting values.

Chronostratigraphic event definition. The program quantifies all important processes as a function of time, and the basin development is thus defined in terms of chronostratigraphic units valid for the entire area (model layers or events). The number of events in the geological model, including periods of deposition, non-deposition and erosion, must be chosen in such a way that all major changes can be described and related to existing stratigraphy while avoiding excessive calculation times.

A total of 45 events was chosen to describe the geological development within the study area from Cambrian time until the present (a time step of 250 ka and a depth step of 25 m). The duration of the events is shorter in the Cenozoic interval to describe the rapid changes in this period. The lithologies used in the modelling were kept constant for the same event if no geological information dictated otherwise. The main lithotypes for the events are: Cenozoic events 30–45 (excluding Danian time; 60–0 Ma): sand and shale with occasional coals, more shaly towards the base of the succession; Late Cretaceous–Danian events 26–29 (96–60 Ma): Chalk; Early Cretaceous events 24–25 (129–96 Ma): mixture of silt, marl and shale; Late Jurassic events 20–23 (152–129 Ma): shale and siltstone; Mid-Jurassic events 17–19 (178–152 Ma): mixed sandstone, siltstone and shale; Early Jurassic events 13–16 (210–178 Ma): shale, silt and sandy shale; Triassic events 10–12 (250–210 Ma): sandstone, sandy shale to shale with carbonate; locally salt; Late Permian event 9 (256–250 Ma): clean salt if the layer is thick, otherwise a mixture of shale, anhydrites and salt; Cambrian–Early Permian events 1–8 (570–256 Ma): shale and sandstone.

Sparse temperature and vitrinite reflectance data are available below uppermost Triassic level, and even fewer below lowermost Permian level. The model for each well was extrapolated to basement by the use of seismic data where no well data were available.

Palaeo-surface-temperatures. The palaeo-surface-temperature is the average temperature at the sediment–water interface during a particular period. Estimates of palaeo-surface-temperature were modified from Buchardt (1978) (Fig. 8).

Heat-flow model. The heat-flow history was constrained by two datasets only: (1) the present-day temperature in the wells, which combined with the given lithologies (thermal conductivities) defines the present-day heat flow; (2) values of vitrinite reflectance, which define the heat flow at the time of maximum burial for a given formation.

A simple heat-flow model that honours these values is estimated for each well based on an assumption of a constant heat flow of 1 Heat Flow Unit (HFU) from Cambrian time (event 1) until the beginning of Oligocene time (event 34) (1 HFU = 42 mWm^{-2}). This value is typical for the constrained part of the heat-flow history, but the exact choice is of minor importance because only the heat flow at maximum burial affects the vitrinite reflectance and most of the study area has been affected by erosion. The heat-flow model for the Oligocene–Recent time interval was modified to match present-day temperature and vitrinite reflectance data, but also to result in smooth heat-flow variations in time and space (Fig. 8).

The heat flow used in the model is the heat flow at the base of the sedimentary succession. In contrast to the background heat flow from the upper mantle, the heat flow at this level is affected by transient effects from sedimentation and erosion (Vik & Hermanrud 1993) and by local variations in the geometry and lithology of the sedimentary units (e.g. salt domes). Transient

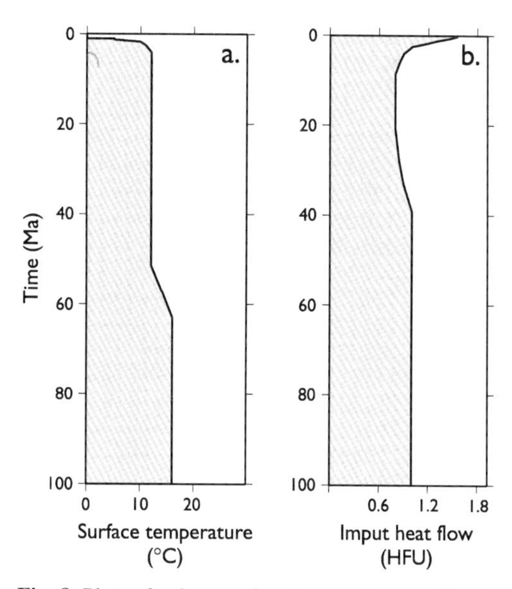

Fig. 8. Plots of palaeo-surface-temperatures and heat-flow history used in the modelling of Hyllebjerg-1 well. A heat flow of 1 HFU has been used until Oligocene time, after which heat flow was allowed to change smoothly to match calibration data. The palaeo-surface-temperatures are the same for all wells in the study (Buchardt, 1978). (See the corresponding calibration plot in Fig. 9.)

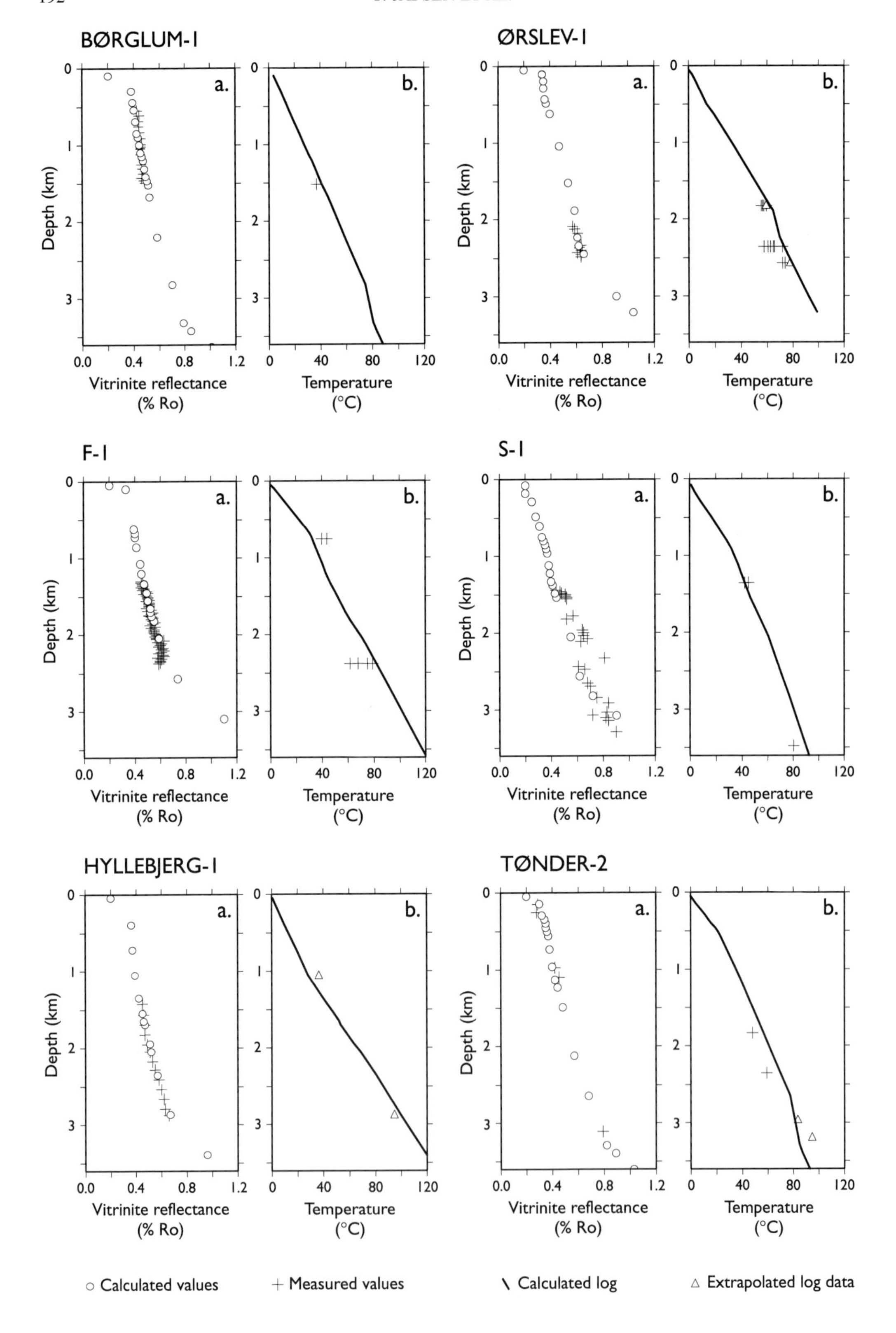

BØRGLUM-1
ØRSLEV-1
F-1
S-1
HYLLEBJERG-1
TØNDER-2
a.
b.
Depth (km)
Vitrinite reflectance
(% Ro)
Temperature
(°C)
0.0
0.4
0.8
1.2
0
40
80
120
○ Calculated values
+ Measured values
\ Calculated log
△ Extrapolated log data

heat-flow variation from 0.9 to 1.1 HFU in the sedimentary section was modelled by assuming deposition (1 km per 5 Ma) followed by erosion (also 1 km per 5 Ma) on top of a sedimentary succession of 5 km and assuming constant heat flow of 1 HFU at the base of a 50 km thick basement section (using the PetroMod 2D software, version 6.1).

Erosion model. Vitrinite reflectance is more sensitive to temperature than to time, and therefore depends mainly on maximum temperature and not on the timing of maximum temperature. In this study, therefore, assessment of the timing of erosion was based on the existing stratigraphy and on general geological considerations because no fission-track or (U–Th)/He data were available. A model of erosion starting in late Miocene time and continuing until late Quaternary time was used in the basin modelling throughout the area (see the discussion in the section 'Timing of maximum burial and subsequent erosion').

Model calibration and results

Basin modelling was performed for 35 wells where the heat-flow and erosion model could be calibrated against temperature and vitrinite data (Figs 2 and 9). The estimated heat-flow model varies smoothly with time for individual wells (e.g. Fig. 8), and in space for different time intervals (Fig. 10). To match the steep vitrinite reflectance gradients observed in this area (Fig. 9), low heat-flow values have been used for central and northern Jylland during maximum burial (partly as a result of transient effects caused by sedimentation before maximum burial). To match present-day temperatures, high heat-flow values have been used for the most recent development in northern Jylland (partly because of transient effects reflecting late erosion).

Calibration of the model assumed the deposition of overburden sediments and their subsequent removal in all wells studied, apart from the L-1 and S-1 wells (Fig. 11). The thickness of the missing section can only be estimated within a range of possible solutions, and this range is constrained by the data quality (e.g. vitrinite data), by the lateral consistency of the model of erosion and heat flow, and by the general geological understanding of the area. The uncertainties on the estimates of erosion are of the order of 100–200 m.

Quantification of late Cenozoic erosion in Denmark

Estimates of missing section based on different data

Estimates of the missing section based on Chalk sonic data and on basin modelling are similar; the correlation coefficient is 0.81 for 24 wells in common and the mean difference between the estimates is 30 m, standard deviation 130 m (Fig. 12a). The estimates of the missing section based on Chalk and Triassic sonic data are also rather similar, but the scatter is greater; the correlation coefficient is 0.72 for the 19 wells in common and the mean difference between the estimates is 10 m, standard deviation 210 m (Fig. 12c). The estimates of the missing section based on Chalk sonic data are generally smaller than those based on Lower Jurassic sonic data; the correlation coefficient is 0.71 for the 27 wells in common and the mean difference between the estimates is 150 m, standard deviation 270 m (Fig. 12b). This difference is suggested to be due to lithological variations within the Lower Jurassic sequence as discussed below.

Only in NE Denmark does the magnitude of the burial anomaly based on sonic data for the pre-Chalk section generally exceed the estimates from Chalk sonic data and from basin modelling; e.g. differences of 500–1000 m between anomalies based on sonic data for the Chalk and the Lower Jurassic units. A possible interpretation of this difference could be that the Lower Jurassic units in that area experienced maximum burial before the deposition of the Chalk (see Japsen 2000). It is not, however, possible to identify a major hiatus in the stratigraphic record corresponding to a deep erosional event that could explain the difference between the Chalk and the pre-Chalk burial anomalies. Maximum burial of the Mesozoic succession is thus suggested to have occurred during Cenozoic time in all wells studied.

Lateral lithological variation within the Lower Jurassic sequence is a likely cause for the high

Fig. 9. Plots of calculated and measured values of (**a**) vitrinite reflectance and (**b**) temperature for the wells Børglum-1, F-1, Hyllebjerg-1, Ørslev-1, S-1 and Tønder-2. The slow increase of vitrinite reflectance with depth for, for example, the Børglum-1 well should be noted. This is suggested to be partly due to a transient thermal effect related to deposition followed by erosion during Cenozoic time.

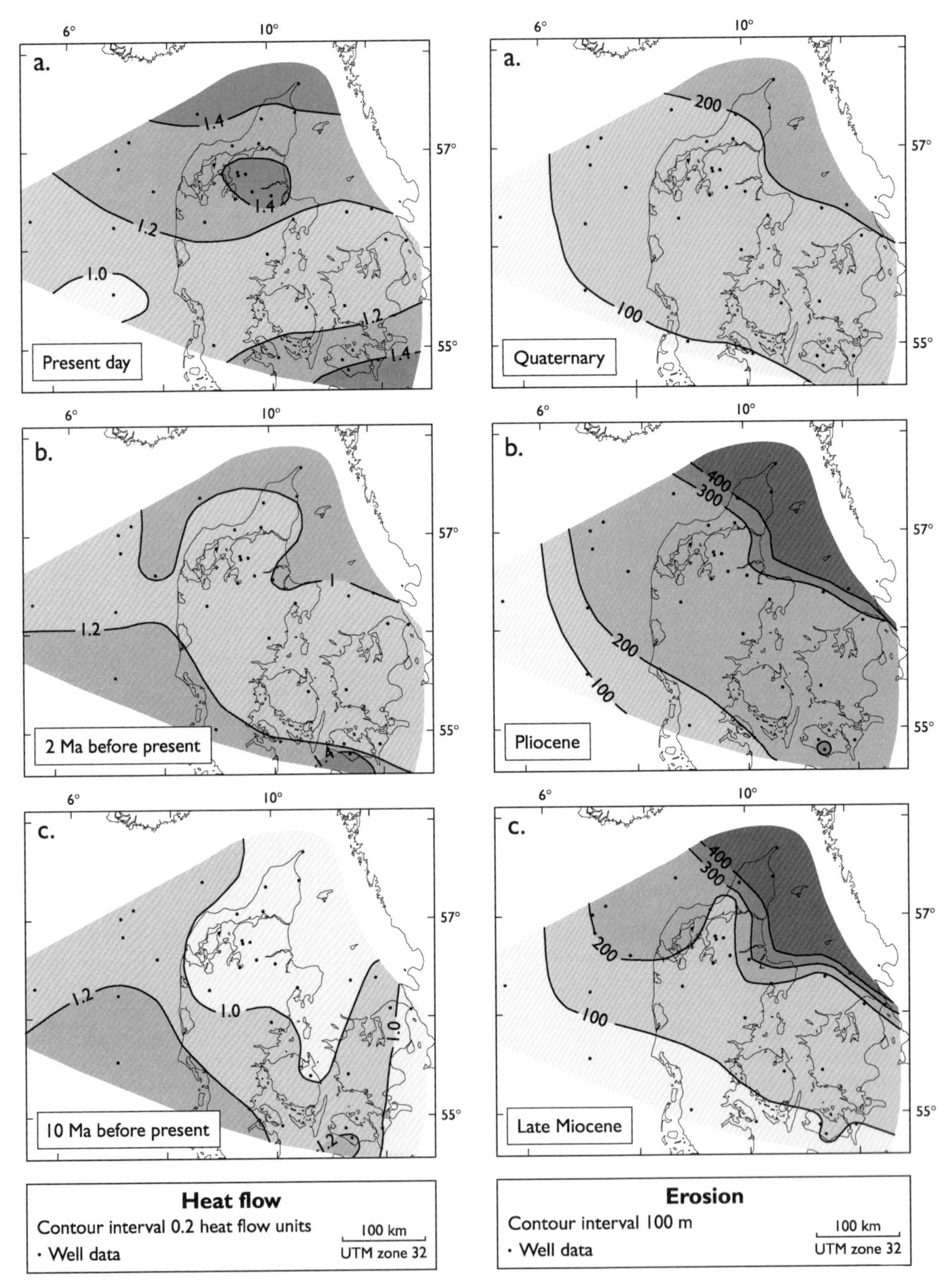

Fig. 10. Maps of heat flow at three times during the late Cenozoic period resulting from the basin modelling. The heat flow is assumed to be constant until Oligocene time (1 HFU), but throughout Oligocene–Recent time the heat flow is allowed to vary smoothly to match calibration data. (**a**) Present day; (**b**) 2 Ma before present; (**c**) 10 Ma before present.

Fig. 11. Maps of erosion at three intervals during late Cenozoic time resulting from the basin modelling. The existing stratigraphy constrains the temporal development of the erosion. (**a**) Quaternary time; (**b**) Pliocene time; (**c**) late Miocene time. A model of erosion starting in late Miocene time has been used in the basin modelling throughout the area.

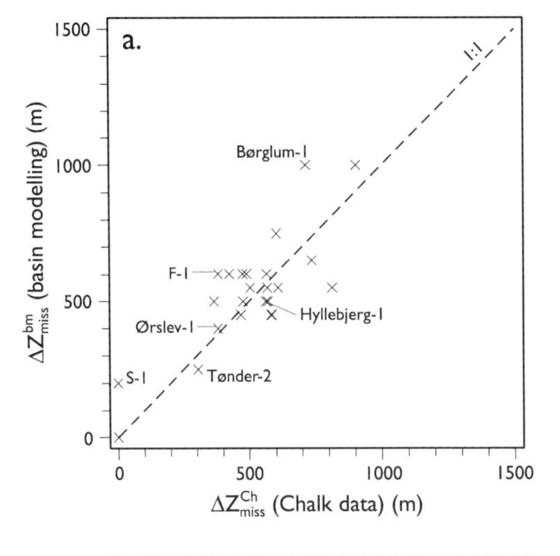

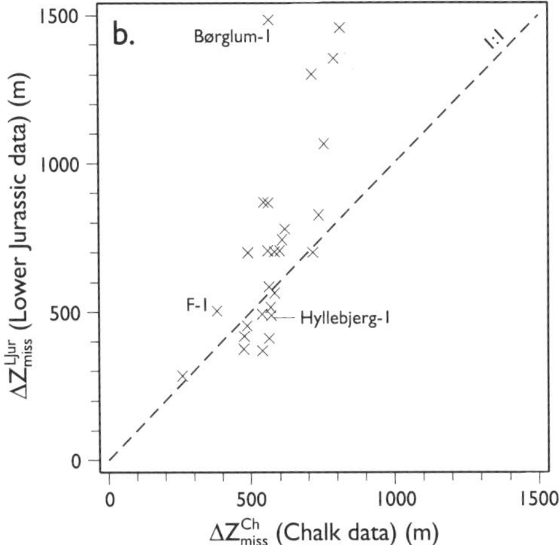

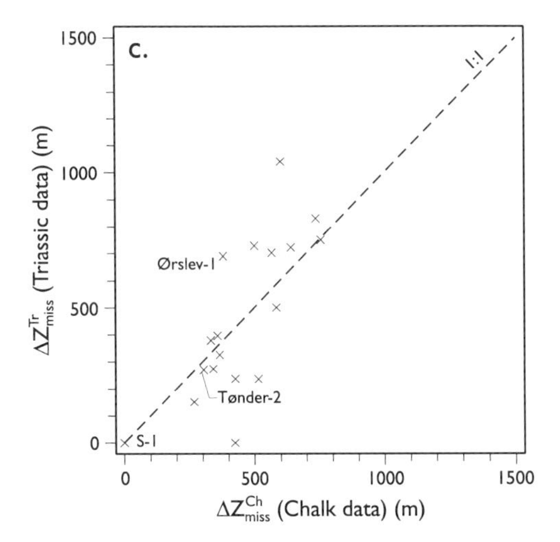

Fig. 12. Correlation between estimates of missing section. (**a**) Estimates based on Chalk sonic data, ΔZ^{Ch}_{miss}, v. estimates from basin modelling, ΔZ^{bm}_{miss}. (**b**) Estimates based on Chalk sonic data, ΔZ^{Ch}_{miss}, v. estimates based on sonic data for Lower Jurassic F-I Member, ΔZ^{LJur}_{miss}. (**c**) Estimates based on Chalk sonic data, ΔZ^{Ch}_{miss}, v. estimates based on sonic data for Lower Triassic Bunter Sandstone and Bunter Shale, ΔZ^{Tr}_{miss}. The good correlation between estimates based on Chalk velocities and on basin modelling should be noted. Estimates based on data for Lower Jurassic shale are overestimated relative to estimates from Chalk data in NE Denmark because of lithological variations in the shale. The lines illustrating the 1:1 relationship between the estimates are shown. Well names are given for wells with results of basin modelling shown in Fig. 9.

estimates of erosion based on sonic data from these sediments compared with estimates from both basin modelling and from Chalk sonic data. During earliest Jurassic time a shallow marine environment prevailed close to the exposed basement of the Scandinavian Shield, and correspondingly a high content of sand and kaolin has been reported for the Lower Jurassic shale in this area (Pedersen 1985; Lindgreen 1991). The relatively high velocity of the Lower Jurassic sequence in NE Denmark could thus be due to the high content of both sand and kaolin (see the section 'Derivation of normal velocity–depth trends' and Appendix). Furthermore, the general agreement between the erosional estimates based on Chalk sonic data and those based on basin modelling in northern Jylland, as in the rest of Denmark, suggests that estimates based on Chalk data are to be preferred to those based on Lower Jurassic data.

The high velocities of the F-I Member have previously been suggested to reflect maximum burial before Late Cretaceous–Paleogene inversion within the Sorgenfrei–Tornquist Zone and the subsequent removal of 1 km of Chalk (Fig. 1) (Japsen 1993; Michelsen & Nielsen 1993). Removal of a thick Chalk section during the inversion is, however, unlikely because seismic reflectors within the Chalk section onlap an anticlinal structure along the Sorgenfrei–Tornquist Zone in the western parts of the Kattegat (Liboriussen *et al.* 1987). Syndepositional growth of the inversion structure, with peak movements during mid-Cretaceous times, implies that only a thin Chalk section was deposited in the inversion zone.

Magnitude of late Cenozoic erosion

The section removed by late Cenozoic erosion has been estimated for 68 Danish wells on the

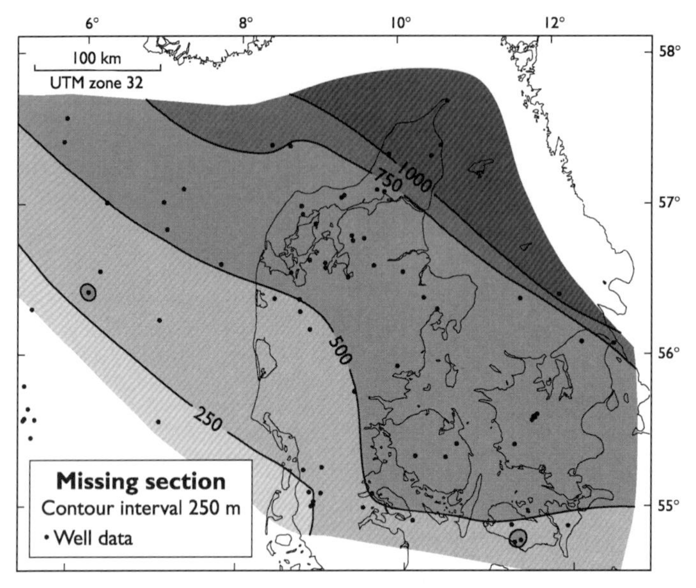

Fig. 13. Map of the section missing as a result of late Cenozoic erosion based on estimates from basin modelling and sonic data from 68 Danish and three Norwegian wells. A succession of about 500 m of post-Chalk sediments is missing from large parts of the area. Towards the NE, *c.* 1000 m are missing where the Chalk is absent or deeply eroded on and along the Skagerrak–Kattegat Platform.

basis of a combination of results from basin modelling in 35 wells with compaction studies based on velocity–depth data in 60 wells (Figs 13 and 14) (Japsen & Bidstrup 1999).

(1) The estimated missing section reaches 1000 m in a number of wells in NE Denmark, on or close to the Skagerrak–Kattegat Platform, where the Chalk is deeply truncated or absent (Fig. 1). For the Frederikshavn-1 well, both methods indicate a missing section of *c.* 1000 m, whereas the maximum based on basin modelling alone is 1200 m in the Hans-1 well, and the maximum from sonic data alone is 1300 m in the Sæby-1 well. The latter value is, however, based on data from the Lower Jurassic sequence, and appears to be overestimated when compared with results from neighbouring wells.

(2) The missing section is found to be just over 500 m in many Danish wells located in a broad band from NW to SE where the Upper Cretaceous–Danian Chalk is preserved.

(3) Missing sections between 250 and 500 m are mapped in the western and southern part of the study area. Erosion estimates of only 100–250 m are found in southwesternmost Jylland. Estimates of 250 m are close to the accuracy of the methods, and the Mesozoic succession is thus probably close to normal compaction in this area. The two easternmost Danish wells with Mesozoic sediments found to be at maximum burial are the L-1 and S-1 wells. The absence of upper Pliocene sediments in the L-1 indicates non-deposition rather than an episode of erosion (Laursen 1992).

The magnitude of erosion increases clearly across the NE side of the Sorgenfrei–Tornquist Zone, where estimates of erosion are as high as on the Skagerrak–Kattegat Platform to the north, *c.* 1000 m. In the southern part of the inversion zone, estimates of erosion are similar to values found south of the zone, *c.* 500 m. This difference indicates stronger tectonic movements on the Skagerrak–Kattegat Platform than in the area to the south.

Along the SW edge of the Skagerrak–Kattegat Platform, the missing section of 1000 m is suggested to be *c.* 500 m of Cenozoic sediments as in the Danish Basin to the south plus a Chalk section of *c.* 500 m. Farther north on the Platform, the missing Cenozoic section is suggested to be *c.* 250 m, and the missing Chalk section *c.* 750 m. On the basis of these assumptions a profile along Jylland has been reconstructed to the situation before Neogene uplift and erosion (Fig. 14). The reconstructed profile reveals the known Chalk depocentre (thickness <2 km) south of the Sorgenfrei–Tornquist Zone as well as one NE of the Zone (reconstructed thickness <1 km) corresponding

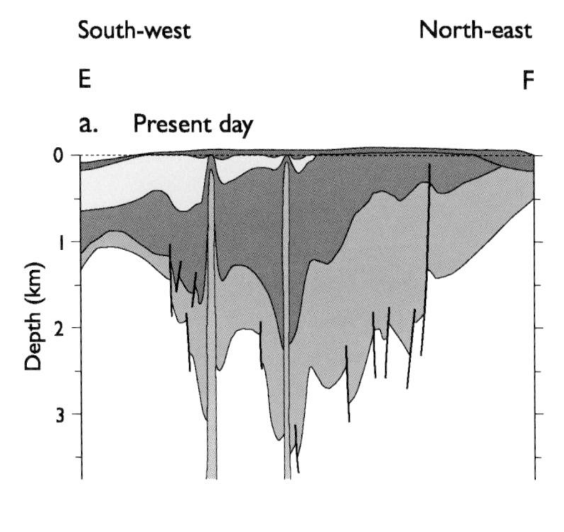

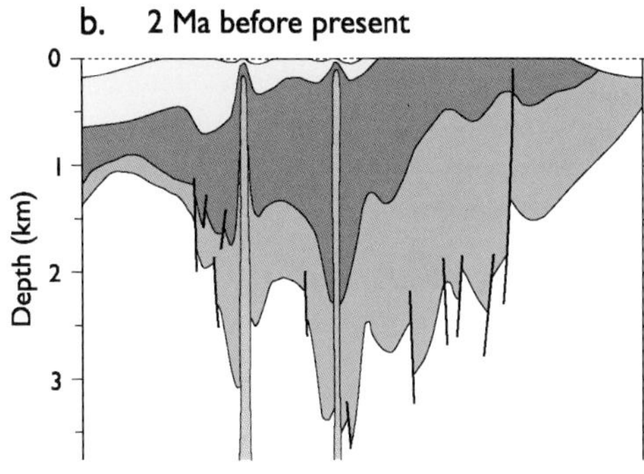

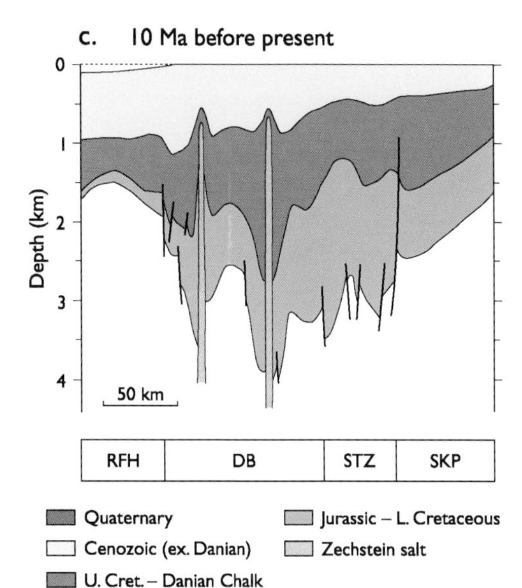

Fig. 14. (**a**) Profile of the post-Triassic succession in Jylland, and reconstructions at (**b**) 2 Ma and (**c**) 10 Ma before present (before uplift and erosion increasing to 1 km towards the NE). The (now partly eroded) Chalk depocentre on the Skagerrak–Kattegat Platform should be noted. DB, Danish Basin; RFH, Ringkøbing–Fyn High; SKP, Skagerrak–Kattegat Platform; STZ, Sorgenfrei–Tornquist Zone. Location shown in Fig. 2. (Compare Japsen (1993).)

to a prediction based on thermo-mechanical modelling (Gemmer 2002).

We have estimated the thickness of sediments deposited within the study area during the Cenozoic epochs from the known stratigraphy, the estimation of the total missing section and from the above assumptions about removed chalk cover: no Pliocene sediments are found to have been deposited within the study area apart from the SW part where thicknesses are estimated to have reached 300 m; a Miocene depocentre <1100 m thick is found to have been located west of Jylland with thicknesses gradually decreasing to 100 m towards the NE and east and 300 m towards the SE; an Oligocene depocentre <600 m thick is found to have been located NW of Jylland; in the rest of the area thicknesses range from 50 to 100 m; Eocene deposits are found to have covered the entire area, except for the SW part of the North Sea, with thicknesses ranging from 50 to 150 m; Paleocene (excluding the Danian) deposits are found to have covered the entire area with thicknesses of 50–100 m, locally 150 m.

Timing of maximum burial and subsequent erosion

Along the margins of the North Sea Basin, the Chalk was at maximum burial before Neogene erosion (Japsen 1998). This conclusion applies to the majority of wells in the present study where Chalk is overlain by Paleogene sediments, and where sonic data for the Chalk indicate a previous greater depth of burial. Thus maximum burial of the Chalk must have occurred during Cenozoic time after early Cenozoic burial.

(1) The timing of erosion can be further detailed in southern and central Jylland and offshore west of Jylland. Here erosion must postdate the deposition of offshore to shoreface sediments of the upper Miocene Gram Formation; for example, during Plio-Pleistocene time (Rasmussen 1961; Rasmussen, pers. comm.). This erosional event thus matches the basin-wide hiatus at the base of the Plio-Pleistocene deposits that are younger than 2.4 Ma (the late Cenozoic succession is complete only in a narrow zone in the central North Sea) (e.g. Zagwijn 1989; see Japsen 1998).

(2) The timing of erosion is less constrained on and along the Skagerrak–Kattegat Platform, where the Chalk is deeply eroded or absent (Fig. 1). The pronounced movements of the Platform documented here may, however, lead to the suggestion that erosion on the Platform was initiated earlier than in the rest of the area; for example, by uplift during mid-Miocene time. This suggestion is further supported by the shift in sediment transport direction during mid-Miocene time observed in the northeastern North Sea Basin as discussed below (Fig. 1) (Clausen *et al.* 1999).

The sedimentary deposits in the study area were thus eroded during late Cenozoic time subsequent to their maximum burial, which generally must have taken place during Neogene time. The erosion is likely to have been initiated in two phases, an early phase of ?Miocene age and a better constrained later phase that began in the Pliocene. The late event has affected a vast area that reaches far into the North Sea, whereas the extent of the early phase must be restricted to areas close to Scandinavia where no Miocene is present, e.g. the Skagerrak–Kattegat Platform and southern Sweden.

The late Cenozoic erosion was followed by Quaternary reburial, which in onshore Denmark occurred later than *c.* 0.3 Ma, the age of oldest Quaternary sediments (Knudsen 1995). The age of the oldest Quaternary deposits is progressively older farther into the North Sea, and in the westernmost part of the Danish sector, sedimentation was almost continuous during late Cenozoic time (see Konradi 1995; Japsen 1998).

Comparison with other studies

Three studies from the early 1990s found the missing section to be substantially greater along the Sorgenfrei–Tornquist Zone than suggested here (Jensen & Schmidt 1992; Japsen 1993; Michelsen & Nielsen 1993). The studies by Japsen (1993) and by Michelsen & Nielsen (1993) were based on data from the Lower Jurassic F-I Member, which, as suggested in the previous section, may lead to overestimated erosion towards the NE because of lithological variations within the unit. Furthermore, the difficulty in defining the absolute level of baselines for different lithologies contributes to the uncertainty of the estimated burial anomalies.

Jensen & Schmidt (1992, 1993) overestimated erosion by an average of *c.* 450 m relative to the results presented here for the 14 Danish wells in common; the maximum overestimate was 1000 m. The burial anomalies of Jensen & Schmidt (1992, 1993) were estimated from vitrinite reflectance, density and sonic data.

Japsen (1993) overestimated erosion by an average of *c.* 250 m relative to the results presented here for the 31 wells in common; the maximum overestimate was 1000 m. The estimates of erosion were based on the same data for the F-I Member as used here, and were calculated relative to the linear velocity–depth trend for Lower Jurassic shale in the NW German Basin determined by Scherbaum (1982). This trend deviates less than 100 m from the shale trend applied here for the relevant velocity interval (Eq. 4).

Michelsen & Nielsen (1993) overestimated erosion by *c.* 650 m relative to the results presented here for the seven wells in common; the maximum overestimate was 1300 m. The burial anomalies were calculated for the F-I Member relative to a simple exponential transit time–depth trend established by those workers. This trend results in overestimates of erosion of up to 300 m for the relevant velocity interval relative to the trend applied here (Eq. 4).

Japsen (1998) overestimated erosion by an average of *c.* 200 m relative to the present study for the 51 wells in common; the maximum overestimate was 500 m. In the present study, Chalk burial anomalies are calculated relative to the revised Chalk baseline (Eq. 3), which, for the main velocity interval, is shifted *c.* 200 m towards more shallow depths relative to the Chalk trend applied by Japsen (1998).

Huuse *et al.* (2001) found erosion to be several hundred metres less when estimated from stratal geometries along a north–south seismic section west of Jylland than when estimated from the maximum burial studies of Jensen & Schmidt (1993) and Japsen (1998). This conclusion regarding the S-1 well to the south of the section is in agreement with the present study, where the drilled section in this well is found to be at maximum burial today. This interpretation, which is supported by vitrinite data, is also a consequence of the revision of the Chalk baseline as discussed in the Appendix (Fig. 4). We find, however, that a missing section of *c.* 500 m of Miocene sediments is compatible with the stratigraphy around the northern F-1 well, where no Miocene sediments are present today.

Neogene uplift of southern Scandinavia and of the South Swedish Dome

Lidmar-Bergström (1999) discussed the geomorphological evidence concerning the uplift history of the three surface domes in Scandinavia. The

Northern and Southern Scandes are the most prominent of the domes. The Southern Scandes constitute the main part of southern Norway and have a maximum elevation of 2540 m above sea level (a.s.l.). There is a general consensus that major uplift of the Southern Scandes occurred in Neogene time (Peulvast 1985; Jensen & Schmidt 1992; Rohrman *et al.* 1995; Riis 1996). This timing of uplift is in accordance with a Neogene onset of erosion in northern Denmark. The Southern Scandes were also affected by a Paleogene uplift phase (Clausen *et al.* 2000; Lidmar-Bergström *et al.* 2000).

Denudation surfaces of the South Swedish Dome

The South Swedish Dome, which is the smallest of the Scandinavian domes, culminates NE of Kattegat, just south of lake Vättern, and has a maximum elevation of 380 m a.s.l. (Figs 1 and 15) (Lidmar-Bergström 1988, 1996). On its northern and eastern flanks, the sub-Cambrian peneplain can be traced up to its summits (see Fig. 16). This surface is extremely flat and was formed during a period of major denudation over all of the Baltic Shield in late Proterozoic time (e.g. Högbom 1910; Högbom & Ahlström 1924). It is still well preserved in parts of eastern and south–central Sweden (Rudberg 1954; Lidmar-Bergström 1996). In contrast, the sub-Cambrian peneplain has been destroyed on the southern and western flanks of the South Swedish Dome, where the Precambrian basement was re-exposed during the warm and humid Mesozoic climate (Lidmar-Bergström 1989, 1995). Deep weathering took place and thick kaolinitic saprolites were formed, and, after partial erosion of the saprolites, an undulating, hilly relief developed. This sub-Cretaceous hilly relief can be seen up to 125 m a.s.l. today as a result of preservation under a long-lasting Cretaceous cover (Lidmar-Bergström 1982, 1996). Where the sub-Cambrian peneplain is well preserved, it must have been protected during Mesozoic time by a Palaeozoic cover. Consequently, the latest uplift of the South Swedish Dome with re-exposure of the sub-Cambrian peneplain did not occur until some

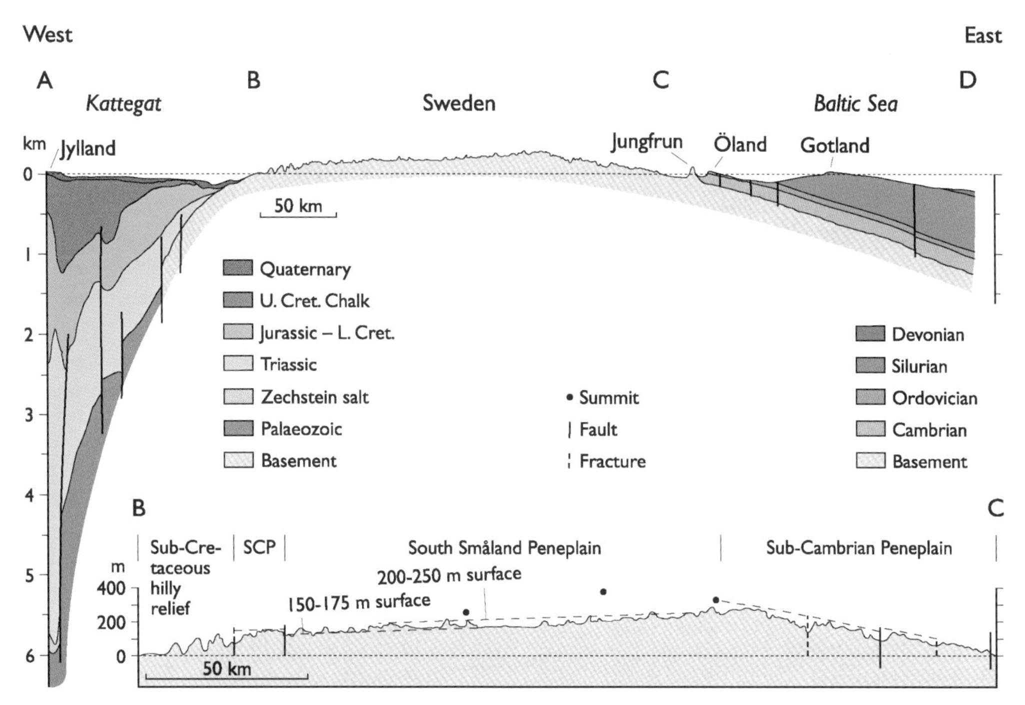

Fig. 15. Profile of cover rocks and basement topography and across southern Sweden from the Kattegat to the Baltic. A first phase of Neogene uplift and erosion led to the formation of the South Småland Peneplain and a second phase to the re-exposure of the sub-Cretaceous hilly relief (i.e. during mid-Miocene and Pliocene time). Location shown in Fig. 2. SCP, Sub-Cambrian Peneplain. Modified from Kornfält & Larsson (1987), Britze & Japsen (1991), Japsen & Langtofte (1991a, 1991b), Lykke-Andersen (1991), Lidmar-Bergström (1995) and Vejbæk (1997).

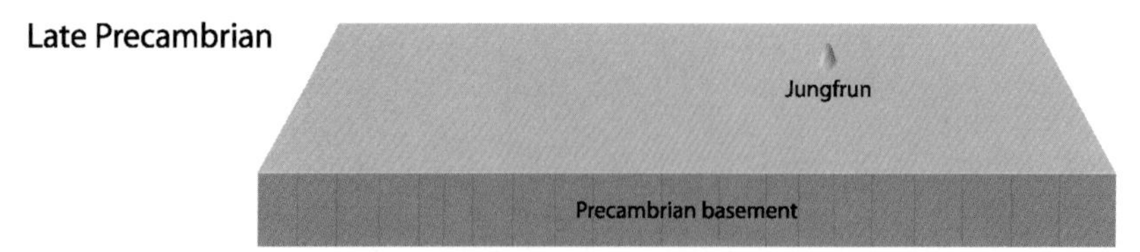

The surface of the Precambrian bedrock was denuded to an extremely flat surface, the sub-Cambrian peneplain. Residual hills only occurred as exceptions.

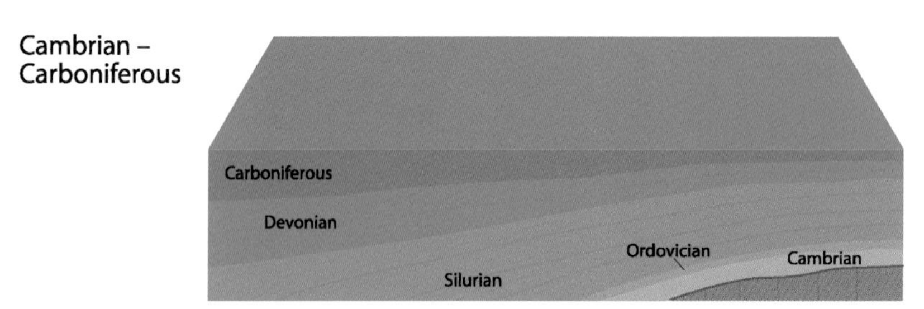

The sea transgressed the peneplain and Cambrian rocks were deposited on the flat surface and were succeeded by Ordovician–Carboniferous strata. These cover rocks protected the Precambrian basement in southern Sweden from further erosion for a long time. In the Kattegat area, a thick Palaeozoic cover accumulated in the Caledonian foreland basin.

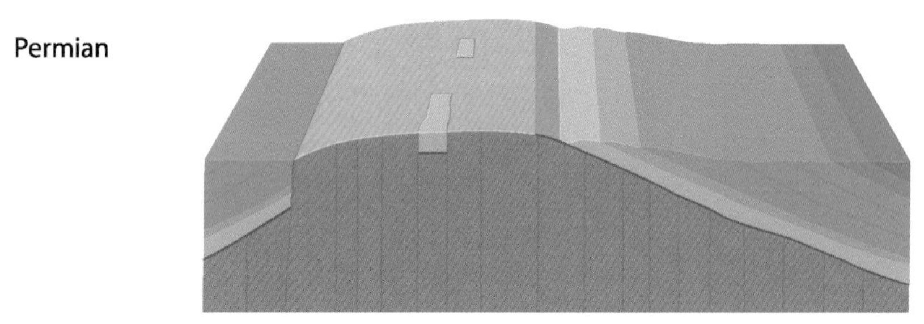

Inversion of the Caledonian foreland basin removed the Palaeozoic cover rocks in parts of the Kattegat area and probably in most of south-western Sweden.

Fig. 16. Block diagrams illustrating the development of bedrock relief in southern Sweden between the Kattegat and the Baltic (see Figs 2 and 15). Neogene uplift and erosion is assumed to have been initiated in two phases (?mid-Miocene, Pliocene). The final diagram illustrates how the South Swedish Dome is composed of surface facets from widely different periods. Surface features are exaggerated and Quaternary deposits are not shown in the final diagram. SCP, Sub-Cambrian Peneplain; SCr, Sub-Cretaceous hilly relief; SSP, South Småland Peneplain. Modified from Nielsen & Japsen (1991), Fredén (1994), Buchardt *et al.* (1997) and Vejbæk (1997).

time during the Cenozoic period (Lidmar-Bergström 1991).

The South Småland Peneplain is an almost horizontal and very flat erosion surface with only few, low, residual hills, which extends SW of the crestal part of the South Swedish Dome (Fig. 1) (Lidmar-Bergström 1988, 1996). The peneplain cuts off the inclined sub-Cretaceous hilly relief, and therefore developed during Cenozoic time by erosion by rivers flowing towards the south and west down to a base level that corresponds to the present level of about 125 m a.s.l. (Figs 15 and 16) (Lidmar-Bergström 1982). The formation of the South Småland Peneplain destroyed the sub-Cretaceous hilly relief. The peneplain must have continued across Cretaceous rocks in the west where the Kattegat is found today. Towards the NE, the South Småland Peneplain was cut into the sub-Cambrian peneplain,

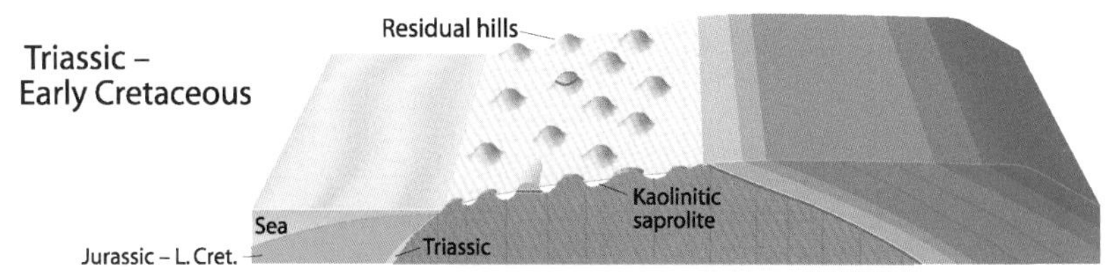

Uplift and erosion continued. The climate became humid and kaolinitic weathering penetrated deep into the basement along fracture zones (etching). Thick kaolinitic weathering mantles (saprolites) were produced. By alternating etching and erosion of the saprolites (stripping) the landscape developed an undulating hilly relief. Locally Palaeozoic remnants still occurred on down-faulted blocks.

The sea transgressed Denmark and large parts of southern Sweden and the area was covered with Upper Cretaceous – mid-Miocene sediments.

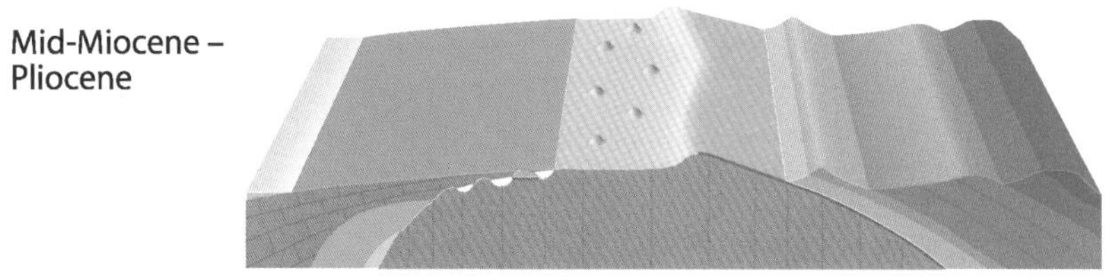

After a Neogene rise followed by erosion of Upper Cretaceous – mid-Miocene cover rocks, the South Småland Peneplain developed as a gently inclined flat rock surface with few residual hills (a pediplain). It probably formed in late Miocene dry climates. South- and west-facing escarpments were formed along the elevated rim of the sub-Cambrian peneplain.

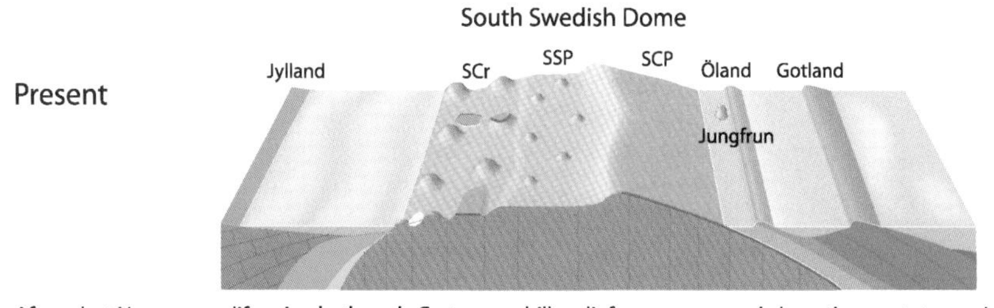

After a last Neogene uplift episode, the sub-Cretaceous hilly relief was re-exposed along the coasts toward south-east and west, while the flat sub-Cambrian peneplain re-appeared at surface in the northern and eastern part of South Sweden. Locally sub-Cambrian facets reappeared in western Sweden.

Fig 16. – *continued*

leaving south- and west-facing erosional scarps and high residual hills (150 m) in the basement surface.

In the crestal part of the South Swedish Dome, several occurrences of weathering products suggested as being of Plio-Pleistocene age have been described (Lidmar-Bergström *et al.* 1997). These gravelly saprolites differ markedly from the kaolinitic saprolites associated with the pre-Cenozoic weathering. The stripping of these weathering mantles appears to be responsible for much of the hilly relief incised in the uplifted sub-Cambrian peneplain. Thus the erosion of the last remnants of Palaeozoic rocks may not have occurred until Neogene time.

Timing of the Cenozoic uplift of the South Swedish Dome

We propose a new theory for the Cenozoic development of southern Scandinavia based on a combination of these observations from the exposed basement in southern Sweden and the geological evidence from the sedimentary cover in the adjacent areas in eastern Denmark, e.g. Sjælland, where Paleocene sediments are preserved and a Cenozoic cover of *c.* 500 m has been removed (Figs 1 and 13). Neogene uplift of the whole of southern Scandinavia, centred around the South Swedish Dome, may account for the erosion of the missing Cenozoic cover in eastern Denmark, and be in agreement with the geomorphological evidence in Sweden.

The formation of the South Swedish Dome started in Late Palaeozoic time with continuation in Mesozoic time (Cederbom 2001; Fig. 16). The present form arose after the Neogene uplift with subsequent formation of a new erosion surface (the South Småland Peneplain) and re-exposure of sub-Cambrian and sub-Cretaceous surfaces. In Denmark, the Neogene uplift of the South Swedish Dome led to removal of Cenozoic cover rocks in large parts of the territory, where the Skagerrak–Kattegat Platform in particular suffered deep erosion (Figs 1 and 13).

A Neogene uplift of the South Swedish Dome is in agreement with interpretation of fission-track data, which indicate the removal of about 650 m of Upper Cretaceous–Paleogene sediments from SW Sweden and of about 1000 m of such sediments from SE Sweden during Neogene time (Cederbom 2002). An Upper Cretaceous–Palaeogene cover over large parts of southern Sweden is in agreement with evidence for a vast Cretaceous cover in south Sweden (Lidmar-Bergström 1982, 1995) and with the report of redeposited Eocene diatoms in lakes in Småland at about 200 m a.s.l. (Cleve-Euler 1941). The suggested Plio–Pleistocene age of the weathering products found on the crestal part of the South Swedish Dome is consistent with Neogene uplift of the Dome (Lidmar-Bergström *et al.* 1997).

The re-exposure of the sub-Cretaceous hilly relief at altitudes below 125 m a.s.l. occurred after the formation of the South Småland Peneplain, when the remaining Cretaceous cover was eroded. Consequently, the Neogene rise of the South Swedish Dome must have taken place in two phases: a ?Miocene and Pliocene phase would be in agreement with the erosional pattern in Denmark (see the section 'Timing of maximum burial and subsequent erosion'.

A rise of the South Swedish Dome during Neogene time is also consistent with the clockwise shift in sediment transport directions that is observed in the NE North Sea Basin. The transport directions changed from southwards in Oligocene–Early Miocene time to southwestwards in mid- and late Miocene time and finally westwards in Pliocene time (Fig. 1) (Clausen *et al.* 1999). The Oligocene transport direction from the north in the eastern North Sea Basin corresponds to a Late Paleogene uplift phase of the Southern Scandes (Lidmar-Bergström *et al.* 1997, 2000; Clausen *et al.* 2000).

The late Cenozoic sediments found in SW Denmark are thus expected to reflect the erosion of Palaeozoic–Paleogene cover rocks in Sweden as well as the re-exposure of a Mesozoic surface on Precambrian rocks and the subsequent formation of the South Småland Peneplain. Several observations agree with these suggestions, for example, the occurrence of silicified Lower Palaeozoic fossils in Miocene sand in Jylland (Spjeldnæs 1975), and the occurrence of gibbsite in the Gram Formation, which indicates erosion of re-exposed tropical soil during late Miocene time and a short fluvial transport (Rasmussen & Larsen 1989). Larsen & Dinesen (1959) found an increasing content of amphibole in Miocene sediments in the Danish area, and suggested that the considerable amount of immature weathering material was derived from exposed basement and cover rocks on the Scandinavian Shield.

Discussion and conclusions

Using a combination of basin modelling and analysis of sonic data from different stratigraphic levels to estimate erosion allows identification of anomalous values and extends the areal coverage relative to that achieved by the application of a single method. We estimate erosion in the eastern North Sea Basin to be on average 200–600 m lower than suggested in previous studies. In particular, we find erosion estimates based on Chalk velocities to be in good agreement with estimates based on basin modelling, and this is interpreted as being due to the homogeneous composition of the Chalk, the large thickness of the Chalk section over which the mean velocity is calculated in most wells, and the stress dependence of Chalk compaction. Basin modelling predicts low heat flow during Oligocene–Pliocene time, to account for observed steep vitrinite reflectance gradients and present-day temperatures in central and northern Jylland. We suggest this to be partly due to transient thermal effects induced by deposition followed by erosion, but further studies of this matter are needed.

Comparison of results from different methods indicates that erosion is overestimated when based on sonic data from Lower Jurassic shale in NE Denmark; this could be due to variations in lithology, but further studies are needed to fully understand these variations. It is concluded that maximum burial of the Mesozoic succession throughout the region occurred before Neogene erosion where Paleogene–Neogene strata are preserved in large parts. A previous suggestion of deep erosion in the Sorgenfrei–Tornquist Zone during the Late Cretaceous–Paleogene inversion is rejected.

The thickness of the missing section removed by late Cenozoic erosion increases from the eastern North Sea towards the Norwegian and Swedish coasts. We estimate the erosion to be *c.* 500 m in a broad zone across Denmark from NW to SE. This zone largely conforms to the area where Paleocene–Miocene deposits subcrop the Quaternary, and the eroded sediments must thus have been of Paleocene–Miocene age. Erosion decreases towards zero in the western and southern part of the Danish North Sea. The age of the removed sediments must be progressively younger in this direction, as for the age of the Quaternary subcrop. To the north, the missing upper Cretaceous–Danian section reaches *c.* 1000 m on and along the Skagerrak–Kattegat Platform. The deeper erosion on and along the Skagerrak–Kattegat Platform relative to the Danish Basin cannot be explained by glacial erosion or by a drop in sea level, and this provides a further argument for tectonic uplift of the Platform during Neogene time. Finally, we have demonstrated that the predicted thickness of the missing strata corresponds to a likely Cenozoic depositional history.

The pattern of late Cenozoic erosion in Denmark agrees with a Neogene uplift of south Norway centred around the Southern Scandes and of southern Scandinavia centred around the South Swedish Dome. The formation of the South Swedish Dome started in Late Palaeozoic time, but the present form arose after Neogene uplift with subsequent formation of the South Småland Peneplain and re-exposure of sub-Cambrian and sub-Cretaceous surfaces. The re-exposure of the sub-Cretaceous hilly relief at altitudes below 125 m a.s.l. occurred after the formation of the South Småland Peneplain, when the remaining Cretaceous cover was eroded. Consequently, the Neogene rise of the South Swedish Dome must have taken place in two phases. We suggest that these phases correspond to the erosional episodes of ?Miocene and Plio-Pleistocene age that affected the sedimentary cover in Denmark. The areal extent of the ?Miocene episode appears restricted to southern Sweden and the Skagerrak–Kattegat Platform, whereas the effects of the later episode reach from the Baltic and far into the North Sea. A Neogene rise of the Dome is consistent with the clockwise shift in sediment transport directions observed in the NE North Sea Basin, changing from southwards in Oligocene–Early Miocene time to southwestwards in mid- and late Miocene time and finally westwards in Pliocene time (Clausen *et al.* 1999). The Oligocene transport direction from the north corresponds to a Late Paleogene uplift phase of the Southern Scandes. Apart from erosion induced by tectonic uplift, glacial erosion was also important, as witnessed by the large volume of Pleistocene sediments in the central North Sea Basin (e.g. Japsen 1998). A drop in sea level during late Cenozoic time may also have increased erosion (e.g. Nielsen *et al.* 2001).

The Neogene uplift along the eastern margin of the North Sea Basin and of the South Swedish Dome fits into the general pattern of late Cenozoic vertical movements around the North Atlantic (Japsen & Chalmers 2000). A model explaining these phenomena must be constrained by observations of the magnitude and timing of uplift and erosion based on independent methods as we have aimed at demonstrating here, and it must thus separate the effects of Paleogene uplift of plate boundaries from those of Neogene intraplate uplift.

Appendix: Normal velocity–depth trends for homogeneous formations

A revised normal velocity–depth trend for the North Sea Chalk

Japsen (1998) published a normal velocity–depth trend for the Chalk Group based on an analysis of data from 845 wells throughout the North Sea Basin and ODP data. For the shallowest part of the trend, no data representing normal compaction were found for the Chalk of the North Sea Basin, so sonic log data from Eocene to Recent ooze and chalk deposits from a stable platform were used to guide the trend (Urmos *et al.* 1993). At intermediate depths, Japsen (1998) applied qualitative arguments to identify North Sea data representing normal compaction along the lower bound for velocity–depth data for which the effect of overcompaction as a result of erosion is minimal. At greater depths, data representing normal compaction were identified along the upper bound where the effect of undercompaction as a result of overpressuring is a minimum.

Later, however, Japsen (2000) found additional geological constraints to refine the identification of reference data at intermediate depths where the influence of erosion and overpressuring is difficult to ascertain. Because the sonic method identifies deviations from maximum burial, post-erosional reburial of a formation will reduce its observable burial anomaly, e.g. a pre-Quaternary erosion of 500 m will be masked by a subsequent Quaternary reburial of 500 m (Eq. 2). This implies that, where the Quaternary sequence is thick, even minor deviations from maximum burial as a result of Neogene erosion may correspond to a substantial missing section. Deep erosion is, however, not likely where the base-Quaternary hiatus is minor, for example, where the Quaternary sequence is underlain by Neogene sediments.

Normally compacted Chalk is thus likely to be found in areas where the Quaternary sequence is thick, Neogene deposits are present, and pressure is hydrostatic. Consequently, the normal velocity–depth trend for the North Sea Chalk should follow the upper bound for data from such areas, whereas data representing undercompaction as a result of overpressuring should plot below the trend (Fig. 4). A revised baseline was thus defined by Japsen (2000) using such maximum velocity data for $900 < z < 1700$ m. The revised trend lines up with the maximum velocity data used to define the original trend for $z > 2000$ m, and with a velocity at the surface of $1550\,\mathrm{m\,s^{-1}}$:

$$
\begin{aligned}
V_N^{Ch} &= 1550 + 1.3z, & z &< 900\,\mathrm{m} \\
V_N^{Ch} &= 920 + 2z, & 900 &< z < 1471\,\mathrm{m} \\
V_N^{Ch} &= 1950 + 1.3z, & 1471 &< z < 2250\,\mathrm{m} \\
V_N^{Ch} &= 2625 + z, & 2250 &< z < 2875\,\mathrm{m}.
\end{aligned} \tag{3}
$$

The fourth of the above segments is unchanged from the original trend, in which the upper three segments were expressed as $1600 + z$, $500 + 2z$, and $937.5 + 1.75z$ (Japsen 1998). The revised trend is shifted towards shallower depths by a mean of 160 m for the velocity interval affected by the revision and where North Sea data are found, $2100 < V < 4875\,\mathrm{m\,s^{-1}}$; the maximum shift is 210 m for $2920 < V < 3920\,\mathrm{m\,s^{-1}}$.

The shift towards higher velocities for the revised baseline results in a reduction in estimates of erosion by up to 210 m, and an increase in estimates of overpressure by up to 2 MPa for data points that plot above and below the line, respectively (overcompaction as a result of overpressure: $dZ_B/100 = 210/100\,\mathrm{MPa} \approx 2\,\mathrm{MPa}$; see Japsen 1998). The increased overpressure that the revised model predicts is an improvement relative to the original model, which explained only 80% of the observed overpressure in the Chalk for 52 wells in the central North Sea located away from diapirs and where the overpressure exceeded 4 MPa (Japsen 1998). The corresponding percentage based on the revised baseline is 91%. This improvement is particularly clear for data from relatively shallow depth or moderate overpressure, e.g. the Danish Dan field where overpressure is 7.3 MPa, and for which the overpressure prediction from velocity data has been increased from 4.3 to 6.5 MPa.

A baseline for marine shale of the Lower Jurassic F-I Member

Japsen (2000) formulated a constrained baseline, V_N^{Sh}, for marine shale dominated by smectite–illite based on velocity–depth data for the Lower Jurassic F-I Member from 31 Danish wells of which 28 have data for the Chalk (Fig. 6):

$$1/V_N^{Sh} = tt_N^{Sh} = 460\,e^{-z/2175} + 185. \tag{4}$$

The baseline was reconstructed by correcting present formation depths for the effect of late Cenozoic erosion as estimated from the velocity of the overlying Chalk in these wells relative to the revised Chalk trend (Eq. 3). The corrected depths correspond to the burial of the formation before erosion when the sediments were at maximum burial at more locations than today. The baseline can thus be traced more easily in a plot of velocity versus the corrected depths, and is well defined at great depth where velocity–depth data for normally compacted shale at maximum burial can be difficult to identify ($2.1 < z < 3.8$ km). This formulation is a constrained, exponential transit time–depth model that fulfils reasonable boundary conditions at the surface and at infinite depth: $V_0 = 1550\,\mathrm{m\,s^{-1}}$ and $V_\infty = 5405\,\mathrm{m\,s^{-1}}$; maximum velocity–depth gradient $0.6\,\mathrm{m\,s^{-1}\,m^{-1}}$ for $z = 2.0$ km. The shale trend given by Eq. 4 corresponds closely to baselines for marine shale found by other workers (Scherbaum 1982; Hansen 1996; see discussion by Japsen 1999).

A baseline for the Lower Triassic Bunter Shale

Japsen (2000) formulated a segmented, linear baseline, V_N^{BSh}, for the Lower Triassic Bunter Shale based on velocity–depth data from 142 British and Danish wells of which 91 have velocity–depth data for the Chalk (Fig. 7):

$$V_N^{BSh} = 1550 + 0.6z, \quad 0 < z < 1393\,\text{m}$$
$$V_N^{BSh} = -400 + 2z, \quad 1393 < z < 2000\,\text{m}$$
$$V_N^{BSh} = 2600 + 0.5z, \quad 2000 < z < 3500\,\text{m} \quad (5)$$
$$V_N^{BSh} = 3475 + 0.25z, \quad 3500 < z < 5300\,\text{m}.$$

The trend indicates a pronounced variation of the velocity gradient with depth. The gradient is only $0.5\,\text{m}\,\text{s}^{-1}\,\text{m}^{-1}$ in the upper part, and increases to $1.5\,\text{m}\,\text{s}^{-1}\,\text{m}^{-1}$ for depths around 2 km, from where it decreases gradually with depth to 0.5 and then $0.25\,\text{m}\,\text{s}^{-1}\,\text{m}^{-1}$. The decline of the gradient with depth reflects that velocity approaches an upper limit.

The Bunter Shale baseline was reconstructed by applying the same procedure as for the Lower Jurassic shale by correcting present formation depths for the effect of late Cenozoic erosion as estimated from Chalk velocities. The trend was constructed to predict likely values near the surface ($V_0 = 1550\,\text{m}\,\text{s}^{-1}$), and is based on reference data with corrected depths from 1600 to 5600 m (Japsen 2000).

Rather than proposing a specific baseline for the Lower Triassic Bunter Sandstone, Japsen (2000) found that the trend derived for the Bunter Shale was a reasonable approximation for a dataset from 133 British and Danish wells of which 87 have velocity–depth data for the Chalk (see Fig. 7). Data from shale are preferable to those from sandstone in studies of maximum burial for several reasons: shale porosity is less affected by diagenetic processes, shale does not act as an aquifer with the consequent porosity variations, and shale may be more uniform with regard to both grain size and mineralogy. Burial anomalies for the Bunter Sandstone can thus be used to place an upper limit on estimates of erosion based on Bunter Shale data.

The dominance of smectite–illite in the distal parts of the Fjerritslev Formation (Lindgreen, pers. comm.), and of kaolin in the continental Bunter Shale was suggested by Japsen (2000) to be a possible explanation of why baselines for these two formations diverge, and why those for Bunter Shale and Bunter Sandstone converge at depth. Alternatively, the similarity of the baselines for the Bunter Shale and the Bunter Sandstone may be due to the dominance of quartz in both formations, and, correspondingly, a high content of quartz in the Fjerritslev Formation close to the exposed Scandinavian basement may explain the relatively high velocity observed in wells in that area.

We wish to thank reviewers D. Issler and J. Turner, as well as U. Gregersen, A. Mathiesen, C. Pulvertaft, E. S. Rasmussen and O. Vejbæk (all GEUS) for constructive comments that improved the manuscript considerably.

References

BAILEY, S.W. 1980. Structure of layer silicates. *In*: BRINDLEY, G.W. & BROWN, G. (eds) *Crystal Structures of Clay Minerals and their X-ray Identification*. Mineralogical Society, London, Monograph, **1**, 1–124.

BERTELSEN, F. 1980. Lithostratigraphy and Depositional History of the Danish Triassic. *Geological Survey of Denmark Series B*, **4**. Geological Survey of Denmark, Copenhagen.

BJØRLYKKE, K. & EGEBERG, P.K. 1993. Quartz cementation in sedimentary basins. *AAPG Bulletin*, **77**, 1538–1548.

BORRE, M. & FABRICIUS, I. 1998. Chemical and mechanical processes during burial diagenesis of chalk: an interpretation based on specific surface data of deep-sea sediments. *Sedimentology*, **45**, 755–769.

BRITZE, P. & JAPSEN, P. 1991. *Geological Map of Denmark 1:400 000. The Danish Basin. 'Top Zechstein' and the Triassic; Two-way Traveltime and Depth, Thickness and Interval Velocity*. Geological Survey of Denmark Map Series, **31**.

BUCHARDT, B. 1978. Oxygen isotope paleotemperatures from the Tertiary period in the North Sea area. *Nature*, **275**, 121–123.

BUCHARDT, B., NIELSEN, A.T. & SCHOVSBO, N. 1997. Alunskiferen i Skandinavien. *Geologisk Tidsskrift*, **1997/3**, 1–30.

BULAT, J. & STOKER, S.J. 1987. Uplift determination from interval velocity studies, UK, southern North Sea. *In*: BROOKS, J. & GLENNIE, K.W. (eds) *Petroleum Geology of North West Europe*. Graham & Trotman, London, 293–305.

CEDERBOM, C. 2001. Phanerozoic, pre-Cretaceous thermotectonic events in southern Sweden revealed by fission track thermochronology. *Earth and Planetary Science Letters*, **188** (1–2), 199–209.

CEDERBOM, C. 2002. The thermotectonic development of southern Sweden during Mesozoic and Cenozoic time. *In*: DORÉ, A.G., CARTWRIGHT, J.A., STOKER, M.S., TURNER, J.P. & WHITE, N. (eds) *Exhumation of the North Sea Margin: Timing, Mechanisms and Implications for Petroleum Exploration*. Geological Society, London, Special Publications, **196**, 169–182.

CLAUSEN, O.R., GREGERSEN, U., MICHELSEN, O. & SØRENSEN, J.C. 1999. Factors controlling the

Cenozoic sequence development in the eastern parts of the North Sea. *Journal of the Geological Society, London*, **156**, 809–816.

CLAUSEN, O.R., NIELSEN, O.B., HUUSE, M. & MICHELSEN, O. 2000. Geological indications for Palaeogene uplift in the eastern North Sea Basin. *Global and Planetary Change*, **24**, 175–187.

CLEVE-EULER, A. 1941. Alttertiäre Diatomeen und Silicioglagellanten im inneren Schwedens. *Palaeontographica*, **92A**, 165–208.

FREDÉN, C. 1994. *Geology. National Atlas of Sweden*. SNA, Stockholm.

GEMMER, L., NIELSEN, S.B., HUUSE, M. & LYKKE-ANDERSEN, H. 2002. Post-mid-Cretaceous eastern North Sea evolution inferred from 3-D dynamic modelling. *Tectonophysics*, **350**, 315–342.

HANSEN, S. 1996. Quantification of net uplift and erosion on the Norwegian Shelf south of 66°N from sonic transit times of shale. *Norsk Geologisk Tidsskrift*, **76**, 245–252.

HILLIS, R.R. 1995. Quantification of Tertiary exhumation in the United Kingdom southern North Sea using sonic velocity data. *AAPG Bulletin*, **79**, 130–152.

HÖGBOM, A.G. 1910. Precambrian geology of Sweden. *Bulletin of the Geological Institution of the University of Uppsala*, **10**, 1–80.

HÖGBOM, A.G. & AHLSTRÖM, N.G. 1924. Über die sub-kambrische Landfläche am fusse vom Kinnekulle. *Bulletin of the Geological Institution of the University of Uppsala*, **19**, 55–88.

HUUSE, M., LYKKE-ANDERSEN, H. & MICHELSEN, O. 2001. Cenozoic evolution of the eastern Danish North Sea. *Marine Geology*, **177**, 243–269.

ILIFFE, J.E. & DAWSON, M.R. 1996. Basin modelling history and predictions. *In*: GLENNIE, K.W. & HURST, A. (eds) *AD1995: NW Europe's Hydrocarbon Industry*. Geological Society, London, 83–105.

JAPSEN, P. 1993. Influence of lithology and Neogene uplift on seismic velocities in Denmark; implications for depth conversion of maps. *AAPG Bulletin*, **77**, 194–211.

JAPSEN, P. 1998. Regional velocity–depth anomalies. North Sea Chalk: a record of overpressure and Neogene uplift and erosion. *AAPG Bulletin*, **82**, 2031–2074.

JAPSEN, P. 1999. Overpressured Cenozoic shale mapped from velocity anomalies relative to a baseline for marine shale, North Sea. *Petroleum Geoscience*, **5**, 321–336.

JAPSEN, P. 2000. Investigation of multi-phase erosion using reconstructed shale trends based on sonic data, Sole Pit axis, North Sea. *Global and Planetary Change*, **24**, 189–210.

JAPSEN, P. & BIDSTRUP, T. 1999. Quantification of late Cenozoic erosion in Denmark based on sonic data and basin modelling. *Bulletin of the Geological Society of Denmark*, **46**, 79–99.

JAPSEN, P. & CHALMERS, J.A. 2000. Neogene uplift and tectonics around the North Atlantic: overview. *Global and Planetary Change*, **24**, 165–173.

JAPSEN, P. & LANGTOFTE, C. 1991a. *Geological Map of Denmark 1:400 000. The Danish Basin. 'Base Chalk' and the Chalk Group, Two-way Traveltime and Depth, Thickness and Interval Velocity*. Geological Survey of Denmark Map Series, **29**.

JAPSEN, P. & LANGTOFTE, C. 1991b. *Geological map of Denmark 1:400 000. The Danish Basin. 'Top Triassic' and the Jurassic–Lower Cretaceous, Two-way Traveltime and Depth, Thickness and Interval Velocity*. Geological Survey of Denmark Map Series, **30**.

JENSEN, L.N. & SCHMIDT, B.J. 1992. Late Tertiary uplift and erosion in the Skagerrak area; magnitude and consequences. *Norsk Geologisk Tidsskrift*, **72**, 275–279.

JENSEN, L.N. & SCHMIDT, B.J. 1993. Neogene uplift and erosion offshore South Norway; magnitude and consequences for hydrocarbon exploration in the Farsund Basin. *In*: SPENCER, A.M. (ed.) *Generation, Accumulation, and Production of Europe's Hydrocarbons; III*. Springer, Berlin, 79–88.

KNUDSEN, K.L. 1995. Kvartæret. *In*: NIELSEN, O.B. (ed.) *Danmarks geologi fra Kridt til i dag*. Aarhus Universitet, Aarhus, 247–269.

KONRADI, P. 1995. Foraminiferal biostratigraphy of the post mid-Miocene in two boreholes in the Danish North Sea. *In*: MICHELSEN, O. (ed.) *Proceedings of the 2nd Symposium on Marine Geology*. Danmarks Geologiske Undersøgelse Series C, **12**, 101–112.

KORNFÄLT, K.A. & LARSSON, K. 1987. *Geological Maps and Cross-sections of Southern Sweden*. Svensk Kärnbränslehantering (SKB) Technical Report, **87-24**.

LARSEN, G. & DINESEN, A. 1959. *Vejle Fjord Formationen ved Brejning*. Danmarks Geologiske Undersøgelse II. Række, **82**.

LAURSEN, G.V. 1992. Foraminifera of the eastern North Sea. *In*: LAURSEN, G.V., HEILMANN-CLAUSEN, C. & THOMSEN, E. (eds) *Cenozoic Biostratigraphy of the Eastern North Sea based on Foraminifera, Dinoflagellates, and Calcareous Nannofossils*. Geologisk Institut, Aarhus, 1–68.

LIBORIUSSEN, J., ASHTON, P. & TYGESEN, T. 1987. The tectonic evolution of the Fennoscandian Border Zone in Denmark. *Tectonophysics*, **137**, 21–29.

LIDMAR-BERGSTRÖM, K. 1982. *Pre-Quaternary Geomorphological Evolution in Southern Fennoscandia*. Sveriges Geologiska Undersökning C, **785**.

LIDMAR-BERGSTRÖM, K. 1988. Preglacial weathering and landform evolution in Fennoscandia. *Geografiska Annaler, 70A*, 273–276.

LIDMAR-BERGSTRÖM, K. 1989. Exhumed Cretaceous landforms in south Sweden. *Zeitschrift für Geomorphologie, Neue Folge, Supplement Band*, **72**, 21–40.

LIDMAR-BERGSTRÖM, K. 1991. Phanerozoic tectonics in southern Sweden. *Zeitschrift für Geomorphologie, Neue Folge, Supplement Band*, **82**, 1–16.

LIDMAR-BERGSTRÖM, K. 1995. Relief and saprolites through time on the Baltic Shield. *Geomorphology*, **12**, 45–61.

LIDMAR-BERGSTRÖM, K. 1996. Long term morphotectonic evolution in Sweden. *Geomorphology*, **16**, 33–59.

LIDMAR-BERGSTRÖM, K. 1999. Uplift histories revealed by landforms of the Scandinavian domes. *In*: SMITH, B.J., WHALLEY, W.B. & WARKE, P.A. (eds) *Uplift, Erosion and Stability: Perspectives on Long-term Landscape Development*. Geological Society, London, Special Publications, **162**, 85–91.

LIDMAR-BERGSTRÖM, K., OLLIER, C.D. & SULEBAK, J.C. 2000. Landforms and uplift history of southern Norway. *Global and Planetary Change*, **24**, 211–231.

LIDMAR-BERGSTRÖM, K., OLSSON, S. & OLVMO, M. 1997. Palaeosurfaces and associated saprolites in southern Sweden. *In*: WIDDOWSON, M. (ed.) *Palaeosurfaces; Recognition, Reconstruction and Palaeoenvironmental Interpretation*. Geological Society, London, Special Publications, **120**, 95–124.

LINDGREEN, H. 1991. Elemental and structural changes in illite/smectite mixed-layer clay minerals during diagenesis in Kimmeridgian–Volgian (–Ryazanian) clays in the Central Trough, North Sea and the Norwegian–Danish Basin. *Bulletin of the Geological Society of Denmark*, **39**, 1–82.

LYKKE-ANDERSEN, H. 1991. Nogle hovedetræk af Kattegats kvartærgeolgi—foreløbige resultater af en seismisk undersøgelse 1988–1991. *Dansk Geologisk Forening, Årsskrift for 1990–91*, 57–65.

MARIE, J.P.P. 1975. Rotliegendes stratigraphy and diagenesis. *In*: WOODLAND, A.W. (ed.) *Petroleum and the Continental Shelf of North-west Europe*. Applied Science, London, 205–211.

MICHELSEN, O. 1989. *Revision of the Jurassic Lithostratigraphy of the Danish Subbasin*. Geological Survey of Denmark Series A, **24**.

MICHELSEN, O. & NIELSEN, L.H. 1993. Structural development of the Fennoscandian border zone, offshore Denmark. *Marine and Petroleum Geology*, **10**, 124–134.

NIELSEN, L.H. & JAPSEN, P. 1991. *Deep Wells in Denmark 1935–1990. Lithostratigraphic Subdivision*. Geological Survey of Denmark Series A, **31**.

NIELSEN, S.B., PAULSEN, G.E., HANSEN, D.L. *et al.* 2002. Paleocene initiation of Cenozoic uplift in Norway. *In*: DORÉ, A.G., CARTWRIGHT, J.A., STOKER, M.S., TURNER, J.P. & WHITE, N. (eds) *Exhumation of the North Sea Margin: Timing, Mechanisms and Implications for Petroleum Exploration*. Geological Society, London, Special Publications, **196**, 45–65.

PEDERSEN, G.K. 1985. Thin, fine-grained storm layers in a muddy shelf sequence: an example from the Lower Jurassic in the Stenlille 1 well, Denmark. *Journal of the Geological Society, London*, **142**, 357–374.

PEULVAST, J.-P. 1985. Post-orogenic morphotectonic evolution of the Scandinavian Caledonides during the Mesozoic and Cenozoic. *In*: GEE, D.G. & STURT, B.A. (eds) *The Caledonide Orogen—Scandinavia and Related Areas*. Wiley, Chichester, 979–996.

RASMUSSEN, E.S. & LARSEN, O. 1989. *Mineralogi og geokemi af det Øvre Miocæne Gram ler*. Danmarks Geologiske Undersøgelse, Serie D, **7**.

RASMUSSEN, L.B. 1961. *De miocæne formationer i Danmark*. Danmarks Geologiske Undersøgelse IV. Række, **4**.

RIIS, F. 1996. Quantification of Cenozoic vertical movements of Scandinavia by correlation of morphological surfaces with offshore data. *Global and Planetary Change*, **12**, 331–357.

RIIS, F. & JENSEN, L.N. 1992. Introduction; measuring uplift and erosion; proposal for a terminology. *Norsk Geologisk Tidsskrift*, **72**, 223–228.

ROHRMAN, M., VAN DER BEEK, P., ANDRIESSEN, P. & CLOETINGH, S. 1995. Meso-Cenozoic morphotectonic evolution of southern Norway: Neogene domal uplift inferred from apatite fission track thermochronology. *Tectonics*, **14**, 700–714.

RUDBERG, A. 1954. Västerbottens berggrundsmorfologi. *Geographica*, **25**, 1–457.

SCHERBAUM, F. 1982. Seismic velocities in sedimentary rocks; indicators of subsidence and uplift. *Geologische Rundschau*, **71**, 519–536.

SHIPBOARD SCIENTIFIC PARTY, *et al.* 1991. Site 807. *In*: KROENKE, L.W., BERGER, W.H. & JANACEK, T.R. (eds) *Proceedings of the Ocean Drilling Program, Initial Reports, 130*. Ocean Drilling Program, College Station, TX, 369–493.

SPJELDNÆS, N. 1975. Palaeogeography and facies distribution in the Tertiary of Denmark and surrounding areas. *Norges Geologiske Undersøgelse Bulletin*, **316**, 289–311.

URMOS, J., WILKENS, R.H., BASSINOT, F., LYLE, M., MARSTERS, J.C., MAYER, L.A. & MOSHER, D.C. 1993. Laboratory and well-log velocity and density measurements from the Ontong Java Plateau: new in-situ corrections to laboratory data for pelagic carbonates. *In*: BERGER, W.H., KROENKE, L.W., MAYER, L.A. & JANECEK, T.R. (eds) *Proceedings of the Ocean Drilling Program, Scientific Results, 130*. Ocean Drilling Program, College Station, TX, 607–622.

VAN OLPHEN, H. 1966. Collapse of potassium montmorillonite clays upon heating—'potassium fixation'. *In*: BAILEY, S.W. (ed.) *Clays and Clay Mineralogy*. Pergamon, Oxford, 393–405.

VEJBÆK, O.V. 1997. Dybe strukturer i danske sedimentære bassiner. *Geologisk Tidsskrift*, **1997/4**, 1–31.

VEJBÆK, O.V. & BRITZE, P. 1994. *Geological Map of Denmark 1:750 000 Top Pre-Zechstein (Two-way Traveltime and Depth)*. Geological Survey of Denmark Map Series, **45**.

VIK, E. & HERMANRUD, C. 1993. Transient thermal effects of rapid subsidence in the Haltenbanken area. *In*: DORÉ, A.G., AUGUSTSON, J.H., HERMANRUD, C., STEWART, D.J. & SYLTA, O. (eds) *Basin Modelling: Advances and Applications*. Norwegian Petroleum Society Special Publication, **3**, 107–117.

YÜKLER, M.A. 1978. One-dimensional model to simulate geologic, hydrodynamic and thermodynamic development of a sedimentary basin. *Geologische Rundschau*, **67**, 960–979.

ZAGWIJN, W.H. 1989. The Netherlands during the Tertiary and the Quaternary: a case story of coastal lowland erosion. *Geologie en Mijnbouw*, **68**, 107–120.

Cenozoic uplift and denudation of southern Norway: insights from the North Sea Basin

MADS HUUSE[1,2]

[1]*Department of Earth Sciences, University of Aarhus, Aarhus, Denmark*

[2]*Present address: Department of Earth Sciences, Cardiff University, Cardiff CF10 3YE, UK*

(e-mail: m.huuse@abdn.ac.uk)

Abstract: The Cenozoic evolution of the North Sea Basin is described, drawing on subsurface data and a series of palaeogeographical maps compiled from a variety of published studies, mainly emphasizing the development of the eastern part of the basin. A model that accounts for the sedimentation history of the North Sea Basin and the topography (including maximum and mean surface elevation) of southern Norway is proposed. The model involves regional plume-related uplift of an initial low-elevation peneplain in early Paleogene time followed by repeated episodes of climatic deterioration and eustatic fall, most notably at the Eocene–Oligocene transition, in late Mid-Miocene time, and eventually culminating with the development of full glacial conditions in southern Norway in Plio-Pleistocene time. These episodes correspond to periods of accelerated sediment supply from southern Norway that reflect increased rates of incision (dissection) of the source area. It is argued that the present-day elevation of >2 km of mountain peaks in southern Norway adjacent to deep valleys and fjords could have been caused by isostatic uplift in response to dissection of a high-elevation peneplain. Hence it may not be necessary to invoke late Cenozoic tectonic uplift events to explain the present-day topography of southern Norway.

This paper describes the Cenozoic infill history of the North Sea Basin and its implications for understanding how the topography of southern Norway was created. At present, the North Sea Basin is filled to its brim. It comprises the shallow North Sea, the low-relief areas of Denmark, northern Poland, northern Germany, the Netherlands and SE England. This low-relief area is surrounded by the topographically high areas of southern Norway to the NE, Scotland to the NW and the Central European Massif to the south (Fig. 1). During Cenozoic time the margins of the North Sea Basin became exhumed whereas the centre of the basin subsided more than 3 km (Fig. 2). The subsidence history of the basin is fairly well constrained by the preserved sedimentary record, although the details are still the subject of debate (McKenzie 1978; Nielsen *et al.* 1986; Vinken 1988; Cloetingh *et al.* 1990, 1992; Ziegler 1990; Joy 1992; Galloway *et al.* 1993; Jordt *et al.* 1995; Liu & Galloway 1997; Japsen 1998; Michelsen *et al.* 1998; Huuse 2002; Nielsen *et al.* 2002). In contrast, the magnitude, mechanisms and exact timing of uplift and subsequent denudation of the basin margins and the hinterland mountains has remained enigmatic to this day (Torske 1972; Doré 1992; Jensen & Michelsen 1992; Jensen & Schmidt 1993; Jordt *et al.* 1995; Rohrman *et al.* 1995; Hansen 1996; Riis 1996; Solheim *et al.* 1996; Japsen 1998; Michelsen *et al.* 1998; Doré *et al.* 1999; Lidmar-Bergström 1999; Chalmers & Cloetingh 2000; Nøttvedt 2000; Huuse, 2002). It is often argued that uplift and denudation occurred in two phases, one synrift and one post-rift (e.g. Riis & Fjeldskaar 1992; Eyles 1996; Riis 1996; Rohrman & van der Beek 1996; Lidmar-Bergström *et al.* 2000), although the two phases can be difficult to separate (Japsen & Chalmers 2000).

The magnitude of uplift and denudation may be estimated from proxies such as geothermometry (fission tracks, vitrinite reflectance, etc.), compaction estimates, structural trends, sedimentary geometries and geomorphology. These methods are, however, all associated with rather large uncertainties, as indicated by the discrepancies observed when comparing uplift estimates derived using different methods (compare Jensen & Schmidt 1993; Japsen & Bidstrup 1999; Huuse 2002). It is even more speculative to assess the mechanisms and the exact timing of uplift, and the amount and timing of denudation that followed. The uplift and denudation history of southern Norway is an intriguing problem in its own right, but is also of significant interest to

From: Doré, A.G., Cartwright, J.A., Stoker, M.S., Turner, J.P. & White, N. 2002. *Exhumation of the North Atlantic Margin: Timing, Mechanisms and Implications for Petroleum Exploration*. Geological Society, London, Special Publications, **196**, 209–233. 0305-8719/02/$15.00

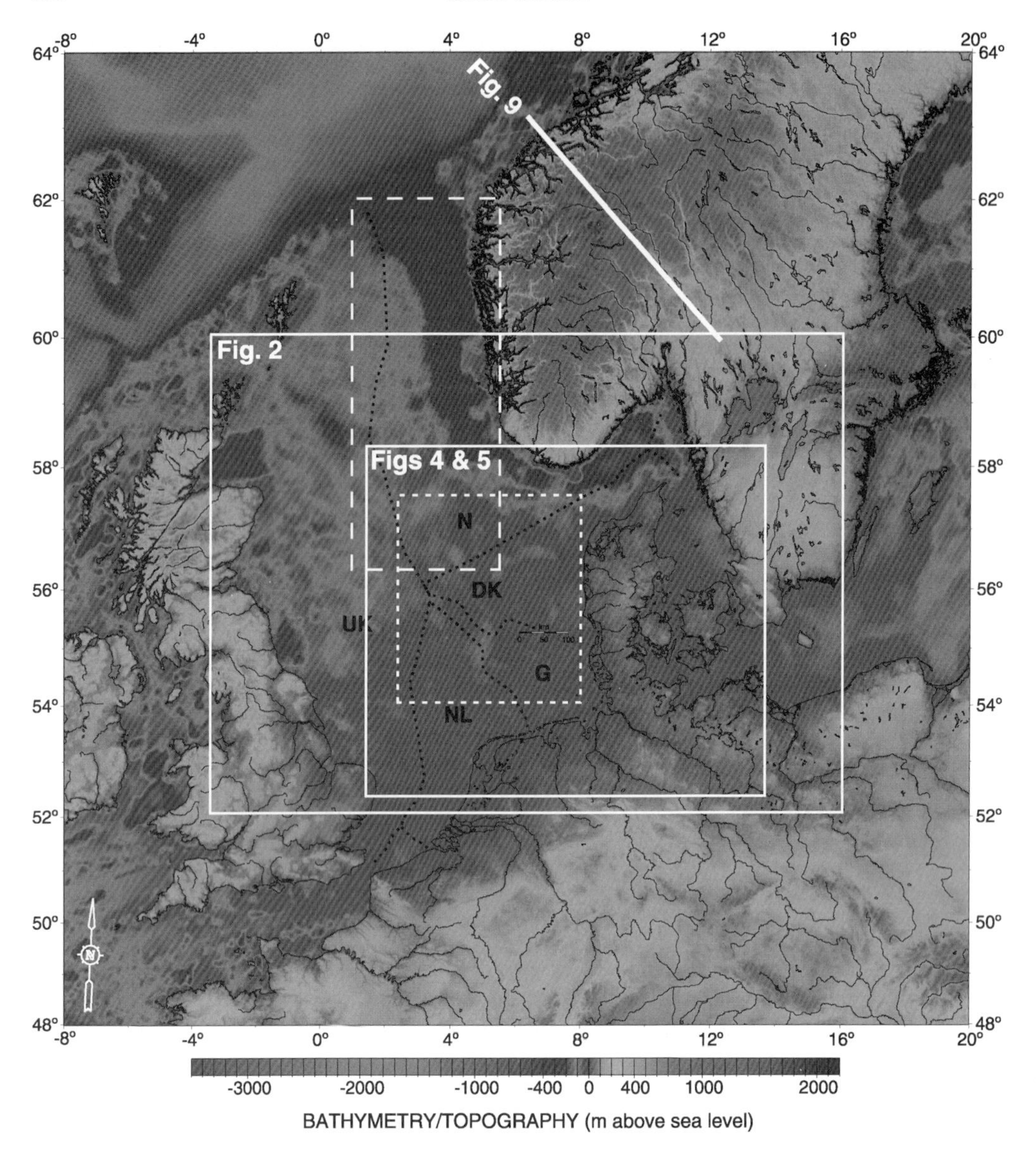

Fig. 1. Present-day topography and bathymetry of NW Europe. The study areas of Jordt *et al.* (1995) (long dash), and of Michelsen *et al.* (1998) (short dash) and locations of Figs 2, 4, 5 and 9 are shown for reference.

the oil industry, as uplift and denudation may have both positive and negative effects on the hydrocarbon potential of the NW European margin (e.g. Sales 1992; Jensen & Schmidt 1993; Doré & Jensen 1996; Doré *et al.* 1997, 1999).

Uplift and denudation

Definition

When discussing the uplift history of any area of the Earth it is extremely important to properly define what is actually meant by the term 'uplift', i.e. is it regional surface uplift or local uplift of mountain peaks ('uplift of rocks'). It is equally important to properly define the reference level to which uplift is measured. In this paper the term 'surface uplift' is used to describe uplift of the Earth's surface with respect to the geoid (mean global sea level) averaged over an area of *c.* $10^4\,km^2$, as this is the relevant scale for studying vertical movements of the lithosphere (England & Molnar 1990). The local inversion of former normal faults is thus not considered here.

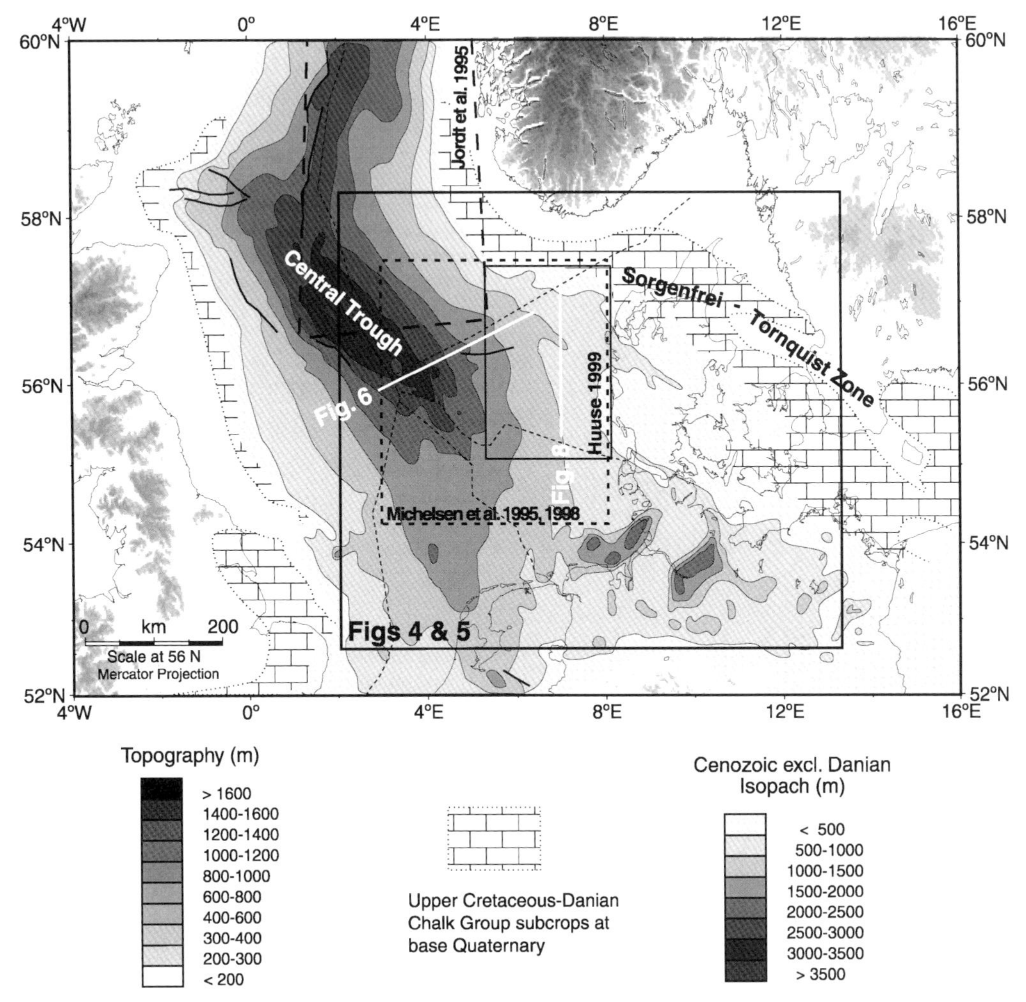

Fig. 2. Depth to the Upper Cretaceous–Danian limestone (Ziegler 1990; Japsen 1998). The study areas of Jordt *et al.* (1995) (long dash), of Michelsen *et al.* (1998) (short dash) and of Huuse (1999) (continuous line) and locations of Figs 4, 5, 6 and 8 are shown for reference.

Surface uplift over large areas requires upward displacement of rocks with respect to the geoid, i.e. work against gravity, and thus requires an active tectonic mechanism of large magnitude (England & Molnar 1990; Summerfield & Brown 1998; Nielsen *et al.*, 2002). When discussing relatively small amounts (some hundreds of metres) of surface uplift, it is important to consider eustatic changes (England & Molnar 1990; Huuse, 2002). In particular, when discussing Cenozoic uplift events, the long-term eustatic fall since mid-Eocene time of *c.* 200 m (e.g. Haq *et al.* 1987) corresponds to (nontectonic) surface uplift of the same magnitude. Finally, it is important to note that 'surface uplift', 'uplift of rocks' and 'denudation' are related in the following way (England & Molnar 1990): surface uplift = uplift of rocks − denudation.

Hence, the implicit assumption often made that rock uplift is equal to surface uplift is true only on the rare occasions when denudation is zero (England & Molnar 1990). The term 'denudation' is used here instead of 'exhumation' as the latter is generally used to describe the re-exposure of a buried land surface (see Summerfield & Brown 1998). It should be noted also that surface uplift in itself does not cause increased denudation, as it is the local relief rather than average elevation that governs denudation rates (Ahnert 1970; Summerfield 1991). Moreover, denudation rates will usually be less than rates of surface uplift, as otherwise there would be no mountains. Indeed, it has been shown that in some cases the denudational response may lag several tens of million years behind an uplift event (Summerfield & Brown 1998).

Tectonic mechanisms

A large number of active tectonic mechanisms have been proposed to explain the denudation record and present-day topography of the North Atlantic margins, in particular the Norwegian and Scottish highlands and the east Greenland margin (e.g. Doré 1992; Japsen & Chalmers 2000). Crustal shortening is a well-known mechanism for generating surface uplift, but the amount of shortening observed in NW Europe is nowhere near enough to explain the amount of exhumation inferred from vitrinite reflectance profiles and fission-track analyses (Brodie & White 1995). Crustal shortening is observed in the shape of elongate domes along the NW European Atlantic margin (Doré & Lundin 1996; Boldreel & Andersen 1998; Doré *et al.* 1999), but these domes are significantly smaller than, for example, the south Norwegian dome, which must be explained by other means. Other active tectonic mechanisms capable of generating regional surface uplift include increased buoyancy by heated lithosphere (White 1992), mantle upwelling (Rohrman & van der Beek 1996; Nadin *et al.* 1997), magmatic underplating of the lower crust (McKenzie 1984; White & Lovell 1997), and delamination of the lower lithosphere (Bird 1979; Molnar *et al.* 1993). These mechanisms could all be related to the impingement of the Iceland Plume onto the North Atlantic and subsequent North Atlantic rifting at the Paleocene–Eocene transition (*c.* 54 Ma). Thus they are all plausible agents of surface uplift (see discussion by Nielsen *et al.* 2002). However, it has been argued that no single mechanism seems capable of explaining all of the observed (and inferred) 'ups and downs' around and within the North Atlantic (Doré *et al.* 1999).

In late Cenozoic time the NW European margin was far removed from the Iceland Hotspot (Lawver & Müller 1994; Doré *et al.* 1999). Because of the increased distance to the Iceland Plume and the lack of extensive crustal shortening, active tectonic surface uplift of late Cenozoic age is difficult to invoke without jumping to rather exotic explanations. The various tectonic mechanisms capable of causing early Paleogene surface uplift around the North Atlantic are discussed in a companion paper (Nielsen *et al.* 2002), which argues that early Paleogene delamination of the lithosphere is the most plausible mechanism. This mechanism causes several hundred metres of almost instantaneous uplift followed by decelerating uplift rates through the remainder of the Cenozoic.

To keep the model developed in the present paper as simple as possible and independent of the mechanism of the tectonic uplift, it is simply assumed that significant (500–1000 m) surface uplift occurred in relation to the arrival of the plume and rifting of the North Atlantic in early Paleogene time. Also, it is assumed that for the case of southern Norway this uplift was permanent, as opposed to transient, uplift caused by increased buoyancy of heated lithosphere.

Isostatic uplift response to denudation

Gravity data indicate that the mean surface topography of southern Norway is isostatically compensated at depth (Balling 1980; Rohrman & van der Beek 1996), suggesting that Cenozoic denudation was compensated by regional isostatic uplift of the crust. It has been argued that intense localized denudation (dissection) of an initially flat topography of high elevation could theoretically cause mountain peaks to reach elevations of twice or more the height of the initial topography (see Molnar & England 1990; Gilchrest *et al.* 1994). It should be noted that in this case uplift is a consequence of denudation and not vice versa. This is possible because of the regional isostatic response to dissection, which is governed by the isostatic response function (i), which equals ρ_c/ρ_m, where ρ_c is the density of the material eroded from the top of the crust, ρ_m is the density of the mantle at the depth of compensation ($\rho_m \approx 3.3\,g\,cm^{-3}$), and i is the amount of isostatic uplift per unit mean depth of dissection (Gilchrest *et al.* 1994). In the central parts of southern Norway, where the removed material is likely to have been mainly crystalline rock ($\rho_c \approx 2.7$–$3.0\,g\,cm^{-3}$), the compensation is relatively high (approaching 0.8–0.9). In marginal areas, where a greater proportion of the rocks removed are likely to have been of sedimentary origin ($\rho_c \approx 2.3$–$2.7\,g\,cm^{-3}$) it is somewhat lower (*c.* 0.7–0.8).

Climatic and eustatic change

The denudation history of southern Norway and the infill patterns of the North Sea Basin were modulated by climatic and eustatic changes (Spjeldnæs 1975; Doré 1992; Jordt *et al.* 1995; Eyles 1996; Solheim *et al.* 1996; Michelsen *et al.*

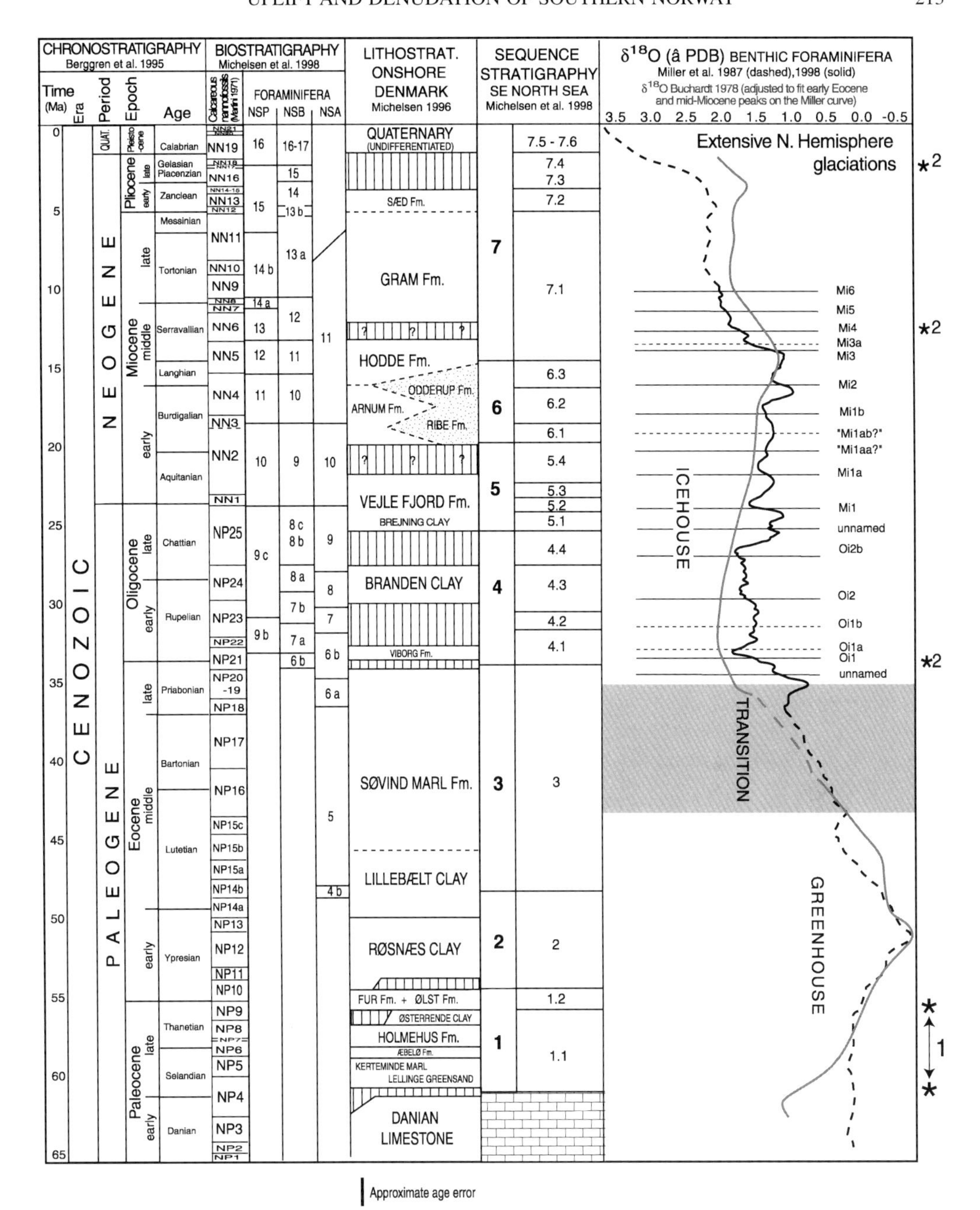

Fig. 3. Cenozoic stratigraphy of the eastern North Sea Basin. Time scale according to Berggren *et al.* (1995). The composite $\delta^{18}O$ curve is compiled from Miller *et al.* (1987, 1998). The Miller *et al.* (1987) curve was shifted −0.25‰ at 10 Ma and +0.20‰ at 37 Ma to match values of the Miller *et al.* (1998) curve. Correlation of North Sea sequences with the Berggren *et al.* (1995) time scale and the $\delta^{18}O$ curve is based on the calcareous nannofossil zonation of Martini (1971). The North Sea $\delta^{18}O$ curve of Buchardt (1978), adjusted to fit early Eocene and mid-Miocene peaks on the Miller curve, is shown for reference. *Episodes of marked increase in the supply of coarse clastic sediments to the North Sea Basin. 1, North Atlantic rifting and volcanism; 2, major ice-sheet expansion (Lear *et al.* 2000). The mid-Paleocene event roughly coincides with the commencement of rift-related uplift of areas bordering the North Atlantic rift. The latest Eocene–earliest Oligocene, the late Mid-Miocene and the Plio-Pleistocene events all coincide with major cooling events and eustatic lowerings as a result of increased continental ice volume (Lear *et al.* 2000). Foraminifera: NSP, North Sea Planktic; NSB, North Sea Benthic; NSA, North Sea Agglutinated.

1998; Clausen *et al.* 1999, 2000; Huuse 2002; Nielsen *et al.* 2002).

Most researchers agree that eustasy peaked in response to relatively large spreading rates at mid-ocean ridges and greenhouse climate during mid-Cretaceous time. The long-term trends of global sea level (eustasy) may be estimated from spreading rates of mid-ocean ridges (e.g. Pitman 1978; Kominz 1984), seismic stratigraphy (Vail *et al.* 1977; Haq *et al.* 1987), flexural backstripping (Pekar & Miller 1996; Steckler *et al.* 1999) and from stable isotopes (Fig. 3; $\delta^{18}O$, Sr, Mg/Ca; Matthews 1984; Miller *et al.* 1987, 1996, 1998; Abreu & Anderson 1998; Lear *et al.* 2000). From these studies a general consensus has emerged that long-term global sea level has fallen some 150–250 m during Cenozoic time. This, of course, corresponds to a surface uplift of the same magnitude relative to sea level.

A long-term trend of climatic cooling started in late Mid-Eocene time, both globally (Savin 1977; Wolfe 1978; Miller *et al.* 1987, 1998; Zachos *et al.* 1992; Lear *et al.* 2000) and in the North Sea region (Buchardt 1978; Collinson *et al.* 1981). In association with the change towards a colder climate, the continental ice sheet on Antarctica expanded, causing a pronounced eustatic lowering, culminating at the Eocene–Oligocene transition (*c.* 34 Ma; Miller *et al.* 1998; Lear *et al.* 2000). A similar episode of cooling and eustatic lowering occurred once again in late Mid-Miocene time, following an early to mid-Miocene warm period (Molnar & England 1990; Miller *et al.* 1998; Lear *et al.* 2000), and in Plio-Pleistocene time, eventually causing full glacial conditions in NW Europe (Eyles 1996; Solheim *et al.* 1996; Lear *et al.* 2000). With the exception of the Plio-Pleistocene glaciations, the role of climate change has largely been neglected in previous studies of the North Atlantic margins. This may be due to the notion that climate as a mechanism acts on a time scale an order of magnitude shorter than that of tectonics (see Vail *et al.* 1991) or simply because of lack of attention to Cenozoic climate changes. However, it is important to bear in mind that a major long-term change in climate may significantly affect denudation rates (Summerfield & Brown 1998).

Major episodes of climatic cooling during the Cenozoic generally corresponded to major ice-sheet expansions on Antarctica that caused major eustatic lowerings (Lear *et al.* 2000). Thus it is possible that the effects of major eustatic falls and stepwise climatic deterioration could cause effects similar to those widely attributed to regional surface uplift, i.e. accelerated denudation of topography and increased sediment supply to adjacent basins (Donnelly 1982; Molnar & England 1990; Huuse 2002).

Rationale

From the above discussion it should be clear that the following factors should be considered when attempting to account for the observed topography of southern Norway and the sedimentary record of the North Sea Basin: (1) plume-related surface uplift along Atlantic margins in early Paleogene time; (2) episodic inversion tectonics driven by Alpine compression and Atlantic ridge push; (3) stepwise climatic deterioration since mid-Eocene time as documented by stable isotope records (Fig. 3); (4) eustatic lowering of *c.* 250 m since mid-Cretaceous time (*c.* 200 m since mid-Eocene time); (5) passive (flexural?) isostatic response to denudation and deposition. These are all relatively well-documented phenomena, although there may be some uncertainty about the exact mechanisms of early Paleogene uplift and the magnitude and effects of climatic deterioration and eustatic fall (see Doré *et al.* 1999; Nielsen *et al.* 2002).

Another factor that must be addressed in an account of the genesis of present-day topography is palaeo-topography, i.e. the topography of the so-called ‘pre-uplift peneplain’ (Stuevold & Eldholm 1996) or ‘palaeic surface’ of Norway (Gjessing 1967; Lidmar-Bergström *et al.* 2000). It is assumed here that the present-day elevation of the palaeo-peneplain roughly coincides with the summit envelope (see Doré 1992). The purity of the Late Cretaceous–Danian chalks of the North Sea Basin indicates that any siliciclastic source area must have been close to a peneplain by late Cretaceous time (Hancock 1975). However, the presence of upper Cretaceous siliciclastic wedges along the Atlantic margin (Knott *et al.* 1993; Doré *et al.* 1999) demonstrates that there must have been some topography above sea level. This topography probably coincided with the Shetland Platform and the present-day Norwegian mainland. Some topography was probably also generated along the late Cretaceous and early Paleogene inversion zones (Ziegler 1990; Gemmer *et al.* 2002). The elevation and relief of the topography of southern Norway and Shetland is difficult to constrain because of the absence of late Mesozoic and Cenozoic sediments, but a maximum elevation of the order of a few hundred metres above (palaeo-) sea level, gently sloping towards the shoreline with only minor local relief seems plausible. The shoreline was probably close to or slightly inboard of the present shorelines of southern Norway, whereas all of Denmark and southern Sweden was submerged (Spjeldnæs 1975; Ziegler 1990; Stuevold & Eldholm 1996). This combination of maximum elevation and shoreline positions would correspond to an average surface slope of about 0.1°, which does not seem unrealistic for a peneplain across a former mountain range.

Against this background, it will be assessed whether the interaction of the above-mentioned mechanisms can account for the present-day topography of southern Norway and the stratigraphic record of the adjacent basins. If this were the case, then there would be no need to invoke late Cenozoic tectonic uplift events. Conversely, if the combination of the above factors does not support the hypothesis, one may begin to speculate about late Cenozoic tectonic events. The scenario given here is based mainly on qualitative evidence and simple isostatic calculations, and should therefore be regarded as a

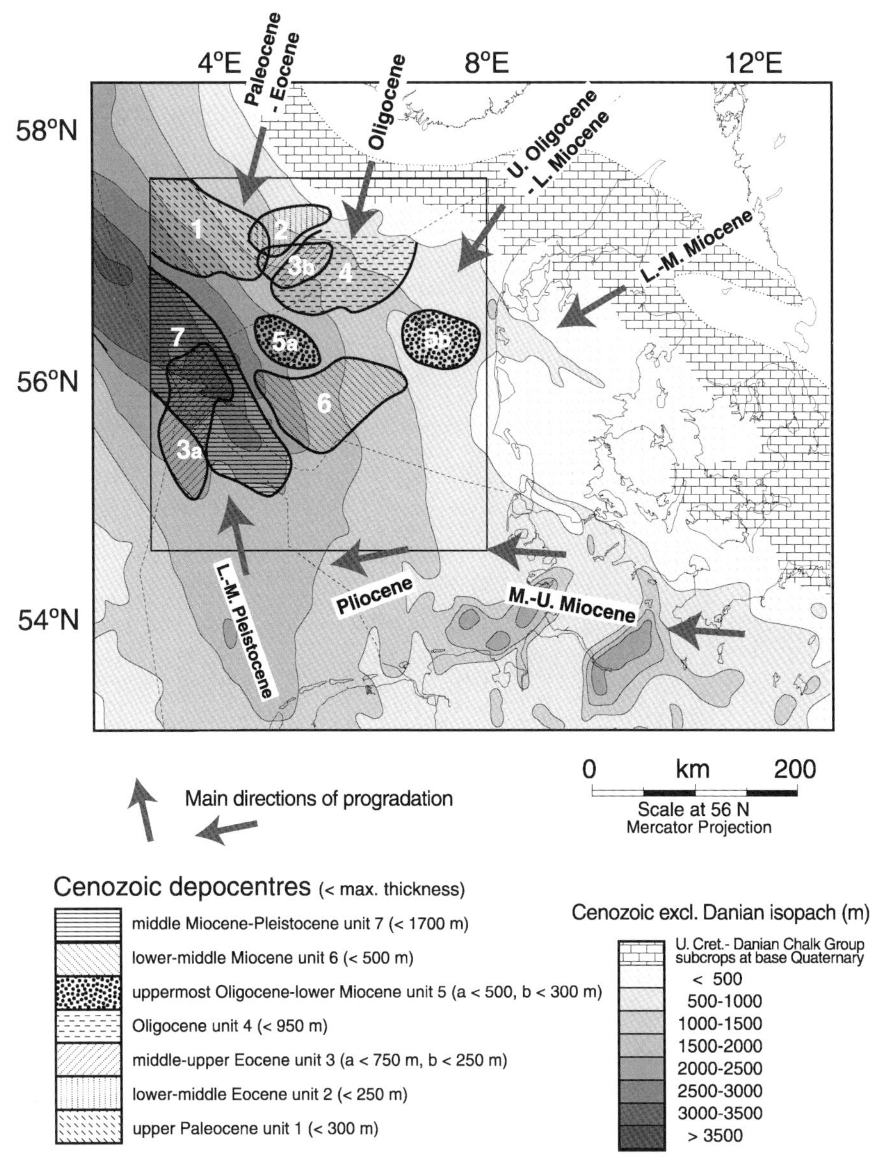

Fig. 4. Post-Danian depocentres in the eastern North Sea superimposed on depth contours of the Upper Cretaceous–Danian limestone. Depocentres were mapped by Bidstrup (1995) and Michelsen *et al.* (1998).

somewhat subjective perception of how present-day topography and sedimentary thicknesses of the North Sea area may be accounted for. The importance of global (allocyclic) mechanisms (climate, eustasy) is stressed, as opposed to local or regional (autocyclic) mechanisms (e.g. tectonics). On the basis of geodynamic modelling, Nielsen *et al.* (2002) have provided a quantitative account of a similar scenario.

The ultimate test of any qualitative scenario of the development of source areas and their basins would be to carry out mass-balanced palaeogeographical reconstructions such as that performed for the Mississippi catchment and the associated depocentre in the Gulf of Mexico (Hay *et al.* 1989). Such an approach requires a detailed integration of regional denudation estimates from onshore areas with the extensive offshore database of seismic and well data and thus calls for international co-operation and integration of large regional databases.

Cenozoic evolution of the North Sea Basin

Database

The Cenozoic evolution of the North Sea Basin has been pieced together from a large database that covers various parts of the basin. Seismic data, well data, outcrops, structure and isopach maps (Fig. 4; Nielsen *et al.* 1986; Bidstrup 1995; Jordt *et al.* 1995; Michelsen *et al.* 1995, 1998; Joy 1996; Sørensen *et al.* 1997; Huuse & Clausen 2001), seismic and sedimentary facies maps (Joy 1996; Mudge & Bujak 1996; Danielsen *et al.* 1997; Sørensen *et al.* 1997), patterns of clinoform breakpoint migration (Figs 5 and 6; Sørensen *et al.* 1997; Clausen *et al.* 1999), and palaeogeographical compilations (Gramann & Kockel 1988; Kockel 1988; Ziegler 1990) have been compiled to yield palaeogeographical (palaeobathymetric) maps of the entire North Sea Basin (Fig. 7). Huuse, 2002 has provided a complete list of references for each of the maps (Fig. 7b–g) and an account of their compilation.

Palaeogeographical development

It is generally accepted that the overall infill of the North Sea Basin was dominated by westerly source areas during Paleocene and Eocene time, whereas easterly source areas dominated during the remainder of Cenozoic time (e.g. Jordt *et al.* 1995, 2000; Joy 1996; Mudge & Bujak 1996; Michelsen *et al.* 1998). However, this picture may be severely biased if one merely looks at the preserved sediments. For example, regional cross-sections of the northern North Sea (e.g. Jordt *et al.* 1995, Fig. 3), show a completely preserved Paleocene–Eocene succession prograding from the Shetland Platform, whereas similar age progradational deposits are truncated

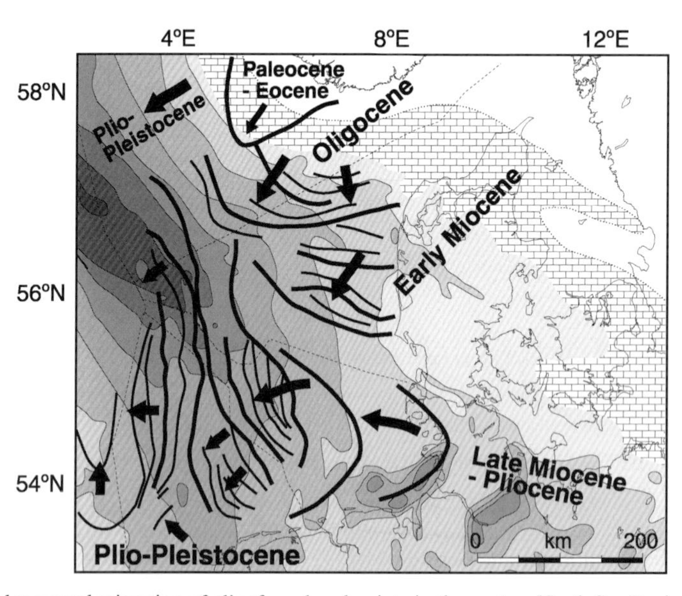

Fig. 5. Spatial and temporal migration of clinoform breakpoints in the eastern North Sea Basin superimposed on depth contours of the Upper Cretaceous–Danian limestone (clinoform breakpoints after Funnell 1996; Clausen *et al.* 1999). The clockwise infill from the Oligocene time onwards should be noted. This caused the northern part of the area to be filled to base level some 15–20 Ma before the southern part.

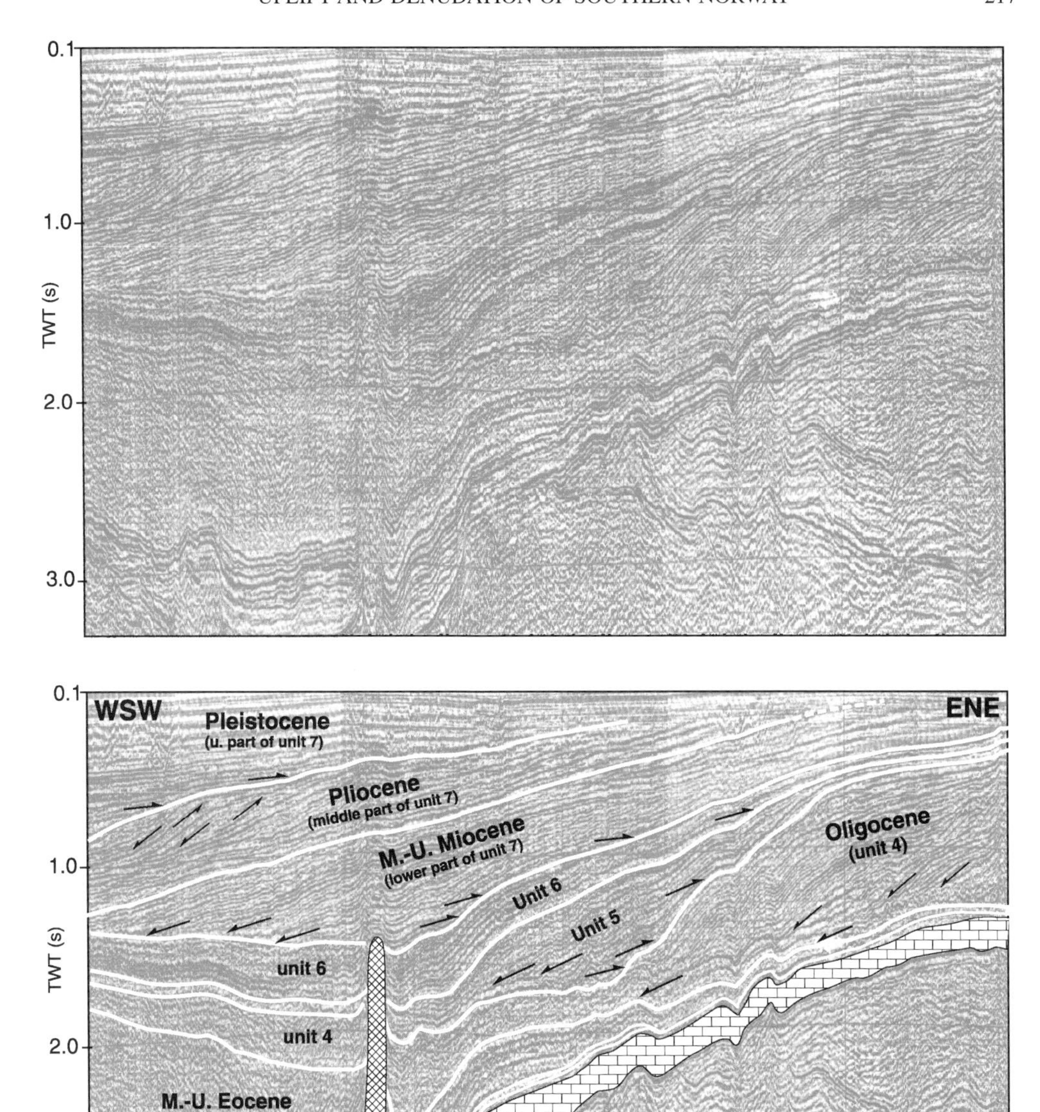

Fig. 6. ENE–WSW-oriented seismic profile (RTD81-22) showing large-scale depositional geometries along the Norwegian–Danish sector boundary. (For location, see Figs 2 and 7.) In the Central Graben area a condensed upper Paleocene to lower Middle Eocene succession is overlain by thick upper Middle–Upper Eocene smectitic clays with chaotic internal structure onlapping towards the east. The overlying succession of Oligocene silty clays interbedded with thick sand units exhibits markedly southwestward progradational geometries. The Miocene deposits are less markedly progradational along this profile and consist mainly of silty clay with thin sand stringers. High-angle progradational geometries are also observed in the upper Pliocene succession to the WSW. The final phase of infill in Pleistocene time is characterized by regional onlap of mainly shallow-water sediments supplied from the SSE. The height of the Oligocene–Miocene clinoforms indicates that palaeo-water depths in the Central Graben area were substantial (500–1000 m).

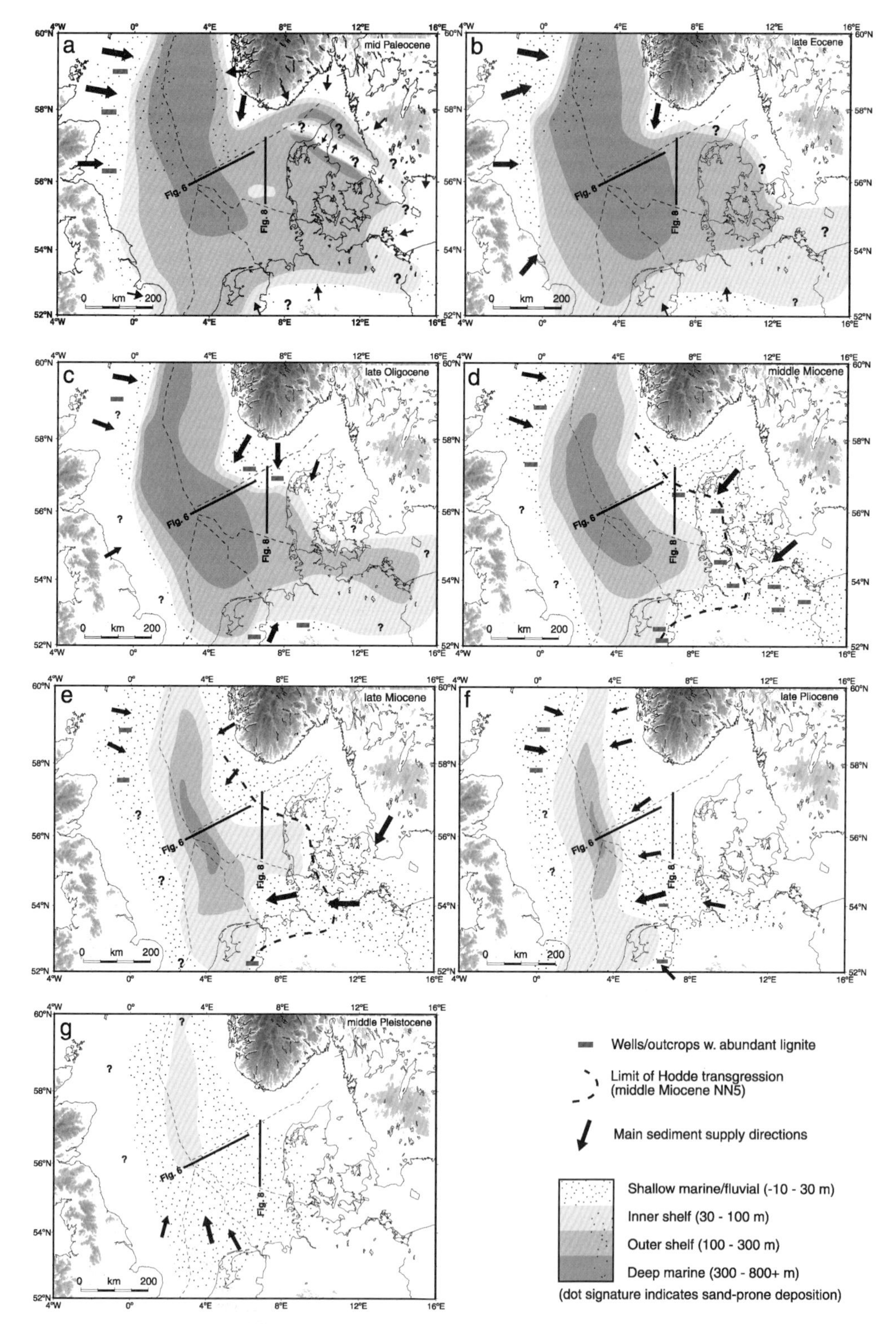

Fig. 7. Palaeogeographical development of the eastern North Sea Basin (modified after Huuse 2002); at (**a**) mid-Paleocene, (**b**) late Eocene, (**c**) late Oligocene, (**d**) mid-Miocene (**e**) late Miocene, (**f**) late Pliocene, and (**g**) mid-Pleistocene time. The maps are based on observations from the eastern North Sea Basin integrated with an extensive number of published data. See Huuse (2002) for complete listing of data and references used.

at a high angle towards the Norwegian coast. These progradational wedges are probably the remnants of a larger system of progradational lower Paleogene deposits originating from southern Norway. It is thus likely that significant amounts of sediment were supplied from easterly source areas during Paleocene and Eocene time (Jordt *et al.* 2000). This is also indicated by sand-prone depocentres of late Paleocene to early Eocene age SW of Norway (Fig. 4). Possible causes for the differential preservation of the Paleocene–Eocene succession on either side of the northern North Sea will be discussed later in this paper.

The large middle to upper Eocene depocentre in the Central Trough (unit 3, Fig. 4; Michelsen *et al.* 1998) consists mainly of smectite-dominated clay (Thyberg *et al.* 2000). The deposits contain little, if any, evidence for palaeo-transport directions and comprise an abnormally thick succession of hemipelagic clays deposited far away from potential source areas.

In the eastern North Sea, the Eocene–Oligocene transition is characterized by a massive increase in the amount of coarse clastic sediments supplied from southern Norway, resulting in an almost 1 km thick depocentre of markedly progradational sand-prone deltaic sediments of Oligocene age in the Norwegian–Danish Basin (Figs 4–6). This shift from hemipelagic clays and marls to silty and sandy clays was probably an effect of the late Eocene climatic deterioration (Buchardt 1978; Collinson *et al.* 1981), which led to increased seasonality and eustatic lowering (Ivany *et al.* 2000; Lear *et al.* 2000). The lower temperatures and increased seasonality probably increased the amount of precipitation and caused significant changes in vegetation (Spjeldnæs 1975; Collinson *et al.*, 1981), thus increasing the erosivity of the geomorphological system (see Summerfield & Brown 1998).

The clockwise rotation of the direction of progradation (Fig. 5; Clausen *et al.* 1999) shows how the basin was filled by sediments prograding from the NE, east, SE and finally south during Oligocene to Pleistocene time. As the most proximal parts of the basin were filled during Oligocene time (Fig. 7c; Danielsen *et al.* 1997), sediments simply bypassed the Oligocene depocentre during Miocene time, filling up the easternmost part of the basin coinciding with the central parts of Denmark (Figs 4–6 and 7d). Following the Hodde transgression (Koch 1989), sediments started prograding from the east across the northern part of Germany into the relatively deep waters of the German Bight of the southern North Sea Basin (Figs 5 and 7e; Kockel 1988; Jürgens 1996). The distal toes of these late Mid- to Upper Miocene deltas of the southern North Sea obliquely onlap the Oligocene–lower Mid-Miocene depocentres farther north (see Figs 6 and 8). In Pliocene time the eastern parts of the basin had been filled and the pattern of progradation was east to west along most of the North Sea Basin (Figs 5 and 7f; Sørensen *et al.* 1997). The final phase of infill during latest Pliocene–mid-Pleistocene time was mainly sourced from the SSE by the ancestors to the large NW European rivers of the present-day landscape (Figs 5 and 7g; Gibbard 1988; Zagwijn 1989), whereas sediments from southern Norway made their way into the northern North Sea and the North Atlantic (Jordt *et al.* 1995, 2000; Riis 1996; Evans *et al.* 2000).

The general picture of infill is thus comparable with that of passive margins, which generally show an overall basinward progradation of successive sedimentary wedges. However, the last phase of infill is remarkable in that the lower to middle Pleistocene sediments are regionally very extensive compared with previous units. In fact, they attain thicknesses of up to more than 500 m in areas where accommodation had previously been filled by the Miocene and Pliocene deltas (see western part of Fig. 6), thus indicating that additional accommodation was being created in early to mid-Pleistocene time. Apart from the central parts of the basin, the lower to middle Pleistocene sediments are generally of relatively shallow-water origin, suggesting that sedimentation rates kept up with the increased subsidence rates (Huuse 2002).

Correlation with regional tectonic events

Apart from regional plume-related uplift in Paleocene–early Eocene time there is only minor sign of tectonic activity in Cenozoic time in the North Sea Basin. This is mainly in the form of local inversion of old fault systems such as the Sorgenfrei–Tornquist Zone and the Central Graben (Vejbæk & Andersen 1987, 2002; Ziegler 1990). The inversion of these structures was probably caused by the combined effects of Atlantic ridge push and Alpine compression (e.g.Vejbæk & Andersen 2002) rather than by Alpine compression alone. The effect of Atlantic ridge push is also seen as large inversion domes along the NW Atlantic margin (Doré & Lundin 1996; Boldreel & Andersen 1998; Doré *et al.* 1999).

It is unlikely that intra-plate compression could cause significant uplift of cratonic source

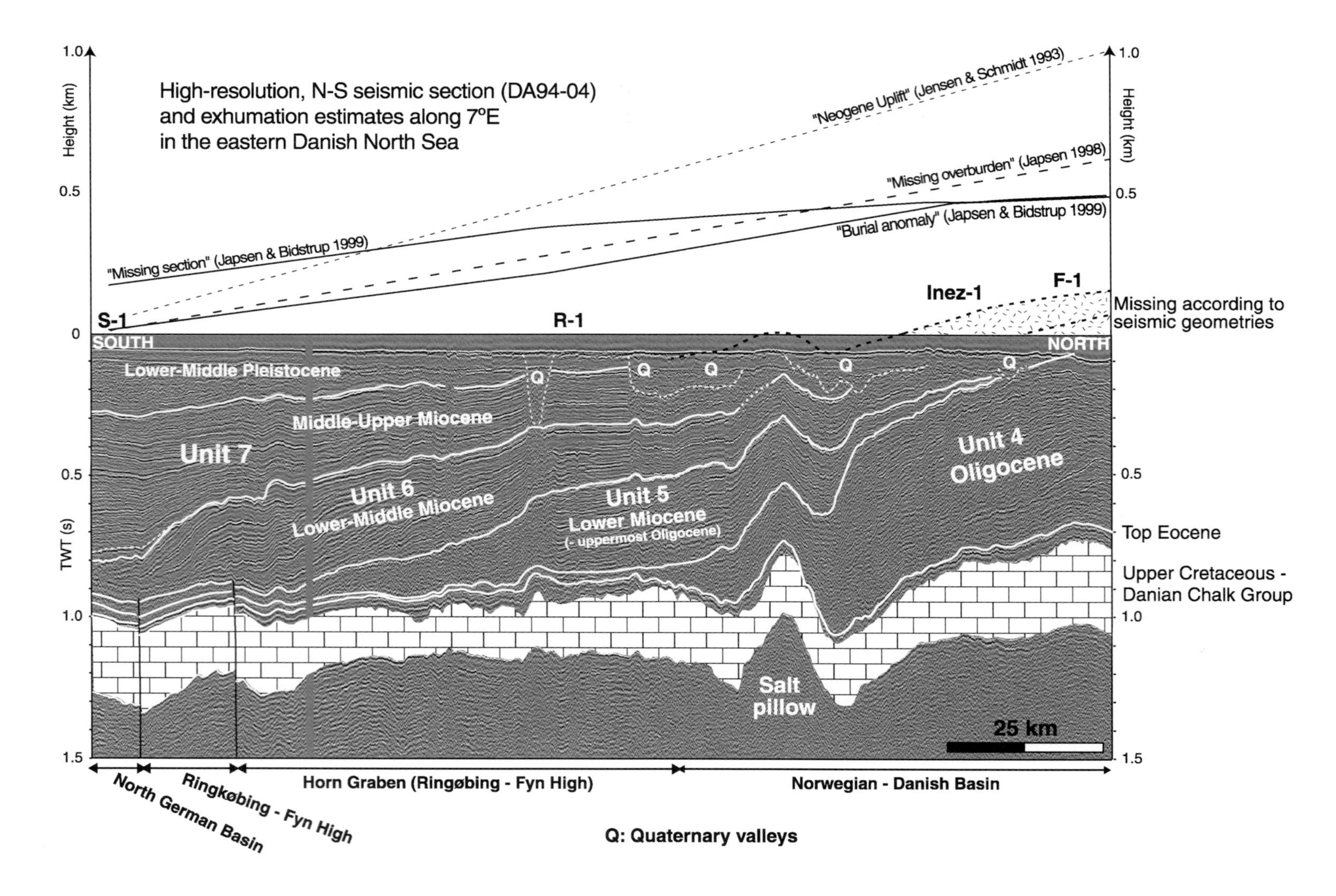

High-resolution, N-S seismic section (DA94-04)
and exhumation estimates along 7°E
in the eastern Danish North Sea
Height (km)
1.0
0.5
0
TWT (s)
0.5
1.0
1.5
"Neogene Uplift" (Jensen & Schmidt 1993)
"Missing overburden" (Japsen 1998)
"Burial anomaly" (Japsen & Bidstrup 1999)
"Missing section" (Japsen & Bidstrup 1999)
Missing according to seismic geometries
Top Eocene
Upper Cretaceous - Danian Chalk Group
S-1
R-1
Inez-1
F-1
SOUTH
NORTH
Lower-Middle Pleistocene
Middle-Upper Miocene
Unit 7
Unit 6
Lower-Middle Miocene
Unit 5
Lower Miocene
(- uppermost Oligocene)
Unit 4
Oligocene
Q
Salt pillow
25 km
North German Basin
Ringkøbing - Fyn High
Horn Graben (Ringøbing - Fyn High)
Norwegian - Danish Basin
Q: Quaternary valleys

areas such as southern Norway (Rohrman & van der Beek 1996; Doré *et al.* 1999). On the other hand, it may be possible that intra-plate compression could create enough disturbance along old fault zones to cause avulsion of major rivers and thus result in major changes of sediment input directions such as observed in mid-Miocene time (see Figs 5 and 7d and e).

It has been suggested that abnormally rapid early Pleistocene subsidence could have been caused by intra-plate compression (e.g. Cloetingh *et al.* 1990, 1992). However, the absence of evidence for early Pleistocene compressional faulting in the North Sea area makes it less likely that extensive compression was the cause of rapid early Pleistocene subsidence.

Another effect to consider is the load-induced subsidence caused by the last phase of infill of the North Sea Basin. By late Pliocene time only a narrow seaway of relatively great water depth existed in the central North Sea. The infill of a narrow (say 400 m) deep basin remaining at the beginning of Pleistocene time could cause some degree of flexural downwarping of the margins. The extent to which the sediment-loaded subsidence would be distributed laterally is, however, strongly dependent on the flexural strength of the lithosphere, and numerical modelling of the loading effect needs to be carried out to quantify this effect.

Correlation with global climate and sea level

The supply of coarse clastic sediment to the North Sea Basin accelerated several times during Cenozoic time, most notably in late Paleocene time, at the Eocene–Oligocene transition, in late Mid-Miocene time and in Plio-Pleistocene time (Fig. 3). It appears straightforward that the abruptly increased supply of siliciclastic sediments in late Paleocene time was caused by uplift of source areas in relation to the arrival of the plume and rifting of the North Atlantic. Because of the substantial time lag (>20 Ma), direct plume-related effects cannot satisfactorily explain the increases in sediment supply at the Eocene–Oligocene transition, and in late Mid-Miocene and Plio-Pleistocene times. Hence, it should be investigated whether these increases could have been caused by other well-documented phenomena such as major long-term climatic and eustatic changes. Although plume-related effects cannot account for abrupt changes in sediment supply, the effect of delamination of the lithosphere may be part of the explanation by providing decelerating but continued uplift after the initial uplift pulse (Nielsen *et al.* 2002).

Previous studies have noted a conspicuous correlation between major North Sea sequence boundaries and major $\delta^{18}O$ increases (Jordt *et al.* 1995; Huuse & Clausen 2001; Huuse 2002). Huuse (2002) also noted that large-scale sedimentation patterns of the eastern North Sea Basin are comparable with sedimentation patterns observed along many continental margins (see Donnelly 1982; Bartek *et al.* 1991; Cameron *et al.* 1993; Miller *et al.* 1998; Séranne 1999; Huuse & Clausen 2001). Moreover, the periods of increased sediment supply at the Eocene–Oligocene transition, and in late Mid-Miocene and Plio-Pleistocene times roughly correlate with episodes of climatic deterioration and major increases in continental ice volume (see Buchardt 1978; Molnar & England 1990; Lear *et al.* 2000). Other evidence of significant increases in denudation comes from geomorphological studies, which have indicated that rates of stream incision accelerated in late Cenozoic time, both in southern Norway and in the eastern USA (Lidmar-Bergström *et al.* 2000; Mills 2000). Such virtually contemporaneous increases in denudation indicate a global rather than regional or local cause (Molnar & England 1990).

Fig. 8. North–south-oriented seismic profile (DA94-04) showing a conformable upper Paleocene to Eocene succession consisting of hemipelagic clay and marl (units 1–3) overlying the Upper Cretaceous–Danian Chalk Group. The condensed Paleocene–Eocene succession is overlain by a thick progradational succession of Oligocene to mid-Miocene age (units 4–6). The post-middle Miocene sequence (unit 7) is characterized by a thick shallowing-upwards aggradational succession of marine clay and silt, onlapping the mid-Miocene unconformity. P-wave velocities from four wells located along the profile (S-1, R-1, Inez-1, F-1) indicate that the velocity of the post-Chalk Group is very close to 2 km s^{-1} (i.e. 1 s two-way travel time (TWT) $\approx$ 1 km). This relationship is used to directly compare geometries on the seismic with inferred amounts of 'missing section' or 'Neogene uplift' based on sonic-derived compaction trends of the Chalk Group (Japsen 1998) and of Jurassic shales (Jensen & Schmidt 1993), respectively. Estimates of 'missing section' and 'missing overburden' based on integration of chalk compaction and vitrinite reflectance data (Japsen & Bidstrup 1999) are also shown. The amount of uplift estimated from the seismic geometries is significantly lower than that based on compaction trends. Q, Quaternary valley. (For location, see Figs 2 and 7.)

Cenozoic uplift and denudation of southern Norway

Constraints on magnitude

Apatite fission-track thermochronology (AFTT) has become a popular proxy for inferring uplift. However, AFTT relates to cooling histories of the rocks analysed, i.e. to denudation and not directly to uplift (Brown 1991). This is important, as denudation may lag uplift by several tens of million years (Summerfield & Brown 1998). Moreover, AFTT is sensitive only to temperatures of minimum *c.* 60°C (depths >2–3 km) and is thus generally not sensitive to late Cenozoic effects (Gunnell 2000), although inversion of the AFTT data may hint at late Cenozoic cooling histories (e.g. Rohrman *et al.* 1995). The latter indicate a maximum denudation of 2 ± 0.5 km in the inner fjords of southern Norway decreasing outward to <0.5 km at the present-day coastline, whereas the amount of denudation at the highest mountain peaks such as Jotunheimen is below the resolution of AFTT inversion (Rohrman *et al.* 1995; Fig. 8). The rather large uncertainties of denudation estimates based on the AFTT method call for additional constraints on denudation of southern Norway. The most powerful method for producing reliable denudation estimates is probably (U–Th)–He thermochronometry, coupled with investigations of cosmogenic isotope studies to date exposed landforms, and it is recommended that such investigations be carried out in the future.

The denudation estimates cited above are almost opposite to the estimates of Riis & Fjeldskaar (1992), which show greatest average amounts of denudation at the coastline (their fig. 11). This study relies on geomorphological characteristics of the landscape and extrapolation of well-dated offshore surfaces over land areas (Doré 1992; Riis & Fjeldskaar 1992; Riis 1996). The use of this method is somewhat problematic when datable rocks are absent in the areas of maximum uplift and the age of the surface onshore is thus poorly constrained. It is possible, however, to define relatively coherent palaeosurfaces in even severely denuded areas using geomorphological criteria, although the age of the surface will be conjectural (Gjessing 1967; Riis & Fjeldskaar 1992; Riis 1996; Lidmar-Bergström *et al.* 2000). The use of palaeosurfaces thus may yield some idea about the magnitude of uplift, but with rather poor constraints on timing, and may, in some cases, agree poorly with results of AFTT.

A widely used method of estimating uplift and denudation is to compare regional compaction trends of shales and chalks relative to a defined 'normal trend' (Jensen & Schmidt 1993; Hansen 1996; Japsen 1998). The estimates from compaction-trend methods are highly variable, even using the same wells (Fig. 8) and it would appear that the regional *trends* of over- (and under-) compaction are more useful than the absolute values. The estimates of 'missing section' may be further constrained by integrating compaction analyses with vitrinite reflectance studies (Fig. 8; Japsen & Bidstrup 1999). However, although this integration yields even lower estimates of uplift and denudation, they are still of the order of 300 m above those based on seismic geometries (Fig. 8: compare with 'burial anomaly' of Japsen & Bidstrup 1999). Japsen & Bidstrup inferred that late Neogene erosion is responsible for the 'burial anomaly' shown in Fig. 8. If this is the case, then a 300–400 m thick wedge of Pliocene sediments must have been deposited between the (complete) Middle–Upper Miocene succession and the Lower–Middle Pleistocene unit in Fig. 8. However, in Pliocene time the eastern Danish North Sea was mainly bypassed by sediments, which were deposited in large deltas 100 km farther to the west and SW (Fig. 7e and f). Hence, the notion of rapid late Neogene deposition and erosion in the eastern Danish North Sea inferred from compaction-based exhumation estimates does not agree with the palaeogeographical evolution of the area.

As demonstrated by the above example, in areas where the bulk of the Cenozoic succession is preserved, it is possible to use large-scale depositional geometries observed in seismic profiles to estimate the amount of tilting and denudation that has occurred within the basin during Cenozoic time. It should be noted that it is the amount of tilt of previously horizontal surfaces that indicates differential uplift or subsidence; the volume of sediment of any given age reflects only the amount of denudation.

Geometrical denudation estimates and 'uplift' and/or denudation estimates based on compaction trends all show similar trends of denudation increasing northwards, but the amplitudes vary widely, with the geometrical estimate being the lowest (Fig. 8). It is beyond the scope of this study to scrutinize the methods and assumptions behind compaction analyses, but if the seismic geometries are to be relied upon, it appears that compaction methods overestimate the amount of uplift and denudation of the eastern North Sea. Moreover, it is remarkable that the geometrical estimate is in accord with an estimate of 'overburial' of the Chalk Group based on 3D basin modelling (S.B. Nielsen pers. comm. 2000). The modelling study incorporates a thermally subsiding North Sea Basin, subject to long-term eustatic fall, varying sediment input and sedimentary loading, without the influence of late Cenozoic tectonics. Hence, it appears that the large-scale stratal geometries observed on seismic data (Figs 6 and 8) reflect the infill of a thermally subsiding basin during generally falling sea level.

Constraints on timing

Paleogene and Neogene sediments are almost exclusively located around the fringes of the uplifted areas in Scandinavia and Britain and there are no sediments preserved in the most elevated parts. Hence, until high-resolution fission-track data become available, the timing of uplift and denudation has to be inferred from the sedimentary record of the adjacent basins. This task requires regional data coverage to filter out local variations in sediment supply, which could cause potentially misleading sedimentation patterns. When trying to establish uplift and denudation histories from the sedimentary record it is extremely important to bear in mind that although a volume of sediment is directly related to denudation it tells us very little about uplift. Also, it is important to bear in mind that isopach maps show only the present distribution of the erosional products. Hence, it is possible that sediments recording early denudation have now been removed as a result of later uplift (or base-level fall) and associated erosion. Thus, the present distribution of sediments may be dominated by the most recent episodes of denudation and the associated isostatic response. This is especially true for the earliest proximal sediments deposited along the margins of the source area.

It seems likely that the apparent lack of proximal sediments of Paleocene and Eocene age off Norway is due to cannibalization and redeposition. This is indicated by the truncation of thick, highly progradational sediment wedges off the west coast of Norway (see Jordt *et al.* 1995; Fig. 3). The less marked truncation around the Shetland Platform could be due to the narrowness of the Shetland topography (about one-third the width of the south Norwegian dome) causing less uplift as a result of erosional unloading. However, variations in thermal uplift and subsidence (with the Shetland Platform experiencing the largest of both) may also have influenced this pattern.

From the sedimentary record it appears that denudation accelerated at least four times during Cenozoic time: in late Paleocene time, at the Eocene–Oligocene transition, in late Mid-Miocene time and in Plio-Pleistocene time. The first episode was probably a response to regional uplift associated with plume activity and rifting in the North Atlantic region. The remaining episodes appear to correlate with similar responses on continental margins far from the North Atlantic domain, thus indicating the influence of global factors such as climate and eustasy. Hence, it seems that the latter three episodes of accelerated denudation may reflect climatic and eustatic changes rather than tectonic events.

Isostatic response to localized denudation (dissection)

The width of the south Norwegian dome is of the order of 300–400 km along any cross-section (Figs 1 and 9). The area of the dome is therefore *c.* 10^5 km^2 and removal of material from the surface of the dome is thus likely to be isostatically compensated by addition of mantle material at depth (England & Molnar 1990). This is in agreement with gravity data (Balling 1980), which indicate that the topography of southern Norway is isostatically compensated at depth. Because the lithosphere has a certain flexural strength there are likely to be flexural effects, seen as subdued uplift (or subsidence) response to erosion (or sedimentation), along the margins of the dome.

A schematic illustration of the formation of dissected topography from an initial low-elevation peneplain is shown in Fig. 10 and described below, leaving out flexural effects at the margins of the rock column. The dimensions of the rock column are comparable with those of the central parts of the south Norwegian dome. Removal of crustal material ($\rho_c \approx 2.7\,\mathrm{g\,cm^{-3}}$) from the top of the 200 km wide columns is isostatically compensated by addition of mantle material ($\rho_m \approx 3.3\,\mathrm{g\,cm^{-3}}$) at depth. The isostatic compensation ($i = \rho_c/\rho_m$) for denudation is thus *c.* 0.8. Hence, 1 km (mean) denudation would decrease the surface elevation by only 0.2 km as a result of the isostatic response, which would cause 0.8 km uplift of the entire rock column.

Regional studies of active orogens (European Alps, Andes, Himalayas) indicate that dissection is rarely fully developed in their central parts and that only about half of the peak height at the centre of orogens can be explained by the isostatic response to denudation (Gilchrest *et al.* 1994). Looking at a regional topographic cross-section of southern Norway (Fig. 9) it appears that denudation is unevenly distributed, with

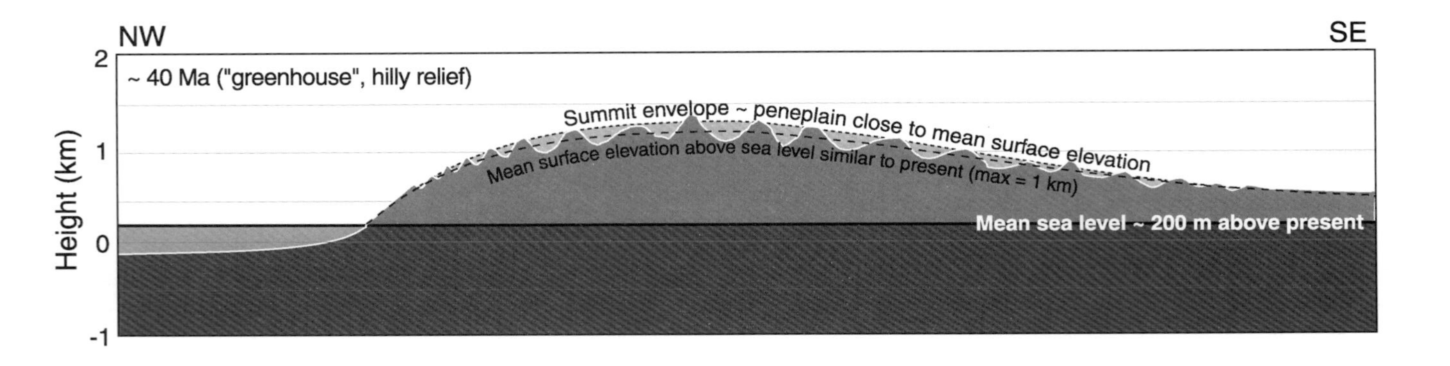

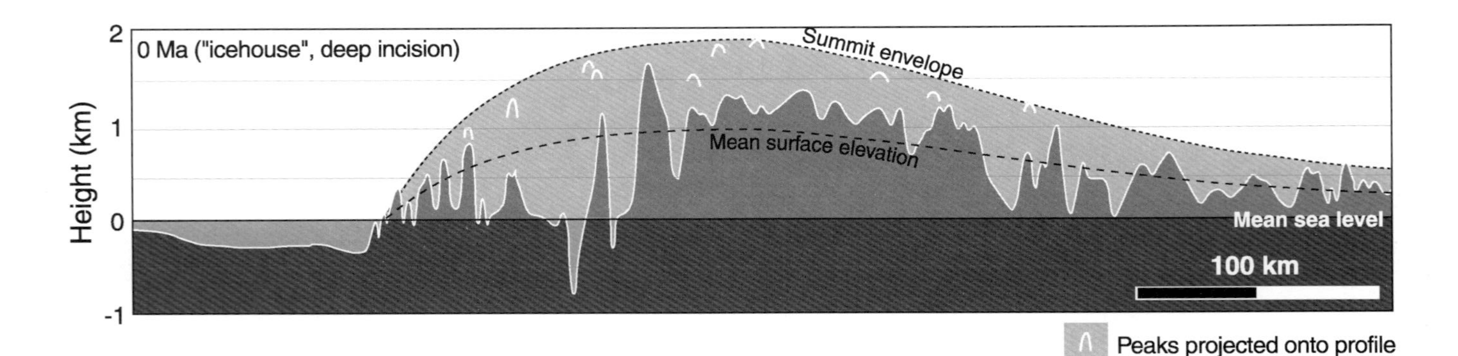

Fig. 9. Topographic cross-sections of the south Norwegian dome. Location is shown in Fig. 1. The upper profile shows a hypothetical cross-section of the dome at mid-Eocene time (*c*. 40 Ma). The lower profile shows the present topography and summit envelope (after Torske 1972) and the mean surface elevation averaged over 50–100 km. The mean surface elevation is a qualitative estimate of the isostatically compensated topography envelope. In mid-Eocene time (*c*. 40 Ma) a hilly relief had developed in response to weathering of a Mesozoic peneplain uplifted to 1–1.5 km elevation in earliest Eocene time. A warm climate, dense vegetation and a low-gradient local relief probably caused low rates of denudation. It should be noted that sea level was *c*. 200 m higher at 40 Ma than at present. The present-day deeply incised relief probably developed in response to repeated episodes of climatic deterioration and eustatic lowering during mid- to late Cenozoic time, culminating with full glacial conditions and extreme rates of incision in Plio-Pleistocene time. Deep dissection of the former high-elevation peneplain caused mountain peaks to rise to approximately twice their initial elevation. As a result of the eustatic fall of *c*. 200 m since mid-Eocene time, the mean surface elevation (with respect to sea level) has remained at roughly the same level throughout.

maximum denudation of 1–1.5 km in a zone stretching *c.* 100 km inboard of the coastline. Local denudation is of the order of 0.7–0.8 km in the area of highest mean elevation and decreases to 0.5 km farther to the SE. Average denudation (defined as summit envelope minus mean surface elevation) is highest (*c.* 1 km) above the central and northwestern parts of the dome (Fig. 9). This picture is, of course, strongly dependent on the length scale over which denudation is averaged and on the 3D distribution of valley incision, but it seems to indicate that the central parts of the dome have been affected by the combined effects of local denudation (causing surface lowering) and isostatic uplift in response to deep incision of neighbouring areas.

Early Cenozoic uplift and late Cenozoic denudation; a hypothetical model

A regional cross-section of the south Norwegian dome is shown in Fig. 9. The upper panel shows a hypothetical low-relief landscape after regional uplift of a peneplain, before the development of the deeply incised valleys and fjords that characterize the present-day topography, illustrated by the lower profile.

A hypothetical model of Cenozoic uplift and denudation of the central parts of southern Norway is shown in Fig. 10. At the end of Mesozoic time, southern Norway was probably worn down to a peneplain of some mean elevation above sea level. The exact elevation of the peneplain is poorly constrained and here it is assumed that the mean height of the central parts was close to 200 m elevation, although the highest peaks in southern Norway may be remnants of inherited topography (Riis & Fjeldskaar 1992).

Late Paleocene–early Eocene plume- and rift-related tectonics uplifted the peneplain to *c.* 1 km elevation (including pre-uplift elevation), but incision was initially minimal in the central parts of the dome, as a result of the low local relief, and a warm humid climate favouring dense vegetation and limited runoff. The bulk of sediment supplied to the basin at this time (40–60 Ma) probably derived from the margins of the uplifted dome (not shown in Fig. 10).

In late Eocene–early Oligocene time, climatic deterioration (increased seasonality) and eustatic lowering caused increased rates of stream incision. Climate recovered during late Oligocene–early Mid-Miocene time, thus stabilizing incision rates.

In late Mid-Miocene time another phase of climatic deterioration and eustatic lowering caused accelerated rates of stream incision and thus increased denudation rates. Finally, in Plio-Pleistocene time, full glacial conditions in the high-lying parts of southern Norway caused extreme rates of incision, eventually carving out the dissected topography we observe today (Fig. 9).

In the scenario described here, the only tectonic uplift of the surface (and of the rock column) occurred in early Paleogene time. Deep dissection of the uplifted peneplain and resultant isostatic uplift of the rock column caused the remaining part of the rock uplift, whereas the mean surface was lowered by *c.* 20% of the amount of mean denudation. It should be noted that the average denudation of 1 km (surface lowering of 0.2 km) is compensated by a eustatic lowering of the same magnitude, thus maintaining the mean surface elevation (with respect to sea level) at *c.* 1 km.

The landscape evolution as depicted in Figs 9 and 10 is in agreement with denudation as recorded by the offshore stratigraphic record (Figs 4–7) and with geomorphological studies, which indicate accelerated rates of incision in late Cenozoic time (Lidmar-Bergström *et al.* 2000). A late Cenozoic increase in the rates of stream incision was also found in a regional study of the eastern USA (Mills 2000), supporting the notion of an allogenic control on incision rates, i.e. climate and eustasy rather than regional tectonics. The Plio-Pleistocene increase in the rates of incision is also reflected by the occurrence of thick Plio-Pleistocene sediment wedges all along the NW European Atlantic margin (Riis & Fjeldskaar 1992; Riis 1996; Evans *et al.* 2000) and in the central North Sea (Figs 4 and 6). The concept of mountain building by isostatic response to dissection of an uplifted peneplain is also backed by the apatite fission-track analyses of Rohrman *et al.* (1995). These results indicate rapid late Cenozoic denudation at the base of the deepest valleys (fjords) and only minor denudation (below the detection threshold) on the mountain peaks. However, acquisition of high-resolution fission-track data is needed to further constrain the Cenozoic denudation history of southern Norway.

The model proposed here is not in agreement with the interpretations by Riis (1996), who found that eastern Denmark has suffered more than 1 km of Plio-Pleistocene denudation. This estimate was probably driven by an attempt to honour the pattern of late Neogene erosion inferred from compaction analyses. As argued above, such estimates probably overestimate the amount of exhumation of the eastern North Sea Basin by several hundred metres. Moreover, the

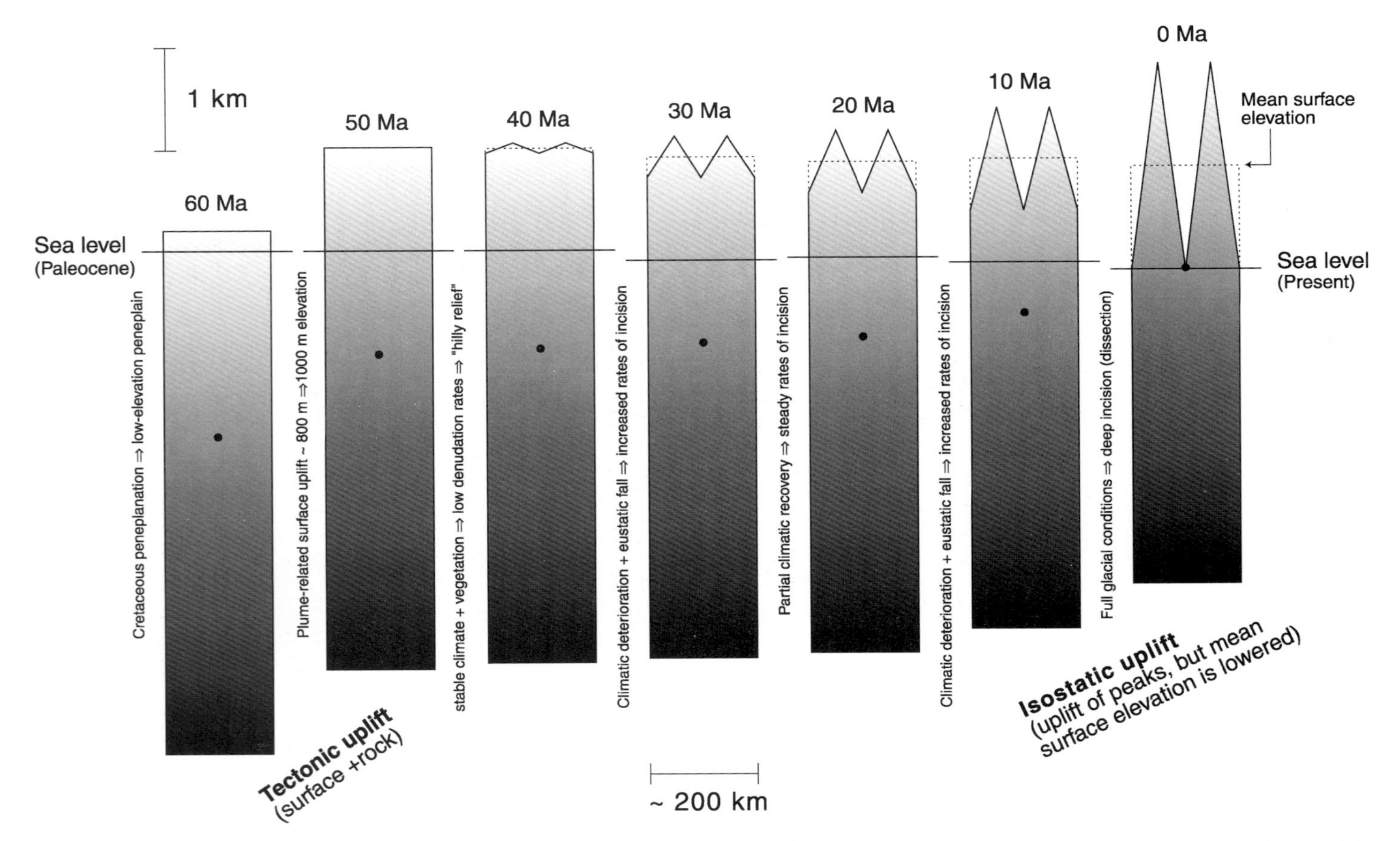

Fig. 10. Schematic illustration of uplift and bisection of an initial low-elevation peneplain. The dimensions are comparable with those of the central parts of the south Norwegian dome. (See text for discussion.)

study relied on extrapolation of key surfaces over several hundred kilometres, obviously a somewhat tricky discipline when Cenozoic sediments are absent over most of the area. In particular, the occurrence of late Early Eocene diatoms in northern Finland (Tynni 1982; Fenner 1988) has been used to infer a Paleogene episode of low-elevation peneplanation and submergence (Rohrman *et al.* 1995; Riis 1996). The locations of the diatom finds are, however, all of relatively low elevation (<300 m) and could thus have been submerged during the early Eocene highstand of sea level (Fig. 3) without the need for Paleogene peneplanation of southern Norway. The subsequent emergence of the diatom locations could easily be explained by the combined effects of post mid-Eocene eustatic lowering of *c.* 200 m and flexural uplift as a result of denudation (dissection) and isostatic uplift (of rocks) of the adjacent northern Norwegian highland.

Modelling studies (e.g. Riis & Fjeldskaar 1992; Stuevold & Eldholm 1996) generally fall some hundred metres short of explaining the present topography of southern Norway, leading the researchers to suggest that active tectonic mechanisms such as mantle phase changes or palaeo-topographic relief are required to explain present-day topography of southern Norway. However, these studies assumed a constant global sea level through Cenozoic time as opposed to the 200 m eustatic lowering that occurred since mid-Eocene time. Neglecting this fall would introduce 200 m surface uplift of presumed tectonic origin into the modelling results.

Differences between southern Norway and the Shetland Platform

Differential preservation of upper Paleocene and Eocene sediments across the northern North Sea could be due to a number of factors such as: (1) different amounts of transient uplift caused by heating of the lithosphere beneath the Shetland Platform and southern Norway; (2) different amounts of permanent uplift caused by underplating or delamination or other non-transient mechanisms; (3) presence v. absence of extrusive volcanic rocks in Scotland v. Norway; (4) different (flexural) isostatic response to denudation; (5) palaeobathymetric variations; (6) width of the topography (full isostatic or flexural isostatic behaviour).

Plate tectonic reconstructions (Skogseid *et al.* 2000) indicate that the Shetland Platform was closer to the Iceland Plume at the time of rifting (*c.* 54 Ma) and thus experienced a larger thermal (transient) uplift than southern Norway. It should thus be expected that the depocentres adjacent to the Shetland Platform subsequently experienced greater thermal subsidence and thus had greater preservation potential than those at the same latitude off southern Norway.

Possible mechanisms responsible for non-transient uplift of southern Norway and other uplifted areas were discussed by Nielsen *et al.* (2002), who invoked delamination of the lithosphere as the uplift mechanism for southern Norway. Other areas, e.g. Scotland, may not have been affected to the same degree or may even have experienced underplating, leading to a different uplift and subsidence history.

Early Paleogene extrusive volcanic rocks are abundant in western Scotland, but absent in Norway. Extrusive volcanic rocks are readily erodible and volcanic activity contributes to a high local relief, both of which may have contributed to a larger sediment supply from Scotland in early Paleogene time (see Hall 1991).

The south Norwegian dome has a much larger areal extent than Scotland and the Shetlands (Fig. 1). It is possible that this could cause differences in the amount of (flexural) isostatic response to denudation (and deposition) between the two areas. The relatively large extent of the south Norwegian dome probably causes most of the loading and unloading of material across southern Norway to be isostatically compensated, except around the fringes of the dome, where flexural strength may have an effect. In the case of Scotland and Shetlands, the narrowness of the topography may cause a relatively large part of the loading and unloading of material across the platform to be supported by the flexural strength of the lithosphere. Hence, the amount of rock uplift as a result of denudation may be significantly less for Scotland and the Shetlands than for southern Norway. The effects of loading by deposition adjacent to the uplifted area would also have a greater effect on the Shetlands by depressing any isostatic uplift response to denudation.

Palaeogeographical reconstructions (Fig. 7a; Ziegler 1990, fig. 54) indicate that sediments deposited east of Scotland and the Shetlands were fed across a narrow shelf into relatively deep water whereas the sediments on the Norwegian side were deposited on a wide shelf. Such differences in palaeobathymetry could have caused the greater thickness and better preservation of depocentres observed east of Scotland and the Shetlands.

The 'self-perpetuating' model of uplift and denudation of southern Norway invoked here (Fig. 10) works only until maximum equilibrium of valley slopes is reached; then mountain tops will commence eroding and regional down-wearing will prevail (Ahnert 1984; Gilchrest *et al.* 1994). Because of their narrowness, the Shetlands might already have reached this stage in Eocene time whereas it has not yet been reached in the central parts of southern Norway (Fig. 9). This could explain why sediment supply from the northern part of the British Isles slowed down in Neogene time.

Conclusions

The Cenozoic evolution of the North Sea Basin has been summarized based on a compilation of detailed studies from the eastern North Sea Basin integrated with published regional studies.

The first pulses of coarse clastic sediment into the basin occurred in response to regional surface uplift coinciding with increased plume activity and the onset of North Atlantic rifting in late Paleocene and earliest Eocene time. The preserved upper Paleocene and Eocene sediments mainly originate from the west, whereas progradational siliciclastic wedges of similar age have been truncated towards the south Norwegian dome.

In early Oligocene time, sediment supply from Scotland and the Shetlands diminished, probably as a result of thermal subsidence and regional down-wearing of topography in Scotland and the Shetlands. The supply of coarse clastic material from southern Norway accelerated at the Eocene–Oligocene transition, and from Oligocene time onwards the central, eastern and southern North Sea Basin was filled by large deltas prograding in a clockwise fashion from the NE (Oligocene–early Miocene), east (late Miocene–Pliocene), and finally SSE during Pleistocene time (Figs 5 and 7).

It is argued here that seismic geometries can be used with confidence to infer uplift and denudation of areas where Cenozoic sediments are preserved. Such geometric uplift estimates have been compared with estimates based on compaction trends along a profile in the eastern Danish North Sea. All methods show similar trends of increasing uplift towards southern Norway, but the comparison indicates that compaction-based estimates may overestimate the amount of uplift by several hundred metres. Recent uplift estimates integrating chalk compaction and vitrinite reflectance data (Japsen & Bidstrup 1999) are lower than previous compaction-based estimates, but still *c.* 300 m too high compared with the geometric estimates. The geometric estimates are similar to estimates of overburial based on 3D basin modelling of a passively subsiding basin filled during falling sea level (S.B. Nielsen pers. comm. 2000), indicating that late Cenozoic tectonic events may not be required to explain the sedimentation and erosion patterns of the North Sea Basin.

A model describing the development of topography in southern Norway is developed (Fig. 10) based on the evidence provided by the Cenozoic succession of the North Sea Basin and the present-day topography of the adjacent land areas. The model involves late Paleocene–earliest Eocene uplift of a low-elevation peneplain developed during late Mesozoic and early Paleocene time in an area roughly coinciding with present-day southern Norway. The uplift episode was followed by *c.* 15 Ma of warm climatic conditions favouring a dense vegetation and low amounts of runoff. In combination with low local relief and high sea level, this caused relatively low rates of denudation during Eocene time. Subsequently, repeated episodes of climatic deterioration and eustatic fall caused increased rates of incision, most markedly at the Eocene–Oligocene transition, and in late Mid-Miocene and Plio-Pleistocene times. The development of full glacial conditions in southern Norway in Plio-Pleistocene time was probably instrumental for carving out the main parts of the deep valleys and fjords of southern Norway, i.e. dissecting the initial high-elevation peneplain and causing uplift of the adjacent mountain peaks to twice the initial elevation of the uplifted peneplain. This model is in agreement with the large-scale sedimentation patterns of the North Sea Basin, with geomorphological studies (Lidmar-Bergström *et al.* 2000) and with apatite fission-track modelling (Rohrman *et al.* 1995). It also accounts for the occurrence of thick Plio-Pleistocene wedges along the NW European Atlantic margin (Riis 1996; Evans *et al.* 2000).

The rise of the mountain peaks is facilitated by the regional isostatic response to dissection of an initially flat surface of some elevation (Molnar & England 1990; Gilchrest *et al.* 1994), which will cause an uplift of the rock column *c.* 0.8 times the amount of mean denudation. The amount of denudation averaged over the south Norwegian dome approximates 1 km (Riis & Fjeldskaar 1992). This would correspond to *c.* 200 m of mean surface lowering. However, the eustatic fall of *c.* 200 m since mid-Eocene time has maintained the mean surface elevation of central southern Norway at approximately the level of the uplifted peneplain.

To test the model advocated here and to further constrain the uplift and denudation history of southern Norway and other uplifted areas there is a need for an integrated approach, leading to mass-balanced palaeogeographical reconstructions. This requires integration of well-dated sediment volumes (mainly offshore) with detailed denudation estimates from both onshore and offshore regions. In particular, the onshore regions are in need of more detailed studies, e.g. based on apatite (U–Th)–He thermochronometry, to resolve the Cenozoic denudation history in sufficient detail. Ideally, such an integrated database would be fed into 3D geodynamic models incorporating isostatic and flexural effects of loading and unloading of the crust.

The contents of this paper derive from a PhD study carried out at the Department of Earth Sciences, University of Aarhus, co-sponsored by the Danish Natural Science Research Council (Grants 9401161 and 9502760) and the Faculty of Science, University of Aarhus. The paper was written while in receipt of funding by EFP-2000 (Project ENS-1313/00-0001). Discussions with N. Balling, O. R. Clausen, P. Japsen and S. B. Nielsen are greatly appreciated. I thank A. Hurst, G. E. Paulsen and the referees for comments, which helped improve the manuscript.

References

ABREU, V.S. & ANDERSON, J.B. 1998. Glacial eustasy during the Cenozoic: sequence stratigraphic implications. *AAPG Bulletin*, **82**, 1385–1400.

AHNERT, F. 1970. Functional relationships between denudation, relief, and uplift in large mid-latitude drainage basins. *American Journal of Science*, **268**, 243–263.

AHNERT, F. 1984. Local relief and the height limits of mountain ranges. *American Journal of Science*, **284**, 1035–1055.

BALLING, N. 1980. The land uplift in Fennoscandia, gravity field anomalies and isostasy. *In*: MÖRNER, N.A. (ed.) *Earth Rheology, Isostasy and Eustasy*. Wiley, Chichester, 297–321.

BARTEK, L.R., VAIL, P.R., ANDERSON, J.B., EMMET, P.A. & WU, S. 1991. Effect of Cenozoic ice sheet fluctuations in Antarctica on the stratigraphic signature of the Neogene. *Journal of Geophysical Research*, **96** (B4), 6753–6778.

BERGGREN, W.A., KENT, D.V., SWISHER, C.C. III & AUBRY, M.-P. III 1995. A revised Cenozoic geochronology and chronostratigraphy. *In*: BERGGREN, W.A., KENT, D.V., AUBRY, M.-P. & HARDENBOL, J. (eds) *Geochronology, Time Scales and Global Stratigraphic Correlation*. Society of Economic Paleontologists and Mineralogists, Special Publications, **54**, 129–212.

BIDSTRUP, T. 1995. *Seismic sequence stratigraphy of the Tertiary in the Danish North Sea; incl. 7 structural maps of the main sequences, and 35 isochore maps of main sequences and smaller units*. Report of the EFP-92 project: Basin development of the Tertiary of the Central Trough with emphasis on possible hydrocarbon reservoirs, **5**. Danish Energy Agency.

BIRD, P. 1979. Continental delamination and the Colorado Plateau. *Journal of Geophysical Research*, **B84**, 7561–7571.

BOLDREEL, L.O. & ANDERSEN, M.S. 1998. Tertiary compressional structures on the Faroe–Rockall Plateau in relation to northeast Atlantic ridge-push and Alpine foreland stresses. *Tectonophysics*, **300**, 13–28.

BRODIE, J. & WHITE, N. 1995. The link between sedimentary basin inversion and igneous underplating. *In*: BUCHANAN, J.G. & BUCHANAN, P.G. (eds) *Basin Inversion*. Geological Society, London, Special Publications, **88**, 21–38.

BROWN, R.W. 1991. Backstacking apatite fission-track 'stratigraphy': a method for resolving the erosional and isostatic rebound components of tectonic uplift histories. *Geology*, **19**, 74–77.

BUCHARDT, B. 1978. Oxygen isotope palaeotemperatures from the Tertiary period in the North Sea area. *Nature*, **275**, 121–123.

CAMERON, T.D.J., BULAT, J. & MESDAG, C.S. 1993. High resolution seismic profile through a Late Cenozoic delta complex in the southern North Sea. *Marine and Petroleum Geology*, **10**, 591–599.

CHALMERS, J.A. & CLOETINGH, S. (eds) (2000) *Neogene Uplift and Tectonics around the North Atlantic. Global and Planetary Change*, **24**, 165–318.

CLAUSEN, O.R., GREGERSEN, U., MICHELSEN, O. & SØRENSEN, J.C. 1999. Factors controlling the Cenozoic sequence development in the eastern parts of the North Sea. *Journal of the Geological Society, London*, **156**, 809–816.

CLAUSEN, O.R., NIELSEN, O.B., HUUSE, M. & MICHELSEN, O. 2000. Geological indications for Palaeogene uplift in the eastern North Sea Basin. *Global and Planetary Change*, **24**, 175–187.

CLOETINGH, S., GRADSTEIN, F.M., KOOI, H., GRANT, A.C. & KAMINSKI, M. 1990. Plate reorganization: a cause of rapid late Neogene subsidence and sedimentation around the North Atlantic? *Journal of the Geological Society, London*, **147**, 495–506.

CLOETINGH, S., REEMST, P., KOOI, H. & FANAVOLL, S. 1992. Intraplate stresses and the post-Cretaceous uplift and subsidence in northern Atlantic basins. *Norsk Geologisk Tidsskrift*, **72**, 229–235.

COLLINSON, M.E., FOWLER, K. & BOULTER, M.C. 1981. Floristic changes indicate a cooling climate in the Eocene of southern England. *Nature*, **291**, 315–317.

DANIELSEN, M., MICHELSEN, O. & CLAUSEN, O.R. 1997. Oligocene sequence stratigraphy and basin development in the Danish North Sea sector based on log interpretations. *Marine and Petroleum Geology*, **14**, 931–950.

DONNELLY, T. 1982. Worldwide continental denudation and climatic deterioration during the late Tertiary: evidence from deep-sea sediments. *Geology*, **10**, 451–454.

DORÉ, A.G. 1992. The Base Tertiary Surface of southern Norway and the northern North Sea. *Norsk Geologisk Tidsskrift*, **72**, 259–265.

DORÉ, A.G. & JENSEN, L.N. 1996. The impact of late Cenozoic uplift and erosion on hydrocarbon exploration: offshore Norway and some other uplifted basins. *Global and Planetary Change*, **12**, 415–436.

DORÉ, A.G. & LUNDIN, E.R. 1996. Cenozoic compressional structures on the NE Atlantic margin: nature, origin and potential significance for hydrocarbon exploration. *Petroleum Geoscience*, **2**, 299–311.

DORÉ, A.G., LUNDIN, E.R., BIRKELAND, Ø., ELIASSEN, P.E. & JENSEN, L.N. 1997. The NE Atlantic Margin: implications for late Mesozoic and Cenozoic events for hydrocarbon prospectivity. *Petroleum Geoscience*, **3**, 117–131.

DORÉ, A.G., LUNDIN, E.R., JENSEN, L.N., BIRKELAND, Ø., ELIASSEN, P.E. & FICHLER, C. 1999. Principal tectonic events in the evolution of the northwest European Atlantic margin. *In*: FLEET, A.J. & BOLDY, S.A.R. (eds) *Petroleum Geology of Northwest Europe: Proceedings of the 5th Conference*. Geological Society, London, 41–61.

ENGLAND, P. & MOLNAR, P. 1990. Surface uplift, uplift of rocks, and exhumation of rocks. *Geology*, **18**, 1173–1177.

EVANS, D., MCGIVERSON, MCNEIL, A.E., HARRISON, Z.H., ØSTMO, S.R. & WILD, J.B.L. 2000. Plio-Pleistocene deposits on the mid-Norwegian margin and their implications for late Cenozoic uplift of the Norwegian mainland. *Global and Planetary Change*, **24**, 233–237.

EYLES, N. 1996. Passive margin uplift around the North Atlantic region and its role in Northern Hemisphere late Cenozoic glaciation. *Geology*, **24**, 103–106.

FENNER, J. 1988. Occurrences of pre-Quaternary diatoms in Scandinavia reconsidered. *Meyniana*, **40**, 133–141.

FUNNELL, B.M. 1996. Plio-Pleistocene palaeogeography of the southern North Sea Basin (3.75–0.60 Ma). *Quaternary Science Reviews*, **15**, 391–405.

GALLOWAY, W.E., GARBER, J.L., LIU, X. & SLOAN, B.J. 1993. Sequence stratigraphic and depositional framework of the Cenozoic fill, Central and Northern North Sea Basin. *In*: PARKER, J.R. (eds) *Petroleum Geology of Northwest Europe: Proceedings of the 4th Conference*. Geological Society, London, 33–43.

GEMMER, L., HUUSE, M., CLAUSEN, O.R. & NIELSEN, S.B. 2002. Mid-Paleocene palaeogeography of the eastern North Sea Basin: integrating geological evidence and 3D geodynamic modelling. *Basin Research*, **14**(3), in press.

GIBBARD, P.L. 1988. The history of the great northwest European rivers during the past three million years. *Philosophical Transactions of the Royal Society of London, Series B*, **318**, 559–602.

GILCHREST, A.R., SUMMERFIELD, M.A. & COCKBURN, H.A.P. 1994. Landscape dissection, isostatic uplift, and the morphologic development of orogens. *Geology*, **22**, 963–966.

GJESSING, J. 1967. Norway's paleic surface. *Norges Geografiske Tidsskrift*, **21**, 69–132.

GRAMANN, F. & KOCKEL, F. 1988. Palaeogeographical, lithological, palaeoecological and palaeoclimatic development of the Northwest European Tertiary Basin. *In*: VINKEN, R. (ed.) *The Northwest European Tertiary Basin*. Geologisches Jahrbuch, **A100**, 428–441.

GUNNELL, Y. 2000. Apatite fission track thermochronology: an overview of its potential and limitations in geomorphology. *Basin Research*, **12**, 115–132.

HALL, A.M. 1991. Pre-Quaternary landscape evolution in the Scottish Highlands. *Transactions of the Royal Society of Edinburgh: Earth Sciences*, **82**, 1–26.

HANCOCK, J.M. 1975. The petrology of the Chalk. *Proceedings of the Geologists' Association*, **86**, 499–535.

HANSEN, S. 1996. Quantification of net uplift and erosion on the Norwegian Shelf south of 66°N from sonic transit times of shale. *Norsk Geologisk Tidsskrift*, **76**, 245–252.

HAQ, B.U., HARDENBOL, J. & VAIL, P. 1987. Chronology of fluctuating sea levels since the Triassic. *Science*, **235**, 1156–1167.

HAY, W.W., SHAW, C.A. & WOLD, C.N. 1989. Mass-balanced paleogeographic reconstructions. *Geologische Rundschau*, **78**, 207–242.

HUUSE, M. 1999. *Cenozoic evolution of the eastern North Sea Basin—new evidence from high-resolution and conventional seismic data*. PhD thesis, University of Aarhus.

HUUSE, M. 2002. Late Cenozoic palaeogeography of the eastern North Sea Basin: climatic vs. tectonic forcing of basin margin uplift and deltaic progradation. *Bulletin of the Geological Society of Denmark*, **49**, in press.

HUUSE, M. & CLAUSEN, O.R. 2001. Morphology and origin of major Cenozoic sequence boundaries in the eastern Danish North Sea Basin: top Eocene, near top Oligocene and the mid-Miocene unconformity. *Basin Research*, **13**, 17–41.

IVANY, L.C., PATTERSON, W.P. & LOHMANN, K.C. 2000. Cooler winters as a possible cause of mass extinctions at the Eocene/Oligocene boundary. *Nature*, **407**, 887–890.

JAPSEN, P. 1998. Regional velocity–depth anomalies, North Sea Chalk: a record of overpressure and Neogene uplift and erosion. *AAPG Bulletin*, **82**, 2031–2074.

JAPSEN, P. & BIDSTRUP, T. 1999. Quantification of late Cenozoic erosion in Denmark based on sonic data and basin modelling. *Bulletin of the Geological Society of Denmark*, **46**, 79–99.

JAPSEN, P. & CHALMERS, J.A. 2000. Neogene uplift and tectonics around the North Atlantic: overview. *Global & Planetary Change*, **24**, 165–173.

JENSEN, L.N. & MICHELSEN, O. 1992. Tertiææ̈r hævning og erosion I Skagerrak, Nordjylland og Kattegat. *Dansk Geologisk Forening, Årsskrift*, **1990–1991**, 159–168.

JENSEN, L.N. & SCHMIDT, B.J. 1993. Neogene uplift and erosion offshore south Norway: magnitude and consequences for hydrocarbon exploration in the Farsund Basin. *In*: SPENCER, A.M. (ed.) *Generation, Accumulation and Production of Europe's*

Hydrocarbons III. European Association of Petroleum Geoscientists Special Publication, **3**, 79–88.

JORDT, H., FALEIDE, J.I., BJØRLYKKE, K. & IBRAHIM, M.T. 1995. Cenozoic sequence stratigraphy of the central and northern North Sea Basin: tectonic development, sediment distribution and provenance areas. *Marine and Petroleum Geology*, **12**, 845–879.

JORDT, H., THYBERG, B.I. & NØTTVEDT, A. 2000. Cenozoic evolution of the central and northern North Sea with focus on differential movements of the basin floor and surrounding clastic source areas. *In*: NØTTVEDT, A. (ed.) *Dynamics of the Norwegian Margin*. Geological Society, London, Special Publications, **167**, 219–243.

JOY, A.M. 1992. Right place, wrong time: anomalous post-rift subsidence in sedimentary basins around the North Atlantic Ocean. *In*: STOREY, B.C., ALABASTER, T. & PANKHURST, R.J. (eds) *Magmatism and the Causes of Continental Break-up*. Geological Society, London, Special Publications, **68**, 387–393.

JOY, A.M. 1996. Controls on Eocene sedimentation in the central North Sea Basin: results of a basinwide correlation study. *In*: KNOX, R.W.O'B., CORFIELD, R.M. & DUNAY, R.E. (eds) *Correlation of the Early Paleogene in Northwest Europe*. Geological Society, London, Special Publications, **101**, 79–90.

JÜRGENS, U. 1996. Mittelmiozäne bis pliozäne Randmeer-Sequenzen aus dem deutschen Sektor der Nordsee. *Geologisches Jahrbuch*, **A146**, 217–232.

KNOTT, S.D., BURCHELL, M.T., JOLLEY, E.J. & FRASER, A.J. 1993. Mesozoic to Cenozoic plate reconstructions of the North Atlantic and hydrocarbon plays of the Atlantic margins. *In*: PARKER, J.R. (eds) *Petroleum Geology of Northwest Europe: Proceedings of the 4th Conference*. Geological Society, London, 953–974.

KOCH, B.E. 1989. *Geology of the Søby–Fasterholt area*. Geological Survey of Denmark, **A22**.

KOCKEL, F. 1988. The palaeogeographical maps. *In*: VINKEN, R. (ed.) *The Northwest European Tertiary Basin*. Geologisches Jahrbuch, **A100**, 423–427.

KOMINZ, M.A. 1984. Oceanic ridge volumes and sea-level change—an error analysis. *In*: SCHLEE, J.S. (ed.) *Interregional Unconformities and Hydrocarbon Accumulation*. American Association of Petroleum Geologists, Memoir, **36**, 109–127.

LAWVER, L.A. & MÜLLER, R.D. 1994. Iceland hot spot track. *Geology*, **22**, 311–314.

LEAR, C.H., ELDERFIELD, H. & WILSON, P.A. 2000. Cenozoic deep-sea temperatures and global ice volumes from Mg/Ca in benthic foraminiferal calcite. *Science*, **287**, 269–272.

LIDMAR-BERGSTRÖM, K. 1999. Uplift histories revealed by landforms of the Scandinavian domes. *In*: SMITH, B.J., WHALLEY, W.B. & WARKE, P.A. (eds) *Uplift, Erosion and Stability: Perspectives on Long-term Landscape Development*. Geological Society, London, Special Publications, **162**, 85–91.

LIDMAR-BERGSTRÖM, K., OLLIER, C.D. & SULEBAK, J.R. 2000. Landforms and uplift history of southern Norway. *Global and Planetary Change*, **24**, 211–231.

LIU, X. & GALLOWAY, W.E. 1997. Quantitative determination of Tertiary sediment supply to the North Sea Basin. *AAPG Bulletin*, **81**, 1482–1509.

MARTINI, E. 1971. Standard Tertiary and Quaternary calcareous nannoplankton zonation. In: *Proceedings of the II Planktonic Conference, Roma*, 739–785.

MATTHEWS, R.K. 1984. Oxygen isotope record of ice-volume history: 100 million years of glacio-eustatic sea-level. *In*: SCHLEE, J.S. (ed.) *Interregional Unconformities and Hydrocarbon Accumulation*. American Association of Petroleum Geologists, Memoir, **36**, 97–107.

MCKENZIE, D. 1978. Some remarks on the development of sedimentary basins. *Earth and Planetary Science Letters*, **40**, 25–32.

MCKENZIE, D. 1984. A possible mechanism for epeirogenic uplift. *Nature*, **307**, 616–618.

MICHELSEN, O. 1996. Late Cenozoic basin development of the eastern North Sea Basin. *Bulletin of the Geological Society of Denmark*, **43**, 9–21.

MICHELSEN, O., DANIELSEN, M., HEILMANN-CLAUSEN, C., JORDT, H., LAURSEN, G.V. & THOMSEN, E. 1995. Occurrence of major sequence stratigraphic boundaries in relation to basin development in Cenozoic deposits of the southeastern North Sea. *In*: STEEL, R.J., FELT, V.L., JOHANNESSEN, E.P. & MATHIEU, C. (eds) *Sequence Stratigraphy of the Northwest European Margin*. Norwegian Petroleum Society Special Publication, **5**, 415–427.

MICHELSEN, O., THOMSEN, E., DANIELSEN, M., HEILMANN-CLAUSEN, C., JORDT, H. & LAURSEN, G.V. 1998. Cenozoic sequence stratigraphy in the eastern North Sea. *In*: HARDENBOL, J., DE GRACIANSKY, P.C., JACQUIN, T. & VAIL, P.R. (eds) *Mesozoic–Cenozoic Sequence Stratigraphy of Western European Basins*. Society of Economic Paleontologist and Mineralogists, Special Publication, **60**, 91–118.

MILLER, K.G., FAIRBANKS, R.G. & MOUNTAIN, G.S. 1987. Tertiary oxygen isotope synthesis, sea level history, and continental margin erosion. *Paleoceanography*, **2**, 1–19.

MILLER, K.G., LIU, C. & FEIGENSON, M.D. 1996. Oligocene to middle Miocene Sr-isotopic stratigraphy of the New Jersey continental slope. *In*: MOUNTAIN, G.S., MILLER, K.G., BLUM, P., POAG, C.W. & TWITCHELL, D.C. (eds) *Proceedings of the Ocean Drilling Program, Scientific Results, 150*. Ocean Drilling Program, College Station, TX, 97–114.

MILLER, K.G., MOUNTAIN, G.S., BROWNING, J.V., KOMINZ, M., SUGARMAN, P.J., CHRISTIE-BLICK, N., KATZ, M.E. & WRIGHT, J.D. 1998. Cenozoic global sea level, sequences, and the New Jersey Transect: results from coastal plain and continental slope drilling. *Reviews of Geophysics*, **36**, 569–601.

MILLS, H.H. 2000. Apparent increasing rates of stream incision in the eastern United States during the late Cenozoic. *Geology*, **28**, 955–957.

MOLNAR, P. & ENGLAND, P. 1990. Late Cenozoic uplift of mountain ranges: chicken or egg? *Nature*, **346**, 29–34.

MOLNAR, P., ENGLAND, P. & MATINOD, J. 1993. Mantle dynamics, uplift of the Tibetan Plateau, and the Indian Monsoon. *Reviews of Geophysics*, **31**, 357–396.

MUDGE, D.C. & BUJAK, J.P. 1996. An integrated stratigraphy for the Paleocene and Eocene of the North Sea. *In*: KNOX, R.W.O'B., CORFIELD, R.M. & DUNAY, R.E. (eds) *Correlation of the Early Paleogene in Northwest Europe*. Geological Society, London, Special Publications, **101**, 91–113.

NADIN, P.A., KUZNIR, N.J. & CHEADLE, M.J. 1997. Early Tertiary plume uplift of the North Sea and Faroe–Shetland Basins. *Earth and Planetary Science Letters*, **148**, 109–127.

NIELSEN, O.B., SØRENSEN, S., THIEDE, J. & SKARBØ, O. 1986. Cenozoic differential subsidence of North Sea. *AAPG Bulletin*, **70**, 276–298.

NIELSEN, S. B., PAULSEN, G. E., HANSEN, D. L. *et al.* 2002. Paleocene initiation of Cenozoic uplift in Norway. *In*: DORÉ, A.G., CARTWRIGHT, J.A., STOKER, M.S., TURNER, J.P. & WHITE, N. (eds) *Exhumation of North Atlantic Margins: Timing, Mechanisms and Implications for Petroleum Exploration*. Geological Society, London, Special Publications, **196**, 45–65.

NØTTVEDT, A., *et al.* 2000. Integrated basin studies—dynamics of the Norwegian Margin: an introduction. *In*: NØTTVEDT, A. (ed.) *Dynamics of the Norwegian Margin*. Geological Society, London, Special Publications, **167**, 1–14.

PEKAR, S. & MILLER, K.G. 1996. New Jersey Oligocene 'Icehouse' sequences (ODP Leg 150X) correlated with global $\delta^{18}O$ and Exxon eustatic records. *Geology*, **24**, 567–570.

PITMAN, W.C. III 1978. Relationship between eustasy and stratigraphic sequences of passive margins. *Geological Society of America Bulletin*, **89**, 1389–1403.

RIIS, F. 1996. Quantification of Cenozoic vertical movements of Scandinavia by correlation of morphological surfaces with offshore data. *Global and Planetary Change*, **12**, 331–357.

RIIS, F. & FJELDSKAAR, W. 1992. On the magnitude of the Late Tertiary and Quaternary erosion and its significance for the uplift of Scandinavia and the Barents Sea. *In*: LARSEN, R.M., BREKKE, H., LARSEN, B.T. & TALLERAAS, E. (eds) *Structural and Tectonic Modelling and its Application to Petroleum Geology*. Norwegian Petroleum Society Special Publication, **1**, 163–185.

ROHRMAN, M. & VAN DER BEEK, P. 1996. Cenozoic postrift domal uplift of North Atlantic margins: an asthenospheric diapirism model. *Geology*, **24**, 901–904.

ROHRMAN, M., VAN DER BEEK, P., ANDRIESSEN, P. & CLOETINGH, S. 1995. Meso-Cenozoic morphotectonic evolution of southern Norway: Neogene domal uplift inferred from apatite fission track thermochronology. *Tectonics*, **14**, 704–718.

SALES, J.K. 1992. Uplift and subsidence in northwestern Europe: causes and influence on hydrocarbon entrapment. *Norsk Geologisk Tidsskrift*, **72**, 253–258.

SAVIN, S. 1977. The history of the Earth's surface temperature during the past 100 million years. *Annual Review of Earth and Planetary Sciences*, **5**, 319–355.

SÉRANNE, M. 1999. Early Oligocene stratigraphic turnover on the west Africa continental margin: a signature of the Tertiary greenhouse-to-icehouse transition? *Terra Nova*, **11**, 135–140.

SKOGSEID, J., PLANKE, S., FALEIDE, J.I., PEDERSEN, T., ELDHOLM, O. & NEVERDAL, F. 2000. NE Atlantic continental rifting and volcanic margin formation. *In*: NØTTVEDT, A. (ed.) *Dynamics of the Norwegian Margin*. Geological Society, London, Special Publications, **167**, 295–326.

SOLHEIM, A., RIIS, F., ELVERHØI, A., FALEIDE, J.I., JENSEN, L.N. & CLOETINGH, S. (eds) 1996. *Impact of Glaciations on Basin Evolution: Data and Models from the Norwegian Margin and Adjacent Areas*. Global and Planetary Change, **12**, 1–450.

SPJELDNÆS, N. 1975. Palaeogeography and facies distribution in the Tertiary of Denmark and surrounding areas. *Geological Survey of Norway Bulletin*, **316**, 289–311.

STECKLER, M.S., MOUNTAIN, G.S., MILLER, K.G. & CHRISTIE-BLICK, N. 1999. Reconstruction of Tertiary progradation and clinoform development on the New Jersey passive margin by 2-D backstripping. *Marine Geology*, **154**, 399–420.

STUEVOLD, L.M. & ELDHOLM, O. 1996. Cenozoic uplift of Fennoscandia inferred from a study of the mid-Norwegian margin. *Global and Planetary Change*, **12**, 359–386.

SUMMERFIELD, M.A. 1991. Sub-aerial denudation of passive margins: regional elevation versus local relief models. *Earth and Planetary Science Letters*, **102**, 460–469.

SUMMERFIELD, M.A. & BROWN, R.W. 1998. Geomorphic factors in the interpretation of fission-track data. *In*: VAN DEN HAUTE, P. & DE CORTE, F. (eds) *Advances in Fission-Track Geochronology*. Kluwer Academic, Dordrecht, 269–284.

SØRENSEN, J.C., GREGERSEN, U., BREINER, M. & MICHELSEN, O. 1997. High-frequency sequence stratigraphy of Upper Cenozoic deposits in the central and southeastern North Sea areas. *Marine and Petroleum Geology*, **14**, 99–123.

THYBERG, B.I., JORDT, H., BJØRLYKKE, K. & FALEIDE, J.I. 2000. Relationships between sequence stratigraphy, mineralogy and geochemistry in Cenozoic sediments of the northern North Sea. *In*: NØTTVEDT, A. (ed.) *Dynamics of the Norwegian Margin*. Geological Society, London, Special Publications, **167**, 245–272.

TORSKE, T. 1972. Tertiary oblique uplift of western Fennoscandia; crustal warping in connection with rifting and break-up of the Laurasian continent. *Norges Geologiske Undersøkelse*, **273**, 43–48.

TYNNI, R. 1982. *The reflection of geological evolution in Tertiary and interglacial diatoms and silicoflagellates in Finnish Lapland*. Geological Survey of Finland **320**.

VAIL, P.R., MITCHUM, R.M. JR & THOMPSON, S. III 1977. Seismic stratigraphy and global changes of sea level, Part 3: Relative changes of sea level from coastal onlap. *In*: PAYTON, C.E. (ed.) *Seismic Stratigraphy—Applications to Hydrocarbon Exploration*. American Association of Petroleum Geologists, Memoir, **26**, 63–81.

VAIL, P.R., AUDEMARD, F., BOWMAN, S.A., EISNER, P.N. & PEREZ-CRUZ, C. 1991. The stratigraphic signatures of tectonics, eustacy and sedimentology—an overview. *In*: EINSELE, G., RICKEN, W. & SEILACHER, A. (eds) *Cycles and Events in Stratigraphy*. Springer, Berlin, 617–659.

VEJBÆK, O.V. & ANDERSEN, C. 1987. Cretaceous–Early Tertiary inversion tectonism in the Danish Central Trough. *Tectonophysics*, **137**, 221–238.

VEJBÆK, O.V. & ANDERSEN, C. 2002. Post mid-Cretaceous inversion tectonics in the Danish Central Graben. *Bulletin of the Geological Society of Denmark*, **49**, in press.

VINKEN, R. (ed.) *The Northwest European Tertiary Basin*. Geologisches Jahrbuch, **A100**.

WHITE, N. & LOVELL, B. 1997. Measuring the pulse of a plume with the sedimentary record. *Nature*, **387**, 888–891.

WHITE, R.S. 1992. Magmatism during and after continental break-up. *In*: STOREY, B.C., ALABASTER, T. & PANKHURST, R.J. (eds) *Magmatism and the Causes of Continental Break-up*. Geological Society, London, Special Publications, **68**, 1–16.

WOLFE, J.A. 1978. A paleobotanical interpretation of Tertiary climates in the northern hemisphere. *American Scientist*, **66**, 694–703.

ZACHOS, J.C., BREZA, J.R. & WISE, S.W. 1992. Early Oligocene ice-sheet expansion on Antarctica: stable isotope and sedimentological evidence from Kerguelen Plateau, southern Indian Ocean. *Geology*, **20**, 569–573.

ZAGWIJN, W.H. 1989. The Netherlands during the Tertiary and Quaternary: a case history of Coastal Lowland evolution. *Geologie en Mijnbouw*, **68**, 107–120.

ZIEGLER, P.A. 1990. *Geological Atlas of Western and Central Europe*. Shell Internationale Petroleum Maatschappij BV, The Hague.

Tectonic impact on sedimentary processes during Cenozoic evolution of the northern North Sea and surrounding areas

JAN INGE FALEIDE[1], RUNE KYRKJEBØ[2], TOMAS KJENNERUD[3], ROY H. GABRIELSEN[2], HENRIK JORDT[1,4], STEIN FANAVOLL[3,5] & MORTEN D. BJERKE[1,6]

[1]*Department of Geology, University of Oslo, P.O. Box 1047, Blindern, N-0316 Oslo, Norway (e-mail: j.i.faleide@geologi.uio.no)*

[2]*Geological Institute, University of Bergen, Allègaten 41, N-5007 Bergen, Norway*

[3]*SINTEF Petroleum Research, N-7465 Trondheim, Norway*

[4]*Present address: Aarhus Amt, Lyseng Alle 1, DK-8270, Højbjerg, Denmark*

[5]*Present address: IPRES International Ltd., Nedre Vollgt. 4, N-0158 Oslo, Norway*

[6]*Present address: PGS, Strandveien 4, N-1366 Lysaker, Norway*

Abstract: This paper focuses on the Cenozoic evolution of the northern North Sea and surrounding areas, with emphasis on sediment distribution, composition and provenance, as well as on timing, amplitude and wavelength of differential vertical movements. Quantitative information about palaeo-water depth and tectonic vertical movements has been integrated with a seismic stratigraphic framework to better constrain the Cenozoic evolution. The data and modelling results support a probable tectonic control on sediment supply and on the formation of regional unconformities. The sedimentary architecture and breaks are related to tectonic uplift of surrounding clastic source areas, thus the offshore sedimentary record provides the best age constraints on Cenozoic exhumation of the adjacent onshore areas. Tectonic subsidence accelerated in Paleocene time throughout the basin, with uplifted areas to the east and west sourcing prograding wedges, which resulted in large depocentres close to the basin margins. Subsidence rates outpaced sedimentation rates along the basin axis, and water depths in excess of 600 m are indicated. In Eocene times progradation from the East Shetland Platform was dominant and major depocentres were constructed in the Viking Graben area, with deep water along the basin axis. At the Eocene–Oligocene transition, southern Norway and the eastern basin flank became uplifted. The uplift, in combination with prograding units from both the east and west, gave rise to a shallow threshold in the northern North Sea, separating deeper waters to the south and north. The uplift and shallowing continued into Miocene time when a widespread hiatus formed in the northern North Sea, as indicated by biostratigraphic data. The Pliocene basin configuration was dominated by outbuilding of thick clastic wedges from the east and south. Considerable late Cenozoic uplift of the eastern basin flank is documented by the strong angular relationship and tilting of the complete Tertiary package below the Pleistocene unconformity. Cenozoic exhumation is documented on both sides of the North Sea, but the timing is not well constrained. Two major uplift phases in early Paleogene and late Neogene times are related to rifting, magmatism and break-up in the NE Atlantic and isostatic response to glacial erosion, respectively. Additional uplift events may be related to mantle processes and the episodic behaviour of the Iceland plume.

The northern North Sea rift basin (Fig. 1) has been affected by two major episodes of rifting since Devonian time. The two events took place in Permian to earliest Triassic and late Mid-Jurassic to earliest Cretaceous times, each followed by periods of post-rift thermal relaxation and subsidence (Badley *et al.* 1988; Ziegler & van Hoorn 1989; Gabrielsen *et al.* 1990; Roberts *et al.* 1990; Ziegler 1990, 1992; Yielding *et al.* 1992; Milton 1993; Rattey & Hayward 1993; Nøttvedt *et al.* 1995; Færseth 1996; Færseth *et al.* 1997). Most of the post-rift tectonic subsidence related to the Late Jurassic rift phase had ceased at the end of Cretaceous

From: DORÉ, A.G., CARTWRIGHT, J.A., STOKER, M.S., TURNER, J.P. & WHITE, N. 2002. *Exhumation of the North Atlantic Margin: Timing, Mechanisms and Implications for Petroleum Exploration*. Geological Society, London, Special Publications, **196**, 235–269. 0305-8719/02/$15.00

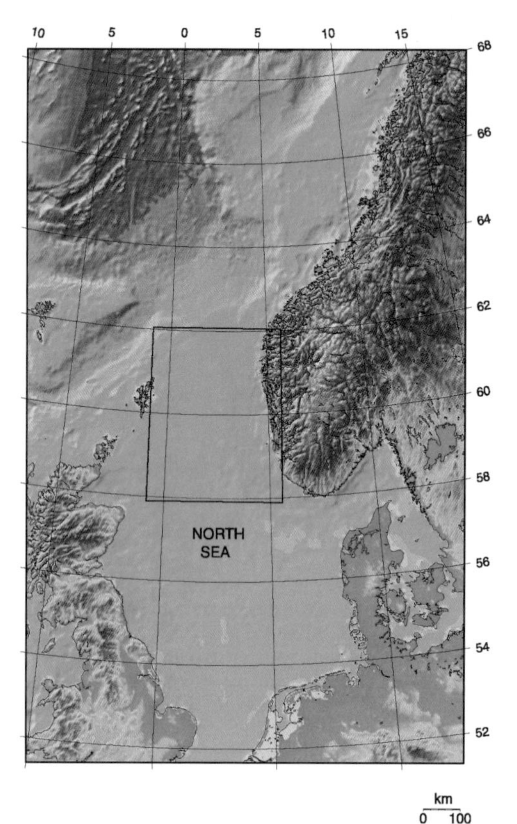

Fig. 1. Regional setting and location of study area in the northern North Sea.

time and the basin had become a wide area of deposition with low relief. Hence it is suggested that thermal equilibrium was reached in latest Cretaceous time (Gabrielsen *et al.* 2001).

The Cenozoic northern North Sea basin forms a wide sagged depocentre containing up to *c.* 2500 m of sediments. Shifts in depocentre locations, outbuilding directions and sediment composition have been related to differential vertical movements of the basin floor and surrounding clastic source areas (Nielsen *et al.* 1986; Rundberg 1989; Rundberg & Smalley 1989; Galloway *et al.* 1993; Jordt *et al.* 1995, 2000; Michelsen *et al.* 1995, 1998; Clausen *et al.* 1999b; Martinsen *et al.* 1999; Thyberg *et al.* 2000). Basin modelling (both backward and forward) has been applied to quantify the syn- and post-rift evolution of the northern North Sea basin (e.g. Joy 1992, 1993; Roberts *et al.* 1993; White & Latin 1993; Hall & White 1994; Nadin & Kusznir 1995, 1996). However, and as pointed out by these workers, the input parameters, and especially the assessed palaeo-water depth, are associated with uncertainty and may have resulted in erroneous subsidence histories.

This paper focuses on the Cenozoic evolution of the northern North Sea and surrounding areas, with emphasis on the timing, amplitude and wavelength of differential vertical movements. Quantitative information about palaeo-water depth and tectonic vertical movements has been integrated with a seismic stratigraphic framework to better constrain the Cenozoic evolution. We will also briefly discuss the main mechanisms responsible for the tectonic evolution of the area.

The main study area of the northern North Sea is restricted to 1°W–5°E and 58–62°N (Figs 1 and 2). However, when discussing the Cenozoic sediment distribution and provenance we have to include the surrounding land areas (Fig. 1). Cenozoic exhumation is documented on both sides of the North Sea, mainly from geomorphological and apatite fission-track studies (Green 1986, 1989; Bray *et al.* 1992; Lewis *et al.* 1992; Holliday 1993; Rohrman *et al.* 1995; Riis 1996; Lidmar-Bergstrom *et al.* 2000. At least two significant episodes of Cenozoic exhumation have been suggested, but the timing is not well constrained and in many places it is difficult to separate the two (Japsen 1997; Japsen & Chalmers 2000). The Cenozoic development of the study area is also linked to the plate tectonic evolution of the North Atlantic (Talwani & Eldholm 1977; Eldholm *et al.* 1990; Doré *et al.* 1999).

The work has been carried out as part of the project 'Tectonic impact on sedimentary processes in the post-rift phase—improved models', which focused on the Cretaceous–Cenozoic succession filling in the structural relief resulting from Late Jurassic–earliest Cretaceous rifting in the northern North Sea (Fig. 3). This paper summarizes the Cenozoic part of this study, and Gabrielsen *et al.* (2001) have given a summary of the Cretaceous post-rift development in the northern North Sea.

Seismic mapping

In this study we have interpreted a grid of regional high-quality seismic reflection profiles tied to key wells. The main database comprised eight regional seismic reflection surveys, providing the best data coverage in the Norwegian part of the northern North Sea. In addition, four regional deep seismic reflection lines were used. Regional crustal transects (Fig. 2) were constructed along these lines by combining them with conventional seismic lines and gravity and magnetic data. The four transects were depth-converted using velocity information from wells,

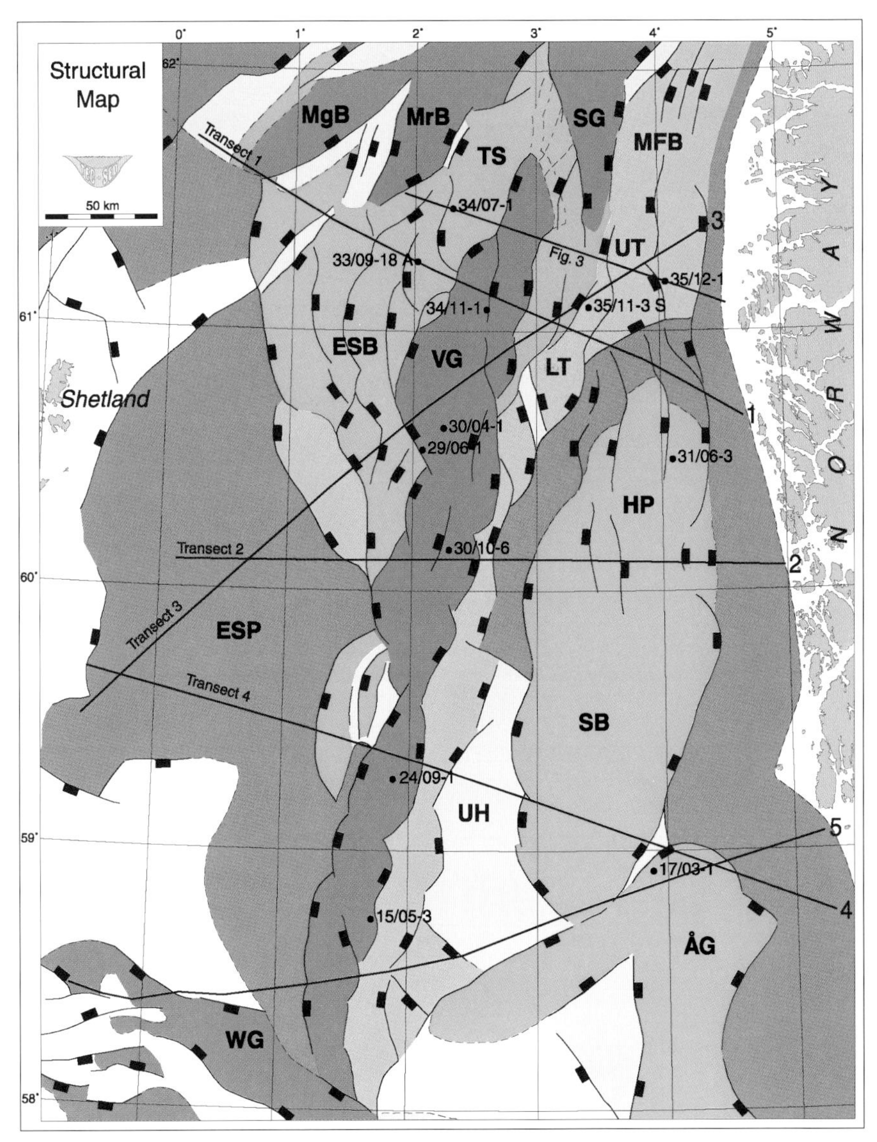

Fig. 2. Structural map of study area in the northern North Sea and location of regional transects and key wells. ESB, East Shetland Basin; ESP, East Shetland Platform; HP, Horda Platform; LT, Lomre Terrace; MFB, Måløy Fault Blocks; MgB, Magnus Basin; MrB, Marulk Basin; SB, Stord Basin; SG, Sogn Graben; TS, Tampen Spur; UH, Utsira High; UT, Uer Terrace; VG, Viking Graben; WG, Witchground Graben;ÅG,Åsta Graben.

interval velocities from stacking velocities and velocities from deep seismic refraction data. The transects were used in the modelling.

Twelve wells (Fig. 2) were selected to be principal sources for analysis on the basis of their location at or close to the regional transects and as representatives of different structural positions. The key wells were analysed to obtain information about age and lithological composition of the main Cenozoic sequences, to recognize stratigraphical breaks, and to estimate palaeo-water depths. In addition, biostratigraphic data and well logs from more than 60 wells were used to calibrate the interpretation of the seismic data.

The Cenozoic seismic stratigraphic framework is based on the work byJordt *et al.* (1995, 2000). The Cenozoic succession is subdivided into 10 seismic sequences (CSS-1 to CSS-10) (Fig. 4). The sequences have been dated using biostrati-

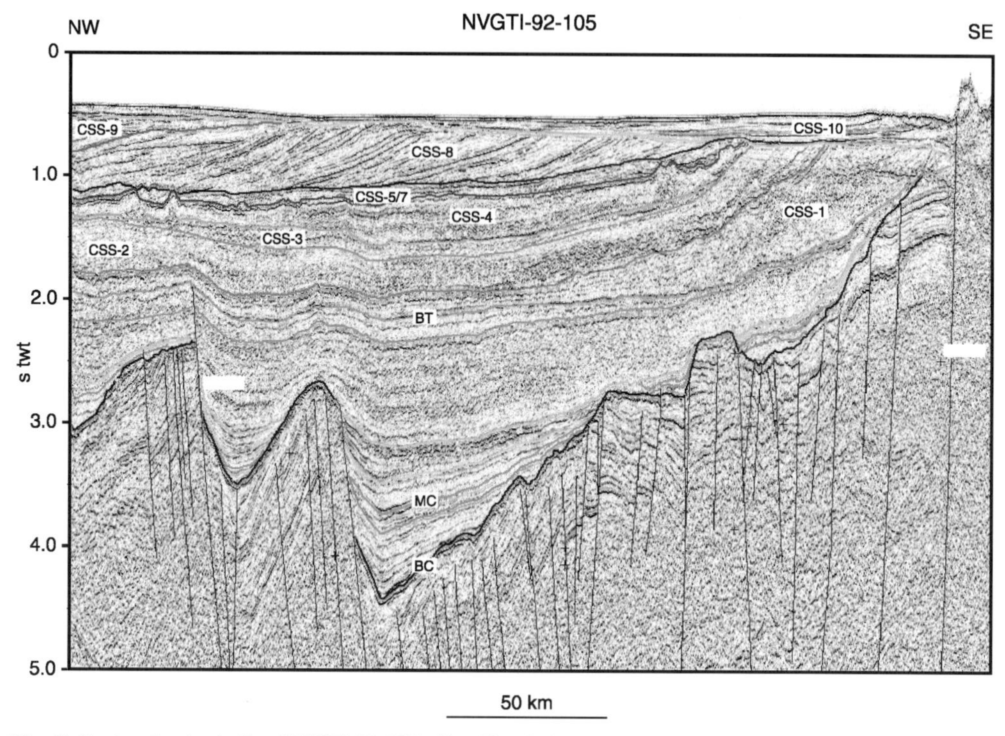

Fig. 3. Regional seismic line NVGTI-92-105. (See Fig. 2 for line location and Fig. 4 for Cenozoic seismic stratigraphy.) BT, Base Tertiary; MC, mid-Cretaceous; BC, Base Cretaceous; twt, two-way travel time.

Ma	Period	Epoch	Group	Formation	Sequence
0	Quaternary	Pleistocene	Nordland		CSS-9, CSS-10
	TERTIARY	Pliocene	Nordland		CSS-8
10	TERTIARY	Miocene	Nordland	Utsira	CSS-7
20	TERTIARY	Miocene	Hordaland	Skade	CSS-6, CSS-5
30	TERTIARY	Oligocene	Hordaland		CSS-4, CSS-3
40–50	TERTIARY	Eocene	Hordaland	Grid, (2.2), Frigg (2.1)	CSS-2
60	TERTIARY	Paleocene	Rogaland	Balder (1.2), Sele, Lista, Våle	CSS-1

Fig. 4. Cenozoic seismic stratigraphic framework, based on Jordt *et al.* (1995, 2000).

graphic data from key wells published by Steurbaut *et al.* (1991), Eidvin & Riis (1992), van Veen *et al.* (1994), Gradstein & Bäckström (1996), Martinsen *et al.* (1999) and Eidvin *et al.* (1999, 2000), in addition to the 12 key wells of the TecSed project (Fig. 2). The seismic stratigraphic framework was related to the time scale of Gradstein & Ogg (1996) and formal lithostratigraphy of the northern North Sea (Isaksen & Tonstad 1989; Knox & Holloway 1992).

The Paleogene succession comprises four seismic sequences (CSS-1, CSS-2, CSS-3 and CSS-4). The CSS-1 sequence is of Late Paleocene–earliest Eocene age and its top corresponds to the top of the Balder tuffs. The CSS-2 sequence is of Eocene age. The seismic boundary at the top of CSS-2 correlates with the Eocene–Oligocene transition, which is associated with a hiatus, in particular along the basin flanks. The CSS-3 sequence covers a narrow period in Early Oligocene time. An upward change in seismic signature from progradation to marked aggradation and onlap occurs against the top of CSS-3 in the northern North Sea. A mid-Early Oligocene age is inferred for this sequence boundary. The top of sequence CSS-4 is of latest Oligocene age.

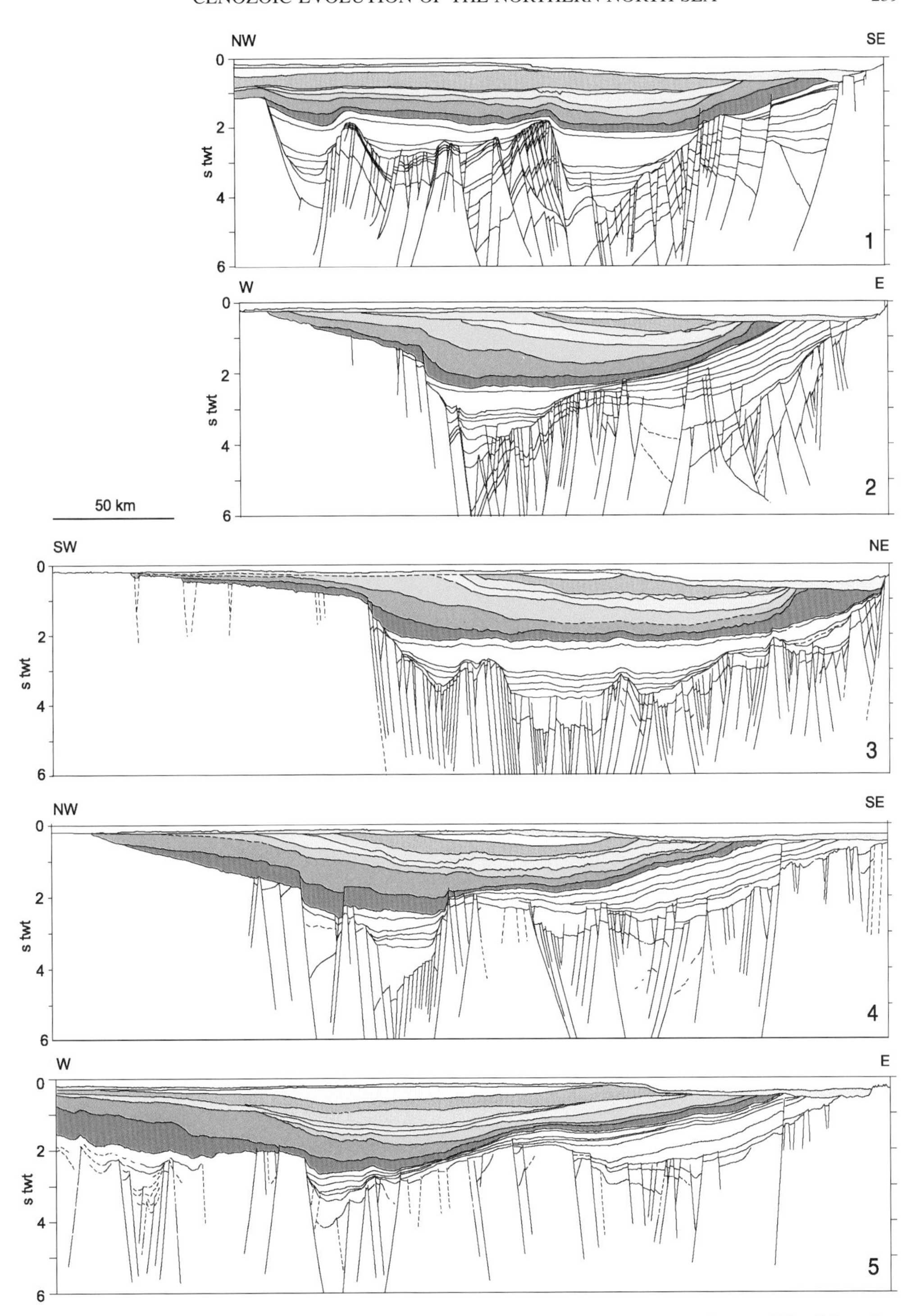

Fig. 5. Regional profiles 1–5 across the northern North Sea. (See Fig. 2 for profile locations and Fig. 4 for colour codes of the Cenozoic seismic sequences CSS-1 to CSS-10.)

The Miocene succession is divided into three seismic sequences (CSS-5, CSS-6 and CSS-7), and the boundaries separating these are dated in the central North Sea (Jordt *et al.* 1995; Michelsen *et al.* 1995). Miocene sediments are also present in the northern North Sea, but it is difficult to make this subdivision here because some of the sequences thin below seismic resolution. Most workers agree on a prominent mid-Miocene hiatus; however, the biostratigraphy is equivocal (Steurbaut *et al.* 1991; Eidvin & Riis 1992; Gradstein & Bäckström 1996; Martinsen *et al.* 1999; Eidvin *et al.* 2000; see Jordt *et al.* 2000 for discussion).

The Plio-Pleistocene sequences (CSS-8, CSS-9 and CSS-10) consist mostly of glacial sediments. The CSS-8 sequence is of Pliocene age (Eidvin & Riis 1992; Gradstein & Bäckström 1996; Eidvin *et al.* 2000) whereas the CSS-9 and CSS-10 sequences are of Pleistocene age.

The seismic interpretation focused on identification of sequence geometries, location of depocentres, outbuilding directions, recognition of tectonic influence and establishment of a general Cenozoic framework for further analyses and basin modelling. Seismic sequence geometries and outbuilding directions provide information about changes in the basin topography, and they are related to underlying structures, shifts in provenance area and changes in relative sea level and sediment accumulation rates. For each seismic sequence we have constructed time–thickness maps.

Palaeo-water depth

Temporal and spatial variations in palaeo-water depth are crucial parameters in basin analysis, as changes in palaeobathymetry detail the amount of sediment underfill during basin evolution (Gradstein & Bäckström 1996). In forward or backward basin modelling, palaeo-water depth is an important input parameter that controls the measured subsidence or uplift and thus the

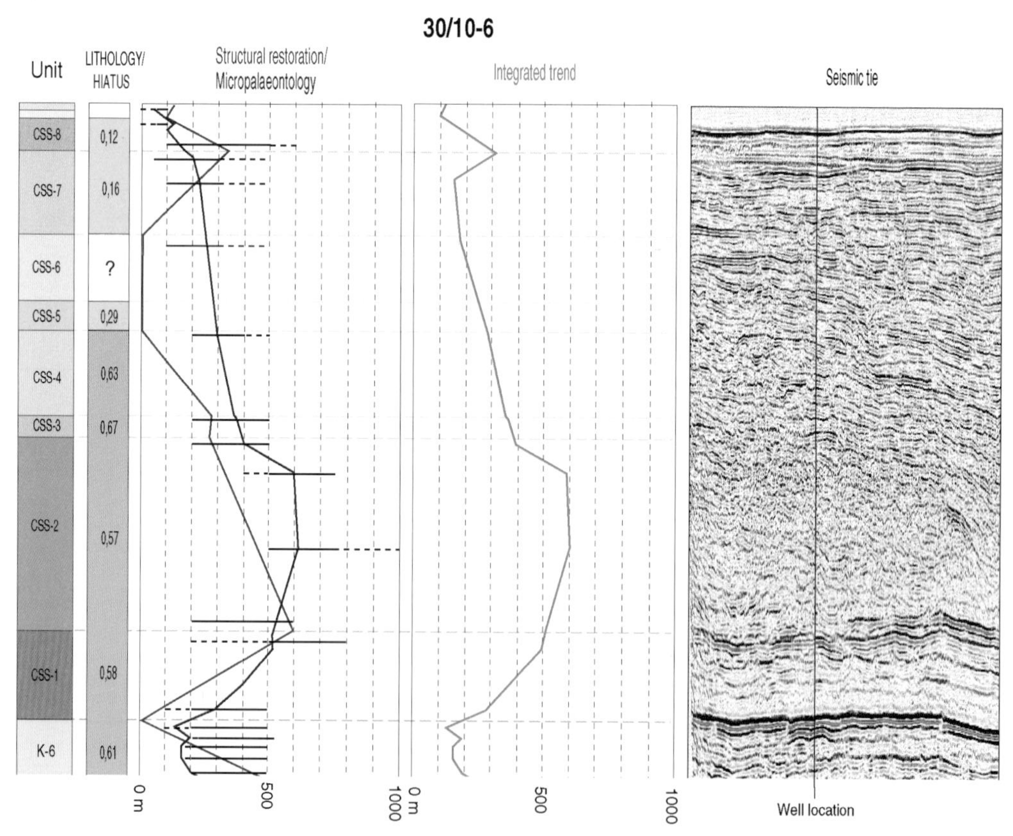

Fig. 6. Palaeo-water depth summary for the Cenozoic sequence of well 30/10-6 (Viking Graben). Modified from Kyrkjebø *et al.* (2001). Dominant lithologies: shales–mudstones in green; sandstones in yellow. Sedimentation rates in mm ka^{-1} are also shown. (See Fig. 4 for Cenozoic seismic stratigraphy.) K-6, Maastrichtian.

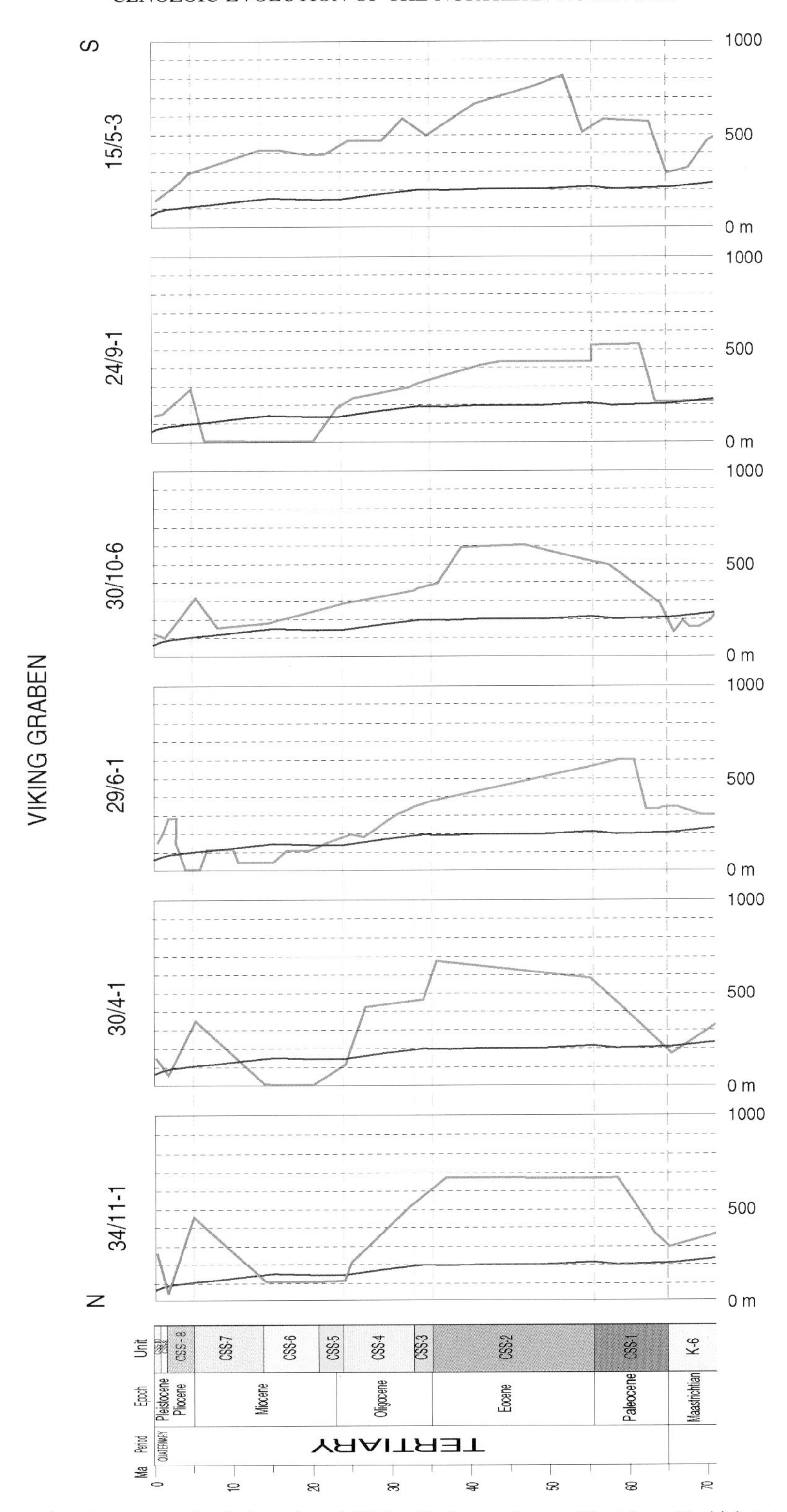

Fig. 7. Cenozoic palaeo-water depth for selected Viking Graben wells, modified from Kyrkjebø *et al.* (2001). Long-term eustatic curve from Haq *et al.* (1987, 1988) is shown in magenta for comparison. (See Fig. 2 for location of wells.)

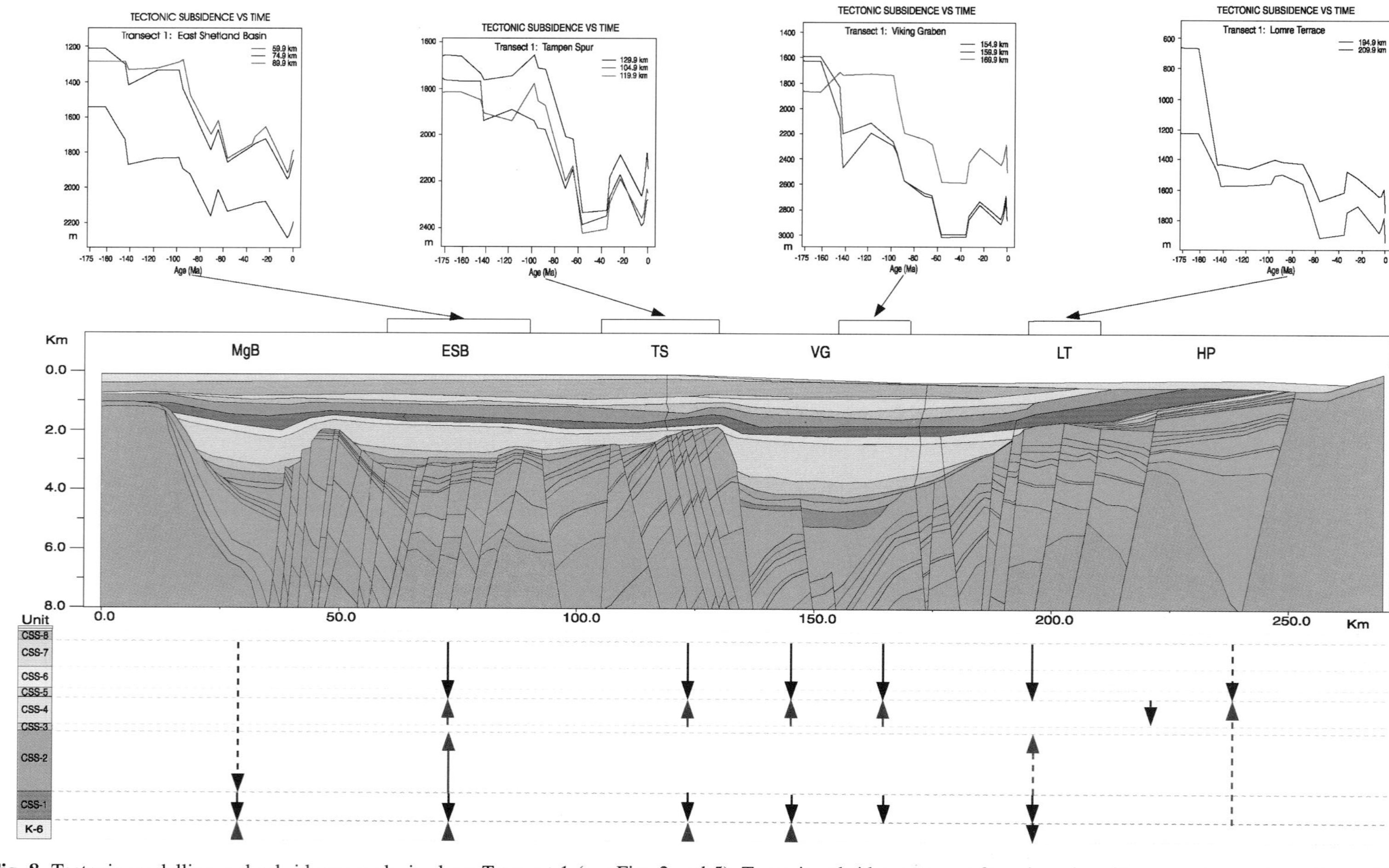

Fig. 8. Tectonic modelling and subsidence analysis along Transect 1 (see Figs 2 and 5). Tectonic subsidence curves for selected positions are shown above the profile. Relative vertical movements indicated by arrows (red, uplift; blue, subsidence). MgB, Magnus Basin; ESB, East Shetland Basin; TS, Tampen Spur; VG, Viking Graben; LT, Lomre Terrace; HP, Horda Platform. Modified from Kyrkjebø *et al.* (1999).

recognition of tectonic events. However, palaeo-water depth is difficult to determine, and depends largely on the quality and quantity of data from boreholes and seismic data (Bertram & Milton 1989; Joy 1992, 1993, 1996; Hall & White 1994; Jones & Milton 1994; Nadin & Kusznir 1995, 1996).

By carefully integrating seismic stratigraphic observations with palaeo-water depth estimates from structural restoration and micropalaeontological data, changes in accommodation space throughout Cretaceous–Tertiary times can be documented on a regional scale in the northern North Sea (Figs 6 and 7) (Kyrkjebø *et al.* 2001).

The palaeorelief was restored along the regional transects (Figs 2 and 5) using the depositional geometries, indications of zero or near-zero water depth (such as subaerial unconformities and coals) and fault restoration (Kjennerud *et al.* 2001). The method provides estimates of palaeorelief along the transects at the base of each seismic unit rather than absolute palaeo-water depth. In most cases this basin relief could be characterized as a minimum water depth. Subaerial topography cannot be determined by this method. Prograding sequences, which are characteristic for parts of the Cenozoic development in the northern North Sea, are associated with the least uncertainty in the restorations.

The 12 key wells from the Norwegian part of the northern North Sea (Fig. 2) were studied in great detail to assess palaeo-water depths. The micropalaeontological analysis (Gillmore *et al.* 2001) gave a range estimate of palaeo-water depth and in some cases additional maxima and minima are included (Fig. 6). Next, a most likely trend through time was determined for each well by integrating the results from the structural restoration and the micropalaeontological analysis (Fig. 6). As it is not possible to determine exactly the palaeo-water depth, we have focused on determining most likely depth intervals, and identifying the principal shallowing and deepening trends. In the seismic interpretation it was not possible to differentiate between the Miocene seismic units CSS-5, CSS-6 and CSS-7 in the northern North Sea. Therefore, palaeo-water depth estimates were not obtainable for each Miocene unit by structural restoration. Furthermore, only a few samples of Miocene age have been available for micropalaeontological analysis. The sensitivity for shallowing or deepening trends is considered to be no better than 100 m (Kyrkjebø *et al.* 2001).

The inferred Cenozoic deepening or shallowing trends from the investigated wells are generally in good agreement with each other on a regional scale, especially when the tectonic position within the basin is taken into account (Fig. 7). The inferred general trends in Cenozoic time are (Kyrkjebø *et al.* 2001): (1) deepening in Early to Late Paleocene time; (2) shallowing from Late Eocene to Late Miocene time; (3) deepening from Late Miocene to Early Pliocene time; (4) shallowing during Pliocene time.

Most workers agree on falling global eustatic sea level throughout Cenozoic times (Pitman 1978; Watts & Steckler 1979; Haq *et al.* 1987, 1988). However, the suggested amplitudes vary between about 100 and 300 m, which are less than the suggested amplitudes of the deepening or shallowing trend (Fig. 7). The Late Eocene to Late Miocene shallowing correlates with the long-term eustatic curve of Haq *et al.* (1987, 1988), but the general shallowing trend was probably amplified by tectono-thermal effects. The deepening events in Early to Late Paleocene and Late Miocene to Pliocene times cannot be explained by the eustatic sea-level curve, and must therefore be explained by purely tectono-thermal events (Kyrkjebø *et al.* 2001). From Mid-Eocene time, there is a reasonable correlation between eustatic curves derived from sequence stratigraphic studies (e.g. Haq *et al.* 1987, 1988) and the composite oxygen isotope record reflecting variations in global climate and glaciations (Abreu & Anderson 1998).

Modelling

The Cenozoic deepening or shallowing trends summarized above clearly point towards tectono-thermal events affecting Cenozoic basin evolution. One of the main objectives of our study was to constrain the amplitude and wavelength of differential vertical movements. Together with timing, these are critical factors to a discussion of possible mechanisms for Cenozoic vertical movements.

To assess tectonic subsidence and uplift in a more quantitative way, we have carried out modelling along the four regional transects (Figs 2, 5 and 8) (Kyrkjebø *et al.* 1999). Conventional backstripping techniques (e.g. Steckler & Watts 1978; Watts *et al.* 1982; Allen & Allen 1990; Roberts *et al.* 1993, 1998; Kuznir *et al.* 1995; Nadin & Kusznir 1995, 1996) were applied to the transects after key parameters such as age, lithology, porosity–depth and palaeo-water depth for each seismic sequence had been constrained from analysis of the seismic and well data. The lithospheric response to loading or unloading of sediment and water was assumed to be compensated for by local Airy isostasy.

Vertical movements different from those predicted by thermal contraction and sediment and water loading were regarded as potential tectonic signals. Kyrkjebø *et al.* (1999) regarded events to be truly tectonic when the amplitude of the vertical movements exceeds the uncertainties related to changes in palaeo-water depth adjusted for eustatic sea-level changes according to Haq *et al.* (1987, 1988). Figure 9 shows how changes in palaeo-water depth are crucial for detection of tectonic events other than thermal cooling.

The tectonic subsidence analysis of the northern North Sea points to the following Cenozoic trends (Kyrkjebø *et al.* 1999) (Figs 8 and 9): (1) accelerated tectonic subsidence in Paleocene time; (2) standstill to uplift in Eocene time followed by uplift in Oligocene time; (3) tectonic subsidence during Miocene time followed by uplift in Pliocene time. Kyrkjebø *et al.* (1999) emphasized that these trends are general, and that local variations occur. The general lack of palaeo-water depth information for the Miocene period affects the resolution in these results. On the basis of evidence discussed below, we believe that the change from uplift to subsidence occurred in Mid-Miocene time and the subsequent change from subsidence to uplift probably took place in Late Pliocene time.

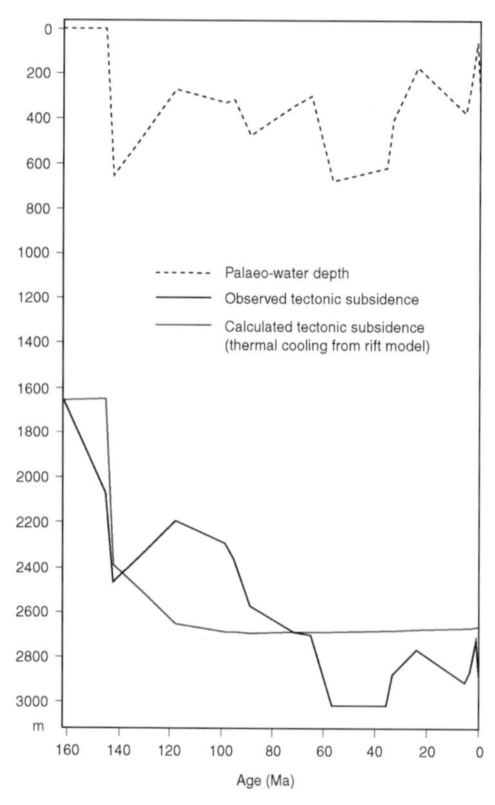

Fig. 9. Typical palaeo-water depth and tectonic subsidence curves in the northern North Sea compared with standard rift model (post-rift thermal cooling).

Cenozoic evolution

Here we integrate the results of the studies described above and summarize the Cenozoic evolution of the northern North Sea and surrounding areas using seismic sections, time–thickness maps and palaeo-water depth maps.

Paleocene time

The CSS-1 sequence (Upper Paleocene–lowermost Eocene units) is typified by prograding wedges that built out from the East Shetland Platform and from southern Norway. Lowermost CSS-1, marking the beginning of the Tertiary sequence, was probably characterized by shallow water depth and little relief (Fig. 10a). The lowermost Paleocene break (66–62 Ma) is interpreted to represent both an erosional vacuity and a hiatus. Part of the break is probably a marine condensation of sedimentation before progradation of the Paleocene depositional systems from the east and the west (Martinsen *et al.* 1999).

The prominent depocentres along the western basin margin (Fig. 10c) were mainly sourced from the uplifted East Shetland Platform and the Scottish Highlands. The prograding shelf-slope system is in part fairly sand rich and several phases of sand deposition were related to tectonic uplift and erosion of the source area. The sequence geometries reflect that some of the Mesozoic graben faults, particularly along the western margin of the East Shetland Basin, were reactivated and that differential compaction took place over deeper-seated Mesozoic fault blocks. The syndepositional faulting was not related to a new phase of rifting, but probably to regional subsidence. The shape of the depocentre was defined by a combination of increased subsidence towards the basin centre, and differential compaction of the Mesozoic sediments in the graben relative to those on the platform areas (Milton *et al.* 1990).

A depocentre in the northeastern part of the North Sea (Fig. 10c) was probably sourced from mainland Norway. It has a progradational stacking pattern that thickens pronouncedly eastward although it has been subjected to later erosion (Fig. 11). This unit is characterized by more fine-grained lithologies. The clay mineral distribution shows an increasing kaolinite–

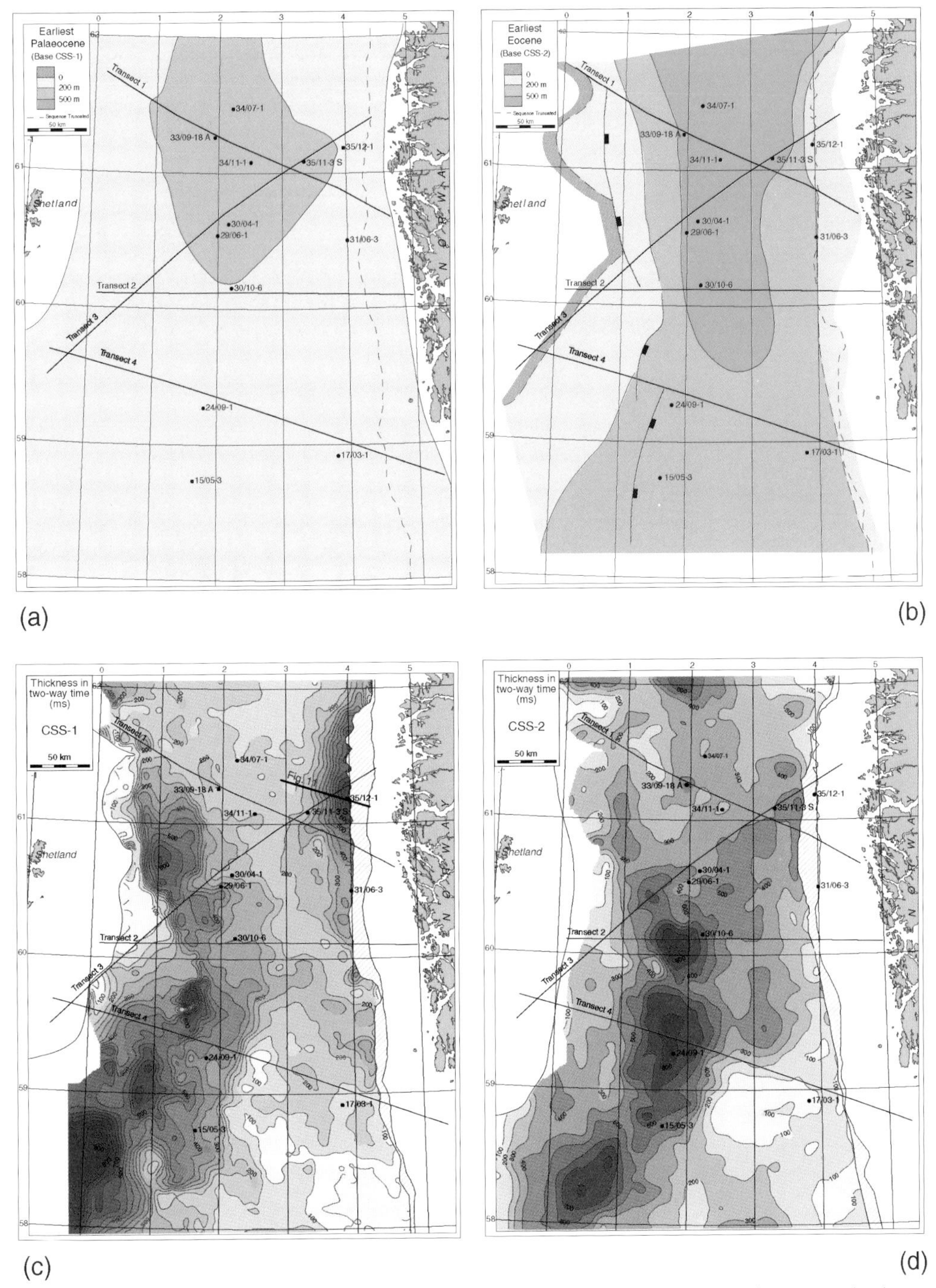

Fig. 10. (**a**) Palaeo-water depth map for earliest Paleocene time (lowermost CSS-1); (**b**) palaeo-water depth map for earliest Eocene time (lowermost CSS-2); (**c**) time–thickness map for CSS-1 (Upper Paleocene–lowermost Eocene sequence); (**d**) time–thickness map for CSS-2 (Eocene sequence).

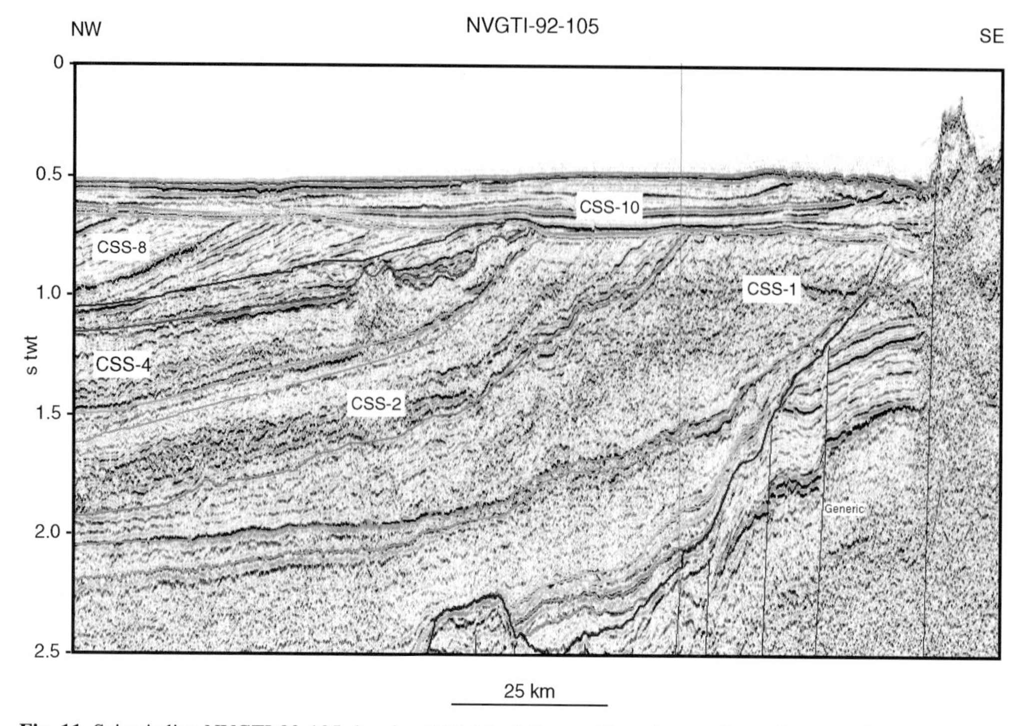

Fig. 11. Seismic line NVGTI-92-105 showing CSS-1 building out from the east in northernmost North Sea. (See Fig. 10c for line location and Fig. 4 for Cenozoic seismic stratigraphy.)

smectite to illite ratio towards the depocentre (Rundberg 1989) indicating that these sediments were derived from the east (Thyberg *et al.* 2000).

A relative sea-level rise in latest Paleocene–earliest Eocene time is indicated by aggradation and even sediment thickness in the uppermost part of CSS-1, which includes the Balder Formation, comprising interbedded shales and tuff layers. Explosive volcanism in the British and the Faeroe–Greenland Tertiary volcanic provinces has been regarded as the source for the widespread volcanic ashes found in the North Sea (Malm *et al.* 1984; Knox & Morton 1988) and onshore Denmark (Spjeldnæs 1975; Nielsen & Heilmann-Clausen 1988). The higher smectitic content in the upper part of the CSS-1 sequence indicates weathering of a basaltic source rock. The most likely provenance area for these sediments is the Late Paleocene–Early Eocene basalt province along the North Atlantic rift zone (Fig. 12) in addition to the widespread tuff deposits. Uppermost CSS-1 probably indicates a condensed section related to periods of low clastic sediment supply and a starved depositional environment.

The sequence geometries and thickness distribution of CSS-1 indicate that deep marine conditions existed along the Viking Graben and towards the continental margin in the north. A deepening during Paleocene time, reaching water depths of about 800 m in the deepest parts of the basin, has been inferred from biostratigraphic data (Gradstein *et al.* 1994; Gradstein & Bäckström 1996; Gillmore *et al.* 2001; Kyrkjebø *et al.* 2001) and structural restoration (Kjennerud *et al.* 2001) (Figs 7 and 10b). As the inherited post-Cretaceous bathymetry along the North Sea basin margin areas was not significant (Fig. 10a), the accommodation and development of the thick Upper Paleocene wedges probably resulted from subsidence of the basin floor combined with increased sediment supply related to tectonic uplift and subsequent erosion of the adjacent land areas. Shallow marine conditions probably prevailed along the basin margins in the west and east. A delta succession characterized by the presence of coal and *in situ* lignites, as well as freshwater floras and faunas, has been reported in the Moray Firth (Andrews *et al.* 1990).

Uplift above sea level of the Hebrides–Shetland axis to the NW of the North Sea basin was accompanied by the outbreak of extensive volcanicity in the areas surrounding the present NE Atlantic Ocean. The uplift established a wholly new geography (Fig. 12). An easterly to southeasterly flowing drainage system became

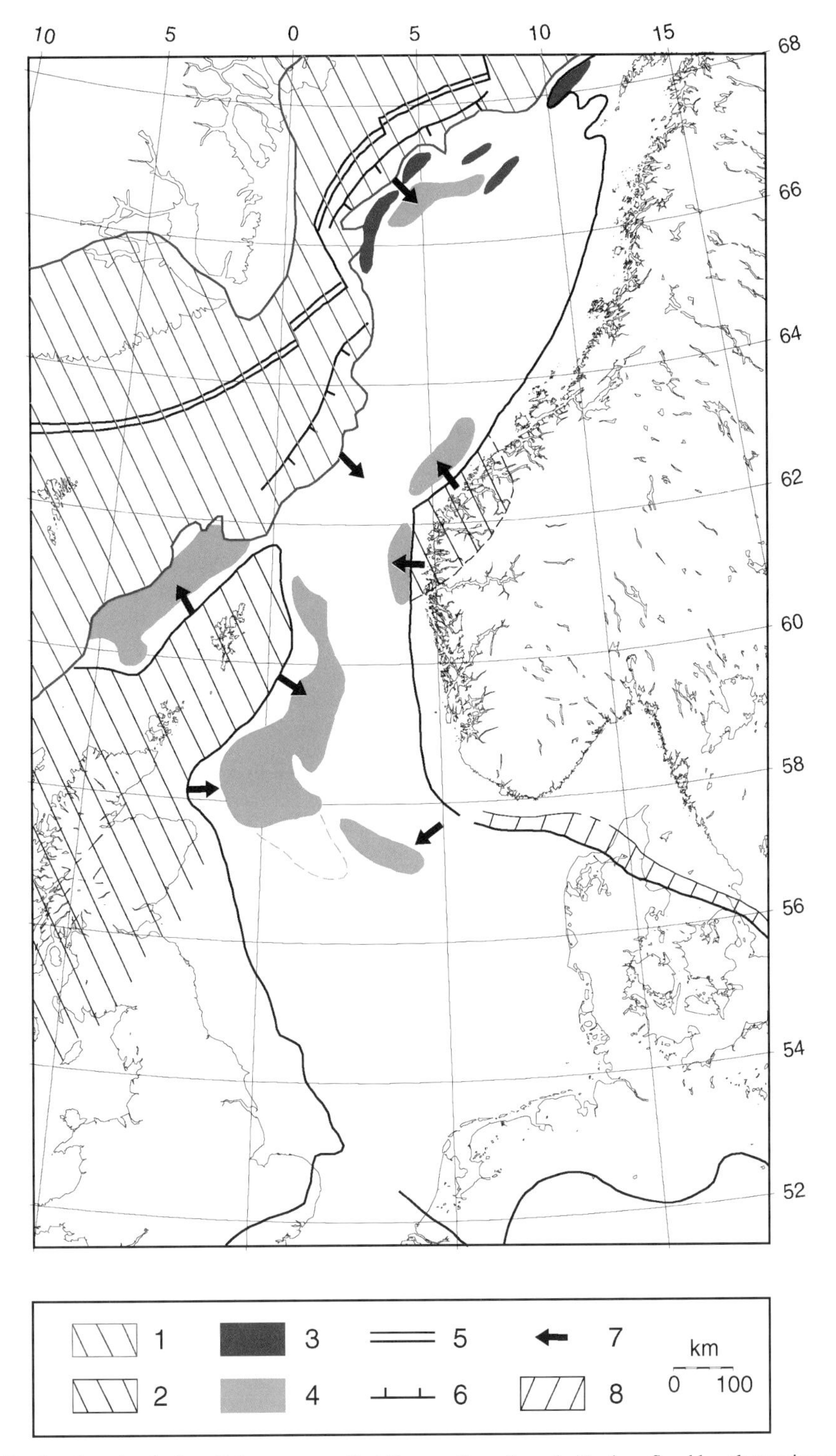

Fig. 12. Regional setting in Late Paleocene–earliest Eocene time. 1, early Tertiary flood basalt province; 2, areas of uplift; 3, intrabasinal highs; 4, main depocentres; 5, plate boundary–line of breakup; 6, Vøring and Faeroe–Shetland escarpments; 7, outbuilding directions; 8, Sorgenfrei–Tornquist Zone–Fennoscandian Border Zone.

established on the Orkney–Shetland Platform and Scottish Highlands, resulting in the reworking of the sedimentary cover over the newly emergent terrains, and dispersal of their detritus into the North Sea basin (Johnson *et al.* 1993; Jones & Milton 1994).

The northwestern corner of southern Norway (Fig. 12), on trend with the Hebrides–Shetland axis, was also uplifted. The uplifted area was bounded by the Øygarden and Møre–Trøndelag fault complexes towards the west and NW, respectively. It extended southwards to about the Sognefjorden area. Northeastwards the uplifted area was probably bounded by the landward extension of the Jan Mayen Lineament. Early Tertiary volcanism (*c.* 56 Ma) is documented within this area just off the coast of Norway (Vestbrona Formation; Bugge *et al.* 1980; Torske & Prestvik 1991; Prestvik *et al.* 1999). Clasts of Danian chalk have been found in Pliocene sediments building out from Norway in this area (Thyberg *et al.* 2000), indicating that no clastic source area existed here before the Late Paleocene uplift.

Most of the sediments derived from the uplifted area on the Norwegian side were deposited in the northeastern North Sea and southeastern margin of the Møre Basin (Fig. 12). However, some sediments may have been transported southwards, where only a small portion reached the North Sea in the Norwegian–Danish Basin. Here, a Late Paleocene depocentre (Fig. 12) reflects outbuilding from the NE. The source area for the CSS-1 sequence in this area may also have included the uplifted Sorgenfrei–Tornquist Zone (STZ)–Fennoscandian Border Zone (FBZ), which was inverted in response to Alpine compression. The uplifted STZ–FBZ must have been a barrier for any sediments coming from southern Scandinavia. Erosional valleys and karst topography suggest that the uppermost chalk was subaerially exposed in the eastern North Sea in mid-Paleocene times (Huuse 1999; Clausen & Huuse 1999). The fossil content in the Danish Fur Formation indicates that the lowermost Eocene sediments were deposited relatively close to a coastline in the Norwegian–Danish Basin.

Erosional products (sands) were also transported northwestward from the uplifted Hebrides–Shetland area into the Faeroe–Shetland Basin (Fig. 12). Renewed tectonic movement allowed deposition adjacent to major faults, but Late Paleocene and Early Eocene events were dominated by regional uplift and massive volcanism associated with North Atlantic rifting. Towards the end of Paleocene times, transfer faulting had largely ceased, and the northwestern margin of the Faeroe–Shetland Basin was inundated by subaerially emplaced lavas of the Faeroes lower series, and to the NE the Erlend and West Erlend volcanoes were erupting (Stoker *et al.* 1993).

Late Paleocene–earliest Eocene depocentres in the western Møre and Vøring basins were sourced from an uplifted area in the west along the incipient plate boundary (Fig. 12). On the Møre Marginal High, Paleocene deposits are thin, typically 200–350 m, and lithologies consist mainly of lavas and shallow-water sediments. A thick wedge in the Møre Basin was sourced from the west.

Aggradation of sequence CSS-1.2 (Jordt *et al.* 1995) indicates that sediment supply from Norway was significantly reduced in earliest Eocene time, and that the land areas further east were subjected to a marine transgression. Reduced runoff caused by climatic change, changes in land vegetation, tectonic quiescence or erosional levelling of southern Norway, are possible mechanisms that could explain a sudden reduction in sediment supply from the east in the very earliest Eocene time (Jordt *et al.* 2000).

Eocene time

There is a composite break in the Early Eocene deposition corresponding to the CSS-1–CSS-2 boundary. The biostratigraphy indicates that the break spans the period 55–52 Ma. The lower part of the break (*c.* 55–54 Ma) is laterally extensive. The upper part of the break (*c.* 54–52 Ma) seemingly has a limited lateral extent, and in the North Sea basin is accompanied by a shift of sedimentation in the basinwards direction, and subsequent onlap onto the basin margin. Thus, the break increases in extent towards the east. This Early Eocene break is interpreted to represent both an erosional vacuity (although evidence is limited) and a hiatus (Martinsen *et al.* 1999).

The Eocene (CSS-2) depocentres (Fig. 10d) are mainly located in the central and western parts of the northern North Sea basin and indicate outbuilding into the basin from the uplifted East Shetland Platform. CSS-2 infills and drapes the topography of top CSS-1 and the depocentre is located basinwards of the former CSS-1 shelf edge. Eocene deposition was dominated by deep-water and slope processes including turbidity currents. Deep water existed throughout most of Eocene times (Figs 10b and 13a). Water depths of about 1000 m have been estimated for the Lower Eocene Frigg Formation, which was deposited as a submarine fan in a deep-sea environment (Heritier *et al.* 1979, 1981).

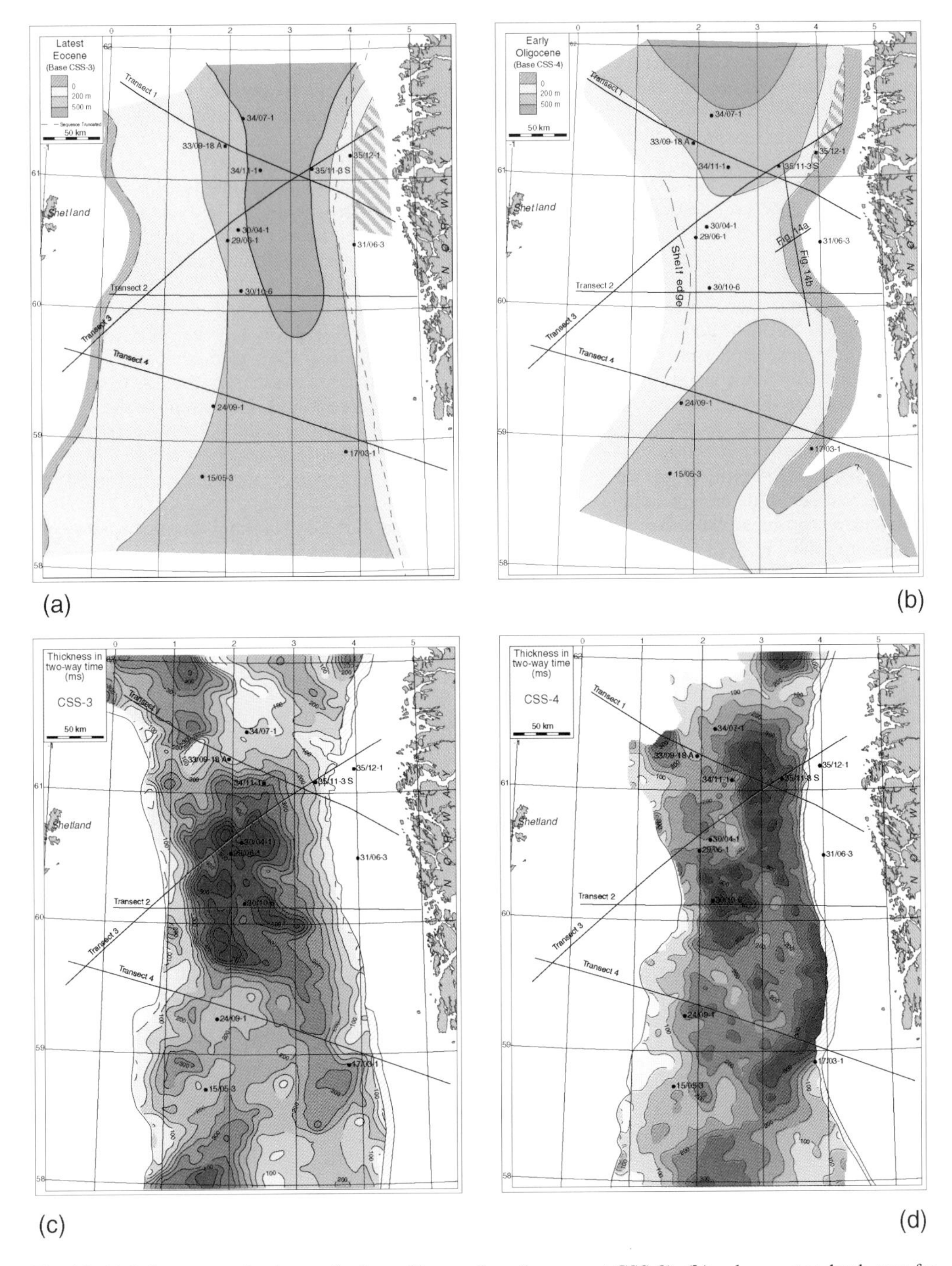

Fig. 13. (**a**) Palaeo-water depth map for latest Eocene time (lowermost CSS-3); (**b**) palaeo-water depth map for Early Oligocene time (lowermost CSS-4); (**c**) time–thickness map for CSS-3 (Lower Oligocene sequence); (**d**) time–thickness map for CSS-2 (Upper Oligocene sequence).

CSS-2 consists of stacked, thickly bedded immature sandstones with interbeds of clay- and siltstone at the margin of the East Shetland Platform and in the Moray Firth Basin and generally of claystones in the Viking Graben and Central Graben areas (Mudge & Bujak 1994). Lignite layers penetrated in well 8/27a-1 point to the existence of swampy coastal lakes landward of the inner shelf (Veeken 1997).

The Eocene succession thins eastwards and is almost absent in some sections close to the eastern basin margin. However, a minor Early Eocene (CSS-2.1 of Jordt *et al.* 1995) depocentre off Sognefjorden, in the same area as for the underlying CSS-1 sequence, comprises a progradational unit that downlaps onto the uppermost Balder surface. The outbuilding of the lower part of sequence CSS-2 in the northeastern North Sea and deposition of sand along the eastern basin margin indicate renewed tectonic uplift and sediment supply from the east (Rundberg 1989; Jordt *et al.* 1995, 2000).

From mid-Eocene time onwards, Tertiary sedimentation in the central areas of the North Sea basin was dominated by monotonous sequences of marine muds and silts, although sands were deposited at the periphery. The fine-grained CSS-2 sediments of low rates of non-volcanic clastic sediment supply indicate a relatively deep marine facies. More or less starved sedimentation conditions, as a result of a relative rise in sea level, favoured the enrichment in both volcaniclastic and organic matter (Thyberg *et al.* 2000). The uniform lithology and composition of the Eocene sediments (smectitic mudstones) in the northeastern North Sea (Rundberg 1989; Thyberg *et al.* 2000), and the velocity distribution of sequence CSS-2 (Jordt *et al.* 2000), suggests a rather uniform source and sediment supply from an area dominated by basaltic rocks, i.e. the North Atlantic basalt province to the NW (Thyberg *et al.* 2000).

The dominance of fine-grained Middle–Upper Eocene deposits in eastern North Sea wells (Rundberg 1989; Steurbaut *et al.* 1991; Mudge & Bujak 1994; Michelsen *et al.* 1995), and the regional thinning of CSS-2.2 on the Horda Platform and further south, suggests that the Scandinavian topography was too limited to supply sufficient clastic material and to allow progradation from the east, and that large parts of Scandinavia, east of the study area, may have been submerged in Mid–Late Eocene times (Jordt *et al.* 1995).

During Late Eocene time, a lowering of relative sea level (eustatic fall and/or uplift?) promoted the advance of delta systems into the North Sea from the west, with deposition of successions characterized by the presence of *in situ* lignites and freshwater floras and faunas. Biostratigraphical evidence indicates that Upper Eocene strata are absent in much of the northern North Sea and along the basin flanks further south. In these areas, Middle Eocene deposits are overlain unconformably by Lower Oligocene shales (e.g. Gradstein *et al.* 1994).

The boundary between the Lower and Middle Eocene sequences in the Norwegian–Danish Basin is associated with a change in sediment outbuilding direction from the NE (Norway) to the west (British Isles and Shetland Platform) (Michelsen *et al.* 1995). A complete deep marine Eocene succession is preserved in the Norwegian–Danish Basin and in the Central Graben (Heilmann-Clausen *et al.* 1985). Here, the Eocene succession appears unaffected by erosion.

In the Faeroe–Shetland Basin, rifting and volcanism waned through Early Eocene time and was followed by subsidence. Middle to Upper Eocene sediments of the Faeroe–Shetland Basin are generally fine-grained claystones and siltstones with occasional limestones and sandstones. These reflect the return to fully marine conditions after a regressive phase in Late Paleocene and Early Eocene times (Stoker *et al.* 1993).

In the Møre Basin, subsidence during Eocene time allowed a thick sequence to accumulate. Wells drilled on the flank of the Margarita Spur penetrated nearly 1000 m of Lower to Middle Eocene sediments, overlain unconformably by the Oligocene sequence (Stoker *et al.* 1993). The thickness increase of CSS-2 towards the Atlantic continental margin indicates sediment supply from that direction. At the Vøring margin off mid-Norway, uplift created the main western source area for the Paleogene Vøring Basin sediments (Skogseid & Eldholm 1989; Stuevold *et al.* 1992; Stuevold & Eldholm 1996; Hjelstuen *et al.* 1999).

Oligocene time

Biostratigraphical data indicate a hiatus between the Eocene (CSS-2) and Oligocene (CSS-3–CSS-4) sequences (van Veen *et al.* 1994; Gradstein & Bäckström 1996). Martinsen *et al.* (1999) placed a widespread composite break in Early Oligocene time (33–27 Ma), of variable extent depending on location.

During Oligocene time (CSS-3 and CSS-4) the basin configuration changed. The thickest Oligocene strata in the northern North Sea basin are located over the Norwegian sector of the North Viking Graben, where up to 900 m accumulated. Isopachs of Oligocene deposits indicate a general north–south trend, with thinning both to the west and east (Fig. 13c and d). The concentration of CSS-3 sediments in the centre of the northern North Sea basin and contemporaneous erosion on the flanks indicate an increased topographic relief, which was probably caused by uplift along the eastern and the western basin margins in Early Oligocene time.

The sequence geometries reflect uplift of the eastern basin flank (Fig. 14). Minor progradation occurred as a consequence of tectonic uplift activity in the east initiated in Early Oligocene time. An increased rate of input of coarser-grained material occurred in Early Oligocene (CSS-3) time, probably in response to uplift and erosion of soft Eocene and Paleocene sediments. However, the volcanic input is still distinctive (Thyberg *et al.* 2000). New ashes from Iceland and reworked Eocene sediments from Norway may have contributed to the high smectite content.

Parts of the basin were uplifted together with Norway so that CSS-4 (Upper Oligocene sequence) locally overlies CSS-2 (Eocene sequence) (Figs 13b and 14). A minor drop in sea level, leading to truncation of the Eocene–Lower Oligocene sequence(s), was followed by a renewed rise that resulted in onlap of CSS-4 deposits. Late Oligocene deposition was characterized by basinal accumulation of aggrading silts and clays. Locally CSS-4 builds out from uplifted source area(s) within the basin. The CSS-4 depocentres are mainly located along the eastern basin margin (Fig. 13d).

Clausen *et al.* (1999a) studied intraformational faults within the Upper Oligocene sequence (CSS-4) in the Troll area on the northern Horda Platform. The faults have a strongly dominant NW–SE trend indicating SW–NE extension, and formed above an intra-Oligocene unconformity that was established as a result of uplift of Fennoscandia and a sea-level fall.

Outbuilding continued from the East Shetland Platform during both CSS-3 and CSS-4 times, resulting in a thick depocentre in the Viking Graben. In the transgressive phase following the intra-Oligocene lowstand of sea level, a major shelf–deltaic foreset unit up to 500 m thick prograded from the East Shetland Platform into the central part of the northern North Sea basin.

The Early Oligocene glacio-eustatic sea-level fall that was caused by expansion of ice sheets in Antarctica (Miller *et al.* 1987; Zachos *et al.* 1992; Abreu & Anderson 1998) was overprinted by local vertical tectonic movements in the North Sea area. Palaeo-water depth data indicate shallowing (Fig. 7) and at the end of Oligocene time the northern North Sea formed a narrow and shallow basin separating deeper basins to the south and north (Fig. 13b) (Thyberg *et al.* 1999).

In the Norwegian–Danish Basin, the marked shift from Late Eocene distal deep-water sedimentation to shore progradation with sediment transport from north and NE in earliest Oligocene time indicates rapid tectonic uplift of southern Norway (Fig. 15) (Jordt *et al.* 1995, 2000; Michelsen *et al.* 1995; Danielsen *et al.* 1997; Clausen *et al.* 2000). The geometry and thickness of CSS-3 indicate that water depths in the Central North Sea may have exceeded 600 m in Early Oligocene time (Jordt *et al.* 1995; Michelsen *et al.* 1995). The presence of reworked Paleocene and Eocene nannofossils in Oligocene sediments is interpreted to be the result of denudation of Paleocene–Eocene sediments exposed at the basin margin to the east (Clausen *et al.* 2000).

The Lower Oligocene sequence forms an extensive depositional unit that extends parallel to the mid-Norwegian coast from Møre to Lofoten (Rokoengen *et al.* 1995). The sediments are interpreted as deltaic and coastal deposits, probably formed in a wave-dominated environment with extensive longshore drift. Seismic lines across this unit show that it is represented by an interval of steeply dipping reflectors, which are interpreted to be a set of prograding foresets. The base of the succession is seismically identified as a downlap surface. The western boundary is a morphological ramp formed by the distal foresets (Eidvin *et al.* 1998).

Outbuilding from Norway observed from the southeastern North Sea to Lofoten (Fig. 15) reflects the onset of regional tectonic uplift of Scandinavia at the Eocene–Oligocene transition. Southern Norway was uplifted, erosion accelerated and sediment transport towards the west became more important.

Early Oligocene outbuilding is also observed at the edge of the Hebrides Shelf (Stoker *et al.* 1993). In the Faeroe–Shetland Channel, the Oligocene sequence is thin.

During Early Oligocene time, the Greenland–Svalbard gap began opening, allowing greater communication between the Atlantic and Arctic oceans, and causing an influx of colder water into the North Sea. This change is evidenced by the replacement of warm-water fish by boreal forms, and a drop of 12 °C in the North Sea bottom temperatures (Buchardt 1978). This change was coincident with shallowing of the basin as a result of sediment infilling and tectonic uplift.

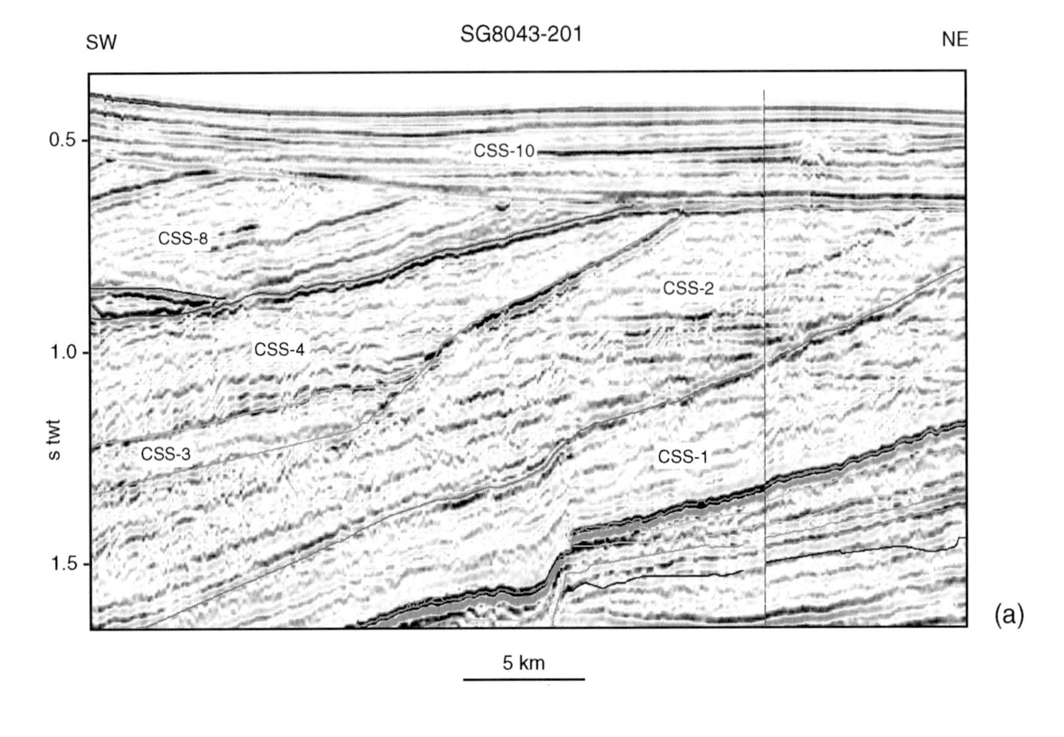

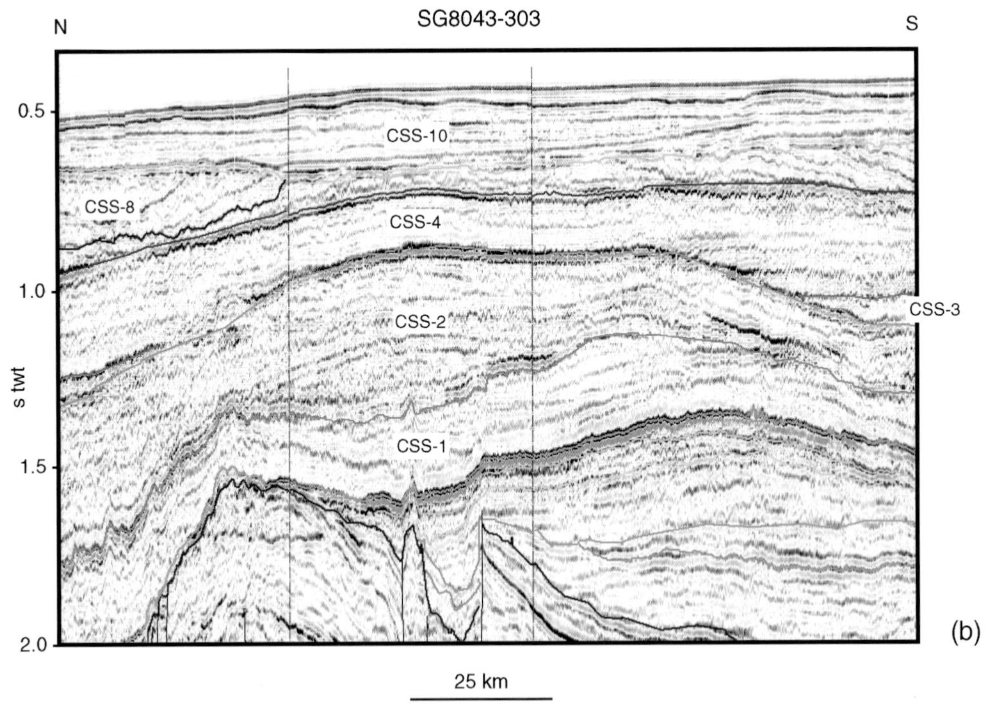

Fig. 14. Seismic lines showing pinchout of CSS-3 towards area at the eastern basin flank uplifted at the Eocene–Oligocene transition. (**a**) Seismic line SG8043-201; (**b**) seismic line SG8043-303. (See Fig. 13b for line locations and Fig. 4 for Cenozoic seismic stratigraphy.)

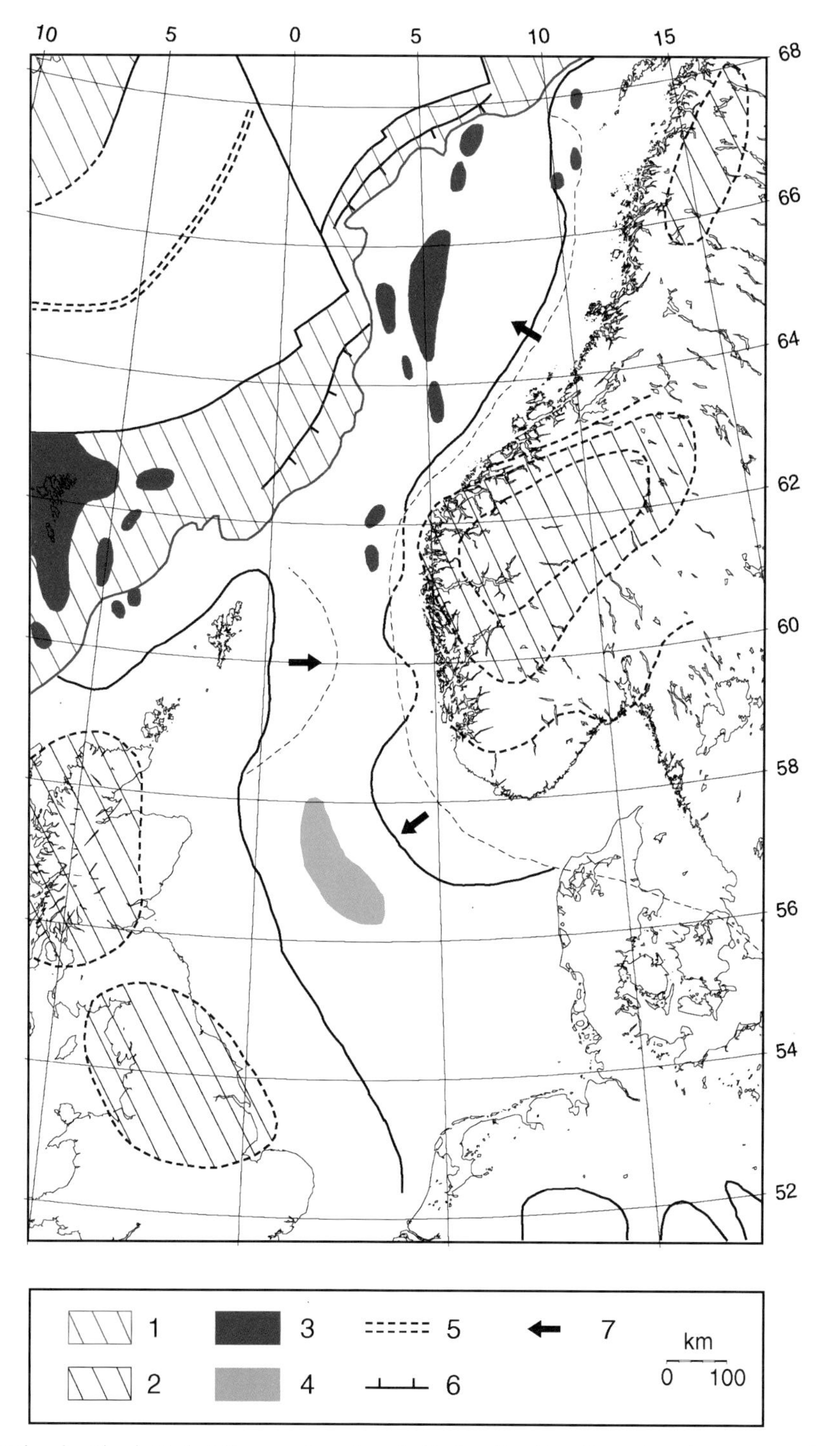

Fig. 15. Regional setting in Early Oligocene time. 1, early Tertiary flood basalt province; 2, areas of regional uplift (Rohrman *et al.* 1995; Japsen 1997); 3, domes or anticlines (Doré & Lundin 1996); 4, main depocentre; 5, extinct spreading axis; 6, Vøring and Faeroe–Shetland escarpments; 7, outbuilding directions.

Miocene time

A prominent hiatus is present in the Miocene sequence but the biostratigraphy is equivocal (Steurbaut *et al.* 1991; Eidvin & Riis 1992; Gradstein & Bäckström 1996; Martinsen *et al.* 1999; Eidvin *et al.* 2000). According to Martinsen *et al.* (1999), this break extends from latest Oligocene (*c.* 25 Ma) to Late Miocene time (*c.* 8–9 Ma), interrupted by sedimentary units of Late Oligocene and Early to Mid-Miocene age. The Miocene break is locally characterized by erosion of the underlying Oligocene sequence.

During Miocene time the northern North Sea constituted a shallow marine basin, which connected the deeper central North Sea and the Norwegian–Greenland Sea (Fig. 16a). The main Miocene depocentres (CSS-5, CSS-6 and CSS-7) are located south of 60°N (Fig. 16b). Farther north some of the sequences are thin or missing, so that it is not possible to tie their boundaries within the seismic grid.

The CSS-5 sequence (Lower Miocene sequence) built out into the northern North Sea from a basin margin to the west. It comprises a seismically well-defined sequence reaching about 200 m in thickness in the basin centre. The base is marked by onlap onto the irregular uppermost Oligocene sequence boundary, whereas the top is identified as a prominent reflector representing the base of the sandy Utsira Formation. To the north, the Lower Miocene sequence pinches out at about 61°30'N (Eidvin *et al.* 2000). Deposition during Early–Mid-Miocene time was dominated by low accumulation rates.

The overlying sequence (CSS-6) thins northward and may be absent in large parts of the study area north of 60°N. A mid-Miocene fall in glacio-eustatic sea level (Miller *et al.* 1987; Haq *et al.* 1987, 1988; Abreu & Anderson 1998) in combination with regional uplift gave rise to erosion and formation of a prominent unconformity, in particular along the eastern basin margin. A Miocene prograding wedge observed in the Møre Basin (Martinsen *et al.* 1999) probably represents outbuilding from uplifted and eroded areas in southern Norway and the northern North Sea (Fig. 17). A Middle Miocene sequence boundary in the Norwegian–Danish Basin represents a shift from a prograding reflection pattern below to an aggrading pattern above (Michelsen *et al.* 1995).

In quadrant 35 of the Norwegian northern North Sea there is evidence of incision into Oligocene strata (Fig. 18) (Rundberg *et al.* 1995; Gregersen 1998; Martinsen *et al.* 1999). The formation of the incised valleys (submarine or

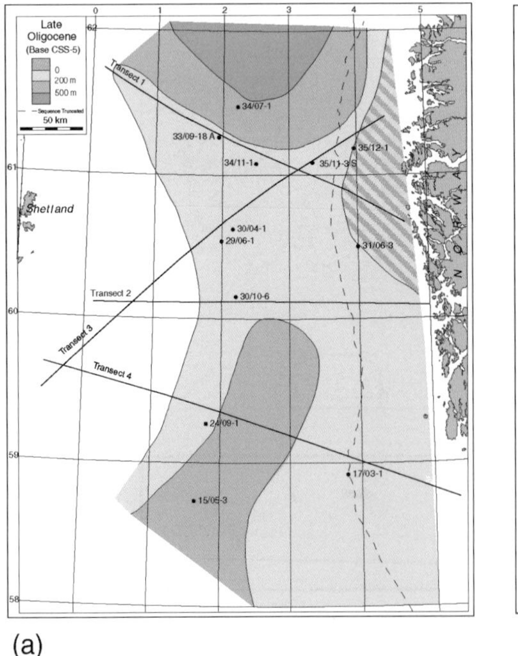

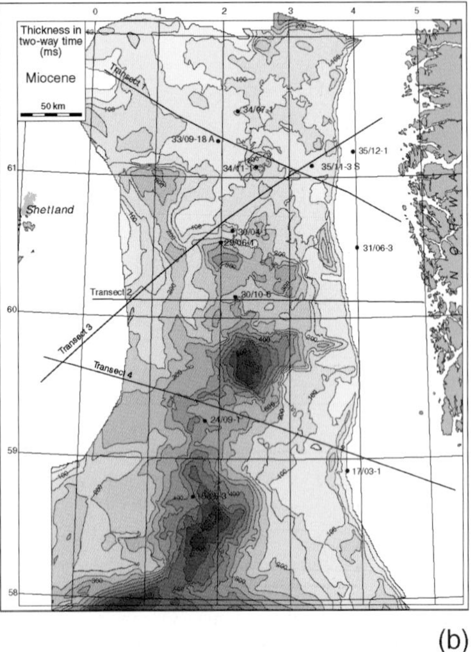

Fig. 16. (**a**) Palaeo-water depth map for Late Oligocene time (lowermost CSS-5); (**b**) time–thickness map for the Miocene units (CSS-5 to CSS-7).

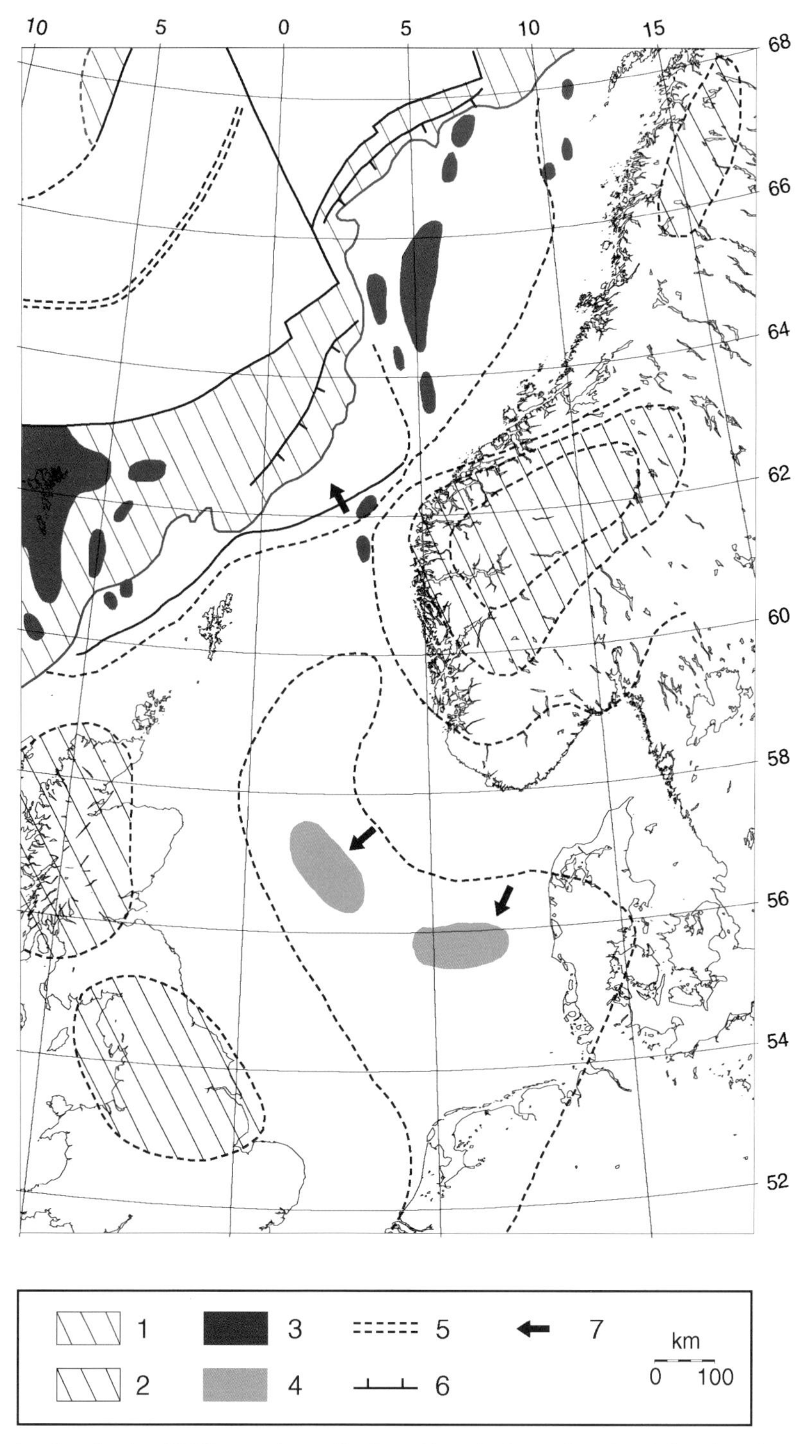

Fig. 17. Regional setting in mid-Miocene time reflecting a hiatus in the northern North Sea. 1, early Tertiary flood basalt province; 2, areas of regional uplift (Rohrman *et al.* 1995; Japsen 1997); 3, domes or anticlines (Doré & Lundin 1996); 4, main depocentres; 5, extinct spreading axis; 6, Vøring and Faeroe–Shetland escarpments; 7, outbuilding directions.

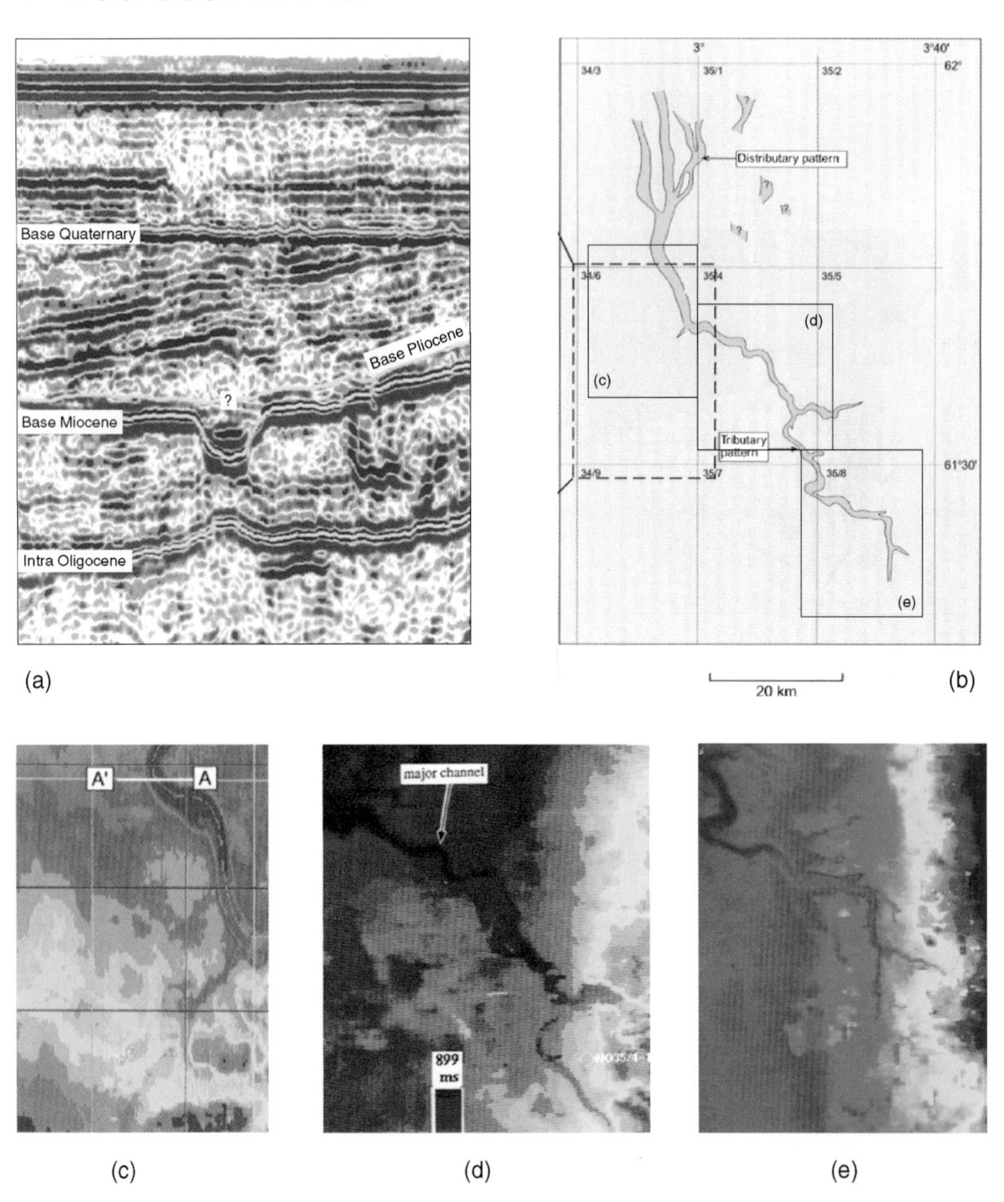

Fig. 18. Neogene incised valleys in the northern North Sea. (**a**) Seismic example across incised valley; (**b**) semi-regional map showing distribution of incised valleys (modified from Martinsen *et al.* (1999); (**c**) earliest Miocene time map, Block 34/6 (Martinsen *et al.* 1999); (**d**) earliest Miocene time map, Block 35/4 (Gregersen 1998); (**e**) earliest Miocene time map, Block 35/8.

subaerial?) and the timing or age are critical questions in constraining the history of vertical movements. Martinsen *et al.* (1999) interpreted the incised features as subaerial, incised valleys because of the morphological patterns. A subaerial origin for the incised features implies considerable tectonic uplift followed by Late Miocene–Pliocene subsidence. The incision probably took place during formation of the mid-Miocene unconformity. However, according

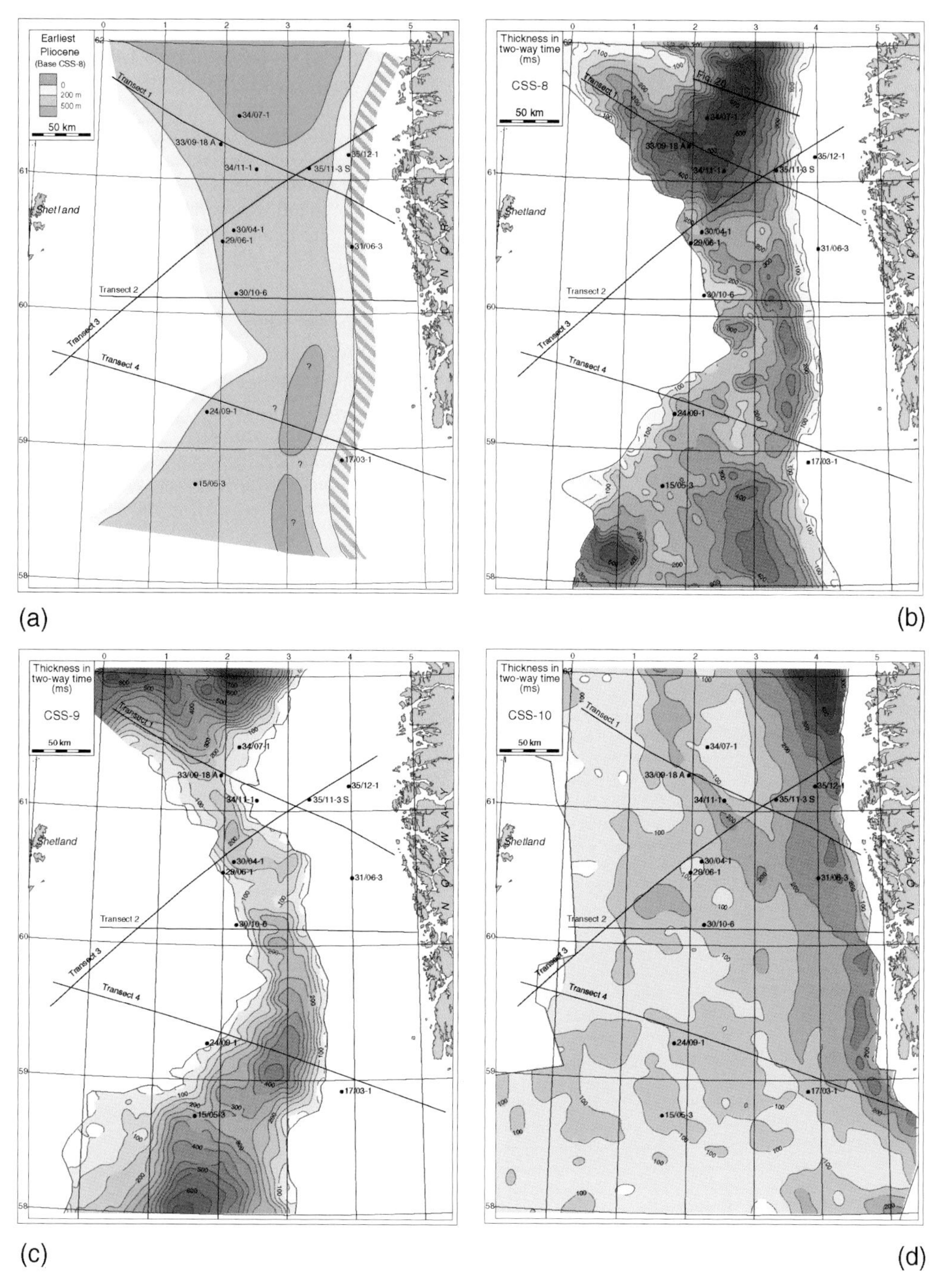

Fig. 19. (**a**) Palaeo-water depth map for earliest Pliocene time (lowermost CSS-8); (**b**) time–thickness map for CSS-8; (**c**) time–thickness map for CSS-9; (**d**) time–thickness map for CSS-10.

to Eidvin *et al.* (2000) the incision is Pliocene in age (see discussion below).

The Utsira Formation corresponds to part of CSS-7 and was deposited on top of the mid-Miocene unconformity in the northern North Sea. The formation accumulated in two main depocentres along the basin centre, south and north of 60°N, respectively. The Utsira

Formation is interpreted as a shallow-marine deposit, shed mainly from the eastern basin margin during late Mid–Late Miocene rising relative sea level (Isaksen & Tonstad 1989; Rundberg 1989; Galloway *et al.* 1993; Gregersen *et al.* 1997; Martinsen *et al.* 1999). The Utsira Formation is sand rich, but towards the western basin margin it fines considerably. The sediments were mainly derived from Scandinavia (Rundberg 1989), but sediment influx from the Shetland Platform or Scotland is also proposed (Gregersen *et al.* 1997). Although the Utsira sandstones are likely to have been supplied from the basin margin, they were probably significantly reworked by basinal currents (Galloway *et al.* 1993). The basin is unlikely to have been more than 150–200 m deep, so that such currents could have existed. The water depth increased and culminated in earliest Pliocene time with condensed deposits on which the succeeding Pliocene deposits downlapped.

Late Miocene uplift of south central Norway promoted further tilting of the Horda Platform area, and hence, erosion of Miocene sediments (Clausen *et al.* 1999a). The continued uplift of Scandinavia forced the Late Miocene and Pliocene deposits to prograde westward, downlapping onto the mid-Miocene unconformity and the underlying Lower–Middle Miocene deposits and filling in the more central parts of the northern North Sea. Deeper-water starved conditions existed in the Norwegian–Danish Basin and the Møre Basin in Late Miocene time contemporaneous with deposition of the Utsira Formation in the northern North Sea.

The Mid-Miocene and younger sedimentation in the southern North Sea was dominated by the expansion of massive delta systems on its eastern seaboard associated with former Baltic rivers that drained from the Fennoscandian Shield (Bijlsma 1981; Gibbard 1988; Cameron *et al.* 1993).

Plio-Pleistocene time

A major depocentre of mainly Upper Pliocene (CSS-8) glacial sediments derived from uplifted Norway is located in the northernmost North Sea (Fig. 19). Sequence geometries characterized by westward prograding clinoforms (Fig. 20) show that the dominant sediment transport directions are from the Norwegian shelf margin towards the west and NW, but occasional inputs from the Shetland Platform are also observed. The regional downlap surface at the base of the Pliocene sequence is interpreted as reflecting starved sedimentation probably caused by a relative sea-level rise (Gregersen *et al.* 1997). The clinoforms within sequence CSS-8 indicate deep water in excess of 500 m (Figs 19a and 20). An age of 2.75 Ma is taken as the maximum age of the Upper Pliocene section, corresponding to a large increase in the supply of ice-rafted material

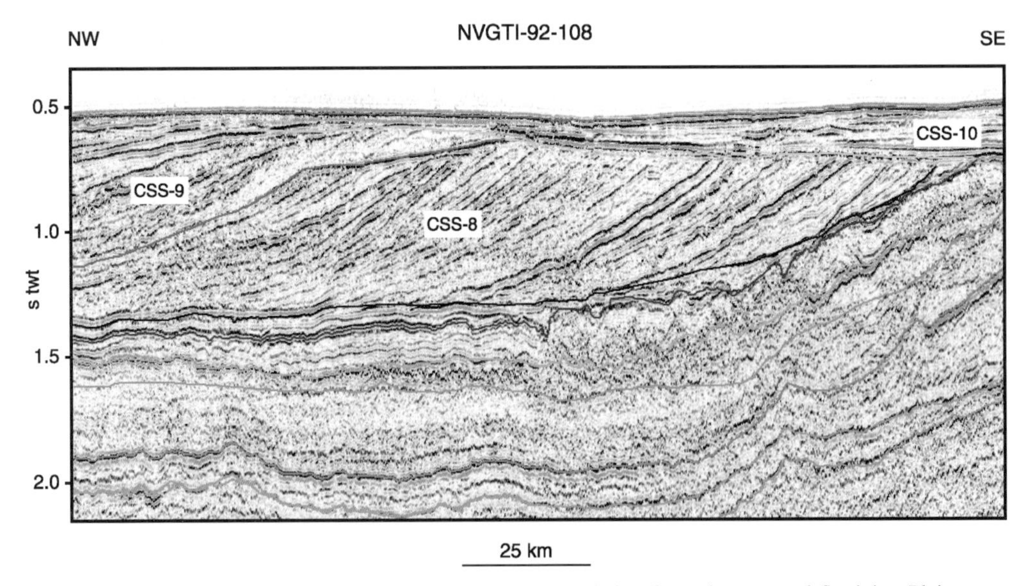

Fig. 20. Seismic line NVGTI-92-108 showing Pliocene progradation from the east and flat-lying Pleistocene sediments above angular unconformity. (See Fig. 19b for line location and Fig. 4 for Cenozoic seismic stratigraphy.)

related to a marked expansion of northern European glaciers (Eidvin *et al.* 2000).

A period of transgression in Early Pliocene time resulted in strongly reduced rates of deposition, and sediments of this age are preserved mainly in the central North Sea (Eidvin *et al.* 2000). Most of the Lower Pliocene sediments, which may have existed, were probably eroded in the subsequent period during an extensive relative fall in the global sea level (4.1–2.9 Ma). A period of regression in earliest Late Pliocene time probably resulted in erosion of most of the Norwegian continental shelf with the exception of the deeper areas of the Central and Viking grabens. If the incised valleys described above (Fig. 18) were formed at this time (earliest Late Pliocene time), and they formed above sea level, it implies extremely rapid Late Pliocene subsidence to create the water depth inferred from the clinoform geometries within the CSS-8 sequence (>500 m). This period was immediately followed, in the later part of Late Pliocene time, by rapid deposition of glacially derived sediments prograding along the entire shelf.

In general, the Pleistocene development is a continuation of the Late Pliocene evolution, but is marked by more extensive erosion of the inner shelf (Eidvin *et al.* 2000). Flat-lying Pleistocene beds lie with an angular unconformity on more or less progradational Upper Pliocene deposits (Fig. 20). The lower part of the Pleistocene sequence and uppermost part of the Upper Pliocene sequence are eroded over large areas of the continental shelf. The base of the Pleistocene section is dated to 1.2 Ma (Sejrup *et al.* 1995). This dating coincides with a marked intensification of glacial activity, as is observed in the deep-sea record (Ruddiman *et al.* 1986; Berger & Jansen 1994). Repeated glaciations (Sejrup *et al.* 1995) eroded the marginal parts of the northern North Sea and the associated glacio-eustatic sea-level falls and the overall lowstand have resulted in major channel cuts or incised valleys in the marginal parts of the North Sea basin (Sejrup *et al.* 1991).

Differences in Pliocene and Pleistocene depositional patterns are probably the result of changes in glaciation cycles that occurred at *c.* 1.1 Ma. During the period before this, the Fennoscandian ice cap probably extended only to the present coastline (Jansen & Sjøholm 1991). Subsequent to *c.* 1.1 Ma, glaciers periodically extended over the continental shelf and transported sediments over greater distances (Sejrup *et al.* 1995, 2000; King *et al.* 1996).

Large-scale Pliocene progradation is also observed offshore mid-Norway (Fig. 21) (Stuevold & Eldholm 1996; Hjelstuen *et al.* 1999; Eidvin *et al.* 2000).

The narrow–elongated depocentre along the basin axis in the central North Sea shows that the basin flanks were uplifted or exhumed together with mainland Norway and the British Isles (Fig. 21) (Hillis 1995a, 1995b; Hansen 1996; Japsen 1998, 1999, 2000). The Great European delta, fed mainly from the east and south, continued to expand northwards into the southern and central North Sea (Zagwijn 1989; Cameron *et al.* 1993; Funnell 1996).

The regional uplift of Scandinavia is believed to be the main control on sediment supply to the prograding Pliocene sequences. Oscillations of the eustatic sea level punctuated the tectonically controlled progradation and affected variations in the accommodation space, and thus created the high-frequency sequences (Sørensen *et al.* 1997). An upward increasing number of Pliocene sequences in the central North Sea has been related to an increasing Neogene uplift of Scandinavia and adjacent areas, combined with a general lowered glacio-eustatic sea level during Pliocene time. Biostratigraphic studies by Seidenkrantz (1992) support a shallowing upward and gradually colder environment during Late Pliocene time. The gradual cooling (Buchardt 1978) and ice-cap growth caused a general glacio-eustatic sea-level lowering (Haq *et al.* 1988) through Pliocene time, which, together with the uplift during Neogene time (Ghazi 1992; Jensen & Schmidt 1992, 1993; Hansen 1996), led to erosion of the exposed shelf areas and increased sediment influx.

Possible mechanisms causing vertical movements

The main mechanisms suggested in the literature to cause Cenozoic vertical movements along the NE Atlantic margin and adjacent areas include: (1) arrival of the Iceland plume and the subsequent lateral spreading of the plume head; (2) mantle processes and thermal regime, and the episodic behaviour of the Iceland plume; (3) emplacement of magma in and at the base of the crust (magmatic underplating) leading to isostatic uplift; (4) metamorphic phase changes in the mantle; (5) intra-plate stress related to ridge-push from the North Atlantic plate boundary and/or to Alpine compression; (6) erosion and isostatic rebound; (7) flexural effects.

It is beyond the scope of this paper to test all these mechanisms quantitatively. Here, some of them will be briefly discussed in light of our new constraints on timing, amplitude and wavelength

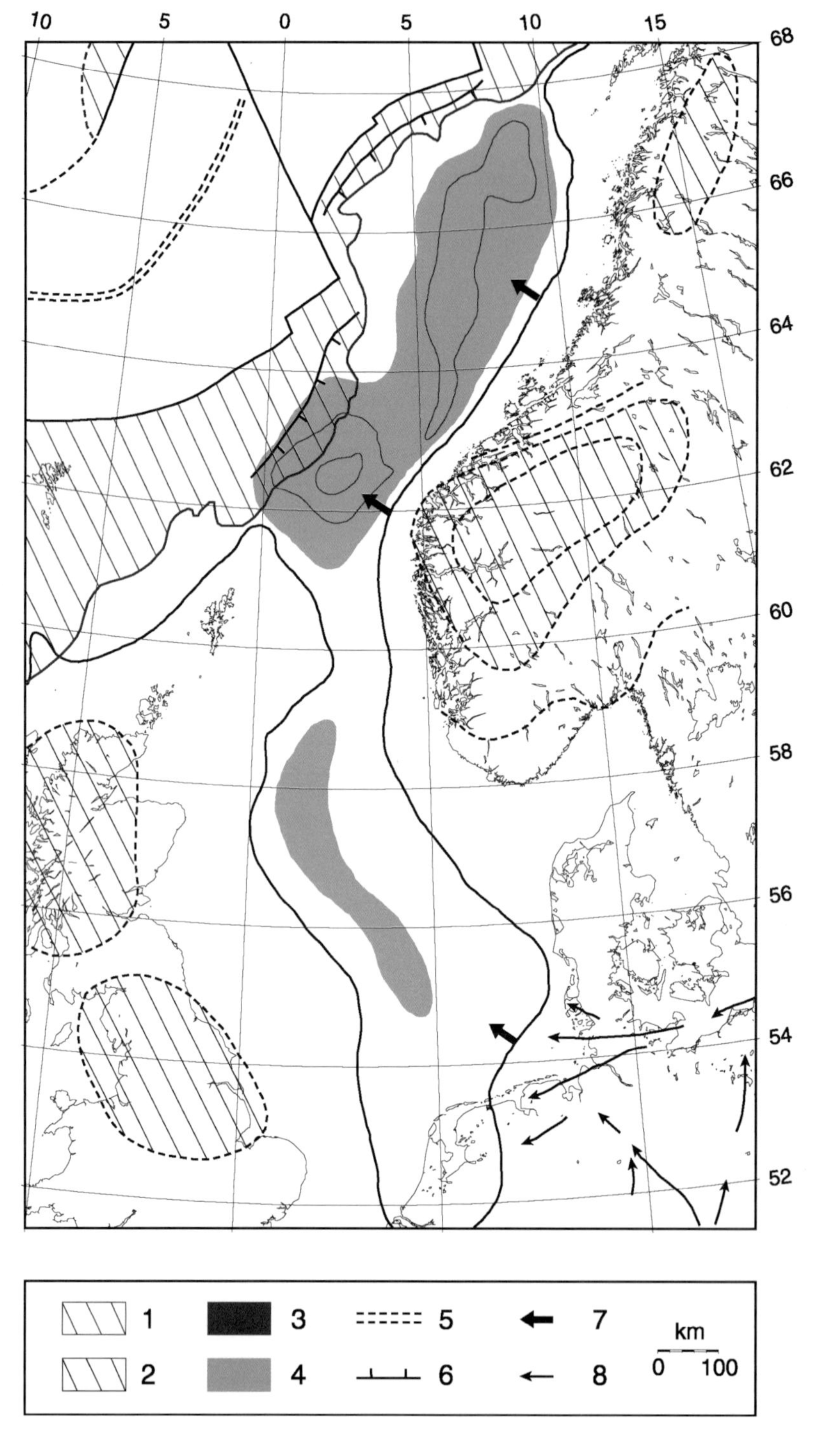

Fig. 21. Regional setting in Late Pliocene time. 1, early Tertiary flood basalt province; 2, areas of regional uplift (Rohrman *et al.* 1995; Japsen 1997); 4, main depocentres; 5, extinct spreading axis; 6, Vøring and Faeroe–Shetland escarpments; 7, outbuilding directions; 8, major river system (Gibbard 1988).

of Cenozoic differential vertical movements. For each of the main regional events we will summarize the main observations or facts that any proposed mechanism has to explain.

Late Paleocene–Early Eocene vertical movements

In simple terms, the Late Paleocene–Early Eocene uplift is referred to as 'marginal uplift' related to the rifting and break-up of the NE Atlantic. However, there are large variations in the wavelength and amplitude of the uplift observed along the NE Atlantic margins, and possibly also in timing.

Important observations that have to be explained include: (1) uplift of the Hebrides–Shetland axis and northwestern corner of southern Norway; (2) uplift of the area along the incipient plate boundary, with subaerial sea-floor spreading; (3) accelerated subsidence in the North Sea.

The accelerated tectonic subsidence in the North Sea basin in Paleocene time and uplift of surrounding areas was probably associated with arrival of the Icelandic plume and crustal break-up in the North Atlantic. There appear to have been two major phases or pulses of Early Tertiary magmatism within the North Atlantic igneous province, with a peak at 59 Ma and at 55 Ma (White & Lovell 1997; Ritchie *et al.* 1999).

Several uplift mechanisms can be attributed to this regional tectonomagmatic event: (1) transient thermal uplift generated by buoyancy of lithosphere heated by hot plume material (Sleep 1990; Clift & Turner 1998); (2) transient dynamic uplift generated by mantle fluid flow driven by the ascending hot plume material (Nadin *et al.* 1995, 1997); (3) permanent, isostatically compensated uplift generated by crustal thickening caused by igneous underplating associated with mantle plume activity (Brodie & White 1994, 1995; White & Lovell 1997; Clift & Turner 1998; Clift 1999).

It is likely that one or more of these mechanisms contributed to the regional uplift of the northern British Isles and adjacent areas (surrounding the British Tertiary igneous province). The main development of Paleocene sandstone reservoirs along the axis of the Faeroe–Shetland Basin appears to have been synchronous with the phases of thermal uplift along the basin margin and pulsed volcanism (White & Lovell 1997; Naylor *et al.* 1999). A similar relation between outbuilding of sandy deposits and the episodic nature of the plume-related uplift has been suggested for the southern North Sea area (Knox 1996).

However, it is more difficult to see how they can have contributed to the formation of the regional dome in southern Norway. We suggest that the uplifted area related to the Paleocene–Early Eocene phase on the Norwegian side is much narrower, covering the northwestern corner of southern Norway, including the only area where early Tertiary volcanism has been reported (Vestbrona Formation, Bugge *et al.* 1980; Prestvik *et al.* 1999). Magmatic underplating added to relatively thick continental crust and/or buoyancy from heated lithosphere can both explain the uplift.

The Late Paleocene–Early Eocene accelerated subsidence in large parts of the North Sea basin is also difficult to explain. White & Latin (1993) and Hall & White (1994) claimed that a period of crustal extension occurred in early Tertiary time. Minor normal faulting is observable in the basin to support this hypothesis.

Skogseid *et al.* (2000) related Paleocene vertical movements in the North Atlantic region to the arrival of the Iceland mantle plume to lithospheric levels and the subsequent lateral spreading of the plume head. The thickness of the plume body is both a function of the distance from the plume centre and the structure of the overlying lithosphere. The hot plume material preferentially fills traps at the base of the lithosphere created either by previous plate deformations or by continuing lithospheric thinning (Sleep 1996, 1997). In their dynamic model the central rift zone with the thinnest lithosphere and the thickest plume body is uplifted and eroded. In regions where the lithosphere is thick with respect to the thickness of the plume body (e.g. the North Sea) significant and rapid subsidence is expected during the plume emplacement. This subsidence should theoretically be followed by a rapid rebound, resulting in net uplift as the plume movement ceases (Skogseid *et al.* 2000).

Rapid Eocene subsidence has been related to a decrease in dynamic uplift caused by a reduction in plume activity (Nadin *et al.* 1997). Rapid decay of uplift is attributed to a temperature decrease of the plume in Early Eocene time (*c.* 55 Ma), resulting in a decrease in dynamic uplift. This event coincided with a decrease of the plume activity in the British Tertiary and Greenland igneous provinces, and the initiation of sea-floor spreading between Greenland and NW Europe.

Oligocene–Miocene domal uplift of southern Norway

Important observations that have to be explained include: (1) Early Oligocene onset of outbuilding from Norway–Denmark to Lofoten; (2) contemporaneos uplift of the eastern basin flank (offshore areas); (3) a break in sedimentation at the Eocene–Oligocene transition; (4) low-velocity or hotter upper mantle beneath the domes in southern and northern Norway.

The domal uplift of the Norwegian mainland probably started in Late Eocene to Early Oligocene time, and also contributed to uplift of the shelf areas. The uplift and shallowing in the northern North Sea basin continued into Miocene time. Several mechanisms have been suggested to explain the Neogene domal uplift of southern Norway and similar uplifted areas around the North Atlantic.

Rohrman & van der Beek (1996) proposed a model based on simple fluid dynamics and linked this to the elevated North Atlantic upper-mantle thermal regime surrounding the Iceland hotspot. In their model, an anomalous hot, buoyant asthenospheric layer or lens is present at the base of the lithosphere. On interaction of the hot (low-viscosity) asthenosphere layer with the cold (higher-viscosity) shield lithosphere, diapirs will start to form as a result of convective instabilities. The diapirism was probably triggered by lithospheric stress patterns that caused the plate reorganization at *c.* 30 Ma. The model implies that uplift is transient and will change to subsidence when the thermal anomaly decays. However, diapirs penetrating the lithosphere could generate partial melting and produce underplating, hereby generating permanent uplift. The areas of uplift are associated with strong negative Bouguer gravity anomalies and reduced lithospheric P- and S-wave velocities, suggesting an anomalous mantle structure and temperature underneath the domes (Bannister *et al.* 1991).

Plio-Pleistocene glacial erosion and uplift

Important observations that have to be explained include: (1) Upper Pliocene and older sequences tilted away from the dome in south Norway; (2) Pleistocene sediments, nearly flat-lying above an angular unconformity; (3) the angular unconformity separating Pleistocene and Pliocene sediments becoming less pronounced towards both the central North Sea and the shelf offshore mid-Norway; (4) accelerated subsidence of basin centres adjacent to the uplifted landmasses.

The seismic lines show that the post-Eocene sequences seem to be uniformly tilted away from the Scandinavian dome, indicating that a significant part of the Neogene uplift of south Norway occurred in latest Pliocene and earliest Pleistocene times. Isostatic response to unloading caused by glacial erosion contributed significantly to the Neogene uplift (Riis & Fjeldskaar 1992). However, unloading cannot be the only operating mechanism for the south Norway Neogene uplift (Riis 1996).

Intra-plate stress: compressional deformation

Tertiary epeirogeny is often attributed to compression that is assumed to be related in a general sense to Alpine mountain building. However, to remove *c.* 3 km of sedimentary rock from a basin *c.* 100 km wide requires >15 km of shortening (Brodie & White 1994). Minor Tertiary compression is observed all over the continental shelf, but nowhere is it sufficient to account for the required amount of uplift and erosion. In addition, exhumation dramatically increases from south to north, whereas the observed compression decreases markedly in the same direction.

Compressional structures of Cenozoic age, including simple domes or anticlines, reverse faults and broad-scale inversion, are widespread in the North Sea and along the NE Atlantic margin (Figs 15 and 17) (Doré & Lundin 1996). A multiphase growth history has been reported for some of the structures at the Norwegian margin (Mid-Eocene to Early Oligocene and Miocene times; Doré & Lundin 1996) and in the Faeroe region (Late Eocene to Early Oligocene and mid-Miocene times; Boldreel & Andersen 1998; Andersen *et al.* 2000). Vågnes *et al.* (1998), on the other hand, reported a surprisingly constant growth rate for the Ormen Lange Dome from earliest Eocene time to the present. These studies all relate the Cenozoic compressional deformation to two main sources: (1) far-field effects reflecting episodes of deformation in the Alpine Orogeny and (2) ridge-push from the North Atlantic mid-ocean ridge system.

Late Miocene and Pliocene differential vertical movements in the North Sea basin (increased subsidence rates at basin centre and relative uplift along basin edges) have also been related to changes in intra-plate stress correlated with plate tectonic adjustments in the Alpine hinterland and the Atlantic spreading system (Cloetingh *et al.* 1990; Galloway *et al.* 1993).

Summary and conclusions

This paper has focused on the Cenozoic evolution of the northern North Sea and surrounding areas, with emphasis on sediment distribution, composition and provenance as well as on timing, amplitude and wavelength of differential vertical movements. The data and modelling results support a probable tectonic control on sediment supply and on the formation of the regional unconformities. The sedimentary architecture and breaks are related to tectonic uplift of surrounding clastic source areas; thus, the offshore sedimentary record provides the best age constraints on the Cenozoic exhumation of the adjacent onshore areas. However, climatic and glacio-eustatic changes also played a role and for some events it is difficult to separate the effects of tectonics and eustasy.

Cenozoic exhumation is documented on both sides of the North Sea, but the timing is not well constrained. Two major uplift phases are clearly reflected by the available data: (1) an early Paleogene (Late Paleocene–Early Eocene) phase and (2) a late Neogene (Plio-Pleistocene) phase. The first is related to rifting, magmatism and break-up in the NE Atlantic associated with the arrival of the Iceland plume and the subsequent lateral spreading of the plume head. The latter is related to the isostatic response to unloading caused by glacial erosion during the widespread Northern Hemisphere glaciations. However, other uplift events interposed between these episodes, in particular on the Scandinavian side during Oligocene and Miocene times. Some of these may also be related to mantle processes and the episodic behaviour of the Iceland plume. Intra-plate stress related to ridge-push from the North Atlantic plate boundary and/or to Alpine compression also contributed to differential vertical movements, but cannot be the main uplift mechanism for the long-wavelength domal uplifts.

The paper is based on work within the project 'Tectonic impact on sedimentary processes in the post-rift phase—improved models'. The project was supported by the Research Council of Norway through grant 32842/211. The authors would like to thank the companies participating in the project (Amoco Norway Oil Company, den Norske Stats Oljeselskap (Statoil), Mobil Exploration Norway Inc., Norsk Agip A/S, Norsk Hydro ASA, Phillips Petroleum Company Norway, Saga Petroleum ASA). The seismic data were kindly made available by TGS-NOPEC. We are also grateful to D. Mudge and K. G. Røssland for reviewing the paper and for suggesting improvements.

References

ABREU, V.S. & ANDERSON, J.B. 1998. Glacial eustasy during the Cenozoic; sequence stratigraphic implications. *AAPG Bulletin*, **82**, 1385–1400.

ALLEN, P.A. & ALLEN, J.R. 1990. *Basin Analysis; Principles and Applications*. Blackwell Scientific, Oxford.

ANDERSEN, M.S., NIELSEN, T., NIELSEN, T., SØRENSEN, Aa.B., BOLDREEL, L.O. & KUIJPERS, A. 2000. Cenozoic sediment distribution and tectonic movements in the Faeroe region. *Global and Planetary Change*, **24**, 239–259.

ANDREWS, I.J., LONG, D., RICHARDS, P.C., THOMSON, A.R., BROWN, S., CHESHER, J.A. & MCCORMAC, M. 1990. *United Kingdom Offshore Regional Report: the Geology of the Moray Firth*. HMSO for the British Geological Survey, London.

BADLEY, M.E., PRICE, J.D., RAMBECH DAHL, C. & AGDESTEIN, T. 1988. The structural evolution of the northern Viking Graben and its bearing upon extensional modes of basin formation. *Journal of the Geological Society, London*, **145**, 455–472.

BANNISTER, S.C., RUUD, B.O. & HUSEBYE, E.S. 1991. Tomographic estimates of sub-Moho seismic velocities in Fennoscandia and structural implications. *Tectonophysics*, **189**, 37–53.

BERGER, W.H. & JANSEN, E. 1994. Mid-Pleistocene climate shift; the Nansen connection. *In*: JOHANNESSEN, O.M., MUENCH, R.D. & OVERLAND, J.E. (eds) *The Polar Oceans and their Role in Shaping the Global Environment; the Nansen Centennial Volume*. Geophysical Monograph, American Geophysical Union, **85**, 295–311.

BERTRAM, G.T. & MILTON, N.J. 1989. Reconstructing basin evolution from sedimentary thickness; the importance of palaeobathymetric control, with reference to the North Sea. *Basin Research*, **1**, 247–257.

BIJLSMA, S. 1981. Fluvial sedimentation from the Fennoscandian area into the North-West European Basin during the late Cenozoic. *Geologie en Mijnbouw*, **60**, 337–345.

BOLDREEL, L.O. & ANDERSEN, M.S. 1998. Tertiary compressional structures on the Faroe–Rockall Plateau in relation to Northeast Atlantic ridge-push and Alpine foreland stresses. *Tectonophysics*, **300**, 13–28.

BRAY, R.J., GREEN, P.F. & DUDDY, I.R. 1992. Thermal history reconstruction using apatite fission track analysis and vitrinite reflectance; a case study from the UK East Midlands and southern North Sea. *In*: HARDMAN, R.P.F. (ed.) *Exploration Britain; Geological Insights for the Next Decade*. Geological Society, London, Special Publications, **67**, 2–25.

BRODIE, J. & WHITE, N. 1994. Sedimentary basin inversion caused by igneous underplating; Northwest European continental shelf. *Geology*, **22**, 147–150.

BRODIE, J. & WHITE, N. 1995. The link between sedimentary basin inversion and igneous underplating. *In*: BUCHANAN, J.G. & BUCHANAN, P.G.

(eds) *Basin Inversion*. Geological Society, London, Special Publications, **88**, 21–38.

BUCHARDT, B. 1978. Oxygen isotope palaeotemperatures from the Tertiary period in the North Sea area. *Nature*, **275**, 121–123.

BUGGE, T., PRESTVIK, T. & ROKOENGEN, K. 1980. Lower Tertiary volcanic rocks off Kristiansund; Mid Norway. *Marine Geology*, **35**, 277–286.

CAMERON, T.D.J., BULAT, J. & MESDAG, C.S. 1993. High resolution seismic profile through a Late Cenozoic delta complex in the southern North Sea. *Marine and Petroleum Geology*, **10**, 591–599.

CLAUSEN, O.R. & HUUSE, M. 1999. Topography of the Top Chalk surface on- and offshore Denmark. *Marine and Petroleum Geology*, **16**, 677–691.

CLAUSEN, J.A., GABRIELSEN, R.H., REKSNES, P.A. & NYSÆTHER, E. 1999*a*. Development of intraformational (Oligocene–Miocene) faults in the northern North Sea; influence of remote stresses and doming of Fennoscandia. *Journal of Structural Geology*, **21**, 1457–1475.

CLAUSEN, O.R., GREGERSEN, U., MICHELSEN, O. & SØRENSEN, J.C. 1999*b*. Factors controlling the Cenozoic sequence development in the eastern parts of the North Sea. *Journal of the Geological Society, London*, **156**, 809–816.

CLAUSEN, O.R., NIELSEN, O.B., HUUSE, M. & MICHELSEN, O. 2000. Geological indications for Palaeogene uplift in the eastern North Sea basin. *Global and Planetary Change*, **24**, 175–187.

CLIFT, P.D. 1999. The thermal impact of Paleocene magmatic underplating in the Faeroe–Shetland–Rockall region. *In*: FLEET, A.J. & BOLDY, S.A.R. (eds) *Petroleum Geology of Northwest Europe; Proceedings of the 5th Conference*. Geological Society, London, 585–593.

CLIFT, P.D. & TURNER, J. 1998. Paleogene igneous underplating and subsidence anomalies in the Rockall–Faeroe–Shetland area. *Marine and Petroleum Geology*, **15**, 223–243.

CLOETINGH, S., GRADSTEIN, F.M., KOOI, H., GRANT, A.C. & KAMINSKI, M. 1990. Plate reorganization; a cause of rapid late Neogene subsidence and sedimentation around the North Atlantic. *Journal of the Geological Society, London*, **147**, 495–506.

DANIELSEN, M., MICHELSEN, O. & CLAUSEN, O.R. 1997. Oligocene sequence stratigraphy and basin development in the Danish North Sea sector based on log interpretations. *Marine and Petroleum Geology*, **14**, 931–950.

DORÉ, A.G. & LUNDIN, E.R. 1996. Cenozoic compressional structures on the NE Atlantic margin; nature, origin and potential significance for hydrocarbon exploration. *Petroleum Geoscience*, **2**, 299–311.

DORÉ, A.G., LUNDIN, E.R., JENSEN, L.N., BIRKELAND, Ø., ELIASSEN, P.E. & FICHLER, C. 1999. Principal tectonic events in the evolution of the northwest European Atlantic margin. *In*: FLEET, A.J. & BOLDY, S.A.R. (eds) *Petroleum Geology of Northwest Europe: Proceedings of the 5th Conference*. Geological Society, London, 41–61.

EIDVIN, T. & RIIS, F. 1992. *En biostratigrafisk og seismo-stratigrafisk analyse av tertiære sedimenter i nordlige deler av Norskerenna, med hovedvekt på øvre pliocene vifteavsetninger*. Norwegian Petroleum Directorate Contribution, **32**.

EIDVIN, T., BREKKE, H., RIIS, F. & RENSHAW, D.K. 1998. Cenozoic stratigraphy of the Norwegian Sea continental shelf, 64 degrees N–68 degrees N. *Norsk Geologisk Tidsskrift*, **78**, 125–151.

EIDVIN, T., JANSEN, E., RUNDBERG, Y., BREKKE, H. & GROGAN, P. 2000. The upper Cainozoic of the Norwegian continental shelf correlated with the deep sea record of the Norwegian Sea and the North Atlantic. *Marine and Petroleum Geology*, **17**, 579–600.

EIDVIN, T., RIIS, F. & RUNDBERG, Y. 1999. Upper Cainozoic stratigraphy in the central North Sea (Ekofisk and Sleipner fields). *Norsk Geologisk Tidsskrift*, **79**, 97–128.

ELDHOLM, O., SKOGSEID, J., SUNDVOR, E. & MYHRE, A.M. 1990. The Norwegian–Greenland Sea. *In*: GRANTZ, A., JOHNSON, L. & SWEENEY, J.F. (eds) *The Arctic Ocean Region, The Geology of North America, Series L*. Geological Society of America, Boulder, CO, 351–364.

FÆRSETH, R.B. 1996. Interaction of Permo-Triassic and Jurassic extensional fault-blocks during the development of the northern North Sea. *Journal of the Geological Society, London*, **153**, 931–944.

FÆRSETH, R.B., KNUDSEN, B.E., LILJEDAHL, T., MIDBØ, P.S. & SØDERSTRØM, B. 1997. Oblique rifting and sequential faulting in the Jurassic development of the northern North Sea. *Journal of Structural Geology*, **19**, 1285–1302.

FUNNELL, B.M. 1996. Plio-Pleistocene palaeogeography of the southern North Sea basin (3.75–0.60 Ma). Quaternary stratigraphy and palaeoecology; festschrift in honour of Richard West. *Quaternary Science Reviews*, **15**, 391–405.

GABRIELSEN, R.H., FAERSETH, R.B., STEEL, R.J., IDIL, S. & KLØVJAN, O.S. 1990. Architectural styles of basin fill in the northern Viking Graben. *In*: BLUNDELL, D.J. & GIBBS, A.D. (eds) *Tectonic Evolution of the North Sea Rifts*. International Lithosphere Program Publication, **181**, 158–179.

GABRIELSEN, R.H., KYRKJEBØ, R., FALEIDE, J.I., FJELDSKAAR, W. & KJENNERUD, T. 2001. The Cretaceous post-rift development in the northern North Sea. *Petroleum Geoscience*, **7**, 137–154.

GALLOWAY, W.E., GARBER, J.L., LIU, X. & SLOAN, B.J. 1993. Sequence stratigraphic and depositional framework of the Cenozoic fill, central and northern North Sea basin. *In*: PARKER, J.R. (ed.) *Petroleum Geology of Northwest Europe; Proeedings of the 4th Conference*. Geological Society, London, 33–43.

GHAZI, S.A. 1992. Cenozoic uplift in the Stord Basin area and its consequences for exploration. *Norsk Geologisk Tidsskrift*, **72**, 285–290.

GIBBARD, P.L. 1988. The history of the great northwest European rivers during the past three million years. *Philosophical Transactions of the Royal Society of London, Series B*, **318**, 559–602.

GILLMORE, G.K., KJENNERUD, T. & KYRKJEBØ, R. 2001. The reconstruction and analysis of palaeowater depths: a new approach and test of micropalaeontological approaches in the post-rift (Cretaceous to Quaternary) interval of the Northern North Sea. *In*: MARTINSEN, O. & DREYER, T. (eds) *Sedimentary Environments Offshore Norway—Palaeozoic to Recent*. Norwegian Petroleum Society (NPF) Special Publication, **10**, 365–381.

GRADSTEIN, F. & BÄCKSTRÖM, S. 1996. Cainozoic biostratigraphy and paleobathymetrty, northern North Sea and Haltenbanken. *Norsk Geologisk Tidsskrift*, **76**, 3–32.

GRADSTEIN, F.M. & OGG, J. 1996. A Phanerozoic time scale. *Episodes*, **19**, 3–6.

GRADSTEIN, F. M., KAMINSKI, M. A., BERGGREN, W. A., KRISTIANSEN, I. L. & D'IORIO, M. 1994. Cainozoic Biostratigraphy of the North Sea and Labrador Sea. *Micropaleontology*, **40** (Supplement).

GREEN, P.F. 1986. On the thermo-tectonic evolution of northern England; evidence from fission track analysis. *Geological Magazine*, **123**, 493–506.

GREEN, P.F. 1989. Thermal and tectonic history of the East Midlands shelf (onshore UK) and surrounding regions assessed by apatite fission track analysis. *Journal of the Geological Society, London*, **146**, 755–774.

GREGERSEN, U. 1998. Upper Cenozoic channels and fans on 3D seismic data in the northern Norwegian North Sea. *Petroleum Geoscience*, **4**, 67–80.

GREGERSEN, U., MICHELSEN, O. & SØRENSEN, J.C. 1997. Stratigraphy and facies distribution of the Utsira Formation and the Pliocene sequences in the northern North Sea. *Marine and Petroleum Geology*, **14**, 893–914.

HALL, B.D. & WHITE, N. 1994. Origin of anomalous Tertiary subsidence adjacent to North Atlantic continental margins. *Marine and Petroleum Geology*, **11**, 702–714.

HANSEN, S. 1996. Quantification of net uplift and erosion on the Norwegian Shelf south of 66 degrees N from sonic transit times of shale. *Norsk Geologisk Tidsskrift*, **76**, 245–252.

HAQ, B.U., HARDENBOL, J. & VAIL, P.R. 1987. Chronology of fluctuating sea-levels since the Triassic. *Science*, **235**, 1156–1167.

HAQ, B.U., HARDENBOL, J. & VAIL, P.R. 1988. Mesozoic and Cenozoic chronostratigraphy and cycles of sea-level changes. *In*: WILGUS, C.K., HASTINGS, B.S., POSAMENTIER, C.G.St.C., ROSS, C.A. & VAN WAGONER, J.C. (eds) *Sea Level Changes: an Integrated Approach*. Society of Economic Paleontologists and Mineralogists, Special Publication, **42**, 71–108.

HEILMANN-CLAUSEN, C., NIELSEN, O.B. & GERSNER, F. 1985. Lithostratigraphy and depositional environments in the Upper Paleocene and Eocene of Denmark. *Bulletin of the Geological Society of Denmark (Meddelelser fra Dansk Geologisk Forening)*, **33**, 287–323.

HERITIER, F.E., LOSSEL, P. & WATHNE, E. 1979. Frigg Field; large submarine-fan trap in Lower Eocene rocks of North Sea Viking graben. *AAPG Bulletin*, **61**, 1999–2020.

HERITIER, F.E., LOSSEL, P. & WATHNE, E. 1981. The Frigg gas field. *In*: ILLING, L.V. & HOBSON, G.D. (eds) *Petroleum Geology of the Continental Shelf of North-West Europe; Proceedings of the Second Conference*. Heyden and Son, London, 380–391.

HILLIS, R.R. 1995*a*. Quantification of Tertiary exhumation in the United Kingdom southern North Sea using sonic velocity data. *AAPG Bulletin*, **79**, 130–152.

HILLIS, R.R. 1995*b*. Regional Tertiary exhumation in and around the United Kingdom. *In*: BUCHANAN, J.G. & BUCHANAN, P.G. (eds) *Basin Inversion*. Geological Society, London, Special Publications, **88**, 167–190.

HJELSTUEN, B.O., ELDHOLM, O. & SKOGSEID, J. 1999. Cenozoic evolution of the northern Vøring margin. *Geological Society of America Bulletin*, **111**, 1792–1807.

HOLLIDAY, D.W. 1993. Mesozoic cover over northern England; interpretation of apatite fission track data. *Journal of the Geological Society, London*, **150**, 657–660.

HUUSE, M. 1999. Detailed morphology of the Top Chalk surface in the eastern Danish North Sea. *Petroleum Geoscience*, **5**, 303–314.

ISAKSEN, D. & TONSTAD, K. 1989. A Revised Cretaceous and Tertiary Lithostratigraphy Nomenclature for the Norwegian North Sea. *Norwegian Petroleum Directorate Bulletin*, 5.

JANSEN, E. & SJØHOLM, J. 1991. Reconstruction of glaciation over the past 6 Myr from ice-borne deposits in the Norwegian Sea. *Nature*, **349**, 600–603.

JAPSEN, P. 1997. Regional Neogene exhumation of Britain and the western North Sea. *Journal of the Geological Society, London*, **154**, 239–247.

JAPSEN, P. 1998. Regional velocity–depth anomalies, North Sea Chalk; a record of overpressure and Neogene uplift and erosion. *AAPG Bulletin*, **82**, 2031–2074.

JAPSEN, P. 1999. Overpressured Cenozoic shale mapped from velocity anomalies relative to a baseline for marine shale, North Sea. *Petroleum Geoscience*, **5**, 321–336.

JAPSEN, P. 2000. Investigation of multi-phase erosion using reconstructed shale trends based on sonic data; Sole Pit axis, North Sea. *Global and Planetary Change*, **24**, 189–210.

JAPSEN, P. & CHALMERS, J.A. 2000. Neogene uplift and tectonics around the North Atlantic; overview. *Global and Planetary Change*, **24**, 165–173.

JENSEN, L.N. & SCHMIDT, B.J. 1992. Late Tertiary uplift and erosion in the Skagerrak area; magnitude and consequences. *Norsk Geologisk Tidsskrift*, **72**, 275–279.

JENSEN, L.N. & SCHMIDT, B.J. 1993. Neogene uplift and erosion offshore South Norway; magnitude and consequences for hydrocarbon exploration in the Farsund Basin. *In*: SPENCER, A.M. (ed.) *Generation, Accumulation, and Production of Europe's Hydrocarbons; III*. Special Publication of the European Association of Petroleum Geoscientists, **3**, 79–88.

JOHNSON, H., RICHARDS, P.C., LONG, D. & GRAHAM, C.C. 1993. *United Kingdom Offshore Regional*

Report: the Geology of the Northern North Sea. HMSO for the British Geological Survey, London.

JONES, R.W. & MILTON, N.J. 1994. Sequence development during uplift; Palaeogene stratigraphy and relative sea-level history of the Outer Moray Firth, UK North Sea. *Marine and Petroleum Geology*, **11**, 157–165.

JORDT, H., FALEIDE, J.I., BJØRLYKKE, K. & IBRAHIM, M.T. 1995. Cenozoic sequence stratigraphy of the central and northern North Sea basin; tectonic development, sediment distribution and provenance areas. *Marine and Petroleum Geology*, **12**, 845–879.

JORDT, H., THYBERG, B.I., NØTTVEDT, A. 2000. Cenozoic evolution of the central and northern North Sea with focus on differential vertical movements of the basin floor and surrounding clastic source areas. *In*: NØTTVEDT, A. (ed.) *Dynamics of the Norwegian Margin*. Geological Society, London, Special Publications, **167**, 219–243.

JOY, A.M. 1992. Estimation of Cenozoic water depths in the western Central Graben, UK North Sea, by subsidence modelling. *In*: HARDMAN, R.F.P. (ed.) *Exploration Britain; Geological Insights for the Next Decade*. Geological Society, London, Special Publications, **67**, 107–125.

JOY, A.M. 1993. Comments on the pattern of post-rift subsidence in the central and northern North Sea Basin. *In*: WILLIAMS, G.D. & DOBB, A. (eds) *Tectonics and Seismic Sequence Stratigraphy*. Geological Society, London, Special Publications, **71**, 123–140.

JOY, A.M. 1996. Controls on Eocene sedimentation in the central North Sea Basin; results of a basinwide correlation study. *In*: KNOX, R.W.O'B., CORFIELD, R.M. & DUNAY, R.E. (eds) *Correlation of the Early Paleogene in Northwest Europe*. Geological Society, London, Special Publications, **101**, 79–90.

KING, E.L., SEJRUP, H.P., HAFLIDASON, H., ELVERHØI, A. & AARSETH, I. 1996. Quaternary seismic stratigraphy of the North Sea Fan; glacially-fed gravity flow aprons, hemipelagic sediments, and large submarine slides. *Marine Geology*, **130**, 293–315.

KJENNERUD, T., FALEIDE, J.I., GABRIELSEN, R.H., GILLMORE, G.K., KYRKJEBØ, G.K., LIPPARD, S.J. & LØSETH, H. 2001. Structural restoration of Cretaceous–Cainozoic palaeobathymetry in the northern North Sea Basin. *In*: MARTINSEN, O. & DREYER, T. (eds) *Sedimentary Environments Offshore Norway—Palaeozoic to Recent*. Norwegian Petroleum Society (NPF) Special Publication, **10**, 347–364.

KNOX, R.W.O'B. 1996. Correlation of the early Paleogene in Northwest Europe; an overview. *In*: KNOX, R. W. O'B., CORFIELD, R.M. & DUNAY, R.E. (eds) *Correlation of the Early Paleogene in Northwest Europe*. Geological Society, London, Special Publications, **101**, 1–11.

KNOX, R. W. O'B. & HOLLOWAY, S. 1992. Paleogene of the central and northern North Sea. *In*: KNOX, R. W. O'B. & CORDEY, W.G. (eds) *Lithostratigraphic Nomenclature of the UK North Sea*. British Geological Survey, Keyworth.

KNOX, R. W. O'B. & MORTON, A.C. 1988. The record of early Tertiary N Atlantic volcanism in sediments of the North Sea Basin. *In*: MORTON, A.C. & PARSON, L.M. (eds) *Early Tertiary Volcanism and the Opening of the NE Atlantic*. Geological Society, London, Special Publications, **39**, 407–419.

KUZNIR, N.J., ROBERTS, A.M. & MORLEY, C.K. 1995. Forward and reverse modelling of rift basin formation. *In*: LAMBIASE, J.J. (ed.) *Hydrocarbon Habitat in Rift Basins*. Geological Society, London, Special Publications, **80**, 33–56.

KYRKJEBØ, R., FJELDSKAAR, W., FALEIDE, J. I. & GABRIELSEN, R. H. 1999. The post-rift (Cretaceous–Tertiary) vertical tectonic movements in the northern North Sea as obtained by 2D backstripping. In: Kyrkjebø, R. *The Cretaceous–Tertiary of the northern North Sea: thermal and tectonic influences in a post-rift setting*. Dr. Scient. thesis, University of Bergen, Norway.

KYRKJEBØ, R., KNENNERUD, T., GILLMORE, G.K., FALEIDE, J.I. & GABRIELSEN, R.H. 2001. Cretaceous–Tertiary palaeo-bathymetry in the northern North Sea; integration of palaeo-water depth estimates obtained by structural restoration and micropalaeontological analysis. *In*: MARTINSEN, O. & DREYER, T. (eds) *Sedimentary Environments Offshore Norway—Palaeozoic to Recent*. Norwegian Petroleum Society (NPF) Special Publication, **10**, 321–345.

LEWIS, C.L.E., GREEN, P.F., CARTER, A. & HURFORD, A.J. 1992. Elevated K/T paleotemperatures throughout Northwest England; three kilometres of Tertiary erosion? *Earth and Planetary Science Letters*, **112**, 131–145.

LIDMAR-BERGSTROM, K., OLLIER, C.D. & SULEBAK, J.R. 2000. Landforms and uplift history of Southern Norway. *Global and Planetary Change*, **24**, 211–231.

MALM, O.A., CHRISTENSEN, O.B, FURNES, H., LØVLIE, R., RUSELÅTTEN, H., ØSTBY, K.L., *et al.* 1984. The Lower Tertiary Balder Formation: an organogenic and tuffaceous deposit in the North Sea region. *In*: SPENCER, A.M. *et al.* (eds) *Petroleum Geology of the North European Margin*. Graham & Trotman, London, 149–170.

MARTINSEN, O.J., BØEN, F., CHARNOCK, M.A., MANGERUT, G. & NØTTVEDT, A. 1999. Cenozoic development of the Norwegian margin 60–64°N: sequences and sedimentary response to variable basin physiography and tectonic setting. *In*: FLEET, A.J. & BOLDY, S.A.R. (eds) *Petroleum Geology of NW Europe, Proceedings of the 5th Conference*. Geological Society, London, 293–304.

MICHELSEN, O., DANIELSEN, M., HEILMANN-CLAUSEN, C., JORDT, H., LAURSEN, G.V. & THOMSEN, E. 1995. Occurrence of major sequence stratigraphic boundaries in relation to basin development in Cenozoic deposits of the southeastern North Sea. *In*: STEEL, R.J., FELT, V.L., JOHANNESEN, E.P. & MATHIEU, C. (eds) *Sequence Stratigraphy on the Northwest European Margin*. Norwegian Petroleum Society Special Publication, **5**, 415–427.

MICHELSEN, O., THOMSEN, E., DANIELSEN, M., HEILMANN-CLAUSEN, C., JORDT, H. & LAURSEN, G.V. 1998. Cenozoic sequence stratigraphy in the eastern North Sea. *In*: DE GRACIANSKY, P.-C., HARDENBOL, J., JACQUIN, T. & VAIL, P.R. (eds) *Mesozoic and Cenozoic Sequence Stratigraphy of European Basins*. Special Publication, Society for Sedimentary Geology, **60**, 91–118.

MILLER, K.G., FAIRBANKS, R.G. & MOUNTAIN, G.S. 1987. Tertiary oxygen isotope synthesis, sea level history, and continental margin erosion. *Paleoceanography*, **2**, 1–19.

MILTON, N.J. 1993. Evolving depositional geometries in the North Sea Jurassic rift. *In*: PARKER, J.R. (ed.) *Petroleum Geology of Northwest Europe; Proceedings of the 4th Conference*. Geological Society, London, 425–442.

MILTON, N.J., BERTRAM, G.T. & VANN, I.R. 1990. Early Paleogene tectonics and sedimentation in the central North Sea. *In*: HARDMAN, R.F.P. & BROOKS, J. (eds) *Proceedings of Tectonic Events Responsible for Britain's Oil and Gas Reserves*. Geological Society, London, Special Publications, **55**, 339–351.

MUDGE, D.C. & BUJAK, J.P. 1994. Eocene stratigraphy of the North Sea Basin. *Marine and Petroleum Geology*, **11**, 166–181.

NADIN, P.A. & KUSZNIR, N.J. 1995. Palaeocene uplift and Eocene subsidence in the northern North Sea basin from 2D forward and reverse stratigraphic modelling. *Journal of the Geological Society, London*, **152**, 833–848.

NADIN, P.A. & KUZNIR, N.J. 1996. Forward and reverse stratigraphic modelling of Cretaceous–Tertiary post-rift subsidence and Paleogene uplift in the Outer Moray Firth Basin, central North Sea. *In*: KNOX, R. W. O'B., CORFIELD, R.M. & DUNAY, R.E. (eds) *Correlation of the Early Paleogene in Northwest Europe*. Geological Society, London, Special Publications, **101**, 43–62.

NADIN, P.A., KUSZNIR, N.J. & CHEADLE, M.J. 1997. Early Tertiary plume uplift of the North Sea and Faeroe–Shetland basins. *Earth and Planetary Science Letters*, **148**, 109–127.

NADIN, P.A., KUZNIR, N.J. & TOTH, J. 1995. Transient regional uplift in the early Tertiary of the northern North Sea and the development of the Iceland Plume. *Journal of the Geological Society, London*, **152**, 953–958.

NAYLOR, P.H., BELL, B.R., JOLLEY, D.W., DURNALL, P. & FREDSTED, R. 1999. Palaeogene magmatism in the Faeroe–Shetland Basin; influences on uplift history and sedimentation. *In*: FLEET, A.J. & BOLDY, S.A.R. (eds) *Petroleum Geology of Northwest Europe; Proceedings of the 5th Conference*. Geological Society, London, 545–558.

NIELSEN, O.B. & HEILMANN-CLAUSEN, C. 1988. Palaeogene volcanism; the sedimentary record in Denmark. *In*: MORTON, A.C. & PARSON, L.M. (eds) *Early Tertiary Volcanism and the Opening of the NE Atlantic*. Geological Society, London, Special Publications, **39**, 395–405.

NIELSEN, O.B., SØRENSEN, S., THIEDE, J. & SKARBØ, O. 1986. Cenozoic differential subsidence of North Sea. *AAPG Bulletin*, **70**, 276–298.

NØTTVEDT, A., GABRIELSEN, R.H. & STEEL, R.J. 1995. Tectonostratigraphy and sedimentary architecture of rift basins, with reference to the northern North Sea. *Marine and Petroleum Geology*, **12**, 881–901.

PITMAN, W.C. 1978. III Relationship between eustacy and stratigraphic sequences of passive margins. *Geological Society of America Bulletin*, **89**, 1389–1403.

PRESTVIK, T., TORSKE, T., SUNDVOLL, B. & KARLSSON, H. 1999. Petrology of early Tertiary nephelinites off mid-Norway; additional evidence for an enriched endmember of the ancestral Iceland Plume. *Lithos*, **46**, 317–330.

RATTEY, R.P. & HAYWARD, A.B. 1993. Sequence stratigraphy of a failed rift system; the Middle Jurassic to Early Cretaceous basin evolution of the central and northern North Sea. *In*: PARKER, J.R. (ed.) *Petroleum Geology of Northwest Europe; Proceedings of the 4th Conference*. Geological Society, London, 215–249.

RIIS, F. 1996. Quantification of Cenozoic vertical movements of Scandinavia by correlation of morphological surfaces with offshore data. *Global and Planetary Change*, **12**, 331–357.

RIIS, F. & FJELDSKAAR, W. 1992. On the magnitude of the late Tertiary and Quaternary erosion and its significance for the uplift of Scandinavia and the Barents Sea. *In*: LARSEN, R.M., BREKKE, H., LARSEN, B.T. & TALLERAAS, E. (eds) *Structural and Tectonic Modelling and its Application to Petroleum Geology*. Norwegian Petroleum Society (NPF) Special Publication, **1**, 163–185.

RITCHIE, J.D., GATLIFF, R.W. & RICHARDS, P.C. 1999. Early Tertiary magmatism in the offshore NW UK margin and surrounds. *In*: FLEET, A.J. & BOLDY, S.A.R. (eds) *Petroleum Geology of Northwest Europe; Proceedings of the 5th Conference*. Geological Society, London, 573–584.

ROBERTS, A.M., KUSZNIR, N.J., YIELDING, G. & STYLES, P. 1998. 2D flexural backstripping of extensional basins; the need for a sideways glance. *Petroleum Geoscience*, **4**, 327–338.

ROBERTS, A.M., YIELDING, G. & BADLEY, M.E. 1990. A kinematic model for the orthogonal opening of the Late Jurassic North Sea rift system, Denmark–mid Norway. *In*: BLUNDELL, D.J. & GIBBS, A.D. (eds) *Tectonic Evolution of the North Sea Rifts*. International Lithosphere Program Publication, **181**, 180–199.

ROBERTS, A.M., YIELDING, G. & BADLEY, M.E. 1993. Tectonics and bathymetric controls on stratigraphic sequences within evolving half-graben. *In*: WILLIAMS, G.D. & DOBB, A. (eds) *Tectonics and Seismic Sequence Stratigraphy*. Geological Society, London, Special Publications, **71**, 87–121.

ROHRMAN, M. & VAN DER BEEK, P. 1996. Cenozoic postrift domal uplift of North Atlantic margins; an asthenospheric diapirism model. *Geology*, **24**, 901–904.

ROHRMAN, M., VAN DER BEEK, P., ANDRIESSEN, P. & CLOETINGH, S. 1995. Meso-Cenozoic morphotectonic evolution of southern Norway; Neogene domal uplift inferred from apatite fission track thermochronology. *Tectonics*, **14**, 704–718.

ROKOENGEN, K., RISE, L., BRYN, P., FRENGSTAD, B., GUSTAVSEN, B., NYGAARD, E. & SAETTEM, J. 1995. Upper Cenozoic stratigraphy on the mid-Norwegian continental shelf. *Norsk Geologisk Tidsskrift*, **75**, 88–104.

RUDDIMAN, W.F., RAYMO, M. & MCINTYRE, A. 1986. Matuyama 41,000-year cycles; North Atlantic Ocean and Northern Hemisphere ice sheets. *Earth and Planetary Science Letters*, **80**, 117–129.

RUNDBERG, Y. 1989. *Tertiary sedimentary history and basin evolution of the Norwegian North Sea between 60°N and 62°N. An integrated approach.* PhD thesis, University of Trondheim.

RUNDBERG, Y. & SMALLEY, P.C. 1989. High-resolution dating of Cenozoic sediments from northern North Sea using $^{87}Sr/^{86}Sr$ stratigraphy. *AAPG Bulletin*, **73**, 298–308.

RUNDBERG, Y., OLAUSSEN, S. & GRADSTEIN, F. 1995. Incision of Oligocene strata; evidence for northern North Sea Miocene uplift and key to the formation of the Utsira sands. *Geonytt*, **22**, (abstract).

SEIDENKRANTZ, M.-S. 1992. Plio-Pleistocene foraminiferal paleoecology and stratigraphy in the northernmost North Sea. *Journal of Foraminiferal Research*, **22**, 363–378.

SEJRUP, H.P., AARSETH, I. & HAFLIDASON, H. 1991. The Quaternary succession in the northern North Sea. *Marine Geology*, **101**, 103–111.

SEJRUP, H.P., AARSETH, I., HAFLIDASON, H., LØVLIE, R., BRATTEN, Å., TJOSTHEIM, G., FORSBERG, C.F. & ELLINGSEN, K.L. 1995. Quaternary of the Norwegian Channel; glaciation history and palaeoceanography. *Norsk Geologisk Tidsskrift*, **75**, 65–87.

SEJRUP, H.P., LARSEN, E., LANDVIK, J., KING, E.L., HAFLIDASON, H. & NESJE, A. 2000. Quaternary glaciations in southern Fennoscandia; evidence from southwestern Norway and the northern North Sea region. *Quaternary Science Reviews*, **19**, 667–685.

SKOGSEID, J. & ELDHOLM, O. 1989. Vøring Plateau continental margin; seismic interpretation, stratigraphy, and vertical movements. *In*: ELDHOLM, O., THIEDE, J., TAYLOR, E. *et al.* (eds) *Proceedings of the Ocean Drilling Program, Scientific Results, 104*. Ocean Drilling Program, College Station, TX, 993–1030.

SKOGSEID, J., PLANKE, S., FALEIDE, J.I., PEDERSEN, T., ELDHOLM, O., NEVERDAL, F., *et al.* 2000. NE Atlantic continental rifting and volcanic margin formation. *In*: NØTTVEDT, A. (ed.) *Dynamics of the Norwegian Margin*. Geological Society, London, Special Publications, **167**, 111–222.

SLEEP, N.H. 1990. Hotspots and mantle plumes; some phenomenology. *Journal of Geophysical Research (B)*, **95**, 6715–6736.

SLEEP, N.H. 1996. Lateral flow of hot plume material ponded at sublithospheric depths. *Journal of Geophysical Research (B)*, **101**, 28065–28083.

SLEEP, N.H. 1997. Lateral flow and ponding of starting plume material. *Journal of Geophysical Research (B)*, **102**, 10001–10012.

SØRENSEN, J.C., GREGERSEN, U., BREINER, M. & MICHELSEN, O. 1997. High-frequency sequence stratigraphy of upper Cenozoic deposits in the central and southeastern North Sea areas. *Marine and Petroleum Geology*, **14**, 99–123.

SPJELDNÆS, A. 1975. Palaeogeography and facies distribution in the Tertiary of Denmark and surrounding areas. *In*: WHITEMAN, A., ROBERTS, D. & SELLEVOLL, M. (eds) *Petroleum Geology and Geology of the North Sea and Northwest Atlantic Continental Margin*. Norges Geologiske Undersøkelse Bulletin, **316**, 289–311.

STECKLER, M.S. & WATTS, A.B. 1978. Subsidence of the Atlantic-type continental margin off New York. *Earth and Planetary Science Letters*, **41**, 1–13.

STEURBAUT, E., SPIEGLER, D., WEINELT, M. & THIEDE, J. 1991. *Cenozoic Erosion and sedimentation on the Northwest European Continental Margin*. Geomar Research Centre for Marine Geosciences, Christian-Albrechts Universität, Kiel.

STOKER, M.S., HITCHEN, K. & GRAHAM, C.C. 1993. *United Kingdom Offshore Regional Report: the Geology of the Hebrides and West Shetland Shelves, and Adjacent Deep-water Areas*. HMSO for the British Geological Survey, London.

STUEVOLD, L.M. & ELDHOLM, O. 1996. Cenozoic uplift of Fennoscandia inferred from a study of the mid-Norwegian margin. *Global and Planetary Change*, **12**, 359–386.

STUEVOLD, L.M., SKOGSEID, J. & ELDHOLM, O. 1992. Post-Cretaceous uplift events on the Voring continental margin. *Geology*, **20**, 919–922.

TALWANI, M. & ELDHOLM, O. 1977. Evolution of the Norwegian–Greenland Sea. *Geological Society of America Bulletin*, **88**, 969–999.

THYBERG, B.I., JORDT, H., BJØRLYKKE, K., FALEIDE, J.I., *et al.* 2000. Relationships between sequence stratigraphy, mineralogy and geochemistry in Cenozoic sediments of the northern North Sea. *In*: NØTTVEDT, A. (ed.) *Dynamics of the Norwegian Margin*. Geological Society, London, Special Publications, **167**, 111–222.

THYBERG, B.I., STABELL, B., FALEIDE, J.I. & BJØRLYKKE, K. 1999. Upper Oligocene diatomaceous deposits in the northern North Sea; silica diagenesis and paleogeographic implications. *Norsk Geologisk Tidsskrift*, **79**, 3–18.

TORSKE, T. & PRESTVIK, T. 1991. Mesozoic detachment faulting between Greenland and Norway; inferences from Jan Mayen fracture zone system and associated alkalic volcanic rocks. *Geology*, **19**, 481–484.

VÅGNES, E., GABRIELSEN, R.H. & HAREMO, P. 1998. Late Cretaceous–Cenozoic intraplate contractional deformation at the Norwegian continental shelf; timing, magnitude and regional implications. *Tectonophysics*, **300**, 29–46.

VAN VEEN, P., SKOLD, L.J. & RYSETH, A. 1994. *A high-resolution stratigraphic framework for the Paleogene in the Northern North Sea. A contribution to the IBS-DNM Project*. Norsk Hydro Research Centre, Bergen.

VEEKEN, P.C. 1997. The Cenozoic fill of the North Sea Basin (UK sector 56–62 degrees N), a seismic stratigraphic study with emphasis on Paleogene massflow deposits. *Geologie en Mijnbouw*, **75**, 317–340.

WATTS, A.B. & STECKLER, M.S. 1979. Subsidence and eustasy at the continental margin of eastern North America. *In*: TALWANI, M., HAY, W. & RYAN, W.B.F. (eds) *Deep Drilling Results in the Atlantic Ocean; Continental Margins and Paleoenvironment*. Maurice Ewing Series, American Geophysical Union, **3**, 218–234.

WATTS, A.B., KARNER, G.D. & STECKLER, M.S. 1982. Lithospheric flexure and the evolution of sedimentary basins. *Philosophical Transactions of the Royal Society of London, Series A*, **305**, 249–281.

WHITE, N. & LATIN, D. 1993. Subsidence analyses from the North Sea 'triple-junction'. *Journal of the Geological Society, London*, **150**, 473–488.

WHITE, N. & LOVELL, B. 1997. Measuring the pulse of a plume with the sedimentary record. *Nature*, **387**, 888–891.

YIELDING, G., BADLEY, M.E. & ROBERTS, A.M. 1992. The structural evolution of the Brent Province. *In*: MORTON, A.C., HASZELDINE, R.S. & BROWN, S. (eds) *Geology of the Brent Group*. Geological Society, London, Special Publications, **61**, 27–43.

ZACHOS, J.C., BREZA, J.R. & WISE, S.W. 1992. Early Oligocene ice-sheet expansion on Antarctica; stable isotope and sedimentological evidence from Kerguelen Plateau, southern Indian Ocean. *Geology*, **20**, 569–573.

ZAGWIJN, W.H. 1989. The Netherlands during the Tertiary and the Quaternary; a case history of coastal lowland evolution. *Geologie en Mijnbouw*, **68**, 107–120.

ZIEGLER, P.A. 1990. Tectonic and palaeogeographic development of the North Sea rift system. *In*: BLUNDELL, D.J. & GIBBS, A.D. (eds) *Tectonic Evolution of the North Sea Rifts*. International Lithosphere Program Publication, **181**, 1–36.

ZIEGLER, P.A. 1992. North Sea rift system. *Tectonophysics*, **208**, 55–75.

ZIEGLER, P.A. & VAN HOORN, B. 1989. Evolution of North Sea rift system. *Extensional Tectonics and Stratigraphy of the North Atlantic Margins. In*: TANKARD, A.J. & BALKWILL, H.R. (eds) American Association of Petroleum Geologists, Memoir, **46**, 471–500.

Scotland's denudational history: an integrated view of erosion and sedimentation at an uplifted passive margin

ADRIAN HALL[1,2] & PAUL BISHOP[3]

[1]*Department of Geography, University of Edinburgh, Drummond Street, Edinburgh EH8 9XP, UK*

[2]*Fettes College, Carrington Road, Edinburgh EH4 1QX, UK (e-mail: am.hall@fettes.com)*

[3]*The CRUST Project, Department of Geography and Topographic Science, University of Glasgow, Glasgow G12 8QQ, UK*

Abstract: Denudational history is commonly reconstructed from basin sediments derived from the denuded source area, and less frequently from the source area itself. Northern Britain is an important source area for the surrounding sedimentary basins and this paper reviews the erosional history of Scotland from Devonian time to the present using evidence both from onshore geology and geomorphology and from patterns of sedimentation in surrounding basins. Cover rocks were extensive in Scotland during late Palaeozoic time but the persistence of sediment source areas within the upland areas of Scotland makes it unlikely that basement highs were ever completely buried, and depths of post-Devonian erosion of basement have been correspondingly modest (< 1–2 km). During Mesozoic time, Scotland experienced several major erosional cycles, beginning with uplift, reactivation of relief and stripping of cover rocks, followed by progressive reduction of relief through etchplanation and culminating in extensive marine transgressions in Late Triassic, Late Jurassic and Late Cretaceous time. Mid-Paleocene pulses of coarse sediment to the Moray Firth Basin coincided with major uplift. This uplift was associated with major differential tectonics within the Highlands, with warping and faulting along the margins of the Minch and the inner Moray Firth Basins. Tectonic activity was renewed on a lesser scale in late Oligocene time and continued into Late Neogene time. Differential weathering and erosion under the warm to temperate humid climates of Neogene time created the major elements of the preglacial relief, with formation of valleys, basins, scarps and inselbergs, features often closely adjusted to lithostructural controls and, in some cases, with precursors that can be traced back to Devonian time. The history that can be 'read' from the onshore region complements the source area history interpreted from sedimentary basins derived from these areas.

The nature and rate of deposition of sedimentary basin sequences depend on many factors, including rates of basin subsidence, sea-level history and source area characteristics, such as climate history, lithology and uplift rates. Source area uplift is generally interpreted to be associated with an essentially instantaneous 'basin' signal of high rates of flux of sediments that have experienced limited chemical weathering. Although in some cases such a response may be demonstrable (e.g. Copeland & Harrison 1990), the degree to which major uplift is signalled by a sedimentary pulse depends on the extent to which the potential energy associated with high elevation can be converted into the kinetic energy (and hence the stream power) necessary to detach, entrain and transport high volumes of sediment. Two interrelated factors determine the extent to which potential energy provided by uplift can be converted to kinetic energy, namely, the geomorphological character of the source area, and the extent to which the source area 'knows' about the uplift event. A high-elevation, uplifting source area that is efficiently connected to base level will generate high volumes of sediment via a combination of river incision, mass movement from steep valley sides and efficient evacuation of sediment. The southwards draining rivers of the Himalayas offer excellent examples of such systems (e.g. Burbank *et al.* 1996; Hancock *et al.* 1998), which contrast markedly with the systems draining northwards from the Himalayas to the high-elevation, but low-energy and internally drained, Tibetan plateau (Summerfield & Brown 1998).

From: Doré, A.G., Cartwright, J.A., Stoker, M.S., Turner, J.P. & White, N. 2002. *Exhumation of the North Atlantic Margin: Timing, Mechanisms and Implications for Petroleum Exploration*. Geological Society, London, Special Publications, **196**, 271–290. 0305-8719/02/$15.00

The most rapidly uplifting areas, such as the Southern Alps (Hovius 2000; Tippett & Hovius 2000), Himalayas (Fielding 2000), Japan (Ohmori 2000) and Taiwan (Lin 2000), are associated with plate convergence zones. On the other hand, passive margin uplands, such as the southern African uplands (Brown *et al.* 2000), the Western Ghats in India (Gunnel & Fleitout 2000), the SE Australian highlands (Bishop & Goldrick 2000), and the Scottish Highlands portion of the western European Atlantic margin, are commonly associated with low to very low rates of denudation and sediment flux. Fleming *et al.* (1999) and Cockburn *et al.* (2000), for example, have reported very low rates of denudation from the southern African passive margin highlands, of the same order of magnitude as those reported from the SE Australian uplands from mass balance, geomorphological and thermochronological studies (Bishop 1985; Bishop & Goldrick 2000). Bishop & Goldrick (2000) also reported widespread and persistent disequilibria in SE Australian river long profiles, reflecting the low gradients and low stream power of this margin's drainage systems. These long profile disequilibria can be attributed to passive denudational rebound (Bishop & Brown 1992; Bishop & Goldrick 2000), with the margin evidently not having experienced active tectonic uplift during Cenozoic time apart from temporary uplift events related to transient thermal effects at the central volcanoes that mark eastern Australia's Cenozoic passage over mantle hotspots (Wellman & McDougall 1974; Wellman 1986; McDougall & Duncan 1988; Sun *et al.* 1989).

The tectonic character and histories of most of the passive margins descibed above have been reconstructed largely from subaerial terrestrial data, with relatively little reliance on the sedimentary basin record. The sedimentary record has been used mainly for mass balance studies of these margins to determine rates of source area subaerial denudation (e.g. Bishop 1985) or as a guide to the evolution of the river systems (e.g. Rust & Summerfield 1990). Rates of source area denudation may also be determined more directly from the source area itself using geomorphological studies (e.g. Bishop 1985; Nott *et al.* 1996), cosmogenic isotope analysis (Fleming *et al.* 1999; Cockburn *et al.* 2000), and low-temperature thermochronological techniques, such as apatite fission-track analysis (Gleadow & Brown 2000). Conflicts, which are not yet fully resolved, are often apparent, however, between geomorphological interpretations of source area history and thermochronological approaches to source area denudation (Kohn & Bishop 1999).

Reconstruction of the evolution of the Scottish Highlands (Fig. 1) on the Western European continental margin has relied on both offshore and onshore data, with often much greater emphasis on the offshore record, no doubt because of the wealth and quality of these data (see Jones *et al.* 2002). There is a corresponding wealth of data from onshore areas, and in this paper we re-examine the post-Palaeozoic geomorphological history of the Scottish Highlands as a source area for surrounding basins, especially the main sediment receiving area, the North Sea Basin. We have two aims: (1) to summarize critically the onshore data on the evolution of the Scottish Highlands, for a readership that might not be fully aware of the literature on this topic; (2) to assess the extent to which source area uplift, denudation and geomorphological development, and landscape antiquity, can be 'read' from the source area itself, thereby complementing the offshore record.

An outline of the history of the Highlands source region from Palaeozoic time to the present

The disposition and provenance of (often thin) remnants of Devonian sediments show that many key morphotectonic elements of the current Highlands relief were already established by the end of Devonian time (Fig. 2). These include the main Grampian watershed, the linear depression of the Great Glen, the large basins of NE Scotland and major valley systems draining NE towards the Moray Firth along the Caledonian fracture zones. The Caledonian mountains had been eroded, exposing many late Caledonian Newer Granites, together with some older intrusions (Watson 1985). The Caledonian granites were intruded into already stabilized crust or their intrusion completed the stabilization process (Leake & Cobbing 1993). Reconstruction of the sub-Devonian relief around the inner Moray Firth implies surfaces of high relief, with fault-bounded half-basins and fault-guided valleys partly infilled with conglomerates and sandstones. These Devonian fills were largely removed between late Palaeozoic time and the present so that the current level of erosion lies close to that at the end of Devonian time (see Leake & Cobbing 1993). A similar equivalence exists in the NW Highlands, where the present terrain lies at the same general elevation as the base of the Torridonian sequence (Watson 1985).

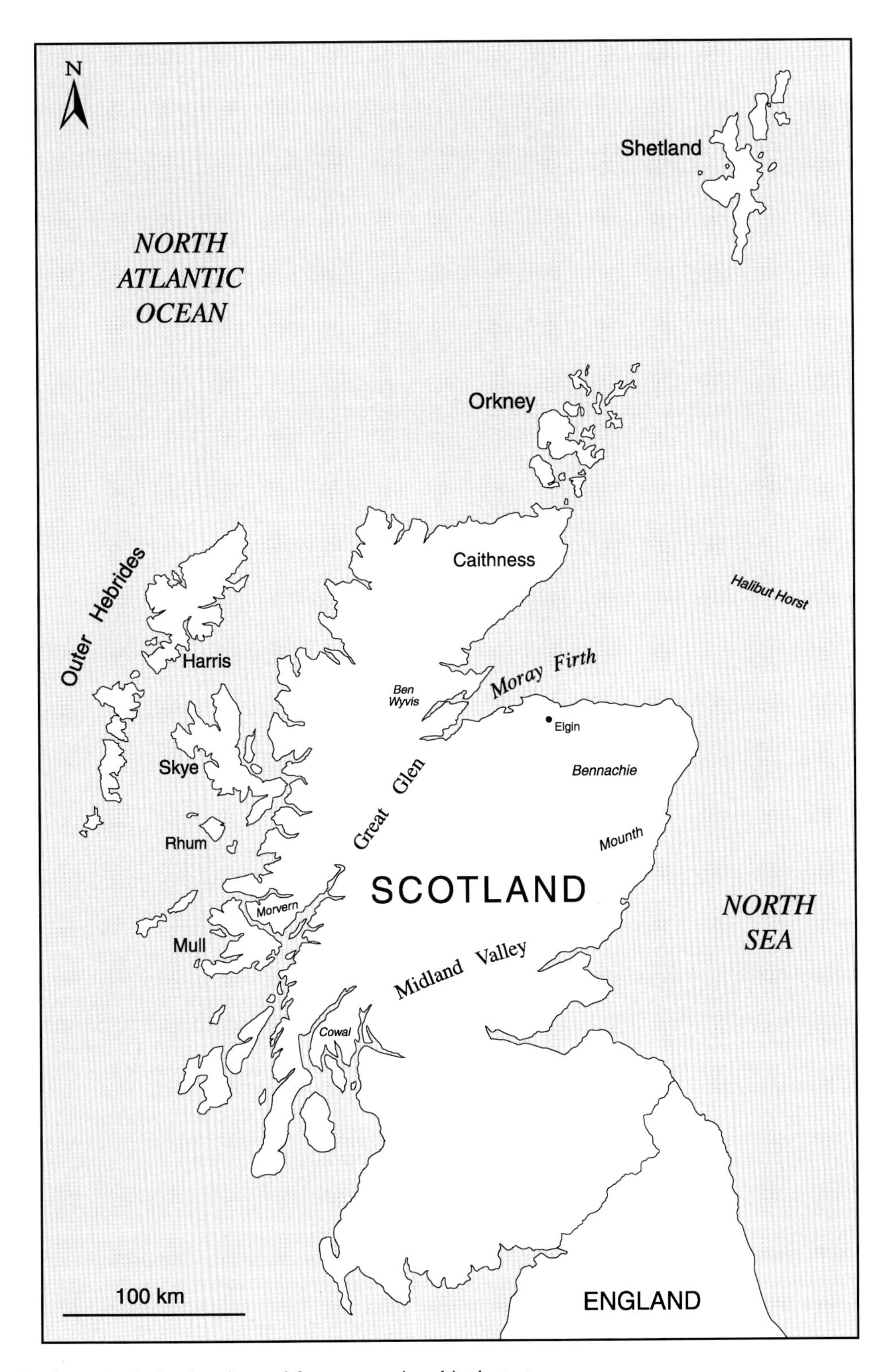

Fig. 1. Scotland, showing sites and features mentioned in the text.

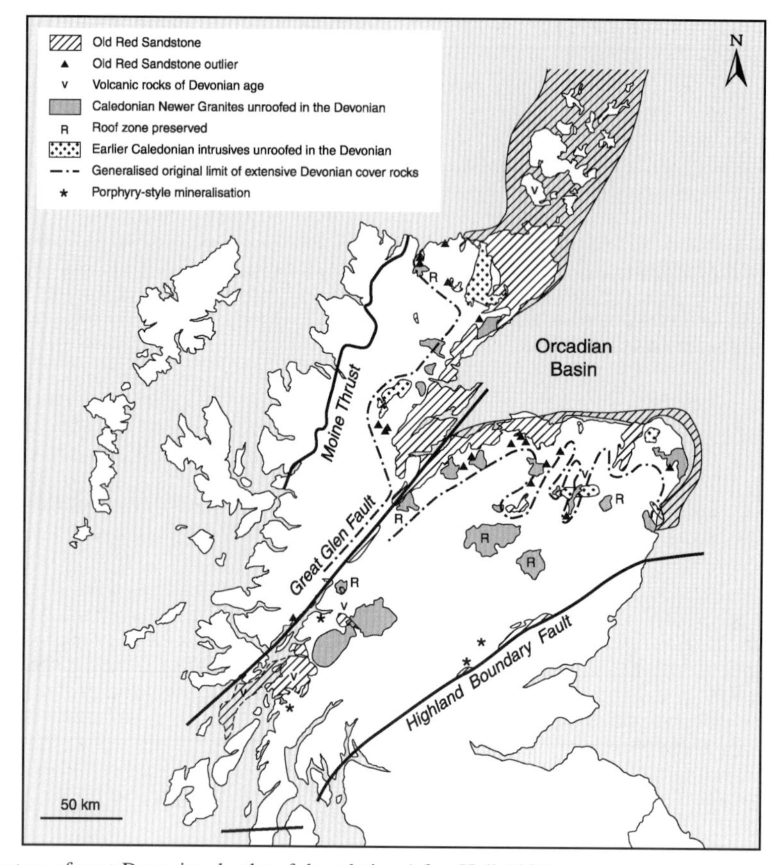

Fig. 2. Indicators of post-Devonian depths of denudation (after Hall 1991).

There is little direct evidence of events in the Highlands during the Carboniferous period, as rocks of this age are restricted to the marginal basins of the Midland Valley and the Moray Firth. The considerable thicknesses of Carboniferous sediments, with up to 4 km in the Midland Valley (Francis 1991) and 1.5 km in the outer Moray Firth (Andrews *et al.* 1990), were largely sourced from the Highlands. The Westphalian outliers in Morvern resting on the Moine sequence imply a former cover of Carboniferous rocks in parts of the SW Highlands (Francis 1991). A reworked Carboniferous microflora is present in Jurassic rocks as far west as the onshore outcrops in the inner Moray Firth Basin (Andrews *et al.* 1990).

The Highlands are generally shown as an emergent and exposed basement area in palaeogeographical maps of the Carboniferous period (Guion *et al.* 2000) but the extent and depth of Carboniferous denudation in the Highlands remain unclear. Recent apatite fission-track studies imply that as much as 3 km of late Palaeozoic cover rocks have been removed from the Highlands area (Thomson *et al.* 1999) but deep Late Palaeozoic erosion appears incompatible with the widespread survival of near-surface volcanic and intrusive rocks of late Carboniferous age (Watson 1985; Hall 1991). Volcanic activity continued throughout Permian time and is represented in the Highlands by the camptonite–monchiquite dyke swarms of Orkney and the Western Highlands (Francis 1991). Associated uplift was probably limited as the total volume of magma was small (Watson 1985). This accords with the survival of Carboniferous deep weathering mantles in northern Scotland that acted as a major source of kaolinitic detritus for Jurassic sediments in the inner Moray Firth (Hurst 1985a).

Alternating periods of moderate uplift, reduction of relief and marine transgression affected the Highlands during Mesozoic time (Hall 1991). Triassic sediments are up to 500 m thick against the Great Glen Fault but thin to 150 m around Elgin (Frostick *et al.* 1988). By the end of the Triassic period uplift had ceased and relief was considerably reduced, and the outer Moray Firth formed part of an extensive continental plain of low relief (Andrews *et al.* 1990). Calcretes and silcretes formed and the Rhaetic Sea transgressed close to the present margins of the inner Moray Firth. Continued transgression in Early Jurassic time saw the deposition of fluviatile sand around the margins of the Moray Firth. Sand and clay mineralogy suggests derivation dominantly from Devonian and Carboniferous cover rocks to the north (Hurst 1985a) and from Moinian chloritic metasediments to the south (Hurst 1985b). Thermal doming in Mid-Jurassic time in the Moray Firth Basin caused deep truncation of Early Jurassic and older sediments and sediment transfer to the Viking Graben and inner Moray Firth. Crustal collapse in the central North Sea in Callovian time was accompanied by rapid sedimentation in the inner Moray Firth and synsedimentary movements along the Helmsdale Fault (Anderton *et al.* 1979). Marginal marine sands overstepped the current basin margins and may have covered the axis of the Great Glen (Hallam & Sellwood 1976: Wignall & Pickering 1993). The faults controlling sedimentation in the inner Moray Forth appear to have also been active on the adjacent land area (Roberts & Holdsworth 1999).

Tectonic activity was renewed at the Jurassic–Cretaceous boundary in the Moray Firth Basin. Uplift of the Halibut Horst led to erosion of Carboniferous sandstones. Fault scarps along the northern margin of the inner Moray Firth generated coarse mass flow deposits (Anderton *et al.* 1979). Early Cretaceous sediments later overstepped the Helmsdale Fault north of Helmsdale to overlie Jurassic and Devonian sediments (Chesher & Lawson 1983). In Morvern, Cretaceous greensands rest on Moinian schists (George 1966). A small outlier of late Hauterivian–early Barremian glauconitic sandstone rests on Devonian and basement rocks in eastern Buchan (Hall & Jarvis 1994).

By Late Cretaceous time the Highlands had been reduced to an area of relatively low relief. Cretaceous sequences along the eastern margin of the Hebrides basin are thin, implying limited sediment supply, and lie close to sea level, implying tectonic stability (Hancock 2000). Terrigenous sedimentation ceased in the Moray Firth with the deposition of thick chalk sequences. On land, the sub-Cenomanian surface, before transgression, carried deep kaolinitic weathering mantles, later reworked to form the highly quartzose sands of Lochaline (Humphries 1961) and the kaolinitic Paleocene sands and muds of the inner Moray Firth (Carman & Young 1981). The extent of marine transgression in Late Cretaceous time is unclear but deposition of the chalk in depths of several hundred metres of water (Hancock 1975) suggests that only a small area of the Highlands can have escaped submergence.

Around 60 Ma, the passage of the Iceland plume was accompanied by major magmatic activity in western Scotland (Bell & Jolley 1997). Magmatism involved emplacement of igneous centres, extrusion of flood basalts well beyond the present outcrop and injection of regional dyke forms to form the Tertiary Igneous Province. The period of magmatism was brief, concentrated between 61 and 55 Ma (Jolley 1997), and in individual igneous centres volcanism was largely confined to single palaeomagnetic polarity intervals of 0.4–3 Ma (Musset 1984). Accelerated sand accumulation in the Moray Firth Basin (Liu & Galloway 1997) can be linked via sediment routeways and provenance to erosion of uplifted source areas on the Orkney–Shetland Platform and in the Highlands (Jones & Milton 1994). Sediment flux reached a maximum in Late Paleocene time and declined into Eocene time (Joy 1993; White & Lovell 1997).

Small outliers of thin Cretaceous sequences occur on both the western (Hancock 2000) and the eastern (Hall & Jarvis 1994) margins of the Highlands. As any emergent areas of the Highlands had been reduced to low relief by the end of Cretaceous time (Hall 1991), patterns of Tertiary uplift can be reconstructed using the present summit topography of the Highlands (Fig. 3). The distribution of summits above 800 m defines a zone of maximum uplift; terrain that now forms the main watersheds of the NW Highlands and of the Grampian Mountains. In Northern Scotland, high summits overlook the sedimentary basins of The Minch and the innermost Moray Firth, basins with margins that retain attenuated and localized sequences of Mesozoic sediments (Fig. 4). Major differential tectonics is implied between the Highlands and the surrounding basins. Early Tertiary reactivation of the Helmsdale Fault produced a major fault scarp, now marked by the line of hills between Ben Wyvis and Helmsdale. Another major escarpment existed in Early Tertiary time on the west coast of the Northern Highlands, stretching from the Cuillins to Cape Wrath. This escarpment is now dissected into a chain of isolated hills and hill groups, including the inselbergs of Suilven and Quinag. Its alignment runs parallel to the edge of the Minch Basin but it is not fault controlled, implying that uplift of the NW Highlands was associated with significant

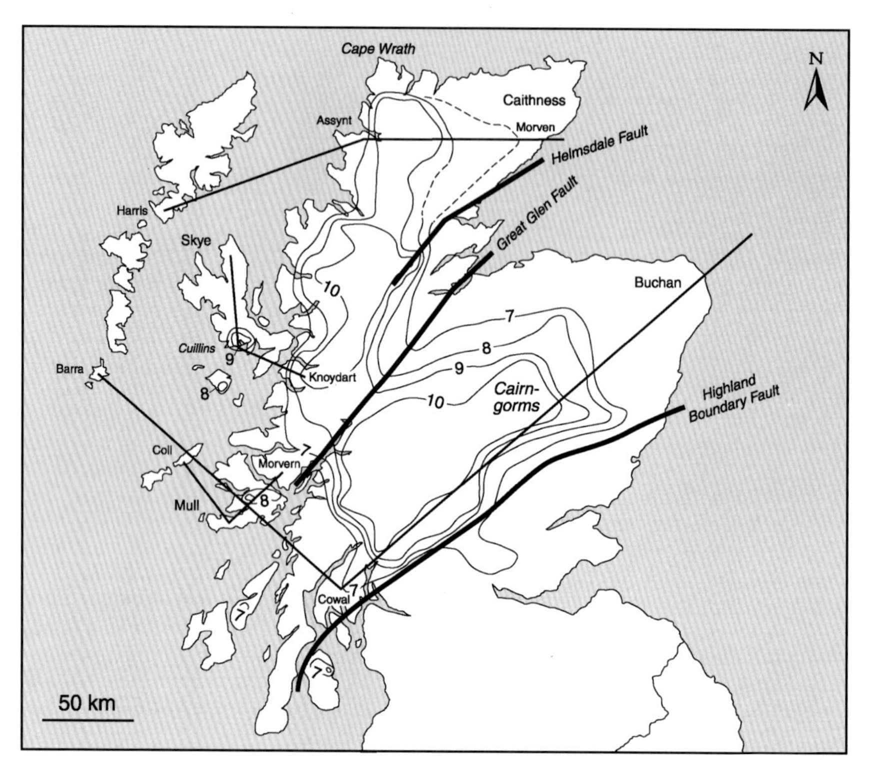

Fig. 3. Patterns of Tertiary uplift implied by Highland summit heights (in hundreds of metres). Fine lines give locations of diagrammatic cross-sections in Fig. 4 (longer section lines) and Fig. 5 (shorter section lines).

warping on its western margin. South of the Great Glen, there is also evidence of differential uplift. In Buchan the preservation of Cretaceous chalk flints and greensand demonstrates modest Tertiary uplift yet the Cairngorms, only 50 km to the west, even today reach 1300 m (Fig. 5). Differential movements are required, with possible downwarping towards the east in the Dalradian belt between Ballater and Keith (Ringrose & Migon 1997) and dislocation at the eastern edge of the Mounth and the Hill of Fare (Hall 1987). On the southwestern edge of the Western Grampians lies a zone of lower summits, centred on Cowal, where the presence of vesicular dykes (Gunn *et al.* 1897) suggests relative proximity to the Early Tertiary land surface (Fig. 4). The preservation on the Lorne Plateau of a small Carboniferous outlier at Bridge of Awe, resting on Devonian lavas (Johnstone 1966), is noteworthy, as are the fragments of sub-Triassic surfaces (Godard 1965) found in Morvern. These occurrences together imply that post-Caledonian vertical movements of the Cowal peninsula and adjacent areas have been modest when compared with those that have affected the main area of the SW Grampians.

In early Eocene time the Highland area foundered as it moved away from the Iceland plume (Nadin & Kusznir 1995) and sedimentation rates dropped in the North Sea (Liu & Galloway 1997). This coincided with the onset of a period, *c.* 20 Ma in duration, of humid and initially subtropical conditions, and apparently limited uplift. Deep kaolinitic weathering covers probably developed widely in association with extensive erosion surfaces (Hall 1991). Tectonic activity was resumed throughout NW Europe in Late Oligocene time, with the onset of major uplift of Fennoscandia (Rohrman *et al.* 1995) and basin development throughout western Britain, including the Hebridean region (Fyfe *et al.* 1993). The presence of depositional hiatuses west of the Shetlands (Ridd 1981), deltaic and lignitic sands east of the Shetlands (Johnson *et al.* 1993) and unconformities in the central North Sea (Gatliff *et al.* 1994) indicate significant uplift in the Scottish area and associated erosion and enhanced sediment supply (Liu & Galloway 1997).

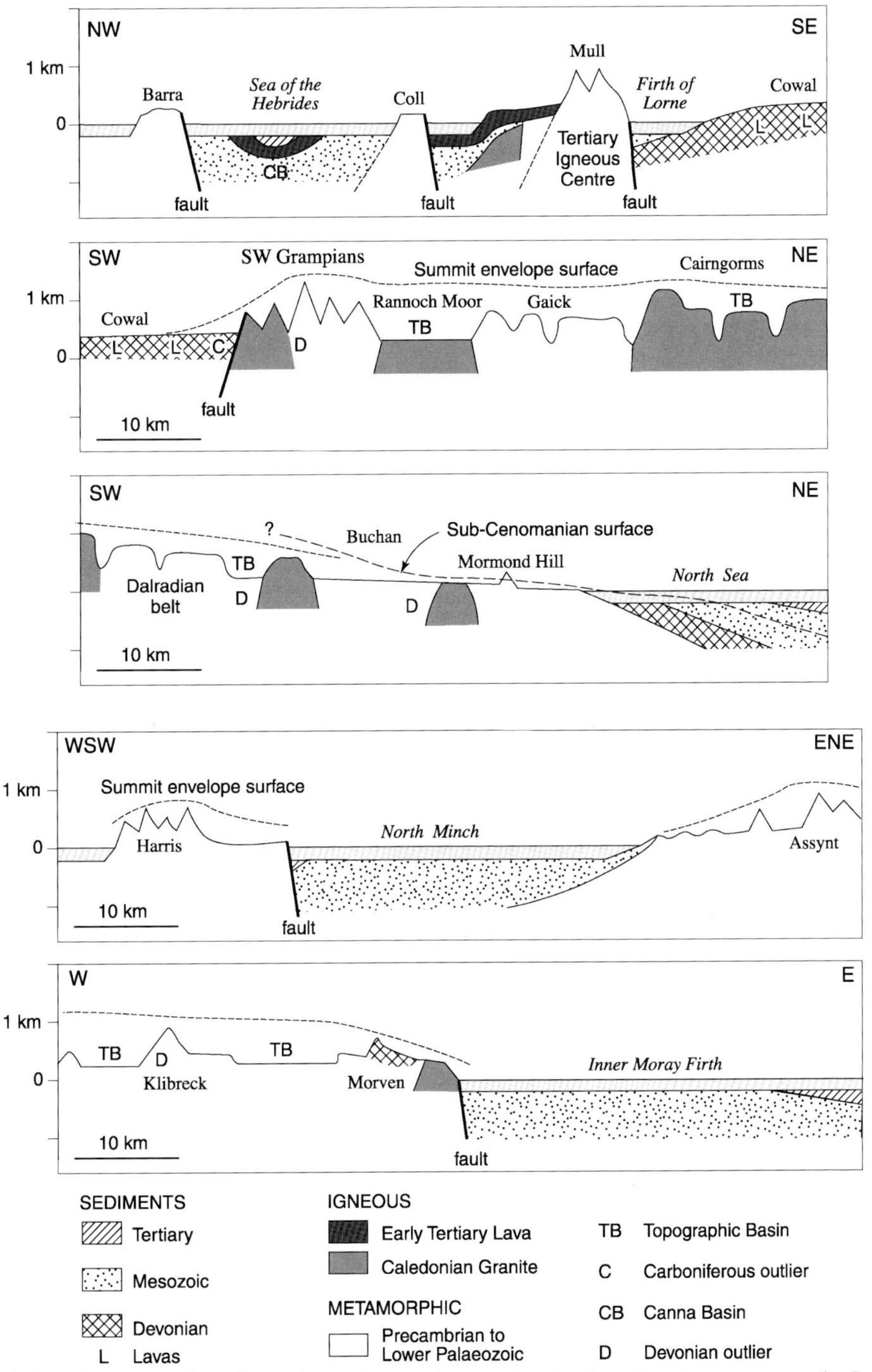

Fig. 4. Schematic cross-sections illustrating major morphotectonic units along two traverses across the Scottish Highlands.

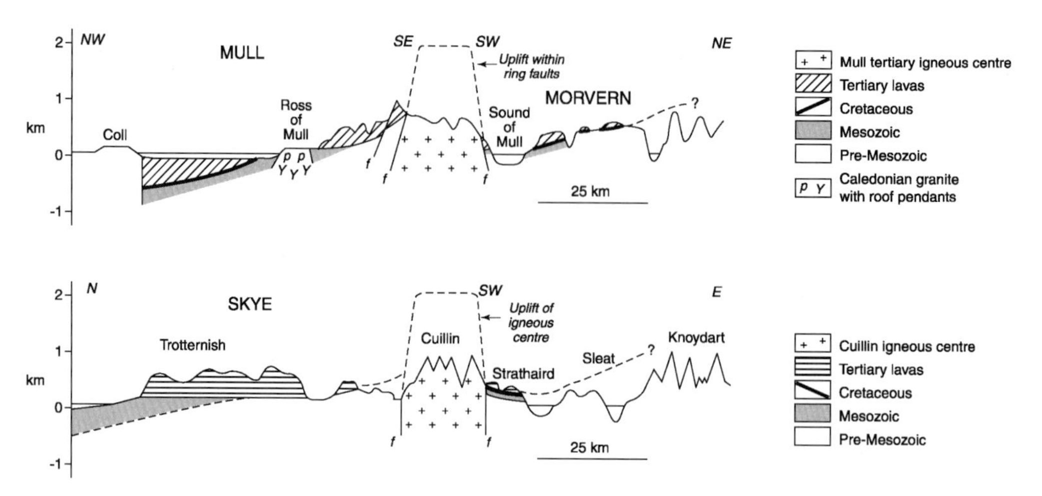

Fig. 5. Schematic cross-sections illustrating differential tectonics and deundation in the Tertiary Igneous Province: Skye and Mull (from various data sources).

Early to mid-Miocene time was the second long period of relative tectonic stability in the Tertiary period, lasting for over 10 Ma (Le Coeur 1999). Kaolinitic weathering resumed under humid, warm to temperate conditions (Berstad & Dypvik 1982) and it is possible that this was an important period for the formation, extension and remodelling of erosion surfaces by etch processes. Miocene marine clays, up to 8 m thick, together with Cretaceous sandstone, are reported from Leavad, Caithness, as part of a large glacial erratic transported from the Moray Firth (Crampton & Carruthers 1914). The occurrence appears to demonstrate Miocene marine sedimentation in the inner Moray Firth.

A marked change occurs in the clay mineralogy of North Sea sediments in Late Miocene time. An increasing content of feldspar, chlorite and illite (Karllson *et al.* 1979; Berstad & Dypvik 1982) reflects climatic cooling and the input of immature terrigenous material. This material was increasingly sourced first from Fennoscandia and later from the Rhine and Baltic river systems, reflecting uplift of source areas. The significance of Neogene uplift in the morphogenesis of northern Britain has long been recognized (George 1966), although its scale and pattern are as yet uncertain. Exhumation of the Chalk from beneath *c.* 1 km in the inner Moray Firth appears to have been achieved largely in Neogene time (Japsen 1997; Japsen & Chalmers 2000) and implies contemporaneous uplift of the Scottish Highlands. Geomorphological evidence for Late Tertiary tectonics is provided by the apparent warping of mid-Tertiary erosion surfaces in northern Scotland (Godard 1965), the uplift, warping and dislocation of Late Tertiary surfaces in western Scotland (Le Coeur 1988, 1999) and the widespread evidence of valley incision and deepening of topographic basins set into mid-Tertiary erosion surfaces throughout the Highlands at this time (Hall 1991).

The influx of ice-rafted material to the Hebridean margin at 2.5 Ma marks the onset of mid-latitude climatic deterioration (Stoker *et al.* 1994). Episodic mountain glaciation is likely to have occurred thereafter (Clapperton 1997) but the first ice sheets reached the North Sea Basin only after 1 Ma (Andrews *et al.* 1990). Multiple glaciation of the Highlands and the adjacent shelves during the Quaternary period brought the transfer of sediment from current land and nearshore area to the axial area of the North Sea Basin and to the continental shelves (Clayton 1996). On land, each glaciation tended to remove the deposits of its predecessor so that only the deposits of the last (Late Devensian) ice sheet are usually preserved.

The impact of the Quaternary glaciations has varied in time and space. The volume of sediment deposited in the central North Sea indicates that sedimentation, and hence denudation of the adjacent land masses and shelves, doubled between the dominantly non-glacial conditions of Pliocene and early Pleistocene time and the conditions of episodic ice sheet glaciation over the last 1 Ma. The last ice sheets during Late Quaternary time were also less effective agents of erosion and transportation than those of mid-Quaternary time, reflecting the earlier removal of pre-glacial weathered rock, the progressive adaptation of the glacier bed to the efficient evacuation of ice, and the greater thickness and extent of the Elsterian and Saalian ice sheets (Glasser & Hall 1997). The Scottish Highlands also exhibit a wide range of glacial landscapes, from the deeply dissected terrain of the western Highlands to the zones of selective linear erosion of the Cairngorms and the limited erosion of the Buchan lowlands (Linton 1959; Clayton 1974). The average depth of glacial erosion across Britain is estimated at 76 m, with 175 m in mountainous zones of intense erosion and as little as 15 m in zones of slow-moving or cold-based ice (Clayton 1996). This is equivalent to a volume less than the total amount of Quaternary sediment on the shelves surrounding Britain, implying that a significant component has been derived from the deep erosion of material from the inner shelves (Clayton 1996), including the inner Moray Firth and The Minch. Mass transfer on this scale must have caused isostatic uplift in the glaciated mountain areas of western and northern Britain. The depth of some of the west coast fjords may reflect glacial incision into the still-rising edge of the NW Highlands.

Former cover rocks in the Scottish Highlands

Palaeogeographical maps of the Highlands have tended to show the area as a persistent topographic high (e.g. Anderton *et al.* 1979; Ziegler 1981), but the extent and thickness of former cover rocks in the Scottish Highlands remain controversial. The region is routinely seen as a source area for sediments that have accumulated in the surrounding basins. This seems consistent with evidence of relatively modest depths (<1–2 km) of post-Devonian erosion of basement rocks (Watson 1985; Hall 1991). Yet in recent years, apatite fission-track thermochronology (AFTT), vitrinite reflectance and compaction studies have suggested that the Scottish Highlands have supported much greater thicknesses of overburden than previously thought and that depths of Tertiary erosion were as much as 3 km across northern Britain. The modelling of rates of cooling and the removal of overburden has the potential to provide valuable insights into the history of not only the offshore basins but also the adjacent source regions, yet it is currently often difficult to reconcile these models with onshore regional geology and geomorphology (Cope 1994; Holliday 1993; McCallan 1994; Smith *et al.* 1994). These are precisely the issues identified in parallel geomorphological and thermochronological studies of SE Australia (see Kohn & Bishop 1999).

Current models of Highland source area evolution based on AFTT (Thomson *et al.* 1999) suggest that: (1) late Palaeozoic cover rocks up to 3 km thick formerly covered the basement rocks of the Highlands; (2) Tertiary erosion has removed 1–2 km of rock from above the current topography.

There is little doubt that late Palaeozoic rocks once covered considerably larger areas of the Highlands massif than at present. Devonian outliers occur widely around the coastal rim of the Moray Firth and reach thicknesses of several hundred metres in fault-bounded basins. Carboniferous rocks formerly extended across part of the SW Highlands (George 1960) and reach a thickness of over 1 km in the inner Moray Firth (Thomson *et al.* 1999). Yet it is unlikely that late Palaeozoic rocks once covered all of the Highlands and hence pre-Mesozoic overburden thicknesses of 2–3 km seem unreasonable. The Lower and Middle Old Red Sandstones around the Orcadian Basin were largely derived from mountains sited in the area of the current Eastern Grampians and Western Highlands (Mykura 1983) and there is little obvious sign that these mountain areas were eventually worn down sufficiently to be buried. Typically, a regional unconformity separates the Upper Old Red Sandstone from older sediments and this is ascribed to regional tectonics (Mykura 1983). In Morayshire, on the margin of the Orcadian Basin where uplift might be expected to be limited, the Upper Old Red Sandstone rests in places directly on the Moine sequence (Horne 1923), indicating removal of the Lower and Middle Old Red Sandstone before deposition. A similar situation occurs in the Midland Valley, where the Upper Old Red Sandstone, including conglomerates with pebbles of metamorphic rocks derived from the Southern Highlands, rests with marked angular unconformity on the Lower Old Red Sandstone (Francis *et al.* 1970). By implication, substantial removal of the Lower and Middle Old Red Sandstone had been achieved before the end of Devonian time.

The Highlands also acted as a source region for Carboniferous sediments in the Midland Valley (Francis 1991; Guion *et al.*, 2000) and Namurian–Westphalian sandstones in the Stirling district contain heavy minerals ultimately derived from low-grade metamorphic rocks north of the Highland Boundary Fault (Francis *et al.* 1970). Along the eastern margin of the Minch Basin numerous small outliers of Permo-Trias occur. Only at Inninmore on the Sound of Mull is Carboniferous sediment found underlying Trias units and here the Carboniferous rocks are only

100–160 m thick. Elsewhere the Permo-Trias sequence rests on older rocks (Johnstone & Mykura 1989). By implication, the Carboniferous rocks were removed before the start of the Trias period or were never laid down to any significant thickness along the western edge of the Northern Highlands.

The thick sequences of Mesozoic clastic sediments in the central North Sea, which might be taken to indicate deep erosion of the Highlands, derive only in part from the Scottish area. A major contribution of material from Fennoscandia occurred during Triassic and early Jurassic time (Ziegler 1981). Erosion of intrabasinal highs also provided sediment (Andrews *et al.* 1990). Yet the presence of igneous and metamorphic debris in sediments at the margins of the Highlands implies continuing erosion and thus exposure of the basement of the Highlands area (Hudson 1964; Hurst 1985a, 1985b). The volume of debris also implies a significant, but unknown depth of erosion. In the Minch Basin, it is possible to quantify depths of Permian to Jurassic erosion of the terrain surrounding the Sea of the Hebrides Trough. Preserved sediment volumes are close to the originally deposited volumes (Steel 1978: Fyfe *et al.* 1993) and suggest that *c.* 280 m of rock was removed in Permo-Trias time and *c.* 210 m in Jurassic time from the main contributing area of the southern Outer Hebrides Platform. These relatively low values are consistent with the removal of only thin cover from the Highlands.

Depths of Tertiary erosion from the source area of the Moray Firth Basin in the Highlands can be quantified using sediment volumes in the North Sea. The North Sea acted as a sediment trap throughout Tertiary time (Liu & Galloway 1997), with only a narrow and shallow connection to the Norwegian–Greenland Sea (Nielsen *et al.* 1986). In early Tertiary time the dominant sediment source was the Scottish Highlands and the Orkney–Shetland Platform but throughout Neogene time sediment was increasingly derived from Scandinavia and the great river systems of NW Europe. Rough calculations indicate the removal of 600–800 m of rock from the contributing area east of the main Scottish watershed during Tertiary time (Hall 1991). As only 25% of the Scottish land area now lies above 300 m (Haynes 1983), it appears that the summit envelope surface of the Highlands at 900–1200 m (Fig. 3) lies only a few hundred metres below the uplifted sub-Cretaceous land surface. On these estimates, there appears to be a good fit between the volume of rock removed by erosion from the Highlands and that received in the North Sea Basin. In the inner Moray Firth, sonic velocities in the Kimmeridge Clay indicate removal of around 1 km of overburden (Hillis *et al.* 1994; Thomson & Hillis 1995) and this seems generally compatible with the existence of hills, such as Ben Rinnes, at up to 800 m elevation on both the northern and southern margins of the basin. In contrast, the removal of over 1.1 km of rock from above the present terrain of Caithness and Sutherland (Thomson *et al.* 1999) seems excessive. On Morvern, AFTA data indicate *c.* 1.7 km of Tertiary erosion on the Strontian Granodiorite (Thomson *et al.* 1999), yet the current land surface nearby retains thin sequences of Mesozoic sediments buried beneath Early Tertiary lavas (Johnstone & Mykura 1989).

Further judgement on these issues must await continuing work using AFTT and (U–Th)/He thermochronology. None the less, onshore geological and geomorphological data from the passive margins of southern Africa (B.J. Bluck, pers. comm) and southeastern Australia (Kohn & Bishop 1999; Bishop & Goldrick, 2000) seem to suggest that AFTT may overestimate the depths of former cover rocks and of denudation. Of course, the regional pattern of AFTT data, and the denudation that they imply, cannot be taken to provide a detailed history of any particular locality. The lavas of 100 Ma age on the south coast of New South Wales, for example, are inconsistent with deep denudation having occurred along and across the whole of the coastal strip below the SE Australian escarpment, an interpretation that has been implicit in much of the AFTT discussion. Likewise, there are many areas throughout the SE Australian highlands, as in the Scottish Highlands, where ancient landscape elements (dating from Mesozoic time in the SE Australian case) have been identified (Young 1981; Bird & Chivas 1989; Twidale 1994; Twidale & Campbell 1995; Hill 1999). Hill (1999) has argued that variations in relief in the SE Australian highlands and across the coastal strip below the escarpment, and differential preservation of Mesozoic landscape elements, may explain the apparent conundrum in SE Australia of kilometre-scale denudation in Late Mesozoic time and the preservation of Mesozoic landscape elements. Where local, detailed AFTT data are available in SE Australia, particularly in areas where the regional structure is dominated by individual fault blocks, differential fault block movement is clearly indicated by the AFTT data (Kohn & Bishop 1999; Kohn *et al.* 1999). Elevated geothermal gradients would also assist in minimizing the amounts of denudation required for Late Mesozoic fission-track ages to crop out at the present ground surface in SE Australia, but there is currently a

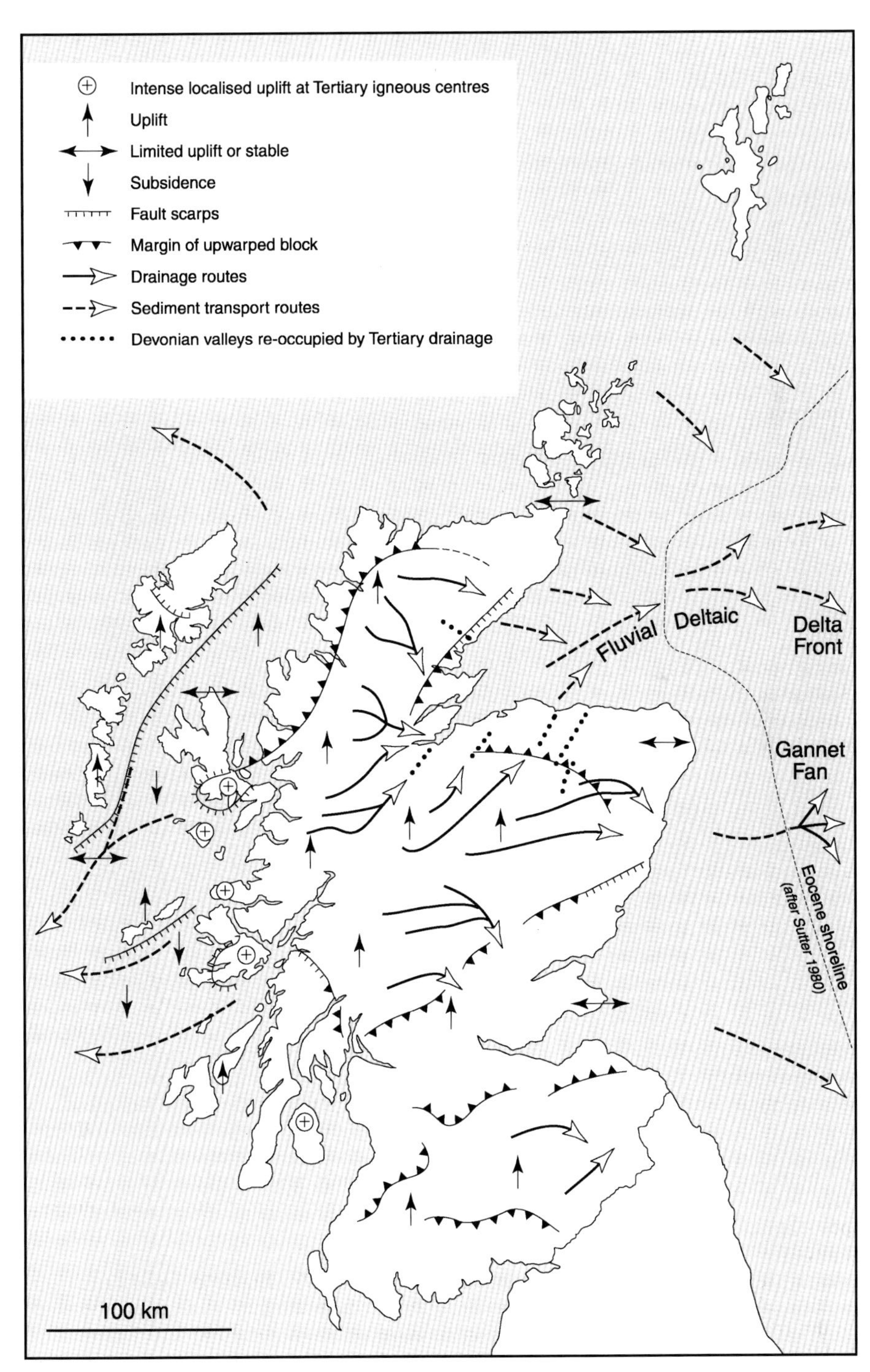

Fig. 6. Early Tertiary differential tectonics and sediment transport (from various data sources).

strong consensus among AFTT researchers that geothermal gradients along the SE Australian continental margin were probably not significantly greater than 25–30 °C km^{-1} in Late Mesozoic time (Dumitru *et al.* 1991; Brown *et al.* 1994). These matters will almost certainly emerge as important for the Scottish Highlands as more detail becomes available.

Cenozoic uplift and denudation of the Scottish Highlands

The timing, amount and pattern of uplift in the Highlands through Cenozoic time are gradually becoming clearer. Uplift appears to have been intermittent, with the main phases centred on Late Paleocene and Late Oligocene time and a later phase of continuing uplift starting in Late Miocene time (Hall 1991; Liu & Galloway 1997). In terms of sediment supply to the Moray Firth, the first of these phases is the most important (Nielsen *et al.* 1986; Liu & Galloway 1997), implying that this was the period of maximum Tertiary uplift in the Highlands. The timings of these vertical movements match those in western Norway (Riis 1996). Indeed, the four major unconformities dating from mid-Paleocene, mid- to late Oligocene, mid-Miocene and latest Pliocene time are remarkably consistent in development across the central and northern North Sea and the Norwegian Shelf (Huuse *et al.*, 2001). It appears therefore that the Scottish–Norwegian section of the NE Atlantic margin has reacted contemporaneously to crustal break-up in Paleocene time, intraplate deformation in Neogene time (Stuevold & Eldholm 1996) and to the onset of regional glaciation. The magnitude of these events, however, may well have varied spatially. The maximum uplift of southern Fennoscandia appears to have occurred from late Oligocene to Pliocene time (Doré *et al.* 1998; Japsen & Chalmers 2000) and thus postdates the main uplift phase in the Scottish Highlands. The reconstructed burial history of the Chalk in the western North Sea (Japsen 1997) implies that parts of eastern Scotland may have experienced maximum Tertiary uplift in Neogene time.

Major uplift occurred at the Paleocene-Eocene boundary via a combination of permanent and transient (dynamic) uplift (Jones *et al.* 2002). This uplift event is consistent with the evidence of rapid local denudation in the Hebridean Igneous Province. As much as 2 km of roof and cover rocks are missing from above the igneous centres of Skye, Mull and Arran (George 1966), but this unroofing is likely to have been spatially variable (Fig. 6) (see foregoing discussion of SE Australia). Holness (1999), for example, has argued that the metamorphic grade of the Torridonian arkose intruded by the Rhum Igneous Complex implies 500–550 m of overburden, which is less than the current relief on Rhum, and 'points to a topography at the time of melting [i.e. metamorphism] as very similar to that of today' (p. 538). This erosion of the Hebridean Igneous Province took place remarkably quickly. Some 2 km of basalt was removed from Mull between 58 and 56 Ma and the unroofing of the Western Granophyre of Rhum took place within 3 Ma (Emeleus 1983). Sedimentary interbeds in the Rhum Central Complex demonstrate that the complex was already unroofed and was undergoing active erosion as the lavas were being extruded (Emeleus & Forster 1979; Holness 1999). Palynological studies of former interbasaltic vegetation suggest that during the 0.24 Ma of existence of the Skye Lava Field there were two major subsidence phases and one uplift phase. Thermal doming related to the emplacement of the Cuillin centre caused elevation to altitudes in excess of 1200 m a.s.l. (above sea level; Jolley 1997). Lava erupted mainly from fissures spread out over an area of *c.* 40 000 km^2 stretching from Harris to the Firth of Clyde over the axis of the dyke swarms (Preston 1982). By the end of the magmatic phase many igneous centres were reduced to close to present levels, as shown by the late lavas resting on the Western Granophyre of Rhum (Emeleus 1983) and the deeply denuded basalts lying beneath the Sgurr of Eigg pitchstone (Dickin & Jones 1983). This deep erosion implies that the lava fields were also largely stripped in early Tertiary time, together with any underlying Palaeozoic and Mesozoic sediments. As Permian sandstones on Lewis and Jurassic sandstones on Skye are associated with maximum burial depths of only 1.5–2 km (Carter *et al.* 1995), it is likely that the Paleogene uplift event was a key phase in the stripping of Paleozoic and Mesozoic cover rocks throughout the Inner Hebrides.

The Leavad clays probably relate to early Miocene marine transgression (Crampton & Carruthers 1914), although a modern faunal study of these intriguing deposits is required. If correct, an early Miocene age implies that deep erosion of the Jurassic and lower Cretaceous sequence in the inner Moray Firth (Hillis *et al.* 1994) was largely completed between Late Paleocene and Late Oligocene time.

Jones *et al.* (2002) interpreted gravity data from the NE Atlantic in terms of Neogene uplift but because of the absence of later Tertiary rocks from much of the Highlands area it is not easy to assess amounts of subsequent uplift and denudation. The small fault-bounded Late Oligocene basins of The Minch and the NW Scottish shelf provide important evidence of a phase of tectonic activity that can be traced throughout western Britain and may be related to minor plate reorganization (Evans *et al.* 1991, 1997). It may be significant that the Late Oligocene floodplain and swamp deposits in these basins do not rest on Eocene sediments but on kaolinized basement, implying prolonged weathering under humid conditions in Eocene time. On the NW Scottish shelf, the Oligocene sediments are succeeded by Miocene shallow marine sands (Evans *et al.* 1997). The sequence implies that despite Eocene crustal collapse The Minch and the adjacent shelf remained above sea level until marine transgression in Miocene time.

Crustal movement resumed in Late Miocene time (Stoker *et al.* 1994). Japsen (1997) considered that the amount of exhumation during Neogene uplift was equivalent in Britain to that achieved in Late Paleocene–Early Eocene time (see also Japsen & Chalmers, 2000). This seems unlikely, as Neogene sediments in the North Sea are largely sourced from Scandinavia and NW Europe, rather than Britain (Jordt *et al.* 1995). Moreover, the large volumes of coarse clastic sediments of Late Paleocene and early Eocene age in the Moray Firth contrast markedly with the more restricted sequences of Neogene muds (Liu & Galloway 1997). The increasing resolution of the Neogene succession in the central and northern North Sea and on the Norwegian shelf helps to constrain the timing of this late uplift, with pulses of terrigenous sediments evident in Late Miocene time and, coincident with the first Scandinavian ice sheets, in Late Pliocene time (Eidvin *et al.* 2000).

The thick Neogene sequences in the Faeroe–Shetland basin system (Stoker *et al.* 1993) are more difficult to account for. Likewise, Jones *et al.* (2002) reported that the volume of Paleocene sediment in the Faeroe–Shetland basins cannot be accounted for by denudation of a source area in NW Scotland. They suggested that Faeroe–Shetland basin sediments may also have been derived from source areas other than NW Scotland, such as the Faeroe Islands region or more widely in northern and western Scotland, and a similar explanation may have to be invoked to reconcile the thickness of the Neogene sequence in the Faeroe–Shetland basin systems and the apparently minor Neogene uplift of Scotland.

The signature of these Neogene vertical movements is recorded by elements of the Highland topography. In the Northern Highlands, Godard (1965) recognized three major erosion surfaces: a surface between 400 and 550 m, which cuts across parts of the Tertiary igneous centres and so is post-Paleocene and possibly Eocene in age; an intermediate and extensive Scottish Surface at *c.* 300 m of possible Oligocene age; the low coastal plateau of the Niveau Pliocène at *c.* 100 m. Assuming that these erosion surfaces originated close to sea level, it has been estimated that uplift in the NW Highlands is of the order of 400 m in the last 40 Ma (Le Coeur 1999).

The long-term trend has been towards the progressive tilting or downwarping of the Highlands towards the inner Moray Firth Basin. Despite a degree of glacial diversion and disruption the main drainage routes continue to flow towards the Moray Firth, just as in Early Tertiary and Late Devonian time. The most extensive erosion surface recognized south of the Moray Firth is the Eastern Grampians Surface, comprising the ramp-like interfluves of the Monadliath, the Dee–Don watershed and the Mounth. These surfaces slope from their inner margins around the Cairngorms at around 800 m towards the inner Moray Firth and the North Sea, dropping to elevations of *c.* 500 m at the outer margins (Hall 1991).

Recent morphometric analysis has confirmed that regional tilting was far from uniform. Erosion surfaces between 200 and 600 m on the Monadliath are strongly influenced by the Great Glen Fault and the Ericht–Laidon Fault. This may imply a significant tectonic event in mid- to late Tertiary time in which an extensive medium-level erosion surface was disrupted by block movement (Ringrose & Migon 1997). Ringrose & Migon also identified a possible zone of flexure located in the Dalradian belt between the high tops of the Cairngorms and the lowlands of Buchan. Late Neogene faulting has also been proposed along the SE coasts of Rhum and Coll along the Camasunary–Skerryvore Fault (Le Coeur 1988) and in the Elgin area (Hall 1991) and may represent part of a continuing patterns of neotectonic activity (Muir Wood 1989).

Denudation and landscape evolution

The main morphotectonic units in the Scottish area were already in existence by the end of the Palaeozoic era. The Orkney–Shetland Platform, the Highlands and Southern Uplands massifs have remained above sea level for most of post-Palaeozoic time and have shed sediment to surrounding basins. The persistence of differential movements between basement massifs and sedimentary basins has allowed conservation, during surface lowering, of major topographic features from as long ago as Devonian time, including watershed zones, drainage patterns and, especially, the scarps at the margins of the buoyant basement massifs. The Helmsdale Fault, for example, has been intermittently active throughout Mesozoic and Cenozoic time (Andrews *et al.* 1990).

Superimposed on this physiographic framework are the main Tertiary landforms. The magnitude of Paleocene uplift and denudation means that these Tertiary landforms are of mid- to late Tertiary age. Only in areas of minimal displacement, such as Buchan and Caithness, is the preservation of extensive Mesozoic landforms feasible.

Exhumed terrains are of restricted extent in Scotland. They include the rugged sub-Torridonian surface in NW Scotland (Godard 1957; Stewart 1972) and the equally irregular sub-Devonian surface exposed beneath the outliers around the margins of the Moray Firth Basin (Godard 1965). Aside from areas such as these, the oldest landforms recognized in Scotland occur in the lowlands of Buchan. This is an area of long-term relative stability where typical Highland rocks are associated with surfaces of low relief (Clayton & Shamoon 1999). It is also an area of very limited glacial erosion. The antiquity of the landscape is demonstrated by the Cretaceous residues in the form of chalk flints within the quartzite- and kaolin-rich Buchan Gravels (Hall 1987) and an outlier of Lower Cretaceous greensand (Hall & Jarvis 1994). These residues rest on weathered igneous and metamorphic rocks, known in boreholes to extend to depths of many tens of metres (Hall 1986). Largely by correlation with clay minerals in North Sea sediments, the spatially restricted kaolin-rich weathering profiles have been assigned to pre-Pliocene time and the less mature, but still deep sandy weathering profiles to Pliocene and Pleistocene time (Hall 1985; Hall *et al.* 1989). Given the apparent stability of the area, it is conceivable that some of the kaolinitic weathered materials are older, surviving from Paleogene time or exhumed from beneath Late Cretaceous cover rocks (Hall 1993).

Deep weathering is of fundamental importance in understanding the nature and evolution of the pre-Quaternary relief throughout Scotland and NW Europe (Godard 1965; Thomas 1989; Migon & Lidmar-Bergstrom 2001). Its presence has been used widely as an indicator of the preservation or limited modification of preglacial forms (Linton 1951; Godard 1961; Hall & Sugden 1987). There is often a close correspondence between pre-glacial morphology, levels of rock resistance to chemical weathering (Godard 1962) and deep weathering patterns (Hall 1986). The Tertiary period in Scotland was a time of warm to temperate humid climates when most, if not all, of the country stood above base level, conditions favouring the deep penetration of weathering.

The sustained etching out of contrasts in rock resistance by chemical processes led to the formation of major landforms of differential weathering and erosion. These include valley systems, such as the pre-glacial headwaters of the Spey, Don and Dee, where there is pervasive litho-structural controls on valley alignment (Threlfall 1981). Deep topographic basins, some floored by Devonian sediments, are strung out along many of the valleys that drained east from the main watersheds in the NW Highlands and Grampian Highlands towards the Moray Firth (Fig. 4). The largest examples include the Rannoch, Atholl and Naver basins, with areas of more than 500 km^2 (Linton 1951). In NE Scotland, the basin floors are preferentially located on biotite-bearing granite and gabbro and boreholes show widespread deep weathering, reaching depths of as much as 50 m (Hall 1986, 1991). These susceptible rocks have provided the foci for weathering and erosion as the surrounding terrain was uplifted. In counterpoint stand inselbergs. These isolated hills include those of resistance, notably quartzite hills such as Schiehallion. Inselbergs of position also occur (Godard 1965), where the isolation of the hill mass appears to be a result of backwearing of slopes. A few exhumed hills occur, notably the sub-Devonian inselberg of Scaraben in Caithness (Crampton & Carruthers 1914). The quartzite inselberg of Mormond Hill in Buchan may be a Mesozoic relic, as it is associated with deep kaolinization and lies close to or at the level of the sub-Cenomanian surface (Hall 1987).

The landforms of differential weathering and erosion occur as mesoscale features as part of major erosion surfaces. The Buchan Surface, at an elevation of *c.* 100 m, includes most of the lowlands of NE Scotland, apart from the glacially modified coastal strip. It is an etch surface, where subtle differences in rock resistance give rise to hills and depressions and deep weathering is widespread. The origins of the eastern part of the surface date back to late Mesozoic time, for Late Cretaceous greensand and chalk were deposited on its surface, but it has a long history and its various elements are of different age. The preservation of unworn flints at the base of the flint gravels in Buchan testifies to the proximity of the sub-Cenomanian surface (Bridgland *et al.* 1997; Merritt *et al.* 2002). On the high ground of central Buchan at 100–150 m highly kaolinitic saprolites and the flint gravels themselves have been ascribed a pre-Pliocene age (Hall 1985). The lower tiers of the terrain support deep sandy weathering covers and appear to be of Plio-Pleistocene age.

It is likely that high-level erosion surfaces are also polycyclic and spatially diachronous. The surfaces at 800 m in the Gaick Forest (Hall & Mellor 1988) and at high elevations in the Cairngorms (Hall 1996) retain pockets of deep sandy weathering dating from the latest Plio-Pleistocene phase of etching. Within the Cairngorms mountains are a range of major paleic forms which have a longer history, including high-level basins and open valleys (Hall 1996) and the major depression of the Upper Avon Embayment (Linton 1950). The precursors of the major Tertiary rivers of NE Scotland, the Dee and the Don, were already established in Paleocene time and fed material to the Gannet Fan in the North Sea (Morton 1979). The Cairngorms and the Eastern Grampians have been an area of positive relief since at least the start of Tertiary time and these headwater erosion surfaces have evolved far above sea level. Phases of valley incision, indicated by benches on valley and basin sides, are likely to have been driven by changes in local base levels rather than in response to regional uplift.

Discussion and conclusions

Variations in the character and rate of supply of sediments to the offshore region has generally been interpreted in terms of source area responses to Cenozoic regional tectonics. Indeed, Liu & Galloway (1997) were explicit about the link between uplift and erosion in the North Sea Basin: 'Tectonic uplift of source areas exercises the commanding role in modulating long term sediment supply to the North Sea' (p. 1506). The precise process linkages between uplift and enhanced denudation are, however, not always as clear as is implied by the assumed link between uplift and enhanced sediment flux (see Introduction).

White & Lovell (1997) acknowledged that a simple process link between an increase in offshore sedimentation rate and the uplift of Scotland cannot be assumed, and suggested that uplift may be linked closely in time to increases in offshore sedimentation via an uplift-driven fall in relative sea level and an associated mobilization of sediments that were in storage in the nearshore area and on the continental shelf. It seems clear, none the less, that parts of Scotland, for example, the Tertiary Igneous Province, have responded rapidly to presumed uplift events. This is probably because these western areas are composed of small catchments that are well connected to base level and that, therefore, would be expected to respond rapidly to a fall in relative sea level. The preservation of ancient landscape elements throughout Scotland demonstrates, however, that such rapid response cannot be assumed for all areas, and that not all areas are equally sensitive to base-level changes. Variations in lithology will also be expected to be associated with variations in landscape sensitivity (see Brunsden & Thornes 1979); lithological variations in the response to Cenozoic weathering regimes in the Scottish Highlands point to a variant of such sensitivity. In the case of the Tertiary Igneous Province, it is also worth noting that construction of the Province's central volcanic edifices and the extrusion of the regional lava fields must have caused a relative elevation of the land surface, in turn triggering incision and enhanced denudation. The relative magnitudes of this latter effect and uplift remain to be quantified.

Many problems remain in understanding long-term landscape development in Scotland and yet there are many promising lines of enquiry. There is an urgent need for detailed morphometric analysis of the relief comparable with that available in southern Fennoscandia (e.g. Lidmar-Bergström *et al.* 2000). The scattered outliers of sedimentary rocks of Devonian to possible Miocene age that occur on land in Scotland are important archives of information regarding the evolution of the terrain, both via provenance studies and by investigation of burial histories. The dyke systems of late Caledonian, Carboniferous, Permian and Paleogene age that criss-cross much of Scotland have yet to be examined in detail with regard to depths of emplacement beneath contemporary land surfaces. Saprolites occur throughout Scotland, including buried and formerly buried pre-Cenozoic weathering mantles and Cenozoic deep weathering profiles. These saprolites await detailed mineralogical study and dating (Hall 1993), using techniques such as K–Ar dating of mica clays (Sturt *et al.* 1979) and D–H analysis of kaolins (Gilg 2000). Finally, there are opportunities to tie in the uplift and denudational histories of regional source areas within Scotland with sub-basins and fans in the offshore area, such as the Tertiary Barra Fan NW of The Minch (Stoker *et al.* 1993) and the Early Tertiary Gannet Fan fed by rivers draining the Eastern Grampians (Gatliff *et al.* 1994).

The preservation of ancient landscape remnants of varying spatial extent in the Scottish Highlands points to the types of spatial variability of denudation that have been identified in more detailed studies in southeastern Australia. Jones *et al.* (2002) has called for improved quantitative understanding of the Cenozoic denudation of northern Britain and it is therefore timely for thermochronologically based studies of the denudation of Scotland: (1) to move beyond regional denudational studies to more focused AFTT studies and (U–Th)/He analysis in apatite (Zeitler *et al.* 1987) to identify spatial variability in denudation; (2) to use AFTT and (U–Th)/He analysis to assess the extent of differential movements of crustal blocks; (3) to use the (U–Th)/He system to assess the age(s) of major landscape elements features (e.g. House *et al.* 1998); (4) to use cosmogenic isotope analysis to assess in more detail the spatial and temporal variations in Quaternary denudation.

Despite the advent of these various recently developed techniques, the major challenges of dating landscapes and landforms, and of determining the timing of uplift and denudation, remain. So far, the excellent temporal resolution available in the offshore record eludes us in studying the onshore, source area record, but this shortcoming does not justify, for example, oversimplified assumptions concerning the relationships between uplift and denudation.

A.M.H. thanks the Carnegie Trust for the Universities of Scotland for financial support of fieldwork. P.B. gratefully acknowledges the support of the Australian Research Council. The CRUST project (Constraining Regional Uplift, Sedimentation and Thermochronology) is supported by a Scottish Higher Education Council Research Development Grant. We thank K. Lidmar-Bergström, J. Cartwright and the Editor for their reviews and suggestions to improve this contribution, and M. Shand for preparing the figures.

References

ANDERTON, R., BRIDGES, P.H., LEEDER, M.R. & SELLWOOD, B.W. 1979. *A Dynamic Stratigraphy of the British Isles*. Allen and Unwin, London.

ANDREWS, I.J., LONG, D., RICHARDS, P.C., THOMSON, A.R., BROWN, S., CHESHER, J.A. & MCCORMAC, M. 1990. *The Geology of the Moray Firth*. British Geological Survey, Keyworth.

BELL, B.R. & JOLLEY, D.W. 1997. Application of palynological data to the chronology of the Palaeogene lava fields of the British Province: implications for magnetic stratigraphy. *Journal of the Geological Society, London*, **154**, 701–708.

BERSTAD, S. & DYPVIK, H. 1982. Sedimentological evolution and natural radioactivity of Tertiary sediments from the central North Sea. *Journal of Petroleum Geology*, **5**, 77–88.

BIRD, M.I. & CHIVAS, A.R. 1989. Geomorphic and palaeoclimatic implications of an oxygen-isotope chronology for Australian deeply weathered profiles. *Australian Journal of Earth Sciences*, **40**, 345–358.

BISHOP, P. 1985. Southeast Australian late Mesozoic and Cenozoic denudation rates: a test for late Tertiary increases in continental denudation. *Geology*, **13**, 479–482.

BISHOP, P. & BROWN, R. 1992. Denudational isostatic rebound of intraplate highlands: The Lachlan River valley, Australia. *Earth Surface Processes and Landforms*, **17**, 345–360.

BISHOP, P. & GOLDRICK, G. 2000. Geomorphological evolution of the East Australian continental margin. *In*: SUMMERFIELD, M.A. (ed.) *Geomorphology and Global Tectonics*. Wiley, Chichester, 225–254.

BRIDGLAND, D.R., SAVILLE, A. & SINCLAIR, J.M. 1997. New evidence for the origin of the Buchan Ridge Gravel, Aberdeenshire. *Scottish Journal of Geology*, **33**, 43–50.

BROWN, R., GALLAGHER, K., GLEADOW, A.J.W. & SUMMERFIELD, M.A. 2000. Morphotectonic evolution of the South Atlantic margins of Africa and South America. *In*: SUMMERFIELD, M.A. (ed.) *Geomorphology and Global Tectonics*. Wiley, Chichester, 255–281.

BROWN, R.W., SUMMERFIELD, M.A. & GLEADOW, A.J.W. 1994. models of long-term landscape evolution. *In*: KIRBY, M.J. (ed.) *Process Models and Theoretical Geomorphology*. Wiley, London, 23–53.

BRUNSDEN, D. & THORNES, J.B. 1979. Landscape sensitivity and change. *Transactions of the Institute of British Geographers, NS*, **4**, 463–484.

BURBANK, D.W., LELAND, J., FIELDING, E., ANDERSON, R.S., BROZOVIC, N., REID, M. & DUNCAN, C. 1996. Bedrock incision, rock uplift and threshold slopes in the northwestern Himalayas. *Nature*, **379**, 505–510.

CARMAN, G.J. & YOUNG, R. 1981. Reservoir geology of the Forties Field. *In*: ILLING, L.V. & HOBSON, G. (eds) *Petroleum Geology of the Continental Shelf of NW Europe*. Heyden, London, 371–379.

CARTER, A., YELLAND, A., BRISTOW, C. & HURFORD, A.J. 1995. Thermal histories of Permian and Triassic basins in Britain derived from fission track analysis. *In*: BOLDY, S.A.R. (ed.) *Permian and Triassic Rifting in Northwest Europe*. Geological Society, London, Special Publications, **91**, 41–56.

CHESHER, J.A. & LAWSON, D. 1983. *Geology of the Moray Firth*. Institute of Geological Sciences, London.

CLAPPERTON, C.M. 1997. Greenland ice cores and North Atlantic sediments: implications for the last glaciation in Scotland. *In*: GORDON, J.E. (ed.) *Reflections on the Ice Age in Scotland*. Scottish Natural Heritage, Edinburgh, 45–58.

CLAYTON, K. 1996. Quantification of the impact of glacial erosion on the British Isles. *Transactions of the Institute of British Geographers, NS*, **21**, 124–140.

CLAYTON, K.M. 1974. Zones of glacial erosion. *In*: BROWN, E.H. & WATERS, R. S. (eds) Institute of British Geographers Special Publication, **7**, 163–176.

CLAYTON, K. & SHAMOON, N. 1999. A new approach to the relief of Great Britain III. Derivation of the contribution of neotectonic movements and exceptional regional denudation to the present relief. *Geomorphology*, **27**, 173–189.

COCKBURN, H.A.P., BROWN, R.W., SUMMERFIELD, M.A. & SEIDL, M.A. 2000. Quantifying passive margin denudation and landscape development using a combined fission-track thermochronology and cosmogenic isotope analysis approach. *Earth and Planetary Science Letters*, **179**, 429–435.

COPE, J.C.W. 1994. A latest Cretaceous hotspot and the southeasterly tilt of Britain. *Journal of the Geological Society, London*, **151**, 904–908.

COPELAND, P. & HARRISON, T.M. 1990. Episodic rapid uplift in ther Himalaya revealed by $^{40}Ar/^{39}Ar$ analysis of detrital K-feldspar and muscovite, Bengal fan. *Geology*, **18**, 354–357.

CRAMPTON, C.B. & CARRUTHERS, R.G. 1914. *The Geology of Caithness*. Geological Survey of Scotland, Edinburgh.

DICKIN, A.P. & JONES, N.W. 1983. Isotopic evidence for the age and origin of pitchstones and felsites, Isle of Eigg, North-West Scotland. *Journal of the Geological Society, London*, **140**, 691–700.

DORÉ, A.G., LUNDIN, E.R., JENSEN, L.N., BIRKELAND, Ø., ELIASSEN, Ø. & FICHLER, C. 1998. Principal tectonic events in the evolution of the northwest European Atlantic margin. *In*: FLEET, A.J. & BOLDY, S.A.R. (eds) *Petroleum Geology of Northwest Europe*. Geological Society, London, 41–61.

DUMITRU, T.A., HILL, K.C., COYLE, D.A. & 7 OTHERS 1991. Fission track thermochronology: application to continental rifting of south-eastern Australia. *APEA Journal*, **31**, 131–142.

EIDVIN, T., JANSEN, E., RUNDBERG, Y., BREKKE, H. & GROGAN, P. 2000. The upper Cainozoic of the Norwegian continental shelf correlated with the deep sea record of the Norwegian Sea and the North Atlantic. *Marine and Petroleum Geology*, **17**, 579–600.

EMELEUS, C.H. 1983. Tertiary igneous activity. *In*: CRAIG, G.Y. (ed.) *Geology of Scotland*. Scottish Academic Press, Edinburgh, 357–397.

EMELEUS, C.H. & FORSTER, R.M. 1979. *Field Guide to the Tertiary Igneous Rocks of Rhum*. Inner Hebrides Nature Conservancy Council (Geology and Physiography Section), Newbury.

EVANS, D., HALLSWORTH, C., JOLLEY, D.W. & MORTON, A.C. 1991. Late Oligocene terrestrial sediments from a small basin in the Little Minch. *Scottish Journal of Geology*, **27**, 33–40.

EVANS, D., MORTON, A.C., WILSON, S., JOLLEY, D. & BARREIRO, B.A. 1997. Palaeoenvironmental significance of marine and terrestrial Tertiary sediments on the NW Scottish Shelf in BGS borehole 77/7. *Scottish Journal of Geology*, **33**, 31–42.

FIELDING, E.J. 2000. Morphotectonic evolution of the Himalayas and Tibetan Plateau. *In*: SUMMERFIELD, M.A. (ed.) *Morphotectonic evolution of the Himalayas and Tibetan Plateau*. Wiley, Chichester, 202–222.

FLEMING, A., SUMMERFIELD, M.A., STONE, J.O., FIFIELD, L.K. & CRESSWELL, R.G. 1999. Denudation rates for the southern Drakensberg escarpment, SE Africa, derived from *in-situ*-produced cosmogenic ^{36}Cl: initial results. *Journal of the Geoogical Society, London*, **156**, 209–212.

FRANCIS, E.A. 1991. Carboniferous. *In*: CRAIG, G.Y. (ed.) *Geology of Scotland*. Geological Society, London, 347–392.

FRANCIS, E.A., FORSYTH, I.H., READ, W.A. & ARMSTRONG, M. 1970. *The Geology of the Stirling District. Memoir of the Geological Survey of Great Britain*. HMSO, Edinburgh.

FROSTICK, L., REID, I., JARVIS, J. & EARDLEY, H. 1988. Triassic sediments in the inner Moray Firth, Scotland: early rift deposits. *Journal of the Geological Society, London*, **145**, 235–248.

FYFE, J.A., LONG, D. & EVANS, D. 1993. *United Kingdom Offshore Regional Report: The Geology of the Malin–Hebrides Sea Area*. HMSO, London.

GATLIFF, R. W., RICHARDS, P. C., SMITH, K. & 9 OTHERS (1994). *The Geology of the Central North Sea*. HMSO, London.

GEORGE, T.N. 1960. The stratigraphical evolution of the Midland Valley. *Transactions of the Geological Society of Glasgow*, **24**, 32–107.

GEORGE, T.N. 1966. Geomorphic evolution in Hebridean Scotland. *Scottish Journal of Geology*, **2**, 1–34.

GILG, H.A. 2000. D–H evidence for the timing of kaolinization in northeast Bavaria, Germany. *Chemical Geology*, **170**, 5–18.

GLASSER, N.F. & HALL, A.M. 1997. Calculating Quaternary erosion rates in North East Scotland. *Geomorphology*, **20**, 29–48.

GLEADOW, A.J.W. & BROWN, R.W. 2000. Fission-track thermochronology and the long-term denudational response to tectonics. *In*: SUMMERFIELD, M.A. (ed.) *Geomorphology and Global Tectonics*. Wiley, Chichester, 57–75.

GODARD, A. 1957. La surface prétorridonienne en Écosse. *Revue de Géographie Alpine*, **45**, 135–153.

GODARD, A. 1961. L'efficacité de l'érosion glaciare en Écosse du Nord. *Revue de Géomorphologie Dynamique*, **12**, 32–42.

GODARD, A. 1962. Essais de corrélation entre l'altitudes des reliefs et les caractères pétrographiques des roches dans les socles de l'Écosse du nord. *Compte Rendus de l'Académie des Sciences*, **255**, 139–141.

GODARD, A. 1965. *Recherches en Géomorphologie en Écosse du Nord-Ouest*. Masson, Paris.

GUION, P.D., GUTTERIDGE, P. & DAVIES, S.J. 2000. Carboniferous sedimentation and volcanism on the Laurussian margin. *In*: WOODCOCK, N. &

STRACHAN, R. (eds) *Geological History of Britain and Ireland*. Blackwell, Oxford, 227–270.

GUNN, W., CLOUGH, C.T. & HILL, J.B. 1897. *The Geology of Cowal. Memoir of the Geological Survey of Scotland*. HMSO, Edinburgh.

GUNNEL, Y. & FLEITOUT 2000. Morphotectonic evolution of the Western Ghats, India. *In*: SUMMERFIELD, M.A. (ed.) *Geomorphology and Global Tectonics*. Wiley, Chichester, 321–338.

HALL, A.M. 1985. Cenozoic weathering covers in Buchan, Scotland, and their significance. *Nature*, **315**, 392–395.

HALL, A.M. 1986. Deep weathering patterns in north-east Scotland and their geomorphological significance. *Zeitschrift für Geomorphologie*, **30**, 407–422.

HALL, A.M. 1987. Weathering and relief development in Buchan, Scotland. *In*: GARDINER, V. (ed.) *International Geomorphology 1986*. Wiley, Chichester, 991–1005.

HALL, A.M. 1991. Pre-Quaternary landscape evolution in the Scottish Highlands. *Transactions of the Royal Society of Edinburgh: Earth Sciences*, **82**, 1–26.

HALL, A.M. 1993. Deep weathering in Scotland: a review. *In*: HALL, A.M. (ed.) *Scottish Geographical Studies*. Universities of Dundee and St. Andrews, St. Andrews, 37–46.

HALL, A.M. 1996. The paleic relief of the Cairngorm Mountains. *In*: GLASSER, N.F. & BENNETT, M.R. (eds) *The Quaternary of the Cairngorms*. Quaternary Research Association, London, 13–27.

HALL, A.M. & JARVIS, J. 1994. A concealed Lower Cretaceous outlier at Moss of Cruden, Grampian Region. *Scottish Journal of Geology*, **30**, 163–166.

HALL, A.M. & MELLOR, T. 1988. The characteristics and significance of deep weathering in the Gaick area, Grampian Highlands, Scotland. *Geografiska Annaler, 70A*, 309–314.

HALL, A.M. & SUGDEN, D.E. 1987. Limited modification of mid-latitude landscapes by ice sheets: the case of north-east Scotland. *Earth Surface Processes and Landforms*, **12**, 531–542.

HALL, A.M., MELLOR, T. & WILSON, M.J. 1989. The clay mineralogy and age of deeply weathered rock in north-east Scotland. *Zeitscrift für Geomorphologie, Supplement Bund*, **72**, 97–108.

HALLAM, A. & SELLWOOD, B. 1976. Middle Mesozoic sedimentation in relation to tectonics in the British area. *Journal of Geology*, **84**, 302–321.

HANCOCK, G.S., ANDERSON, R.S. & WHIPPLE, K.X. 1998. Beyond power: bedrock incision process and form. *In*: TINKLER, K.J. & WOHL, E.E. (eds) *Rivers over Rock: Fluvial Processes in Bedrock Channels*. Geophysical Monograph, American Geophysical Union, **107**, 25–36.

HANCOCK, J.M. 1975. The petrology of the Chalk. *Proceedings of the Geologists' Association*, **86**, 499–536.

HANCOCK, J.M. 2000. The Gribun Formation: clues to the latest Cretaceous history of western Scotland. *Scottish Journal of Geology*, **36**, 137–142.

HAYNES, V. 1983. Scotland's landforms. *In*: CLAPPERTON, C.M. (ed.) *Scotland: a New Study*. David and Charles, Newton Abbott, 28–63.

HILL, S.M. 1999. Mesozoic regolith and palaeo-landscape features in southeastern Australia: significance for interpretations of the evolution of the eastern highlands. *Australian Journal of Earth Sciences*, **46**, 217–232.

HILLIS, R.R., THOMSON, K. & UNDERHILL, J.R. 1994. Quantification of Tertiary erosion in the Inner Moray Firth by sonic velocity data from the Chalk and Kimmeridge Clay. *Marine and Petroleum Geology*, **11**, 282–293.

HOLLIDAY, D.W. 1993. Mesozoic cover over northern England: interpretation of fission track data. *Journal of the Geological Society, London*, **150**, 657–660.

HOLNESS, M.B. 1999. Contact metamorphism and anatexis of Torridonian arkose by minor intrusions of the Rum Igneous Complex. Inner Hebrides, Scotland. *Geological Magazine*, **136**, 527–542.

HORNE, J. 1923. *The Geology of the Lower Findhorn and Lower Strath Nairn. Memoir of the Geological Survey of Scotland*. HMSO, Edinburgh.

HOUSE, M.A., WERNICKE, B.P. & FARLEY, K.A. 1998. Dating topography of the Sierra Nevada, Cali-formia, using apatite (U–Th)/He ages. *Nature*, **396**, 66–69.

HOVIUS, N. 2000. Macroscale process systems of mountain belt erosion. *In*: SUMMERFIELD, M.A. (ed.) *Geomorphology and Global Tectonics*. Wiley, Chichester, 77–105.

HUDSON, J.D. 1964. The petrology of the sandstones of the Great Estuarine Series and the Jurassic palaeogeography of Scotland. *Proceedings of the Geologists' Association*, **75**, 499–528.

HUMPHRIES, D.W. 1961. The Upper Cretaceous White Sandstone of Loch Aline, Argyll, Scotland. *Proceedings of the Yorkshire Geological Society*, **33**, 47–76.

HURST, A. 1985*a*. The implications of clay mineralogy to palaeoclimate and provenance during the Jurassic in north-east Scotland. *Scottish Journal of Geology*, **21**, 143–160.

HURST, A. 1985*b*. Mineralogy and diagenesis of Lower Jurassic sediments of the Lossiemouth borehole, north-east Scotland. *Proceedings of the Yorkshire Geological Society*, **45**, 189–197.

HUUSE, M., LYKKE-ANDERSEN, H. & MICHELSEN, O. 2001. Cenozoic evolution of the eastern Danish North Sea. *Marine Geology*, **177**, 243–269.

JAPSEN, P. 1997. Regional Neogene exhumation of Britain and the western North Sea. *Journal of the Geological Society, London*, **154**, 239–247.

JAPSEN, P. & CHALMERS, J.A. 2000. Neogene uplift and tectonics around the North Atlantic: overview. *Global and Planetary Change*, **24**, 165–173.

JOHNSON, H., RICHARDS, P.C., LONG, D. & GRAHAM, C.C. 1993. *United Kingdon Offshore Regional Report: the Geology of the Northern North Sea*. HMSO, London.

JOHNSTONE, G.S. 1966. *The Grampian Highlands: British Regional Geology*. HMSO, London.

JOHNSTONE, G.S. & MYKURA, W. 1989. *The Northern Highlands of Scotland: British Regional Geology*. HMSO, London.

JOLLEY, D.W. 1997. Palaeosurface palynofloras on the Skye lava field, and the age of the British Tertiary volcanic province. *In*: WIDDOWSON, M. (ed.) *Palaeosurfaces: Recognition, Reconstruction and Palaeoenvironmental Interpretation*. Geological Society, London, 67–94.

JONES, R.W. & MILTON, N.J. 1994. Sequence development during uplift: Palaeogene stratigraphy and relative sea-level history of the Outer Moray Firth, UK North Sea. *Marine and Petroleum Geology*, **11**, 157–165.

JONES, S.M., WHITE, N., CLARKE, B.J., ROWLEY, E. & GALLAGNER, K. 2002. Present and past influence of the Iceland Plume on sedimentation. *In*: DORÉ, A.G., CARTWRIGHT, J.A., STOKER, M.S., TURNER, J.P. & WHITE, N. (eds) *Exhumation of the North Atlantic Margin: Timing, Mechanisms and Implications for Petroleum Exploration*. Geological Society, London, Special Publications, **196**, 13–25.

JORDT, H., FALEIDE, J.I., BJØRLYKKE, K. & IBRAHIM, M. 1995. Cenozoic sequence stratigraphy of the central and northern North Sea Basin: tectonic development, sediment distribution and provenance areas. *Marine and Petroleum Geology*, **12**, 845–879.

JOY, A.M. 1993. Comments on the pattern of post-rift subsidence in the Central and Northern North Sea Basin. *In*: WILLIAMS, G.D. & DOBB, A. (eds) *Tectonics and Seismic Sequence Stratigraphy*. Geological Society, London, Special Publications, **71**, 123–140.

KARLLSON, W., VOLLSET, J., BJØRLYKKE, K. & JÖRGENSEN, P. 1979. Changes in the mineralogical composition of Tertiary sediments from North Sea wells. *Proceedings of 6th International Clay Conference*, **27**, 281–289.

KOHN, B. & BISHOP, P. 1999. Apatite fission track thermochronology and geomorphology. (Thematic issue.). *Australian Journal of Earth Sciences*, **46**, 155–233.

KOHN, B.P., GLEADOW, A.J.W. & COX, S.J.D. 1999. Denudation history of the Snowy Mountains: constraints from apatite fission track thermochronology. *Australian Journal of Earth Sciences*, **46**, 181–198.

LE COEUR, C. 1988. Late Tertiary warping and erosion in western Scotland. *Geografiska Annaler, 70A*, 361–368.

LE COEUR, C. 1999. Rythmes de dénudation tertiaire et quaternaire en Écosse occidentale. *Géomorphologie*, **4**, 291–304.

LEAKE, B.E. & COBBING, J. 1993. Transient and long-term correspondence of erosion level and the tops of granite plutons. *Scottish Journal of Geology*, **29**, 177–182.

LIDMAR-BERGSTRÖM, K., OLLIER, C.D. & SULEBAK, J.R. 2000. Landforms and uplift history of southern Norway. *Global and Planetary Change*, **24**, 211–231.

LIN, J.-C. 2000. Morphotectonic evolution of Taiwan. *In*: SUMMERFIELD, M.A. (ed.) *Geomorphology and Global Tectonics*. Wiley, Chichester, 135–146.

LINTON, D. 1951. Problems of the Scottish scenery. *Scottish Geographical Magazine*, **67**, 65–85.

LINTON, D.L. 1950. The scenery of the Cairngorm Mountains. *Journal of the Manchester Geographical Society*, **55**, 1–14.

LINTON, D.L. 1959. Morphological contrasts between eastern and western Scotland. *In*: MILLER, R. & WATSON, J.W. (eds) *Geographical Essays in Memory of Alan G. Ogilvie*. Nelson, Edinburgh, 16–45.

LIU, X. & GALLOWAY, W.E. 1997. Quantitative determination of Tertiary sediment supply to the North Sea Basin. *AAPG Bulletin*, **81**, 1482–1509.

MCCALLAN, A.A. 1994. Discussion on Mesozoic cover in N England: interpretation of apatite fission track data. *Journal of the Geological Society, London*, **151**, 735–736.

MCDOUGALL, I. & DUNCAN, R.A. 1988. Age progressive volcanism in the Tasmantid seamounts. *Earth and Planetary Science Letters*, **89**, 207–220.

MERRITT, J., AUTON, C.A., CONNELL, E.R., HALL, A.M. & PEACOCK, J.D. 2002. *Quaternary Geology and Landscape Evolution of north-east Scotland*, Memoir of the British Geological Survey. NERC, Nottingham, in press.

MIGON, P. & LIDMAR-BERGSTROM, K. 2001. Weathering mantles and their significance for geomorphological evolution of central and northern Europe since the Mesozoic. *Earth-Science Reviews*, **56**, 285–324.

MORTON, A.C. 1979. The provenance and distribution of the Palaeocene sands of the North Sea. *Journal of Petroleum Geology*, **2**, 11–21.

MUIR WOOD, R. 1989. Fifty million years of 'passive margin' deformation in North West Europe. *In*: GREGERSEN, S. & BASHAM, P.W. (eds) *Earthquakes at North Atlantic Passive Margins: Neotectonics and Postglacial Rebound*. Kluwer, Dordrecht, 7–36.

MUSSET, A.E. 1984. Timing and duration of Tertiary igneous activity of Rhum and adjacent areas. *Scottish Journal of Geology*, **20**, 273–280.

MYKURA, W. 1983. *Old Red Sandstone*. Scottish Academic Press, Edinburgh.

NADIN, P.A. & KUSZNIR, N.J. 1995. Palaeocene uplift and Eocene subsidence in the northern North Sea Basin from 2D forward and reverse stratigraphic modelling. *Journal of the Geological Society, London*, **152**, 833–848.

NIELSEN, O.B., SORENSEN, S., THIEDE, J. & SKARBO, O. 1986. Cenozoic differential subsidence of the North Sea. *AAPG Bulletin*, **70**, 276–298.

NOTT, J., YOUNG, R. & MCDOUGALL, I. 1996. Wearing down, wearing back, and gorge extension in the long-term denudation of a highland mass: Quantitative evidence from the Shoalhaven catchment, southeast Australia. *Journal of Geology*, **104**, 224–232.

OHMORI, H. 2000. Morphotectonic evolution of Japan. *In*: SUMMERFIELD, M.A. (ed.) *Geomorphology and Global Tectonics*. Wiley, Chichester, 147–166.

PRESTON, J. 1982. Eruptive volcanism. *In*: SUTHERLAND, D.S. (ed.) *Igneous Rocks of the British Isles*. Wiley, London, 351–368.

RIDD, M.F. 1981. Petroleum geology west of the Shetlands. *In*: ILLING, L.V. & HOBSON, G. (eds) *Petroleum Geology of the Continental Shelf of NW Europe*. Heyden, London, 414–425.

RIIS, F. 1996. Quantification of Cenozoic vertical movements of Scandinavia by correlation of morphological surfaces with offshore data. *Global and Planetary Change*, **12**, 331–357.

RINGROSE, P.S. & MIGON, P. 1997. surfaces. *In*: WIDDOWSON, M. (ed.) *Palaeosurfaces: Recognition, Reconstruction and Palaeoenvironmental Interpretation*. Geological Society, London, 25–36.

ROBERTS, A.M. & HOLDSWORTH, R.E. 1999. Linking onshore and offshore structures: Mesozoic extension in the Scottish Highlands. *Journal of the Geological Society, London*, **156**, 1061–1064.

ROHRMAN, M., BEEK, P.v.d., ANDRIESSEN, P. & CLOETINGH, S. 1995. Meso-Cenozoic morphotectonic evolution of southern Norway: Neogene domal uplift inferred from apatite fission track thermochronology. *Tectonics*, **14**, 704–718.

RUST, D.J. & SUMMERFIELD, M.A. 1990. Isopach and borehole data as indicators of rifted margin evolution in southwestern Africa. *Marine and Petroleum Geology*, **7**, 277–287.

SMITH, K., GATLIFF, R.W. & SMITH, N.J.P. 1994. Discussion on the amount of Tertiary erosion in the UK using sonic velocity analysis. *Journal of the Geological Society, London*, **151**, 1041–1044.

STEEL, R.J. 1978. Triassic rift basins of Northwest Scotland—their configuration, infilling and development. *In*: FINSTAD, K.G. & SELLEY, R.C. (eds) *Mesozoic Northern North Sea Symposium*. Norwegian Petroleum Society, Stavanger, Paper 7.

STEWART, A.D. 1972. Precambrian landscapes in northwest Scotland. *Geological Journal*, **8**, 111–124.

STOKER, M.S., HITCHEN, K. & GRAHAM, C.C. 1993. *United Kingdom Offshore Regional Report: the Geology of the Hebrides and the West Shetland Shelves, and Adjacent Deep Water Areas*. HMSO, London.

STOKER, M.S., LESLIE, A.B., SCOTT, W.D. & 6 OTHERS 1994. A record of late Cenozoic stratigraphy, sedimentation and climate change from the Hebrides Slope, NE Atlantic Ocean. *Journal of the Geological Society, London*, **151**, 235–249.

STUEVOLD, L.M. & ELDHOLM, O. 1996. Cenozoic uplift of Fennoscandia inferred from a study of the mid-Norwegian margin. *Global and Planetary Change*, **12**, 359–386.

STURT, B.A., DALLAND, A. & MITCHELL, J.L. 1979. The age of the sub-Jurassic tropical weathering profiles of Andøya, northern Norway, and the implications for the Late Palaeozoic paleogeography of the North Atlantic region. *Geologische Rundschau*, **68**, 523–542.

SUMMERFIELD, M.A. & BROWN, R.W. 1998. Geomorphic factors in the interpretation of fission-track data. *In*: VAN DEN HAUTE, P. & DE CORTE, F. (eds) *Advances in Fission-Track Geochronology*. Kluwer, Dordrecht, 269–284.

SUN, S.-S., MCDONOUGH, W.F. & EWART, A. 1989. Four component model for Australian basalts. *In*: JOHNSON, R.W. (ed.) *Intraplate Volcanism in Eastern Australia and New Zealand*. Cambridge University Press, Cambridge, 333–347.

SUTTER, A.A. 1980. *Palaeogene sediments from the U.K. sector of the central North Sea*. PhD thesis, University of Aberdeen.

THOMAS, M.F. 1989. The role of etch processes in landform development. *Zeitschrift für Geomorphologie, NF*, **33**, 129–142.

THOMSON, K. & HILLIS, R.R. 1995. Tertiary structuration and erosion of the inner Moray Firth. *In*: SCRUTTON, R.A., STOKER, M.S., SHIMMIELD, G.B. & TUDHOPE, A.W. (eds) *The Tectonics, Sedimentation and Palaeooceanography of the North Atlantic Region*. Geological Society, London, Special Publications, **90**, 249–269.

THOMSON, K., UNDERHILL, J.R., GREEN, P.F., BRAY, R.J. & GIBSON, H.J. 1999. Evidence from apatite fission track analysis for the post-Devonian burial and exhumation history of the northern Highlands, Scotland. *Marine and Petroleum Geology*, **16**, 27–39.

THRELFALL, W.F. 1981. Structural framework of the central and northern North Sea. *In*: ILLING, L.V. & HOBSON, G. (eds) *Petroleum Geology of the Continental Shelf of NW Europe*. Heyden, London, 98–103.

TIPPETT, J.M. & HOVIUS, N. 2000. Geodynamic processes in the Southern Alps. *In*: SUMMERFIELD, M.A. (ed.) *Geomorphology and Global Tectonics*. Wiley, Chichester, 109–134.

TWIDALE, C.R. 1994. Gondwanan (Late Jurassic and Cretaceous) palaeosurfaces of the Australian craton. *Palaeogeography, Palaeoclimatology, Palaeoecology*, **112**, 157–186.

TWIDALE, C.R. & CAMPBELL, E.M. 1995. Pre-Quaternary landforms in the low latitude context: the example of Australia. *Geomorphology*, **12**, 17–35.

WATSON, J. 1985. Scotland as an Atlantic–North Sea divide. *Journal of the Geological Society, London*, **142**, 221–243.

WELLMAN, P. 1986. Intrusions beneath large alkaline intraplate volcanoes. *Exploration Geophysics*, **17**, 135–139.

WELLMAN, P. & MCDOUGALL, I. 1974. Cainozoic ingenous activity in eastern Australia. *Tectonophysics*, **23**, 49–65.

WHITE, N. & LOVELL, B. 1997. Measuring the pulse of a plume with the sedimentary record. *Nature*, **387**, 888–891.

WIGNALL, P.B. & PICKERING, K.T. 1993. Palaeoecology and sedimentology across a Jurassic fault scarp, NE Scotland. *Journal of the Geological Society, London*, **150**, 323–340.

YOUNG, R.W. 1981. Denudational history of the south–central uplands of New South Wales. *Australian Geographer*, **15**, 77–88.

ZEITLER, P.K., HERCZIG, A.L., MCDOUGALL, I. & HONDA, M. 1987. U–Th–He dating of apatite: a potential thermochronometer. *Geochimica et Cosmochimica Acta*, **51**, 2865–2868.

ZIEGLER, P.A. 1981. *Evolution of Sedimentary Basins of North-West Europe*. Heyden, London.

Cenozoic evolution of the Faroe Platform: comparing denudation and deposition

MORTEN SPARRE ANDERSEN[1], AAGE BACH SØRENSEN[1], LARS OLE BOLDREEL[2] & TOVE NIELSEN[1]

[1]*Geological Survey of Denmark and Greenland (GEUS), Thoravej 8, DK-2400 Copenhagen NV, Denmark (e-mail: msa@geus.dk)*

[2]*University of Copenhagen, Department of Geology, Øster Voldgade 10, DK-1350, Copenhagen K, Denmark*

Abstract: Throughout Paleocene and Eocene time the Faroe–Shetland Channel and the eastern part of the Faroe Platform was a subsiding marine basin. In Early Paleocene time, basin-floor fans of a British provenance were deposited in the eastern part of the basin. In Late Paleocene time, *c.* 6 km of basalt entered the basin from the west and north, and the basin was constricted by the large volumes of basalt that entered the basin, creating the Faroe–Shetland Escarpment. In Eocene time subsidence continued in the basinal areas. Again, sediments of a dominantly eastern provenance were deposited. Throughout Eocene time, erosion products from the Faroe Platform were possibly deposited in the Faroe Bank Channel and the Norwegian Sea Basin, but only to a limited degree in the Faroe–Shetland Channel. The oldest sediments of documented western provenance on the eastern margin of the Faroe Platform are of Early Oligocene age. During a compressional phase commencing in Mid–Late Miocene time some basinal areas emerged and erosion took place on the top of emerged anticlines. However, denudation throughout Late Miocene and Early Pliocene time was apparently rather limited compared with a Late Pliocene phase of denudation. During this phase of denudation, a large progradational wedge was deposited on the eastern margin of the Faroe Platform. On the basis of a structural analysis of the Faroe Platform, the amount of basalt removed from it during Cenozoic time is estimated to be *c.* 46 000 km^3 ($131\,100 \times 10^{12}$ kg). Using 2900 kg m^{-3} as the density of basalt and 2300 kg m^{-3} as sediment density the estimated amount of removed basalt is in fair agreement with the estimate of the volume of sediments derived from the platform (*c.* 56 000 km^3, $114\,800 \times 10^{12}$ kg). The greatest deposition rates on the eastern Faroe Platform and in the Faroe–Shetland Channel apparently occurred after two distinct inversion or compression events in Mid-Eocene and Mid–Late Miocene time. However, uplift of the Faroe Platform could have been forced by denudation rather than endogenous processes.

In this paper we present a preliminary attempt to quantify the denudational and depositional history of the Faroe region during Cenozoic time (Fig. 1). The stratigraphic subdivision and discussion of the depositional history is mainly based on published work (Nielsen & Van Weering 1998; Andersen *et al.* 2000; Stoker *et al.* 2002).

Structural elements and regional setting

In a broader context the Faroes are located close to the line of the Late Paleocene break-up between Europe and Greenland, centrally in the area affected by the Icelandic mantle plume during Paleocene time, and on the margin of documented Mesozoic rifting in NW Europe (Fig. 2).

The most important Cenozoic structural elements within the Faroe region are shown in Fig. 1. Most of the structural elements are known from previous work (e.g. Boldreel & Andersen 1994; Lamers & Carmichael 1999). The Foinaven Basin is the Paleocene–Eocene depocentre in the southern part of the Faroe–Shetland Channel. Geographically it is equivalent to the Foinaven Sub-basin of Lamers & Carmichael (1999). The Fugloy Basin, introduced here, is a broad Eocene–Oligocene depocentre in the northern part of the Faroe–Shetland Channel. Upper Eocene–Oligocene sediments, which are thin or absent in the Foinaven basin, thicken northward in the Fugloy Basin (Fig. 3).

From: DORÉ, A.G., CARTWRIGHT, J.A., STOKER, M.S., TURNER, J.P. & WHITE, N. 2002. *Exhumation of the North Atlantic Margin: Timing, Mechanisms and Implications for Petroleum Exploration*. Geological Society, London, Special Publications, **196**, 291–311. 0305-8719/02/$15.00

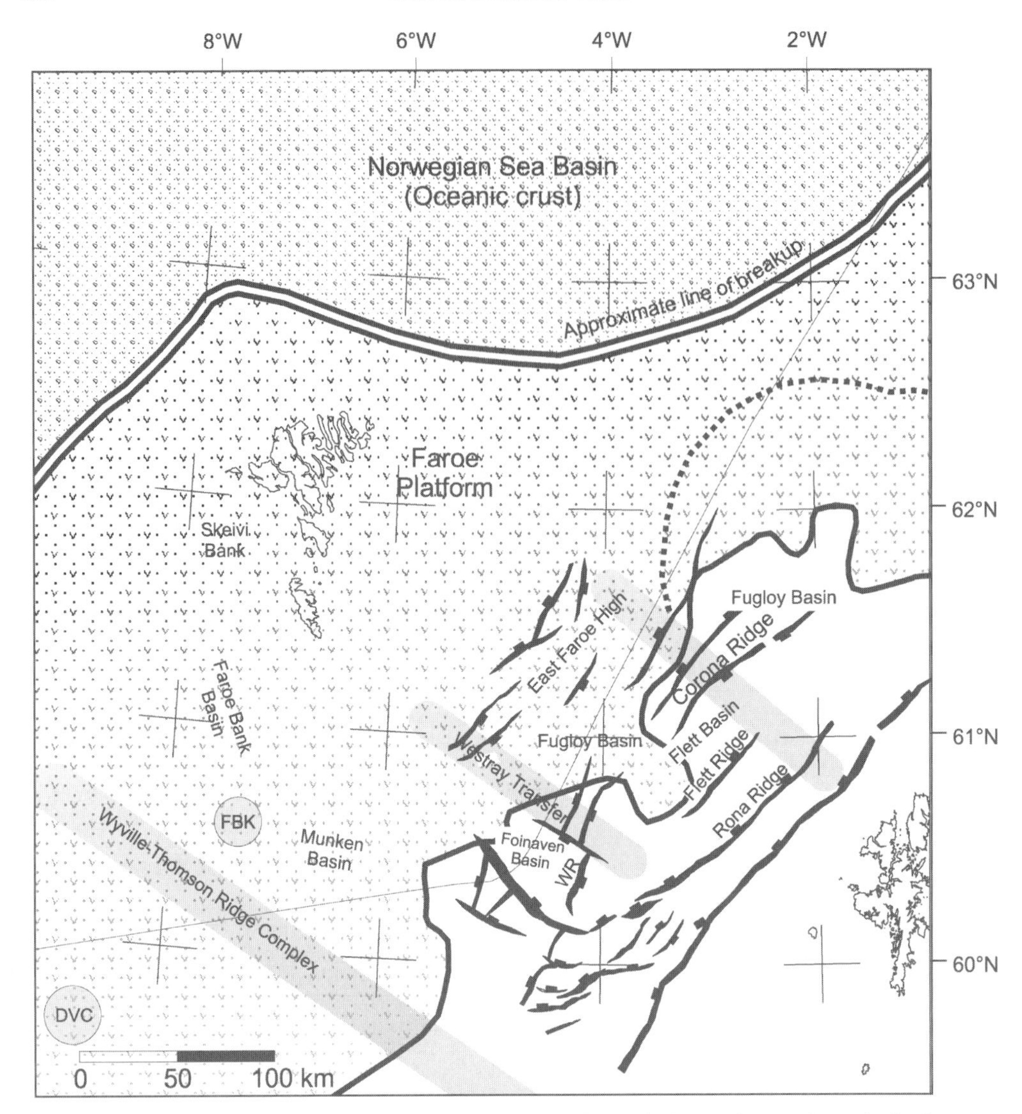

Fig. 1. Structural elements of the study area. Interpreted and inferred faults are shown schematically. Important transfer zones are shown with shading. Fugloy Basin is introduced as a new term representing the northern Eocene–Oligocene depocentre in the Faroe–Shetland Channel. It overlies the Paleocene Flett Basin, the Corona Ridge and a poorly described depocentre west of Corona Ridge. Foinaven Basin is a depocentre in the southern part of the Faroe–Shetland Channel throughout Paleocene–Eocene time. Other elements have been described elsewhere (e.g. Andersen *et al.* 2000). FBK, Faroe Bank Knoll; DVC, Darwin volcanic centre; WR, Westray Ridge.

Geographically, the Fugloy Basin overlies the Paleocene Flett Basin (Hitchen & Ritchie 1987), the Corona Ridge (e.g. Rumph *et al.* 1993) and a relatively poorly constrained Paleocene depocentre west of the Corona Ridge. The Westray Transfer Zone (e.g. Lamers & Carmichael 1999), including the anticlinal structure the Westray Splay, separates the Foinaven Basin from the Flett and Fugloy basins and the Corona Ridge.

During Paleocene time, normal faulting occurred along approximately NNE–SSW- and east–west-trending faults producing subsidence of the Foinaven and Flett basins (e.g. Dean *et al.* 1999; Lamers & Carmichael 1999). To some extent, this may reflect compaction of a thick underlying succession of Cretaceous sediments (Ebdon *et al.* 1995). However, the Paleocene depocentres are offset and slightly rotated

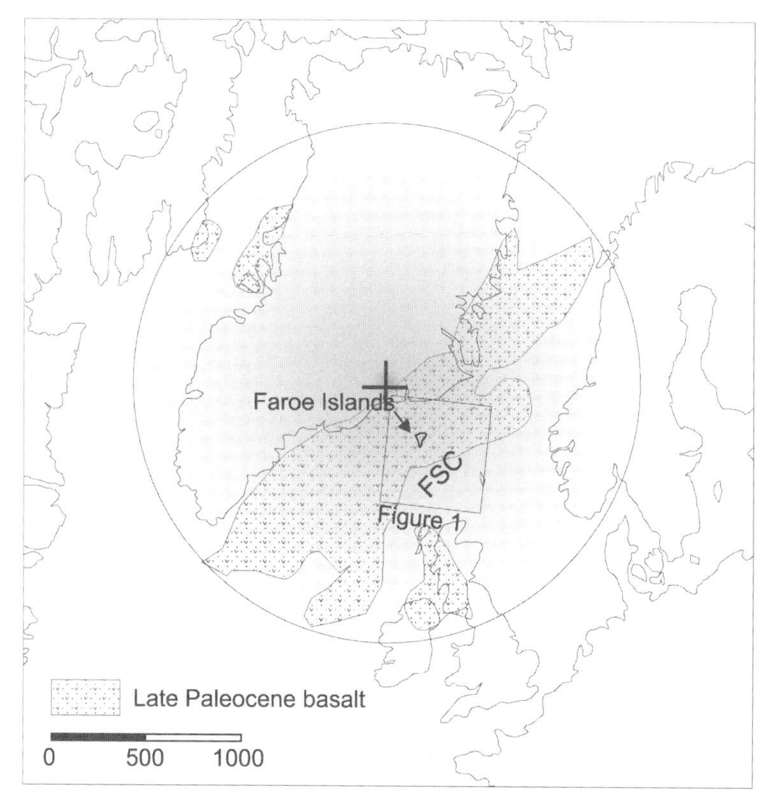

Fig. 2. The NE Atlantic region in earliest Eocene time (magnetic chron C23). The approximate extent of Paleocene pre-break-up and syn-break-up basalts is shown. The approximate location of the centre of the Icelandic plume and its extent is shown according to White & McKenzie (1989). FSC, Faroe–Shetland Channel.

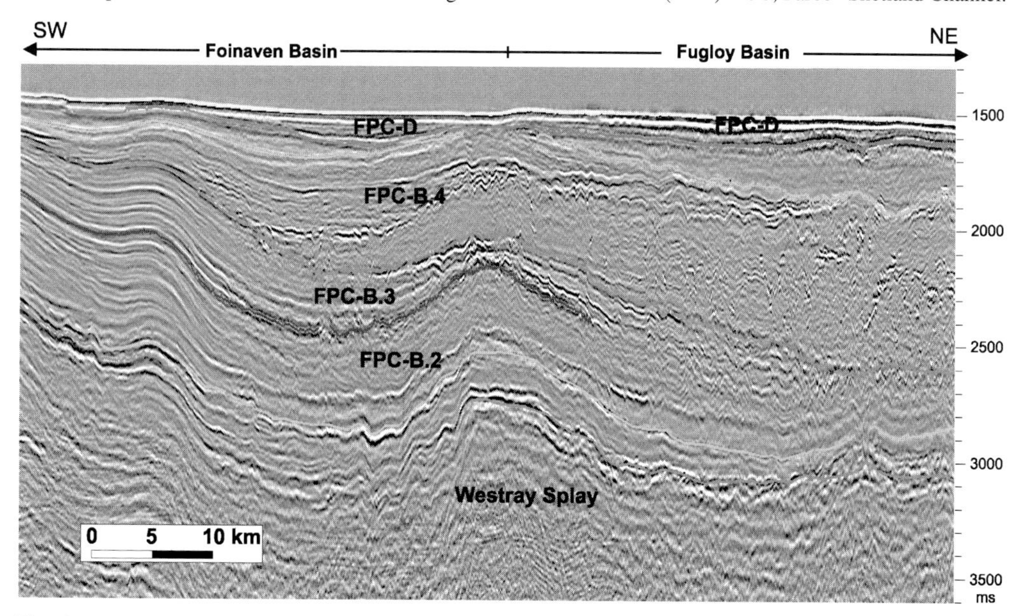

Fig. 3. Seismic section from Foinaven Basin into Fugloy Basin. The seismic character of FPC-B changes dramatically as the section crosses the Westray Splay. Location of profile is shown in Fig. 7. By courtesy of Veritas.

relative to the Late Cretaceous depocentres (Dean *et al.* 1999). In addition, a strike-slip component along some east–west fault planes may possibly be indicated by abrupt change of the throw along strike (e.g. Judd Fault). This suggests an important Paleocene rift event controlled by a stress system rotated relative to the mid-Cretaceous system (Andersen 1999; Dean *et al.* 1999). Cenozoic subsidence curves calculated for wells in the West Shetland area could support the concept of Paleocene rifting in the Faroe–Shetland area. However, Paleocene–Recent subsidence in basinal areas of the northern North Sea and Paleocene denudation in large areas of the British Isles are considered arguments for the existence of a large mantle plume, the Iceland plume, in the area during Paleocene time (Hall & White 1994). The Iceland plume would have even stronger influence in the Faroe–Shetland area than in the North Sea, but in the Faroe region separation of rift-induced tectonic movements from plume-induced tectonic movements has not been possible.

Following the Paleocene rift episode (or late in the rift period) the Faroe Platform was, within a short time span (59–55.5 Ma), covered by a thick succession of basalts (stratigraphic thickness exceeds 5 km: Hald & Waagstein 1984). Since Eocene time, basinal areas of the Faroe region subsided about 2500 m (e.g. Boldreel & Andersen 1993; Ritchie *et al.* 1999). During this period both the Shetland region and the Faroe Platform supplied sediments to basinal areas (e.g. Andersen *et al.* 2000).

Stratigraphic correlation

Work towards a unified mid–late Cenozoic stratigraphy across the Faroe–Shetland Channel is currently in progress under the auspices of the EU-funded STRATAGEM project (Evans 2000). This work should provide a significantly improved stratigraphic model for the Faroe Platform and its margins. However, this stratigraphic model is not finished, and the stratigraphic subdivision used in this paper (Table 1) is adapted from Boldreel & Andersen (1995), Nielsen & Van Weering (1998) and Andersen *et al.* (2000). Revised age assignments and correlation of the Neogene section across the Faroe–Shetland Channel are primarily based on Cloke *et al.* (2000) and Stoker *et al.* (2002). The stratigraphic correlations in Table 1 are considered an improvement relative to those presented by Andersen *et al.* (2000). However, the chronostratigraphic correlations are still mostly tentative. The following changes have been introduced relative to Andersen *et al.* (2000).

A progradational wedge in unit FPC-D.2 on the Faroe Platform's eastern margin (Fig. 4) (Andersen *et al.* 2000) is similar to a Late Pliocene–early Pleistocene progradational

Table 1. *Seismic units mapped on the Faroe Platform and its margins*

Approximate age	Bounded units on Faroe Platform		Seismic horizons		Lithostratigraphy	
	Eastern margin	Northern margin	Faroes East and South	West Shetland BGS		
	FPC-D.3	Unit 9			Nordland Group	N3
mid-Pleistocene			CN-050	GU		
	FPC-D.2	Sequence 3			Nordland Group	N2
early Pliocene			CN-040	INU		
	FPC-D.1	Sequence 2			Nordland Group	N1
Mid-Miocene/Late Miocene			CN-030			
	FPC-C	Sequence1				
latest Oligocene/early Miocene			CN-010	LOEMU		
	FPC-B.4	Unit A			Westray Group	
Mid-Eocene/Late Eocene			CP-100			
	FPC-B.3	Sequence 0			Stronsay Group	
Early Eocene/Mid-Eocene			CP-060			
	FPC-B.2					
intra-Ypresian			CP-030			
	FPC-B.1				Moray Group	
base Eocene			CP-010			
	FPC-A	Basalt basement				

The seismic subdivision used on the eastern margin of the Platform (FPC-A–FPC-D) has also been recognized on the SW margin of the platform and in the Faroe Bank Basin. The correlation with the Neogene seismic stratigraphy on the West Shetland margin (Stoker 1999; Stoker *et al.* 2002) is discussed in the text. The table also shows a tentative correlation with the seismic stratigraphy on the northern margin of the platform (Nielsen & Van Weering 1998). Further discussion of this correlation has been given by Andersen *et al.* (2000).

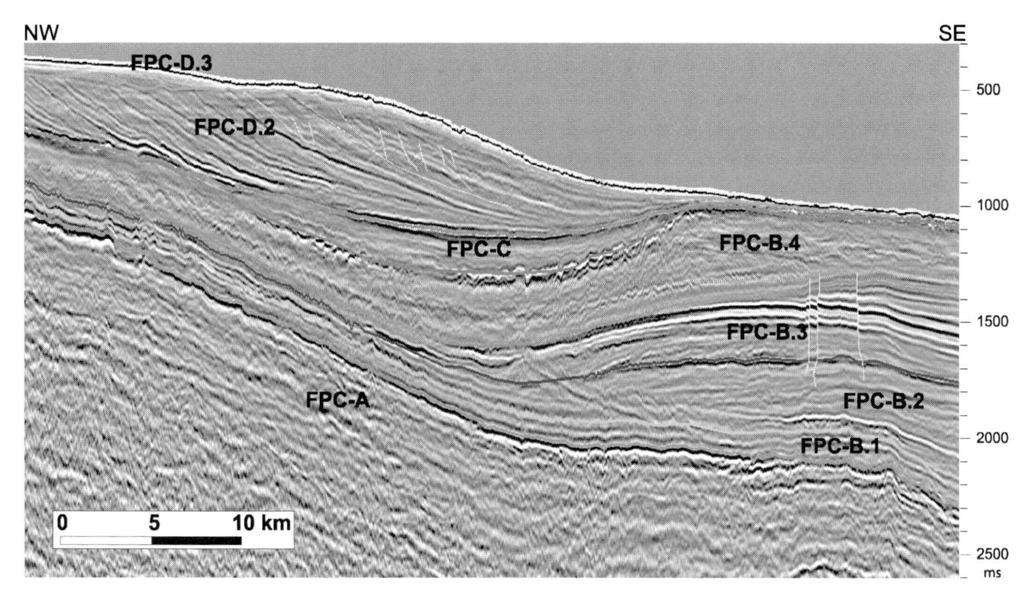

Fig. 4. Seismic section across the NW part of the Foinaven Basin and the SE margin of the Faroe Platform. The distal part of FPC-B.1 continues from the basin onto the margin. Location of profile is shown in Fig. 7. By courtesy of Western Geophysical.

wedge between the 'Glacial Unconformity' (GU) and the 'Intra Neogene Unconformity' (INU) on the West Shetland and Hebrides margin (Stoker 1999) and to a Late Pliocene–Pleistocene wedge off Mid-Norway (Henriksen & Vorren 1996). Therefore the base of FPC-D.2 is tentatively correlated with the INU (Stoker *et al.* 2002), and the base of FPC-D.3 is considered equivalent to the GU (Stoker *et al.* 2002).

Recent biostratigraphic work in exploration wells offshore West Shetland has indicated that Neogene basin inversion and contraction in the UK sector of the Faroe–Shetland Channel commenced approximately at the boundary between Mid- and Late Miocene time (Cloke *et al.* 1999, 2000; Cloke, pers. comm.). We have adopted this age for the base of FPC-D.1. However, distinct along-strike variations in the

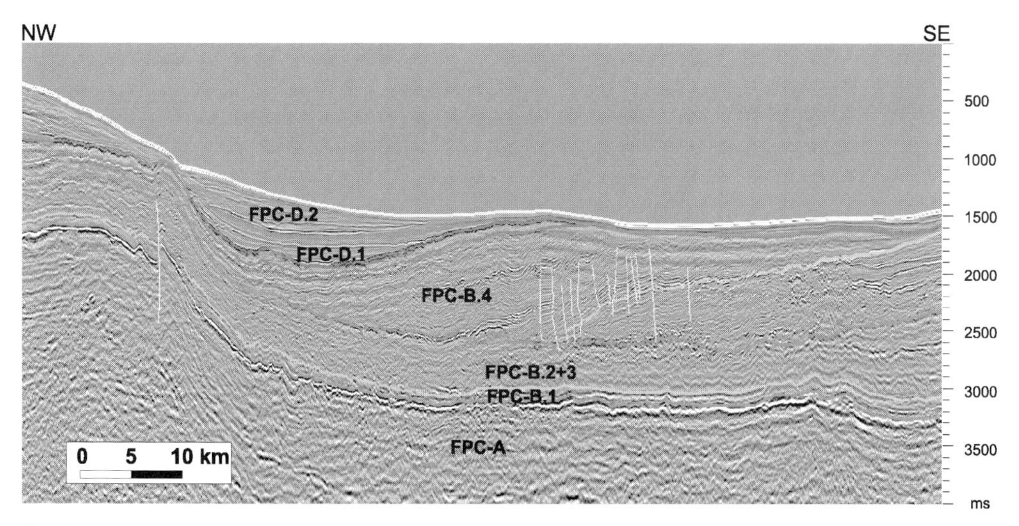

Fig. 5. Seismic section across the Fugloy Basin. Onlap from west onto FPC-B.4 should be noted. Location of profile is shown in Fig. 7. By courtesy of Western Geophysical.

timing of Neogene inversion and contraction in the Faroe–Shetland Channel were recorded by Cloke *et al.* (2000).

The base of FPC-C is tentatively correlated with the 'Late Oligocene–Early Miocene Unconformity' (LOEMU) (Stoker 1999; Stoker *et al.* 2002), and the base of FPC-C is thus reassigned a Late Oligocene–Early Miocene age. The correlation between the base of FPC-C and the LOEMU is supported by ties to wells 214/28-1 and 204/28-1 on the West Shetland margin and correlation with seismic sections presented by Stoker (1999).

The Cenozoic sequence is thus divided into four main units separated by three distinct seismic marker horizons (Andersen *et al.* 2000):

(1) FPC-A, which on the Faroe margin and in the Faroe Bank Basin, consists of Paleocene plateau basalts (Fig. 4). Further east in the Foinaven and Fugloy basins the T36–T45 sequences (Ebdon *et al.* 1995) are considered time equivalents of FPC-A. On the northern margin of the Faroe Platform FPC-A consists of a wedge of seaward-dipping basalt. An outer high characterized by chaotic reflections is found below the lower slope, basinwards from seaward-dipping basalts. According to the emplacement model of Planke *et al.* (2000), the outer high, located seawards of the seaward-dipping basalts, may represent the submergence of the volcanic centre after the break-up between Faroes and Greenland. Further north FPC-A consists of typical oceanic basement formed from subaqueously erupted basalt.

(2) FPC-B consists of a thick succession of Eocene–Oligocene sediments, which in the Faroe–Shetland Channel are mostly of eastern provenance. In the Foinaven Basin FPC-B is subdivided into four subunits FPC-B.1–FPC-B.4, all characterized by fairly good reflection continuity (Fig. 4). This subdivision is not recognized on seismic data from the Fugloy

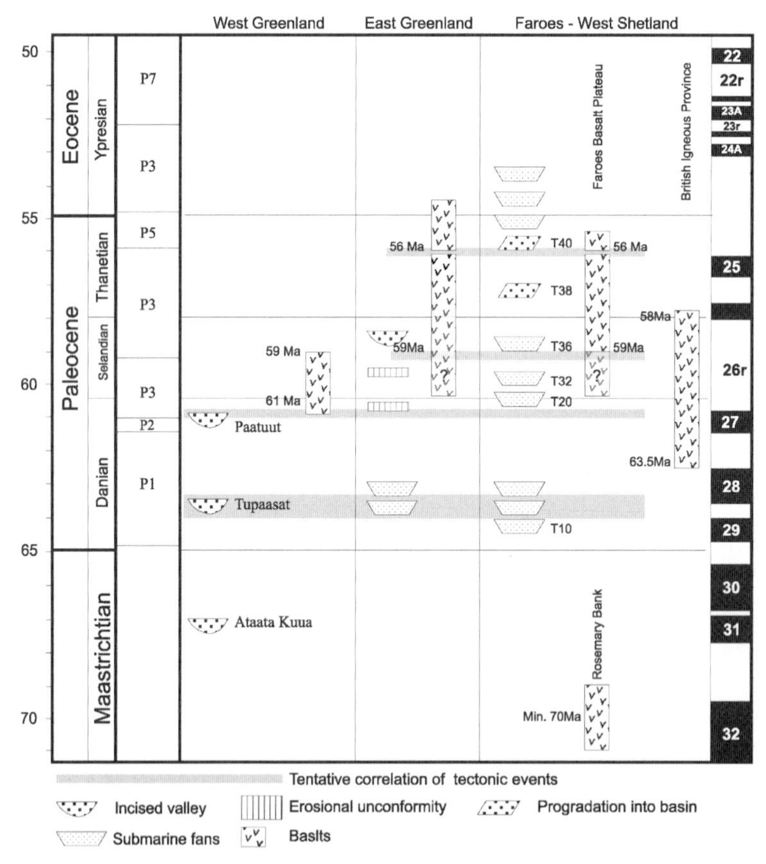

Fig. 6. Correlation chart of Maastrichtian–Paleocene sedimentological and volcanic events in the NE Atlantic region. Four possible regional events are indicated. Modified from White & Lovell (1997), Dam *et al.* (1998), Storey *et al.* (1998) and Larsen *et al.* (1999).

Basin, where FPC-B generally is characterized by poor reflection continuity. However, the shift from sediments of dominantly eastern provenance to sediments of mixed or dominantly western provenance is marked in the Fugloy Basin by a seismic horizon, which approximately correlates with the base of FPC-B.4 (Figs 3 and 5). In the Foinaven Basin, the three lower (Eocene) subunits of FPC-B are of eastern provenance but extend westward on to the Faroe Platform (Fig. 4). In Late Eocene–Oligocene time sediments of a western provenance become abundant, and in the upper part of FPC-B.4 sediments of a western provenance are dominant. In the Faroe Bank Basin FPC-B consists of a seismically homogeneous succession of sediments originating mainly from the Faroe Platform. Sequence 0 (Nielsen & Van Weering

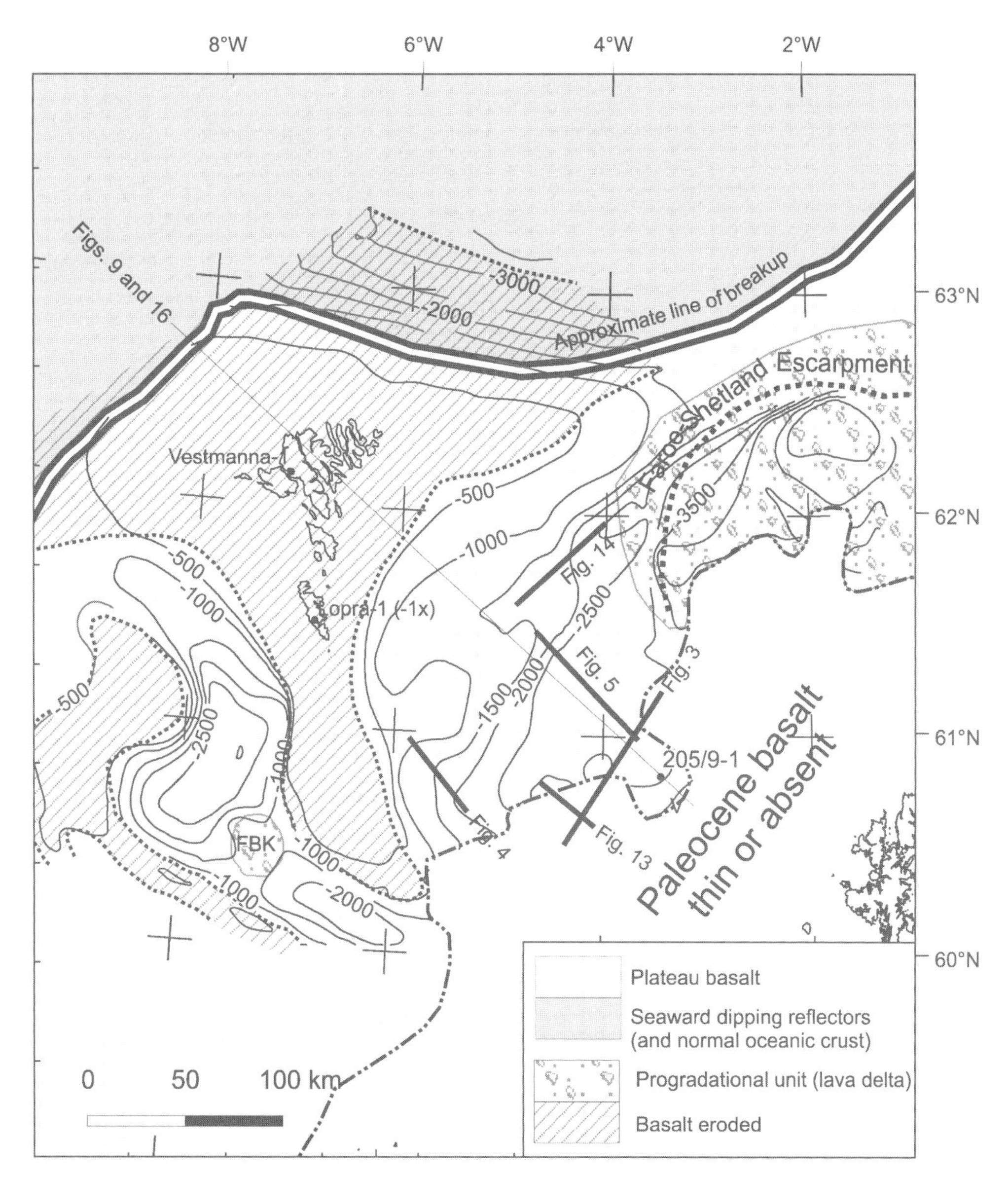

Fig. 7. Distribution of Paleocene basalt in the Faroe region. Contours of the depth to the basalt are based on seismic interpretation. Contour interval 500 m. The approximate extent of erosional truncation of the Paleocene basalt is also shown.

1998), a thick succession of sediments onlapping the basaltic basement on the northern margin of the Faroe Platform, may be equivalent to all or part of unit FPC-B.

(3) The Lower–Middle Miocene unit FPC-C is found on the eastern margin of the Faroe Platform and in the Fugloy Basin, where it thickens northwards. In the Faroe Bank Channel FPC-C is fairly thin (<200 m). The up to 500 m thick Sequence 1 (Nielsen & Van Weering 1998) on the northern margin of the platform may be the equivalent of FPC-C (Andersen *et al.* 2000).

(4) FPC-D (of Late Miocene–Recent age), in the Faroe–Shetland Channel and Faroe Bank Channel, is mostly characterized by good reflection continuity (Fig. 5). The thickness is variable, probably reflecting complex interaction of down-slope and along-slope sediment transport in the deeper part of the basins. FPC-D is divided into three subunits: FPC-D.1 (of Late Miocene–Early Pliocene age), FPC-D.2 (of Late Pliocene–early Pleistocene age) and FPC-D.3 (of late Pleistocene–Recent age).

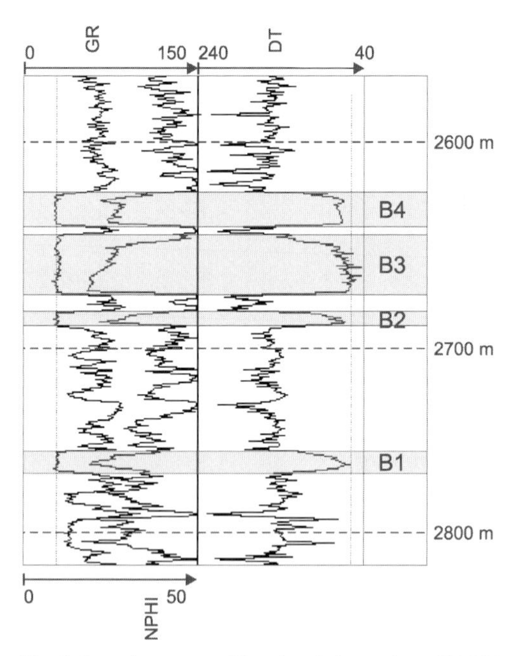

Fig. 8. Log signatures of four basalt layers in well 205/9-1. The constant low values of the gamma-ray log (GR) through each of the four basalt layers should be noted. The top of all four basalt layers is characterized by a porous 'scoria zone' with downward decreasing neutron porosity (NPHI) and increasing velocity (decreasing DT (delta time)). The base of the three upper layers is sharp on three logs. The lower layer, B1, is atypical for subaerially emplaced basalts, as the lower halves of the logs are a mirror image of the upper half. This may indicate rapid cooling from below as well as from above during and immediately after emplacement. Emplacement in water or wet sediments (a bulldozing intrusion, Planke *et al.* 2000) could explain the log signature of layer B4.

Cenozoic tectonics, denudation and deposition in the Faroe region

Paleocene time

Several generations of slope and basin-floor fans were deposited in the Faroe–Shetland Channel (Ebdon *et al.* 1995; Lamers & Carmichael 1999) (Fig. 6). On the basis of well data and seismic interpretation it appears that most if not all of the reasonably well-defined fan deposits in the eastern and central part of the Faroe–Shetland Channel were derived from sediment sources in the area around the Shetland and Orkney Islands or further south (Ebdon *et al.* 1995; Lamers & Carmichael 1999). Further west in the channel the basalt cover prohibits sufficient seismic imaging quality to identify fan systems. No direct evidence of Early Paleocene vertical movements on the Faroe Platform is available (e.g. Kiørboe 1999; Naylor *et al.* 1999).

In Late Paleocene time a thick succession of basalts was emplaced in the Faroe area (Fig. 7). On the Faroe Islands the exposed remnant of this succession is represented by a *c.* 3000 m thick sequence of parallel-bedded plateau basalt (Rasmussen & Noe-Nygaard 1970). An additional *c.* 2000 m of plateau basalt was penetrated in Lopra-1 without reaching the base of the Upper Paleocene basalt succession (Hald & Waagstein 1984). Parallel-bedded plateau basalt has been recognized on seismic reflection profiles as a unit of parallel-bedded reflectors below the top of the basalt throughout the Faroe Platform, the Faroe Bank Channel and the western part of the Faroe–Shetland Channel (Andersen 1988; Boldreel & Andersen 1993). On seismic sections a strong reflection, representing the top of the plateau basalt, can be traced as far east as well 205/91 in the UK sector of the Faroe–Shetland Channel (e.g. Naylor *et al.* 1999; Ritchie *et al.* 1999). In this well the plateau basalts are represented by four distinct basalt layers intercalated with sediments containing terrestrial to marginal marine palynofloras (e.g. Naylor *et al.* 1999). Three of the four basalt layers have log signatures comparable with the log signature of a typical subaerially emplaced basalt flow from the Lopra-1 well (Fig. 8). North and west of the Faroe Platform a wedge of basalts characterized by seaward-dipping reflectors is found (Smythe *et al.* 1983). In the northern part of the Faroe–Shetland Channel, a unit of prograding reflectors is interpreted as the lateral continuation of the parallel-bedded plateau basalt (Gatliff *et al.* 1984; Andersen 1988; Ritchie *et al.* 1999). The eastward termination of the progradational unit is defined by the Faroe–Shetland Escarpment, and it has been suggested that the break of the escarpment formed a Late Paleocene shoreline (e.g. Smythe 1983; Boldreel & Andersen 1994; Ritchie *et al.* 1999).

On the basis of observed lateral thickness changes, we estimate that the surface dip during emplacement of the plateau basalt was *c.* 0.2°. This value is considered reasonable as the largest depositional dips in the plateau basalts, *c.* 0.5°, have been observed on the flanks of individual small shield volcanoes (of the scutulum type) in the upper basalt series (Noe-Nygaard 1968). Volcanic feeder systems along the line of break-up were emphasized in a recent emplacement model for volcanic rifted margins (Planke *et al.* 2000). With 0.2° monotonous dip away from feeder systems along the line of break-up, the Faroe Platform would culminate *c.* 900 m above sea level close to the line of break-up (Fig. 9). Observations of feeder systems along the straits between the Faroe Islands may indicate that the feeder systems were spread out over most of the Faroe Platform (Rasmussen & Noe-Nygaard 1970). The Late Paleocene topography of the platform could thus have been lower, culminating *c.* 500 m above sea level (Fig. 9).

Eocene–Recent denudation of the Paleocene plateau basalts

Since the basalt was emplaced in Late Paleocene time considerable erosion has taken place, and throughout most of the Faroe Platform the top of the basalt is an erosional unconformity (Fig. 7) (Rasmussen & Noe-Nygaard 1970; Boldreel & Andersen 1993; Andersen *et al.* 2000). To estimate the amount of erosion, we have reconstructed the missing section, taking into consideration that the basalts are parallel bedded in most of the study area (Fig. 7). The principle of the reconstruction and the resulting contour map showing the top of the missing basalt section are shown in Fig. 10. To derive a reasonable fit between onshore and offshore geology, we had to assume that the middle and upper basalt series thinned from more than 2000 m on the Northern Islands to about 1000 m on Suðuroy. This is in agreement with detailed observations in the upper and middle series (e.g. Waagstein 1988). This reconstruction also implies that on the northern part of the Faroe Platform *c.* 700 m of basalt was emplaced above the youngest known basalt of the upper series. A detailed correlation between the Faroes basalt series and the Nansen Fjord and Milne Land formations on East Greenland has been established (Larsen *et al.* 1999). A few dykes are found on the Faroes with compositions similar to the basalts of the Geikie Plateau Formation, which overlies the Milne

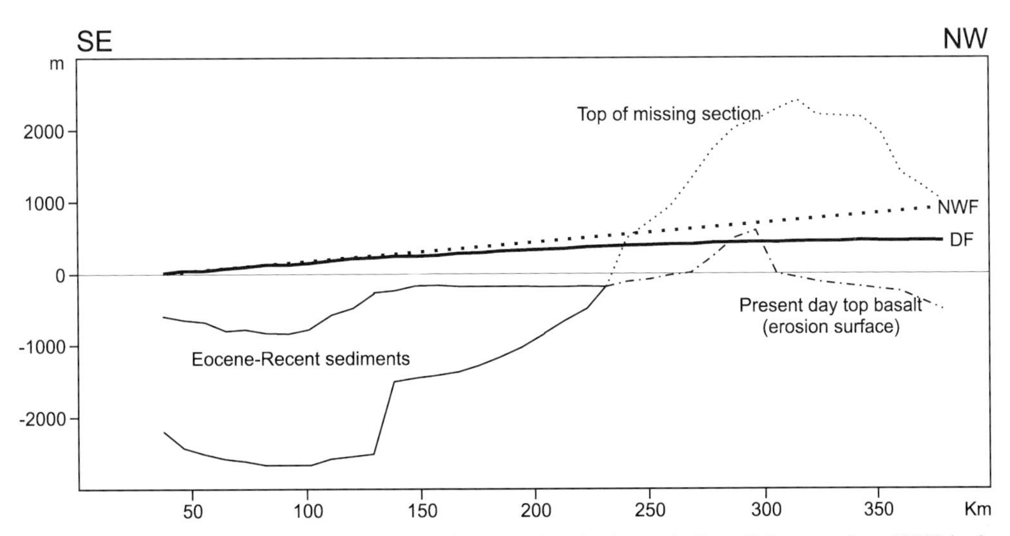

Fig. 9. Two alternative surface topographies of the Faroe basalt plateau in Late Paleocene time. NWF is the topography expected if the volcanic feeders were located in a narrow zone around the line of break-up. DF is the topography expected in the case of distributed feeder systems. (See text for further discussion.) The present-day surface topography, Eocene–Recent sediments in the Faroe–Shetland Channel and the estimated missing section of basalt are shown for comparison. Location of profile is shown in Fig. 7.

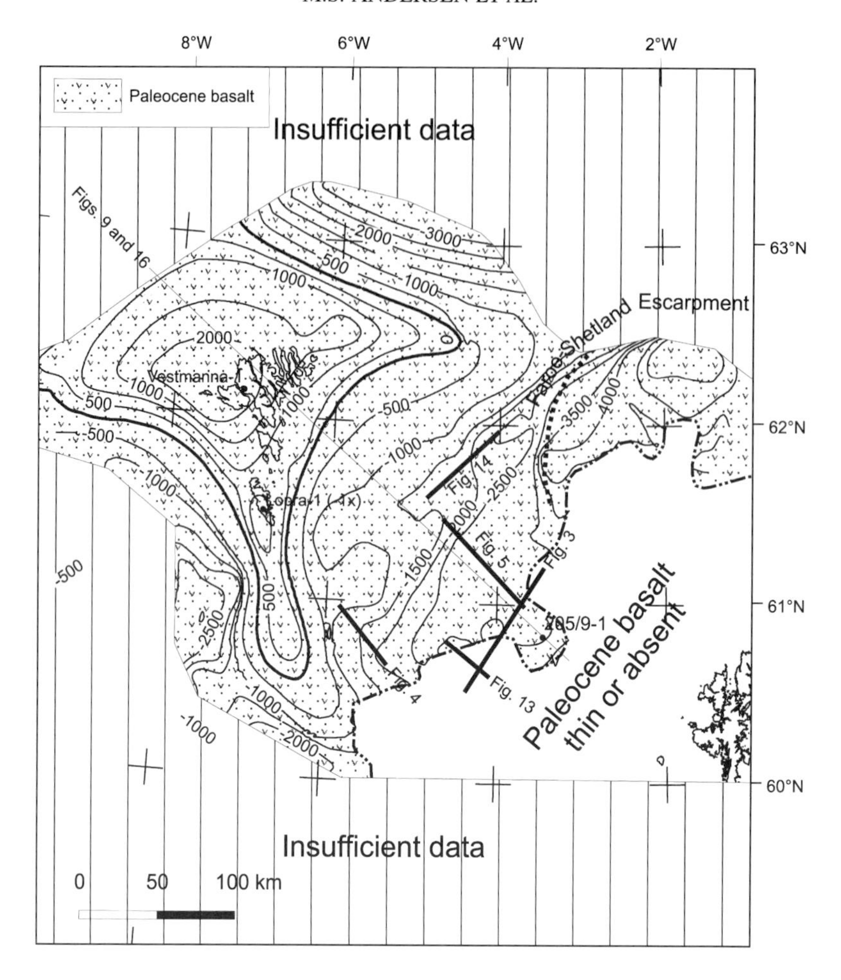

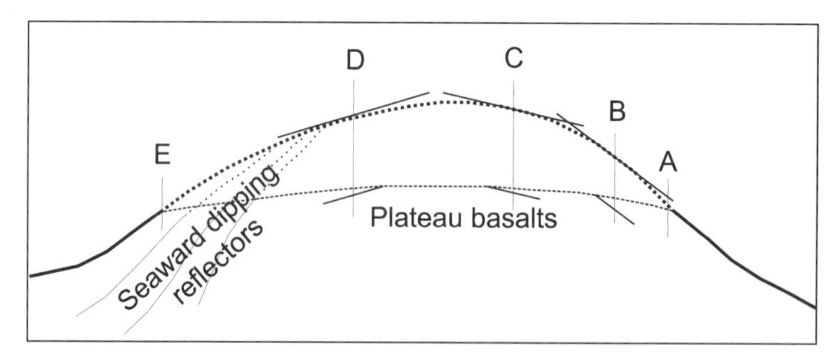

Fig. 10. Reconstruction of the structure of the top of the Faroes Basalt Plateau. The top of the missing section is contoured with 500 m interval. The principles of the reconstruction are shown in the lower part of the figure. Between A and E the top of the basalt is an erosional unconformity. The top of the missing section is drawn as a smooth curve joining the top of the basalt at A and E. The curve is constructed so that its tangents are parallel to bedding planes at points B, C and D. The removed part of the seaward-dipping reflectors is reconstructed by a smooth curve joining the last point within the parallel-bedded basalts, D, to the point of truncation, E. At any location along this part of the curve, the dip of the curve is less than the dip of apparent bedding planes within the seaward-dipping reflectors.

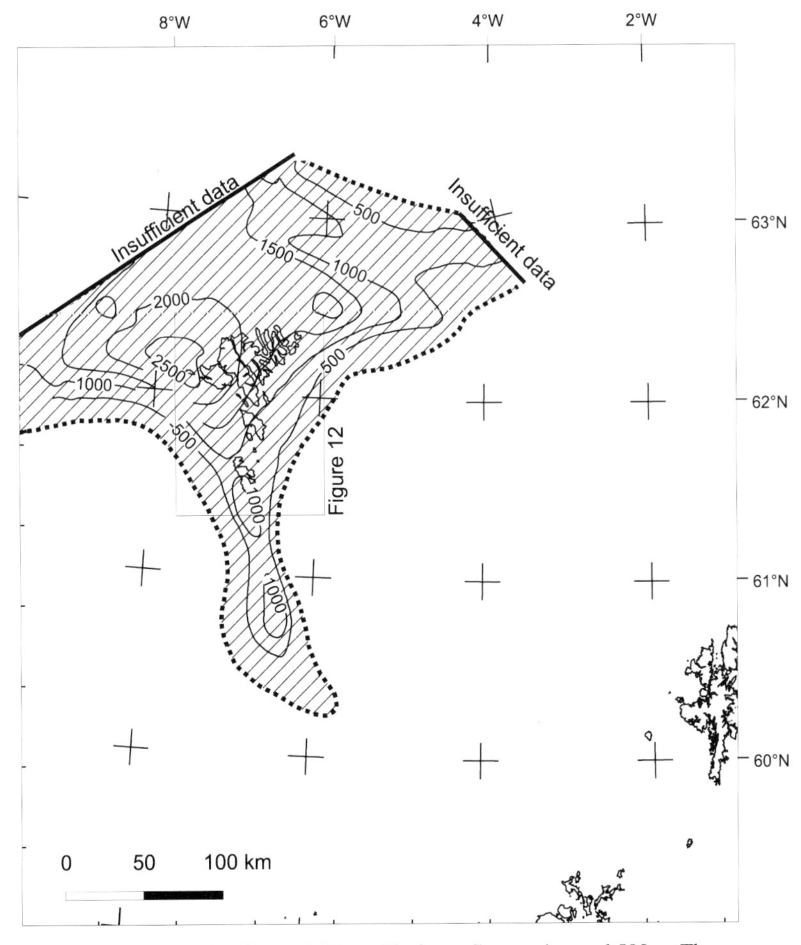

Fig. 11. Estimated denudation on the Cenozoic Faroe Platform. Contour interval 500 m. The map was constructed by subtracting the present top of the basalt (Fig. 7) from the reconstructed structure of the top of the basalt plateau shown as Fig. 10.

Land Formation in East Greenland (Larsen *et al.* 1999). It is thus possible that the 700 m thick missing section is equivalent to the *c.* 1500 m thick Geikie Plateau Formation in East Greenland. An estimate of Eocene–Recent denudation of the Faroe Platform is obtained by subtracting the actual top of the basalt shown in Fig. 7 from the contour map in Fig. 10. A map of estimated Eocene–Recent denudation is presented in Fig. 11. On the basis of this map, *c.* 46 000 km^3 of basalt has been removed from the Faroe Platform since earliest Eocene time.

Studies of zeolite zonation in the Lopra-1 and Vestmanna-1 wells provide independent estimates of the magnitude of denudation at these two locations (Jørgensen 1984). In Lopra-1 the denudation estimate from zeolite zonation obtained by Jørgensen (1984) is comparable with our estimate based on simple structural reconstruction. However, in the 650 m deep Vestmanna-1 well Jørgensen (1984) found a complex crystallization sequence characterized by fossil thermal gradients in the range 67–100 °C km^{-1}. Extrapolation of these gradients to the present mean temperature of 7 °C indicates that the original top of the basalts should be found *c.* 600–800 m above sea level. This implies that the surface of the basalt plateau should coincide approximately with the C-horizon, between the middle and upper basalt. However, during the regional mapping of the Faroes neither an angular unconformity at the base of the upper basalt series nor westward thinning of the upper basalt series was observed (Rasmussen & Noe-Nygaard 1970). A re-evaluation of the zeolite mineralization in the

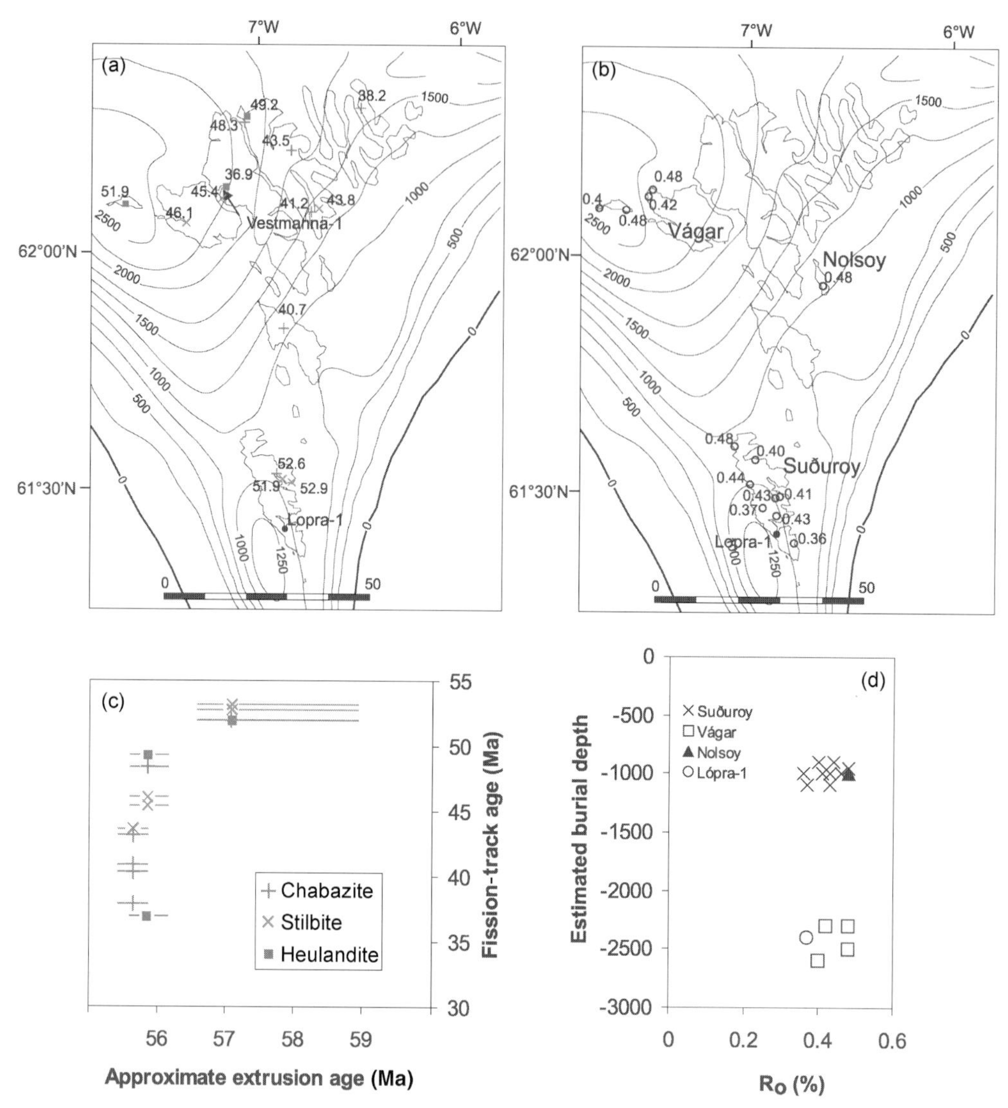

Fig. 12. (**a**) Detail of Fig. 11 showing denudation in the central part of the Faroe Platform. Location of samples taken for measurement of vitrinite reflectance and zeolite fission-track ages from Koul *et al.* (1983) are shown on the map. (**b**) Measurements of vitrinite reflectance values of surface samples from the Faroes. (**c**) Zeolite fission-track ages are plotted against extrusion age of the stratigraphic level (M. Storey cited by Larsen *et al.* 1999). (**d**) Vitrinite reflectance values plotted against estimated depth of denudation at the sample localities.

Vestmanna-1 well is in progress (Jørgensen, pers. comm.). Vitrinite reflectance measurements of coal samples from 14 localities on the Faroes, including cuttings from 1364 m depth in the Lopra-1 well, yield values of R_0 in the range from 0.36 to 0.48% (Fig. 12). However, we have not been able to find systematic variation of the reflectance values that could be used to constrain the estimate of denudation on the Faroe Platform.

Few data are available that give direct indication of the timing of denudation on the Faroe Platform. Because of the absence of apatites of a suitable size, no successful apatite fission-track study of the platform has ever been completed. The only published fission-track study of samples from the Faroes is a reconnaissance study of fission tracks in zeolite minerals (Koul *et al.* 1983). All zeolite fission-track ages from the Faroes are younger than current estimates of the ages of the Faroes basalts (Fig. 12). There is a considerable spread of the fission-track ages within both the middle and upper basalt series. The annealing characteristics of zeolites are poorly constrained and any conclusions based on zeolite fission tracks are thus speculative (H. Andriessen, pers. comm.). However, there is no apparent conflict between the zeolite fission-track ages obtained by Koul *et al.* (1983) and the conclusions in this and following sections.

It appears that the zeolite assemblage in Lopra-1 is the only independent estimate of the thickness of removed basalt on the Faroe Platform, which can be used as a firm constraint on the map of estimated denudation of the Faroe Platform shown in Fig. 11. The calculated volume of removed basalt (46 000 km^3) should thus be considered a first estimate that may be refined when additional constraints become available.

Eocene deposition

Sediment transport from the south and east into the Foinaven Basin continued throughout Eocene time, and sediments were deposited as units FPC-B.1–FPC-B.3. During this period no sediments of any significance entered the Faroe–Shetland Channel from the west (e.g. Andersen *et al.* 2000). Also, farther north in the Fugloy Basin, sediments of eastern provenance were dominant throughout Eocene time. However, the thickness of the Eocene sequence is considerably less than in the Foinaven Basin and the three lower units of FPC-B are not identified as separate units on seismic data. The Westray Transfer Zone thus may have been a significant tectonic boundary in Eocene time (Fig. 1). In the West Shetland area Early Eocene tuff, which presumably originated in the now-subsided volcanic zone around the line of break-up, was interspersed with the mostly paralic sediments of the Balder Formation (e.g. Knox *et al.* 1997). Equivalents of the Balder tuff are also found on the Faroe Platform (Waagstein & Heilmann-Clausen 1995). On the Faroe Platform Ypresian–Lutetian limestone follows the 'Balder tuff' (Waagstein & Heilmann-Clausen 1995). This limestone contains tuffaceous fragments that appear to have been laid down directly in the limestone. However, transported clastic fragments, which would indicate denudation and eastward drainage of the Faroe Platform, are not seen. It was therefore concluded that the Faroe Platform had a fairly low topography throughout most of Eocene time and that northward or westward drainage from a

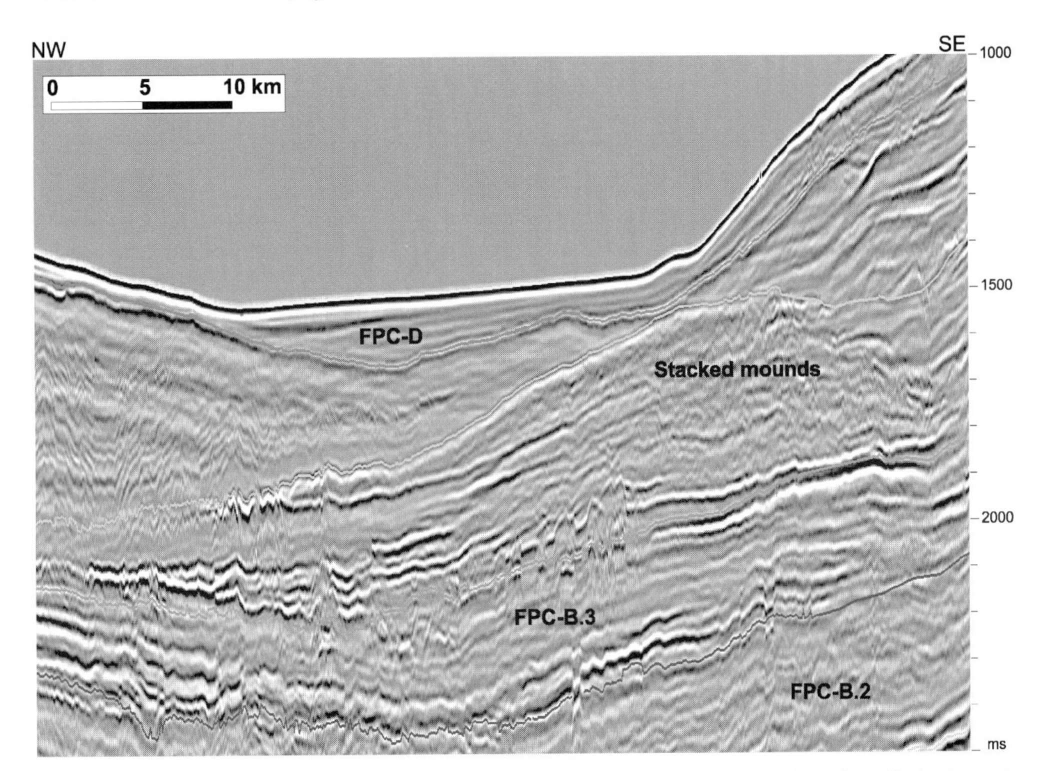

Fig. 13. Seismic section through basin-floor mound in southern Fugloy Basin. Location of profile is shown in Fig. 7. By courtesy of Western Geophysical.

significant part of the plateau was possibly established in Early Eocene time and prevailed without significant modifications into Oligocene time (Andersen *et al.* 2000).

Oligocene time

No Upper Eocene sediments have been encountered on the eastern margin of the Faroe Platform. Lower Oligocene volcaniclastic sandstones are the oldest known sediments originating from denudation of the platform to the west (Waagstein & Heilmann-Clausen 1995). Andersen *et al.* (2000) suggested that the Lower Oligocene volcaniclastic sandstones correlate with seismic unit FPC-B.4 on the outer part of the margin and in the Faroe–Shetland Channel. Abundant reworked dinoflagellates of Early and Mid-Eocene age are found in most of the investigated samples, and reworked dinoflagellates of Mid–Late Eocene age were found in two samples (Waagstein & Heilmann-Clausen 1995). Sediments of Mid–Late Eocene age may thus have been present on the platform (Waagstein & Heilmann-Clausen 1995). It thus appears that uplift of the eastern Faroe Platform occurred in Late Eocene–Early Oligocene time after a period of burial in Early and Mid-Eocene time. On the basis of the seismic interpretation, inversion of the Foinaven Basin commenced in Mid-Eocene time. Therefore, the inversion of the Foinaven Basin apparently started before the Late Eocene–Early Oligocene denudation (and uplift) of large parts of the Faroe Platform.

A mounded basin-floor body, low in unit FPC-B.4, is seen on seismic data from the south-eastern part of Fugloy Basin (Fig. 13). This body is characterized by continuous high-amplitude reflections on its west flank. The central and thickest part of the body is characterized by a fairly chaotic reflection configuration. On the eastern flank a progradational system on the lower West Shetland slope merges into or onlaps the basin-floor body (Fig. 13). We interpret this body as a basin-floor fan system, which was fed by sediments from the south and east. The central part may represent along-slope transport channels whereas the continuous high-amplitude reflections on the flanks may represent more widespread 'overbank deposits'. The tectonic event that caused inversion of the Foinaven Basin and movements along Westray Splay may have caused uplift in the areas that supplied the sediments for the basin floor. Further north and west and above the basin-floor body, FPC-B.4 is characterized by a succession of parallel reflections cut by numerous small faults that mostly terminate downward within unit FPC-B.4. Upwards, the faults mostly terminate in the upper part of FPC-B or in FPC-C. Between the faults, the continuity of the reflections is high, whereas regional continuity is relatively poor (Fig. 5). Westwards, the densely faulted parallel reflectors terminate by downlap onto the top of FPC-B.4 or onto reflectors in the lower part of FPC-B.4. We suggest that the densely faulted parallel reflectors represent slope and basin-floor sediments that were transported westwards from the West Shetland area. These sediments are characterized by syndepositional and early post-depositional deformation. These faults are arranged in a polygonal pattern suggesting volumetric contraction during burial (Davies *et al.* 1999) or gravitational instability of the thick succession of sediments.

The Faroe Bank Knoll (Fig. 1) was emplaced as a central magmatic complex presumably during early Eocene time. A hyaloclastic foreset breccia is interpreted from seismic data. The breccia terminates in a volcanic escarpment on the eastern flank, suggesting that the Faroe Bank Knoll was emplaced in a shallow sea. Throughout most of Eocene time, the Faroe Bank Knoll was a barrier for sediments of eastern provenance. In the Munken Basin, east and NE of the Faroe Bank Knoll, the Eocene–Oligocene evolution was almost identical to the evolution in the Foinaven Basin. Seismic interpretation indicates that the Eocene and Oligocene deposits in the Faroe Bank Basin are up to 1800 m thick. Most of the sediments are presumably derived from the Faroe Platform.

On the northern margin the equivalent of FPC-B is a transgressive unit (Nielsen & Van Weering 1998), and is represented by marine clay- and siltstones at Deep Sea Drilling Project (DSDP) Site 336 (Talwani *et al.* 1976). Andersen *et al.* (2000) have suggested that deposition could have already started in Eocene time. However, no well ties are available that constrain the start of deposition on the northern margin of the Faroe Platform.

Early and Mid-Miocene time

Lower and Middle Miocene sediments are found on the shelf and margin of the Faroe Platform as seismic unit FPC-C. These deposits were previously considered older (Upper Oligocene–Lower Miocene: Andersen *et al.* 2000). The revised age range of this unit is based on the tentative correlation of the base of unit FPC-C with the Late Oligocene–Early Miocene unconformity offshore West Shetland (Stoker *et al.* 2002). On the basis of the geometry and to a lesser degree the internal reflection configuration, we suggest that unit FPC-C on the eastern margin of the Faroe Platform was fed almost exclusively by sediments derived from the platform. In the basinal setting sediments of pelagic and hemipelagic origin may have been dominant.

On the northern margin of the Faroe Platform, down-slope and east of DSDP Site 336, the equivalent of FPC-C was interpreted as a type 2 sequence by Nielsen & Van Weering (1998). The unit reflects deposition after a significant drop in relative sea level following continuous subsidence throughout Eocene–Oligocene times.

Late Miocene–Early Pliocene time

Deposition of FPC-C was followed by mild folding (Andersen *et al.* 2000). Recent biostratigraphic studies in exploration wells in the British sector of the Faroe–Shetland Channel show that deformation occurred close to the Mid–Late Miocene boundary (Cloke *et al.* 2000). This tectonic phase was followed by deposition of unit FPC-D.1 in Late Miocene–Early Pliocene time (Fig. 14). On the eastern margin of the Faroe Platform, sediments of unit FPC-D.1 were deposited in small synclinal basins. The sediments were apparently derived from the anticlinal crests on the Faroe Platform, some of which show indication of erosion (Andersen *et al.* 2000). Along-slope bottom currents apparently influenced Unit FPC-D.1 along the slope and in the Faroe–Shetland Channel. This is especially evident along the northern part of the slope east of the Faroe Platform.

Late Pliocene time

A large progradational wedge, comprising 1900 km^3 of undifferentiated sediments originating on the Faroe Platform, was deposited east of the Faroes during Pliocene and possibly early Pleistocene time as unit FPC-D2 (Figs 4 and 14) (Andersen *et al.* 2000). A similar but much smaller wedge is found SW of the Faroes in the Skeivi Bank area. The total volume of sediments in the Skeivi Bank Wedge is unknown. By analogy with similar wedges offshore Norway and the UK, Andersen *et al.* (2000) have suggested that the wedges east and SW of the Faroes were mostly deposited in Late Pliocene time. A second depocentre within unit FPC-D.2 is found in the northern part of the Faroe–Shetland Channel. This depocentre partially coincides with the depocentre of the underlying FPC-D.1. In the eastern depocentre the dominant

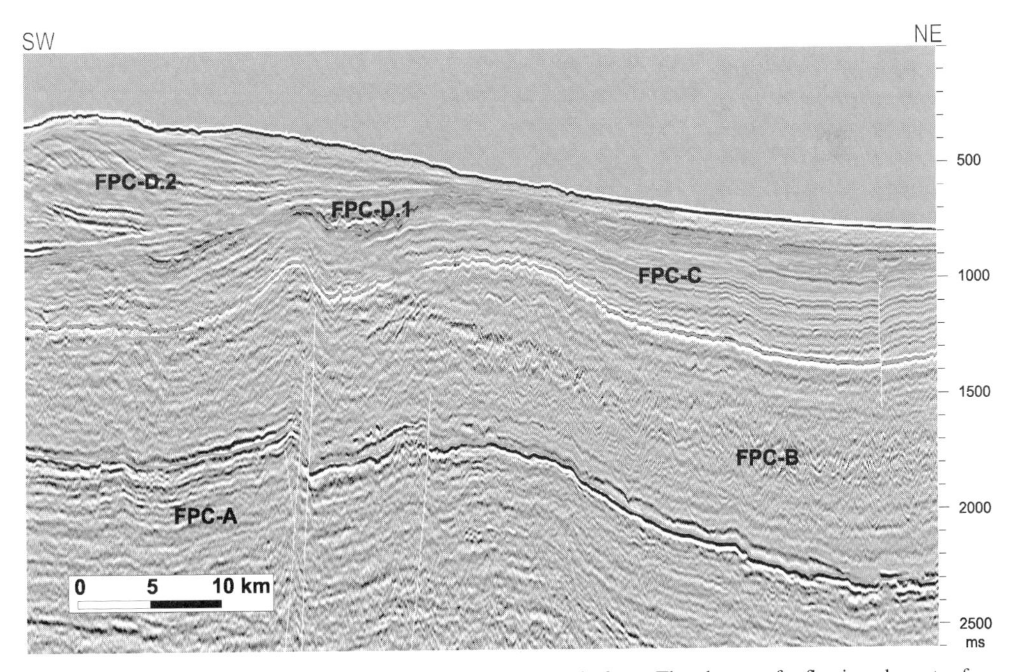

Fig. 14. Seismic section from the eastern margin of the Faroe Platform. The change of reflection character from FPC-B to FPC-C and erosional unconformity on top of FPC-C should be noted. Location of profile is shown in Fig. 7. By courtesy of Western Geophysical.

seismic characteristics are subparallel, continuous and often draping reflectors (Fig. 5). Onlaps onto the underlying unit are common. Therefore, we suggest that the sediments in the eastern depocentre are intercalations of pelagic or hemipelagic sediments and turbidites. In the southern part of the eastern depocentre downslope sedimentation, representing a basinward continuation of the progradational wedge, appears to be dominant.

Pleistocene time

The Faroes were glaciated during Pleistocene time. Study of the glacial landforms on the islands indicates that during the last glaciation the ice cover was limited to a small area around the islands, possibly as three small separate ice caps (Humlum & Christiansen 1998). All of the Faroe Platform experienced erosion. Transverse valleys were cut into the sediments on the outer shelf, marginal valleys that follow the limit of the basalt outcrop around the Faroes were cut into Pliocene and older sediments, and (late) Pleistocene sediments were deposited in these depositional lows. In late Quaternary time mass-flows occurred on the northern and eastern margins of the platform. On the eastern margin the mass-flows have been related to a trough-mouth setting associated with the transverse valleys mentioned above (Kuijpers 2000).

Faroe Platform mass balance

We have made a preliminary calculation of the volume of sediments originating from denudation of the Faroe Platform (Table 2). The volumetric calculations are based on seismic interpretations. The provenance of sediment bodies was determined based on the geometry and internal reflector configuration. This approach is rather crude, as sediment originating outside the Faroe Platform could be hidden within the units, which, on the basis of seismic expression, are interpreted as originating on the platform. For instance, pelagic sediments and hemipelagic sediments derived from other denudation areas are probably included in the calculations. Sediments derived from the Faroe Platform may also have been carried outside the three depositional centres considered. Seismic units that appear to be derived from one side of the Faroe–Shetland Channel may contain veneers derived from the opposite side. In addition to the errors mentioned above, the magnitudes of which are currently unknown, we estimate 10–15% error on the volumes

Table 2. *Estimates of denudation on the Faroe Platform and deposition of sediments derived from the platform*

	FSC	FNM	FBC	FP	Sum
Volume (km^3)					
Removed Paleocene basalt				−46000	−46000
Eocene–Oligocene (FPC-B)	9900[a,b]				
Lower–Middle Miocene (FPC-C)	4450				
Upper Miocene–Lower Pliocene (FPC-D.1)	2250				
Upper Pliocene–Recent (FPC-D.2–3)	4500				
Total	21100	25000[c]	9900[c]	−46000	10000
Mass balance (mass in kg × 10^{12})[d]					
Removed Paleocene basalt				−131100	−131100
Oligocene (FPC-B)	20295				53864
Lower–Middle Miocene (FPC-C)	9123				24211
Upper Miocene–Lower Pliocene (FPC-D.1)	4613				12242
Upper Pliocene–Recent (FPC-D.2–3)	9225				24483
Total	43255	51250	20295	−131100	−16300

FSC, Faroe–Shetland Channel; FNM, northern margin of Faroe Platform; FBC, Faroe Bank Channel; FP, Faroe Platform. Positive numbers indicate deposition; negative numbers indicate denudation.
a Excluding Eocene–Oligocene sediments that are removed later.
b Including some Upper Eocene sediments.
c Bulk volume, all Eocene–Recent sediments included.
d Assuming dry density of basalt is 2850 kg m^{-3} (porosity *c.* 10%), and of sediments is 2050 kg m^{-3} (porosity *c.* 35%).

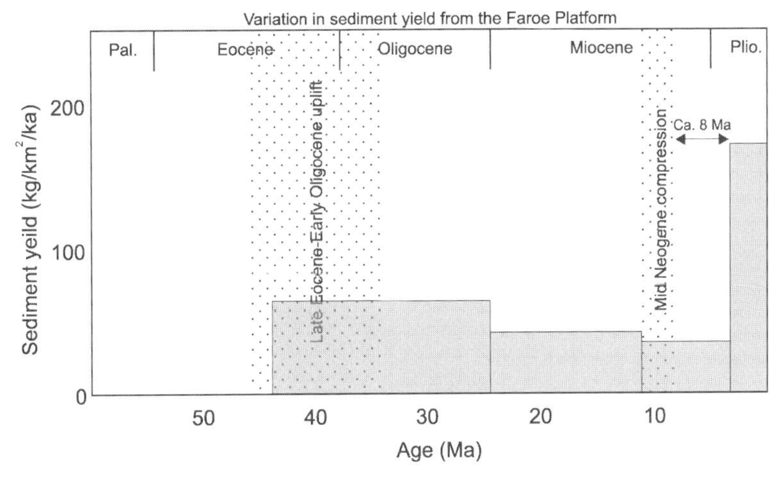

Fig. 15. Estimated Eocene–Recent sediment yield from the Faroe Platform. A peak in sediment production in Late Pliocene time is seen *c.* 8 Ma after the start of the Neogene compression phase. A less pronounced peak partly overlaps the Late Eocene–Oligocene deformation event.

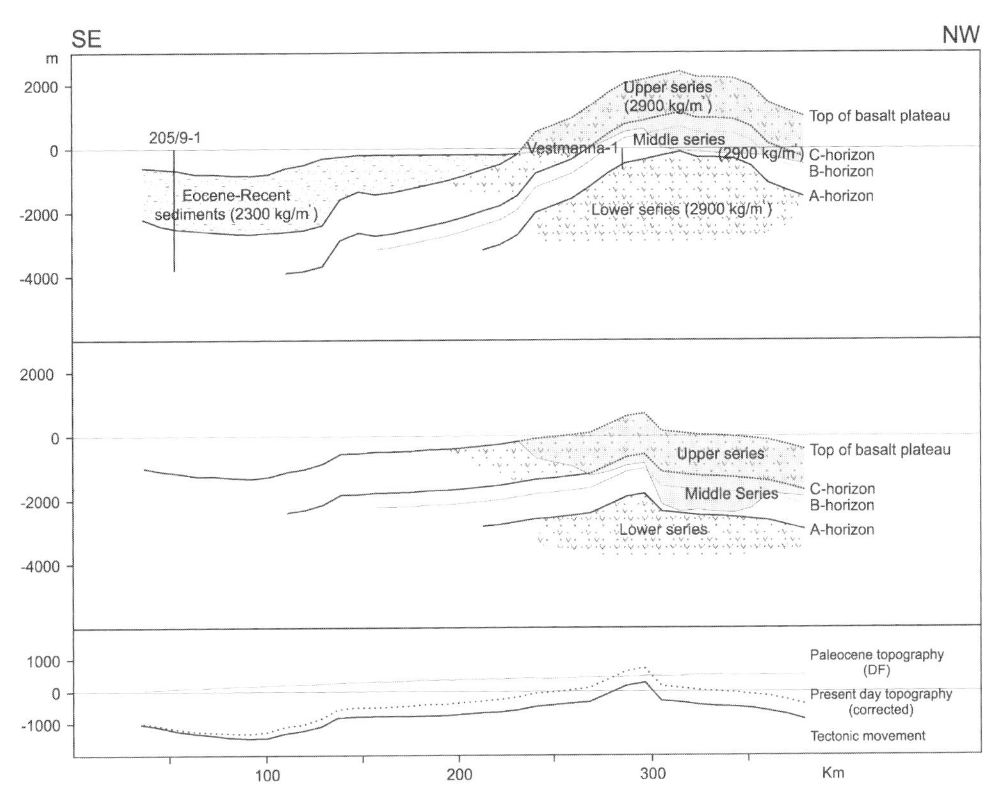

Fig. 16. Cross-sections showing aspects of the structural evolution of the Faroe Platform. (**a**) Present-day configuration of the platform; removed basalt is indicated by grey shading. (**b**) Load effects of Eocene–Recent sediments and the removed basalt section have been removed assuming Airy isostasy and a mantle density of 3100 kg m^{-3}. (**c**) A possible model for the configuration of the platform in latest Paleocene time. The model presented assumes volcanic feeder systems close to the line of incipient break-up. Location of profile is shown in Fig. 7.

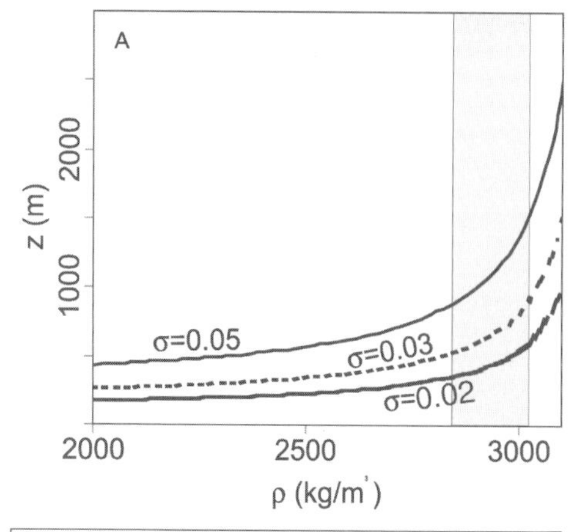

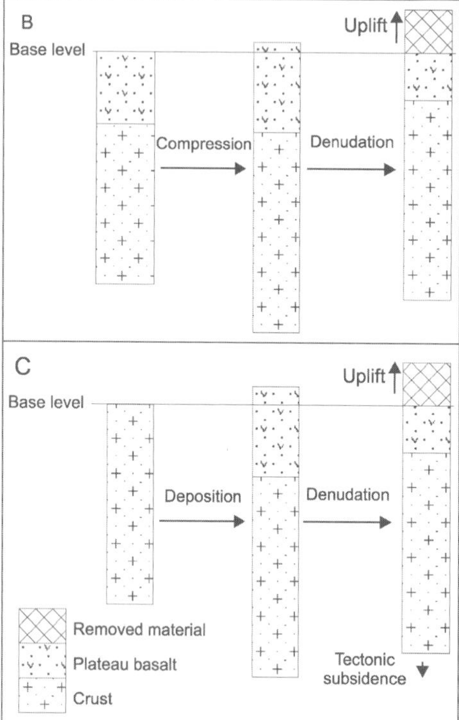

Fig. 17. (**a**) The denudation depth of a 20 km thick crust as a function of upper-crustal density for three compressional strains (2%, 3% and 5% shortening). Erosion to base level is assumed. (**b**) and (**c**) show two conceptual models, which could explain the uplift and denudation of the Faroe Platform. (**b**) A small part of the basalt plateau is lifted above base level during a compressional event. The plateau is then eroded until the top of the plateau again is at base level. (**c**) Basalt is emplaced above base level. The crust will subside to retain isostatic equilibrium, and only a small part of the basalt will be above base level. As a result of thermal subsidence during erosion of the plateau, the erosion surface will reach base level before the plateau is completely removed.

calculated for both removed basalt and deposited sediments.

Basalt is generally less porous than sediments. We have thus recalculated the volumes of denudation and deposition to dry mass assuming the porosity of basalt is *c.* 10% and the porosity of volcaniclastic sediments is *c.* 30%. The estimated total amount of sediments deposited in the basins (56 000 km^3, 114 800 × 10^{12} kg) is reasonably close to the estimated amount of basalt removed from the Faroe Platform (46 000 km^3, 131 100 × 10^{12} kg) (Table 1). Overall the mean specific sediment yield based on the total Cenozoic denudation of the Faroe Platform and deposition in the Faroe–Shetland Channel is approximately 55 × 10^3 kg km^{-2} a^{-1}. This is within the normal range of sediment yield from large rivers (Hovius 1998). On the basis of the depositional data from the eastern margin of the Faroe Platform and the Faroe–Shetland Channel it can be seen that denudation rates changed over time (Fig. 15). It appears that the denudation rate in Late Eocene–Oligocene time was rather high (*c.* 70 × 10^3 kg km^{-2} a^{-1}) possibly reflecting uplift of the Faroe Platform. In Early–Mid-Miocene time it was decreased to *c.* 40 × 10^3 kg km^{-2} a^{-1}, and in Late Miocene time the denudation rate was *c.* 35 × 10^3 kg km^{-2} a^{-1}. This was followed by a sudden increase in denudation rate in mid-Pliocene time to *c.* 175 × 10^3 kg km^{-2} a^{-1}. A more detailed analysis of the sediment succession around the Faroe Platform would presumably reveal significant fluctuations of denudational and depositional rates. Peaks in depositional rates occurred after the two distinct inversion or compression events during Late Eocene–Early Oligocene and Mid–Late Miocene time. However, only the Late Pliocene depositional peak is well constrained in time, and it occurred *c.* 8 Ma after the mid-Neogene deformation event.

Discussion

A simplified profile across the Faroe–Shetland Channel from well 205/9-1 in the UK sector is shown in Fig. 16a. It has been suggested that the uplift and denudation of the Faroe Platform to some extent could be attributed to compressional deformation (Boldreel *et al.* 1998; Andersen *et al.* 2000). Simple considerations concerning isostatic balance indicate that denudation induced by tectonic uplift depends on the density of the removed rock (Fig. 17). Unless the removed rock is relatively dense, small initial uplift of tectonic origin would give rise to only relatively insignificant additional uplift and denudation. However, removal of heavy material such as basalt will give rise to significant denudation even in the case of small tectonic strain. As can be seen in Fig. 17, about 2–3% crustal shortening could account for the average denudation of the Faroe Platform. Crustal shortening of the Faroe Platform, as a result of compression, is indicated both by dip-slip analysis of fault planes on the Faroe Islands (Geoffroy 1993, 1994) and by interpretation of offshore seismic data (e.g. Boldreel *et al.* 1998). However, horizon balancing based on the regional profile in Fig. 16a indicates that the accumulated crustal shortening of the platform since the emplacement of the Late Paleocene basalts presumably is less than 1% and may be insignificant.

White & Lovell (1997) proposed that fluctuating intensity of the Icelandic plume was responsible for episodic Paleocene uplift on the British Isles. Thermal fluctuations of the plume are evidenced by V-shaped ridges arranged symmetrically around the Reykjanes Ridge (White *et al.* 1995). However, the V-shaped ridges were formed where thermal pulses of the plume intersected the spreading ridge, and there is apparently no documentation for off-axis uplift connected to the Neogene thermal pulses. It is thus unlikely that thermal pulses are responsible for the Neogene uplift and denudation of the Faroe Platform.

Assuming Airy isostasy, we may remove the approximate effect of the load of sediments deposited in the Faroe–Shetland Channel and of basalts removed from the Faroe Platform since Late Paleocene time and obtain the profile in Fig. 16b. Comparing the profile in Fig. 16b with an estimate of the Late Paleocene topography, we obtain an approximation of the accumulated tectonic vertical movements along the profile (Fig. 16c). The largest tectonic subsidence (*c.* 1600 m) occurred in the centre of the Faroe–Shetland Channel. On the Faeroe Platform the tectonic subsidence is 400–800 m. Northwest of the Faroe Platform the tectonic subsidence is greater than 800 m.

This distribution of estimated tectonic subsidence may be explained partly by thermal contraction of the head of the Icelandic plume, which would have been present below the entire Faroe region (e.g. White & McKenzie 1989). Reduced dynamic support from the plume may also be involved (e.g. Nadin *et al.* 1997). However, the larger tectonic subsidence in the Faroe–Shetland Channel and NW of the platform indicates an additional component of tectonic subsidence caused by post-rift thermal contraction. Mid-Cretaceous rifting of the Faeroe–Shetland Channel with $\beta \approx 2$ has been proposed (e.g. Nadin *et al.* 1997). Less pronounced Paleocene rifting (e.g. Dean *et al.* 1999) may also be involved. On the Faroe Platform, tectonic subsidence has been more than compensated by isostatic uplift caused by the removal of Paleocene basalt. The low tectonic subsidence around the Faroe Islands (260–300 km in Fig. 16) is presumably an artefact introduced because we assumed that the lithosphere has no flexural strength. Suitable 2D flexural modelling, including a continuation of the profile on the uplifted Shetland Platform, SE of the Faroe–Shetland Channel, should modify the subsidence values, especially around the Faroe Islands.

Flexural strength of the lithosphere is ignored in the discussion above. However, it is still evident from Fig. 16 that Eocene–Recent erosion of the Faroe Platform could be the dominant cause for the Eocene–Recent uplift of the platform. Therefore, we suggest that the Eocene–Recent uplift of the Faroe Platform is primarily caused by isostatic compensation for the erosion of the platform (Fig. 17c). In this context, the widespread Late Pliocene–Pleistocene denudation that provided the sediments found in the large Pliocene wedges around the Faroes is most readily understood in association with the climatic–eustactic changes that affected the area during the Northern Hemisphere glaciation.

We would like to thank W.G. Harrar and two anonymous referees for constructive review of the paper. The vitrinite reflectance measurements were made by C. Guvad and supported financially by the Atlantic Margin Group (Mobil North Sea Limited, Enterprise Oil plc and Statoil UK Limited). We are grateful to Western Geophysical and Veritas DGC for their contribution. We wish to thank the Føroya Jarðfrøðisavn and the Geological Survey of Denmark and Greenland for permission to publish this work.

References

ANDERSEN, M.S. 1988. Late Cretaceous and early Tertiary extension and volcanism around the Faeroe Islands. *In*: MORTON, A.C. & PARSON, L.M. (eds) *Early Tertiary Volcanism and the Opening of the NE Atlantic*. Geological Society, London, Special Publications, **39**, 112–122.

ANDERSEN, M.S. 1999. Structural evolution of the Faroe–Rockall region in the Cenozoicum (abstract). *Geonytt, Norsk Geologisk Vintermøte, 6–8 January 1999*. Norsk Geologisk Forening, Stavanger.

ANDERSEN, M.S., NIELSEN, T., SORENSEN, A.B., BOLDREEL, L.O. & KUIJPERS, A. 2000. Cenozoic sediment distribution and tectonic movements in the Faroe region. *Global and Planetary Change*, **24** (3–4), 239–259.

BOLDREEL, L.O. & ANDERSEN, M.S. 1993. Late Paleocene to Miocene compression in the Faeroe–Rockall area. *In*: PARKER, J.R. (ed.) *Petroleum Geology of Northwest Europe: Proceedings of the 4th Conference*. Geological Society, London, 1025–1034.

BOLDREEL, L.O. & ANDERSEN, M.S. 1994. Tertiary development of the Faeroe–Rockall Plateau based on reflection seismic data. *Bulletin of the Geological Society of Denmark*, **41**, 162–180.

BOLDREEL, L.O. & ANDERSEN, M.S. 1995. The relationship between the distribution of Tertiary sediments, tectonic processes and deep-water circulation around the Faeroe Islands. *In*: SCRUTTON, R.A., STOKER, M.S., SHIMMIELD, G.B. & TUDHOPE, A.W. (eds) *The Tectonics, Sedimentation and Palaeoceanography of the North Atlantic Region*. Geological Society, London, Special Publications, **90**, 145–158.

BOLDREEL, L.O., ANDERSEN, M.S. & KUIJPERS, A. 1998. Neogene seismic facies and deep gateways in the Faroe Bank area, NE Atlantic. *Marine Geology*, **152**, 129–140.

CLOKE, I., DAVIES, R.J., FERRERO, C., LINE, C. & BINGHAM, J. 2000. Inversion: the critical element in determining the success of the petroleum system of the Faroe–Shetland Channel. *Exhumation of Circum-Atlantic Margins: Timing, Mechanisms and Implications for Hydrocarbon Exploration, London, 13–14 June 2000.*

CLOKE, I., DAVIES, R.J., LINE, C., HORNAFIUS, S. and MCLACHLAN, K. 1999. Petroleum system of the Faroe–Shetland Basin. *AAPG International Conference and Exhibition, Birmingham, 12–15 September 1999*, pp. 119–123.

DAM, G., LARSEN, M. & SØNDERHOLM, M. 1998. Sedimentary response to mantle plumes: implications from Paleocene onshore successions, west and east Greenland. *Geology*, **26**, 207–210.

DAVIES, R., CARTWRIGHT, J. & RANA, J. 1999. Giant hummocks in deep-water marine sediments: evidence for large-scale differential compaction and density inversion during early burial. *Geology*, **27** (10), 907–910.

DEAN, K., MCLACHLAN, K. & CHAMBERS, K. 1999. Rifting and development of the Faeroe–Shetland Basin. *In*: FLEET, A.J. & BOLDY, S.A.R. (eds) *Petroleum Geology of Northwest Europe: Proceedings of the 5th Conference*. Geological Society, London, 533–544.

EBDON, C.C., GRANGER, P.J., JOHNSON, H.D. & EVANS, A.M. 1995. Early Tertiary evolution and sequence stratigraphy of the Faeroe–Shetland Basin: implications for hydrocarbon prospectivity. *In*: SCRUTTON, R.A., STOKER, M.S., SHIMMIELD, G.B. & TUDHOPE, A.W. (eds) *The Tectonics, Sedimentation and Palaeoceanography of the North Atlantic Region*. Geological Society, London, Special Publications, **90**, 51–69.

EVANS, D. 2000. EU project on development of the glaciated European Margin. *EOS Transactions, American Geophysical Union*, **81**, 423–425.

GATLIFF, R.W., HITCHEN, K., RITCHIE, J.D. & SMYTHE, D.K. 1984. Internal structure of the Erlend Tertiary volcanic complex, north of Shetland, revealed by seismic reflection. *Journal of the Geological Society, London*, **141**, 555–562.

GEOFFROY, L., ANGELIER, J. & GERGERAT, F. 1993. Sur l'évolution tectonique cassante tertiaire des Iles Féroé, Atlantique Nord; la compression féringienne. *Comptes Rendus de l'Académie des Sciences*, **316**, 975–982.

GEOFFROY, L., BERGERAT, F. & ANGELIER, J. 1994. Tectonic evolution of the Greenland–Scotland ridge during the Paleogene: new constraints. *Geology*, **22**, 653–656.

HALD, N. & WAAGSTEIN, R. 1984. Lithology of a 2-km sequence of Lower Tertiary tholeiitic lavas drilled on the Suduroy Faeroe Islands (Lopra-1). *In*: BERTHELSEN, O., NOE-NYGAARD, A. & RASMUSSEN, J. (eds) *The Deep Drilling Project 1980–1981 in the Faeroe Islands*. Føroya Fródskaparfelag, Tórshavn, 15–37.

HALL, B.D. & WHITE, N. 1994. Origin of anomalous Tertiary subsidence adjacent to North Atlantic continental margins. *Marine and Petroleum Geology*, **11**, 702–714.

HENRIKSEN, S. & VORREN, T.O. 1996. Late Cenozoic sedimentation and uplift history on the mid-Norwegian continental shelf. *Global and Planetary Change*, **12**, 171–199.

HITCHEN, K. & RITCHIE, J.D. 1987. Geological review of the West Shetland area. *In*: BROOKS, J.D. & GLENNIE, K.W. (eds) *Petroleum Geology of North West Europe, Procedings of the Third Conference*. Graham & Trotman, London, 737–749.

HOVIUS, N. 1998. Controls on sediment supply by large rivers. *In*: SHANLEY, K.W. & MCCABE, P.W. (eds) *Relative Role of Eustasy, Climate, and Tectonism in Continental Rocks*. Society of Economic Paleontologists and Mineralogists, Special Publications, **59**, 3–16.

HUMLUM, O. & CHRISTIANSEN, H.H. 1998. Mountain climate and periglacial phenomena in the Faeroe Islands. *Permafrost and Periglacial Processes*, **9**, 189–211.

JØRGENSEN, O. 1984. Zeolite zones in the basaltic lavas of the Faroe Islands. *In*: NOE-NYGAARD, O. & RASMUSSEN, J. (eds) *The Deep Drilling Project 1980–1981 in the Faeroe Islands*. Føroya Fródskaparfelag, Tórshavn, 71–91.

KIØRBOE, L. 1999. Stratigraphic relationships of the Lower Tertiary of the Faeroe Basalt Plateau and the Faeroe–Shetland Basin. *In*: FLEET, A.J. & BOLDY, S.A.R. (eds) *Petroleum Geology of Northwest Europe: Proceedings of the 5th Conference*. Geological Society, London, 559–572.

KNOX, R.W.O.B., HOLLOWAY, S., KIRBY, G.A. & BAILY, H.E. 1997. *Early Paleogene Lithostratigraphy and Sequence Stratigraphy. Stratigraphic Nomenclature of the UK North West Margin, 2.* British Geological Survey, Keyworth.

KOUL, S.L., CHADDERTON, L.T. & BROOKS, C.K. 1983. East Greenland and the Faeroe Islands: a fission track study. *Kongelige Danske Videnskabernes Selskab, Matematisk-fysiske Meddelelser*, **40**, 7–34.

KUIJPERS, A. 2000. Late Quaternary slope instability on the Faeroe margin: mass flow features and timming of events. *Geo-Marine Letters*, 149–159.

LAMERS, E. & CARMICHAEL, M.M. 1999. The Paleocene deepwater sandstone play West of Shetland. *In*: FLEET, A.J. & BOLDY, S.A.R. (eds) *Petroleum Geology of Northwest Europe: Proceedings of the 5th Conference*. Geological Society, London, 645–659.

LARSEN, L.M., WAAGSTEIN, R., PEDERSEN, A. & STOREY, M. 1999. Trans-Atlantic correlation of Palaeogene volcanic successions in the Faeroe Islands and East Greenland. *Journal of the Geological Society, London*, **156**, 1081–1095.

NADIN, P.A., KUZNIR, N.J. & CHEADLE, M.J. 1997. Early Tertiary plume uplift of the North Sea and the Faeroe–Shetland Basins. *Earth and Planetary Science Letters*, **148**, 109–127.

NAYLOR, P.H., BELL, B.R., JOLLEY, D.W., DURNALL, P. & FREDSTED, R. 1999. Palaeogene magmatism in the Faeroe–Shetland Basin: influences on uplift history and sedimentation. *In*: FLEET, A.J. & BOLDY, S.A.R. (eds) *Petroleum Geology of Northwest Europe: Proceedings of the 5th Conference*. Geological Society, London, 545–558.

NIELSEN, T. & VAN WEERING, T.C.E. 1998. Seismic stratigraphy and sedimentary processes at the Norwegian Sea margin northeast of the Faeroe Islands. *Marine Geology*, **152**, 141–157.

NOE-NYGAARD, A. 1968. On extrusion forms in plateau basalts. Shield volcanoes of 'Scutulum' type. *In*: FREDERIKSON, S. (ed.) *Science in Iceland*. Vísindafélag Icelandinga, Reykjavík, 10–13.

PLANKE, S., SYMONDS, P.A., ALVESTAD, E. & SKOGSEID, J. 2000. Seismic volcanostratigraphy of large-volume basaltic extrusive complexes on rifted margins. *Journal of Geophysical Research*, **105** (B8), 19335–19351.

RASMUSSEN, J. & NOE-NYGAARD, A. 1970. *Geology of the Faroe Islands (Pre-Quaternary). Danmarks Geologiske Undersøgelse I. Series, 25*. Reitzels Forlag, Copenhagen.

RITCHIE, J.D., GATLIFF, R.W. & RICHARDS, P.C. 1999. Early Tertiary magmatism in the offshore NW UK margin and surrounds. *In*: FLEET, A.J. & BOLDY, S.A.R. (eds) *Petroleum Geology of Northwest Europe: Proceedings of the 5th Conference*. Geological Society, London, 573–584.

RUMPH, B., REAVES, C.M., ORANGE, V.G. & ROBINSON, D.L. 1993. Structuring and transfer zones in the Faeroe Basin in a regional tectonic context. *In*: PARKER, J.R. (ed.) *Petroleum Geology of Northwest Europe: Proceedings of the 4th Conference*. Geological Society, London, 999–1009.

SMYTHE, D.K. 1983. Faeroe–Shetland Escarpment and continental margin north of the Faroes. *In*: MORTON, A.C. & PARSON, L.M. (eds) *Early Tertiary Volcanism and the Opening of the NE Atlantic*. Geological Society, London, Special Publications, **39**, 109–119.

SMYTHE, D.K., CHALMERS, J.A., SKUCE, A.G., DOBINSON, A. & MOULD, A.S. 1983. Early opening history of the North Atlantic—I. Structure and origin of the Faeroe–Shetland Escarpment. *Geophysical Journal of the Royal Astronomical Society*, **72**, 373–398.

STOKER, M.S. 1999. *Mid- to Late Cenozoic Stratigraphy. Stratigraphic Nomenclature of the UK North West Margin*. British Geological Survey, Keyworth.

STOKER, M.S., NIELSEN, T., VAN WEERING, T.C.E. & KUIJPERS, A. 2002. Towards an understanding of the Neogene tectonostratigraphic framework of the NE Atlantic Margin between Ireland and the Faroe Islands. *Marine Geology*, in–press.

STOREY, M., DUNCAN, R.A., PEDERSEN, A.K., LARSEN, L.M. & LARSEN, H.C. 1998. $^{40}Ar/^{39}Ar$ geochronology of the West Greenland Tertiary volcanic province. *Earth and Planetary Science Letters*, **160** (3/4), 569–586.

TALWANI, M., UDINTSEV, G. *et al.* 1976. *Initial Reports of the Deep Sea Drilling Project, 38*. US Government Printing Office, Washington, DC.

WAAGSTEIN, R. 1988. Structure, composition and age of the Faeroe basalt plateau. *In*: MORTON, A.C. & PARSON, L.M. (eds) *Early Tertiary Volcanism and the Opening of the NE Atlantic*. Geological Society, London, Special Publication, **39**, 225–238.

WAAGSTEIN, R. & HEILMANN-CLAUSEN, C. 1995. Petrography and biostratigraphy of Paleogene volcaniclastic sediments dredged from the Faeroes shelf. *In*: SCRUTTON, R.A., STOKER, M.S., SHIMMIELD, G.B. & TUDHOPE, A.W. (eds) *The Tectonics, Sedimentation and Palaeoceanography of the North Atlantic Region*. Geological Society, London, Special Publications, **90**, 179–197.

WHITE, N. & LOVELL, B. 1997. Measuring the pulse of a plume with the sedimentary record. *Nature*, **387** (6636), 888–891.

WHITE, R.S. & MCKENZIE, D. 1989. Magmatism at rift zones: the generation of volcanic continental margins and flood basalts. *Journal of Geophysical Research, 94B*, 7685–7729.

WHITE, R.S., BOWN, J.W. & SMALLWOOD, J.R. 1995. The temperature of the Iceland plume and origin of outward propagating V-shaped ridges. *Journal of the Geological Society, London*, **152**, 1039–1045.

Late Neogene development of the UK Atlantic margin

M. S. STOKER
British Geological Survey, Murchison House, West Mains Road, Edinburgh EH9 3LA, UK
(e-mail: mss@bgs.ac.uk)

Abstract: The late Neogene (Pliocene–Holocene) interval witnessed a significant change in sedimentation style across the UK Atlantic margin that culminated in its present morphological expression. The onset of change is marked by the creation of a regional, angular, erosional unconformity that can be traced from the Hebrides and West Shetland margins into the adjacent deep-water basins of the Rockall Trough and Faeroe–Shetland Channel. In the Rockall Trough, the unconformity is a submarine erosion surface that is dated as early Pliocene in age, between *c.* 3.85 and 4.5 Ma. On the Hebrides and West Shetland margins, the dating of the unconformity is slightly less well constrained and spans the latest Miocene(?)–early Pliocene (*c.* 5.5–3.8 Ma) interval. The formation of this unconformity may have resulted from the seaward tilting and subsidence of the shelf margin, which may have further modified the oceanographic circulation pattern in the adjacent basins. The sedimentary response to this event was relatively quick in the deep-water basins, which preserve a record of early Pliocene (post 3.85 Ma) to Holocene sediment-drift accumulation, albeit with a shift in the focus of sedimentation relative to the underlying strata. In contrast, major prograding wedges, which have contributed extensively to the construction of the Hebrides and West Shetland margins and to a lesser extent the Rockall Bank, date essentially from late Pliocene time and largely correlate with the influx of ice-rafted material to the margin. However, indications for a restricted lower Pliocene component to the shelf-margin succession suggest that this apparent delay or lag in sedimentation, relative to the basins, may be a natural response to the rate of denudation of the adjacent landmasses. The regional observations off NW Britain support the concept of Neogene tectonic uplift.

It is becoming increasingly apparent that high rates of denudation in the late Neogene interval have strongly influenced the construction of continental margins surrounding the NE Atlantic Ocean. There are indications that this phenomenon resulted from the uplift of adjacent landmasses, although the mechanism driving such a large-scale event remains open to question (Japsen & Chalmers 2000, and references therein). However, there is no doubting that one of the main sedimentary responses to the widespread denudation is preserved in the form of major prograding shelf-margin wedges. Substantial wedges have been described from around the NE Atlantic Ocean, including: the NW British margin (Stoker 1999; Stoker *et al.* 2001, 2002), the east Faeroes margin (Andersen *et al.* 2000), the northern North Sea (Gregersen *et al.* 1997), the SW Norwegian margin (Sejrup *et al.* 1996), the mid-Norwegian margin (Rokoengen *et al.* 1995; Henriksen & Vorren 1996; Evans *et al.* 2000), the Barents Sea margin (Vorren & Laberg 1997), and the east and west Greenland margins (Solheim *et al.* 1998; Chalmers 2000).

Although there is general agreement that circum-NE Atlantic shelf-margin progradation is of Plio-Pleistocene age, an improved stratigraphic resolution of the offshore sequences is necessary to determine whether or not sedimentation is linked directly to the causal (?uplift) mechanism or is a delayed response. There are indications, from both offshore Norway (Eidvin *et al.* 2000) and NW Britain (Stoker *et al.* 1994), that the bulk of the sediment preserved in the prograding wedges is of late Pliocene–Pleistocene age. However, the evidence from offshore NW Britain, including the deep-water oceanographic and sedimentary response, suggests an early Pliocene age for the onset of change as marked by the lower bounding unconformity.

The aim of this paper is to summarize the stratigraphic record and style of sedimentation for the upper Neogene (Pliocene–Holocene) succession off NW Britain, extending from the

From: DORÉ, A.G., CARTWRIGHT, J.A., STOKER, M.S., TURNER, J.P. & WHITE, N. 2002. *Exhumation of the North Atlantic Margin: Timing, Mechanisms and Implications for Petroleum Exploration*. Geological Society, London, Special Publications, **196**, 313–329. 0305-8719/02/$15.00

Hebrides and West Shetland shelves into the adjacent deep-water basins of the Rockall Trough and Faeroe–Shetland Channel (Fig. 1). The paper utilizes and builds upon regional stratigraphic studies undertaken in this region, by the author (e.g. Stoker 1999; Stoker *et al.* 2001, 2002), by focusing in more detail on the nature, distribution and timing of formation of the regionally significant intra-Pliocene unconformity, and the subsequent development of the overlying shelf-margin and deep-water basinal successions. In view of its margin-wide occurrence, the event responsible for the development of this unconformity implies a regional control on continental margin evolution in this region.

Neogene setting off NW Britain

The Neogene succession preserved on the continental margin off NW Britain has been divided into two megasequences of predominantly Miocene (including lower Pliocene in the basins) and Pliocene–Holocene age, which are separated by a regional unconformity of early Pliocene age (Stoker *et al.* 2002) (Fig. 2). A basinal succession of sediment drifts and

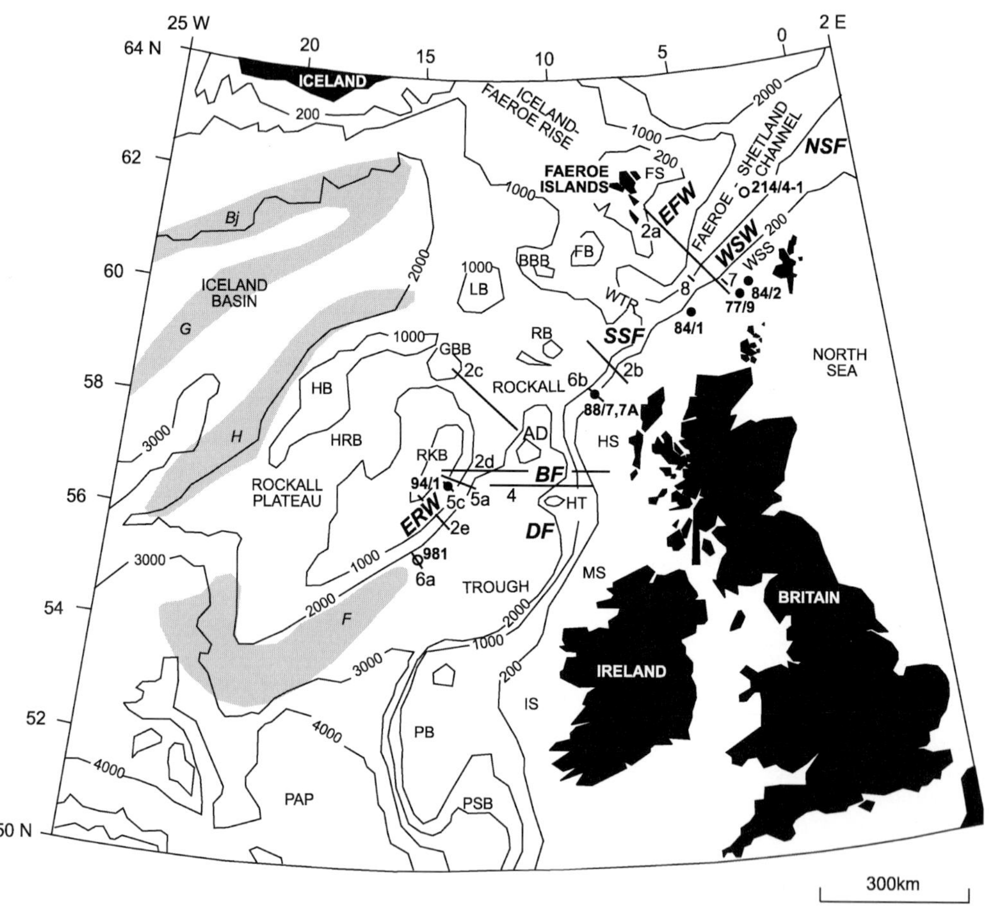

Fig. 1. Bathymetric setting of the UK Atlantic margin (contours in metres), showing locations of BGS boreholes (●), ODP and commercial boreholes (○) and major depocentres referred to in the text, the geoseismic sections illustrated in Fig. 2, and the seismic profiles in Figs 4–8. Depocentres: DF, Donegal Fan; BF, Barra Fan; ERW, East Rockall wedge; SSF, Sula Sgeir Fan; WSW, West Shetland wedge; EFW, East Faeroes wedge; NSF, North Sea Fan. Sediment drifts (shaded): F, Feni Ridge; H, Hatton Drift; Bj, Bjorn Drift; G, Gardar Drift. Geographical locations: WSS, West Shetland Shelf; FB, Faeroe Bank; BBB, Bill Bailey's Bank; WTR, Wyville-Thomson Ridge; LB, Lousy Bank; RB, Rosemary Bank; GBB, George Bligh Bank; AD, Anton Dohrn Seamount; HT, Hebrides Terrace Seamount; RKB, Rockall Bank; HRB, Hatton Rockall Basin; HB, Hatton Bank; FS, Faeroe Shelf; HS, Hebrides Shelf; MS, Malin Shelf; IS, Irish Shelf; PB, Porcupine Bank; PSB, Porcupine Seabight; PAP, Porcupine Abyssal Plain.

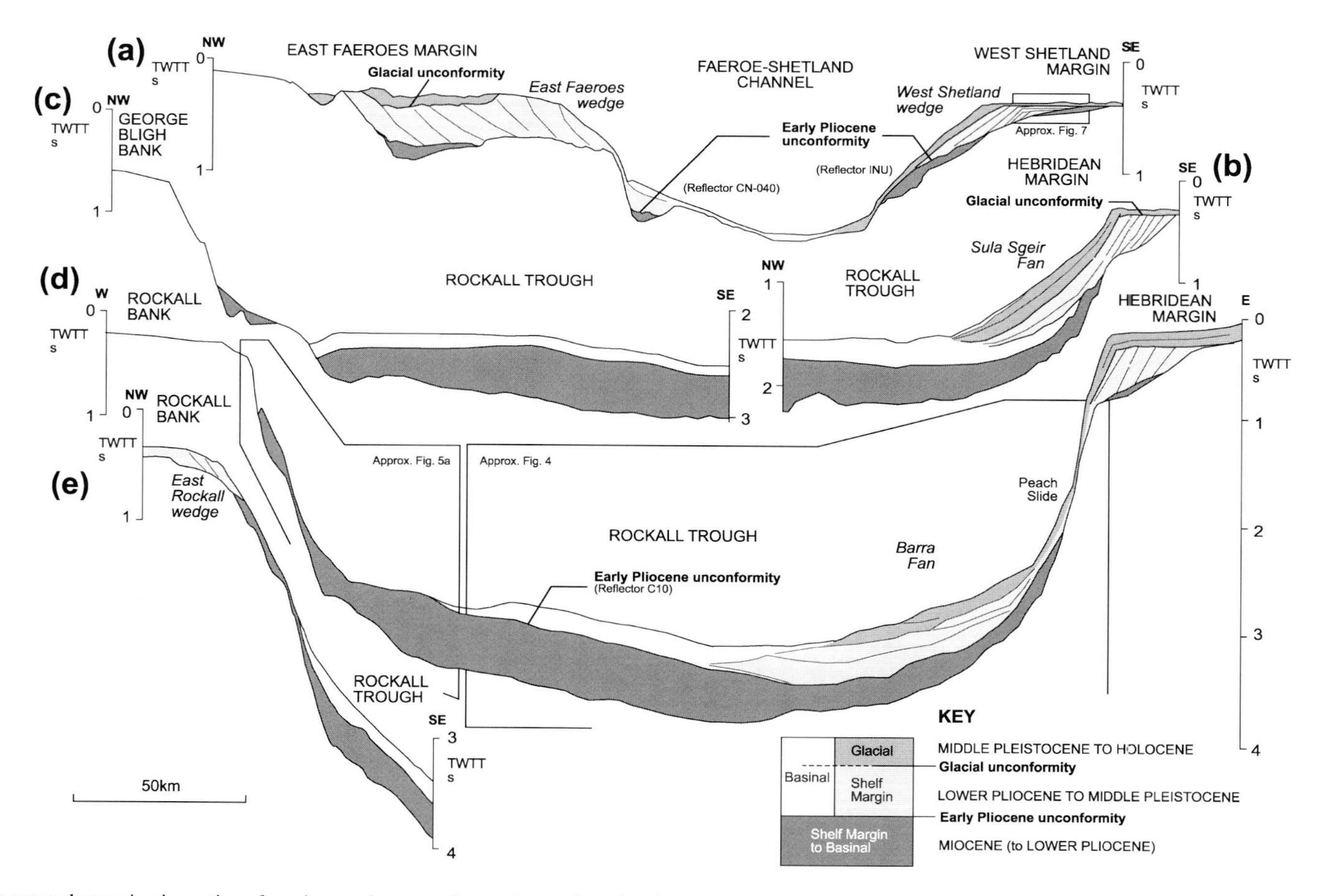

Fig. 2. Interpreted geoseismic sections focusing on the upper Cenozoic stratigraphy. Section (**a**) based on profile OF94-29 (Faeroes margin) and BGS profiles 83/04-29 and -30 (West Shetland margin); section (**b**) based on profiles DG95-8 and -10 (courtesy of Conoco and Digicon) and BGS profile 83/04-6; section (**c**) based on BGS profiles 92/01-38 and -39; section (**d**) based on Mobil line M89-WB-2 acquired by the 'Rockall Continental Margin Consortium' (see Acknowledgements); section (**e**) based on profile WRM96-107 (courtesy of Fugro Geoteam). The variable early Pliocene reflector notation is based on several stratigraphic studies: C10, Stoker *et al.* (2001); INU, Stoker (1999); CN-040, Andersen *et al.* (2000). Lotations of sections are shown in Fig. 1. The boxes indicate the approximate regional context of the seismic profiles illustrated in Figs 4, 5 and 7. TWTT, two-way travel time.

contourites, locally up to 400–500 ms thick (two-way travel time, TWTT), forms the Miocene megasequence. These deposits characteristically display extensive onlap and upslope accretion on both flanks of the Rockall Trough and on the West Shetland Slope (Stoker 1999; Stoker *et al.* 2001). Transgressive shallow-water glauconitic sandstones are locally preserved on the Hebrides and West Shetland shelves (Stoker *et al.* 1993, 1994; Stoker 1999). This megasequence developed in response to the establishment of the modern deep-water exchange between the Arctic and North Atlantic oceans, with deposition of sediment drifts occurring in the Rockall Trough and Faeroe–Shetland Channel as the deep-water currents stabilized.

The indication of a significant change in the development of the continental margin is preserved at the top of this megasequence, which has been extensively eroded both on the shelves and in the adjacent deep-water basins. The overlying Pliocene–Holocene megasequence is interpreted as a third-order composite lowstand systems tract that rests with angular discordance on the Miocene megasequence (Stoker 1999; Stoker *et al.* 2001). Plio-Pleistocene prograding wedges are the most prominent indicator of a change in style of late Neogene sedimentation and margin construction; however, a less well-documented but equally significant change is the modification to the oceanographic circulation that substantially altered the pattern of deep-water sedimentation. It is the development of the Pliocene–Holocene megasequence that forms the basis of the present study.

Late Neogene stratigraphy and style of sedimentation

The geometry and generalized distribution of the Pliocene–Holocene megasequence off NW Britain is depicted in Figs 2 and 3. The most distinctive feature of the megasequence is highlighted by the shelf-margin depocentres that are the prograding wedges. However, these wedges form only one component of a mixed depositional system that involves both downslope and alongslope processes. Sediment drifts and contourites dominate the basinal strata and where they overlap with the prograding wedges an intercalated succession of alongslope and downslope sediments is generally developed. In terms of late Neogene development of the UK Atlantic margin, the deep-water sedimentary response is inextricably linked to the formation of the prograding wedges. These features together with the chronological evidence for the onset of change (the basal unconformity) are described below and summarized in terms of a late Neogene event stratigraphy.

The basal unconformity

Nature of the boundary. The base of the Pliocene–Holocene megasequence is marked by a widespread unconformity. In the Rockall Trough, the C10 reflector of Stoker *et al.* (2001) represents this boundary (Fig. 2b–e), which forms an erosional, angular unconformity truncating strata both on the flanks and in the axis of the Trough (Figs 4–6). Adjacent to the Hebridean margin, this reflector forms the base of the prograding shelf-margin succession that includes the Barra and Sula Sgeir fans (Fig. 2b and d). Figure 4 illustrates the erosive character of the basal unconformity as traced from the basin floor into the Barra Fan on the lower Hebrides Slope. Stratal terminations in the Miocene (to lower Pliocene) basinal deposits are clearly truncated by the unconformity. Although some of this erosion may be associated with the deposition of the basal debris-flow package in the Fan, the removal of a significant section of the Miocene succession beyond the limit of the Fan (Fig. 4 (inset a), and Figs 5 and 6a) implies submarine erosion associated with deep-water bottom currents. On the western flank of the Rockall Trough, this reflector is, itself, commonly eroded and truncated by the present-day sea-bed surface (Figs 2d–e and 5); a further consequence of persistent bottom-current activity (see below). On the Hebridean margin, the unconformity can be traced from the basin floor onto the slope (Fig. 4). On the upper slope, it forms a seaward-tilted erosion surface that has cut into and locally removed shallow-water sediments of mid- to late Miocene age and older (Fig. 6b) (Stoker *et al.* 1993, 1994). This linkage between the deep-water basin and the shelf margin demonstrates the regional significance of this reflector. On the east Rockall Bank, the nature of the base of the East Rockall wedge is less clear, although there is downlap onto older Miocene strata (Fig. 2e).

In the Faeroe–Shetland region, the INU (intra-Neogene unconformity) of Stoker (1999) and the CN-040 reflector of Andersen *et al.* (2000) denote the basal unconformity on, respectively, the West Shetland and East Faeroes margins (Fig. 2a). West of Shetland, the basal unconformity can be traced from the West Shetland Shelf into the Faeroe–Shetland Channel. On the shelf margin, it is a planar, seaward-tilted erosion surface that truncates the underlying upper

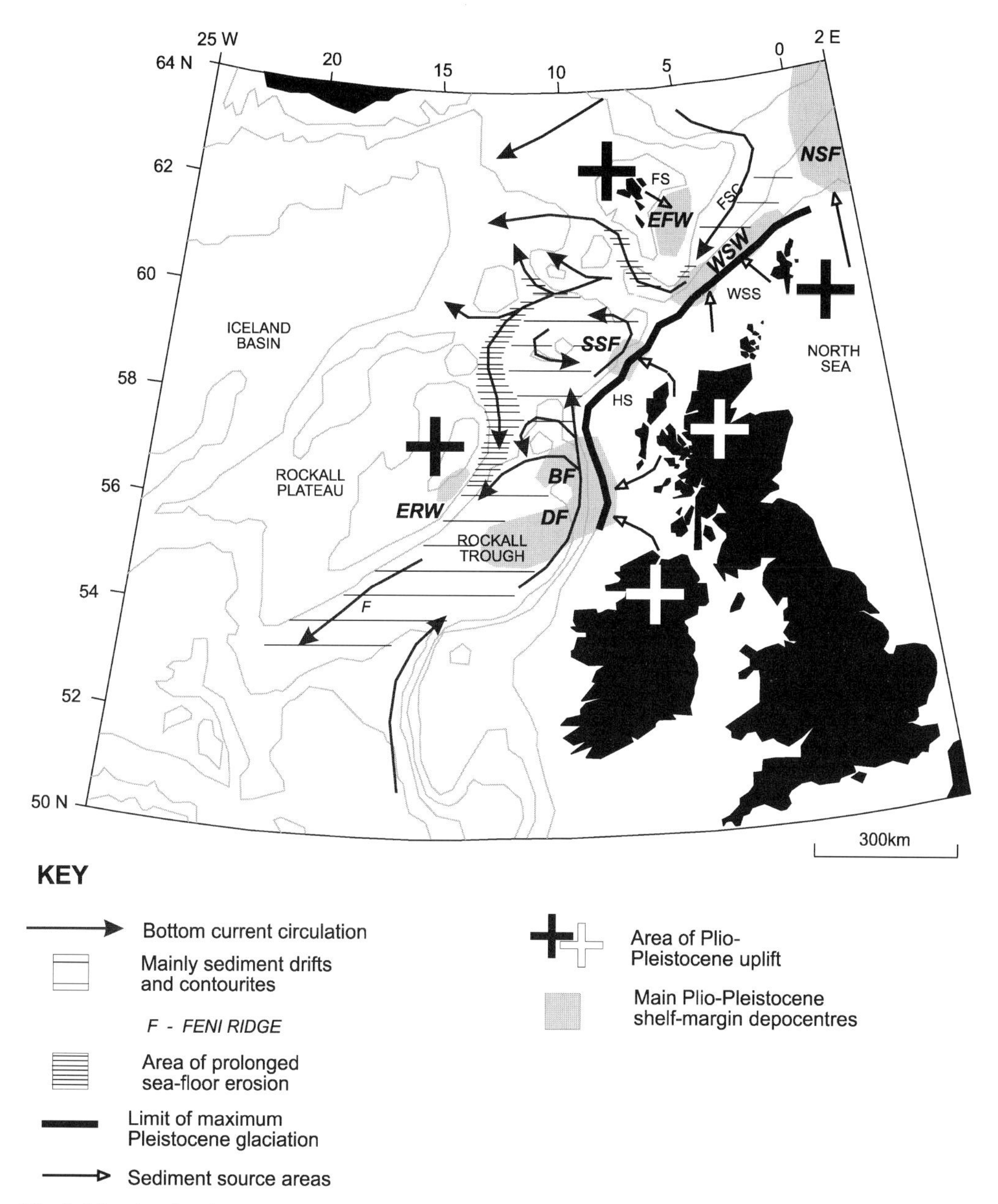

Fig. 3. Map showing the generalized distribution and gross depositional environment of the Pliocene–Holocene megasequence along the Atlantic margin between Ireland and the Faeroe–Shetland region. Abbreviations as in Fig. 1, and FSC, Faeroe-Shetland Channel. Present-day contours (as in Fig. 1) are superimposed on the map to act as an approximate guide to the palaeomorphology of the continental margin. The present-day bottom-circulation pattern is taken from Stoker (1998).

Miocene and older strata (Fig. 7). It is envisaged that the unconformity here formed as a flat-lying erosion surface on the outer shelf, which was tilted seaward before the deposition of the overlying prograding wedge. In the Faeroe–Shetland Channel, the unconformity is a planar to irregular, angular erosion surface. At the narrow SW end of the Channel, the eroded top of a sequence of Miocene sediment-drift deposits infilling a palaeo-erosional deep is clearly

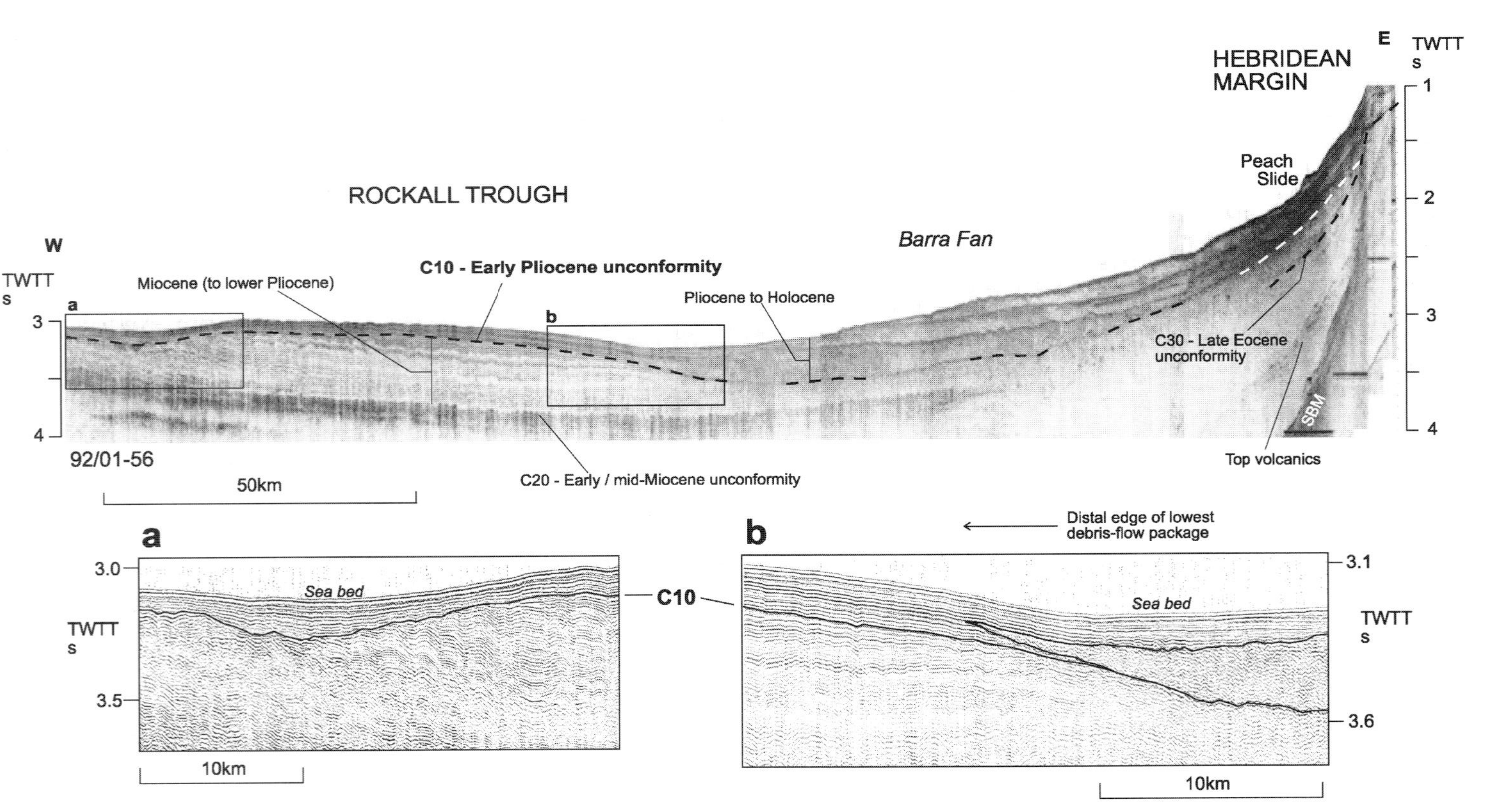

Fig. 4. BGS airgun profile 92/01-56 extends across the lower part of the Barra Fan and into the Rockall Trough and shows the early Pliocene unconformity (C10) truncating the underlying Miocene (to lower Pliocene) strata, and the lateral relationship between the Pliocene–Holocene shelf-margin and basinal deposits. Both insets (**a**) and (**b**) highlight the erosive nature of C10; inset (**b**) also illustrates the interdigitating nature of the debris-flow and deep-marine deposits, and indicates that the lowest debris-flow package is underlain by a thin section of basinal sediment at its distal edge. Cenozoic reflector notation, C10–C30, is from Stoker *et al.* (2001). TWTT, two-way travel time; SBM, sea-bed multiple. Location of profile is shown in Fig. 1 (see also Fig. 2).

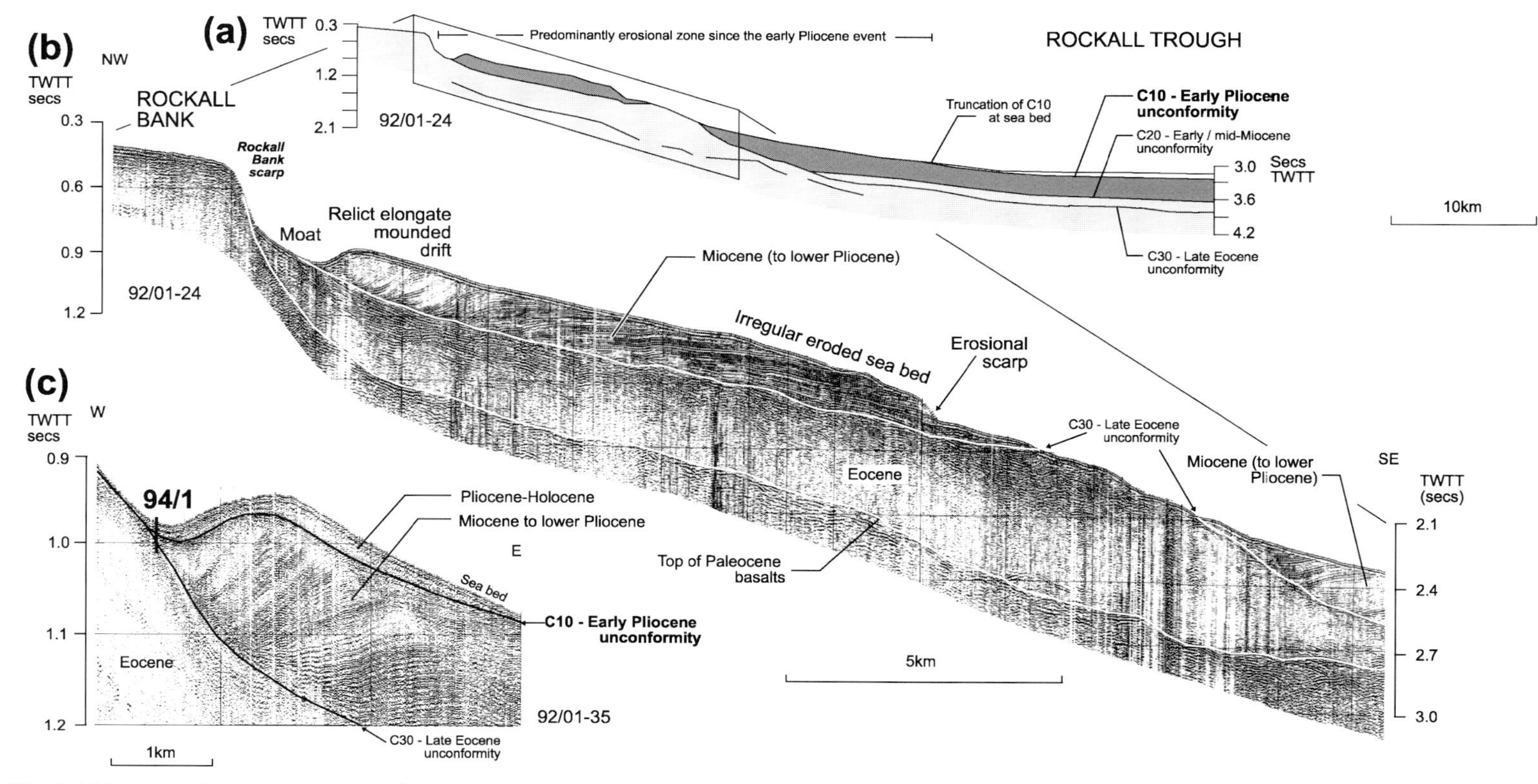

Fig. 5. (**a**) Interpreted geoseismic section of BGS airgun profile 92/01-24 across the western flank of the Rockall Trough. Inset expanded in (**b**) shows detail of erosional sea bed, the relict Miocene (to lower Pliocene) sediment drift, and locally exposed Eocene strata. (**c**) The BGS sparker profile 92/01-35, located SW of profile 24, shows the early Pliocene unconformity (C10) preserved beneath a drape of Pliocene–Holocene sediments, together with the location of BGS borehole 94.1. Cenozoic reflector notation, C10–C30, from Stoker *et al.* (2001). Locations of profiles are shown in Fig. 1 (see also Fig. 2 for (**a**)). TWTT, two-way travel time.

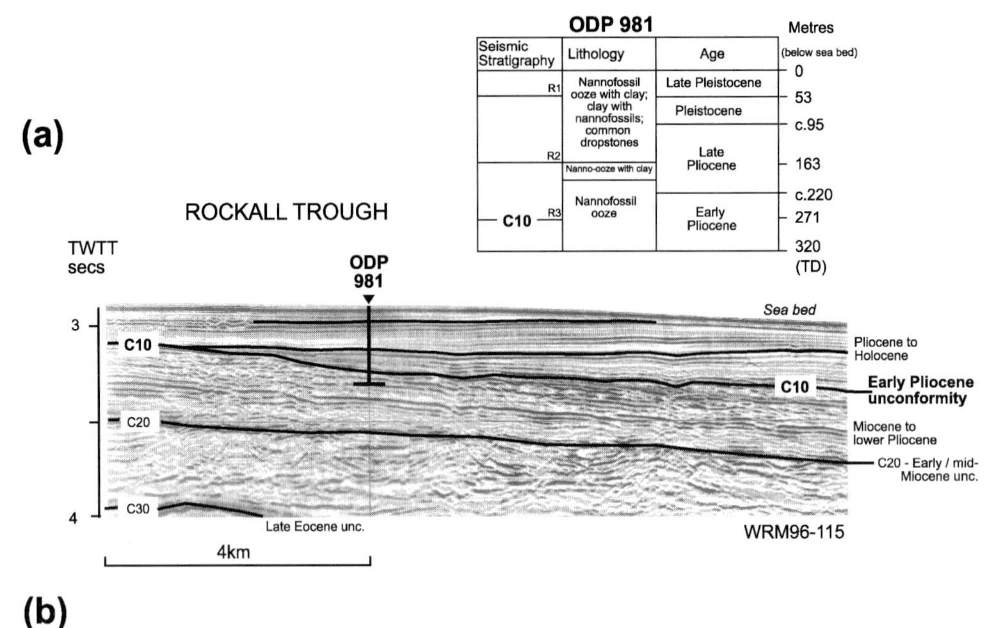

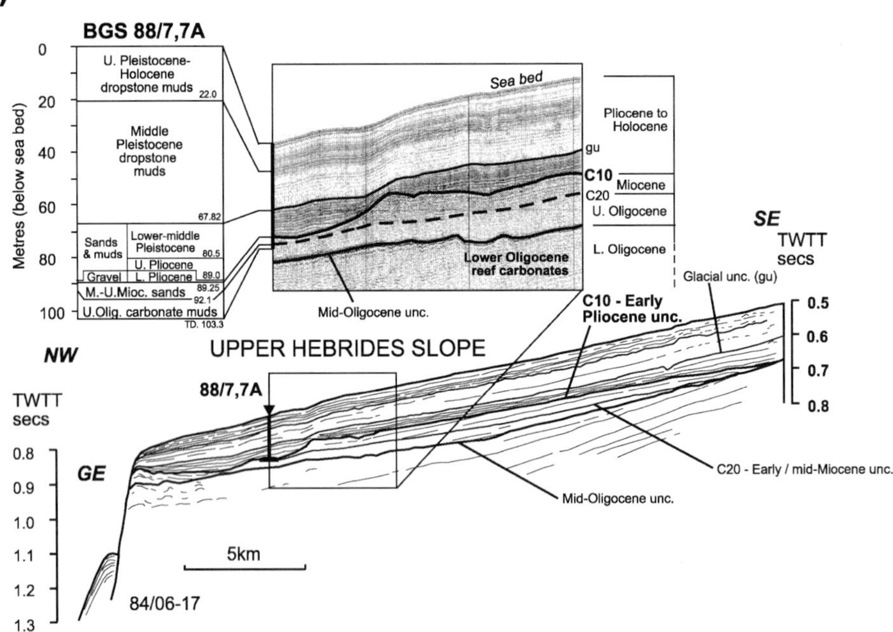

Fig. 6. Seismic-stratigraphic setting of the middle to upper Cenozoic succession in the Hebrides–Rockall region. (**a**) Fugro-Geoteam commercial seismic profile WRM96-115 from the western flank of the Rockall Trough showing the basinal succession calibrated to ODP site 981 (modified from Jansen *et al.* 1996; Stoker *et al.* 2001). (**b**) BGS sparker profile 84/06-17 and interpreted line drawing across the Hebrides Slope showing the shelf-margin succession calibrated to BGS borehole 88/7,7A (modified from Stoker *et al.* 1994, 2001). Locations of profiles are shown in Fig. 1. TWTT, two-way travel time; GE, Geikie Escarpment.

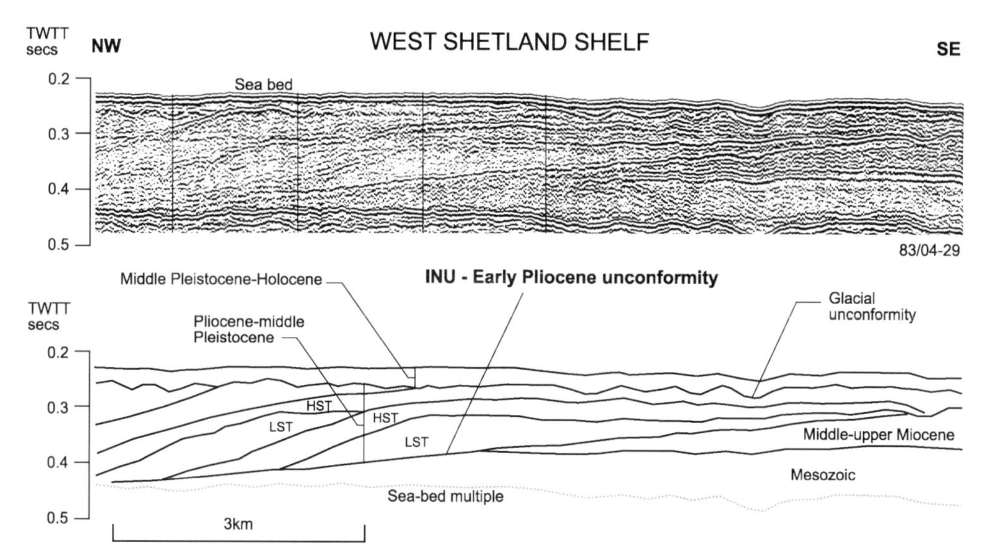

Fig. 7. BGS airgun profile 83/04-29 and interpreted line drawing from the West Shetland Shelf showing the seaward-tilted, shelf-margin expression of the early Pliocene unconformity (INU) and the progradational build-out of the overlying Pliocene–Holocene megasequence. The latter displays several lowstand (LST)–highstand (HST) couplets within an overall third-order composite lowstand systems tract. Reflector notation and stratigraphic data based on Stoker (1999). Location of profile is shown in Fig. 1 (see also Fig. 2). TWTT, two-way travel time.

observed in Fig. 8. Moreover, the reduced thickness of Miocene strata on the flanks of this deep is also partly attributable to erosion associated with the unconformity. Farther to the NE, Davies *et al.* (2001) have described a similar erosional relationship. At the NW end of the profile in Fig. 8, this unconformity is locally truncated by the sea bed, and the erosional deep (part of the Judd Deeps) that here remains open represents a composite erosion surface that has been eroding into the underlying Paleogene strata throughout the Neogene interval (Stoker 1999).

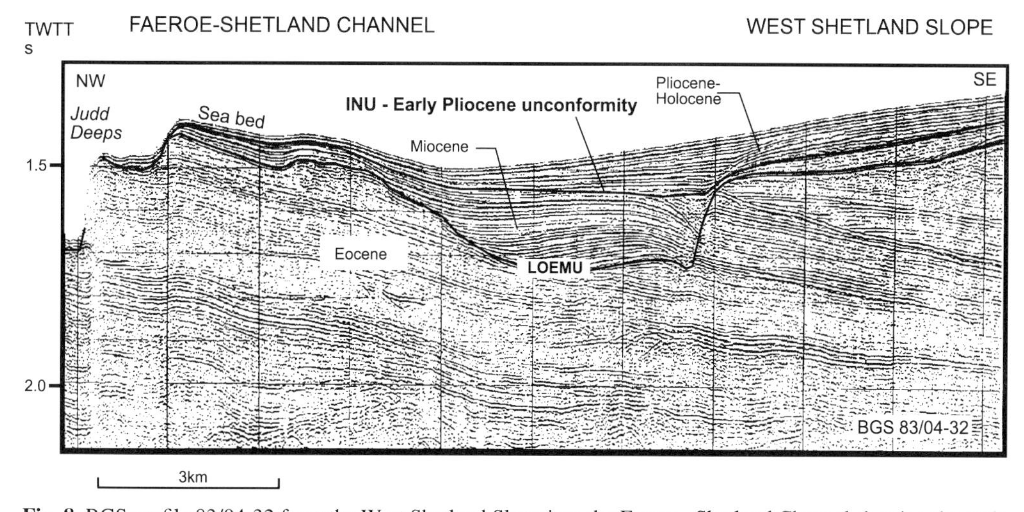

Fig. 8. BGS profile 83/04-32 from the West Shetland Slope into the Faeroe–Shetland Channel showing the early Pliocene unconformity (INU) truncating the underlying Miocene sediment-drift deposits, but being truncated itself by the present-day sea bed that forms a composite erosion surface in the area of the Judd Deeps. Stratigraphic data and notation are from Stoker (1999): LOEMU, latest Oligocene–early Miocene unconformity. Location of profile is shown in Fig. 1. TWTT, two-way travel time.

East of the Faeroe Islands, significant erosion also appears to have occurred at the base of the East Faeroes wedge (Andersen *et al.* 2000).

Age of formation. In the Rockall Trough, the C10 reflector correlates with reflector R3 of Jansen *et al.* (1996) at Ocean Drilling Program (ODP) site 981 (Stoker *et al.* 2001) (Fig. 6a). At this site, the unconformity occurs at *c.* 271 m below sea bed (bsb) (Jansen *et al.* 1996). Several important planktonic foraminifer datum levels (first occurrence, FO; last occurrence, LO) bracket this depth. Above the unconformity, between 207.8 and 215.63 m bsb, the LO of *Globorotalia* cf. *crassula* (3.3 Ma) is recorded, and the FO of *Globorotalia puncticulata* (4.5 Ma) occurs just below the unconformity between 277.7 and 279.2 m bsb (Flower 1999). These datum levels fall within the early Pliocene *Gr. puncticulata* Zone, *c.* 4.2–3.2 Ma, of Weaver & Clement (1986). Additionally, the CN11–12a (Okada & Bukry 1980) or NN15–16 (Martini 1971) calcareous nannoplankton biozone boundary, dated at 3.8 Ma by Berggren *et al.* (1995), is recorded at about 220 m bsb, and the FOs of the nannofossil *Pseudoemiliania lacunosa* (3.7 Ma) and the diatom *Thalassiosira convexa* (3.85 Ma) are recorded at depths of about 244 m bsb and 244–272 m bsb, respectively (Jansen *et al.* 1996). These data suggest an early Pliocene age for the unconformity, between 3.85 and 4.5 Ma. This is consistent with biostratigraphic information from British Geological Survey (BGS) borehole 94/1 that penetrated reflector C10 farther north (Fig. 5c), and proved a mid-Miocene to early Pliocene (NN10–12 to NN15, about 9–4 Ma) age range for the sediments immediately underlying the unconformity (Stoker *et al.* 2001). It is also supported by biostratigraphic data from well 214/4-1 in the NE Faeroe–Shetland Channel, which indicate an intra-early Pliocene age for the INU reflector (Davies *et al.* 2001).

On the Hebrides Slope, BGS borehole 88/7,7A penetrated the entire preserved Pliocene shelf-margin section on the upper Hebrides Slope (Fig. 6b), between the Barra and Sula Sgeir fans (Stoker *et al.* 1994). Although the bulk of the Pliocene section, between 80.5 and 89.0 m bsb, is of late Pliocene age, the basal 0.25 m (89.0–89.25 m) comprised a glauconite-rich lag gravel interpreted to overlie the unconformity, and dated by Foraminifera to the early Pliocene *Gr. puncticulata* Zone. Additionally, the benthonic species *Uvigerina venusta saxonica* was also present in this section; its LO in the North Sea is at the top of the benthonic zone NSB13 of King (1989) dated at *c.* 3.8 Ma. The sediments immediately below the unconformity at this site are dated as late Miocene (about 9.5 Ma) in age; however, the seismic data indicate that younger strata subcropping the unconformity are preserved further upslope from the site (Fig. 6b).

On the West Shetland Shelf, the age of the unconformity is less well constrained and a more general latest Miocene–early Pliocene age has been inferred from a regional study of BGS boreholes and commercial wells (Stoker 1999). However, evidence from BGS borehole 77/9 (Fig. 1), which penetrated the proximal part of the West Shetland wedge, does indicate the presence of the early Pliocene dinoflagellate cysts *Amiculosphaera umbracula* and *Spiniferites splendidus* and the planktonic foraminifer *Globorotalia crassaformis* at the base (61.5 m bsb) of the Pliocene section. These suggest that the sediments immediately above the unconformity are unlikely to be younger than biozone NN13–14, which implies an upper age limit of *c.* 4.2–4.0 Ma (Berggren *et al.* 1995). The sediments immediately below the unconformity are of late Miocene age, dated on the basis of $^{87}Sr/^{86}Sr$ between 5.5 and 8 Ma (Stoker 1999).

Clearly, the resolution of the available data is variable and to some extent dependent upon the geological setting. The most precise dating is the intra-early Pliocene age from the Rockall Trough, supported by the Faeroe–Shetland Channel data, whereas in the generally more erosional setting of the shelf margin a latest Miocene–early Pliocene age is at best defined. On the basis of these data, the unconformity is here assigned an early Pliocene age. Although the possibility of some diachroneity spanning the latest Miocene–early Pliocene (*c.* 5.5–3.8 Ma) interval cannot be discounted, it should be recognized that the cored sediments subcropping the unconformity do not necessarily represent the youngest pre-unconformity strata on the Hebrides and West Shetland shelves. Nevertheless, the seaward tilting of the unconformity on the shelf margins must have occurred between latest Miocene and early Pliocene time.

Prograding wedges

Distribution, thickness and architecture. Along the eastern margin of the Rockall Trough, the discrete depocentres of the Barra–Donegal and Sula Sgeir fans (Fig. 2b and d) form the largest of the prograding sequences off NW Britain, ranging in thickness from 300 to 800 ms TWTT. West of Shetland, the West Shetland wedge (Fig. 2a) forms a more linear depocentre up to about 300 ms TWTT thick, bordering the eastern margin of the Faeroe–Shetland Channel. Through their development, the shelf break off NW Britain has advanced locally by up to 50 km throughout the late Neogene interval (Stoker *et al.* 1993). Farther north, the North Sea Fan (Fig. 3) encroaches into the UK sector but is more related to the development of the Norwegian Channel (see Sejrup *et al.* 1996) and is not considered further in this study.

In general terms, the prograding wedges can be subdivided into two sequences separated by an irregular, shelf-wide, erosional unconformity, which represents the glacial unconformity (Figs 2, 6b and 7) (Stoker *et al.* 1994; Stoker 1995). The development of this boundary marks the onset of extensive shelf glaciation off NW Britain (see below for age).

The seismic-stratigraphic architecture of these wedges indicates that their growth pattern was initially restricted to the outer shelf and upper slope, utilizing the accommodation space created by the seaward tilting of the outer shelf (Figs 2 and 7). Locally, the wedges preserve both a prograding and an aggrading component, representing higher-order lowstand-highstand couplets (Fig. 7) (Stoker 1999). Although the predominant trend is toward progradation, and the development of a composite lowstand prograding wedge, the occurrence of these couplets does imply some degree of higher-order relative sea-level fluctuation. Significantly, the preservation of these couplets is indicative of continued subsidence of the margin throughout the late Neogene interval. As the shelf margin developed, the prograding wedges extended into deep water, downlapping older Miocene (to lower Pliocene) sediment-drift deposits, and becoming intercalated with the basinal strata (Figs 2, 4 and 8). Whereas the Sula Sgeir Fan clearly overlaps slightly older Pliocene basinal strata (Fig. 2b), contemporary basinal strata in the area of the Barra Fan may have been partly eroded away, as the bulk of the fan deposits appear to rest on the early Pliocene unconformity (Figs 2d and 4). Close inspection of the seismic profile in Fig. 4 (inset b) indicates that the distal edge of the basal debris-flow package overlies post-unconformity basinal deposits. However, farther in towards the Hebridean Margin, any basinal sedimentation is likely to have been overwhelmed by the sheer volume of the mass-flow sedimentation. The upper part of the shelf-margin succession has locally been modified by slope failure, such as that associated with the Peach Slide on the Barra Fan (Holmes *et al.* 1998).

Shelf-margin progradation was not exclusive to the Hebridean and West Shetland margins. A significant prograding wedge, the East Faeroes wedge, is preserved off the Faeroe Islands (Andersen *et al.* 2000) (Fig. 2a), which together with the West Shetland wedge indicates a symmetry to margin construction in this region. Farther south, Plio-Pleistocene progradation occurs on the eastern flank of Rockall Bank, where the East Rockall wedge is developed downlapping onto Miocene sediment-drift deposits (Stoker 2002) (Figs 2e and 3). However, the East Rockall wedge is relatively small in comparison with the eastern flank of the Rockall Trough.

Age of the wedges. The Plio-Pleistocene succession has been tested by numerous BGS boreholes and commercial wells on the Hebrides and West Shetland margins (Stoker *et al.* 1993; Stoker 1999), but relatively few of these provide the necessary detail and resolution to rigorously date the sediments. BGS borehole 88/7,7A penetrated the entire Plio-Pleistocene succession (89.25 m thick) preserved on the upper Hebrides Slope between the Barra and Sula Sgeir fans (Fig. 6b). Although lower Pliocene strata were proved in the basal 0.25 m (described above), biostratigraphic and magnetostratigraphic data indicate that the bulk of the overlying 89.0 m section (to sea bed) is of late Pliocene to Pleistocene age, probably no older than *c.* 3 Ma (Stoker *et al.* 1994). It was suggested by Stoker *et al.* (1994) that the lower–upper Pliocene boundary may be lost in a core gap between 89.0 and 88.4 m bsb, thus the nature of this boundary remains uncertain. The gap in the ages recorded by the early and late Pliocene strata at this site suggests that here the boundary may represent a hiatus of up to 0.8 Ma. The Pliocene–Pleistocene boundary (1.64 Ma) occurs at 80.5 m bsb in the borehole, with the glacial unconformity penetrated at 67.82 m bsb and dated to early mid-Pleistocene (about 0.44 Ma) age. It is interesting to note that the influx of ice-rafted detritus (IRD) to the Hebrides Slope occurs at 86.0 m bsb, coinciding more or less with the Gauss–Matuyama polarity transition (2.48 Ma) at 85.8 m bsb (Stoker *et al.* 1994).

On the West Shetland Shelf, BGS borehole 77/9 proved lower Pliocene sediments at the base of the borehole (described above). However, BGS boreholes 84/1 and 84/2 (Fig. 1), which similarly penetrated the proximal (landward) part of the wedge, suggest that the bulk of the West Shetland wedge is of late Pliocene–Pleistocene age (Stoker 1999). In borehole 84/2, the sediments recovered between the basal unconformity and the glacial unconformity are assigned a late Pliocene–early Pleistocene age on the basis of the dinoflagellate cyst species, *Amiculosphaera umbracula* and *Habibacysta tectata* (Harland, pers. comm.). This is supported by borehole 84/1, where the basal part of the wedge correlates with the benthic foraminifer *Cibicides grossa* zone, dated to latest Pliocene–earliest Pleistocene time (Wilkinson & Harland 1999). These data suggest that the main onset of deposition of the prograding wedges occurred in late Pliocene time. Borehole 88/7,7A further suggests that this correlates approximately with the influx of ice-rafted material at *c.* 2.48 Ma.

Deep-water sedimentary response

Style and focus of sedimentation. Contour-following bottom currents have strongly influenced basinal sedimentation in the Rockall Trough and Faeroe–Shetland Channel throughout the Neogene period, and a variety of sediment-drift and associated bedforms have been described (Kidd & Hill 1986; Howe *et al.* 1994; Howe 1996; Stoker 1998; Stoker *et al.* 1998; Masson *et al.* 2002). However, the base of the Pliocene–Holocene megasequence indicates that the deep-water sedimentary regime was modified in two ways: (1) by extensive submarine erosion that truncated older drift deposits, forming the basal (early Pliocene) unconformity; (2) by a major shift in the focus of late Neogene sedimentation relative to the underlying Miocene deposits. Examples of these changes are described below.

In the Rockall Trough, the Miocene (to lower Pliocene) sediments characteristically onlap the flanks of the basins, with elongate mounded sediment drifts displaying significant upslope accretion (Stoker 1998; Stoker *et al.* 1998) (Fig. 5). Following early Pliocene erosion (e.g. Figs 4–6a), sediment drifts continued to accumulate in the basin (e.g. Fig. 6a) and along the eastern flank of the Trough (partly overlapping with the prograding wedges), but the western flank of the basin (north of about 56°30′N) has continued to be subject to erosion to the present day (Stoker *et al.* 2001) (Figs 2c–e and 3). This intense bottom-current activity has restricted Pliocene–Holocene accumulation in this area to a thin drape (e.g. Fig. 5c) or a sea-bed veneer, commonly marked by a gravel-lag contourite (Howe *et al.* 2001), and has eroded the underlying Miocene drift deposits, locally exposing Paleogene strata at the sea bed (Fig. 5a and b). The longevity of the erosion is confirmed where the base of the Pliocene–Holocene megasequence, the C10 reflector, is truncated by the sea bed forming a composite sea-bed erosion surface (Fig. 5). In comparison with Miocene (to lower Pliocene) sedimentation, this represents a major eastward (basinward) shift in the accumulation of sediment-drift deposits on the west and NW flank of the northern Rockall Trough during Plio-Pleistocene time. Farther south, in contrast, basinal sedimentation along the western flank of the central and southern Rockall Trough prevailed following early Pliocene erosion, and sediment drifts, such as the giant elongate Feni Drift (Figs 1 and 3), continued to aggrade (Stoker *et al.* 2001).

In comparison with the Rockall Trough, the Faeroe–Shetland Channel has generally been more an area of sediment export, resulting in the preservation of a thinner Neogene succession (Stoker *et al.* 1998; Stoker 1999). Nevertheless, the early Pliocene unconformity is strongly expressed, especially at the SW end of the Channel, where formerly extensive, thick, Miocene sediment drifts have been eroded and locally removed (Figs 2a and 8). Since early Pliocene time, this part of the channel has remained largely an area of prolonged erosion (Fig. 3), characterized by a thin to locally absent sediment cover. The main focus of Pliocene–Holocene sediment-drift accumulation in the Faeroe–Shetland Channel occurs north of about 60°30′N; here the basinal sediments increase in thickness and onlap onto the lower–middle part of the West Shetland Slope where they overlap with the West Shetland wedge. Andersen *et al.* (2000) reported a similar relationship from the East Faeroes Slope.

Rate of response. On the basis of biostratigraphic data and sedimentation rates, Jansen *et al.* (1996) concluded that ODP site 981 does not contain any significant hiatuses. However, the FOs of *T. convexa* (3.85 Ma) and *Gr. puncticulata* (4.5 Ma) were recorded by Jansen *et al.* (1996) immediately above and below, respectively, the unconformity, which does suggest a hiatus of about 0.75 Ma, although the problem of biostratigraphic resolution is highlighted by their placing of the CN11–12a nannofossil boundary (3.8 Ma) about 50 m above the unconformity (see above). Nevertheless, these data suggest that there is a relatively continuous record of late early Pliocene to Holocene (post 3.85 Ma) sedimentation above the unconformity. This represents an earlier sedimentary response than that proved for the prograding wedges (3 Ma or younger). Moreover, the influx of IRD at ODP site 981 is recorded at a depth of about 138 m bsb, which is just above the LO of the planktonic foraminifer *Neogloboquadrina atlantica* (2.41 Ma), and about 133 m above the unconformity (Jansen *et al.* 1996). The thickness of the pre-IRD post-unconformity sediments clearly contrasts with the prograding wedges, which preserve a far more condensed record of equivalent strata.

Late Neogene development of the UK Atlantic margin: a summary event stratigraphy

From the information presented in the preceding sections, the key elements of the late Neogene event stratigraphy off NW Britain are summarized in Fig. 9, and further encapsulated as a gross palaeo-environmental reconstruction in Fig. 3. In general terms, a four-stage history of development can be established that is applicable to the entire UK Atlantic margin, as follows.

(1) The early Pliocene event. This resulted in the formation of the early Pliocene unconformity between about 3.8 and 4.5 Ma, although some diachroneity from latest Miocene time (about 5.5 Ma) cannot be discounted on the basis of the shelf-margin record.

(2) The early Pliocene (post 3.8 Ma) sedimentary response. This is marked by the continuation of sediment-drift accumulation in the deep-water basins, albeit with a shift in the focus of sedimentation in the northern Rockall Trough and SW Faeroe–Shetland Channel. A more restricted record of early Pliocene sedimentation is preserved on the shelf margins at the base of the prograding wedges.

(3) The late Pliocene–early mid-Pleistocene sedimentary response. This interval was dominated by the deposition of the major prograding wedges between about 3 and 0.44 Ma. In the

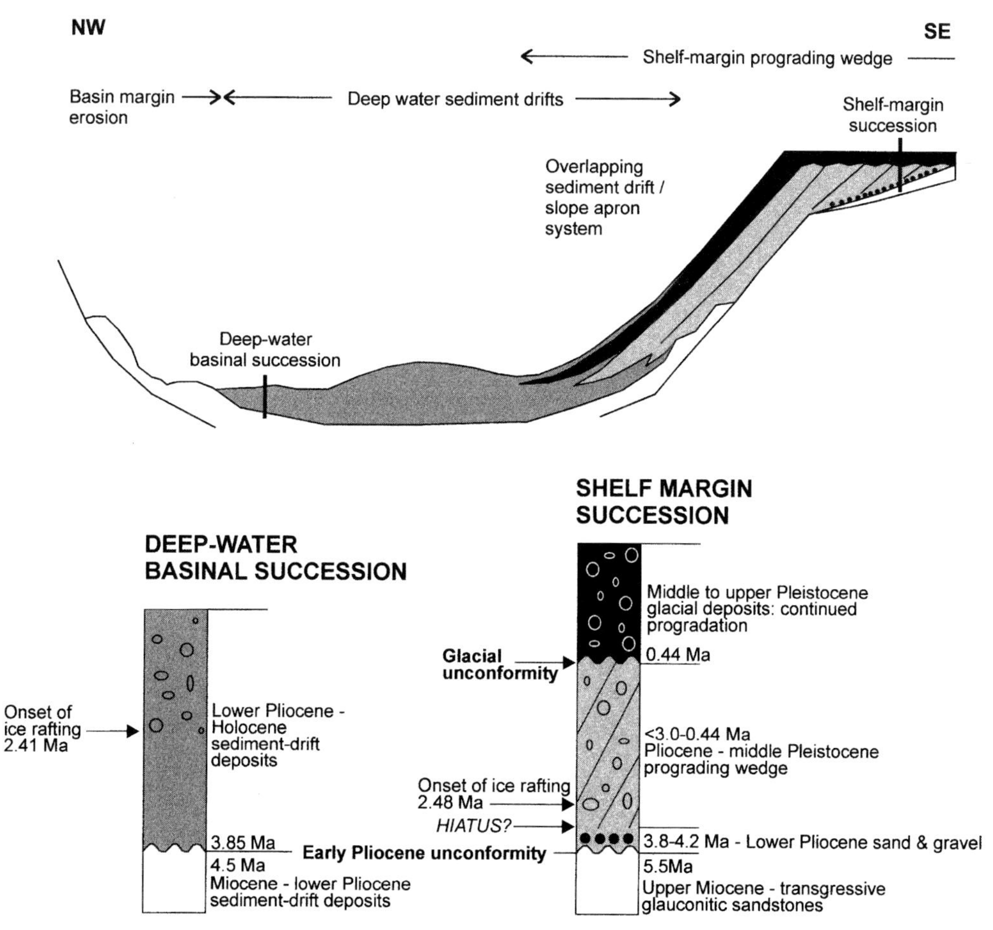

Fig. 9. Schematic diagram summarizing the late Neogene event stratigraphy and development of the UK Atlantic margin (see text for details).

adjacent basins, deep-water sedimentation prevailed and in areas where downslope and alongslope processes interacted overlapping slope apron–sediment-drift systems developed.

(4) Mid- to late Pleistocene shelf-wide glaciation. Ice sheets reached the edge of the Hebrides and West Shetland shelves between 0.44 and *c.* 0.018 Ma and further contributed to shelf-margin progradation (Stoker 1995). Deep-water sedimentation prevailed in the adjacent basins, and bottom-current processes continue to be active at the present day (Howe 1996; Masson *et al.* 2002).

Discussion

The results presented from the UK Atlantic margin raise several interesting aspects of broader relevance concerning the timing and mechanism of change throughout the NE Atlantic region.

Timing of change

Perhaps one of the most significant aspects arising from this study concerns the question of cause and effect and its bearing on the timing of change within late Neogene time. In general, the sedimentary record of shelf-margin progradation around the NE Atlantic region, including NW Britain, appears to indicate that it essentially dates from late Pliocene time. From a comprehensive study of the Norwegian continental margin, Eidvin *et al.* (2000) concluded that there was a link between shelf-margin progradation and the intensification of northern latitude glaciation. Although this climatically related thesis is supported to some extent by the present study, the question remains as to whether this linkage is simply a response to (an effect of) a more fundamental causal event, e.g. uplift. The prograding wedges effectively preserve the record of denudation, be that caused by glacial processes or otherwise, and do not necessarily represent an immediate response to change. As denudation rate is largely a function of local relief, which in turn is related to the degree of fluvial dissection, there is an inherent tendency for the rate of denudation to be initially slower than the rate of uplift (Summerfield 1991).

On the basis of the results of the present study, it is here suggested that the late Neogene change was initiated before the deposition of the bulk of the prograding wedges. The indicator of change is manifest by the regional early Pliocene unconformity, which can be traced unambiguously from the shelf margin into the deep-water basins (Fig. 2). Biostratigraphic data are consistent from both settings and indicate lower Pliocene sediments immediately overlying the unconformity (Fig. 9).

A major question that cannot be resolved from the available data is whether or not there is some degree of diachroneity in the development of the unconformity between the shelf margin and deep-water basin. On the shelf margin, the youngest cored strata from below the unconformity are of latest Miocene (*c.* 5.5 Ma) age, which contrasts with the basinal sediments, where early Pliocene (4.5 Ma) deposits make up the sub-unconformity section (Fig. 9). This disparity may suggest diachroneity of the order of 1 Ma, possibly extending the onset of change on the shelf margin into latest Miocene time, before culminating in the margin-wide early Pliocene event. Further sampling is necessary to better constrain the timing of change on the shelf margin.

Mechanics of change

The mechanism of change in late Neogene time is a problem of very general interest as it has affected a large proportion of the landmasses around the North Atlantic. In a recent review paper, Japsen & Chalmers (2000) summarized the current thinking on Neogene evolution around the North Atlantic that favours some kind of tectonic event, which resulted in uplift of basin margins and subsidence of basinal areas adjacent to the uplifted landmasses. However, the mechanism responsible for uplift remains uncertain, and there is a school of thought that suggests that the so-called 'Neogene uplift' may in fact have begun earlier in Paleogene time (Clausen *et al.* 2000). What is clear is that any general model invoked to try and explain Cenozoic uplift, including both Paleogene and Neogene uplift, must be constrained by observations from the entire NE Atlantic region.

On the UK Atlantic margin, regional observations that must be taken into consideration include (1) the seaward tilting and subsidence of the outer part of the Hebrides and West Shetland margins at some time between latest Miocene and early Pliocene time, (2) the early Pliocene change in the oceanographic circulation pattern, and (3) the predominantly late Pliocene onset of shelf-margin progradation. The regional distribution of the prograding wedges is a further consideration, as west to NW progradation of the NW British margin was accompanied by SE progradation from the Faeroe Islands (Andersen *et al.* 2000) and the Rockall Bank (Fig. 3). One might argue whether or not uplift is necessary to achieve this distribution of wedges; instead, it may be achieved by the eustatic lowering of sea level, for which there is evidence of significant lowering from *c.* 4.1 Ma (Vail & Hardenbol 1979). Presumably, this would also have affected the oceanographic circulation pattern. However, eustacy alone does not explain either the seaward tilting of the continental margin off NW Britain, nor the timing of shelf margin progradation. Instead, the continuity of the early Pliocene unconformity from the shelf margin into deep water suggests that the modification to the oceanographic circulation pattern was inextricably linked to the process responsible for the tilting of the margin.

There is evidence for Neogene uplift of the NW British hinterland from onshore studies in Britain and Ireland (e.g. Japsen 1997; Galewsky *et al.* 1998) and the Irish Sea Basin (Green *et al.* 2001). Uplift has also been invoked to explain the prograding wedge development on the Faeroese margin (Andersen *et al.* 2000). The conjugate development of wedges in the Faeroe–Shetland region implies that the intervening basin must have subsided. Similarly, the development of the East Rockall wedge, facing the Hebridean margin, suggests that not only has the Rockall Bank undergone some degree of uplift concomitant with that of the Hebridean margin, but that the Rockall Trough has probably compensated for this with subsidence. The ensuing change in the shape and palaeobathymetry of the continental margin is also likely to have modified the water circulation pattern. The adjustment and breaching of sills or other barriers and the development or modification of deep-water gateways controlling bottom-current circulation, sedimentation and erosion in the North Atlantic Ocean has been linked by several workers (e.g. Tucholke & Mountain 1986; Eldholm 1990; Thiede & Mhyre 1996; Wright & Miller 1996) to regional tectonic events. Indeed, the triggering of early Neogene deep-water erosion, an event that had margin-wide implications similar to those of the early Pliocene event, and the subsequent widespread accumulation of Miocene (to lower Pliocene) sediment-drift deposits off NW Britain are most probably linked to the plate-tectonic development of the Norwegian–Greenland Sea (Eldholm 1990; Jansen & Raymo 1996; Stoker 1998).

Uplift in combination with climatic cooling may have ultimately been a trigger for the onset of glaciation (Raymo & Ruddiman 1992; Eyles 1996) in late Pliocene time. Glacial processes superimposed on uplift may be the reason for the enhanced denudation that resulted in the prograding wedges, thus accounting, in part, for the delayed shelf-margin sedimentary response to change. The presence of IRD within the wedges supports the concept that the influx of much of this material may be linked to the climatic evolution of the area. The subsequent isostatic response to denudation, including climatically enhanced denudation as glaciation intensified, and sediment loading in the adjacent basin may have continued to drive vertical movements.

Although this explanation of the observations is consistent with the Japsen & Chalmers (2000) view of landmass and basin-margin uplift and basin-centre subsidence, it does not explain the nature of the mechanism responsible for the late Neogene event. However, it does highlight the observations that must be incorporated into any future general model of late Neogene evolution of the NE Atlantic region.

Conclusions

The pattern of observations on and around the UK Atlantic margin, including the evidence of tectonic tilting and subsidence of the shelf margins, the modified deep-water current system, and the distribution of the prograding wedges, support the idea of a late Neogene uplift event. Athough the mechanism remains unknown, it is suggested that this event was initiated in early Pliocene (possibly even latest Miocene) time and has been a major influence in the subsequent development of the UK Atlantic margin. On this basis, it is proposed that the late Neogene event stratigraphy reflects a varied sedimentary response to uplift. The deep-water basins, being open to the NE Atlantic, were sensitive to changes in oceanographic circulation, and the early Pliocene event had a significant, and possibly relatively sudden (of the order of 10–100 ka), effect on patterns of deep-marine sedimentation and erosion. In contrast, it is suggested that the shelf-margin record reflects some delay in the sedimentary response, as the prograding wedges do not begin to fully develop until perhaps 1–2 Ma after the initiation of uplift, albeit enhanced by glacial processes.

I would like to thank D. Evans, K. Hitchen, and the two referees, P. Knutz and J. Cartwright, for their constructive reviews of the paper. I am grateful to I. Walker (Conoco) for providing copies of seismic lines DG95-8 and -10; P. Broad of Fugro Geoteam for allowing access to, and use of data from, the WRM96 survey; and the following oil companies, who, together with the BGS, make up the Rockall Continental Margin Consortium, and without whose support this work could not have been undertaken: Agip, Amerada

Hess, BG, BP-Amoco, Conoco, Enterprise, Exxon/Mobil, Phillips, Statoil, Texaco and TotalFinaElf. The paper is published with the permission of the Director of the British Geological Survey (NERC).

References

ANDERSEN, M.S., NIELSEN, T., SØRENSEN, A.B., BOLDREEL, L.O. & KUIJPERS, A. 2000. Cenozoic sediment distribution and tectonic movements in the Faroe region. *Marine Geology*, **24**, 239–259.

BERGGREN, W.A., HILGEN, F.J., LANGEREIS, C.G. & 5 OTHERS 1995. Late Neogene chronology: new perspectives in high-resolution stratigraphy. *Geological Society of America Bulletin*, **107**, 1272–1287.

CHALMERS, J.A. 2000. Offshore evidence for Neogene uplift in central West Greenland. *Global and Planetary Change*, **24**, 311–318.

CLAUSEN, O.R., NIELSEN, O.B., HUUSE, M. & MICHELSEN, O. 2000. Geological indications for Palaeogene uplift in the eastern North Sea Basin. *Global and Planetary Change*, **24**, 175–187.

DAVIES, R., CARTWRIGHT, J., PIKE, J. & LINE, C. 2001. Early Oligocene initiation of North Atlantic Deep Water formation. *Nature*, **410**, 917–920.

EIDVIN, T., JANSEN, E., RUNDBERG, Y., BREKKE, H. & GROGAN, P. 2000. The upper Cainozoic of the Norwegian continental shelf correlated with the deep sea record of the Norwegian Sea and the North Atlantic. *Marine and Petroleum Geology*, **17**, 579–600.

ELDHOLM, O. 1990. Paleogene North Atlantic magmatic–tectonic events: environmental implications. *Memorie della Società Geologica Italiana*, **44**, 13–28.

EVANS, D., MCGIVERON, S., MCNEILL, A.E., HARRISON, Z.H., ØSTMO, S.R. & WILD, J.B.L. 2000. Plio-Pleistocene deposits on the mid-Norwegian margin and their implications for late Cenozoic uplift of the Norwegian mainland. *Global and Planetary Change*, **24**, 233–237.

EYLES, N. 1996. Passive margin uplift around the North Atlantic region and its role in Northern Hemisphere late Cenozoic glaciation. *Geology*, **24**, 103–106.

FLOWER, B.P. 1999. Data report: Planktonic foraminifers from the subpolar North Atlantic and Nordic seas: sites 980–987 and 907. *In*: RAYMO, M.E., JANSEN, E., BLUM, P. & HERBERT, T.D. (eds) *Proceedings of the Ocean Drilling Program, Scientific Results, 162*. Ocean Drilling Program, College Station, TX, 19–34.

GALEWSKY, J., ALLEN, P.A. & DENSMORE, A.L. 1998. The Tertiary uplift of Ireland: the regional setting. *In*: BOLDREEL, L.O. & JAPSEN, P. (eds) *Neogene Uplift and Tectonics around the North Atlantic, International Workshop, Copenhagen*. Geological Survey of Denmark and Greenland, Copenhagen, 37–38.

GREEN, P.F., DUDDY, I.R., BRAY, R.J., DUNCAN, W.I. & CORCORAN, D. 2001. The influence of thermal history on hydrocarbon prospectivity in the central Irish Sea Basin. *In*: SHANNON, P.M., HAUGHTON, P.D.W. & CORCORAN, D.V. (eds) *The Petroleum Exploration of Ireland's Offshore Basins*. Geological Society, London, Special Publications, **188**, 171–188.

GREGERSEN, U., MICHELSEN, O. & SØRENSEN, J.C. 1997. Stratigraphy and facies distribution of the Utsira Formation and the Pliocene sequences in the northern North Sea. *Marine and Petroleum Geology*, **14**, 893–914.

HENRIKSEN, S. & VORREN, T.O. 1996. Late Cenozoic sedimentation and uplift history on the mid-Norwegian continental shelf. *Global and Planetary Change*, **12**, 171–199.

HOLMES, R., LONG, D. & DODD, L.R. 1998. Large-scale debrites and submarine landslides on the Barra Fan, west of Britain. *In*: STOKER, M.S., EVANS, D. & CRAMP, A. (eds) *Geological Processes on Continental Margins: Sedimentation, Mass-Wasting and Stability*. Geological Society, London, Special Publications, **129**, 67–79.

HOWE, J.A. 1996. Turbidite and contourite sediment waves in the Northern Rockall Trough, North Atlantic Ocean. *Sedimentology*, **43**, 219–234.

HOWE, J.A., STOKER, M.S. & STOW, D.A.V. 1994. Late Cenozoic sediment drift complex, northeast Rockall Trough, North Atlantic. *Paleoceanography*, **9**, 989–999.

HOWE, J.A., STOKER, M.S. & WOOLFE, K.J. 2001. Deep-marine seabed erosion and gravel-lags in the northwestern Rockall Trough, North Atlantic Ocean. *Journal of the Geological Society, London*, **158**, 427–438.

JANSEN, E., RAYMO, M.E., *et al.* 1996. Leg 162: new frontiers on past climates. *In*: JANSEN, E., RAYMO, M.E. & BLUM, P. (eds) *Proceedings of the Ocean Drilling Program, Initial Reports, 162*. Ocean Drilling Program, College Station, TX, 5–20.

JANSEN, E., RAYMO, M.E., BLUM, P., *et al.* 1996. Sites 980/981. *In*: JANSEN, E., RAYMO, M.E. & BLUM, P. (eds) *Proceedings of the Ocean Drilling Program, Initial Reports, 162*. Ocean Drilling Program, College Station, TX, 49–90.

JAPSEN, P. 1997. Regional Neogene exhumation of Britain and the western North Sea. *Journal of the Geological Society, London*, **154**, 239–247.

JAPSEN, P. & CHALMERS, J.A. 2000. Neogene uplift and tectonics around the North Atlantic: overview. *Global and Planetary Change*, **24**, 165–173.

KIDD, R.B. & HILL, P.R. 1986. Sedimentation on mid-ocean sediment drifts. *In*: SUMMERHAYES, C.P. & SHACKLETON, N.J. (eds) *North Atlantic Palaeoceanography*. Geological Society, London, Special Publications, **21**, 87–102.

KING, C. 1989. Cenozoic of the North Sea. *In*: JENKINS, D.G. & MURRAY, J.W. (eds) *Stratigraphical Atlas of Fossil Foraminifera*. 2. Ellis Horwood, Chichester, 419–489.

MARTINI, E. 1971. Standard Tertiary and Quaternary calcareous nannoplankton zonation. *In*: FARINACCI, A. (ed.) *Proceedings of the II Planktonic Conference, Rome, 1969*. Tecnoscienza, Rome, 739–785.

MASSON, D.G., HOWE, J.A. & STOKER, M.S. 2002. Bottom current sediment waves, sediment drifts and contourites in the northern Rockall Trough. *Marine Geology*, in press.

OKADA, H. & BUKRY, D. 1980. Supplementary modification and introduction of code numbers to the low-latitude coccolith biostratigraphic zonation (Bukry 1973; 1975). *Marine Micropalaeontology*, **5**, 321–325.

RAYMO, M.E. & RUDDIMAN, W.F. 1992. Tectonic forcing of late Cenozoic climate. *Nature*, **359**, 117–122.

ROKOENGEN, K., RISE, L., BRYN, P., FRENGSTAD, B., GUSTAVSEN, B., NYGAARD, E. & SAETTEM, J. 1995. Upper Cenozoic stratigraphy on the Mid-Norwegian continental shelf. *Norsk Geologisk Tidsskrift*, **75**, 88–104.

SEJRUP, H.P., KING, E.L., AARSETH, I., HAFLIDASON, H. & ELVERHØI, A. 1996. Quaternary erosion and depositional processes: western Norwegian fjords, Norwegian Channel and North Sea Fan. *In*: DE BATIST, M. & JACOBS, P. (eds) *Geology of Siliciclastic Shelf Seas*. Geological Society, London, Special Publications, **117**, 187–202.

SOLHEIM, A., FALEIDE, J.I., ANDERSEN, E.S. & 5 OTHERS 1998. Late Cenozoic seismic stratigraphy of the East Greenland and Svalbard–Barents Sea continental margins. *Quaternary Science Reviews*, **17**, 155–184.

STOKER, M.S. 1995. The influence of glacigenic sedimentation on slope-apron development on the continental margin off Northwest Britain. *In*: SCRUTTON, R.A., STOKER, M.S., SHIMMIELD, G.B. & TUDHOPE, A.W. (eds) *The Tectonics, Sedimentation and Palaeoceanography of the North Atlantic Region*. Geological Society, London, Special Publications, **90**, 159–177.

STOKER, M.S. 1998. Sediment-drift development on the continental margin off NW Britain. *In*: STOKER, M.S., EVANS, D. & CRAMP, A. (eds) *Geological Processes on Continental Margins: Sedimentation, Mass-Wasting and Stability*. Geological Society, London, Special Publications, **129**, 229–254.

STOKER, M.S. 1999. *Stratigraphic Nomenclature of the UK North West Margin. 3. Mid- to Late Cenozoic Stratigraphy*. British Geological Survey, Edinburgh.

STOKER, M.S. 2001. *Central Rockall Basin 1:500,000*. British Geological Survey, Edinburgh.

STOKER, M.S., AKHURST, M.C., HOWE, J.A. & STOW, D.A.V. 1998. Sediment drifts and contourites on the continental margin off northwest Britain. *Sedimentary Geology*, **115**, 33–51.

STOKER, M.S., HITCHEN, K. & GRAHAM, C.G. 1993. *United Kingdom Offshore Regional Report: the Geology of the Hebrides and West Shetland Shelves, and Adjacent Deep-Water Areas*. British Geological Survey. HMSO, London.

STOKER, M.S., LESLIE, A.B., SCOTT, W.D. & 6 OTHERS 1994. A record of late Cenozoic stratigraphy, sedimentation and climate change from the Hebrides Slope, NE Atlantic Ocean. *Journal of the Geological Society, London*, **151**, 235–249.

STOKER, M.S., VAN WEERING, T.C.E. & SVAERDBORG, T. 2001. A mid- to late Cenozoic tectonostratigraphic framework for the Rockall Trough. *In*: SHANNON, P.M., HAUGHTON, P. & CORCORAN, D. (eds) *The Petroleum Exploration of Ireland's Offshore Basins*. Geological Society, London, Special Publications, **188**, 411–438.

STOKER, M.S., NIELSEN, T., VAN WEERING, T.C.E. & KUIJPERS, A. 2002. Towards an understanding of the Neogene tectonostratigraphic framework of the NE Atlantic margin between Ireland and the Faroe Islands. *Marine Geology*, in press.

SUMMERFIELD, M.A. 1991. *Global Geomorphology: an Introduction to the Study of Landforms*. Longman, Harlow.

THIEDE, J. & MHYRE, A.M. 1996. Introduction to the North Atlantic gateways: plate tectonic–palaeoceanographic history and significance. *In*: THIEDE, J., MHYRE, A.M., FIRTH, J.V., JOHNSON, G.L. & RUDDIMAN, W.F. (eds) *Proceedings of the Ocean Drilling Program, Scientific Results, 151*. Ocean Drilling Program, College Station, TX, 3–23.

TUCHOLKE, B.E. & MOUNTAIN, G.S. 1986. Tertiary palaeoceanography of the western North Atlantic Ocean. *In*: VOGT, P.R. & TUCHOLKE, B.E. (eds) *The Geology of North America, Volume M, The Western North Atlantic Region*. Geological Society of America, Boulder, CO, 631–650.

VAIL, P.R. & HARDENBOL, J. 1979. Sea-level changes during the Tertiary. *Oceanus*, **22**, 245–276.

VORREN, T.O. & LABERG, J.S. 1997. Trough mouth fans—palaeoclimate and ice sheet monitors. *Quaternary Science Reviews*, **16**, 865–881.

WEAVER, P.P.E. & CLEMENT, B.M. 1986. Synchroneity of Pliocene planktonic foraminiferal datums in the North Atlantic. *Marine Micropalaeontology*, **10**, 295–307.

WILKINSON, I.P. & HARLAND, R. 1999. Appendix 1: Mid- to late Cenozoic biostratigraphic markers. *In*: STOKER, M.S. (ed.) *Stratigraphic Nomenclature of the UK North West Margin: 3. Mid- to Late Cenozoic Stratigraphy*. British Geological Survey, Edinburgh, A1–A6.

WRIGHT, J.D. & MILLER, K.G. 1996. Control of North Atlantic Deep Water circulation by the Greenland–Scotland Ridge. *Paleoceanography*, **11**, 157–170.

Quantifying exhumation from apatite fission-track analysis and vitrinite reflectance data: precision, accuracy and latest results from the Atlantic margin of NW Europe

PAUL F. GREEN, IAN R. DUDDY & KERRY A. HEGARTY

Geotrack International Pty Ltd, 37 Melville Road, Brunswick West, Vic. 3055, Australia (e-mail: mail@geotrack.com.au)

Abstract: In areas where significant unconformities are present, palaeotemperatures derived from apatite fission-track analysis (AFTA) and vitrinite reflectance (VR) data through a vertical rock section can be used to estimate palaeogeothermal gradients and (by extrapolation to an assumed palaeo-surface temperature) amounts of exhumation (palaeo-burial). AFTA also provides a direct estimate of the timing of exhumation. These parameters can be used to reconstruct more complete histories than those based purely on the preserved rock record.

Precision and accuracy of these estimates are controlled by a range of theoretical and practical factors, perhaps the most important being the use of appropriate kinetic models. In extracting thermal history information from fission tracks in apatite, it is essential to use models that can describe variation in response between apatite grains within a sample. It is also important to recognize the limitations of the methods. AFTA and VR are dominated by maximum temperatures, preserving no information on events prior to a palaeo-thermal maximum. Recognition of this allows definition of key aspects of the history with greater precision.

Results from NW Europe define a series of regionally synchronous palaeo-thermal episodes, with cooling beginning in Early Cretaceous, Early Tertiary and Late Tertiary times. Latest results show that Early Tertiary palaeo-thermal effects in NW England can be understood as being due to a combination of higher basal heat flow and deeper burial, and emphasize the importance of obtaining data from a vertical sequence of samples. Comparison with similar results from other parts of the world suggests that events at plate margins exert a key influence on the processes responsible for regional exhumation, as recognized through Mesozoic and Cenozoic times across NW Europe.

Over the last 30 years, the importance of exhumation in the sedimentary basins and basement terrains of NW Europe has been increasingly recognized. Here we use 'exhumation' to describe the process by which rock units that were once more deeply buried are brought to shallower depths as a result of removal of overlying rocks. Although debate continues concerning precise definitions (e.g. England & Molnar 1990), most workers appear to be comfortable with this interpretation of the word, used essentially to represent the reverse of 'burial'.

The process of exhumation plays an important role in petroleum systems of many regions of NW Europe (e.g. Green *et al.* 1997; Duncan *et al.* 1998; Doré *et al.* 1999). Recognition of the process raises many questions in terms of tectonic mechanisms, etc. (e.g. Hillis 1992; Brodie & White 1994). Quantification of the effects of exhumation is therefore important for many reasons, and a variety of techniques have been applied to quantify various aspects of exhumation (see, e.g. Riis & Jensen 1992). Here we focus on palaeo-thermal methods, i.e. those based on increase of temperature with depth, specifically within the context of sedimentary basins, although the principles employed are equally applicable in basement terrains. We review the nature of the information obtained from these methods, discuss the various practical factors that affect precision and accuracy in estimating both the magnitude and timing of exhumation, and finally review latest results from the Atlantic margin of NW Europe, including comparison with other areas and some speculation on possible mechanisms.

From: Doré, A.G., Cartwright, J.A., Stoker, M.S., Turner, J.P. & White, N. 2002. *Exhumation of the North Atlantic Margin: Timing, Mechanisms and Implications for Petroleum Exploration*. Geological Society, London, Special Publications, **196**, 331–354. 0305-8719/02/$15.00

Quantifying the magnitude of exhumation using palaeo-thermal methods

Because of the progressive increase in temperature with depth within the lithosphere, palaeothermal indicators such as apatite fission-track analysis (AFTA®) and vitrinite reflectance (VR), which provide estimates of the maximum temperature attained by a rock sample at some time in the past, can also be used to assess former burial depths. Sedimentary units are progressively heated as they are buried, and begin to cool at the initiation of exhumation. AFTA and VR provide quantitative estimates of the temperature of individual samples at the palaeo-thermal maximum, immediately before the onset of cooling (as explained, e.g. by Bray *et al.* 1992; Duddy *et al.* 1994; Green *et al.* 1995). Whereas VR values provide discrete estimates of the maximum post-depositional palaeotemperature, AFTA may provide either lower or upper limits or a range of values for the maximum palaeotemperature in one, two or rarely three separate episodes (for more details, see e.g. Bray *et al.* 1992).

No other palaeo-thermal indicators are currently understood in sufficient detail to allow quantitative estimation of maximum palaeotemperatures. For some techniques, published conversions to equivalent VR values, e.g. Thermal Alteration Index (TAI), Conodont Alteration Index (CAI) and T_{max} (Waples 1985), Spore Colour Index (SCI) (Fisher *et al.* 1980), allow estimation of palaeotemperatures using the kinetics of VR response, although uncertainties in these calibrations may introduce additional error.

As described in detail by Bray *et al.* (1992), a series of palaeotemperature estimates over a range of depths in a well or borehole (or elevations of outcrop samples in mountainous terrains) allows determination of the palaeogeothermal gradient. Extrapolation of the fitted palaeotemperature profile to an assumed palaeosurface temperature then provides an estimate of the amount by which the section was once more deeply buried, i.e. the amount of section removed during exhumation (Fig. 1). (In the absence of direct palaeogeothermal gradient constraints, a range of realistic values may be assumed.)

A palaeotemperature profile can be characterized by a single value of palaeogeothermal gradient only when the profile is linear (from the surface to the base of the section); in situations where palaeotemperature profiles are markedly non-linear (see below), the analysis illustrated in Fig. 1 is not valid. However, it is important to recognize that palaeotemperatures determined from AFTA and/or VR in individual samples are independent of these considerations, and can therefore be used to constrain possible thermal history models regardless of any assumptions about the form of the palaeotemperature profile.

Factors affecting the accuracy of the magnitude of exhumation

Basic assumptions

The analysis in Fig. 1 depends critically on the assumptions that the palaeotemperature–depth profile is linear through the preserved section and that the profile can be linearly extrapolated through the removed section to the assumed palaeo-surface temperature. Explicit estimation of removed section from palaeo-thermal methods requires such assumptions to reduce the problem to a level where formal estimation of palaeogeothermal gradients and amounts of exhumation or removed section is possible. More importantly, these assumptions also allow determination of the associated uncertainties ($\pm 95\%$ confidence limits), thereby allowing rigorous and objective assessment of the range of scenarios that are consistent with the data.

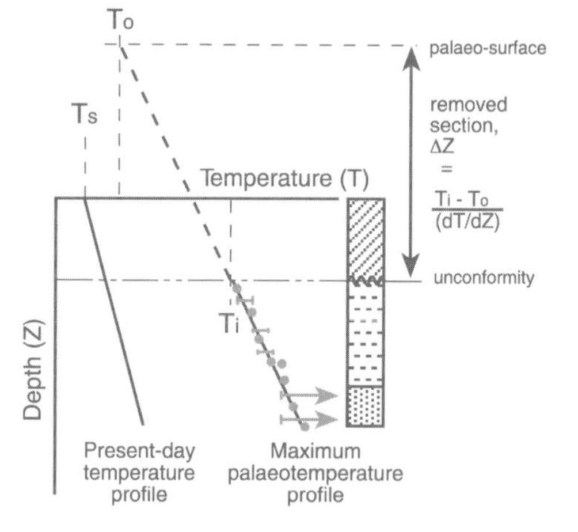

Fig. 1. Where heating is due to deeper burial, possibly combined with elevated heat flow, amounts of exhumation (or deeper burial) can be estimated by fitting a linear palaeogeothermal gradient to a series of down-hole palaeotemperature constraints and then extrapolating to an assumed palaeo-surface temperature, as shown. This approach depends critically on certain assumptions, as discussed in the text.

Although this approach has been criticized as over-simplified (e.g. Holliday 1993; Smith *et al.* 1994), results in well-controlled situations are usually highly consistent with estimates of former burial depths from other sources. For example, in the Fresne-1 well in the Taranaki Basin of New Zealand (Kamp & Green 1990), extrapolation of a linear profile fitted to palaeotemperature constraints from AFTA and VR data gives an estimate of section removed during Late Miocene basin inversion that is highly consistent with values derived from extrapolation of truncated seismic reflectors into the inversion structure (Green *et al.* 1995).

In detail, of course, the variation of temperature with depth is not exactly linear, but depends on the variation of thermal conductivity through the section, which is related, in turn, to variations in lithology (see, e.g. Gretener 1981, Fig. 2.8-1). Practical experience shows that in most situations, a linear approximation is reasonable (e.g. Deming 1994, Fig. 9.4), probably because small-scale variations in lithology serve to blur any local thermal conductivity contrasts to produce a broadly linear variation of temperature with depth. In addition, the accuracy of typical bottom hole temperature (BHT) values (usually the only available control on present-day temperatures) is often so poor that their detailed variation with depth may have more to do with recording practice, rather than real thermal structure within a sedimentary section. With typical precisions on palaeotemperature estimates of the order of 10 °C (see discussion below), typical variation in true temperatures about a linear profile usually means that a linear approximation introduces negligible additional error to the treatment.

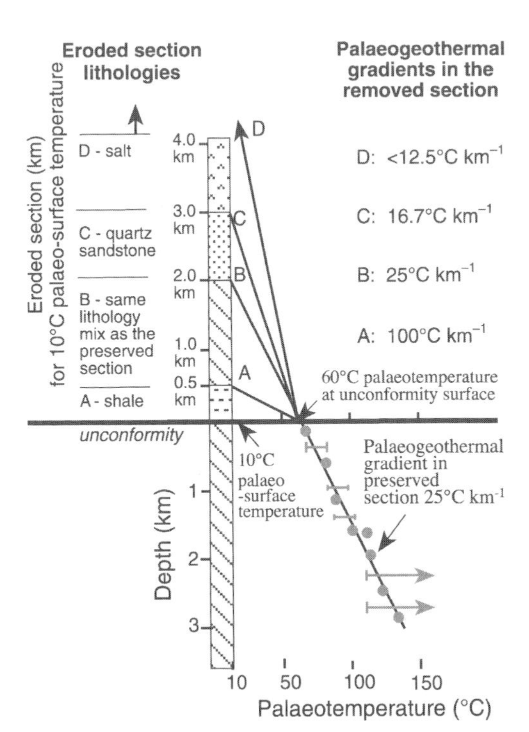

Fig. 2. This plot illustrates the influence of the thermal conductivity of the removed section on the nature of the palaeotemperature profile through that part of the section. Only where the removed and preserved sections are identical will the thermal gradient be the same throughout the entire section, but in practice the assumption of linearity appears to give reliable results (see text).

In summary, although the approach illustrated in Fig. 1 always represents some degree of approximation to the true situation, it has the advantage of providing an objective assessment, with results constrained by the measured data and simplifying assumptions explicitly stated. In situations where the approximation of linear gradients is thought to be inappropriate, the results of this approach may still provide a 'first-pass' assessment of the situation, on which more detailed treatments based on heat flow (see discussion below) can be based.

Non-linear palaeogeothermal gradients

It situations where the palaeotemperature profile is obviously non-linear, the amount of section removed cannot be estimated using the construction shown in Fig. 1. Non-linear palaeotemperature profiles may be expected in two important types of situation. First, large-scale contrasts in thermal conductivity can cause significant non-linearity in the temperature profile within the (removed and/or preserved) section. This is most pronounced when the removed section consisted of very different lithologies compared with the preserved section, as illustrated in Fig. 2. In this case, the analysis illustrated in Fig. 1 will give spurious answers depending on the nature of the eroded lithologies.

The second exception is where observed palaeo-thermal effects are not primarily related to depth of burial, but were caused by enhanced lateral heat flow; for example, as a result of flow of hot fluids in an aquifer system either within the eroded section or shallow in the preserved section. Heating caused by the passage of hot fluids can produce a variety of non-linear palaeotemperature profiles, with different forms depending on the time scale of heating (see, e.g. Ziagos & Blackwell 1986; Duddy *et al.* 1994). Most importantly, longer time scales result in a linear temperature profile below the aquifer, which is parallel to the present-day temperature profile. Erosion of the shallow strata in such cases may leave a palaeotemperature profile that mimics the effects of deeper burial, and that would give erroneously high estimates of former burial depths.

Heating effects owing to minor igneous intrusions can produce purely local anomalies as a result of contact effects, or may be more widespread if the intrusions cause circulation of heated fluids on a regional scale (e.g. Summer & Verosub 1989). Such effects can also mimic the effects of deeper burial, and identification of such effects requires regional data coverage.

Examples of heating that are probably due to the effects of hot fluids, but that mimic the effects of deeper burial, have been discussed, for example, by Green *et al.* (1993a, 1997) in the Irish Sea and by Marshallsea *et al.* (2000) in the Laura Basin of Far North Queensland.

Approaches based on heat flow

Taking into account the spatial variation of thermal conductivity through the section and temporal variation in heat flow can potentially provide more accurate predictions of the variation of temperature with depth and time. However, this procedure is also subject to major uncertainties. For example, thermal conductivity is highly sensitive to lithology (in particular, to porosity and water content), with apparently similar samples giving thermal conductivity values that vary by almost an order of magnitude (e.g. Corrigan 1991a). Such problems are particularly pronounced in sections that have been more deeply buried in the past, as thermal conductivities depend on the degree of compaction, which is not known (without direct measurements, which are available only rarely) until the amount of exhumation is determined. In addition, when considerable section has been removed no information is available on lithologies (and hence thermal conductivities) in the removed section. Therefore, the effects of greater depth of burial are difficult to predict accurately in sequences affected by exhumation when the magnitude of burial is unknown (for further discussion, see Duddy *et al.* 1991; Waples *et al.* 1992).

A further point to note in assessing approaches based on heat flow is that all heat-flow values are derived originally from down-hole temperature measurements, which are combined with thermal conductivities (usually assumed, rarely measured) to produce the reported heat-flow value. In the past, raw temperatures were often discarded, and not reported with the resulting heat-flow values. This allows the possibility that the 'measured' heat flow may then be combined with a different set of thermal conductivity values than those used in their original determination, which will result in a thermal model that does not honour the measured temperatures from which the heat flow was derived. It is not valid to simply take a quoted heat-flow value and combine it with a set of preferred or 'off-the-shelf' thermal conductivities to predict a temperature profile. The quoted heat-flow value is meaningful only when combined with the thermal conductivities from which it was derived, so as to accurately predict the present-day thermal structure.

We should also note that treatments that use a single value of thermal conductivity and heat flow as an approximation to more complex situations (e.g. Gallagher & Brown 1999) are equivalent to assuming linear geothermal gradients, and are subject to exactly the same potential limitations as the approach in Fig. 1 (and possibly more if the combined heat-flow and conductivity values do not match the original temperature data from which the heat flows were derived).

Palaeo-surface temperature

Estimates of removed section from Fig. 1 also depend on the assumed palaeo-surface temperature, independent estimates of which may or may not be available for a particular region. In the absence of such information, either the present-day value can be used, or else calculations can be performed for a range of likely values. Uncertainties associated with this parameter can be easily assessed for any given value of palaeogeothermal gradient. For instance, if the palaeogeothermal gradient was $50\,°C\,km^{-1}$, a $10\,°C$ rise in palaeo-surface temperature requires that the estimated eroded section should be reduced by 200 m.

Heating rates

The thermal history before the onset of cooling from a palaeo-thermal peak cannot be constrained by AFTA data (Green *et al.* 1989a) and as the thermal sensitivity of VR is similar to that of AFTA (Duddy *et al.* 1991, 1994, 1998) the same is true of VR data. In extracting quantitative thermal history information from AFTA and VR, it is therefore necessary to assume a heating rate to estimate a specific temperature. Typical values are usually between 1 and $10\,°C\,Ma^{-1}$. Changing the assumed heating rate by an order of magnitude is equivalent to a change of *c.* 10 °C in the required maximum palaeotemperature for both AFTA and VR (Green *et al.* 1989a) with higher heating rates requiring higher temperatures and vice versa. Thus, if assumed heating rates are systematically high, then amounts of exhumation will also be correspondingly high.

Identifying the appropriate unconformity

In sedimentary sequences, accuracy in estimating amounts of removed section as shown in Fig. 1 depends, in addition to the points discussed so far, on identification of the appropriate unconformity from which the section was removed, as the extrapolation in Fig. 1 is constructed with respect to the depth of that unconformity. In sections containing only a single major unconformity, this is straightforward, but in sections containing multiple unconformities, correct assignment may be more difficult, and erroneous assignment produces a systematic error in estimating the amount of removed section. In such cases, determining the timing of exhumation independently using AFTA (see discussion below) can provide unique insight into which unconformity represents the main phase of exhumation.

Calibration of system response

Given all of the systematic factors discussed to this point, accuracy in estimating amounts of exhumation is controlled primarily by the accuracy of the palaeotemperatures derived from AFTA and VR. This, in turn, depends on using the most reliable quantitative kinetic descriptions of system response, and it is essential to demonstrate that these descriptions are calibrated in geological situations. Estimates of exhumation derived using kinetic models that do not accurately match calibration data from well-understood situations will inevitably introduce systematic errors in estimates of exhumation.

In the studies described below, thermal history information has been extracted from AFTA data using a proprietary 'multi-compositional' kinetic model, which makes full quantitative allowance for the effect of Cl content on annealing rates of fission tracks in apatite (Green *et al.* 1996). This model was derived directly from data in geological conditions, in combination with laboratory data. VR values are converted to maximum palaeotemperatures using the kinetic model developed by Burnham & Sweeney (1989) and Sweeney & Burnham (1990). Good agreement between measured and predicted VR values (at least up to *c.* 1.0%) in sediments that have undergone progressive burial and are now at their maximum post-depositional palaeotemperatures confirms the validity of the kinetic model.

Use of inappropriate system kinetics: apatite fission-track analysis

The first quantitative kinetic model for fission-track annealing in apatite to gain widespread acceptance was that of Laslett *et al.* (1987). This was based purely on laboratory annealing experiments in a single, compositionally uniform apatite. Comparison of predictions based on this model with data from controlled geological conditions (Green *et al.* 1989a) showed that the predictions were broadly consistent with the measured data. But it was also recognized at an early stage in the development of AFTA that apatites of different composition anneal at different rates, with Cl content apparently providing the dominant influence (Green *et al.* 1985). This effect produces a variation in the degree of fission-track annealing in different apatite grains within a single sample, which cannot be described using a kinetic model based on a single apatite species such as the Laslett *et al.* (1987) model. The 'multi-compositional' model (Green *et al.* 1996) introduced in the previous section has the same general form as the Laslett *et al.* (1987) model, but uses constants that vary systematically with Cl content so as to reproduce the observed within-sample variation.

Other workers have adopted an alternative approach to dealing with variation in annealing kinetics between apatite species (Carlson *et al.* 1999; Donelick *et al.* 1999; Ketcham *et al.* 1999). However, the performance of these models in geological conditions has yet to be rigorously demonstrated. A number of other kinetic models based on laboratory annealing of mono-compositional apatites have also been published (Carlson 1990; Crowley *et al.* 1991) but their predictions are not compatible with data from controlled geological situations, which has precluded routine use.

In addition, it is widely accepted that the Laslett *et al.* (1987) model under-predicts the degree of annealing observed at low temperatures (e.g. Vrolijk *et al.* 1992). Because of this, thermal history solutions derived from apatite fission-track data using the Laslett *et al.* (1987) model invariably suggest up to 30 °C or more of Late Tertiary cooling, which is generally regarded as an artefact of the low-temperature behaviour of this kinetic model. An added contribution to this artefactual cooling may well be due to compositional variation, as the majority of apatite grains anneal more rapidly than the Durango apatite on which the Laslett *et al.* (1987) model was derived. Thus, track lengths in such grains will be shortened to a greater degree than expected in apatite of the composition to which the kinetic model directly relates, inevitably resulting in the need for a late, anomalous cooling episode.

The 'multi-compositional' model (Green *et al.* 1996) is based on both laboratory and geological data and provides a much better fit to control data at lower temperatures, as well as incorporating within-sample variation. Thus, thermal histories derived from this model do not suffer from the 'Late Tertiary cooling artefact' that characterizes thermal history solutions derived using the Laslett *et al.* (1987) model.

Despite the problems discussed so far, use of the Laslett *et al.* (1987) model remains widespread in many studies based on analysis of fission tracks in apatite (many of which appear to show Late Tertiary cooling!). Use of this model to estimate palaeotemperatures from fission-track data in apatite will give systematically high temperatures below *c.* 70 °C, and systematically low temperatures at around 100–110 °C. For this reason, use of this model will result in anomalously low palaeogeothermal gradients and anomalously high amounts of removed section. But most importantly, it will fail to adequately describe variation within an apatite population from a single sample, introducing serious errors to thermal history solutions that may be widely erroneous.

To summarize, accuracy of palaeotemperature determination from AFTA is critically dependent on use of kinetic models that describe variation between apatite grains within individual samples, for which incorporation of compositional influences is essential. Further discussion of the kinetic response of the AFTA system is provided below in the section 'Quantifying timing of exhumation using AFTA', as estimates of timing and palaeotemperature obtained from AFTA are inextricably linked.

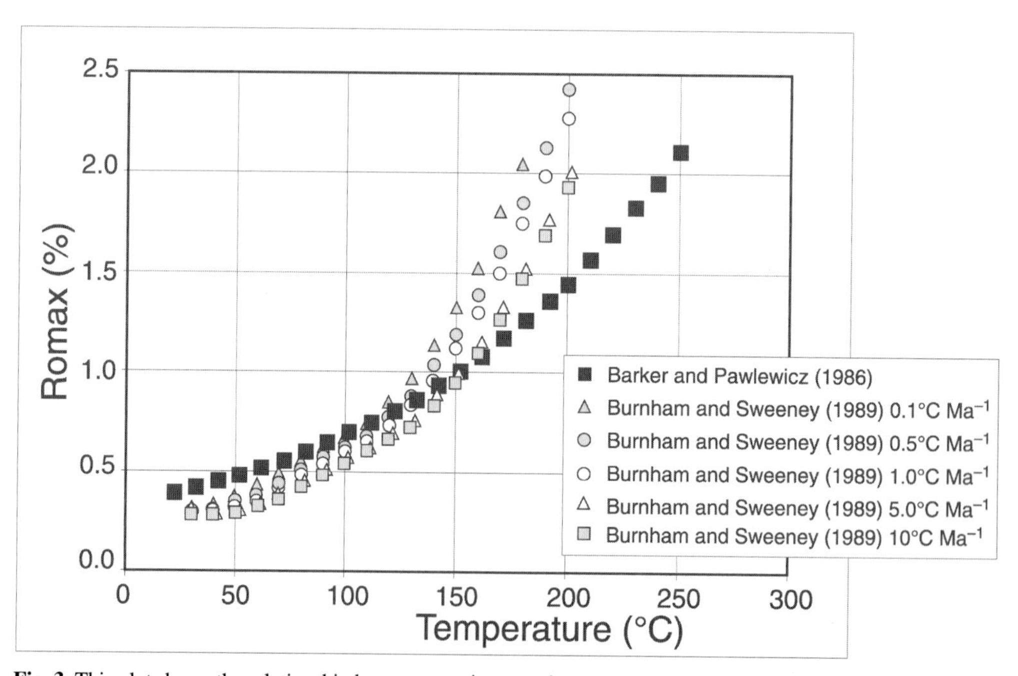

Fig. 3. This plot shows the relationship between maximum palaeotemperatures and VR values predicted from the Burnham & Sweeney (1989) model, using various heating rates, and for the Barker & Pawlewicz (1986) model, which ignores the influence of time and relates reflectance directly to palaeotemperature. The two approaches show very different behaviour, particularly at reflectances above 1% and below 0.5%. For this reason, use of the Barker & Pawlewicz model will generally produce higher palaeogeothermal gradients, and lower amounts of removed section, compared with the Burnham & Sweeney model.

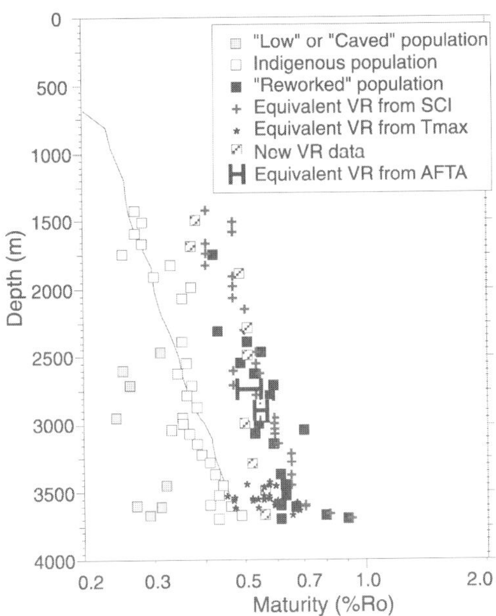

Fig. 4. Various sets of maturity data in an offshore well are plotted against depth (RKB; below Kelly bushing). An earlier VR dataset (grey, white and black squares) plus VR data from new analyses (diagonal stripes) are shown, together with ranges of equivalent VR levels derived from AFTA in two samples from this well, plus equivalent VR values derived from SCI and T_{max} values. In the earlier VR analyses, reflectance measurements were assigned to various populations by the analysts, as indicated in the legend. In the new analyses, identification of indigenous vitrinite was based on petrographic inspection of polished sections. The continuous line shows the VR profile expected if samples throughout the section are currently at their maximum temperature since deposition. There is a clear mismatch between those values from the earlier analyses originally interpreted as representing the indigenous vitrinite population, on one hand, and the new VR values plus the equivalent VR values defined by AFTA, SCI and T_{max} data on the other, which all define a consistently higher trend. Whereas the original data suggest that all units throughout the well are currently at their maximum temperatures, the new VR data and the AFTA, SCI and T_{max} results show that in fact maturity levels are higher than previously thought by *c.* 0.2 or 0.3%, and most units have been hotter in the past. The higher trend is sub-parallel to the predicted profile, suggesting that heating was due to deeper burial, followed by cooling as a result of exhumation. It is also worth noting that the occurrence of three separate sub-parallel trends denoting caved, *in situ* and reworked vitrinite, as in the earlier analyses shown here, is highly unlikely, and more probably arises because of measurement of macerals other than true vitrinite. The occurrence of three sub-parallel trends is usually a reliable sign of maceral misidentification.

Use of inappropriate system kinetics: vitrinite reflectance

In the years since vitrinite reflectance was adopted as a standard measure of maturity for hydrocarbon exploration, a number of kinetic models have been suggested for the evolution of reflectance as a function of temperature and time. The Burnham & Sweeney (1989) model is undoubtedly the most successful, in terms of accurately reproducing observed values in well-controlled situations (see also Morrow & Issler 1993). However, despite the success of this model, other treatments remain in common use, and use of inappropriate kinetic descriptions for VR can provide serious systematic errors in estimating palaeotemperatures and thus amounts of exhumation (or former depths of burial).

At one extreme, understanding of the system kinetics may be ignored completely. For instance, estimates of removed section have been obtained historically by extrapolating VR profiles to a value representing untransformed kerogen as deposited at the surface, typically around 0.2% (e.g. Dow 1977). However, experience shows (Cook, pers. comm.) that untransformed vitrinite, as deposited in a sediment, may have a reflectance as high as 0.32%. Thus, extrapolation to 0.2% should invariably overestimate amounts of removed section.

Problems with another approach are illustrated in Fig. 3, which shows the relationship between maximum palaeotemperatures and VR values predicted from the Burnham & Sweeney (1989) model, using a variety of different heating rates, and from the Barker & Pawlewicz (1986) model, which ignores the influence of time and relates reflectance directly to palaeotemperature. The two approaches show very different behaviour, particularly at reflectances above 1% and below 0.5%. For this reason, use of the Barker & Pawlewicz model will generally produce higher palaeogeothermal gradients, and lower amounts of removed section, compared with the Burnham & Sweeney model.

Integration of results from different methods

If the kinetic algorithm used to extract thermal history information from the AFTA and/or VR data were to be systematically in error, then palaeotemperatures from one method or the other would be consistently either low or high. Comparative results from a wide range of situations (e.g. Duddy *et al.* 1994, 1998; Green *et al.* 1995, 1997) show that the two techniques give highly consistent palaeotemperatures. As the calibration of each system has been carried out independently, this suggests that both sets of values can be regarded as reliable, and that any systematic inaccuracy as a result of errors in system response are not significant.

Analytical problems

Analytical problems in VR analyses are well known, including problems such as suppression or retardation (e.g. Carr 2000). Such problems usually affect a particular horizon (e.g. suppression in hydrogen-rich source-rock facies) and can often be recognized by local departures from an overall trend. A variety of approaches have been proposed for dealing with reflectance suppression, including utilization of fluorescence information (e.g. Wilkins *et al.* 1992; Newman 1997), and development of specific kinetic models (Carr 1999, 2000). Combination of palaeotemperatures derived from different techniques (e.g. AFTA, as discussed above) can also allow detection of anomalous VR values.

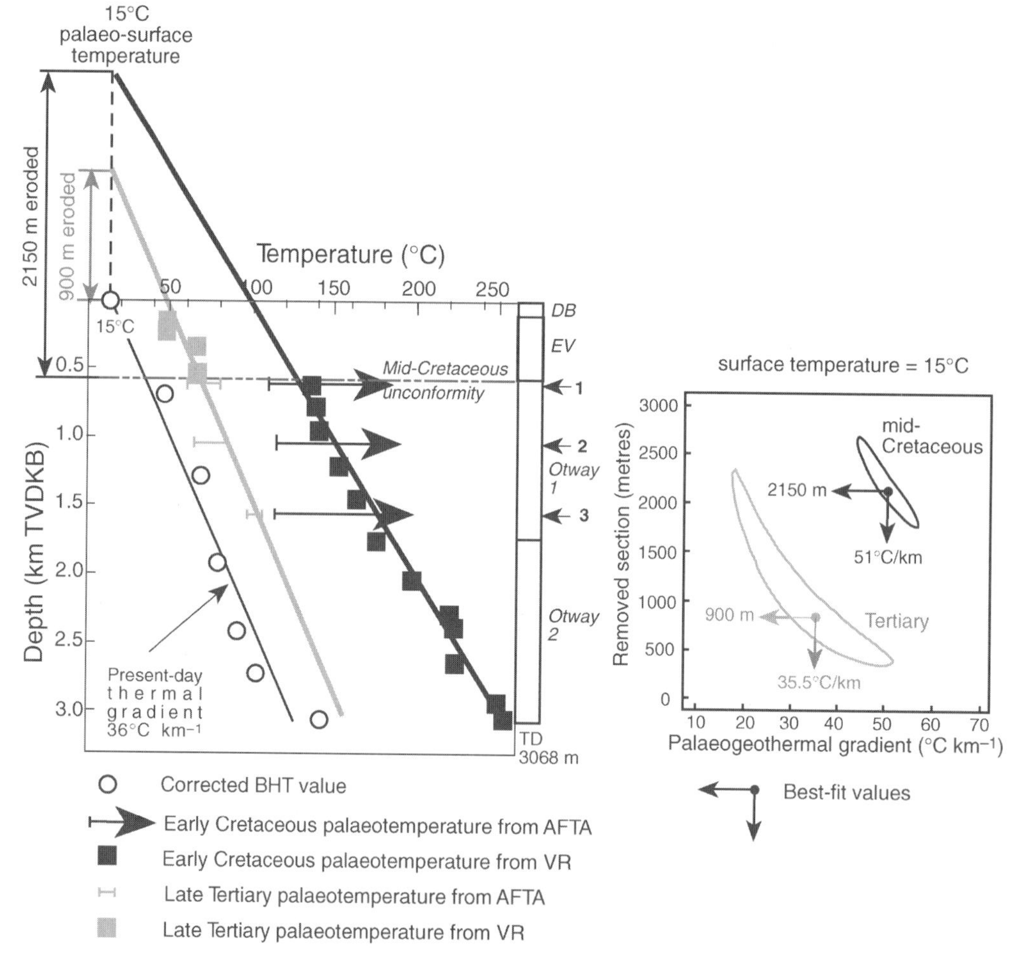

Fig. 5. Palaeotemperature profiles from AFTA and VR data in the Anglesea-1 well. Left: AFTA data clearly reveal two palaeo-thermal episodes, as shown, with palaeotemperature constraints from each method for the two events plotted against depth (RKB). AFTA shows that cooling began in the earlier episode between 110 and 95 Ma and in the later episode between 60 and 10 Ma. Unconformities are present for mid- to Late Cretaceous and Late Tertiary time, suggesting that both episodes can be explained by deeper burial. Palaeotemperature profiles characterizing both episodes are linear, supporting this conclusion. DB, Demons Bluff Formation; EV, Eastern View Formation. Right: statistical analysis of the palaeotemperature constraints defines the range of allowed values of palaeogeothermal gradient and removed section during each episode of exhumation within 95% confidence limits, as shown by the contoured regions, together with the best-fit values.

However, a possibly more common problem, which may be more difficult to recognize, concerns the incorrect assignment of the *in situ* vitrinite population throughout an entire well section. This can result in systematically low estimates of palaeotemperature (and therefore palaeo-burial) and maturity levels. These problems are illustrated in Fig. 4, in which a series of datasets from an offshore well are compared. AFTA data in two samples from this well show consistent evidence of higher temperatures before Miocene cooling, but existing VR data indicated that all units were now at their maximum temperatures since deposition. Results of new VR analyses, carried out to resolve this mismatch, are highly consistent with the equivalent maturity values indicated by the AFTA data, and also with equivalent maturity levels derived from SCI and T_{max} values within the original dataset.

Further investigation of the original VR dataset revealed a number of populations within the measurements, as shown in Fig. 4, including a population of higher reflectances, originally interpreted as 'reworked' vitrinite. Significantly, the new VR values and the equivalent maturity values from AFTA, SCI and T_{max} data are highly consistent with these 'reworked' vitrinite values. All these values define a consistent trend in Fig. 4, consistently higher than those originally attributed to the indigenous population. Thus, these original values are too low and values originally designated as reworked, together with the AFTA, SCI and T_{max} values, provide the most reliable indication of true maturity levels in this well.

Such experience is remarkably common, and illustrates a general tendency amongst many geologists to underestimate the importance of exhumation (or deeper burial). Thus, the reflectance population that falls closest to the values expected on the basis of the preserved section is most commonly identified as the indigenous population. These observations again highlight the importance of combining information on palaeo-thermal effects from different techniques, which provides a more objective assessment.

Factors affecting the precision of the magnitude of exhumation

Availability of palaeotemperature constraints over a range of depths

The main factor affecting precision in estimating amounts of removed section is the range of depths over which palaeotemperature constraints are available, as this controls the degree to which the palaeogeothermal gradient is constrained, which, in turn, controls the precision of the estimate of removed section (Fig. 1). An extreme case of this is where data are available only from outcrop samples, in which case no estimate of the palaeogeothermal gradient is possible and a value must be assumed.

Uncertainties in the fitted palaeogeothermal gradient are magnified because of the extrapolation required to estimate the amount of removed section. This also causes a correlation between allowed values of palaeo-gradient and removed section, such that high palaeo-gradients require smaller amounts of removed section and vice versa. Statistical techniques allow definition of the range of allowed values (within $\pm 95\%$ confidence limits) for each parameter.

These points are illustrated in Fig. 5 using measured data from the Anglesea-1 well located in the Otway Basin of SE Australia (Duddy 1994). Combined AFTA and VR data from this well define two discrete palaeo-thermal episodes. The palaeotemperature profile characterizing the mid-Cretaceous episode is defined by VR data over a depth interval of *c.* 2.5 km (AFTA data provide only minimum limits), which results in relatively tight constraints on the allowed ranges of palaeo-geothermal gradients (43–57 °C km^{-1}) and removed section (1750–2700 m). In contrast, Tertiary palaeotemperature constraints from AFTA and VR are available over a depth of only 1.5 km, and provide much broader ranges of allowed values for each parameter (18–52 °C km^{-1} and 400–2300 m, respectively). These examples typify the levels of precision available from this approach. However, it should be noted that although the ranges of allowed values are broad for both events, for any particular value of gradient the range of allowed values of removed section will be much smaller, typically around 200–500 m in Fig. 5.

Nature of palaeotemperature constraints

Other factors affecting the precision of estimating amounts of exhumation (removed section) include the quality of the VR and AFTA data (poor-quality data will generally provide only broad palaeotemperature constraints, which provide little control on the palaeo-gradient) and the availability of AFTA and VR data through the section, coupled with the nature of the AFTA constraints (which depends to some extent on the nature of the underlying thermal history). For example, if all AFTA samples were totally or near totally annealed before cooling, such that only a lower limit to the maximum palaeotemperature is available from AFTA, useful constraints on the palaeogeothermal gradient will be available only if VR data are also available through the section. Results from the West Newton-1 well (NW England) (Green *et al.* 1997) illustrate this point. Availability of VR data through the Carboniferous section in this well, where AFTA samples were totally annealed before Early Tertiary cooling, results in tight constraints on the palaeo-gradient and the degree of exhumation.

Quantifying timing of exhumation using AFTA

The timing of exhumation events has traditionally been inferred from regional geological evidence, but a major advantage of AFTA is that it provides an independent estimate of the time at which a sample began to cool from its maximum palaeotemperature (or a subsequent peak value). This estimate is derived from the AFTA data alone, and therefore provides an objective measure of timing. If cooling can be attributed to exhumation, then AFTA can define the timing of the onset of exhumation.

In considering the accuracy and precision of timing estimates from AFTA, it is important to recognize how this information is coded in the AFTA data, which in turn requires explanation of the thermal response of fission tracks in apatite. AFTA is based on 'fission-track annealing': the progressive reduction in track length as a function of temperature and time (Green *et al.* 1986, 1989b). This reduction in track length is also manifested as a reduction in fission-track age (Green 1988; Green *et al.* 1989b). As illustrated in Fig. 6, new tracks are produced throughout geological time, as a result of spontaneous fission of uranium impurity atoms within the apatite crystal lattice. In a sample that is heated and then cooled (e.g. Sample 1 in Fig. 6), tracks produced up to the time at which cooling begins will be shortened to a length determined by the maximum palaeotemperature, whereas tracks produced after the onset of cooling will be longer. The time at which cooling

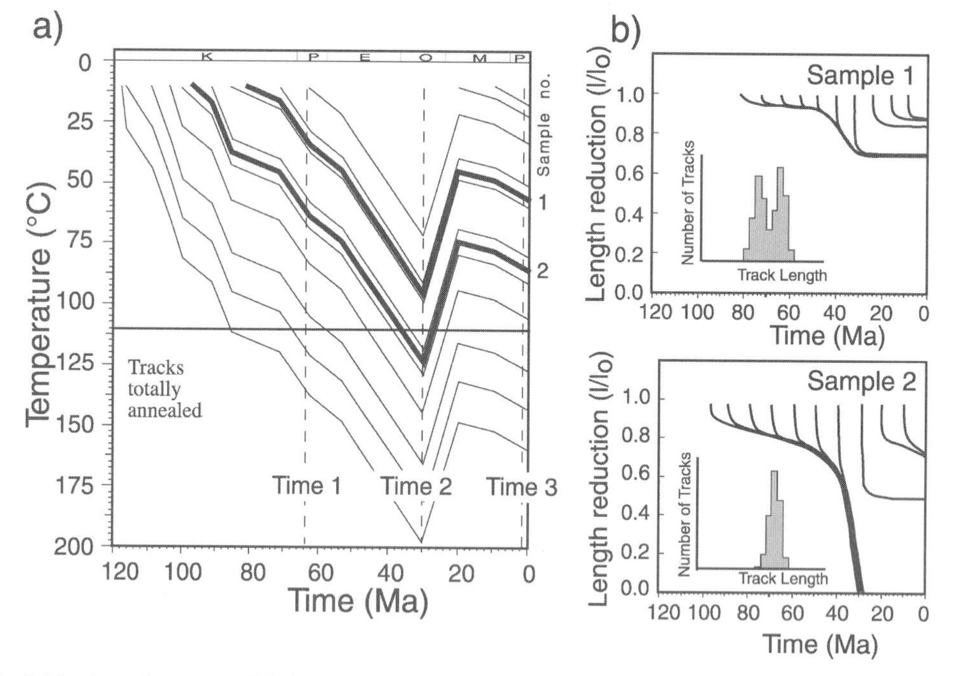

Fig. 6. The thermal response of fission tracks in apatite to geological thermal histories is well understood, based on a series of observations in laboratory and geological conditions (Green *et al.* 1986, 1989a; Laslett *et al.* 1987; Duddy *et al.* 1988). In a thermal history that involves heating and subsequent cooling, followed by minor reheating (**a**), fission tracks in apatite respond in a characteristic way (**b**), which allows the main features of the history to be constrained. Details are discussed in the text. Understanding the way in which fission tracks in apatite respond to heating and cooling allows us to focus on those aspects of the history that can be constrained (i.e. the maximum palaeotemperature and the time of cooling), whereas the history before the onset of cooling cannot be constrained.

begins, in relation to the overall duration of the history, determines the proportion of shorter to longer tracks, and the maximum palaeotemperature determines the length of the shorter peak. Because an apatite may contain tracks at the time of deposition in a sediment, this track length information must be combined with a fission-track age measurement to establish the total duration over which tracks have been retained.

In samples that exceed a critical temperature limit, all tracks are 'totally annealed' (i.e. the track length is reduced to zero, as for Sample 2 in Fig. 6). Such samples retain tracks only after cooling below this limit (which depends on the composition of the apatite), and provide only a minimum estimate of the maximum palaeotemperature. However, such samples usually provide tight constraints on the time of cooling, through the fission-track age.

Determining the time at which cooling begins from AFTA data is thus inextricably linked to extraction of palaeotemperature information, as both aspects of the overall thermal history solution exert critical controls on the measured AFTA parameters.

Factors affecting the accuracy and precision of timing estimates from AFTA

Extracting thermal history solutions from AFTA data involves modelling AFTA parameters (Green *et al.* 1989b) through various thermal history scenarios, based on a detailed knowledge of the kinetics of the annealing process, so as to define the range of maximum palaeotemperatures and timing of cooling for which predictions are consistent with the measured data. Figure 7

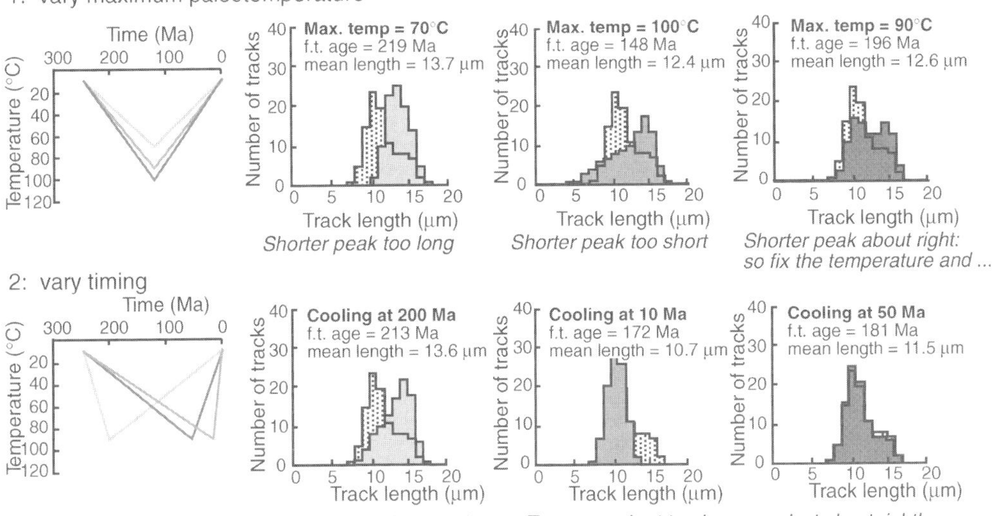

Fig. 7. Extracting thermal history solutions from AFTA data involves modelling AFTA parameters (Green *et al.* 1989a) through various thermal history scenarios, using formal statistical procedures to define the range of maximum palaeotemperatures and timing of cooling for which predictions are consistent with the measured data. This process requires a detailed knowledge of the kinetics of the annealing process. This synthetic example, based on a notional mono-compositional example, for simplicity, illustrates the basic principles involved. Cooling from a maximum palaeotemperature of 90 °C beginning at 50 Ma gives the best fit to the data.

illustrates the basic principles involved, based on a simple mono-compositional example. By modelling the AFTA parameters through likely thermal history scenarios, and comparing predictions with measured data, the range of thermal history solutions compatible with the data within 95% confidence limits can be defined.

In practice, annealing kinetics depends on chlorine content. As illustrated in Fig. 8, thermal history solutions can be extracted from data broken down into discrete compositional groups, using separate kinetics for each group. The final thermal history solution should not only match the pooled data for the whole sample but also the within-sample variation of fission-track age and length with wt % Cl, which provides additional information by which the final thermal history solution can be constrained.

The variation in the track length distributions in Fig. 8 between compositional groups should be compared with that shown by Green (1986) for various outcrop samples from Northern England. Individual compositional groups within the single sample illustrated in Fig. 8 span a

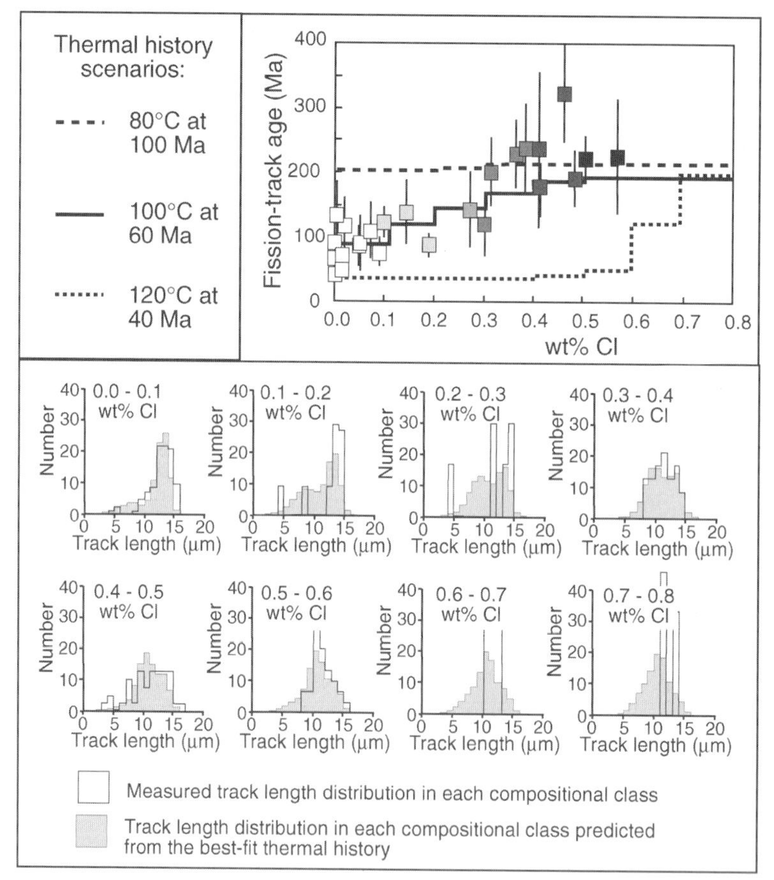

Fig. 8. Influence of Cl content on AFTA data. In detail, fission-track annealing kinetics in apatite depends on chlorine content, and thermal history solutions are extracted from data broken down into discrete compositional groups, using separate kinetics for each group. The final thermal history solution should match not only the pooled data characterizing the sample as a whole but also the variation of fission-track age and length with wt % Cl. The data shown here were measured in a sample of Triassic sandstone from NW England. In the upper plot, fission-track ages of individual apatite grains are plotted against Cl content (measured by electron microprobe). Also shown are predicted patterns of fission-track age and mean track length v. wt % Cl for three different thermal histories, as indicated. A maximum of 100 °C at 60 Ma clearly gives the best fit to the data. The lower plot shows measured length distributions binned into discrete wt % Cl intervals, together with predicted track length distributions corresponding to the best-fit thermal history, which also give a good match to the measured data. It is essential, in extracting thermal history information from fission-track data in apatite, to use kinetic models that incorporate the effect of this within-sample variation.

similar range of annealing shown by outcrop samples over a wide area. Just as it is not acceptable to extract a single thermal history solution from a collection of outcrop samples showing varying degrees of annealing, it is equally unacceptable to attempt to extract such information from data within a single sample showing an equivalent spread.

The accuracy of timing estimates from AFTA depends mainly on use of an appropriate kinetic model, as above, but precision often depends critically on the magnitude of palaeo-thermal effects. In samples that have been heated only to around 70 °C or below, the degree of length reduction in tracks formed before the onset of cooling is relatively small. In such cases, the shorter component can often not be resolved from the longer tracks formed after cooling, and a wide range of timings will be consistent with the data. Conversely, in samples that have been heated to *c.* 90 or 100 °C, the degree of length reduction is severe and the shorter tracks can easily be resolved, allowing much tighter constraints on timing. Samples in which all tracks were totally annealed before cooling (maximum palaeotemperature typically >110–120 °C) generally give the tightest timing constraints, although this depends to some extent on the history after cooling.

Uncertainty in the final estimate of timing can depend to some extent on the timing itself. If cooling (exhumation) began early in the history, the effects may be difficult to discriminate from pre-depositional effects (reflecting the history of sediment source terrains), and a large uncertainty may result. Conversely, if cooling began relatively late in the history, the majority of tracks will be affected and timing can be determined with greater precision.

In practical terms, poor apatite yields and/or low uranium contents (producing low track counts for age determination and small numbers of track lengths) invariably result in poor constraints on timing. The Poissonian uncertainty on track counts (fractional error of $1/N$) provides a basic limit to the final error on the fission-track age of an individual apatite grain and/or sample, which in turn limits the analytical uncertainty in the timing of cooling. Even in the highest quality analyses, it is unusual for more than *c.* 1000 tracks to be counted, and thus uncertainties of several per cent are to be expected in the best case. However, as stressed elsewhere, counting and measuring large numbers of tracks are not as important as incorporating within-sample variation into the analysis.

Limitations

Thermal history resolution

A fundamental characteristic of both the AFTA and VR techniques is that they are dominated by maximum temperatures, as discussed above. For this reason, these techniques are subject to the fundamental limitation that they can provide no information on the approach to the palaeo-thermal maximum. (This is illustrated by the track length shortening trajectories for Sample 1 in Fig. 6, in which all tracks formed before the onset of cooling are shortened to the same degree.) AFTA can provide information on the history following the onset of cooling, because of the continuous production of tracks through time. For example, by reference to Sample 2 in Fig. 6, if tracks formed after the initial cooling were again shortened during a later heating episode and the sample then cooled and retained longer tracks to the present day, the resulting length distribution would preserve evidence of this episode, and data from this sample would thus constrain two discrete events. However, because of the inherent spread in the distribution of track lengths as a function of the degree of annealing (Green *et al.* 1986), there is a limit to the amount of information that can be obtained. In practice, resolution of two episodes of heating and cooling is often the limit allowed by even the highest quality data, although Green *et al.* (2001a) have illustrated a situation where AFTA data from a single sample allow resolution of three discrete episodes. In contrast, VR data are not sensitive to the history after the onset of cooling, except in situations characterized by very slow cooling or where a considerable time is spent at or near the maximum temperature.

For these reasons, the approach described here is based on using AFTA data to rigorously define the maximum or peak palaeotemperature and timing of cooling in one or two (rarely three) discrete episodes of heating and cooling, using assumed heating and cooling rates (see discussion above). AFTA data do not contain sufficient information to allow definition of the entire thermal history of a sample, as attempted by some workers (e.g. Corrigan 1991b; Lutz & Omar 1991; Gallagher 1995; Willett 1997; Ketcham *et al.* 2000), or even the whole of the history after the onset of cooling. Such approaches invariably result in confidence limits that are so wide that the thermal history solutions provide no useful constraints. This arises because the effects of variation within one part of the history can be compensated by events at other times. Only by focusing on those aspects of the thermal history to which the data are sensitive (i.e. maximum or peak palaeotemperature and timing of cooling, under assumed heating rates), and thereby reducing the problem to a manageable number of variables, can useful constraints be obtained.

Onset v. duration of cooling or exhumation

Because of the limitations discussed above, AFTA usually defines only the onset of cooling (as a result, for example, of exhumation) with any precision, and cannot provide very precise constraints on the duration of episodes of cooling or rates of cooling, although broad limits on the magnitude and duration of cooling phases may be possible. Examples are discussed below in the context of results from the NW European Atlantic margin.

Depth resolution v. thermal resolution

Palaeotemperatures from AFTA and VR data are typically accurate to within 10 °C, and precision is usually similar to or better than this. This compares favourably with present-day temperature assessment, which is, at best, probably accurate to no more than ± 10 °C. However, as a consequence of thermal gradients typically between 20 and 60 °C km^{-1}, even under ideal circumstances uncertainties in resulting estimates of eroded section are usually several hundreds of metres. For a palaeogeothermal gradient of *c.* 30 °C km^{-1}, 10 °C is equivalent to 300 m of inherent uncertainty in the estimate of section removed (500 m for a gradient of 20 °C km^{-1}). This emphasizes that thermal history reconstruction is not a particularly precise method of estimating amounts of removed section. However, as illustrated in Fig. 5, greater precision can be obtained by combining data from multiple samples over a range of depths, and allowed ranges can be rigorously defined.

Reburial

A fundamental limitation of all palaeo-thermal techniques that are dominated by maximum temperatures is that the effects of early heating are obscured by later heating when the magnitude of the more recent event exceeds that of the earlier episode. For example, if a sedimentary section is reburied, following an earlier episode of exhumation, the palaeothermal effects associated with cooling during exhumation are progressively 'overprinted'. Therefore, in a section affected by multiple heating episodes only the maximum palaeotemperature event will be clearly revealed by most techniques, including VR. In contrast, AFTA is capable of detecting lesser magnitude events (from the reduction in length of tracks formed before each event), but only if they occur after the maximum palaeotemperature event (and only if they are sufficiently separated in time and temperature).

In areas affected by repeated cycles of burial and exhumation, earlier burial phases involving lower burial depths can be revealed only if the palaeogeothermal gradients at the time were higher than during later episodes. Results from Inner Moray Firth well 12/16-1 (Green *et al.* 1995) provide an example of this situation.

Latest results from the NW European Atlantic margin

Early Tertiary palaeo-thermal effects in NW England

Earliest evidence of the importance of Early Tertiary palaeo-thermal effects in the UK region came from the application of AFTA to outcropping Caledonian basement from the Southern Uplands of Scotland (Hurford 1977) and the Lake District of NW England (Green 1986). As samples were not analysed from vertical sequences, these studies provided no insight into the magnitude of palaeogeothermal gradients and the origin of the observed Early Tertiary palaeotemperatures.

Similar effects were subsequently recognized regionally across Northern England and the Irish Sea (Lewis *et al.* 1992; Green *et al.* 1993b), but although samples were analysed from well sequences, total or near-total annealing of all AFTA samples as a result of the high Early Tertiary palaeotemperatures at sea-bed or outcrop level again precluded any useful constraints on palaeogeothermal gradients. Mainly because of the regional extent of the effect, coupled with the lack of evidence (at that time) for elevated basal heat flow, an explanation in terms of heating primarily caused by deeper burial was considered most likely, with prevailing palaeogeothermal gradients close to present-day values, and with subsequent cooling resulting largely from uplift and erosion (exhumation). On this basis, Lewis *et al.* (1992) calculated that for an assumed palaeogeothermal gradient of 30 °C km^{-1}, the observed palaeotemperatures required *c.* 3 km of section to have been removed over much of NW England.

Such results were in marked contrast to the prevailing consensus view, and provoked considerable comment and criticism. In particular, Holliday (1993) estimated a probable range of 700–1750 m for the amount of former Mesozoic cover over the Lake District and Pennine blocks, and suggested that higher amounts were not consistent with geological information from surrounding regions.

Later studies by Cope (1994) and Chadwick *et al.* (1994) suggested a growing acceptance of the general concept of kilometre-scale Tertiary

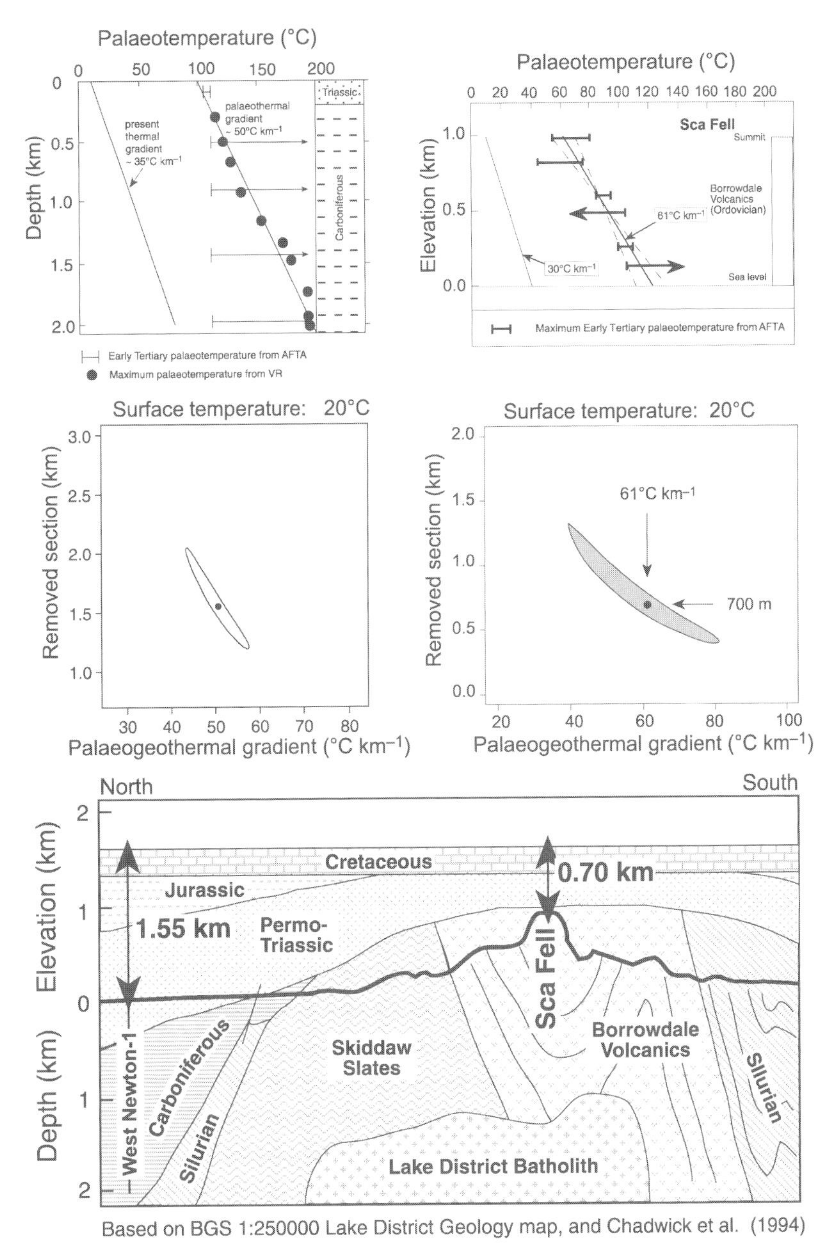

Fig. 9. Summary of Early Tertiary palaeo-thermal effects in NW England. AFTA and VR data from the West Newton-1 well (upper left) and an elevation section from Sca Fell (upper right) allow reconstruction of the region immediately before the onset of exhumation in Early Tertiary time, as shown in the section (lower). For further details, see Green *et al.* (1995, 1999) and Green (2001).

exhumation, but are themselves open to criticism. For example, the concept of a domal erosional event centred in the Irish Sea just north of Anglesey (Cope 1994) is not consistent with published palaeotemperature maps (Green *et al.* 1993b, 1997), which show Early Tertiary palaeothermal effects of maximum magnitude in the Lake District, decreasing southwards into North Wales and westwards to the Isle of Man and Northern Ireland. The erosion map of Cope (1994) also fails to reproduce the observed Early Tertiary palaeotemperature maximum in the Cleveland Basin reported by Green *et al.* (1993b). Also, over most of the Lake District Block, the estimates of removed section reported by Chadwick *et al.* (1994) are outside the limits that Holliday (1993) considered consistent with geological constraints, being >1750 m over all but a small portion in the centre of the region and increasing to the north, west and south to values well outside the range of values considered acceptable. Until recently, therefore, the origin of the Early Tertiary palaeotemperatures revealed by AFTA in Northern England has remained enigmatic, with the nature of the underlying processes unclear.

Subsequent to these studies, an increasing focus on hydrocarbon exploration in the Irish Sea and adjacent regions led to major improvements in definition of thermal history styles in the region, the first signs of which were reported by Green *et al.* (1993a). In addition to recognizing the occurrence of Mesozoic palaeo-thermal episodes, particularly to the west of the region (Green *et al.* 1997), the increasing availability of AFTA and VR data over a range of depths, combined with improved understanding of fission-track annealing kinetics (as discussed above), resulted in much tighter constraints on the nature of Early Tertiary palaeo-thermal effects, revealing a major difference from south to north. Wells from the south of the basin define low palaeogeothermal gradients suggestive of heating related to hot fluid circulation, whereas the northern parts of the basin were characterized by much higher palaeo-gradients, suggesting a major contribution of heating as a result of elevated basal heat flow.

Evidence from the north of the basin are typified by AFTA and VR data from the onshore West Newton-1 well (Green *et al.* 1997, 1999). Assuming that both AFTA and VR data represent the same palaeo-thermal episode, they define an Early Tertiary palaeogeothermal gradient of *c.* 50 °C km^{-1} (Fig. 9), compared with the present-day gradient of *c.* 35 °C km^{-1}, implying that the Early Tertiary heat flow was up to 50% higher than the present-day value. Extrapolating the palaeotemperatures to a palaeo-surface temperature of 20 °C requires around 1.55 km of post-Early Triassic section removed by Tertiary exhumation. However, as explained by Green *et al.* (1997, 1999), data from West Newton-1 would allow an alternative interpretation whereby the VR data represent an earlier episode (perhaps at latest Carboniferous time) in which the Carboniferous units reached their maximum post-depositional palaeotemperatures, whereas Early Tertiary palaeotemperatures in the Carboniferous section were somewhat lower.

Recently, AFTA data in a series of outcrop samples from various elevations in the vicinity of Sca Fell have provided confirmation of elevated palaeogeothermal gradients during Early Tertiary time (Green *et al.* 1999; Green 2001). In this region, characterized by the highest elevations in England at just under 1000 m, and located only *c.* 25 km to the south of the West Newton-1 well, Early Tertiary palaeotemperatures obtained from AFTA data define a palaeogeothermal gradient of 61 °C km^{-1} and require only *c.* 680 m of section removed since Early Tertiary time assuming a palaeo-surface temperature of 20 °C (Fig. 9). The difference of *c.* 870 m in amounts of removed section between the location of the West Newton-1 well and Sca Fell is close to the *c.* 950 m difference in elevation between the (near-coastal) location of the West Newton-1 well and the summit of Sca Fell (particularly bearing in mind typical uncertainties of ±50–100 m).

Thus, these new observations provide a self-consistent framework within which Early Tertiary palaeo-thermal effects in NW England can be understood as being due to a combination of higher basal heat flow and deeper burial, with the amount of section removed during Tertiary exhumation generally varying between *c.* 0.7 km (from mountain peaks) and *c.* 1.6 km (from coastal plains and glacial valleys near sea level) over the region (Fig. 9). Amounts of removed section required to explain the observed Early Tertiary palaeotemperatures are entirely consistent with the conclusions of Holliday (1993) based on regional geological trends. Thus, these latest results provide a geologically plausible mechanism for the origin of the observed Early Tertiary palaeo-thermal effects in NW England (although the underlying cause of the elevated heat flow remains unclear), illustrating the benefits to be gained from analysing data over a range of elevations.

New results from Central England

Green (1989) reported AFTA data in a suite of samples from outcrops and exploration wells on the East Midlands Shelf (EMS) and the Pennine High that revealed Early Tertiary palaeo-thermal effects, interpreted as representing greater depths of burial before Tertiary exhumation. Bray *et al.* (1992) subsequently incorporated VR data with these AFTA data and provided a more quantitative analysis of the palaeotemperature data, which supported an explanation of heating primarily as a result of deeper burial, with amounts of removed section varying between 1 and 2 km across the East Midlands Shelf. Holliday (1993) and Smith *et al.* (1994) suggested that these estimates of section removed from the East Midlands Shelf derived from AFTA were too high by *c.* 1 km.

Further work has been carried out recently to investigate various aspects of these results. This work has included reanalysis of various samples analysed by Green (1989) to incorporate subsequent advances in understanding of annealing kinetics, as described above. New AFTA and VR data from Central England (Green *et al.* 2001*b*) document the transition from inverted basinal regions in the East Midlands Shelf to a stable platform setting in the south (Midland Platform). AFTA reveals two discrete cooling episodes, in Early Tertiary time (beginning between 65 and 60 Ma) and Late Tertiary time (beginning between 25 and 5 Ma). Early Tertiary palaeotemperatures from AFTA and VR in samples collected from outcrop define a consistent increase from $<50\,^{\circ}C$ in Lower Cretaceous and Upper Jurassic units in the SE to around 80–90 °C in Triassic and older units in the NW.

Results from the Rufford-1 well define an Early Tertiary palaeogeothermal gradient of $40.5\,^{\circ}C\,km^{-1}$ ($32-50\,^{\circ}C\,km^{-1}$ at $\pm 95\%$ confidence limits (c.l.), compared with a present-day gradient of *c.* $30\,^{\circ}C\,km^{-1}$), corresponding to deeper burial by 1450 m of additional section, subsequently removed by Tertiary erosion (1.1–2.2 km at $\pm 95\%$ c.l.). Thus, these Early Tertiary palaeotemperatures, as in NW England, also appear to reflect a combination of deeper burial and elevated basal heat flow. Geological considerations suggest a maximum overburden of 800–900 m above the base of the Lias sequence in the vicinity of Rugby where the Early Tertiary palaeotemperature at outcrop is similar to that near the Rufford-1 well site. The discrepancy between stratigraphic and palaeothermal reconstruction of former burial depths, often noted in earlier studies, remains unresolved.

The Late Tertiary episode is much less well constrained, but results from Rufford-1 may require between 910 and 1650 m of eroded section. Thus much of the total amount of removed overburden may have been removed during Late Tertiary time. Results from the Apley Barn Borehole, in the SW of the region, reveal a significantly different thermal history, involving Permian cooling, which probably reflects the protracted effects of Variscan tectonism, and a Late Tertiary episode characterized by a highly non-linear palaeotemperature profile, which probably reflects local heating as a result of passage of hot fluids. Results from this borehole show no evidence of any Early Tertiary effects.

Timing of exhumation

Some workers have explicitly questioned the Early Tertiary timing of cooling revealed by AFTA in NW and Central England (Holliday 1993, 1999; McCulloch, 1994). As documented by the more recent studies described above, recent analyses using latest techniques provide very tight definition of the onset of Early Tertiary cooling in these regions, with the interval 65–60 Ma being the best available estimate.

Although the overall duration of exhumation that began in early Tertiary time cannot be tightly constrained from the data, it is clear that in Central England, samples underwent a considerable amount of Late Tertiary cooling and it remains possible that much of the exhumation actually occurred within the Late Tertiary episode, as suggested by Japsen (1997). Resolution of the effects of discrete episodes of Tertiary exhumation remains a major objective of continuing work in this region.

Mesozoic palaeo-thermal episodes recognized in Ireland and the Central Irish Sea Basin

Integration of AFTA and VR data from onshore Ireland reveals a complex thermal history, characterized by multiple cooling episodes of late Carboniferous, Jurassic, early Cretaceous, early Tertiary and late Tertiary age (Green *et al.* 2000). Peak palaeotemperatures in each episode fall through time to produce an overall long-term cooling trend since late Carboniferous time. Thermal history styles across the region are very similar, although the magnitude of peak palaeotemperatures in individual episodes shows some variation. The regional nature of all these palaeothermal episodes, and their correlation with regionally significant unconformities, suggests that heating was due primarily to greater depth of burial, with subsequent cooling representing the progressive unroofing of the present onshore region since late Carboniferous times. In Northern Ireland, explanations of early Cretaceous and early Tertiary palaeotemperatures in terms of greater depth of burial are more difficult to reconcile with geological evidence, and heating as a result of hot fluid movement appears more likely. This applies particularly to early Tertiary effects, for which the Tertiary Igneous Province provides a ready explanation. Over the entire onshore region, maximum maturity levels in Carboniferous and older units were reached at the end of Carboniferous time, and preservation of hydrocarbons to the present day, through several tectono-thermal episodes, appears unlikely.

Results from wells in the Central Irish Sea Basin (CISB) are dominated by the Early Cretaceous episode, but also reveal Early and Late Tertiary cooling (Duncan *et al.* 1998; Green *et al.* 2001a). Palaeotemperatures in the CISB

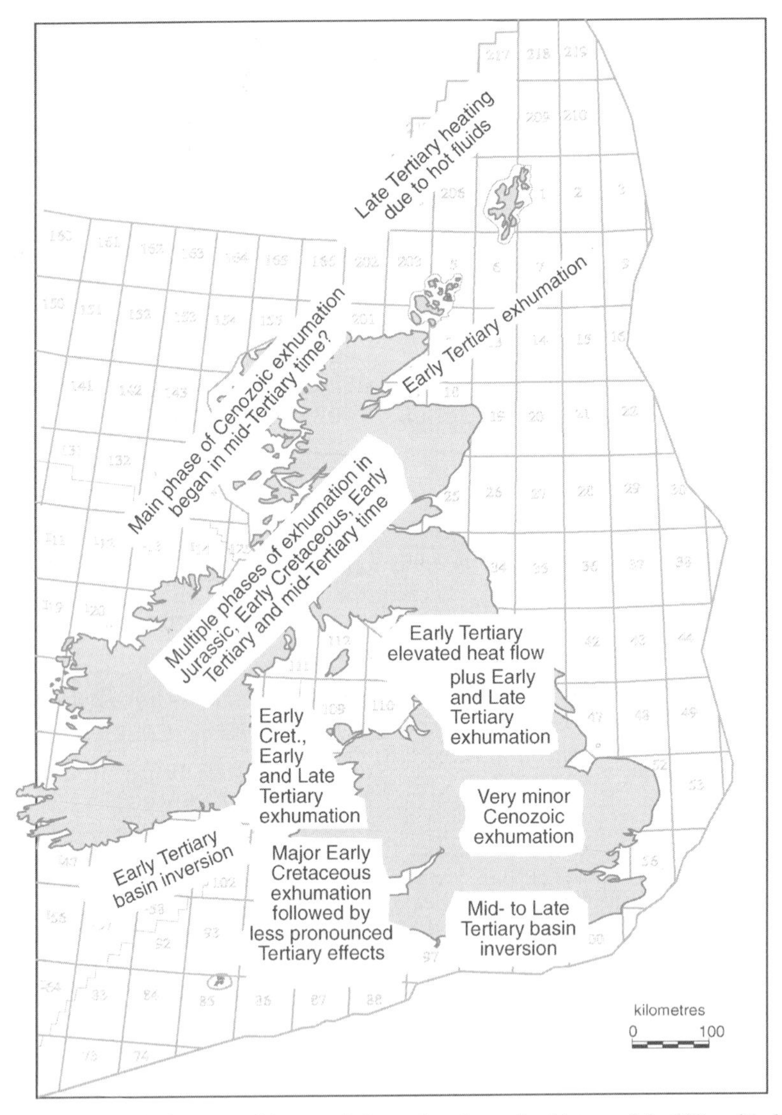

Fig. 10. Summary of the main features of the post-Palaeozoic exhumation history of the UK and Ireland regions, based on work discussed in the text. Several palaeo-thermal episodes of regional extent and variable intensity are recognized. Understanding the timing of these episodes within the regional geological context, and the variation in magnitude of individual episodes across the region is a vital step in understanding the processes involved in producing these palaeo-thermal effects (burial, exhumation, changes in heat flow, fluid circulation, etc.), as well as being a critical aspect of understanding the petroleum systems of the region.

appear to be due almost solely to deeper burial, with up to 3 km of section removed since Early Cretaceous time. However, results from well 42/21-1, located in a SW extension of the St. George's Channel Basin, are dominated by an Early Tertiary onset of exhumation, suggesting some degree of local structural control on discrete phases of exhumation.

Results from other areas

Studies in regions such as the southern North Sea, Forth Approaches and the Moray Firth (some results from which were reported by Green *et al.* (1995)), as well as continuing research work in the Cleveland Basin (preliminary results reported by Green *et al.* (1993b)), all show that a wide region of the UK was affected by Early Tertiary cooling. Results from the Scottish Highlands (Thomson *et al.* 1999b) are very similar to those from onshore Ireland, showing Early Cretaceous, Early Tertiary and Late Tertiary cooling episodes, which presumably again represent the effects of progressive exhumation. Interestingly, towards the Atlantic margin and in the Hebridean Basins, the effects of Early Tertiary exhumation diminish (Green *et al.* 1999) and the mid- to Late Tertiary period appears to represent the main onset of exhumation, with results from the Sea of Hebrides-1 well (aka L134/5-1) defining an onset of cooling between 45 and 20 Ma (Green *et al.* 1999), which may represent a third Tertiary episode, distinct from Early and Late Tertiary episodes discussed thus far.

Figure 10 summarizes the main conclusions that can be drawn regarding the history of Mesozoic and Cenozoic exhumation across the UK region, on the basis of the results discussed here.

Comments on mechanisms of exhumation

Regional events in NW Europe

As discussed above, results from the NW European Atlantic margin show consistent evidence of at least three pulses of Cenozoic exhumation (in Early, mid- and Late Tertiary time), varying in magnitude across the region. In

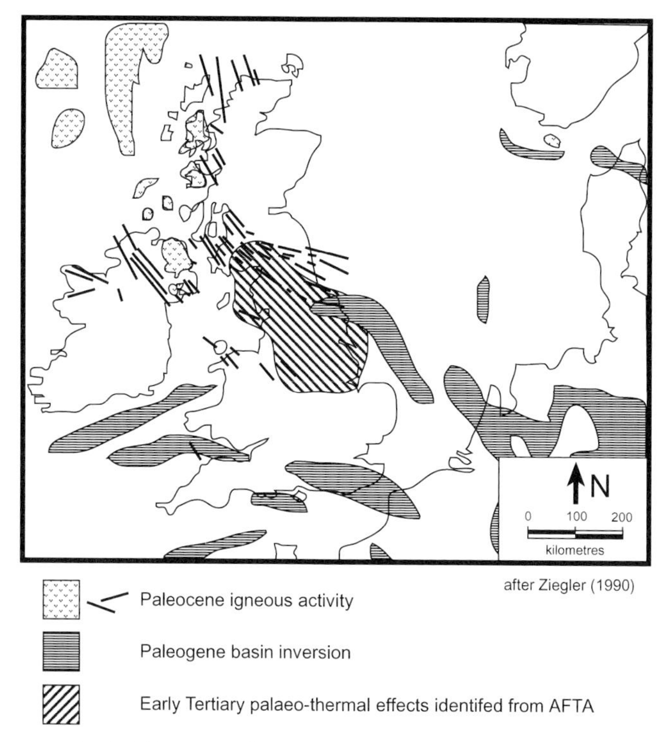

Fig. 11. Palaeogene events in the UK region (after Ziegler 1990). The region where Early Tertiary palaeo-thermal effects are recognized from AFTA appears to link the developing Atlantic margin and Tertiary igneous province in the NW with the region to the SE characterized by the initial stages of Alpine compression and Palaeogene basin inversion. The spatial and temporal relationships between these events is suggestive of some sort of causative link.

considering likely mechanisms, it seems significant that major cooling events that appear to be broadly synchronous with these episodes have also been identified throughout the Arctic region including Svalbard (Blythe & Kleinspehn 1998), Alaska (O'Sullivan *et al.* 1993, 1995) and East Greenland (Thomson *et al.* 1999a). Results from recent unpublished studies in the Barents Sea and the North Slope of Alaska have emphasized the synchroneity of Cenozoic events in these regions and in NW Europe. This suggests that the driving mechanisms are truly regional, and are more likely to be related to events at plate margins rather than more local processes such as igneous underplating.

In particular, as illustrated in Fig. 11, Early Tertiary palaeo-thermal effects identified from AFTA in Northern and Central England and the Irish Sea are broadly synchronous with rifting in the North Atlantic, initial stages of Alpine compression and basin inversion throughout NW Europe. Although the exact relationships between these processes remain to be resolved, the spatial and temporal relationships between the palaeo-thermal effects and these other tectonic processes suggest some sort of genetic relationship.

At a more local level, as detailed by Green *et al.* (2001b), latest results from Central England suggest that at least in the UK region major Early Tertiary exhumation appears to be limited to regions underlain by older Palaeozoic basins whereas regions overlying stable Palaeozoic basement remained dormant until Late Tertiary time, when more regional exhumation occurred. Green *et al.* (2001b) have suggested this reflects the preferential reactivation and progressive locking of the weaker basinal regions as a result of compressional events at plate margins, culminating in regional uplift and erosion. The relationship between the regional Tertiary exhumation discussed here and discrete Tertiary basin inversion events in Southern England (e.g. Bray *et al.*, 1997) also merits consideration, but is beyond the scope of present discussion.

Accelerated burial before uplift

Reconstructed thermal and/or burial and exhumation histories for areas affected by significant exhumation invariably show an acceleration in the rate of burial before the onset of exhumation (see examples given by Hillis 1991; Green *et al.* 1995; Japsen 1997). Although some have suggested that this results from using unreasonable amounts of additional burial, as discussed above, we believe that this is a real effect, and that the accelerated burial and subsequent exhumation are linked. If this is true, any successful model for the underlying mechanism of regional exhumation episodes such as discussed here should explain not only the amounts and regional extent of removed section, but also this prior burial phase.

Repeated cycles of burial and exhumation

Results from onshore Ireland (Green *et al.* 2000), the Central Irish Sea Basin (Green *et al.* 2001a) and the Scottish Highlands (Thomson *et al.* (1999b) strongly suggest the occurrence of repeated cycles of burial followed by exhumation, with the overall magnitude of both processes decreasing over time. Results from other areas show thermal histories that can be interpreted in similar terms (e.g. East Irish Sea Basin, Green *et al.* 1997). In this respect, the phases of accelerated burial discussed in the preceding section become part of a progressive sequence of events, diminishing through time, giving the appearance of a damped oscillation. Although we have no mechanisms for explaining these observations, their ubiquitous recognition suggests some mechanistic control, and we suggest that this provides a fruitful avenue for consideration in developing possible models to explain the sort of effects discussed here.

Comparison with results from other regions

Consideration of data from other regions showing similar styles of palaeo-thermal effects is also useful in considering mechanisms for exhumation. Well-documented examples include Cretaceous effects in Southern Africa and SW Brazil (Gallagher & Brown 1999, and references therein), NE USA (Miller & Duddy 1986) and SE Australia (Moore *et al.* 1986; Dumitru *et al.* 1991; Duddy & Green 1992; O'Sullivan *et al.* 1995b, 1996). In all these areas an association between the onset of exhumation and continental rifting or separation at adjacent plate margins is evident (also in NW Europe, Fig. 11). Although details are yet to be resolved, these empirical observations strengthen the suggestion that events at plate margins exert a key influence on the processes responsible for regional exhumation. Palaeo-thermal effects in these regions also appear to require accelerated rates of burial before the onset of exhumation, further emphasizing previous comments on this topic.

In most of these areas, particularly in Southern Africa and in SE Australia, palaeogeothermal gradients at the palaeo-thermal maximum are not well constrained, because of the lack of data from vertical sections, and therefore amounts of exhumation are still only poorly understood. In this context, it is worth noting that the views expressed by Gallagher & Brown (1999) and Gunnell (2000) that heat flow can generally be treated as constant through time and therefore that all the observed cooling can be interpreted as being due to denudation (exhumation) is at odds with results from NW England (Green *et al.* 1999; Green 2001) and SE Australia (e.g. Duddy 1994, 1997), which reveal consistent evidence for elevated basal heat flow in locations well separated from the sites of rifting at the appropriate time. Amounts of removed section (exhumation or denudation) based on the assumption of constant heat flow should therefore always be regarded as likely to overestimate the true amount, and estimates are likely to be realistic only where estimation is based on vertical sequences of samples that directly constraint the palaeo-heat flow or palaeogeothermal gradient.

AFTA® is a registered trademark of Geotrack International Pty Ltd.

References

BARKER, C.E. & PAWLEWICZ, M.J. 1986. The correlation of vitrinite reflectance with maximum heating in humic organic matter. *In*: BUNTEBARTH, G. & STEGENA, L. (eds) *Palaeogeothermics*. Springer, New York, 79–93.

BLYTHE, A.E. & KLEINSPEHN, K.L. 1998. Tectonically versus climatically driven Cenozoic exhumation of the Eurasian plate margin, Svalbard: fission track analyses. *Tectonics*, **17**, 621–639.

BRAY, R., DUDDY, I.R. & GREEN, P.F. 1997. Multiple heating episodes in the Wessex basin: implications for geological evolution and hydrocarbon generation. *In*: UNDERHILL, J.R. (ed.) *Development, Evolution and Petroleum Geology of the Wessex Basin*. Geological Society, London, Special Publications, **133**, 199–213.

BRAY, R., GREEN, P.F. & DUDDY, I.R. 1992. Thermal history reconstruction in sedimentary basins using apatite fission track analysis and vitrinite reflectance: a case study from the east Midlands of England and the Southern North Sea. *In*: HARDMAN, R.F.P. (ed.) *Exploration Britain: Into the Next Decade*. Geological Society, London, Special Publications, **67**, 3–25.

BRODIE, J. & WHITE, N. 1994. Sedimentary basin inversion caused by igneous underplating: northwest European continental shelf. *Geology*, **22**, 147–150.

BURNHAM, A.K. & SWEENEY, J.J. 1989. A chemical kinetic model of vitrinite reflectance maturation. *Geochimica et Cosmochimica Acta*, **53**, 2649–2657.

CARLSON, W.D. 1990. Mechanisms and kinetics of apatite fission-track annealing. *American Mineralogist*, **75**, 1120–1139.

CARLSON, W.D., DONELICK, R.A. & KETCHAM, R.A. 1999. Variability of apatite fission-track annealing kinetics: I. Experimental results. *American Mineralogist*, **84**, 1213–1223.

CARR, A. 1999. A vitrinite reflectance kinetic model incorporating overpressure retardation. *Marine and Petroleum Geology*, **16**, 355–377.

CARR, A. 2000. Suppression and retardation of vitrinite reflectance, Part 1. Formation and significance for hydrocarbon generation. *Journal of Petroleum Geology*, **23**, 313–343.

CHADWICK, R.A., KIRBY, G.A. & BAILY, H.E. 1994. The post-Triassic structural evolution of north-west England and adjacent parts of the East Irish Sea Basin. *Proceedings of the Yorkshire Geological Society*, **50**, 91–102.

COPE, J.C.W. 1994. A latest Cretaceous hotspot and the southeasterly tilt of Britain. *Journal of the Geological Society, London*, **151**, 905–908.

CORRIGAN, J.D. 1991*a*. Thermal anomalies in the Central Indian Ocean: evidence for de-watering of the Bengal Fan. *Journal of Geophysical Research*, **96**, 14263–14275.

CORRIGAN, J.D. 1991*b*. Inversion of apatite fission track data for thermal history information. *Journal of Geophysical Research*, **96**, 10347–10360.

CROWLEY, K.D., CAMERON, M. & SCHAEFFER, R.L. 1991. Experimental studies of annealing of etched fission tracks in fluorapatite. *Geochimica et Cosmochimica Acta*, **55**, 1449–1465.

DEMING, D. 1994. Overburden rock, temperature and heat flow. *In*: MAGOON, L.B. & DOW, D.G. (eds) *The Petroleum System—From Source to Trap*. American Association of Petroleum Geologists, Memoir, **60**, 165–186.

DONELICK, R.A., KETCHAM, R.A. & CARLSON, W.D. 1999. Variability of apatite fission-track annealing kinetics: II. Crystallographic orientation effects. *American Mineralogist*, **84**, 1224–1234.

DORÉ, A.G., LUNDIN, E.R., JENSEN, L.N., BIRKELAND, Ø., ELIASSEN, P.E. & FICHLER, C. 1999. Principal tectonic events in the evolution of the northwest European Atlantic margin. *In*: FLEET, A.J. & BOLDY, S.A.R. (eds) *Petroleum Geology of North West Europe, Proceedings of the 5th Conference*. Geological Society, London, 41–61.

DOW, W. 1977. Kerogen studies and geological interpretations. *Journal of Geochemical Exploration*, **7**, 79–99.

DUDDY, I.R. 1994. The Otway Basin: thermal, structural, tectonic and hydrocarbon generation histories. *NGMA/PESA Otway Basin Symposium, Extended Abstracts*, **14**, 35–42.

DUDDY, I.R. 1997. Focusing exploration in the Otway Basin: understanding timing of source rock maturation. *APPEA Journal*, **37**, 178–191.

DUDDY, I.R. & GREEN, P.F. 1992. Tectonic development of the Gippsland Basin and Environs: identification of key episodes using apatite fission track analysis (AFTA). *Gippsland Basin Symposium, Joint AusIMM (Melbourne Branch)–PESA*

(Vic/Tas Branch) Energy, Economics and Environment. Australian Institute of Mining and Metallurgy, Melbourne, **3/92**, 111–120.

DUDDY, I.R., GREEN, P.F., BRAY, R.J. & HEGARTY, K.A. 1994. Recognition of the thermal effects of fluid flow in sedimentary basins. *In*: PARNELL, J. (ed.) *Geofluids: Origin, Migration and Evolution of Fluids in Sedimentary Basins*. Geological Society, London, Special Publications, **78**, 325–345.

DUDDY, I.R., GREEN, P.F., HEGARTY, K.A. & BRAY, R.J. (1991). Reconstruction of thermal history in basin modelling using apatite fission track analysis: what is really possible. *Proceedings of the First Offshore Australia Conference (Melbourne)*, III-49–III-61.

DUDDY, I.R., GREEN, P.F., HEGARTY, K.A., BRAY, R.J. & O'BRIEN, G.W. 1998. Dating and duration of hot fluid flow events determined using AFTA® and vitrinite reflectance-based thermal history reconstruction. *In*: PARNELL, J. (ed.) *Dating and Duration of Hot Fluid Flow Events and Fluid–Rock Interaction*. Geological Society, London, Special Publications, **144**, 41–51.

DUDDY, I.R., GREEN, P.F. & LASLETT, G.M. 1988. Thermal annealing of fission tracks in apatite 3. Variable temperature behaviour. *Chemical Geology (Isotope Geoscience Section)*, **73**, 25–38.

DUMITRU, T.A., HILL, K.C., COYLE, D.A. & 7 OTHERS 1991. Fission track thermochronology: application to continental rifting of south-eastern Australia. *Australia Petroleum Explorers Association Journal*, **10**, 131–142.

DUNCAN, W.I., GREEN, P.F. & DUDDY, I.R. 1998. Source rock burial history and seal effectiveness: key facets to understanding hydrocarbon exploration potential in the East and Central Irish Sea Basins. *AAPG Bulletin*, **82**, 1401–1415.

ENGLAND, P. & MOLNAR, P. 1990. Surface uplift, uplift of rocks and exhumation of rocks. *Geology*, **18**, 1173–1177.

FISHER, M.J., BARNARD, P.C. & COOPER, B.S. 1980. Organic maturation and hydrocarbon generation in the Mesozoic sediments of the Sverdrup Basin, Arctic Canada. *Proceedings of the 4th International Palynology Conference, Lucknow (1976–77)*, **2**, 581–588.

GALLAGHER, K. 1995. Evolving temperature histories from apatite fission-track data. *Earth and Planetary Science Letters*, **136**, 421–435.

GALLAGHER, K. & BROWN, R. 1999. The Mesozoic denudation history of the Atlantic margins of southern Africa and southeast Brazil and the relationship to offshore sedimentation. *In*: CAMERON, N.R., BATE, R.H. & CLURE, V.S. (eds) *The Oil and Gas Habitats of the South Atlantic*. Geological Society, London, Special Publications, **153**, 41–53.

GREEN, P.F. 1986. On the thermo-tectonic evolution of Northern England: evidence from fission track analysis. *Geological Magazine*, **123**, 493–506.

GREEN, P.F. 1988. The relationship between track shortening and fission track age reduction in apatite: combined influences of inherent instability, annealing anisotropy, length bias and system calibration. *Earth and Planetary Science Letters*, **89**, 335–352.

GREEN, P.F. 1989. Thermal and tectonic history of the East Midlands shelf (onshore U.K.) and surrounding regions assessed by apatite fission track analysis. *Journal of the Geological Society, London*, **146**, 755–773.

GREEN, P.F. 2001. Early Tertiary palaeo-thermal effects in Northern England: reconciling results from apatite fission track analysis with geological evidence. *Tectonophysics*, **349**, 131–144.

GREEN, P.F., DUDDY, I.R. & BRAY, R.J. 1993*a*. Early Tertiary heating in Northwest England: fluids or burial (or both?) (extended abstract). *In*: PARNELL, J., RUFFELL, A.H. & MOLES, N.R. (eds) *Geofluids '93: Contributions to an International Conference on Fluid Evolution, Migration and Interaction in Rocks*, 119–123.

GREEN, P.F., DUDDY, I.R. & BRAY, R.J. 1995. Applications of thermal history reconstruction in inverted basins. *In*: BUCHANAN, J.G. & BUCHANAN, P.G. (eds) *Basin Inversion*. Geological Society, London, Special Publications, **88**, 148–165.

GREEN, P.F., DUDDY, I.R. & BRAY, R.J. 1997. Variation in thermal history styles around the Irish Sea and adjacent areas: implications for hydrocarbon occurrence and tectonic evolution. *In*: MEADOWS, N.S., TRUEBLOOD, S., HARDMAN, M. & COWAN, G. (eds) *Petroleum Geology of the Irish Sea and Adjacent Areas*. Geological Society, London, Special Publications, **124**, 73–93.

GREEN, P.F., DUDDY, I.R., BRAY, R.J., DUNCAN, W.I. & CORCORAN, D. 2001*a*. The influence of thermal history on hydrocarbon prospectivity in the Central Irish Sea Basin. *In*: SHANNON, P.M., HAUGHTON, P. & CORCORAN, D. (eds) *Petroleum Geology of Ireland's Offshore Basins*. Geological Society, London, Special Publications, **188**, 171–188.

GREEN, P.F., DUDDY, I.R., BRAY, R.J. & LEWIS, C.L.E. 1993*b*. Elevated palaeotemperatures prior to early Tertiary cooling throughout the UK region: implications for hydrocarbon generation. *In*: PARKER, J.R. (ed.) *Petroleum Geology of Northwest Europe: Proceedings of the 4th Conference*. Geological Society, London, 1067–1074.

GREEN, P.F., DUDDY, I.R., GLEADOW, A.J.W. & LOVERING, J.F. 1989*b*. Apatite fission track analysis as a palaeotemperature indicator for hydrocarbon exploration. *In*: NAESER, N.D. & MCCULLOH, T. (eds) *Thermal History of Sedimentary Basins—Methods and Case Histories*. Springer, New York, 181–195.

GREEN, P.F., DUDDY, I.R., GLEADOW, A.J.W., TINGATE, P.R. & LASLETT, G.M. 1985. Fission-track annealing in apatite: track length measurements and the form of the Arrhenius plot. *Nuclear Tracks*, **10**, 323–328.

GREEN, P.F., DUDDY, I.R., GLEADOW, A.J.W., TINGATE, P.R. & LASLETT, G.M. 1986. Thermal annealing of fission tracks in apatite 1. A qualitative description. *Chemical Geology (Isotope Geoscience Section)*, **59**, 237–253.

GREEN, P.F., DUDDY, I.R., HEGARTY, K.A. & BRAY, R.J. 1999. Early Tertiary heat flow along the UK

Atlantic margin and adjacent areas. *In*: FLEET, A.J. & BOLDY, S.A.R. (eds) *Petroleum Geology of North West Europe, Proceedings of the 5th Conference*. Geological Society, London, 348–357.

GREEN, P.F., DUDDY, I.R., HEGARTY, K.A., BRAY, R.J., SEVASTOPULO, G., CLAYTON, G. & JOHNSTON, D. 2000. The post-Carboniferous evolution of Ireland: evidence from thermal history reconstruction. *Proceedings of the Geologists' Association*, **111**, 307–320.

GREEN, P.F., DUDDY, I.R., LASLETT, G.M., HEGARTY, K.A., GLEADOW, A.J.W. & LOVERING, J.F. 1989*a*. Thermal annealing of fission tracks in apatite 4. Quantitative modelling techniques and extension to geological timescales. *Chemical Geology (Isotope Geoscience Section)*, **79**, 155–182.

GREEN, P.F., HEGARTY, K.A. & DUDDY, I.R. 1996. Compositional influences on fission track annealing in apatite and improvement in routine application of AFTA®. *American Association of Petroleum Geologists, San Diego, CA, Abstracts with Program*, A56.

GREEN, P.F., THOMSON, K. & HUDSON, J.D. 2001*b*. Recognising tectonic events in undeformed regions: contrasting results from the Midland Platform and East Midlands Shelf, Central England. *Journal of the Geological Society, London*, **158**, 59–73.

GRETENER, P.E. 1981. *Geothermics: Using Temperature in Hydrocarbon ExplorationAmerican*. Association of Petroleum Geologists, Course Note Series, **17**.

GUNNELL, Y. 2000. Apatite fission-track thermochronology: an overview of its potential and limitations in geomorphology. *Basin Research*, **12**, 115–132.

HILLIS, R.R. 1991. Chalk porosity and Tertiary uplift, Western Approaches Trough, SW UK and NW French continental shelves. *Journal of the Geological Society, London*, **148**, 669–679.

HILLIS, R.R. 1992. A two-layer lithospheric compressional model for the Tertiary uplift of the southern United Kingdom. *Geophysical Research Letters*, **19**, 573–576.

HOLLIDAY, D.W. 1993. Mesozoic cover over northern England: interpretation of apatite fission track data. *Journal of the Geological Society, London*, **150**, 657–660.

HOLLIDAY, D.W. 1999. Palaeotemperatures, thermal modelling and depth of burial studies in northern and eastern England. *Proceedings of the Yorkshire Geological Society*, **52**, 337–352.

HURFORD, A.J. 1977. Fission track ages from two Galloway granites. *Geological Magazine*, **114**, 299–304.

JAPSEN, P. 1997. Regional Neogene exhumation of Britain and the Western North Sea. *Journal of the Geological Society, London*, **154**, 239–247.

KAMP, P.J.J. & GREEN, P.F. 1990. Thermal and tectonic history of selected Taranaki Basin (New Zealand) wells assessed by apatite fission track analysis. *AAPG Bulletin*, **74**, 1401–1419.

KETCHAM, R.A., DONELICK, R.A. & CARLSON, W.D. 1999. Variability of apatite fission-track annealing kinetics: III. Extrapolation to geological timescales. *American Mineralogist*, **84**, 1235–1255.

KETCHAM, R.A., DONELICK, R.A. & DONELICK, M.B. 2000. AFTSolve: a program for multi-kinetic modeling of apatite fission-track data. *Geological Materials Research*, 1–32.

LASLETT, G.M., GREEN, P.F., DUDDY, I.R. & GLEADOW, A.J.W. 1987. Thermal annealing of fission tracks in apatite 2. A quantitative analysis. *Chemical Geology (Isotope Geoscience Section)*, **65**, 1–13.

LEWIS, C.L.E., GREEN, P.F., CARTER, A. & HURFORD, A.J. 1992. Elevated late Cretaceous to Early Tertiary palaeotemperatures throughout Northwest England: three kilometres of Tertiary erosion? *Earth and Planetary Science Letters*, **112**, 131–145.

LUTZ, T.M. & OMAR, G. 1991. An inverse method of modelling thermal histories from apatite fission-track data. *Earth and Planetary Science Letters*, **104**, 181–195.

MARSHALLSEA, S.J., GREEN, P.F. & WEBB, J. 2000. Thermal history of the Hodgkinson and Laura Basins, Far North Queensland: multiple cooling episodes identified from AFTA and VR data. *Australian Journal of Earth Science*, **47**, 779–797.

MCCULLOCH, A.A. 1994. Discussion on Mesozoic cover over Northern England: interpretation of apatite fission track data. *Journal of the Geological Society, London*, **151**, 735–736.

MILLER, D.S. & DUDDY, I.R. 1986. Early Cretaceous uplift and erosion of the northern Appalachian Basin, New York, based on apatite fission track analysis. *Earth and Planetary Science Letters*, **93**, 35–49.

MOORE, M.E., GLEADOW, A.J.W. & LOVERING, J.L. 1986. Thermal evolution of rifted continental margins: new evidence from fission tracks in basement apatites from southeastern Australia. *Earth and Planetary Science Letters*, **78**, 255–270.

MORROW, D.W. & ISSLER, D.R. 1993. Calculation of vitrinite reflectance from thermal histories: a comparison of some models. *AAPG Bulletin*, **77**, 610–624.

NEWMAN, J. 1997. New approaches to the detection and correction of suppressed vitrinite reflectance. *APPEA Journal*, **27**, 524–535.

O'SULLIVAN, P.B., FOSTER, D.A., KOHN, B.P. & GLEADOW, A.J.W. 1996. Multiple postorogenic denudation events: an example from the eastern Lachlan fold belt, Australia. *Geology*, **24**, 563–566.

O'SULLIVAN, P.B., BERGMAN, S.C., DECKER, J., DUDDY, I.R., GLEADOW, A.J.W. & TURNER, D.L. 1993. Multiple phases of Tertiary uplift and erosion in the Arctic National Wildlife Refuge, Alaska, revealed by apatite fission track analysis. *AAPG Bulletin*, **77**, 359–385.

O'SULLIVAN, P.B., HANKS, C.L., WALLACE, W.K. & GREEN, P.F. 1995*a*. Multiple episodes of Cenozoic denudation in the northeastern Brooks Range: fission-track data from the Okpilak Batholith, Alaska. *Canadian Journal of Earth Science*, **32**, 1106–1118.

O'SULLIVAN, P.B., KOHN, B.P., FOSTER, D.A. & GLEADOW, A.J.W. 1995*b*. Fission track data from the Bathurst Batholith: evidence for rapid mid-Cretaceous uplift and erosion within the eastern highlands of Australia. *Australian Journal of Earth Sciences*, **42**, 597–607.

RIIS, F. & JENSEN, L.N. 1992. Introduction: measuring uplift and erosion—proposal for a terminology. *Norsk Geologisk Tidsskrift*, **72**, 223–235.

SMITH, K., GATLIFF, R.W. & SMITH, N.J.P. 1994. Discussion of the amount of Tertiary erosion in the UK estimated using sonic velocity analysis. *Journal of the Geological Society, London*, **151**, 1041–1045.

SUMMER, N.S. & VEROSUB, L.L. 1989. A low temperature hydrothermal maturation mechanism for sedimentary basins associated with volcanic rocks. *In*: PRICE, P.A. (ed.) *Origin and Evolution of Sedimentary Basins and their Economic Potential*. Geophysical Monograph, American Geophysical Union, **48**, 129–136.

SWEENEY, J.J. & BURNHAM, A.K. 1990. Evaluation of a simple model of vitrinite reflectance based on chemical kinetics. *AAPG Bulletin*, **74**, 1559–1570.

THOMSON, K., GREEN, P.F., WHITHAM, A.G., PRICE, S.P. & UNDERHILL, J.R. 1999*a*. New constraints on the thermal history of North-East Greenland from apatite fission-track analysis. *Geological Society of America Bulletin*, **111**, 1054–1068.

THOMSON, K., UNDERHILL, J.R., GREEN, P.F., BRAY, R.J. & GIBSON, H.J. 1999*b*. Evidence from apatite fission track analysis for the post-Devonian burial and exhumation history of the Northern Highlands, Scotland. *Marine and Petroleum Geology*, **16**, 27–39.

VROLIJK, P., DONELICK, R.A., QUENG, J. & CLOOS, M. 1992. Testing models of fission track annealing in apatite in a simple thermal setting: Site 800, Leg 129. *In*: LARSON, R.L., LANCELOT, Y. AND OTHERS (eds) *Proceedings of the Ocean Drilling Program, Scientific Results*, **129**, 169–176. Ocean Drilling Program, College Station, TX.

WAPLES, D.W. 1985. *Geochemistry in Petroleum Exploration*. HRDC, Boston, MA.

WAPLES, D.W., KAMATA, H. & SUIZU, M. 1992. The art of maturity modelling. Part 1: Finding a satisfactory geologic model. *AAPG Bulletin*, **76**, 31–46.

WILKINS, R.W.T., WILMSHURST, J.R., RUSSELL, N.J., HLADKY, G., ELLACOTT, M.V. & BUCKINGHAM, C. 1992. Fluorescence alteration and the suppression of vitrinite reflectance. *Organic Geochemistry*, **18**, 629–640.

WILLETT, S.D. 1997. Inverse modeling of annealing of fission tracks in apatite 1: a controlled random search method. *American Journal of Science*, **297**, 939–969.

ZIAGOS, J.P. & BLACKWELL, D.D. 1986. A model for the transient temperature effect of horizontal fluid flow in geothermal systems. *Journal of Volcanology and Geothermal Research*, **27**, 371–397.

ZIEGLER, P.A. 1990. *Geological Atlas of Western and Central Europe*. Shell International Petroleum Maatschappij, The Hague.

Sonic velocity analysis of the Tertiary denudation of the Irish Sea basin

PHILIP D. WARE[1,2] & JONATHAN P. TURNER[1]

[1]*University of Birmingham, School of Earth Sciences, Birmingham B15 2TT, UK (e-mail: j.p.turner@bham.ac.uk)*

[2]*Present address: Kerr McGee Oil, Crawpeel Road, Aberdeen AB12 3LG, UK*

Abstract: Interaction between uplift related to the Cretaceous–Paleocene opening of the North Atlantic, Neogene shortening (basin inversion) and Pleistocene glacio-isostasy is illustrated by the complex denudation pattern of Britain; such denudation is greatest over the submergent East Irish Sea basin, some 500 km from the Atlantic margin. This paper reports on analysis of sedimentary porosities using sonic velocity logs from 42 wells in the East Irish Sea basin. We present a new map showing the variation in exhumation magnitude at the uppermost Mesozoic unconformity (i.e. thickness of denuded Mesozoic and Cenozoic sedimentary rocks), today buried beneath a thin veneer of Pleistocene sediment. It indicates that exhumation is mostly <1500 m (632–2132 m; mean standard deviation 407 m), less than denudation results obtained from vitrinite reflectance and apatite fission-track data. The map also reveals substantial variation in exhumation over short distances, often between adjacent wells sited on opposing walls of individual faults. This is interpreted in terms of the influence of Neogene basin inversion on the exhumation of the EISB. The role of late Tertiary tectonics in western UK exhumation is therefore discussed.

Estimates of exhumation (i.e. thickness of denuded overburden) vary widely according to the method employed to evaluate it (see Doré & Jensen 1996; Japsen & Chalmers 2000). The main purpose of this paper is to test existing denudation magnitudes obtained from apatite fission-track (AFT) analysis and vitrinite reflectance (VR) data from the East Irish Sea basin (EISB) using sonic velocities (SV) logged in petroleum exploration wells. The long-wavelength (>2 km) form of vertical SV profiles responds chiefly to compaction-driven porosity reduction. Consequently, it is a particularly effective measure of the former maximum burial depth of exhumed sedimentary successions in basins where transient heating episodes elevated thermal burial proxies, such as AFT and VR data, without concomitant burial increases.

The main processes driving exhumation are thermal–isostatic effects related to continental extension, orogenic crustal shortening and epeirogeny – uplift of broad regions of continental interiors driven by plumes (Nadin *et al.* 1995), magmatic underplating (Brodie & White 1994), mantle delamination (Platt & England 1993), post-glacial isostasy (Lambeck 1991), intra-plate stress (Cloetingh 1988), etc. Exhumation differs from uplift, which describes vertical movement relative to the geoid (England & Molnar 1990). Because of the problem in constraining an absolute datum, very few studies are able to measure ancient uplift *per se* (e.g. Abbott *et al.* 1997). Furthermore, in many instances, such as burial history modelling of petroleum source rocks, exhumation is actually a far more useful parameter to constrain because it leads directly to cooling and lithostatic pressure release, with attendant implications for petroleum generation and retention. In the absence of any erosion, uplift on its own results in neither cooling nor pressure decrease.

The EISB is an intra-cratonic sedimentary basin whose Cenozoic evolution records the influence of Cretaceous rifting of the North Atlantic and associated Paleocene break-up, Oligo-Miocene shortening (basin inversion) and Pleistocene glacio-isostasy. Fuller discussion of its geological and tectonic evolution is given by Jackson & Mulholland (1993), Knipe *et al.* (1993), Jackson *et al.* (1995), Cope (1998), Maingarm *et al.* (1999), and references therein.

Denudation analysis has attracted particular interest in the gas-prone EISB, where source rock burial history modelling is hindered by uncertainty over the thickness of the eroded

From: DORÉ, A.G., CARTWRIGHT, J.A., STOKER, M.S., TURNER, J.P. & WHITE, N. 2002. *Exhumation of the North Atlantic Margin: Timing, Mechanisms and Implications for Petroleum Exploration.* Geological Society, London, Special Publications, **196**, 355–370. 0305-8719/02/$15.00

Jurassic–Cretaceous section. Given the primary reliance of hydrocarbon generation on a long-term rise in source rock temperature, cooling (and, to a lesser extent, pressure release) during exhumation leads directly to arrested generation. Conversely, a potential source rock today lying within the appropriate depth window for hydrocarbon generation may not be effective if, before denudation, its generative capacity was exceeded during deeper burial. Exhumation analysis in the EISB has been a particularly lively topic since the work of Lewis *et al.* (1992), whose results from AFT analysis indicated in excess of 3 km of sedimentary section denuded from a broad region centred on the present English Lake District and contiguous Irish Sea. More recent AFT data confirm a major cooling event in the EISB starting at *c.* 60 Ma and decelerating into Neogene time (Green *et al.* 1997; Duncan *et al.* 1998) but they suggest that 3 km exhumation represents an upper bound.

A major difficulty associated with modelling exhumation using VR and AFT data, and the rationale for this work, is their susceptibility to transient hydrothermal effects such as local igneous intrusion and hydrothermal flux (e.g. Green *et al.* 1993; Sibson 1995; White & Morton 1995; Hunt 1996; Green *et al.* 2001). Because of the increased geothermal gradient accompanying these transient heating events, they anneal fission tracks and increase VR without a concomitant increase in burial depth. Consequently, VR and AFT data that have been affected by transient heating events will yield exaggerated denudation magnitudes.

There are two reasons to suppose that this study area was exposed to transient thermal effects during Late Cretaceous and Tertiary time. First, the Iceland Plume, a mantle upwelling that affected a region of the North Atlantic continents of 1000 km width, generated transient heating through (1) extensive volcanism of the North Atlantic region (White & McKenzie 1989) and (2) basaltic underplating of the continental crust conjectured beneath a wide region extending well beyond the area of extrusive volcanism (Brodie & White 1994). Second, hydrothermal fluid flow during Oligo-Miocene basin inversion is implicated in a Late Tertiary (*c.* 20 Ma) cooling event recognized from combined AFT and fluid inclusion studies (Atlantic margins, Green *et al.* 1999; Parnell *et al.* 1999; EISB, Hardman *et al.* 1993; southern England, Green *et al.* 2001). The multiple increments of fault reactivation that typify basin inversion episodes will be characterized by flushing of any overpressured fluids along faults and through their wall rocks during repeated cycles of seal breaching and repair (Sibson 1987). This fault valving can transport extraordinary volumes of hot fluid many times the volume of the pore space in which the fault valving is focused (e.g. Sibson 1995) and it may lead to substantial elevation of wall rock temperature (e.g. Andrews *et al.* 1996).

Regional setting and stratigraphy

The Irish Sea basins comprise a linked system of Mesozoic–Cenozoic depressions that, in the Cardigan Bay basin, attain a post-Carboniferous sedimentary thickness in excess of 12 km. The Mesozoic fill of the Irish Sea basins displays a continental margin-type subsidence signature (Welch & Turner 2000) interrupted by Paleocene epeirogenic uplift, generally linked to the formation of the Iceland Plume (Brodie & White 1994). However, the Irish Sea area also displays clear evidence of shortening of formerly extensional basins (basin inversion) in response to African–European plate collision during late Oligocene and Miocene time (e.g. Ziegler 1987; Roberts 1989). Miocene basin inversion led to only minor modification of the Mesozoic extensional fault geometry (Tucker & Arter 1987). The principal manifestations of inversion in this study area and contiguous basins are transpressional reactivation of steep, NW-trending faults (Turner 1997) and thickening of the Mesozoic basin fill by pure shear (cf. Eisenstadt & Withjack 1995). Although it is difficult to isolate Miocene basin inversion from the effects of the Paleocene Iceland Plume, Cenozoic uplift of the Celtic Sea basins, to the south of the present study area, was responsible for exhumation of between 600 m (Tucker & Arter 1987) and 2500 m (Menpes & Hillis 1995) of Upper Mesozoic and Cenozoic strata.

The petroleum system of the EISB is mainly confined to Triassic rocks and few wells penetrate deeper than this. The Triassic succession can be subdivided into the sandstone-rich Sherwood Sandstone Formation (Scythian; *c.* 1450 m thick in the southern East Irish Sea basin (Jackson & Mulholland 1993)) succeeded by the finer-grained Mercia Mudstone Formation (Anisian–Norian; >1500 m thick in the southern EISB (Jackson & Mulholland 1993)). The Sherwood Sandstone is a distinctive red, fluvial–aeolian arkosic sandstone sequence that constitutes the principal reservoir objective in the Irish Sea. The Mercia Mudstone consists of red mudstone and siltstone with salt units attaining up to 450 m thickness (Jackson & Mulholland

1993). None of the wells used in the analysis reported here drilled through salt of >50 m thickness, which, where encountered, was excluded from the exhumation calculations.

Methods of sonic velocity analysis

Theory and mechanical control over velocity

Until recently, obtaining exhumation from SV has been carried out by using the SV log to plot interval velocity vs. midpoint depth of fine-grained lithofacies units. Through its systematic relation with porosity (Wyllie *et al.* 1956), sonic velocity will exhibit a progressive reduction with mean effective stress ($\sigma_{HYDROSTATIC} - P_F$; where $\sigma_{HYDROSTATIC}$ is hydrostatic stress and P_F is fluid pressure; see Goulty (1998), Giles *et al.* (1998) and references therein). Consequently, in a uniformly pressured sedimentary sequence, plots of SV vs. midpoint depth will highlight anomalously 'fast' (low-porosity) successions where rocks have been exhumed from a formerly greater burial depth. Cross-plotting the exhumation results derived from different parts of the stratigraphy allows for checks on consistency (e.g. Menpes & Hillis 1995). In attempting to compute absolute estimates of denudation magnitude, two problems arise. First, because this approach has no way of defining the 'normal' velocity vs. depth trend for an unexhumed succession, all exhumation estimates will only be relative to the least exhumed log (i.e. that succession which, after the denudation episode(s), remained at or closest to its maximum burial depth). Second, conventional SV analysis relies on individual lithofacies units exhibiting near-identical compaction behaviour across the area of interest (e.g. Issler 1992). In practice, it may be difficult to discriminate lithological variation from short-wavelength changes in exhumation.

In common with other approaches to SV analysis, this study uses the exponential decrease of porosity with burial depth to compute exhumation magnitude:

$$\phi = \phi_0 \exp(-bx) \qquad (1)$$

where ϕ is porosity at depth x, ϕ_0 is surface porosity of uncompacted sediment and b is compaction coefficient per unit lithology (Athy 1930; Rubey & Hubbert 1959). P-wave velocity, as recorded by sonic logs, is a widely used measure of porosity:

$$\Delta t_{log} = \Delta t_{ma}(1 - \phi) + \phi \Delta t_f \qquad (2)$$

where ϕ is porosity, and Δt_{log}, Δt_{ma} and Δt_f are,

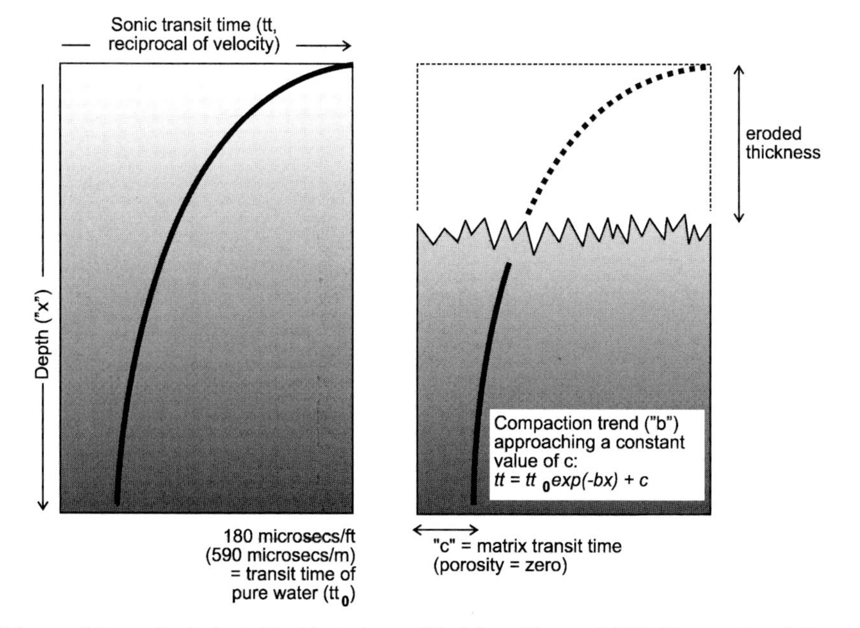

Fig. 1. Scheme of the method adopted in this study, modified from Magara (1976). The pre-denudation condition is given by the combined continuous and dashed line extending, in this study, to 76 m beneath the depositional surface.

respectively, the sonic log (measured), rock matrix and interstitial fluid transit time, the reciprocal of velocity (Wyllie *et al.* 1956). Assuming friction, attenuation and frequency dependence to be negligible, P-wave velocity can be defined in terms of the dynamic elastic moduli (Gassmann 1951). Of these, the shear modulus and rock framework bulk modulus (i.e. bulk

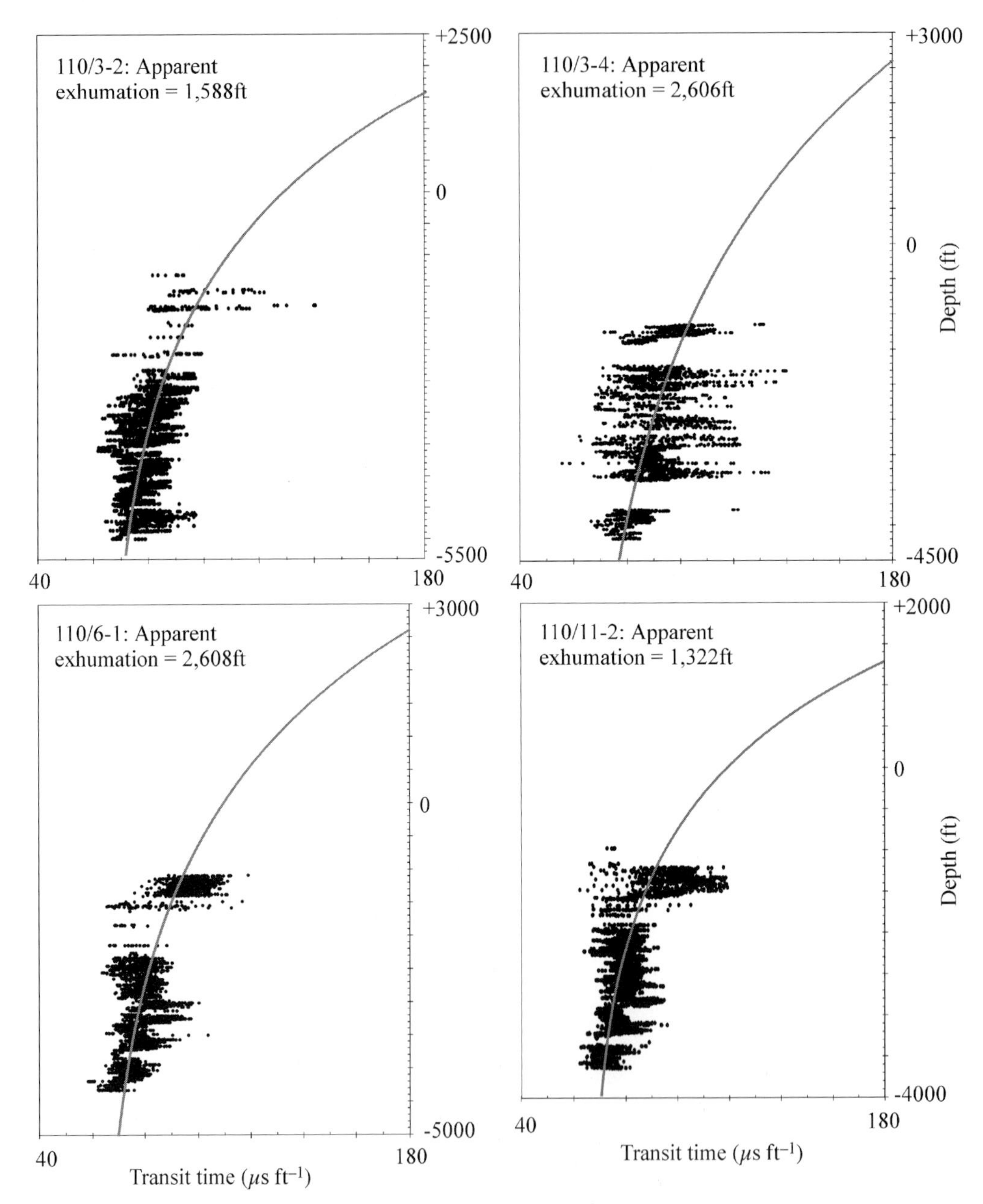

Fig. 2. Data from sonic logs and best-fit transit time vs. depth curves (pale grey) for the four most extensive shale sections in the study area. On the *y*-axis, positive and negative values indicate depths above and below the present sea floor, respectively. Exponential decay constants and shift constants for each well are: for 110/3-2, $-0.00037\,\text{ft}^{-1}$ and 63 μs ft^{-1}; for 110/3-4, $-0.0027\,\text{ft}^{-1}$ and 59 μs ft^{-1}; for 110/6-1, $-0.00035\,\text{ft}^{-1}$ and 61 μs ft^{-1}; for 110/11-2, $-0.00056\,\text{ft}^{-1}$ and 65 μs ft^{-1}.

modulus of the mineral grains) are strongly dependent on porosity and are therefore subject to potentially rapid change with burial depth.

Circumstances in which equation (2) does not work well include: (1) compaction retardation as a result of intensive cementation and/or overpressure (Erickson & Jarrard 1998); (2) porosities greater than c. 25% (Falvey & Middleton 1981; Serra 1984); (3) non-aqueous pore fluid, especially gaseous hydrocarbon (Magara 1978); (4) microcracks (Erickson & Jarrard 1998; Dewhurst *et al.* 1999). Operators' log interpretations were used to identify zones of intensive cementation, overpressure and anomalously high porosity (none encountered), and non-aqueous pore fluid. Data covering these intervals, whose porosity–depth behaviour is likely to be unpredictable, were removed during the initial phase of editing of logs before analysis of the sonic logs.

Account was also taken of evaporitic salt, which was edited out from our dataset before analysis, and poroelasticity during overburden removal. *In situ* stress relaxation generates and opens microcracks such that initial microcrack porosities of <0.5% are sufficient to cause pressure-dependent velocity variations of 5–50% (Erickson & Jarrard 1998). Poroelastic rebound during denudation is a measure of this microcrack porosity change and may cause as

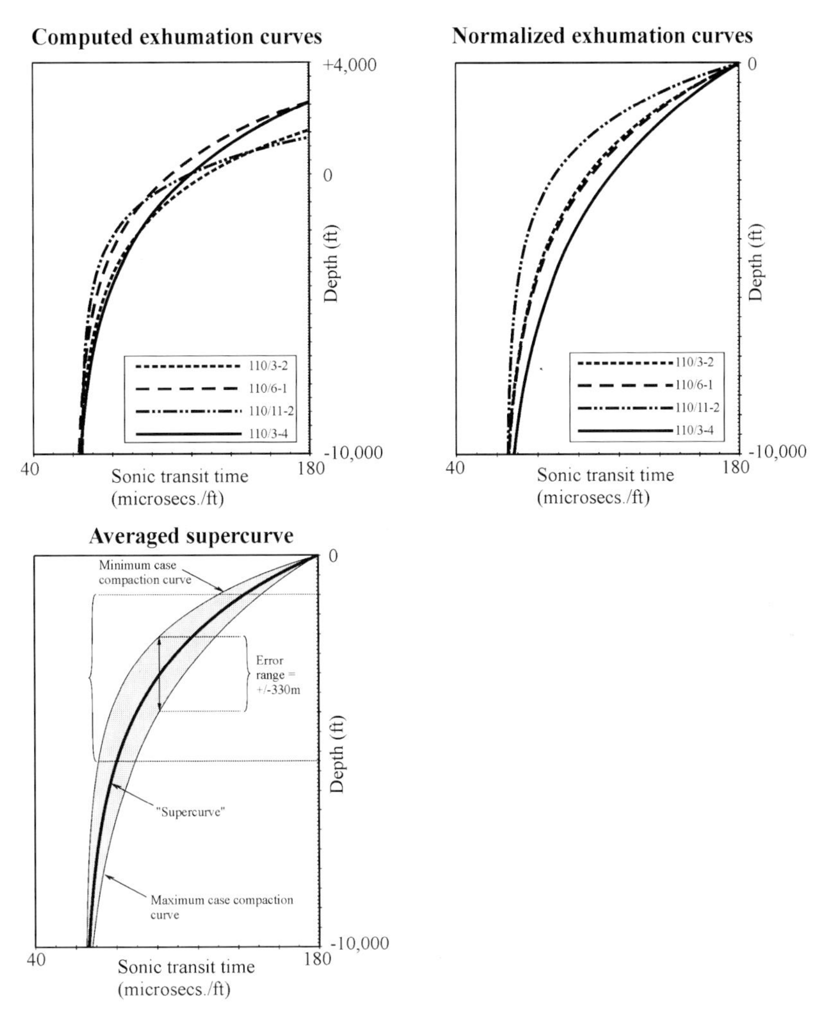

Fig. 3. Comparison of transit time vs. depth curves for the four wells in Fig. 2, before (top left) and after (top right) their normalization. Lower graph summarizes the characteristics of the resultant supercurve derived from averaging the normalized curves shown (supercurve exponential decay constant $-0.00034\ \mathrm{ft}^{-1}$; shift constant $62\ \mu\mathrm{s\ ft}^{-1}$).

Table 1. *Summary of denudation magnitudes and statistical data for wells in the East Irish Sea basin whose locations are given in Figs 4 and 6*

Well number	TVD to unconformity (m)	Water depth (m)	KBE (m TVD)	Minimum E_T before correction (m)	Maximum E_T before correction (m)	SD (m)	E_A (m)	95% CI (m)	Average E_T (m)	Correction (m)	Corrected E_T (m)
110/2-1	64	32	32	132	3341	442	888	18.1	952	140	1092
110/2-2	66	30	36	−85	1695	307	444	15.9	510	142	652
110/2-3	61	29	35	−315	4334	554	670	20.3	731	140	871
110/2-4	62	32	30	−242	4047	454	799	13.8	861	138	999
110/2-5	65	36	30	−259	2953	429	626	16.4	691	141	832
110/2-6	113	34	31	−229	2682	315	522	12.3	635	141	776
110/2-7	69	34	35	−305	5597	703	1003	27.3	1072	146	1218
110/2-8	72	36	37	−333	4324	468	658	17.7	730	148	878
110/3-1	62	29	33	−515	2649	465	498	16.4	560	138	698
110/3-2	80	22	27	−396	5957	425	577	13.1	657	126	783
110/3-3	98	24	37	40	3889	442	722	20.0	820	137	957
110/3-4	22	22	29	−526	4372	380	515	13.2	537	127	664
110/6-1	80	13	34	−65	5347	374	858	11.7	938	123	1061
110/7-1	66	32	34	27	3107	333	811	13.0	877	142	1019
110/7-2	88	37	36	201	2108	295	714	19.4	802	149	951
110/7-3	136	28	37	635	2565	247	1029	16.3	1165	141	1306
110/7-4	91	34	32	−307	2763	363	944	15.4	1035	143	1178
110/7-5	130	42	26	445	3560	351	1051	12.4	1181	145	1326
110/7-6	180	41	31	673	2578	403	1219	23.4	1399	148	1547
110/8-2	99	28	34	87	2751	418	711	16.6	810	139	949
110/9-1	71	10	37	374	7598	467	1314	18.8	1385	123	1508
110/10-1	111	20	35	310	3304	335	909	15.7	1020	132	1152
110/11-1	77	43	34	484	2896	302	1037	23.4	1114	153	1267
110/11-2	75	44	31	217	5698	382	961	12.0	1036	152	1187
110/11-3	37	46	37	−73	5156	399	797	13.1	834	159	993
110/12-1	65	34	31	602	3696	341	1256	13.4	1321	141	1462
110/12-2	71	33	38	746	3372	345	1295	34.8	1366	147	1513
110/12-3	71	38	34	430	1380	204	713	12.9	784	148	932
110/12-4	77	40	37	772	5735	361	1395	11.2	1472	153	1625
110/13-1	61	27	37	237	4461	373	1340	18.8	1401	139	1540
110/13-2	61	33	35	814	5299	515	1663	42.2	1724	144	1868
110/13-5	61	28	29	157	3008	326	1161	13.2	1222	134	1356
110/13-7	61	33	29	599	3171	342	1278	13.0	1339	139	1478
110/13-8	73	29	30	260	4029	413	956	16.2	1029	136	1165
110/13-15	66	26	34	369	5951	738	1519	26.0	1585	136	1721
110/18-1	59	19	34	270	5941	671	1944	21.4	2003	129	2132

Table 1 – *continued*

Well number	TVD to unconformity (m)	Water depth (m)	KBE (m TVD)	Minimum E_T before correction (m)	Maximum E_T before correction (m)	SD (m)	E_A (m)	95% CI (m)	Average E_T (m)	Correction (m)	Corrected E_T (m)
110/20-1	38	5	33	775	2196	223	1249	16.0	1287	114	1401
112/25-1	73	15	38	239	4589	523	754	47.5	827	129	956
112/30-1	83	41	33	297	4513	419	1247	29.1	1330	150	1480
113/26-1	75	42	34	−155	3001	431	672	21.1	747	151	898
113/27-1	101	35	35	422	4742	359	676	41.0	777	146	923
113/27-2	113	28	37	−1	5716	460	950	15.7	1063	141	1204
Mean	257	100	110	162	3954	407	984	19.0	1038	140	1179
Maximum	591	150	125	814	7598	738	1944	48.0	2003	159	2132
Minimum	72	16	86	−526	1380	204	444	11.0	510	114	652

TVD, true vertical depth. Note that in addition to subtracting the drill floor elevation (KBE) and referencing the denudation to the present sea floor, the correction also takes account of a critical porosity depth of 76 m. At depths shallower than this and/or where porosity exceeds 62%, sound waves will travel through interstitial water, hence porosity vs. depth relations are not predictable (see Dickinson 1953; Magara 1976; Pirmez *et al.* 1997).

much as 7% porosity change following exhumation (Hamilton 1976).

The chief control over sonic velocity in undeformed sedimentary successions is therefore burial-controlled mechanical compaction as a result of porosity reduction:

$$\Delta t = \Delta t_0 \exp(-bx) \qquad (3)$$

where Δt is sonic transit time at depth x, and Δt_0 is surface transit time of uncompacted sediment. However, although Athy's Law (1) correctly predicts negligible porosity at depth, sonic velocity in a totally compacted rock will equal that of the rock matrix (Heasler & Kharitonova 1996). Consequently, the correct functional relationship between sonic transit time and depth is

$$\Delta t = \Delta t_0 \exp(-bx) + c \qquad (4)$$

where c is sonic transit time of the rock matrix, or shift constant. Typical sedimentary rock matrix transit times range between 128 μs m^{-1} (c. 7800 m s^{-1}) for dolomites and 223 μs m^{-1} (*c.* 4500 m s^{-1}) for shales (Schlumberger 1989). On a linear plot of sonic transit time vs. depth, the matrix transit time approximates to the asymptote of the curve, which generally flattens out below depths of 2500 m, where porosity approaches zero.

Curve-fitting, geological error and derivation of exhumation

To compute magnitude of denudation from SV data, we modify the scheme of Magara (1976) and Heasler & Kharitonova (1996), in which a compaction curve is fitted through logarithmically transformed transit time vs. depth data (Fig. 1). By varying the shift constant (c in equation (4)), average absolute value and root mean square are used to optimize the fit of the compaction curve to the sonic data. This study follows Heasler & Kharitonova (1996) in assuming that porosity reduction with depth in a heterolithic sedimentary sequence can be described by a single, average compaction coefficient (b in equations (1), (3) and (4); see also Steckler & Watts 1978; Tosaya & Nur 1982; Castagna *et al.* 1985; Han *et al.* 1986; Marion *et al.* 1992).

The best-fit transit time vs. depth curve is extrapolated above the erosion surface (i.e. unconformity or the present basin floor) to the level at which transit time equals that of uncompacted sediment (Fig. 1). Transit times of uncompacted sediment will vary according to

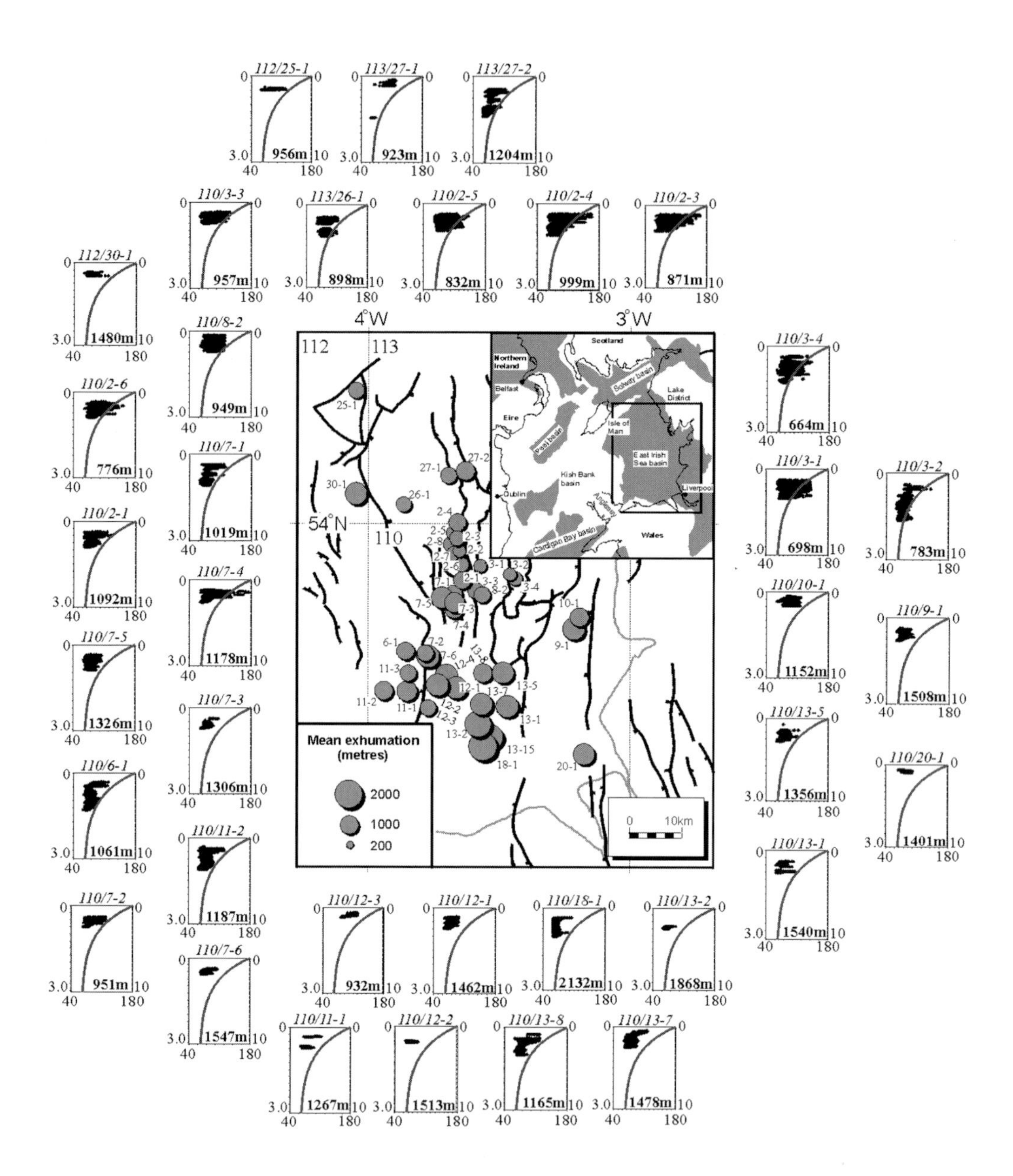

Fig. 4. Variation in denudation relative to the present sea floor at 42 well sites for the East Irish Sea basin, each identified by quad, block and well number. Data for 36 of these wells are shown in the graphs. Bold lines give the traces of principal faults; inset shows the location of the study area (boxed) with exhumed Triassic sedimentary basins shaded grey. The diameter of each circle is proportional to mean total exhumation, given also in bold in the corresponding graph. Sonic velocities are dark grey; pale grey line is the velocity vs. depth supercurve in the unexhumed case. Vertical difference in depth (kilometres on the left-hand *y*-axis, thousands of feet on the right) between the supercurve and its corresponding sonic datapoint indicates exhumation magnitude for that point. Horizontal scatter in sonic datapoints (in $\mu s\ ft^{-1}$) provides a qualitative measure of lithological variation and noise in the data.

bulk sediment porosity, water saltiness, temperature and pressure. Following the approach of Magara (1976), a surface transit time of 591 μs m^{-1} (180 μs ft^{-1}, sonic velocity 1695 m s^{-1}) is adopted.

The geological error is defined here as

$$\text{geological error} = \text{poroelasticity} - \text{temperature effects.}$$

Hamilton (1976) demonstrated that poroelasticity can lead to absolute porosity change of up to 7%. Converting this into transit time gives a reduction of 12.6 μs ft^{-1} for unburied sediment, producing a maximum underestimate of denudation in our data of some 155 m (510 ft), depending on the decay constant of the compaction curve. P-wave velocities in low-porosity sediments increase by up to 1.7% for a 100 °C rise in temperature (Timur 1977). Assuming a geothermal gradient of 33 °C km^{-1}, this temperature-dependent velocity behaviour will lead to underestimation of denudation by up to 40 m (129 ft) in our dataset, again depending on the decay constant. In this study the maximum total geological error is therefore + 115 m (+377 ft). Given that geological errors are positive, all our exhumation values will be underestimates. The small magnitude of geological error with respect to the total denudation values means that they have been ignored in our subsequent interpretation.

Only nine of the wells analysed contain sufficient thickness of shaley section and/or depth of well penetration to produce a statistically acceptable population of sonic data to which a compaction curve could be fitted (Fig. 2). The solution was to compare these data with a 'supercurve', a generic compaction curve for the EISB derived from averaging normalized transit time vs. depth curves from the four most extensive available shale sections. This study

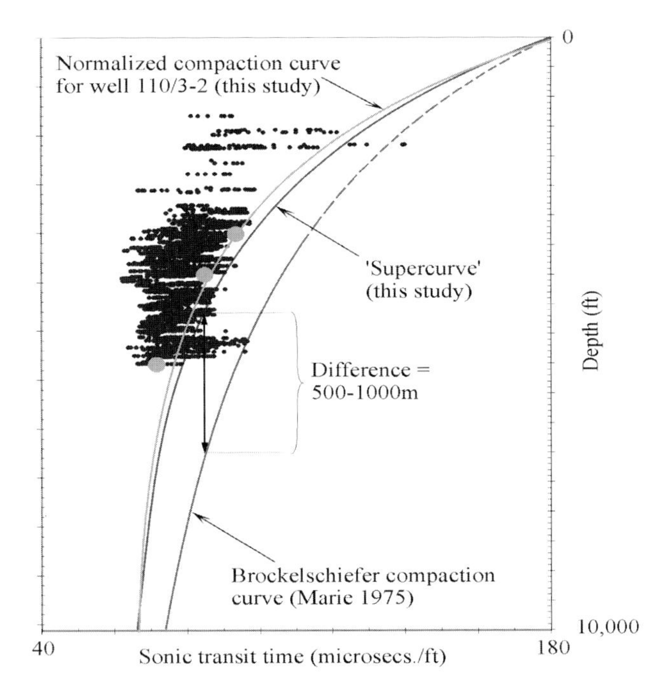

Fig. 5. Comparison of denudation magnitudes computed from well 110/3-2 obtained by comparing sonic velocities in its extensive Mercia Mudstone succession with the supercurve from this study, and transit time vs. depth curves derived from normalized best-fit curves through 110/3-2 and the southern North Sea Rotliegende Brockelschiefer Shale curve of Marie (1975). The pale grey polka-dots indicate the transit time–depth pairs used by Colter (1978) to compute his exhumation magnitude for the East Irish Sea basin.

uses a supercurve derived from sonic logs from wells 110/3-2, 110/6-1, 110/11-2 and 110/3-4 (Fig. 2), each comprising more than 8000 ft (2438 m) of shaley section. Normalization of each curve, such that they intersect the present surface with a transit time of 591 μs m^{-1}, provides us with well-constrained approximations of the form of the unexhumed compaction curve, the average of which yields the supercurve (Fig. 3).

The difference between the supercurve and each datapoint on a sonic log gives a single denudation magnitude, the mean value of which yields apparent mean total exhumation (E_A; Table 1). Because post-exhumational burial can have the effect of returning formerly exhumed rocks to their maximum burial depth, it is necessary to add to E_A the thickness of sediment accumulated since exhumation, to derive mean total exhumation (E_T; Table 1). In this study, we therefore add the thickness of Cenozoic rocks to E_A, to derive E_T relative to the present sea floor at each well location (Fig. 4).

Results

The graphs in Fig. 4 show a large degree of scatter in almost every well, with a similar range of values between wells in which the minimum transit time is generally very close to the asymptote of the velocity–depth curve, and the maximum is close to the supercurve. Comparing exhumation computed from these data (Table 1) with AFT- and VR-derived exhumation, the sonic velocities yield consistently lower values than AFT and VR. Mean exhumation in the EISB ranges between 652 and 2132 m (mean 1179 m, standard deviation 338 m). Contrary to AFT- and VR-based exhumation studies, where estimates of denudation exceed 3000 m (AFT; Lewis *et al.* 1992) and 2500 m (VR; Rowley & White 1998), we compute exhumation of >2000 m in only one well (110/18-1; 2132 m), almost all our values being <1500 m. These results accord with estimates from subsidence modelling (Rowley & White 1998), which, like SV data, are also less susceptible to the distorting effects of transient heating.

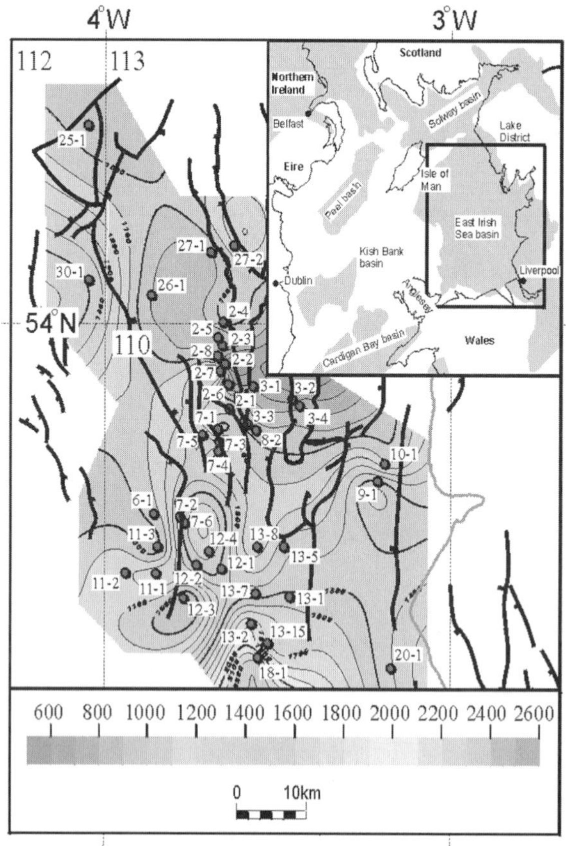

Fig. 6. Denudation contoured from the results presented in Fig. 4 using a standard convergent least-squares gridding algorithm.

In one of the few other studies using SV to compute denudation magnitude in the EISB, Colter (1978) estimated *c*. 2000 m exhumation from well 110/3-2 (this study; $E_T = 783$ m). He compared transit time vs. depth data against a generic compaction curve based on the Rotliegende Brockelschiefer Shale of the southern North Sea (Marie 1975; Fig. 5). Similar results were obtained by Jackson *et al.* (1987). Conversely, our results, especially those from Quad 110 in the southern part of this study area, are in broad agrement with those of Rowley & White (1998), who used VR data to constrain forward models of lithospheric subsidence.

Discussion

Exhumation results from the positive feedback between uplift, erosion and isostasy that brings formerly deeply buried rocks to the surface. Like most major unconformities, the uppermost Mesozoic unconformity in the Irish Sea basins is a composite surface recording several separate uplift and erosion events. Thus, it could be a product of uplift related to North Atlantic opening in Paleocene time, shortening in Oligo-Miocene time, Pleistocene glacio-isostasy or a combination of these. Moreover, hydrothermal fluid flux associated with both the formation of the Atlantic Ocean margin in Paleocene time and shortening during Oligo-Miocene time means that the various proxies used to measure denudation magnitude also require careful interpretation to identify artefacts. As well as the susceptibility of thermal history methods to transient heating episodes, we speculate that the relatively large scatter in the EISB sonic velocity data may record diagenetic effects and differing

i) Cretaceous: Isostatic equilibrium

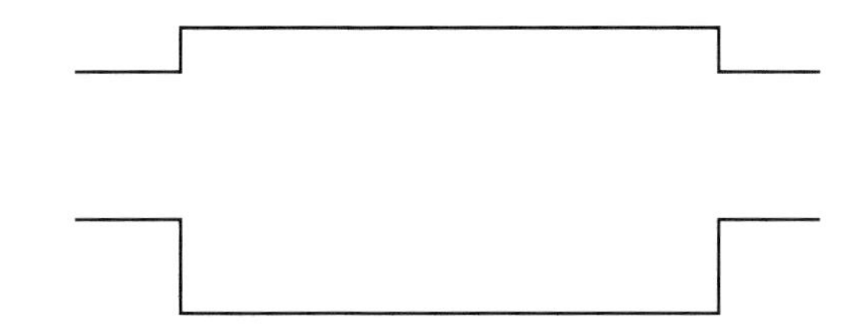

ii) Palaeocene: Magmatic underplating and regionally uniform unconformity development (c. 600m-800m exhumation)

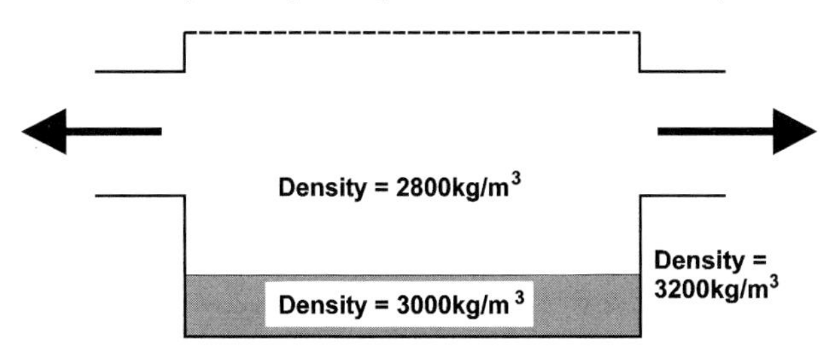

iii) Eocene-Miocene: 'Alpine-Pyrenean' compression and basin inversion (c. 600m -1200m exhumation)

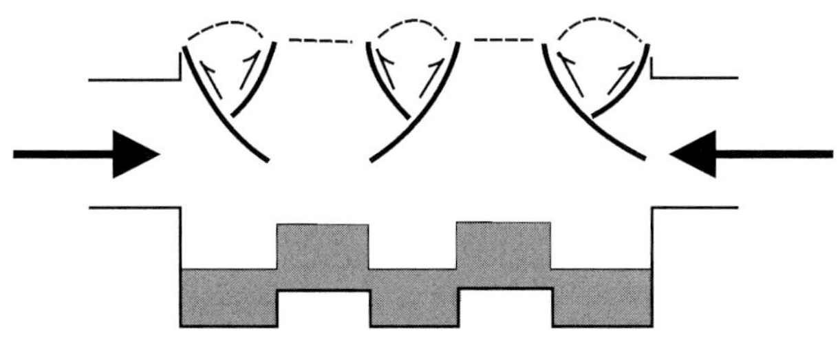

Fig. 7. Model of the uppermost Mesozoic unconformity in the UK Irish Sea, interpreting it as a product of shortening (basin inversion) superimposed on epeirogenesis (here represented by magmatic underplating).

degrees of cementation related directly to Tertiary hydrothermal fluids.

Comparison of overall trends in the SV data with AFT-derived palaeotemperature results reveals no obvious relationship. AFT data show both absolute denudation magnitude and early Tertiary palaeotemperature gradients decreasing southward (Green *et al.* 1997) whereas the SVs indicate denudation increasing southward (Fig. 6). Furthermore, many of the wells whose SV logs are analysed here have AFT-derived early Tertiary (*c.* 60 Ma) palaeotemperatures around 110 °C at depths of 1 km or less (Green, pers. comm.). Hardman *et al.* (1993) reported the apparent discrepancy between VR data from an EISB well, which indicate almost no anomalous palaeotemperatures (and therefore denudation), and apatites from throughout the same well, which were totally annealed. Collectively, these observations are interpreted as a record of early Tertiary hydrothermal circulation within the Collyhurst and Sherwood Sandstone Formation aquifers driven by heating of intra-formational brines during igneous intrusion and/or flushing of

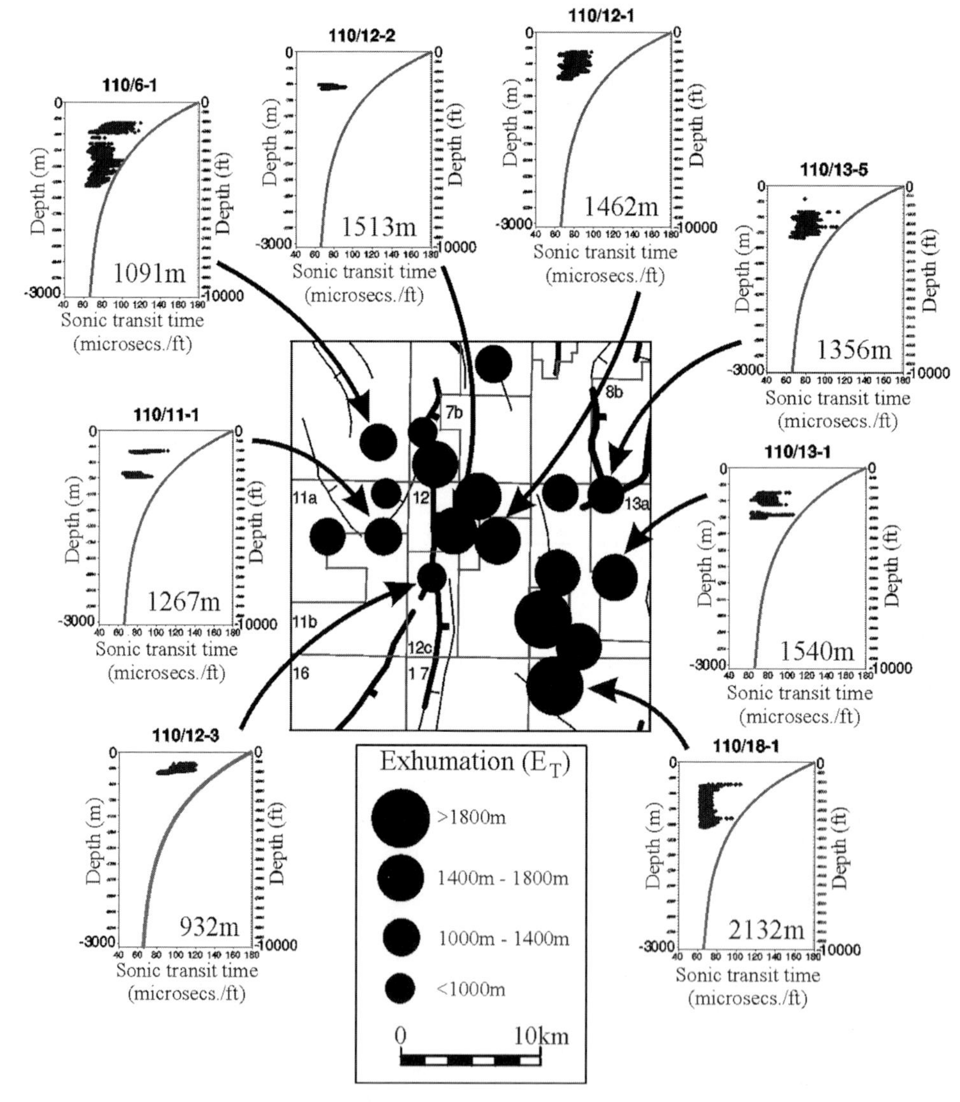

Fig. 8. Detail of the variation in denudation magnitude across faults (shown in bold, ticks in hanging walls) obtained from wells in Quad 110 of the southern East Irish Sea basin. Notations on the graphs are the same as in Fig. 4.

geofluids from deep, overpressured basins (see Iliffe *et al.* 1999).

In the EISB, denudation varies substantially between adjacent fault blocks (Figs 4 and 6), thereby providing indirect support for the contribution of basin inversion to its exhumation. In Fig. 7, this short-wavelength pattern of denudation is interpreted as a record of locally variable shortening superimposed on a uniform peneplain generated by thermal uplift during Atlantic opening (e.g. underplating). This model is consistent with the conclusions of Hardman *et al.* (1993), who described large intra-basinal variation in exhumation magnitude within the EISB, which they interpreted in terms of an early Tertiary denudation episode of *c.* 15 Ma duration, followed by later denudation lasting *c.* 50 Ma. It is also consistent with thermal history analysis of borehole samples from Oxfordshire, where new AFT data with late Tertiary ages record palaeogeothermal gradients up to 90 °C km^{-1}, interpreted as a consequence of flushing of hot fluids during the inversion of the adjacent Wessex basin (Green *et al.* 2001).

Given the above model of superimposed basin inversion and thermal uplift in the EISB, maximum tectonic uplift, and therefore exhumation, would be concentrated along the northern and southern basin margins whose orientation was most favourable for reactivation in the roughly north–south Oligo-Miocene compressive stress field. Thus, denudation increases from the present centre of the basin to the southern basin margin, where the SV analysis records maximum denudation (see also enhanced denudation in the hanging walls of the faults in Fig. 8). Major shortening at basin margins is exemplified by the St. Tudwal's Arch, a Palaeozoic basement high separating the Carnarfon and Cardigan Bay basins, contiguous to this study area. Line-length restoration of the uppermost Lower Jurassic marker across this high indicates that compression in Tertiary time caused bulk shortening of some 15 km across a basin of *c.* 100 km width. We speculate that the present basin-and-high configuration of the Irish Sea area is a product of Neogene compressional reactivation of Palaeozoic basement faults beneath a once-continuous basin system.

Why then is basin inversion such a difficult process to recognize in the EISB area? First, the compressional reactivation of formerly extensional faults means that net reverse fault displacement along an 'inverted' fault will decrease downward. In deeply exhumed basins such as the EISB, the preserved segments of most of the inverted faults will often retain a net normal displacement, even though they may have undergone significant reverse reactivation. Second, not all, or even any, of the shortening during basin inversion is accommodated by fault reactivation. In muddy or overpressured basin fills, shortening will take place by pure shear thickening of the entire basin fill, making it much harder to measure (see Eisenstadt & Withjack 1995).

The significance of pure shear is demonstrated in the inverted St. George's Channel and Cardigan Bay basins, where Oligo-Miocene shortening was accommodated by thickening of the *c.* 10 km thick, fine-grained siliciclastic succession, without much evidence of fault reactivation. Unfortunately, there are few published data on the relative magnitudes of the principal stresses during tectonic shortening and, therefore, it is not possible to comment on the likely change in the magnitude of the mean effective stress during the switch from extension to contraction in the EISB. However, given that a condition for horizontal shortening is that the maximum principal stress is horizontal, a fraction of the denudation magnitudes presented here may be attributable to shortening-induced porosity reduction.

Sonic velocity analysis also provides potentially important information for seismic time-to-depth conversion. Accurate time-to-depth conversion is particularly sensitive to anomalously slow or fast intervals, leading to unreliable estimates of depth to formation tops. This point was demonstrated by EISB well 110/12-3 (Fig. 8), which drilled through a sonically slow mudstone overburden to the reservoir, which was relatively undercompacted compared with the same sequence in surrounding wells. Consequently, the depth to the top of the main reservoir was significantly shallower than had been predicted by the time-to-depth conversion, with attendant ramifications for drilling safety.

Conclusions

(1) Sonic velocity analysis indicates that the EISB underwent major exhumation during which between 652 and 2132 m of overburden was denuded.

(2) SV-derived estimates of thickness of denuded overburden are consistently lower than AFT and VR results. Given the susceptibility of thermal history methods to transient heating, AFT and VR may have overestimated denudation, mainly because of the influence of Tertiary hydrothermal fluid flow.

(3) Contrary to denudation trends from AFT studies, which show a southward decrease, SVs indicate a southward increase in denudation,

toward the faulted margins of the EISB. Together with the distinct short-wavelength variation in intra-basin denudation between adjacent fault compartments, these observations provide indirect support for the importance of basin inversion in the EISB.

We are grateful to Enterprise Oil plc for provision of data and funding for P.D.W.'s MPhil project on which this work is based. Discussions with P. Green, W. Owens, M. Stephenson, R. Swarbrick, K. Thomson, G. Westbrook and N. White were useful and encouraging. We acknowledge helpful reviews by D. James and J. Cartwright.

References

ABBOTT, L.D., SILVER, E.A., ANDERSON, R.S. & 8 OTHERS 1997. Measurement of tectonic surface uplift rate in a young collisional mountain belt. *Nature*, **385**, 501–507.

ANDREWS, J.R., DAY, J. & MARSHALL, J.E.A. 1996. A thermal anomaly associated with the Rusey Fault and its implication for fluid movements. *Proceedings of the Ussher Society*, **9**, 68–71.

ATHY, L.F. 1930. Density, porosity and compaction of sedimentary rocks. *AAPG Bulletin*, **14**, 1–24.

BRODIE, J. & WHITE, N.J. 1994. Sedimentary basin inversion caused by igneous underplating. *Geology*, **22**, 147–150.

CASTAGNA, J.P., BATZLE, M.L. & EASTWOOD, R.L. 1985. Relationships between compressional wave and shear wave velocities in clastic rocks. *Geophysics*, **50**, 571–581.

CLOETINGH, S. 1988. Intraplate stresses: a new element in basin analysis. *In*: KLEINSPEHN, K.L. & PAOLA, C. (eds) *New Perspectives in Basin Analysis*. Springer, New York, 205–230.

COLTER, V.S. 1978. Exploration for gas in the Irish Sea. *Geologie en Mijnbouw*, **57**, 503–516.

COPE, J.C.W. 1998. The Mesozoic and Tertiary history of the Irish Sea. *In*: MEADOWS, N.S., TRUEBLOOD, S.P., HARDMAN, M. & COWAN, G. (eds) *Petroleum Geology of the Irish Sea and Adjacent Areas*. Geological Society, London, Special Publications, **124**, 47–59.

DEWHURST, D.N., YANG, Y. & APLIN, A.C. 1999. Fluid flow through natural mudstones. *In*: APLIN, A.C., FLEET, A.J. & MACQUAKER, J.H.S. (eds) *Muds and Mudstones: Physical and Fluid Flow Properties*. Geological Society, London, Special Publications, **158**, 23–43.

DICKINSON, G. 1953. Geological aspects of abnormal reservoir pressures in Gulf Coast Louisiana. *AAPG Bulletin*, **37**, 410–432.

DORÉ, A.G. & JENSEN, L.N. 1996. The impact of Late Cenozoic uplift and erosion on hydrocarbon exploration; offshore Norway and some other uplifted basins. *Global and Planetary Change*, **12**, 415–436.

DUNCAN, W.I., GREEN, F. & DUDDY, I.R. 1998. Sourcerock burial history and seal effectivness; key facets to understanding hydrocarbon exploration potential in the East and central Irish Sea basins. *AAPG Bulletin*, **82**, 1401–1415.

EISENSTADT, G. & WITHJACK, M.O. 1995. Estimating inversion: results from clay models. *In*: BUCHANAN, J.G. & BUCHANAN, G. (eds) *Basin Inversion*. Geological Society, London, Special Publications, **88**, 119–137.

ERICKSON, S.N. & JARRARD, R.D. 1998. Velocity–porosity relationships for water-saturated siliciclastic sediments. *Journal of Geophysical Research*, **103** (B12), 30385–30406.

FALVEY, D.A. & MIDDLETON, M.F. 1981. Passive continental margins. Evidence of a pre-break-up deep crustal metamorphic subsidence mechanism. *Proceedings of the 26th International Geological Congress. Geology of Continental Margins Symposium, Colloquium*. Oceanologica Acta, Paris, 103–114.

GASSMANN, R. 1951. Elastic waves through a packing of spheres. *Geophysics*, **16**, 673–685.

GILES, M.R., INDRELID, S.L. & JONES, D.M.D. 1998. Compaction—the great unknown in basin modelling. *In*: DUPPENBECKER, S.J. & ILIFFE, J.E. (eds) *Basin Modelling: Practice and Progress*. Geological Society, London, Special Publications, **141**, 15–43.

GOULTY, N.R. 1998. Relationships between porosity and effective stress in shales. *First Break*, **16**, 413–419.

GREEN, P.F., DUDDY, I.R. & BRAY, R.J. 1993. Early Tertiary heating in Northwest England: fluids or burial (or both)? (Extended abstract). *In*: PARNELL, J., RUFFELL, A.H. & MOLES, N.R. (eds) *Contributions to an International Conference on Fluid Evolution, Migration and Interaction in Rocks, Belfast*. 119–123.

GREEN, P.F., DUDDY, I.R. & BRAY, R.J. 1997. Variation in thermal history styles around the Irish Sea and adjacent areas; implications for hydrocarbon occurrence and tectonic evolution. *In*: MEADOWS, N., TRUEBLOOD, S., COWAN, G. & HARDMAN, M. (eds) *Petroleum Geology of the Irish Sea and Adjacent Areas*. Geological Society, London, Special Publications, **124**, 73–93.

GREEN, P.F., DUDDY, I., HEGARTY, K.A. & BRAY, R.J. 1999. Early Tertiary heat flow along the UK Atlantic margin and adjacent areas. *In*: FLEET, A.J. & BOLDY, S.A.R. (eds) *Petroleum Geology of Northwest Europe: Proceedings of the 5th Conference*. Geological Society, London, 349–357.

GREEN, P.F., THOMSON, K. & HUDSON, J.D. 2001. Recognition of tectonic events in undeformed regions: contrasting results from the Midland Platform and East Midlands Shelf, Central England. *Journal of the Geological Society, London*, **158**, 59–73.

HAMILTON, E.L. 1976. Variations of density and porosity with depth in deep-sea sediments. *Journal of Sedimentary Petrology*, **146**, 280–300.

HAN, D., NUR, A. & MORGAN, D. 1986. The effects of porosity and clay content on wave velocities of sandstones. *Geophysics*, **51**, 2093–2097.

HARDMAN, M., BUCHANAN, J., HERRINGTON, P. & CARR, A. 1993. Geochemical modelling of the East Irish Sea Basin. *In*: PARKER, J.R. (ed,) *Petroleum Geology of Northwest Europe: Proceedings of the 4th Conference*. Geological Society, London, 791–808.

HEASLER, H. & KHARITONOVA, N.A. 1996. Analysis of sonic well logs applied to erosion estimates in the Bighorn Basin, Wyoming. *AAPG Bulletin*, **80**, 630–646.

HUNT, J.M. 1996. *Petroleum Geochemistry and Geology*. 2; W. H. Freeman, New York.

ILIFFE, J.E., ROBERTSON, A.G., WARD, G.H.F., WYNN, C., PEAD, S.D.M. & CAMERON, N. 1999. The importance of fluid pressures and migration to the hydrocarbon prospectivity of the Faeroe–Shetland White Zone. *In*: FLEET, A.J. & BOLDY, S.A.R. (eds) *Petroleum Geology of Northwest Europe: Proceedings of the 5th Conference*. Geological Society, London, 601–611.

ISSLER, D.R. 1992. A new approach to shale compaction and stratigraphic restoration, Beaufort–Mackenzie basin and Mackenzie corridor, northern Canada. *AAPG Bulletin*, **76**, 1170–1189.

JACKSON, D.I. & MULHOLLAND, P. 1993. Tectonic and stratigraphic aspects of the East Irish Sea Basin and adjacent areas: contrasts in their post-Carboniferous structural styles. *In*: PARKER, J.R. (ed.) *Petroleum Geology of Northwest Europe: Proceedings of the 4th Conference*. Geological Society, London, 791–808.

JACKSON, D.I., JACKSON, A.A., EVANS, D., WINGFIELD, R.T.R., BARNES, R.P. & ARTHUR, M.J., 1995. *The Geology of the Irish Sea*. United Kingdom Offshore Regional Report. British Geological Survey. HMSO, London.

JACKSON, D.I., MULHOLLAND, P., JONES, S. & WARRINGTON, G. 1987. The geological framework of the East Irish Sea basin. *In*: BROOKS, J. & GLENNIE, K.W. (eds) *Petroleum Geology of Northwest Europe*. Graham and Trotman, London, 191–203.

JAPSEN, P. & CHALMERS, J.A. 2000. Neogene uplift and tectonics around the North Atlantic: overview. *Global and Planetary Change*, **24**, 165–173.

KNIPE, R.J., COWAN, G. & BALENDRAM, V.S. 1993. The tectonic history of the East Irish Sea Basin with reference to the Morecambe Fields. *In*: PARKER, J.R. (ed.) *Petroleum Geology of Northwest Europe: Proceedings of the 4th Conference*. Geological Society, London, 857–866.

LAMBECK, K. 1991. Glacial rebound and sea level change in the British Isles. *Terra Nova*, **3**, 379–389.

LEWIS, C.L.E., GREEN, F., CARTER, A. & HURFORD, A.J. 1992. Elevated K/T palaeotemperatures throughout Northwest England: three kilometres of Tertiary erosion? *Earth and Planetary Science Letters*, **112**, 131–145.

MAGARA, K. 1976. Thickness of removed sedimentary rocks, paleopore pressure and paleotemperature, southwestern part of Western Canada basin. *AAPG Bulletin*, **60**, 554–566.

MAGARA, K. 1978. *Compaction and Fluid Migration. Practical Petroleum Geology*. Elsevier, Amsterdam.

MAINGARM, S., IZATT, C., WHITTINGTON, R.J. & FITCHES, W.R. 1999. Tectonic evolution of the southern–central Irish Sea basin. *Journal of Petroleum Geology*, **22**, 287–304.

MARIE, J.P.P. 1975. Rotliegendes stratigraphy and diagenesis. *In*: WOODLAND, A.W. (ed.) *Petroleum and the Continental Shelf of NW Europe*. Institute of Petroleum, London, 205–211.

MARION, D., NUR, A., YIN, H. & HAN, D. 1992. Compressional velocity and porosity in sand–clay mixtures. *Geophysics*, **57**, 554–563.

MENPES, R.J. & HILLIS, R.R. 1995. Quantification of Tertiary exhumation from sonic velocity data. *In*: BUCHANAN, J.G. & BUCHANAN, G. (eds) *Basin Inversion*. Geological Society, London, Special Publications, **88**, 91–207.

NADIN, A., KUSZNIR, N.J. & TOTH, J. 1995. Transient regional uplift in the Early Tertiary of the northern North Sea and the development of the Iceland Plume. *Journal of the Geological Society, London*, **152**, 953–959.

PARNELL, J., CAREY, P.F., GREEN, P. & DUNCAN, W. 1999. Hydrocarbon migration history, west of Shetland: integrated fluid inclusion and fission track studies. *In*: FLEET, A.J. & BOLDY, S.A.R. (eds) *Petroleum Geology of Northwest Europe: Proceedings of the 5th Conference*. Geological Society, London, 613–625.

PIRMEZ, C., FLOOD, R.D., BAPTISTE, J., HEZHU, Y. & MANLEY, P.L. 1997. Clay content, porosity and velocity of Amazon fan sediments determined from ODP Leg 155 cores and wireline logs. *Geophysical Research Letters*, **24** (3), 317–320.

PLATT, J. & ENGLAND, C. 1993. Convective removal of lithosphere beneath mountain belts: thermal and mechanical consequences. *American Journal of Science*, **293**, 307–336.

ROBERTS, D.G. 1989. Basin inversion in and around the British Isles. *In*: COOPER, M.A. & WILLIAMS, G.D. (eds) *Inversion Tectonics*. Geological Society, London, Special Publications, **44**, 131–153.

ROWLEY, E. & WHITE, N.J. 1998. Inverse modelling of extension and denudation in the East Irish Sea and surrounding areas. *Earth and Planetary Science Letters*, **161**, 57–71.

RUBEY, W.W. & HUBBERT, M.K. 1959. Role of fluid pressure in mechanics of overthrust faulting II. *Geological Society of America Bulletin*, **70**, 167–206.

SCHLUMBERGER, 1989. Sonic logs. *In*: v, v (ed.) *Log Interpretation Principles/Applications*. Schlumberger Educational Services, Houston, TX, 5.1–5.8.

SERRA, O. 1984. *Fundamentals of Well Log Interpretation 1. Acquisition of Logging Data*. Elsevier, Amsterdam.

SIBSON, R.H. 1987. Earthquake rupturing as a mineralizing agent in hydrothermal systems. *Geology*, **15**, 701–704.

SIBSON, R.H. 1995. Selective fault reactivation during basin inversion: potential for fluid redistribution

through fault-valve action. *In*: BUCHANAN, J.G. & BUCHANAN, G. (eds) *Basin Inversion*. Geological Society, London, Special Publications, **88**, 3–21.
STECKLER, M.S. & WATTS, A.B. 1978. Subsidence of the Atlantic-type continental margin off New York. *Earth and Planetary Science Letters*, **42**, 1–13.
TIMUR, A. 1977. Temperature dependence of compressional and shear wave velocities in rocks. *Geophysics*, **42** (5), 950–956.
TOSAYA, C. & NUR, A. 1982. The effects of diagenesis and clays on compressional velocities in rocks. *Geophysics Research Letters*, **9**, 5–8.
TUCKER, R.M. & ARTER, G. 1987. The tectonic evolution of the North Celtic Sea and Cardigan Bay basins with special reference to basin inversion. *Tectonophysics*, **137**, 291–307.
TURNER, J.P. 1997. Strike-slip fault reactivation in the Cardigan Bay basin. *Journal of the Geological Society, London*, **154**, 5–8.
WELCH, M.J. & TURNER, J.P. 2000. Triassic–Jurassic development of the St. George's Channel basin, offshore Wales, UK. *Marine and Petroleum Geology*, **17**, 723–750.
WHITE, R.S. & MCKENZIE, D. 1989. Magmatism at rift zones: the generation of volcanic continental margins and flood basalts. *Journal of Geophysical Research*, **94**, 7685–7729.
WHITE, R.S. & MORTON, A.C. 1995. Edited set of papers on: The Iceland Plume and its Influence on the Evolution of the North Atlantic. *Journal of the Geological Society, London*, **152**, 933–1047.
WYLLIE, M.R.J., GREGORY, J.A.R. & GARDNER, L.W. 1956. Elastic wave velocities in heterogeneous and porous media. *Geophysics*, **21**, 41–70.
ZIEGLER, A. 1987. Late Cretaceous and Cenozoic intraplate compressional deformations in the Alpine foreland—a geodynamic model. *Tectonophysics*, **137**, 389–420.
ENGLAND, P.C. & MOLNAR, P. 1990. Surface uplift, uplift of rocks and the exhumation of rocks. *Geology*, **18**, 1173–1177.

The post-Variscan thermal and denudational history of Ireland

PHILIP A. ALLEN[1,5], STUART D. BENNETT[1], MICHAEL J. M. CUNNINGHAM[1], ANDY CARTER[2], KERRY GALLAGHER[3], ERIC LAZZARETTI[1], JOSEPH GALEWSKY[1], ALEX L. DENSMORE[1, 5], W. E. ADRIAN PHILLIPS[1], DAVID NAYLOR[4,6] & CRISTINA SOLLA HACH[1]

[1]*Department of Geology, Trinity College, Dublin 2, Ireland*

[2]*Research School of Geological and Geophysical Sciences, Birkbeck College and University College London, Gower Street, London WC1E 6BT, UK*

[3]*T. H. Huxley School of Environment, Earth Science and Engineering, Imperial College of Science, Technology and Medicine, RSM Building, Prince Consort Road, London SW7 2BP, UK*

[4]*ERA-Maptec Ltd, 36, Dame Street, Dublin 2, Ireland*

[5]*Present address: Department of Earth Sciences, ETH-Zentrum NO, Sonneggstrasse 5, CH-8092 Zürich, Switzerland (e-mail: philip.allen@erdw.ethz.ch)*

[6]*Present address: Dromreagh, Durrus, Bantry, Co. Cork, Ireland*

Abstract: The thermal and denudational history of Ireland is evaluated using an extensive new apatite fission-track (AFT) dataset derived from surface samples. Modelled thermal histories are used to construct maps of denudation for a number of time slices from Triassic time to 10 Ma using a time-dependent palaeogeotherm. The maps illustrate the spatial variability of denudation and subsidence within each time slice. The patterns of denudation are complex, showing considerable variability at the length scale of 10^1-10^2 km, with especially high denudation rates found over known igneous centres such as the Mournes of County Down. Based on the onshore AFT data alone, there is no definitive signature of an Irish Sea Dome extending significantly across Ireland in Early Tertiary time. The cumulative amount of denudation during Tertiary time varies depending on the AFT annealing model used, but is generally in the region between 1 and 2 km and without clear spatial trends. High amounts of denudation have been mapped over the Tertiary intrusions in County Down, reflecting their unroofing since emplacement in Paleocene time. The cumulative denudation from Triassic time to 10 Ma shows relatively low amounts of denudation (<2 km) in the Irish Midlands and the extreme NE of the island, consistent with the observation that Mesozoic–Tertiary sediments and igneous products are preserved in the Ulster Basin. The western flank of Ireland and the region between Dublin and County Down show high cumulative amounts of denudation (<4 km), the latter being consistent with the high amounts of denudation interpreted for the Irish Sea region. This denudation pattern explains in part the outcrop of Precambrian and Lower Palaeozoic rocks in these areas. The spatial integration of the denudation over the entire landmass gives the average denudation rate and the sediment discharge from Ireland as a function of time. Average denudation rates are moderately high in Triassic time, falling to low values in Cretaceous time, and increasing substantially in Tertiary time. However, the total volumetric discharge of sediment in Tertiary time is an order of magnitude smaller than the preserved solid volume of Tertiary sediment in the basins offshore western Ireland.

A comprehensive understanding of the thermal history of Ireland since the end of the Variscan orogeny is hampered by the scarcity of Mesozoic and Cenozoic rocks exposed on the landmass of the island. Post-Carboniferous geological history can be directly evaluated only from the Ulster Basin in the NE of the island, and from small and scattered outcrops throughout the rest of the landmass. In contrast, the offshore basins (Fig. 1) contain a wealth of information on Mesozoic and Cenozoic sedimentation. A general inference is that the present-day Irish landmass has suffered

From: DORÉ, A.G., CARTWRIGHT, J.A., STOKER, M.S., TURNER, J.P. & WHITE, N. 2002. *Exhumation of the North Atlantic Margin: Timing, Mechanisms and Implications for Petroleum Exploration*. Geological Society, London, Special Publications, **196**, 371–399. 0305-8719/02/$15.00

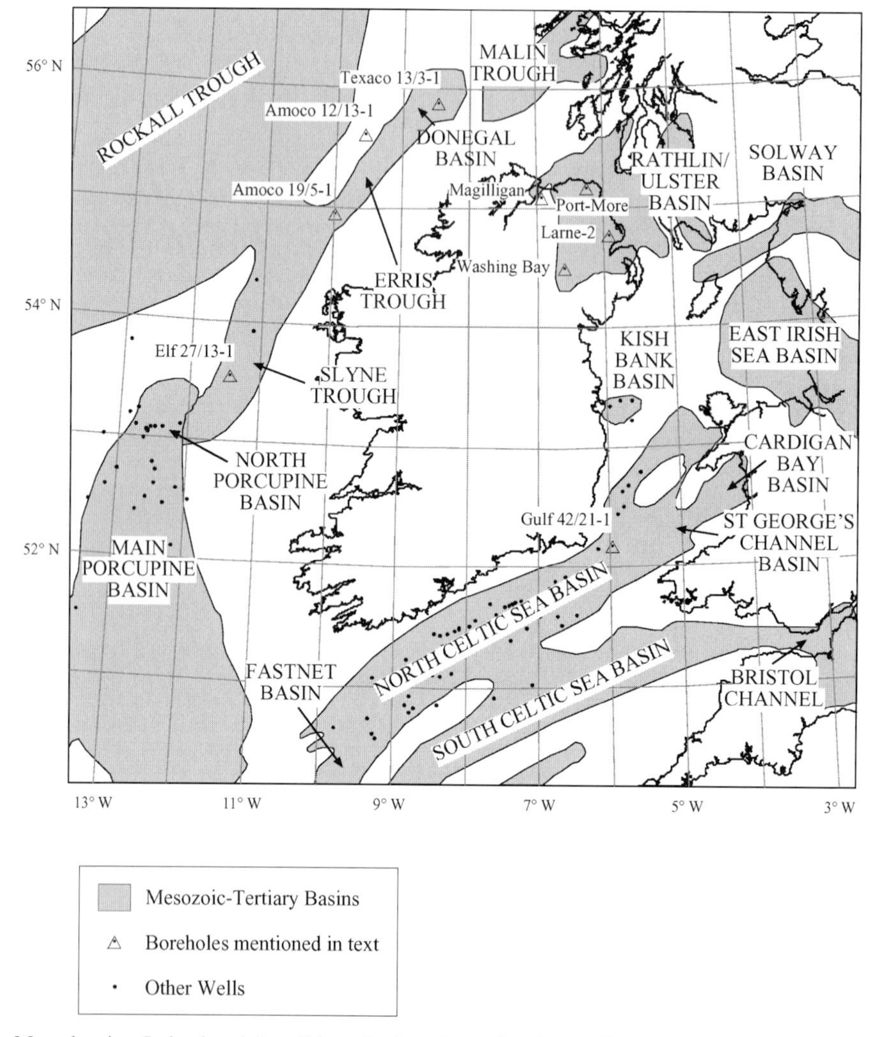

Fig. 1. Map showing Ireland and its offshore basins. ●, exploration wells. Boreholes mentioned in text are highlighted.

extensive uplift and erosion over the last 200 Ma and has served, at least in part, as a source area for clastic detritus now filling the offshore depocentres such as the Porcupine Basin and Rockall margin. A major research goal is therefore to investigate the denudational history of Ireland as a tool for evaluating the likely delivery of clastic sediment to offshore basins in the geological past. By evaluating the denudational history of Ireland, we are also able to provide a jigsaw piece in the broader picture of the uplift and erosional evolution of the hinterlands of the hydrocarbon provinces of the NW European Atlantic continental margin (Doré 1992; Doré *et al.* 1999; Spencer *et al.* 1999) and particularly the British–Irish sector.

The aim of this investigation is to provide a quantitative picture of the thermal and denudational history of Ireland since Triassic time from a new extensive apatite fission-track (AFT) database merging previously published results (McCulloch 1993, 1994; Keeley *et al.* 1993; Green *et al.* 2000) and unpublished results (Gleadow, pers. comm.; McCulloch, unpubl. data; Murphy, pers. comm.) with data collected and analysed during a field campaign in 1998 and 1999. The resulting database contains AFT results from localities throughout Ireland, 139 of which have been modelled in this paper. Individual time–temperature trajectories have been used to contour maps of palaeotemperature as a function of time. These maps are converted to estimates of denudation (presented here) through use of palaeogeothermal gradients constrained by present-day estimates, vitrinite reflectance (VR) profiles and models of early Mesozoic continental stretching. These denudation maps provide vital information on the regional patterns of denudation across the island. By integrating denudation information spatially over the Irish landmass, we are able to make an estimate of the total discharge of sediment since Triassic time. This total discharge history can be compared with the solid volumes of sediment thought to be present in the offshore basins.

The post-Variscan record in Ireland and its offshore basins

The Irish landmass

The geological evolution of Ireland since the end of the Variscan orogeny (Late Carboniferous time) has been summarized by Naylor (1992, 1998). In essence, this long geological history can be directly evaluated only from the stratigraphic and igneous record of the Ulster Basin and from small and scattered occurrences across the island (Fig. 2):

(1) Permo-Triassic and Liassic sedimentary rocks of the Ulster Basin exposed along the shores of Belfast Lough, the coast of Antrim and penetrated in a number of boreholes in the province, and in the small outliers at Kingscourt, County Cavan (Visscher, 1971) and in County

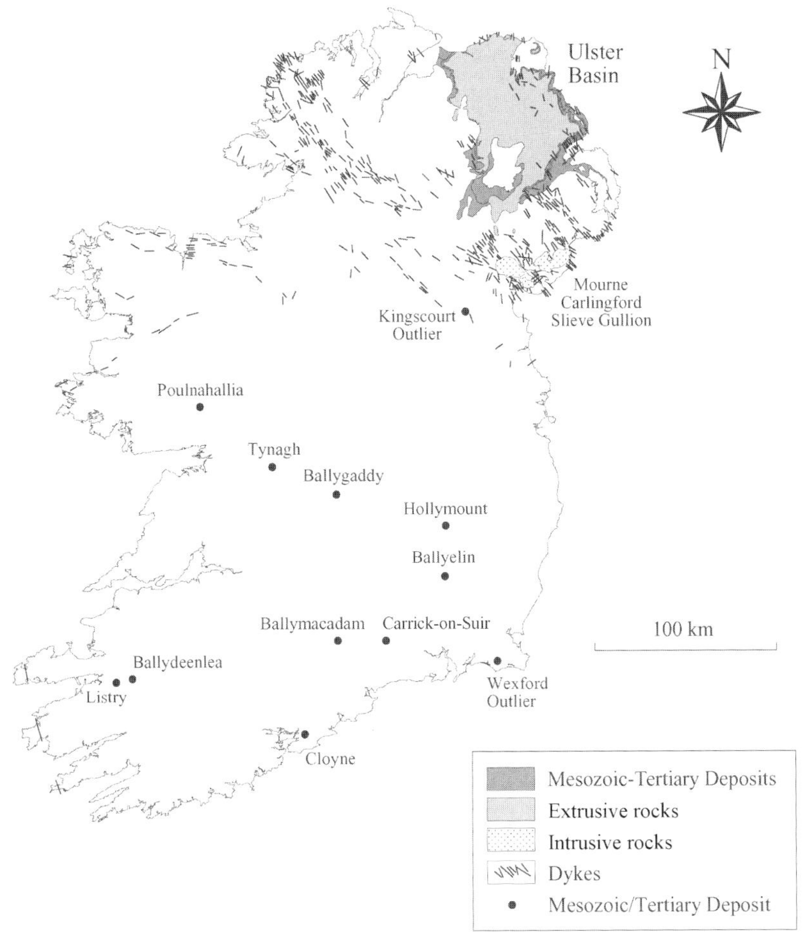

Fig. 2. Simplified map of Ireland showing Mesozoic to Tertiary stratigraphy, Tertiary igneous products and other Mesozoic to Tertiary occurrences discussed in text.

Wexford (Clayton *et al.* 1986). In these areas, the bulk of the post-Variscan succession is made up of Permo-Triassic red beds (1146 m and 1203 m at boreholes near the Antrim coast at Magilligan and Port More, respectively, and 2880 m at Larne 2) (see Fig. 1 for borehole positions). Liassic fine-grained siliciclastic deposits occur at outcrops around the NE Irish coast and in the subsurface, reaching 269 m in thickness in the Port More borehole.

(2) A punctuated and condensed Jurassic to Tertiary succession in the Ulster Basin comprising Upper Cretaceous greensands and chalks (91 m in the Port More borehole), Paleogene basalts (up to 500 m thick) with red lateritic soils and weathered ash falls, and Oligocene clays,

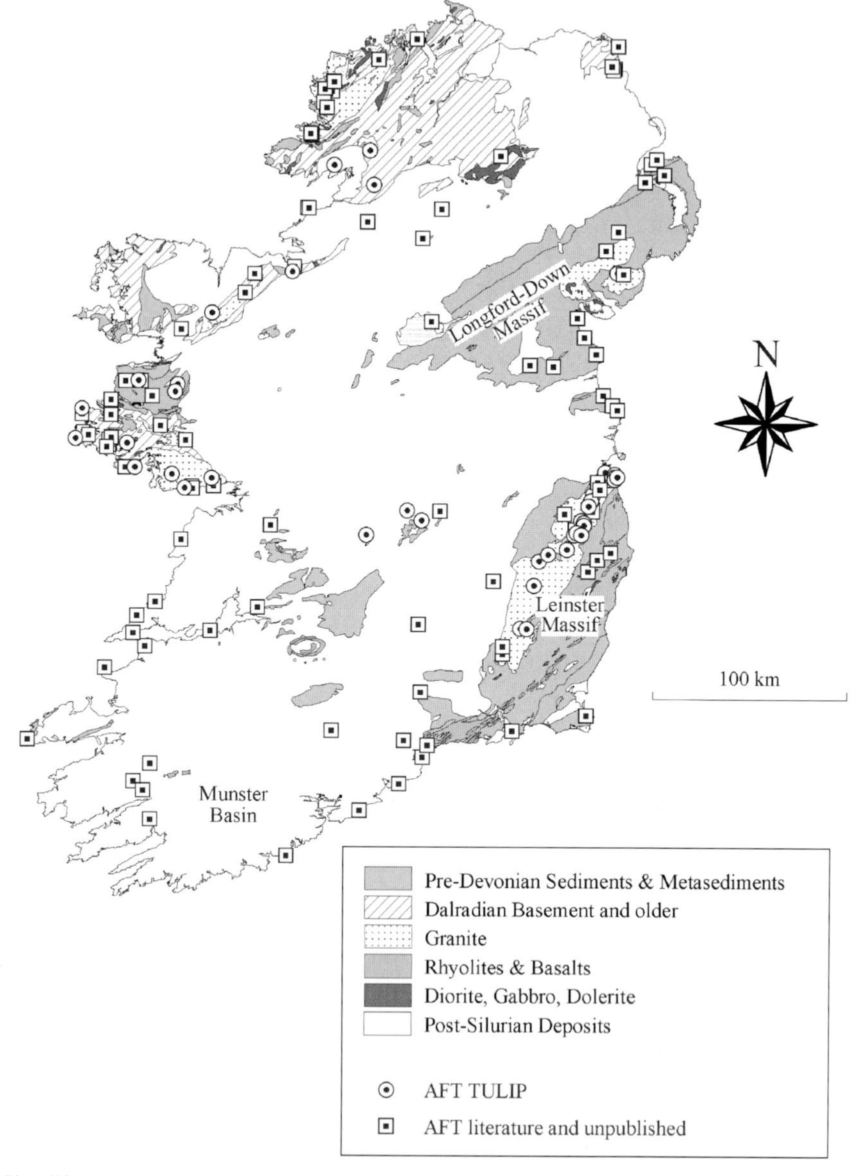

Fig. 3. Simplified map of Ireland showing Lower Palaeozoic and Precambrian rocks, and main igneous plutons, with location of apatite fission-track samples modelled in this study. Circle with dot, TULIP sampling campaign; square with dot, previously published and unpublished samples (Keeley *et al.* 1993; McCulloch 1993, 1994, unpubl. data; Green *et al.* 2000; Gleadow, unpubl. data; Murphy, unpubl. data).

silicliclastic deposits and lignites of the Lough Neagh Group (Parnell *et al.* 1989) (reaching 350 m in the Washing Bay borehole, County Tyrone) (Wilkinson *et al.* 1980).

(3) Isolated pockets of Mesozoic sediment in fault-related depressions, karstic solution hollows or fissure-fills (Fig. 2), such as the chalk breccia of Ballydeenlea, County Kerry (Walsh 1966; Evans & Clayton 1998), Upper Jurassic to Lower Cretaceous clays near Carrick-on-Suir, County Kilkenny (Higgs & Jones 1998) and Middle Jurassic reddened clays at Cloyne, south County Cork (Higgs & Beese 1986).

(4) Sporadic occurrences of supposed Tertiary sediments (Fig. 2) including Oligocene non-marine clays at Ballymacadam (County Tipperary), terrestrial ?Tertiary breccias formed by collapse of karstic limestone caverns at Listry (County Kerry), upper Pliocene dark, non-marine clays at Hollymount (County Carlow), non-marine upper Pliocene–Pleistocene sands within gorges and caves in Carboniferous Limestone bedrock at Poulnahallia (County Galway), a karstic solution pipe in Carboniferous Limestone filled with clay at Ballygaddy (County Offaly), and various Late Tertiary lacustrine records (see summaries and discussions by Davies (1970) and Mitchell (1980)). Deep Tertiary weathering has been described from Tynagh (County Galway). Combined with the fact that the Paleocene basalts of the Antrim Lava Group were erupted subaerially onto a dissected land surface, there is therefore consistent though fragmentary evidence that Ireland has been above sea level throughout Tertiary time.

The onshore stratigraphic record reveals a far from passive post-Palaeozoic history, with multiple periods of relative uplift and denudation separating periods of relative subsidence. In the Ulster Basin of NE Ireland (McCaffrey & McCann 1992), major unconformities, locally demonstrably angular, are present at the base of the Cenomanian–lower Maastrichtian Hibernian Greensand and Ulster White Limestone Formations, as a surface of karstification beneath the subaerially erupted basalts of the Paleocene Antrim Lava Group, below the Oligocene Lough Neagh Group, and below a surface veneer of recent soils, alluvium and Pleistocene tills (McCann 1988, 1990; McCaffrey & McCann 1992).

The geological map of Ireland gives a simple, general impression of the time-integrated denudation of Ireland since the end of Palaeozoic time (Figs 2 and 3). Older rocks emerge from their regional Carboniferous cover in the NW of Ireland from Donegal to County Galway, exposing pre-Dalradian gneisses, Dalradian and Lower Palaeozoic sedimentary and meta-sedimentary rocks, and Caledonian granitic basement (Fig. 3). In eastern Ireland, south of Dublin, the Leinster Massif produces the largest contiguous area of high topography in Ireland and is composed of Caledonian granites intruded into Lower Palaeozoic metasedimentary rocks. This Massif was unroofed in Late Devonian time to provide detritus for the Munster Basin (Penney 1980), again in Jurassic time to feed basins in the Celtic Sea (Petrie *et al.* 1989), and once again in Tertiary time (Davies 1970). The local presence of schist roof pendants demonstrates that the plutons are currently exposed at close to their upper surface. Between Dublin and Belfast, the Longford–Down Massif is a swath of Lower Palaeozoic sedimentary rocks with occasional plutons of Caledonian and Tertiary age, such as the Newry and Mourne granites, respectively. The Tertiary plutons of Counties Down and Armagh are especially informative. SHRIMP U–Pb dates yield 55.5 ± 1.1 Ma and 56.5 ± 1.3 Ma from the Mourne and Slieve Gullion granites, respectively (Meighan *et al.* 1999). As these granites were emplaced at depths of *c.* 3 km in the upper crust, there has evidently been very significant unroofing (*c.* 50 m Ma^{-1} time-averaged) since their crystallization in Paleocene time.

The youngest rocks preserved in Ireland are in Ulster, where Mesozoic strata, including thick Permo-Triassic sequences, are preserved beneath lavas of Paleocene age. The Irish Midlands are a low-elevation, low-relief region of Carboniferous outcrop interspersed with higher-standing inliers of Lower Palaeozoic rocks. As a first approximation therefore, we should suspect that the centre of the island, and the NE in particular, has experienced lower than average time-integrated erosion since the end of Palaeozoic time. Model results presented below support this approximation.

The picture of the denudation of Ireland is aided by consideration of the stratigraphic record of the offshore basins (Fig. 1).

The southern offshore flank

The Fastnet, South Celtic Sea, St. George's Channel and Cardigan Bay basins all contain a thick (<800 m) blanket of Tertiary (Eocene to Miocene or Pliocene) sedimentary rocks above a regional unconformity (Shannon 1991). This base-Tertiary unconformity overlies Cenomanian–Maastrichtian chalk (up to 1 km thick; Naylor & Shannon 1982) throughout most of the Celtic Sea area, but cuts down more deeply into Jurassic sediments in the St. George's Channel and Cardigan Bay basins of the southern Irish Sea (Blundell 1979). On the assumption that Upper Cretaceous chalk formerly extended over the entire southern and eastern offshore flank of Ireland, the southern Irish Sea is inferred to have undergone greater erosion in Early Tertiary time than the Bristol Channel, Celtic Sea and Fastnet basins toward the SW and south. Furthermore, we can conclude from the near-complete absence of Upper Cretaceous chalk and Tertiary deposits in southern Ireland that there was differential uplift and erosion between the Irish landmass and the southern offshore flank both during the Early Tertiary period of unconformity and in the ensuing Eocene–Miocene stages of Tertiary time.

Lower Cretaceous sandstones and shales are found conformably underlying the Upper Cretaceous chalk in the Fastnet, Celtic Sea and Bristol Channel basins. In most exploration wells these Lower Cretaceous rocks are initially continental Wealden, followed by marine sandstones and shales. Around the basin margins, but with the exception of the North Celtic Sea Basin, these successions overlie a base-Cretaceous (Upper Cimmerian) unconformity that cuts out the Middle–Upper Jurassic units. Marine Lower Jurassic shales, marls and limestones occur across the southern offshore flank of Ireland. Lower Jurassic sedimentary rocks (Sinemurian) are thought to have been derived in part from exposed Old Red Sandstone in SW Ireland and from the Leinster Massif in eastern Ireland (Petrie *et al.* 1989), indicating uplift and erosion of these areas during Early Jurassic time. Where Middle Jurassic sediments are present, a Leinster Massif source is also invoked to explain upper Bathonian–lower Callovian red beds prograding into the Celtic Sea area. Upper Jurassic fluvio-lacustrine deposits are found along the northern edge of the North Celtic Sea Basin.

Major inversion along WSW–ENE structures also caused unconformities in the central portion of the North Celtic Sea Basin (Tucker & Arter 1987; Menpes & Hillis 1995; Murdoch *et al.* 1995) in the post-Maastrichtian to pre-mid-Eocene interval.

In summary, the southern flank of Ireland exhibits unconformities of probable latest Jurassic–earliest Cretaceous (Late Cimmerian) and Early Tertiary ages, and locally within Mid-Jurassic time (Mid-Cimmerian; Naylor & Shannon 1982). The Irish landmass was a region of predominantly subaerial exposure and continental deposition during large parts of Mid–Late Jurassic time. This landmass probably became progressively flooded during Early Cretaceous time. There is stratigraphic evidence that whereas the Irish landmass and Irish Sea were undergoing profound erosion in Early Tertiary time, which removed their Upper Cretaceous cover, the southern offshore flank suffered relatively little uplift and denudation except along structural lines of inversion in the North Celtic Sea Basin. This allowed the preservation of chalk beneath a base-Eocene unconformity. During Eocene–Miocene time, the southern flank acted as a depocentre whereas the Irish landmass experienced a long period of subaerial landscape development.

The western offshore flank

Permo-Triassic sedimentary rocks in the Porcupine, Slyne, Erris and Donegal basins are continental (Scotchman & Thomas 1995; Chapman *et al.* 2000). Marine Liassic sedimentary rocks are found throughout the region, passing up into continental and marginal marine Middle Jurassic units. In the main Porcupine Basin, sedimentation was continuous into the mixed continental and shallow marine Upper Jurassic units. Although recognized by the erosional truncation of rotated fault blocks and onlap of Cretaceous sedimentary rocks along the Porcupine Basin margin, the widely recognized uppermost Jurassic–lowermost Cretaceous (Upper Cimmerian) unconformity appears to involve little uplift and erosion in the central part of the Porcupine Basin. Lower Cretaceous sedimentary rocks pass up into *c.* 1 km of Upper Cretaceous chalk along the basin axis of the Porcupine Trough. There is little stratigraphic break at the base of the Tertiary sequence, and a more or less continuous although highly variable succession of Paleocene to Pliocene Tertiary sedimentary rocks (Naylor & Anstey 1987). The stratigraphy is similar in the North Porcupine Basin, but there is a stratigraphic break at the top of Upper Cretaceous chalks (Tate 1993) corresponding to that seen in the onshore sections in the Ulster Basin.

Offshore boreholes in the Donegal and Slyne–Erris basins to the NW of Ireland (Fig. 1) provide a different picture from the Porcupine Trough. In the Donegal Basin (Texaco 13/3-1), Miocene–Pliocene sediments rest directly on Carboniferous units, indicating the merging of the post-Palaeozoic unconformities recognized in the onshore sections into a single stratigraphic break occupying most of Mesozoic and Tertiary time. In contrast, in the NE Erris Trough (Amoco 12/13-1) there is a relatively thick section from Eocene to Aptian rocks, with thin or absent Paleocene units and a thick (817 m) Lower Cretaceous succession resting on Rhaetian units.

A Lower Tertiary unconformity can therefore be recognized in the Erris Trough. The older unconformity is probably due to uplift and erosion during Late Jurassic–Early Cretaceous time, perhaps related to rifting in the Rockall Trough. Further to the SW in Amoco 19/5-1, Lower Cretaceous strata and Upper Cretaceous chalks have been eroded beneath a pre-upper Miocene unconformity that truncates stratigraphy down to Lower Jurassic units. In the Slyne Trough, Elf 27/13-1 penetrated a Neogene–Quaternary cover unconformably overlying a thick (>2 km) Lower–Middle Jurassic succession (Trueblood 1992; Dancer *et al.* 1999).

No exploration wells have yet been drilled in the Irish sector of the Rockall margin. The sedimentary fill of the basin is thought to be a ?Palaeozoic to Jurassic succession that was rifted in Early Cretaceous time and overlain by post-rift Upper Cretaceous to Paleocene and Eocene to Recent sediments (Shannon *et al.* 1995, 1999). A borehole in the UK sector (132/15-1) penetrated Lower Cretaceous sediments resting on crystalline basement (Musgrove & Mitchener 1996), demonstrating considerable erosion of the flank of the Rockall Basin before Early Cretaceous onlap. If present, the base-Tertiary unconformity is poorly developed in the Rockall margin. The region can be thought of as a long-term depocentre since Late Cretaceous time.

The effects of Early Tertiary uplift and erosion were therefore not strongly felt in the main Porcupine Basin or Irish Rockall margin. Instead, these basins continued to subside, allowing thick Cenozoic sediments to accumulate. Likewise, the effects of latest Jurassic–Early Cretaceous (Late Cimmerian) erosion are minor within the main depocentre of the Porcupine Basin. In the Slyne–Erris basins and margins of the Porcupine Basin, however, there is good evidence of a major pre-Early Cretaceous or Late Jurassic–Early Cretaceous erosive event that variably cut down into Middle Jurassic, Lower Jurassic or Triassic sedimentary rocks. The Lower Tertiary unconformity caused by relative uplift and karstification of the Ulster White Limestone beneath the Antrim lavas in the onshore sections of NE Ireland is, however, unrecognized in the western offshore flank basins with the possible exception of the northern Erris Trough. It is likely that a major unconformity beneath the Mio-Pliocene sequences resulting from regional Neogene uplift of the NW flank of Ireland and adjacent continental shelf (Stoker *et al.* 2001) has removed evidence of an earlier Early Tertiary event.

The eastern and northern offshore flanks

Permo-Triassic sedimentary rocks are thought to be extensive in the Irish Sea region. They are overlain by thick (<2700 m), marine Liassic sediments in the Kish Bank Basin (Broughan *et al.* 1989) and Cardigan Bay Basin (Woodland 1971). Middle–Upper Jurassic sedimentary rocks are generally absent, although a more or less complete, entirely marine Jurassic succession is found in the Cardigan Bay Basin. Lower and Upper Cretaceous rocks are also absent from the Irish Sea, indicating considerable erosion, probably during Early Tertiary time. The deep erosion of Mesozoic stratigraphy in the Irish Sea area has been interpreted as due to crustal uplift and denudation caused by igneous underplating (Brodie & White 1994; White & Lovell 1997) or transient dynamic uplift (Cope 1994) related to plume activity centred under the Irish Sea during Paleocene time.

The Rathlin Basin of Ulster extends along a Caledonian trend northeastwards towards Scotland. It contains thick continental Triassic sedimentary rocks overlain by marine transgressive Rhaetian units, but Middle and Upper Jurassic and Cretaceous sedimentary rocks are generally absent (Evans *et al.* 1980; Naylor 1992). A thin basal layer of sandstone and chalk deposited unconformably on the eroded Mesozoic and older rocks of the Inner Hebrides, Malin Shelf and Firth of Clyde correlates with the Hibernian Greensand and Ulster White Limestone of the onshore sections in Ulster and therefore demonstrates a pre-Late Cretaceous erosional event. The subsequent erosional removal of much of this chalk layer records Tertiary denudation.

Palaeothermal and thermochronological data

Although palaeothermal and palaeoburial indices (such as vitrinite reflectance (VR), clay mineral crystallinity, sonic velocities, fluid inclusions) are available to differing degrees on the Irish landmass and adjacent offshore, AFT analysis is potentially extremely valuable in providing information on the time–temperature history of samples, rather than simply a maximum palaeotemperature or burial depth. It has been used as a thermochronological tool extensively in the British Isles region (e.g. Bray *et al.* 1992; Lewis *et al.* 1992; Green *et al.* 1993, 1997; Duncan *et al.* 1998).

Maps of VR show strong heating in the south of Ireland interpreted to be due to a Late

Table 1. *Apatite fission-track results from the TULIP sampling campaign and those of Murphy (unpubl.) (borehole samples and those with low track length populations were not modelled)*

Sample no.	Locality (County)	Easting, Northing	Strat. age (lithology)	Spontaneous ρs^a (Ns)	Induced ρi^a (Ni)	Dosimeter ρd^a (Nd)	$P(\chi^2)^b$ (no. of crystals)	FT central age (Ma) (RE %)[c]
BS1	The Black Gap (Donegal)	204527, 369089	Precambrian (Psammite)	1.547 (1171)	2.114 (1603)	1.235 (3423)	<1 (20)	151 ± 8 (12.5)
F1	Barnesmore Gap (Donegal)	202500, 385400	Devonian (Granite)	1.656 (1772)	1.761 (1884)	1.205 (6679)	99 (22)	189 ± 7 (0)
MT1	Milltown (Donegal)	184980, 378454	Carboniferous (Sandstone)	1.435 (1613)	1.886 (2120)	1.506 (4159)	15 (29)	191 ± 8 (10.2)
F9	Mourne Mts. (Down)	326850, 326277	Silurian (Quartzite)	1.339 (592)	1.366 (604)	1.205 (6679)	40 (12)	197 ± 12 (4)
GG3	nr. Cashel (Galway)	79610, 242400	Devonian (Granite)	0.687 (902)	0.679 (891)	1.168 (6474)	60 (30)	197 ± 10 (0.1)
GG9	Kilkieran (Galway)	82860, 231453	Devonian (Granite)	1.049 (1493)	1.483 (2111)	1.168 (6474)	<1 (30)	138 ± 6 (13.3)
GG12	Bovroughaun (Galway)	101740, 228360	Devonian (Granite)	1.189 (2667)	1.647 (3693)	1.168 (6474)	40 (30)	141 ± 4 (2.9)
GG13	nr. Spiddle (Galway)	109360, 222060	Devonian (Granite)	0.934 (1853)	1.429 (2837)	1.168 (6474)	15 (30)	128 ± 5 (7.8)
GG15	Loughinch (Galway)	122590, 225470	Devonian (Granite)	1.053 (1224)	1.359 (1580)	1.168 (6474)	15 (25)	152 ± 7 (9.2)
IG1	Pollrevagh (Galway)	56490, 244670	Devonian (Granite)	0.734 (1668)	1.013 (2775)	1.168 (6474)	10 (20)	142 ± 6 (9.9)
CL1	Aughrus More (Galway)	56890, 257770	Devonian (Granite)	0.827 (2070)	1.109 (2775)	1.168 (6474)	7 (30)	146 ± 5 (10.6)
LE	Partry Mts. (Mayo)	104900, 267000	Ordovician (Sandstone)	0.859 (243)	0.587 (166)	1.134 (3144)	70 (16)	275 ± 28 (2.9)
GL	Partry Mts. (Mayo)	105900, 270800	Ordovician (Sandstone)	0.533 (153)	0.505 (145)	1.134 (3144)	90 (14)	200 ± 23 (0)
CG1	Laghtaeighter (Mayo)	85170, 274020	Devonian (Granite)	0.837 (847)	0.948 (959)	1.168 (6474)	60 (30)	172 ± 8 (1.4)
OM1	Ox Mts. (Sligo)	164223, 327635	Dalradian (Gneiss)	1.862 (2682)	2.415 (3478)	1.235 (3423)	<1 (22)	159 ± 7 (15.1)
OM2	Pontoon Bridge (Mayo)	123383, 304852	Silurian (Granite)	1.360 (1457)	1.742 (1866)	1.235 (3423)	<1 (25)	159 ± 7 (12.3)
LG2	Carysfort Park (Dublin)	321580, 228571	Devonian (Granite)	4.410 (3527)	5.030 (4023)	1.124 (6234)	82.1 (36)	165 ± 4 (0)
LG3	Sandycove (Dublin)	325700, 228000	Devonian (Granite)	2.569 (2136)	3.516 (2924)	1.124 (6234)	40.2 (28)	138 ± 4 (3.9)
LG5	Dalkey Qu. (Dublin)	326450, 226040	Devonian (Granite)	3.069 (3250)	4.117 (4360)	1.124 (6234)	0 (34)	142 ± 5 (14.9)
LG6	Dalkey Qu. (Dublin)	326247, 226304	Devonian (Granite)	2.117 (2441)	2.975 (3431)	1.124 (6234)	14.3 (34)	133 ± 5 (9.2)
LG7	Killiney Hill (Dublin)	325990, 225550	Devonian (Granite)	1.678 (2240)	2.467 (3293)	1.124 (6234)	28 (32)	127 ± 4 (6.5)
LG8	Dalkey Isl. Hotel (Dublin)	327135, 226371	Devonian (Granite)	2.426 (2132)	3.494 (3070)	1.124 (6234)	94.7 (32)	131 ± 4 (0)
LG10	Church View Rd. (Dublin)	324100, 225459	Devonian (Granite)	1.856 (1201)	2.311 (1496)	1.124 (6234)	69 (23)	151 ± 6 (0.9)
LG11	Ballyedmond Qu. (Dublin)	318739, 223458	Devonian (Granite)	3.883 (5264)	4.407 (5974)	1.124 (6234)	61.3 (34)	166 ± 4 (0.8)
LG12	Barnaculla Qu. (Dublin)	317890, 224230	Devonian (Granite)	3.995 (5454)	0.439 (6059)	1.124 (6234)	37.2 (33)	169 ± 4 (2.1)
LG13	Three Rock Mt. (Dublin)	317600, 223400	Devonian (Granite)	3.661 (3557)	3.808 (3700)	1.124 (6234)	21 (30)	181 ± 5 (1.3)
LG14	Military Rd. (Wicklow)	315090, 218000	Devonian (Granite)	1.710 (2495)	1.817 (2652)	1.124 (6234)	95.9 (40)	177 ± 5 (0.1)
LG15	L. Bray (Wicklow)	314290, 215140	Devonian (Granite)	3.160 (3083)	3.333 (3252)	1.124 (6234)	70 (29)	178 ± 5 (0.1)
LG16	Sally Gap (Wicklow)	312940, 212200	Devonian (Granite)	2.912 (4295)	3.159 (4659)	1.124 (6234)	26.6 (30)	173 ± 4 (2.3)
LG17	Carrigshouk (Wicklow)	309480, 205200	Devonian (Granite)	2.248 (3081)	2.502 (3430)	1.124 (6234)	5.4 (40)	170 ± 5 (9.4)
LG18	Carrigshouk (Wicklow)	309794, 205293	Devonian (Granite)	2.701 (3485)	3.339 (4309)	1.124 (6234)	48.1 (29)	152 ± 4 (0.6)
LG19	Glenmacnass Riv. (Wicklow)	311310, 203994	Devonian (Granite)	2.885 (2911)	3.688 (3721)	1.124 (6234)	31.7 (31)	147 ± 4 (4.9)
LG20	Wicklow Gap Rd. (Wicklow)	310090, 98140	Devonian (Granite)	1.827 (2788)	2.306 (3519)	1.124 (6234)	57.5 (40)	149 ± 4 (0.6)
LG21	L. Nahanagan (Wicklow)	307600, 199200	Devonian (Granite)	4.795 (2645)	5.821 (3211)	1.124 (6234)	32.4 (24)	156 ± 5 (5.8)
LG22	Turlogh Hill (Wicklow)	306500, 198500	Devonian (Granite)	4.121 (3987)	4.388 (4245)	1.124 (6234)	94.9 (30)	177 ± 4 (0)
LG23	Tullow Lowlands (Wicklow)	288770, 185165	Devonian (Granite)	3.952 (3111)	4.399 (3463)	1.124 (6234)	9.1 (29)	169 ± 5 (7.3)

Table 1 – *continued*

Sample no.	Locality (County)	Easting, Northing	Strat. age (lithology)	Spontaneous ρs[a] (Ns)	Induced ρi[a] (Ni)	Dosimeter ρd[a] (Nd)	$P(\chi^2)$[b] (no. of crystals)	FT central age (Ma) (RE %)[c]
LG26	Kelshabeg (Wicklow)	293700, 188250	Devonian (Granite)	2.966 (1858)	3.457 (2166)	1.168 (6474)	<1 (30)	167 ± 7 (11.8)
LG28	Mt. Leinster (Carlow)	282743, 152564	Devonian (Granite)	2.636 (2696)	2.507 (2564)	1.168 (6474)	40 (30)	205 ± 6 (0.9)
LG30	Tomduff (Carlow)	278140, 153114	Devonian (Granite)	2.686 (1452)	3.099 (1675)	1.168 (6474)	30 (30)	169 ± 7 (6.4)
LG32	Tullow Hil (Carlow)	286582, 173267	Devonian (Granite)	0.584 (1098)	0.841 (1581)	1.168 (6474)	5 (30)	136 ± 7 (9.2)
LG36	Lugnaquilla Mt. (Wicklow)	303180, 191200	Devonian (Granite)	3.004 (2590)	4.375 (3772)	1.500 (4159)	2 (20)	174 ± 7 (10.4)
SB1	Slieve Bloom Mts. (Offaly)	223200, 208250	Carboniferous (Sandstone)	1.710 (1989)	1.788 (2079)	1.503 (4159)	40 (28)	240 ± 9 (6.2)
SB2	Slieve Bloom Mts. (Laois)	229750, 204450	Devonian (Sandstone)	2.262 (2144)	2.303 (2183)	1.241 (6882)	<1 (23)	204 ± 9 (11)
SB3	Knockshigowna (Tipperary)	200163, 196360	Carboniferous (Sandstone)	1.342 (1406)	1.507 (1579)	1.504 (4159)	<1 (18)	228 ± 13 (15.7)
MG1[d]	Mourne Mts. (Down)	329979, 325285	Paleocene (Granite)	0.206 (263)	1.365 (1743)	1.461 (4628)	98 (32)	36 ± 3* (0.03)
Unmodelled data								
GA	Killary fjord (Galway)	86500, 262200	Ordovician (Sandstone)	0.459 (233)	0.371 (188)	1.136 (4723)	90 (22)	243 ± 23 (0)
GG4	L. Maumwee (Galway)	97320, 249000	Devonian (Granite)	0.768 (316)	1.105 (455)	1.168 (6474)	<1 (14)	144 ± 17 (28.5)
DB	Lough Mask (Galway)	103700, 261100	Caledonian (QF Porphyry)	0.283 (40)	0.226 (32)	1.136 (4723)	80 (3)	236 ± 57 (4.6)
GS	Lough Mask (Galway)	104000, 260300	Caledonian (QF Porphyry)	0.296 (109)	0.217 (80)	1.124 (6234)	90 (10)	255 ± 38 (0)
DQ1	Doonan Quarry (Fermanagh)	219500, 356600	Devonian (Sandstone)	4.142 (213)	1.175 (604)	1.235 (3423)	5 (13)	73 ± 6 (7.3)
RB1	Western Red Bay (Antrim)	324300, 425500	Triassic (Sandstone)	2.676 (623)	1.499 (349)	1.235 (3423)	10 (10)	369 ± 32 (14.2)
MG2[d]	Mourne Mts. (Down)	330663, 324191	Paleocene (Granite)	0.218 (173)	1.843 (1462)	1.439 (4605)	89 (23)	28 ± 2* (0.18)
NO[e]	Kentstown (NO 1442) (Meath)	295620, 267370	Devonian (Granite)	2.442 (1266)	5.953 (3086)	1.241 (6882)	<1 (21)	84 ± 4 (13.7)
LG38[e]	Quinagh (Carlow)	272642, 173790	Devonian (Granite)	2.163 (1733)	5.246 (4203)	1.501 (4159)	3 (23)	105 ± 4 (11.4)

a Track densities (ρ) are ($\times 10^6$ tracks cm^{-2}); numbers of tracks counted (N) shown in parentheses. Analyses by external detector method using 0.5 for the $4\pi/2\pi$ geometry correction factor.
b $P(\chi^2)$ is probability of obtaining χ^2 value for v degrees of freedom (where v is number of crystals (Nc) − 1) and is the probability that the single grain ages are consistent with one population (<5% denotes failure at the 95% level).
c Fission track (FT) ages calculated using dosimeter glass CN-5 (analyst Carter ζN − 5 = 339 ± 5, *analyst Murphy ζCN − 5 = 330 ± 10), calibrated by multiple analyses of IUGS apatite and zircon age standards (see Hurford 1990). Central age is a modal age, weighted for different precisions of individual crystals (Galbraith & Laslett 1993). RE % is the relative error or age dispersion. Quoted age uncertainties are $\pm 1\sigma$.
d Murphy (unpubl.).
e Boreholes. Sample depths: NO, 472–492 m; LG38, 205 m.

Carboniferous thermal event linked to the Variscan orogeny (Clayton *et al.* 1989). A substantial thickness (5–7 km) of post-Lower Carboniferous rocks is thought to have been removed by erosion to explain the high thermal maturity of Devonian–Carboniferous rocks in southern Ireland (Clayton 1989). Post-Carboniferous rocks, however, have low VR levels. For example, the Campanian chalk breccia at Ballydeenlea has VR values between 0.46 and 0.53% Rm, whereas the underlying Namurian rock has values centred on 4% Rm. The reflectance of chalk clasts indicates that between 1 and 1.5 km of overburden (probably latest Cretaceous and earliest Tertiary rocks) has been removed by erosion from above the chalk breccia (Evans & Clayton 1998). The idea of regional erosion of a significant overburden of latest Cretaceous to earliest Tertiary sediment across Ireland in Early Tertiary time is supported by the occurrence of karstic cave passages and pinnacles along pre-existing joints in the Ulster White Limestone of NE Ireland (Simms 1998), showing that it was well cemented and fractured before the eruption of ashes and lavas of the Antrim Lava Group. This state of induration was probably achieved by burial under a pile of Maastrichtian and possibly lower Paleocene sedimentary rock, subsequently removed by dissolution and erosion.

Although a certain amount of AFT data from Ireland's offshore basins is available in the public domain (Duncan *et al.* 1998), few data are available from onshore Ireland. In the present study, previously published (Keeley *et al.* 1993; McCulloch 1993, 1994; Green *et al.* 2000) and unpublished (Gleadow, pers. comm.; McCulloch, unpubl. data; Murphy, pers. comm.) AFT data have been very substantially augmented by data from a new campaign carried out in 1998 and 1999, part of the continuing TULIP Tertiary Uplift of Ireland Project. New samples were analysed by Carter at the University College London Fission Track Laboratory and modelled by Gallagher to obtain most likely time–temperature trajectories and their uncertainties (see Gallagher 1995; Gallagher & Brown 1999). The published and unpublished data were also modelled or remodelled to provide a fully consistent set of results. Details of procedures involved in the new sampling program and caveats to the interpretation of the results are given below.

Sampling

A wide range of lithologies (igneous, metamorphic and sedimentary) across Ireland were sampled for apatite grains. The best apatite yields were obtained from basement granites. Some lithologies, such as the Hibernian Greensand from Ulster, did not yield sufficient quantities of apatite. In general, thermal aureoles were avoided. We present in tables the data for all samples that were modelled or remodelled in this study, plus those TULIP samples that were not modelled (see below). Analytical data and AFT age results for the TULIP and Murphy (unpubl.). samples, together with location, lithology and stratigraphic interval are given in Table 1. Table 2 contains track length data binned at 1 μm intervals for the same samples. Data for the published and other unpublished AFT samples are given in Table 3. It was deemed inappropriate to model or remodel some of the TULIP and other available published samples because of very low fission-track counts.

Fission-track analytical details (Tables 1 and 2)

Spontaneous fission tracks were revealed by etching polished apatite grain mounts with 5N HNO_3 at 20 ± 1 °C for 20 s. Mounts were irradiated in the thermal facility at the Risø Reactor at the National Research Centre, Roskilde, Denmark (cadmium ratio (thermal/epithermal + fast neutrons) >200), using the external detector method. Fluence was monitored using Corning uranium standard glass dosimeter CN-5. Muscovite mica external detectors were etched after irradiation using 48% HF at 20 ± 1 °C for 55 min. Observation and measurement of fission-track density and length was conducted under transmitted light using a Zeiss axioplan microscope with a total dry magnification of × 1250. Reflected light was used as a tool for discriminating individual tracks. Horizontal confined fission-track lengths were measured using a digitizing tablet calibrated before each analysis against a stage micrometer. Ages were determined using the zeta calibration method and IUGS recommended age standards (Hurford 1990). All data are reported as the central age (modal age weighted for different precisions of individual crystals) together with the percentage age dispersion or relative error (RE %) about the central age (Galbraith & Laslett 1993).

Thermal history and denudation chronology modelling: procedures

Thermal history modelling was undertaken using a search procedure outlined by Gallagher (1995) initially using the annealing method of Laslett *et al.* (1987). This annealing model is derived empirically, based on laboratory annealing experiments performed on a mono-compositional Durango apatite with an F/Cl ratio of *c.* 0.1 (Young *et al.* 1969). This apatite is moderately retentive compared with a pure fluorapatite. Extrapolation of the Laslett *et al.* (1987) Durango apatite model to geological time scales (10^6–10^7 a) predicts a partial annealing temperature range from *c.* 60 °C to *c.* 110 °C with an uncertainty of *c.* 10 °C. However, one of the major limitations of this model is that it does not appear to predict sufficient annealing at temperatures less than *c.* 60 °C. Consequently, thermal history simulations can be biased towards inferring rapid, recent cooling from *c.* 60 °C to surface temperatures, with consequent overestimates of denudation in the most recent time interval. Structure in this temperature range therefore needs to be treated with caution. We also present results from a preliminary new annealing model that removes the effects of excessive denudation at low temperatures (Gunnell *et al.* in prep.). This model is based on the same data as the Laslett *et al.* (1987) model, but uses the Laslett & Galbraith (1996) formulation. The key difference between the two model formulations is the assumption of the error distribution on the data. The Laslett *et al.* (1987) model transforms the data before fitting them, and assumes the variance is constant in the transformed data. The revised model does not make this assumption and fits the observed data directly. Relative to the Laslett *et al.* (1987) model, this revised model leads to more annealing at a given temperature. Consequently, predictions from this model are more consistent with low-temperature (<60 °C) annealing implied by geological data. We compare the two annealing models for cooling and thus denudation during Tertiary time when samples were at temperatures <60 °C as they approached the present surface.

Data from individual AFT samples were modelled independently, and the optimal thermal history model was selected on the basis of the fit to the observed data. The modelling procedure requires the input of stratigraphic age and time–temperature windows through which the modelled thermal history passes (Gallagher 1995). The windows were given broad bounds in temperature and allowed to overlap in time so as not to restrict the modelling process. The same windows were used for all samples for reasons of consistency and are as follows: 500 ± 100 Ma at 80 ± 80 °C; 250 ± 50 Ma at 70 ± 70 °C; 150 ± 50 Ma at 70 ± 70 °C; 100 ± 20 Ma at 70 ± 70 °C; 70 ± 20 Ma at 70 ± 70 °C; 50 ± 20 Ma at 70 ± 70 °C; 30 ± 20 Ma at 70 ± 70 °C; 0 Ma at 20 °C.

The choice of data fit statistic depends on the form of the original data. For new data collected during the TULIP campaign, where we have the raw data (a series of track length measurements and the spontaneous and induced track counts for a suite of crystals), we used a maximum likelihood statistic. For previously unpublished (Gleadow, pers. comm.; McCulloch, unpubl. data; Murphy, pers. comm.) and published datasets (Keeley *et al.* 1993; McCulloch 1993, 1994; Green *et al.* 2000) we had access to mean, pooled or central age, and mean and standard deviation of the track length distributions, as reported in the original papers. A weighted least-squares measure of data fit was used for these data. Gallagher (1995) summarized the various misfit statistics that may be adopted in these different situations.

The form of an individual thermal history is encouraged to show cooling where required to fit the data, but otherwise the modelling approach tries to minimize the structure or variation in the thermal histories. This approach ensures that random variations producing unresolved heating events are minimized, whereas the cooling events required to fit the data are maintained. However, if the modelled palaeotemperature remains more or less constant over a given time interval, this does not mean that progressive heating of the sample to that temperature has not occurred over that time interval. Instead, the modelled thermal history reflects the lack of resolution when heating from a low temperature to a higher temperature occurs. Consequently, a static thermal history may be inferred during a period of burial, with the maximum temperature occurring at the time of maximum burial and little resolution on the earlier, lower-temperature part of the thermal history. This has the effect of damping burial (heating) events in a thermal history. Flat portions of the time–temperature trajectories should therefore be viewed with some caution, but do indicate an upper limit to the palaeotemperature during this interval. In later sections of this paper, we will refer to this effect in the thermal history as deposition, although the strict interpretation is 'no cooling'.

The outcome of the modelling process is a thermal history represented as a series of discrete time–temperature points, with linear interpolation between points adjacent in time. Palaeotemperature maps for a certain time in geological history can then be constructed as described in the following section.

Table 2. *Track length results for TULIP and Murphy (unpubl.) samples*

Sample no.	Number of tracks in μm intervals[a]																Mean track length (μm)[b]	SD[c]	No. of tracks[d]
	2	3	4	5	6	7	8	9	10	11	12	13	14	15	16	17			
BS1	0	0	0	0	0	0	2	7	8	17	16	36	14	8	0	0	12.75 ± 0.16	1.70	108
F1	0	0	0	0	1	1	3	3	13	23	38	30	26	13	4	0	12.84 ± 0.14	1.77	155
MT1	0	4	0	0	2	1	2	3	13	20	38	32	17	6	2	0	12.35 ± 0.20	2.33	140
F4	0	1	0	0	0	1	2	3	4	5	4	6	4	2	1	0	11.93 ± 0.45	2.57	33
F9	0	0	0	0	0	0	0	0	2	1	5	8	8	8	2	0	12.87 ± 0.25	1.47	34
GG3	0	0	1	0	0	1	4	15	24	48	66	26	12	3	0	0	12.98 ± 0.11	1.49	200
GG9	0	0	0	0	0	1	0	0	4	17	15	33	34	13	11	3	13.80 ± 0.15	1.71	131
GG12	0	0	0	0	0	0	0	3	10	30	56	46	40	12	3	0	13.11 ± 0.10	1.37	200
GG13	0	1	0	0	0	1	2	11	15	36	29	53	37	13	5	0	12.82 ± 0.13	1.85	203
GG15	0	0	0	0	1	2	2	4	6	22	21	27	24	17	5	1	13.09 ± 0.17	2.00	132
IG1	0	0	0	0	0	1	3	11	20	47	42	26	38	9	3	0	12.50 ± 0.12	1.75	200
CL1	0	0	0	0	0	0	0	6	21	30	37	47	41	13	5	0	12.98 ± 0.11	1.57	200
LE	0	0	0	0	1	1	2	0	0	0	4	7	5	1	0	1	12.79 ± 0.58	2.64	22
GL	0	0	0	0	0	1	1	3	3	4	11	11	5	2	1	0	12.60 ± 0.30	1.91	42
CG1	0	0	0	0	0	0	1	0	0	3	7	16	21	13	4	2	13.23 ± 0.19	1.51	67
OM1	0	0	0	0	0	1	1	5	5	23	41	46	24	4	0	0	12.79 ± 0.11	1.40	150
OM2	0	0	0	0	0	0	2	0	5	12	38	53	29	8	3	0	13.27 ± 0.11	1.31	150
LG2	0	2	1	1	0	0	6	13	26	25	32	44	38	11	1	0	12.43 ± 0.15	2.11	200
LG3	0	0	0	0	1	2	5	5	19	10	26	40	33	16	2	0	12.91 ± 0.16	1.97	159
LG5	0	0	0	0	0	2	4	8	16	29	30	59	37	12	3	0	12.85 ± 0.12	1.72	200
LG6	0	1	0	0	1	2	8	15	27	19	31	47	36	13	1	0	12.42 ± 0.15	2.09	201
LG7	0	0	0	0	0	1	3	9	16	23	22	38	33	13	2	0	12.85 ± 0.15	1.87	160
LG8	0	1	1	0	0	2	5	17	19	17	32	45	37	18	1	0	12.60 ± 0.15	2.14	195
LG10	0	0	1	1	0	1	0	2	8	11	13	15	11	4	1	0	12.53 ± 0.27	2.20	68
LG11	0	0	0	0	1	2	4	12	17	34	46	47	33	9	0	0	12.55 ± 0.12	1.73	205
LG12	1	0	0	0	0	1	6	7	24	52	37	45	19	9	0	0	12.24 ± 0.12	1.71	201
LG13	0	1	1	1	0	1	1	12	24	42	49	32	28	6	2	0	12.28 ± 0.13	1.85	200
LG14	0	1	0	0	0	0	1	6	20	28	41	50	35	11	0	0	12.78 ± 0.12	1.64	193
LG15	0	0	0	0	0	0	3	5	24	23	40	45	27	13	0	0	12.72 ± 0.12	1.60	180
LG16	2	0	0	2	1	3	5	20	33	47	44	39	5	2	0	0	12.50 ± 0.14	2.02	204
LG17	0	1	0	0	0	0	3	6	13	18	31	26	10	2	1	0	12.22 ± 0.17	1.78	111
LG18	0	0	0	0	1	1	3	6	17	38	36	55	34	10	1	0	12.71 ± 0.11	1.61	202
LG19	0	0	0	0	0	0	1	3	7	17	23	24	19	6	1	0	12.93 ± 0.15	1.54	101
LG20	0	0	0	0	1	1	1	9	17	38	47	43	37	8	1	0	12.66 ± 0.11	1.62	203
LG21	0	0	0	0	0	0	3	13	24	39	57	42	17	4	1	0	12.28 ± 0.10	1.47	200
LG22	0	0	1	0	3	0	0	7	22	42	55	39	27	7	1	0	12.45 ± 0.12	1.68	204
LG23	1	0	1	0	2	1	3	8	17	36	49	51	28	2	2	0	12.38 ± 0.13	1.88	201
LG26	0	0	2	2	0	0	4	16	22	31	28	48	29	14	5	1	12.47 ± 0.15	2.15	200

Table 2 – *continued*

Sample no.	Number of tracks in μm intervals[a]																Mean track length (μm)[b]	SD[c]	No. of tracks[d]
	2	3	4	5	6	7	8	9	10	11	12	13	14	15	16	17			
LG28	1	0	1	0	0	1	5	4	10	44	47	49	24	11	3	0	12.60 ± 0.13	1.83	200
LG30	1	1	2	0	3	1	6	14	40	43	48	27	12	2	0	0	12.56 ± 0.14	1.92	200
LG32	0	0	0	0	1	0	6	11	16	23	55	33	39	14	3	0	12.74 ± 0.13	1.83	201
LG36	0	1	0	0	0	0	0	3	12	39	45	34	18	7	1	0	12.57 ± 0.12	1.50	160
SB1	0	1	0	0	0	0	2	6	12	23	50	34	19	5	0	0	12.52 ± 0.13	1.62	152
SB2	0	0	0	0	0	2	3	4	9	16	53	39	28	5	0	0	12.75 ± 0.12	1.45	159
SB3	0	0	0	1	0	0	0	2	9	8	28	34	15	3	1	0	12.84 ± 0.15	1.47	101
MG1[d]	0	0	0	0	0	0	0	0	1	2	1	3	6	9	6	3	13.98 ± 0.31	1.70	31
Unmodelled samples																			
GA	0	0	0	0	0	0	0	0	0	0	1	1	5	5	2	0	14.94 ± 0.33	1.18	14
GG4	0	0	0	0	0	0	2	1	2	1	4	5	2	1	0	0	12.38 ± 0.47	1.95	18
DQ1	0	0	0	0	0	0	0	0	0	0	0	1	2	2	2	1	14.54 ± 0.41	1.07	8
RB1	0	0	0	0	0	0	0	1	0	2	6	10	7	4	2	0	12.72 ± 0.27	1.51	32
MG2[d]	0	0	0	0	0	0	0	0	0	0	1	2	3	3	0	1	13.90 ± 0.38	1.31	10
NO[e]	0	0	0	0	0	1	2	2	2	7	16	40	21	8	1	0	13.23 ± 0.16	1.57	100
LG38[e]	0	0	1	0	0	0	3	4	10	6	10	32	21	3	0	0	12.82 ± 0.21	1.94	90

a Each track length interval covers the range $n - 1$ to n, where n is the interval heading. No tracks counted <2 μm and >17 μm in length.
b Quoted uncertainties of mean and standard deviation of track length distributions are $\pm 1\sigma$.
c No track length data are available for samples DB and GS owing to low apatite yields and anomalously low uranium concentrations.
d Murphy (unpubl.).
e Borehole samples.

Table 3. *Published and unpublished apatite fission-track results modelled for this study (McCulloch 1993, 1994, unpubl. data; Keeley et al.,* 1993; Green *et al.* 2000; Gleadow (unpubl. data); Murphy (unpubl. data))

Sample no.[a]	Locality	Easting, Northing	Strat. age (lithology)	Fission track age (Ma)[b]	Mean track length (μm)[c]	SD[c]	No. of tracks
Gleadow (unpubl.)							
Gl1	The Rosses (Donegal)	182500, 418000	Devonian (Granite)	188.1 ± 3.6	12.25 ± 0.22	2.15	96
Gl2	The Rosses (Donegal)	183000, 418000	Devonian (Granite)	195.0 ± 7.0	12.05 ± 0.19	1.93	103
Gl3	The Rosses (Donegal)	180000, 416000	Devonian (Granite)	197.1 ± 6.8	11.88 ± 0.22	2.15	96
Gl4	The Rosses (Donegal)	182500, 415800	Devonian (Granite)	200.9 ± 7.1	12.23 ± 0.20	2.04	104
Gl5	SE of Dunglow (Donegal)	179000, 409900	Devonian (Granite)	194.1 ± 8.0	12.22 ± 0.23	2.42	111
Gl6	SE of Dunglow (Donegal)	180000, 407300	Devonian (Granite)	179.9 ± 8.8	12.34 ± 0.23	2.27	97
Gl7	N of Ardara (Donegal)	173000, 393000	Devonian (Granite)	175.5 ± 9.3	12.39 ± 0.18	1.88	109
Gl8	E of Cregganbaun (Mayo)	86000, 274000	Devonian (Granite)	124.1 ± 5.1	12.26 ± 0.24	2.36	97
Gl9	Doonloughan (Galway)	57000, 245500	Devonian (Granite)	119.8 ± 6.1	12.19 ± 0.18	1.77	97
Gl10	nr. Roundstone (Galway)	71000, 239000	Devonian (Granite)	149.7 ± 6.4	12.44 ± 0.21	2.10	100
Gl11	nr. Roundstone (Galway)	73000, 242000	Devonian (Granite)	163.0 ± 7.5	12.39 ± 0.20	1.94	94
Gl12	Carna (Galway)	79000, 232000	Devonian (Granite)	182.7 ± 5.3	11.93 ± 0.22	2.17	97
Gl13	Ballyknockan (Wicklow)	301000, 207500	Devonian (Granite)	200.8 ± 4.6	12.54 ± 0.17	1.81	113
Gl14	Wicklow Gap (Wicklow)	307000, 201000	Devonian (Granite)	208.2 ± 6.4	12.20 ± 0.23	2.30	100
McCulloch (unpubl.)							
Mc1	S of Creeslough (Donegal)	206200, 427800	Devonian (Granite)	159.0 ± 11.0	12.41 ± 0.21	1.65	62
Mc2	Cushendall (Antrim)	324300, 426100	Devonian (Sandstone)	394.0 ± 24.0	12.37 ± 0.20	1.59	63
Mc3	Mannin Bay (Galway)	62000, 245000	Ordovician? (Metagabbro?)	154.0 ± 9.0	12.11 ± 0.34	2.18	41
Keeley et al. (1993)							
659	nr. Hook Head (Wexford)	275300, 101420	Devonian (Sandstone)	226.1 ± 11.1c	12.38 ± 0.15	1.58	113
660	Graiguenamanagh (Kilkenny)	271240, 139420	Devonian (Sandstone)	130.3 ± 7.3c	12.38 ± 0.19	1.67	74
664	Colliganwood (Waterford)	221100, 97380	Devonian (Sandstone)	154.5 ± 13.9c	12.56 ± 0.31	1.55	25
665	Ballycotton (Cork)	199820, 63880	Devonian (Sandstone)	211.8 ± 12.2c	12.10 ± 0.29	1.98	47
666	Ardmore Head (Waterford)	219760, 76520	Devonian (Sandstone)	185.7 ± 9.2c	12.06 ± 0.33	1.92	34
667	Comeragh Mts. (Waterford)	230060, 121300	Devonian (Sandstone)	165.8 ± 7.9c	12.37 ± 0.16	1.58	101
668	Old Head of Kinsale (Cork)	162420, 40690	Devonian (Sandstone)	154.2 ± 5.6c	12.65 ± 0.15	1.44	92
671	Ballyderown (Cork)	184900, 101700	Devonian (Sandstone)	191.6 ± 8.3c	12.68 ± 0.14	1.41	99
672	Creadan Head (Waterford)	231240, 89220	Devonian (Sandstone)	192.1 ± 10.3c	12.63 ± 0.16	1.51	90
McCulloch (1993)							
1	Cregganbaun (Mayo)	81000, 274300	Devonian (Granite)	152.0 ± 14.8p	12.89 ± 0.19	2.00	114
4	Gortamullin (Kerry)	89800, 73000	Devonian (Sandstone)	156.3 ± 26.8m	12.59 ± 0.19	1.45	62
5	Slea Head (Kerry)	31600, 96700	Devonian (Sandstone)	167.6 ± 15.0p	11.80 ± 0.24	2.07	78
7	Caha Mts. (Cork)	93000, 58300	Devonian (Sandstone)	164.6 ± 22.8p	12.61 ± 0.26	1.88	53
8	nr. Tully Cross (Galway)	73000, 262900	Silurian (Sandstone)	170.8 ± 25.4p	13.27 ± 0.23	2.15	92
9	Spiddle (Galway)	113200, 223000	Devonian (Granite)	115.5 ± 13.0p	12.24 ± 0.23	2.13	90
11	Ox Mts. (Sligo)	145000, 323700	Devonian (Granite)	168.7 ± 15.6p	12.49 ± 0.16	1.54	96
12	Ox Mts. (Sligo)	140200, 314500	Devonian (Granite)	154.8 ± 17.6p	12.55 ± 0.16	1.53	94

Table 3 – *continued*

Sample no.[a]	Locality	Easting, Northing	Strat. age (lithology)	Fission track age (Ma)[b]	Mean track length (μm)[c]	SD[c]	No. of tracks
14	Cronagort (Clare)	108500, 194800	Carboniferous (Sandstone)	150.0 ± 47.8m	12.15 ± 0.43	2.30	30
16	Diomond Hill (Galway)	72800, 256000	Precambrian (Granite)	156.4 ± 17.2p	13.22 ± 0.13	1.27	103
17	E of Omey Island (Galway)	57400, 256000	Devonian (Granite)	135.2 ± 15.0p	13.01 ± 0.21	2.07	100
18	Doonloughan (Galway)	56900, 245300	Devonian (Granite)	129.1 ± 11.8p	12.66 ± 0.16	1.61	101
19	Roundstone (Galway)	72300, 240300	Devonian (Granite)	127.3 ± 14.8p	12.77 ± 0.22	2.15	99
20	Spiddle (Galway)	113200, 222200	Devonian (Granite)	115.9 ± 10.2p	12.54 ± 0.20	1.95	101
21	Barna (Galway)	123400, 222500	Devonian (Granite)	119.1 ± 11.0p	12.12 ± 0.17	1.66	105
22	Aughrim (Wicklow)	313100, 179700	Devonian (Granite)	181.9 ± 16.6p	11.81 ± 0.45	2.37	29
23	Ballinacarrig (Wicklow)	318300, 185500	Devonian (Granite)	184.2 ± 24.2m	12.03 ± 0.21	1.91	83
24	Kilmanoge (Wicklow)	325100, 188900	Devonian (Granite)	217.2 ± 19.8m	13.02 ± 0.16	1.61	100
30	N of Ardara (Donegal)	172500, 393300	Devonian (Granite)	180.7 ± 11.6p	12.50 ± 0.16	1.56	101
31	Doonbeg Bay (Clare)	95700, 167500	Carboniferous (Sandstone)	112.7 ± 17.6p	13.20 ± 0.26	1.45	32
32	Mullaghmore (Sligo)	171000, 357900	Carboniferous (Sandstone)	165.7 ± 15.0p	12.45 ± 0.21	1.92	86
39	Crocknaconspody (Fermanagh)	239700, 354700	Carboniferous (Sandstone)	177.4 ± 17.8p	11.48 ± 0.20	1.67	69
43	Cratloe (Clare)	148100, 161600	Devonian (Sandstone)	205.2 ± 27.4 m	12.52 ± 0.18	1.70	89
45	Ballymastocker Bay (Donegal)	225100, 438000	Devonian (Sandstone)	300.8 ± 25.4m	12.85 ± 0.28	2.37	74
46	N of Port Laoise (Laois)	239500, 209000	Devonian (Sandstone)	179.4 ± 16.4p	11.19 ± 0.22	2.15	101
47	Cushendall (Antrim)	324400, 427600	Devonian (Andesite)	267.2 ± 26.2p	12.71 ± 0.16	1.61	101
48	near Enniskillen (Fermanagh)	230200, 341100	Silurian (Sandstone)	176.2 ± 15.0p	12.28 ± 0.20	2.07	100
51	W of Carlow (Laois)	266200, 176700	Carboniferous (Sandstone)	153.7 ± 18.0p	12.14 ± 0.34	2.18	43
54	L. Beltra (Mayo)	107500, 297300	Devonian (Sandstone)	186.5 ± 15.2p	12.05 ± 0.24	2.42	104
55	L. Corrib (Galway)	98100, 249800	Precambrian (Schist)	150.3 ± 11.4p	12.66 ± 0.20	1.94	99
56	Kilkee (Clare)	88400, 159800	Carboniferous (Sandstone)	138.4 ± 22.4m	12.54 ± 0.23	1.36	37
57	NE of Leenane (Mayo)	93500, 265700	Ordovician (Sandstone)	152.2 ± 13.0p	12.62 ± 0.18	1.81	95
60	Oughterard (Galway)	111700, 242600	Devonian (Granite)	129.6 ± 14.2p	12.37 ± 0.25	1.91	60
84	Clonloskan (Cavan)	235200, 300700	Devonian (Granite)	200.7 ± 14.6p	12.13 ± 0.18	1.76	100
64	Belfast Lough (Down)	340900, 380700	Carboniferous (Sandstone)	273.8 ± 26.8p	12.13 ± 0.15	1.54	104
68	Beaghbeg (Tyrone)	268300, 382000	Ordovician (Tuff)	225.2 ± 21.4p	12.40 ± 0.17	1.74	103
75	Ballyvoyle (Waterford)	233600, 94800	Carboniferous (Sandstone)	194.9 ± 12.2p	12.94 ± 0.17	1.67	100
76	Slieveardagh Hills (Tipperary)	228700, 154200	Carboniferous (Sandstone)	147.0 ± 14.4p	12.03 ± 0.26	1.92	54
77	Mouth of Shannon (Kerry)	87600, 144900	Carboniferous (Sandstone)	133.9 ± 28.4m	12.28 ± 0.19	1.85	95
80	Graiguenamanagh (Kilkenny)	270800, 143400	Devonian (Granite)	149.0 ± 11.8p	12.35 ± 0.13	1.76	172
82	Shannon estuary (Clare)	84800, 152400	Carboniferous (Sandstone)	132.9 ± 30.0m	11.77 ± 0.26	2.58	99
83	Kerry Head (Kerry)	69000, 132100	Devonian (Sandstone)	154.3 ± 26.0m	11.56 ± 0.24	2.43	96
86	Tagoat (Wexford)	310300, 111400	Ordovician (Tuff)	219.8 ± 19.4p	12.28 ± 0.19	1.85	95
88	nr. Foynes (Limerick)	124700, 152100	Carboniferous (Sandstone)	143.5 ± 29.6m	12.40 ± 0.21	2.24	113
90	Slieve Aughty Mts. (Galway)	154700, 201300	Devonian (Sandstone)	192.3 ± 16.6p	12.88 ± 0.18	1.71	92
McCulloch (1994)							
2	NW of Roundwood (Wicklow)	316000, 208600	Devonian (Granite)	136.7 ± 13.0p	12.27 ± 0.17	1.71	101
3	NW of Enniskerry (Wicklow)	319200, 219300	Devonian (Granite)	126.5 ± 6.4p	12.27 ± 0.16	1.62	100
4	Three Rock Mt. (Dublin)	317800, 223200	Devonian (Granite)	161.9 ± 33.4m	11.73 ± 0.21	1.96	86
6	N of Slane (Meath)	295500, 279300	Ordovician (Tuff)	212.1 ± 31.6p	12.78 ± 0.24	2.42	101
7	E of Banbridge (Down)	327800, 344500	Devonian (Granite)	134.4 ± 12.2p	11.74 ± 0.15	1.46	101

Table 3 – *continued*

Sample no.[a]	Locality	Easting, Northing	Strat. age (lithology)	Fission track age (Ma)[b]	Mean track length (μm)[c]	SD[c]	No. of tracks
8	Loughshinny (Dublin)	327000, 257700	Carboniferous (Sandstone)	70.2 ± 30.4p	12.56 ± 0.51	2.52	25
9	N of Skerries (Dublin)	325300, 260600	Silurian (Sandstone)	73.4 ± 10.0p	12.20 ± 0.50	2.45	25
11	E of Banbridge (Down)	321000, 334800	Devonian (Granite)	58.3 ± 7.4p	13.94 ± 0.19	1.20	42
12	Dundalk Bay (Louth)	307200, 302600	Silurian (Sandstone)	64.5 ± 14.2m	13.42 ± 0.20	1.30	43
13	S Dundalk Bay (Louth)	311500, 293500	Silurian (Sandstone)	46.0 ± 15.6m	13.72 ± 0.16	1.17	54
14	Clogher Head (Louth)	317100, 284600	Silurian (Sandstone)	50.8 ± 6.4m	13.07 ± 0.25	1.95	60
16	Balbriggan (Dublin)	320300, 264400	Ordovician (Tuff)	128.4 ± 22.8p	13.54 ± 0.38	2.41	40
Green et al. (2000)							
GC345-2	Moll's Gap (Kerry)	86000, 77500	Devonian (Sandstone)	177.0 ± 7.1 p	12.57 ± 0.19	1.90	103
GC345-3	Lough Muckross (Kerry)	93400, 85500	Carboniferous (Sandstone)	150.8 ± 11.1 c	12.87 ± 0.18	1.82	100
GC345-7	Kingscourt (Cavan)	285155, 281582	Carboniferous (Sandstone)	78.5 ± 6.9 c	12.12 ± 0.27	2.72	105
GC458-1	Mullaghmore (Sligo)	171000, 358000	Carboniferous (Sandstone)	164.0 ± 10.0 p	11.76 ± 0.24	2.38	100
GC458-2	Ballysadare (Sligo)	165100, 329600	Carboniferous (Sandstone)	164.2 ± 21.6 c	13.31 ± 0.30	1.49	25
GC458-5	Slisgarrow (Fermanagh)	202525, 351773	Carboniferous (Sandstone)	148.1 ± 13.6 p	11.95 ± 0.30	2.40	65
GC543-63	Strangford Lough (Down)	340477, 369575	Ordovician? (Sandstone)	318.9 ± 25.3 p	12.33 ± 0.17	1.77	108
GC543-64	Craigavad (Down)	342500, 381400	Carboniferous (Sandstone)	321.7 ± 18.7 p	12.62 ± 1.18	1.76	100
GC543-65	Scrabo North Quarry (Down)	347500, 373100	Triassic (Sandstone)	72.9 ± 5.6 p	13.21 ± 0.23	1.51	44
GC543-72	Waterfoot (Antrim)	324700, 425000	Devonian? (Alluvial sands)	312.9 ± 29.2 p	11.89 ± 0.19	1.96	106
GC543-73	Runabay Head (Antrim)	325600, 437000	Precambrian ('Dalradian')	291.9 ± 43.5 p	12.51 ± 0.28	1.50	28

a Sample numbers given as in the literature where applicable.
b Fission track age types given by: c, central age; p, pooled age; m, mean age. McCulloch (unpubl.) and Gleadow (unpubl.) age types unknown. Quoted age uncertainties are $\pm 1\sigma$, except those of McCulloch (1993, 1994) which are $\pm 2\sigma$. McCulloch (unpubl.) age uncertainties unknown.
c Quoted uncertainties of mean and standard deviation of track length distributions are $\pm 1\sigma$.

Palaeotemperatures can be converted to estimates of denudation by choosing a certain time slice of known duration and dividing by an estimate of the palaeogeothermal gradient at each of the sample sites. A physically more appropriate approach is the use of heat flow and thermal conductivity, the latter being dependent on the lithology of the rock section being considered. Here, we avoid the uncertainties inherent in estimating the thermal conductivity of the missing section and assume it is constant. In the absence of information to the contrary, it is common to assume that heat flow or temperature gradients have been constant over time (e.g. Gallagher & Brown 1999). In this situation, it is an assumption that the modelled thermal history reflects only cooling caused by denudation. However, in regions of active tectonics, particularly in regions of lithospheric stretching, heat flow and geothermal gradients are likely to vary strongly as a function of time (McKenzie 1978). Higher palaeogeothermal gradients imply smaller amounts of denudation to explain the cooling. We have therefore mapped denudation with both constant and variable geothermal gradients (see section on 'Evaluation of the geothermal gradient'), but for reasons of space present results only for the variable geotherm case.

Contouring method

Palaeotemperatures as a function of time derived from AFT analysis were contoured. The inverse distance weighted (IDW) interpolator was used to produce a temperature grid for each chosen time slice, at a cell size of 5 km × 5 km. This method determines output cell values using a linearly weighted combination of a set of sample points. The weight is a function of inverse distance. The input number of sample points was limited to 12, using the nearest neighbour method. The power parameter in the IDW interpolation controls the significance of the surrounding points upon the interpolated value, a higher power resulting in a smaller influence from distant points. The power used in the interpolation of the AFT grids was 2, i.e. low. The output value for a cell using IDW is limited to the range of values used to interpolate, that is, the maximum and minimum values constrained by the original palaeotemperature range. As the IDW is a weighted average, the interpolated value cannot be greater than the highest or less than the lowest input (Watson & Philip 1985). The resulting surface is the best estimate of what quantity is on the actual surface for each location. The interpolation preserves the existing data point values on the temperature grid. Although there may be significant gradients and discontinuities in the AFT palaeotemperature data, we have constructed maps of palaeotemperature (and thereby denudation presented here) purely using an IDW interpolator so as to be transparent and unbiased. However, the reader needs to bear in mind that this method necessarily smoothes potential discontinuities in the original data. When presenting the denudation maps we found that the conventional 'equal interval' method of classification smoothed over trends in the data in time slices when much of Ireland underwent relatively similar amounts of denudation. A 'natural breaks' classification scheme was used instead to identify breakpoints between classes using a statistical formula, allowing patterns of denudation to be more readily picked out. No user-defined discontinuities, such as fault block margins, have been incorporated into the contouring procedure.

In contouring palaeotemperature maps, data points were removed from the interpolation exercise when their palaeotemperatures exceeded 120 °C, as higher temperatures are not resolvable with the AFT technique. Consequently, the interpolation is carried out with a variable number of data points, particularly at old time slices. Although this results in a poorer data density, it removes the possibility of bias to the results caused by holding palaeotemperatures at their maximum resolvable level (120 °C) arbitrarily.

Evaluation of the geothermal gradient

A knowledge of palaeogeothermal gradients is important for the correct conversion of palaeotemperature data to estimates of denudation. The available present-day heat-flow data from Ireland are relatively limited (see Brock (1989) and Brock *et al.* (1991)). The typical range is 52–87 mW m^{-2}, with a mean value of 67 mW m^{-2} close to the European average of 64 mW m^{-2} (Cermák 1979). This somewhat limited spatial variation of heat flow, and a lack of evidence to the contrary, justifies our assumption of a constant geothermal gradient spatially.

A number of estimated palaeogeothermal gradients are available from VR profiles in boreholes. This technique relies on the extrapolation of best-fit VR profiles upwards to the appropriate surface temperature. There may be problems in the use of a linear extrapolation where thermal conductivity contrasts characterize the upper part of the basin fill (Holliday 1993; Allen *et al.* 1998), in the parameterization of the temperature–reflectance relation (Burnham &

Sweeney 1989; Rowley & White 1998), and in the recognition and removal of the effects of fluid flow on reflectance values (Green *et al.* 1997). Nevertheless, VR profiles, augmented where possible by data from AFT analysis, have been shown to be useful in the estimation of palaeogeothermal gradients (Corcoran & Clayton 1999).

Palaeogeothermal gradients estimated from VR and AFT profiles vary considerably (e.g. Bray *et al.*, 1992; Green 1986, 1989; Lewis *et al.* 1992; Murdoch *et al.* 1995; Scotchman & Thomas 1995; Green *et al.* 1997; Duncan *et al.* 1998; Corcoran & Clayton 1999) even within relatively small sub-basins. For example, palaeogeothermal gradients of between 10 and 50 °C km^{-1} have been estimated in the East Irish Sea and Solway basins (Green *et al.* 1997). There is therefore some merit in estimating palaeogeothermal gradients for the Irish landmass through a thermal model derived from subsidence history. A large number of boreholes show a period of Permo-Triassic to Early Jurassic stretching followed by thermal relaxation (Rowley & White 1998), with stretch factors of less than 1.4. For example, the subsidence history and VR profile of Gulf 42/21-1 (location in Fig. 1) can be satisfactorily explained using two periods of stretching (in Triassic Sherwood Sandstone and Rhaetian Penarth Group times) with stretch factors (β) of between 1.2 and 1.3 and a radiogenic heat contribution from the upper crust (Allen *et al.* 1998). The palaeogeothermal gradients in this model range from *c.* 40 °C km^{-1} in Triassic time to 22 °C km^{-1} in Late Tertiary time (see captions to Figs 4–8). Similar geotherms are implied (but not directly given) in the much more extensive study by Rowley & White (1998).

In the calculation of denudation in Ireland, therefore, we have used a variable geotherm to capture the early stretching and post-rift history of the region. The geotherm has not, however, been varied to account for possible thermal pulses associated with the onset of widespread igneous activity in the British Tertiary igneous province. Green *et al.* (1999) found no evidence for elevated Early Tertiary heat flows along the UK Atlantic margin. For simplicity, we have applied the time-dependent geotherm uniformly to all of the thermal histories in this study, although acknowledging that the palaeogeothermal gradient is likely to have also been spatially variable in response to different amounts of lithospheric stretching, to the distribution of variably radiogenic crust across Ireland and to the differing thermal conductivities of the sedimentary cover.

Maps of denudation since Triassic time

We present maps corresponding to periods of geological time that are bounded by known

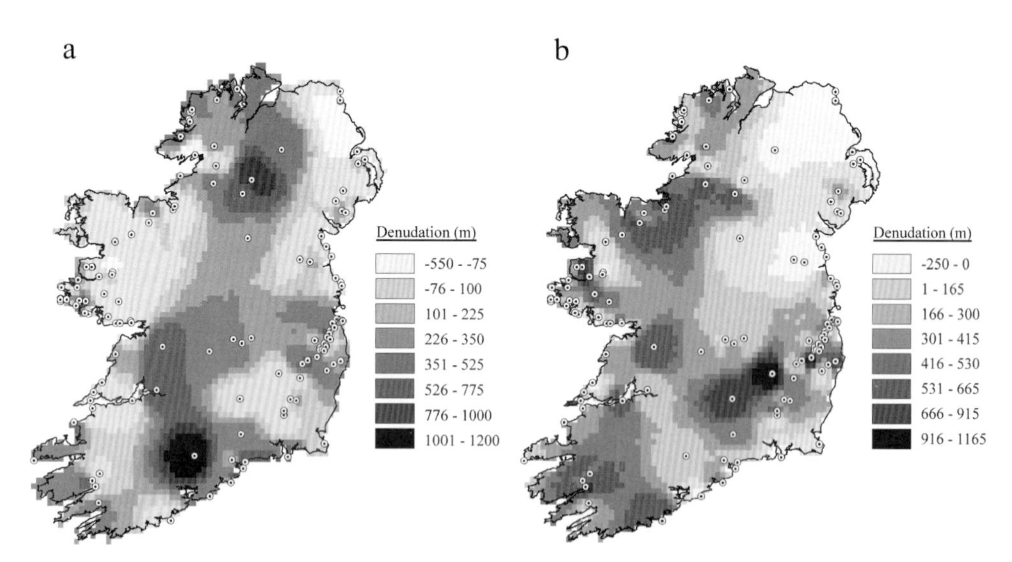

Fig. 4. (**a**) Denudation in the Triassic time slice (250–210 Ma) using the annealing model of Laslett *et al.* (1987) and a geothermal gradient of 40 °C km^{-1}. (**b**) Denudation in the Early Jurassic time slice (210–179 Ma) using the annealing model of Laslett *et al.* (1987) and a geothermal gradient of 36 °C km^{-1}.

stratigraphic 'events' such as unconformities, or corresponding to established biostratigraphic boundaries.

The maps shown in Figs 4–8 illustrate the total amount of denudation during that time slice. Denudation rates can be easily obtained by dividing the denudation by the time period.

Before discussing the results, it is appropriate to consider briefly the precision of the denudation estimates. The precision and resolution of the palaeotemperature vary in space and time, as well as depending on factors such as the uncertainties in the annealing models and the estimation of the geothermal gradient (or heat flow and thermal conductivity). Furthermore, the inferred timing of cooling episodes has an inherent uncertainty. Consequently, it is a non-trivial task to combine these multi-dimensional uncertainties with the results into a visually accessible form. As a guide, for a given sample location, we can consider an uncertainty of 10 °C on the temperature difference between two time points in the thermal history. This, coupled with an uncertainty of 20% on a geothermal gradient of 30 °C, implies a combined uncertainty of about 400–500 m on the denudation estimate. Thus, spatial variations on the maps should be considered with this level of uncertainty in mind. The uncertainties in timing of cooling are less problematical when the results are considered over the relatively long time intervals we adopt here.

Triassic time (Fig. 4a)

The total amount of denudation in Triassic time is low, with most of Ireland experiencing <1 km over the 40 Ma interval. Some areas experienced heating as a result of burial, especially in the NE of the island, corresponding to deposition in the Ulster Basin. The results are therefore consistent with the known Triassic basin development in Ulster. The results also support the view that Permo-Triassic sedimentation did not cover Ireland completely, and was restricted to fault grabens or half-grabens (Naylor 1992). Maximum denudation rates are *c.* 30 m Ma^{-1}, with most of Ireland experiencing rates of 0–20 m Ma^{-1} over the duration of the Triassic time slice.

Jurassic time (Figs 4b and 5a and b)

The amount of denudation continued to be relatively low during Jurassic time. The map for Early Jurassic time shows deposition in the NE and east, which is consistent with known occurrences of Liassic sedimentary rocks in Ulster and the Kish Bank Basin. However, the general pattern across Ireland is one of weak denudation at <20 m Ma^{-1}. This average denudation rate is difficult to reconcile with the notion that the entire area was undergoing subsidence under an extensive Early Jurassic sea. The rates of denudation remain low in

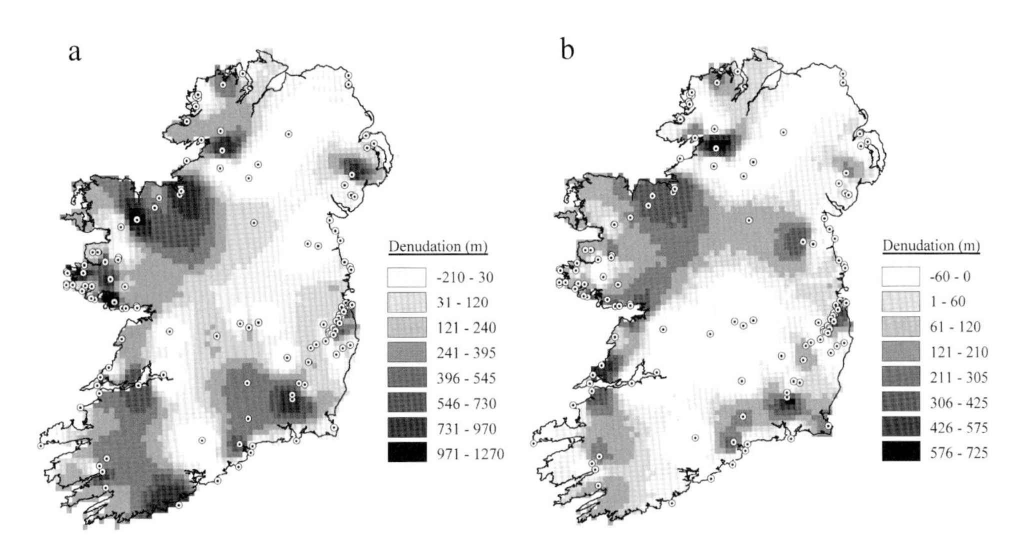

Fig. 5. (**a**) Denudation in the Mid-Jurassic time slice (179–152 Ma) using the annealing model of Laslett *et al.* (1987) and a geothermal gradient of 33 °C km^{-1}. (**b**) Denudation in the Late Jurassic time slice (152–131 Ma) using the annealing model of Laslett *et al.* (1987) and a geothermal gradient of 30 °C km^{-1}.

Mid-Jurassic time, with areas of heating (subsidence) in the north, NE, east and, interestingly, south (Counties Cork and Waterford) of Ireland. The southern part of the Leinster Massif shows denudation rates of 30 m Ma^{-1}, supporting the view that the Massif provided sediment for basins located in the Celtic Sea area. The presence of Middle Jurassic reddened, continental clays at Cloyne, County Cork (Fig. 2), supports the AFT results. The Late Jurassic map shows very low amounts of denudation, with almost all the island experiencing rates of <20 m Ma^{-1}. The AFT results therefore suggest limited realms of subsidence and weak denudation during Jurassic time, with localized areas of uplift such as parts of the Leinster Massif. There is no obvious signature of an early Mid-Jurassic (mid-Cimmerian) erosional event. However, this may be because of the poor resolution in the thermal model of reheating events within an overall cooling trajectory, as described above.

Cretaceous time (Fig. 6a and b)

The map for Early Cretaceous time shows two large zones of no cooling (equivalent to deposition) in the central–south and SE of Ireland and in the north. The extreme west (County Galway) and east of Ireland show denudation rates locally reaching 40 m Ma^{-1} and 30 m Ma^{-1}, respectively. This suggests that basement massifs such as those of Connemara and Leinster were reactivated during Early Cretaceous tectonism, whereas large parts of Ireland were encroached by Early Cretaceous basins. The large swath of Early Cretaceous heating or subsidence in the south of Ireland links well with the presence of thick Lower Cretaceous deposits in the Celtic Sea. The presence of a zone of deposition in the north and NE (along the trend of the Ox Mountains–Fintona block) suggests that the offshore areas of the Ulster Basin and northern Irish Sea were once covered with Lower Cretaceous sediments that have since been removed by erosion before the end of Cretaceous time or during Tertiary time. During Late Cretaceous time, most of Ireland underwent subsidence, with the notable exception of the area of central–east Ireland in Counties Louth and Meath. This area experienced denudation rates of *c.* 50 m Ma^{-1}, and as high as 80 m Ma^{-1} around the shores of Dundalk Bay. The predictions of denudation around the Antrim coast are not consistent with the known occurrences of Campanian to lower Maastrichtian Ulster White Limestone in this region. We suspect that some Early Tertiary cooling has been mapped into the Late Cretaceous time slice during the thermal modelling. Nevertheless, the broad picture is supportive of the idea that with the exception of east–central Ireland, the area underwent flooding under Late Cretaceous chalk seas. This is fully consistent with the occurrence of the chalk breccia at Ballydeenlea, County Kerry (Fig. 2), and with the widespread distribution of flint pebbles in tills across Ireland. The high amount of Cretaceous denudation in east–central Ireland poses the interesting

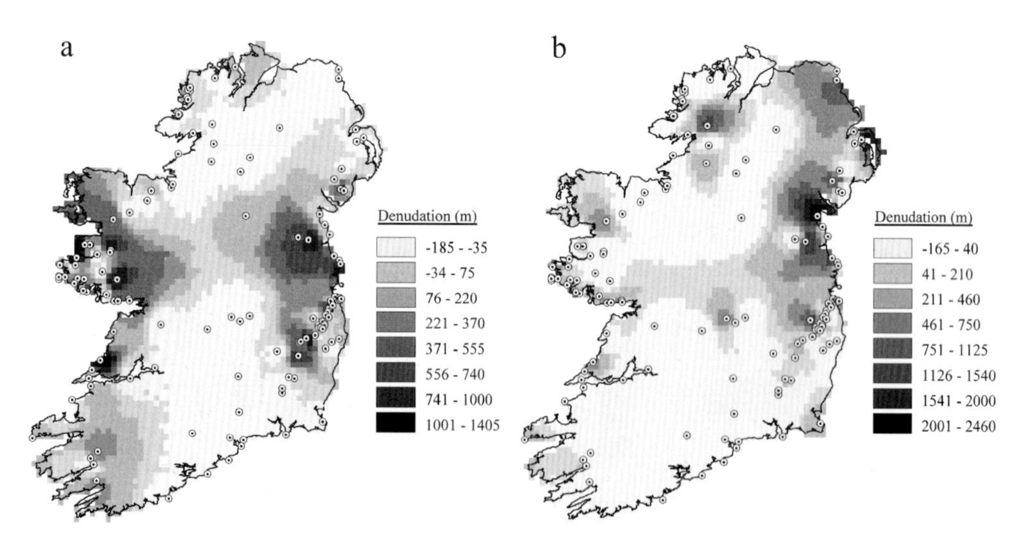

Fig. 6. (**a**) Denudation in the Early Cretaceous time slice (131–96 Ma) using the annealing model of Laslett *et al.* (1987) and a geothermal gradient of 27 °C km^{-1}. (**b**) Denudation in the Late Cretaceous time slice (96–66 Ma) using the annealing model of Laslett *et al.* (1987) and a geothermal gradient of 24 °C km^{-1}.

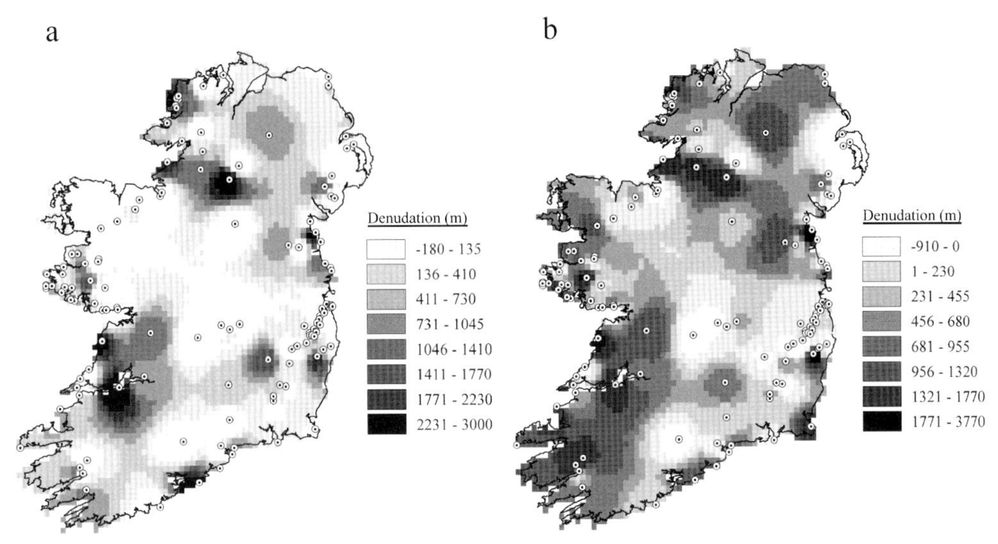

Fig. 7. Denudation in the Paleocene–Eocene time slice (66–36 Ma) with a geothermal gradient of 22 °C km^{-1}, using (**a**) the annealing model of Laslett *et al.* (1987) and (**b**) the annealing model of Gunnell *et al.* (in prep.).

possibility that the absence of Cretaceous rocks in the central Irish Sea may be partly due to Cretaceous erosion and non-deposition rather than solely to erosion during a Paleocene crustal uplift event.

Paleocene–Eocene time (Fig. 7a and b)

The map for Paleocene–Eocene time based on the Laslett *et al.* (1987) annealing model shows a significant change in denudation patterns. It shows a strong spatial variability (even allowing for uncertainties in the estimates), indicative of a patchwork of localized areas of strong denudation (up to 80 m Ma^{-1}) adjacent to areas of weak denudation or subsidence. These variations appear to be maintained by the modified annealing model (Gunnell *et al.* in prep.). Individual fault blocks or geological compartments presumably acted like piano keys, moving relative to adjacent blocks. In some regions, such as the Shannon Estuary and north County Clare and Cork harbour, there appears to be no readily available explanation for the locally high denudation rates inferred from the Laslett *et al.* (1987) model. The fact that these regions become areas of anomalously low denudation in Oligocene–mid-Miocene time suggests that the thermal modelling has generated a strong change across the 36 Ma boundary, which would disappear or be dampened with the choice of a different time slicing. Also, these spatial variations are reduced when the results of the modified model are considered and so should be treated with caution. The centre of Ireland is the only broad zone of no cooling (deposition) during Paleocene–Eocene time. There is no clear indication of an increase in denudation towards a putative Irish Sea Dome (see Cope 1994, 1998) in the Paleocene–Eocene map. The average amount of denudation is 500 m given by the Laslett *et al.* (1987) annealing model and 600 m for the modified model. The latter therefore predicts slightly more denudation (although this is probably at the level of uncertainty) implying samples were not at low enough temperatures during this time slice for the differences between the two models to become apparent.

Oligocene–mid-Miocene time (Fig. 8a and b)

Almost all of Ireland underwent denudation during this time interval, with rates varying between 20 and 50 m Ma^{-1} spread fairly uniformly across the island. A notable localized region of very high denudation is the Mourne region of County Down, where about 3 km of denudation was experienced in this interval, representing a denudation rate of 120 m Ma^{-1}. This is supported by independent evidence of the unroofing of the 56 Ma Slieve Gullion–Mourne Granites to their present-day level. The predominance of denudation over subsidence during Oligocene to mid-Miocene time is consistent with the widespread occurrence of continental and lacustrine Tertiary deposits across the island. There is a suggestion that east–central Ireland (Louth, Meath, Dublin, Wicklow) experienced

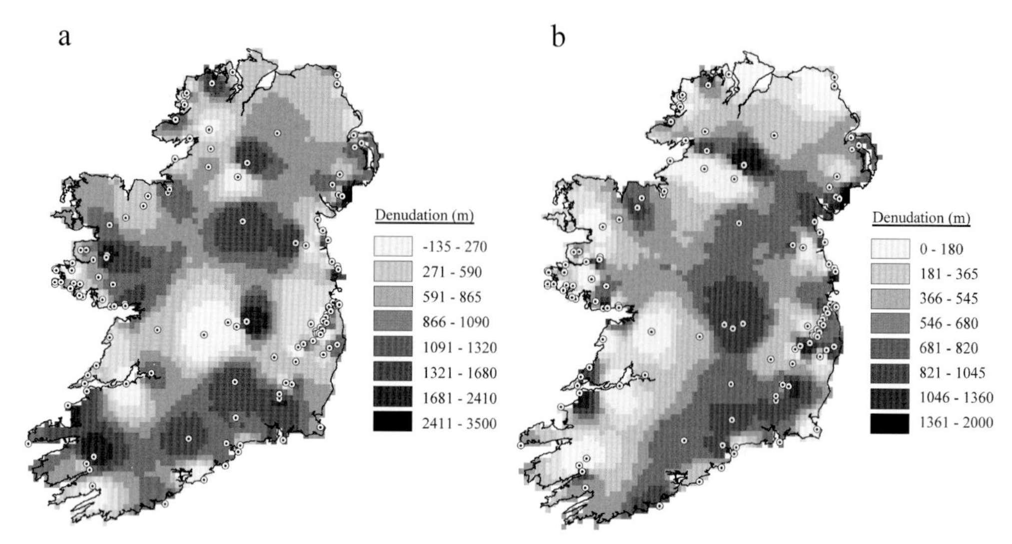

Fig. 8. Denudation in the Oligocene–mid-Miocene time slice (36–10 Ma) with a geothermal gradient of 22 °C km^{-1}, using (**a**) the annealing model of Laslett *et al.* (1987) and (**b**) the annealing model of Gunnell *et al.* (in prep.).

weak denudation and some subsidence, which may link with the preservation of middle to Upper Tertiary sedimentary rocks in Irish Sea basins. It should be borne in mind, however, that the resolution of the thermal model at temperatures below the top of the apatite partial annealing zone is limited, particularly for the Laslett *et al.* (1987) model. This is highlighted by the difference in mean denudations between the Laslett *et al.* (1987) and modified models, 970 m and 560 m, respectively, the former being a likely overestimate.

No map is shown of the 10–0 Ma time interval because of uncertainties in the annealing models at such low temperatures.

The cumulative amount of denudation from Triassic to mid-Miocene time (10 Ma) is shown in Fig. 9, demonstrating the broad pattern of Mesozoic–Tertiary denudation (mean value of 2.7 km). The areas of least denudation are NE Ulster, the Irish Midlands and the extreme SE (Counties Wexford and Waterford). The low values of cumulative denudation (<2 km) over the site of the Ulster Basin is a strong, independent validation of the thermochronology, as this area of Ireland contains preserved Mesozoic–Cenozoic stratigraphy. The areas of maximum denudation from Triassic to mid-Miocene time are eastern Ireland from Dublin Bay to County Down, and to a lesser extent in a tract of western Ireland from Donegal to Kerry, and in the Leinster Massif and its southwestern subsurface extension.

The cumulative amount of denudation in Tertiary time (66–10 Ma) shows less variation than the Triassic to mid-Miocene map (compare Figs 9 and 10). The average denudation for Ireland is c. 1.5 km and c. 1.2 km using the Laslett *et al.* (1987) and modified annealing models, respectively. Both models satisfactorily account for the estimate of <1.5 km Cenozoic denudation at Ballydeenlea, County Kerry, but the match with the new annealing model is better. There is little structure in the map of Tertiary denudation (Fig. 10a and b). There is no convincing regional trend of increasing amounts of denudation to the east that might reflect the effects of a long-wavelength Early Tertiary domal uplift centred on the Irish Sea. However, the high amount of denudation recognized along the coastal tract between Dublin Bay and Dundalk Bay is compatible with the existence of an Irish Sea high to the east.

The average denudation in the last 10 Ma is 650 m as given by the Laslett *et al.* (1987) annealing model and 230 m as given by the modified annealing model.

Total denudational efflux of the Irish landmass

The maps shown in Figs 4–8 portray amounts of denudation during chosen time slices. We are also able to monitor the denudation rate integrated across the entire present-day landmass

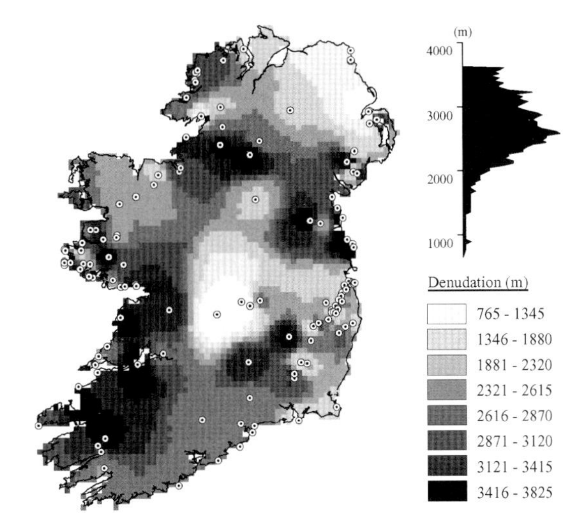

Fig. 9. Cumulative amount of denudation from Triassic time (250 Ma) to 10 Ma using the annealing model of Laslett *et al.* (1987). Inset shows histogram of denudation (in m).

of Ireland as an essentially continuous (time step 1 Ma) function of time. The total net sediment discharge of Ireland as a function of time can be simply calculated by multiplying the average denudation rate per square kilometre by the total land surface area of Ireland.

Two assumptions need to be considered. First, we assume a zero porosity of bedrock undergoing denudation. Second, we assume that all of the denudation of the Irish landmass resulting in delivery of particulate sediment load to neighbouring depocentres is by physical processes. Whereas the first assumption is trivial bearing in mind the order of magnitude estimates being made, the second assumption is potentially important. The ratio of solute to particulate loads in present-day Irish rivers in SE Ireland varies in relation to the bedrock lithologies of the individual catchments. However, the general pattern is that solute load dominates over particulate load (Malone 2001, unpubl. data). Consequently, any estimate of the particulate sediment discharge from Ireland to offshore basins as a function of geological time is likely to

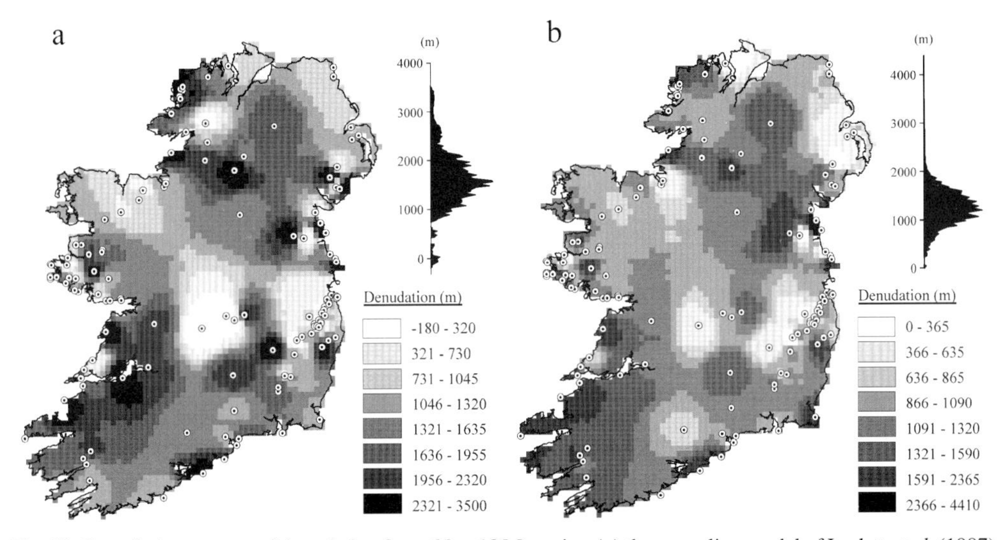

Fig. 10. Cumulative amount of denudation from 66 to 10 Ma, using (**a**) the annealing model of Laslett *et al.* (1987) and (**b**) the annealing model of Gunnell *et al.* (in prep.). Insets show histograms of denudation (in m).

be a significant overestimate. The extent of the overestimation will be a complex function of largely unconstrained climatic, topographic and bedrock variations through geological time in the area now occupied by the Irish landmass.

The sediment yields and discharges of individual ancient sediment routing systems can be calculated with knowledge of the former spatial extent of palaeocatchments. This procedure forms part of continuing research and is not reported here.

The long-term pattern (Figs 11 and 12) is of moderate rates of denudation and discharge during Early to Mid-Jurassic time decreasing through Late Jurassic time to low values through much of Early and Late Cretaceous time. Denudation rates and sediment discharges increase through Late Cretaceous time, with peaks in Maastrichtian–Paleocene, Eocene and Oligocene time. This general pattern is retained with a constant geotherm with time and with a geotherm involving post-stretching thermal relaxation using the annealing model of Laslett *et al.* (1987) (Fig. 11). A second set of curves is shown for the modified annealing model of Gunnell *et al.* (in prep.), which reduces the high cooling rates at low temperatures predicted by the Laslett *et al.* (1987) model (Fig. 12). With the Laslett *et al.* (1987) annealing model, denudation rates and discharges increase further through Neogene time, whereas with the new annealing model, rates fall slightly after a peak in Eocene–Oligocene time. A comparison of these curves also implies that peaks in the denudation chronology inferred from the modified annealing model tend to occur slightly earlier than in the Laslett *et al.* (1987) model. However, these timing differences are potentially within the uncertainties inherent in the modelling.

The results in Figs 11 and 12 can be compared with estimates of solid sediment volumes in the offshore basins. The Porcupine Basin has a surface area of *c.* 20 000 km^2. The total volume of Cenozoic solid sediment in the Porcupine Basin (Jones 2000, unpubl. data) is 54 900 $\pm$ 6300 km^3. The combined Erris Trough and eastern flank of the Irish Rockall Basin has a surface area of 20 000 km^2 (Spencer *et al.* 1999). Approximating the average Cenozoic solid sediment thickness as 3–4 km, the volume of Cenozoic sediment is of the order of 60 000–80 000 km^3. The small surface area (*c.* 5000 km^2) and thin cover of Cenozoic sediment in the Slyne Trough imply that relative to the larger Porcupine and Irish Rockall

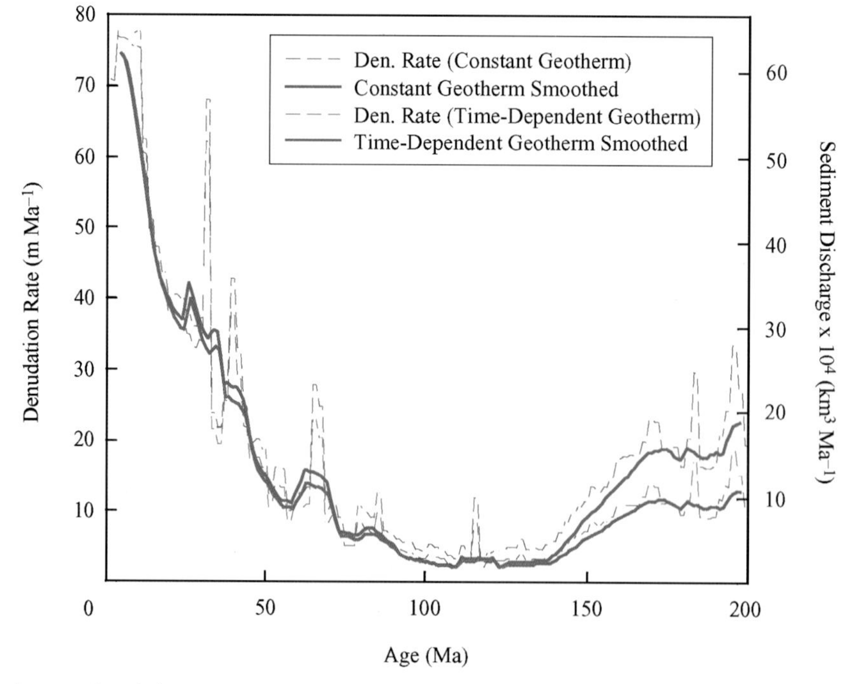

Fig. 11. Average denudation rate and maximum volumetric discharge for the Irish landmass since Triassic time for a constant geothermal gradient (blue lines), and a time-dependent geotherm (red lines).

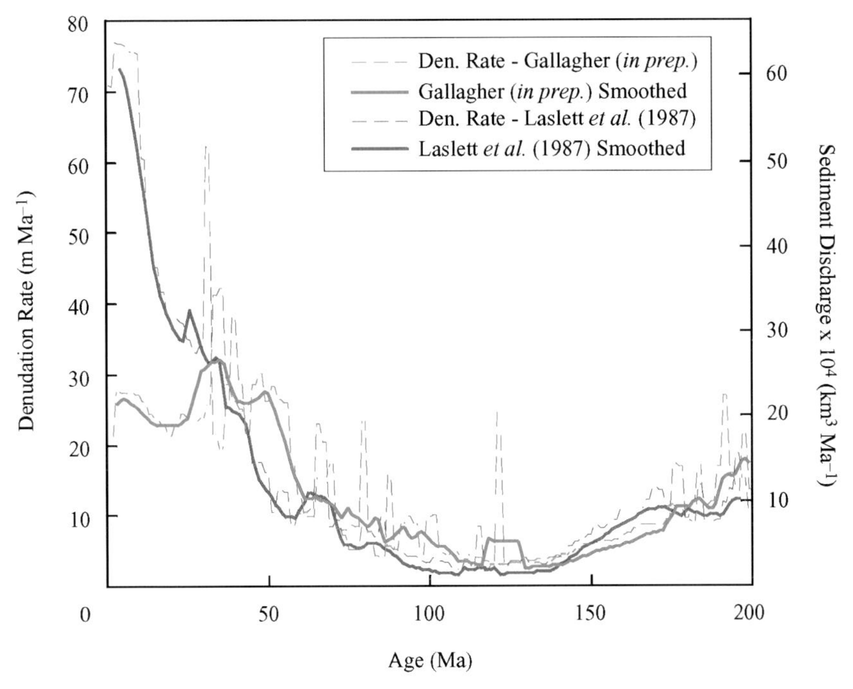

Fig. 12. Average denudation rate and maximum volumetric discharge for the Irish landmass since Triassic time with a new annealing model (green lines) that removes the artefact of high denudation at near-surface temperatures (Gunnell *et al.* in prep.). Results from the Laslett *et al.* (1987) annealing model shown for comparison (red lines). The maximum volume of sediment exported from the area now occupied by the Irish landmass is an order of magnitude smaller than the preserved solid volume of Cenozoic sediment in the Porcupine–Rockall basin system.

depocentres it is unimportant in calculating sediment volumes. The total Cenozoic solid sediment volume in the offshore basins to the west of Ireland is therefore well in excess of 100 000 km^3.

The average denudation rate of Ireland during Cenozoic time derived from the new annealing model is 20–25 m Ma^{-1}, and *c.* 40 m Ma^{-1} using the Laslett *et al.*, 1987 annealing model. Over a period of 65 Ma, with a present-day area of Ireland of 8344 km^2, this gives a maximum total sediment volume of between *c.* 10 000 km^3 and 20 000 km^3, about a factor of 5–10 smaller than the solid volume occupying the Porcupine and Irish Rockall basins. If we were to derive this volume of sediment solely from the Irish mainland, we would need denudation rates of about 180–230 m Ma^{-1}, and this would imply cooling rates of 5–8 °C Ma^{-1} for a 30 °C km^{-1} geotherm. We would then predict all samples to be at temperatures in excess of 120 °C, that is, at total annealing, around 15–25 Ma. We would not expect to measure any fission-track ages older than this. At such high denudation rates, the assumption of a linear geotherm would be invalid, but inferred fission-track ages would be even younger than 25 Ma. Clearly, it is not possible to invoke the current Irish mainland as the dominant source region for the offshore basins on the Atlantic margin. In searching for a source of sediment for the western offshore basins, it is necessary to look far beyond the present-day landmass of Ireland, such as the Irish Sea and the continental shelf bordering the Rockall margin.

A key problem for future research is the delineation of the catchments feeding the offshore basins in Tertiary time. Continuing studies are attempting to map out palaeochannel systems and model the palaeotopographic evolution of Ireland so as to make predictions about Tertiary sediment routing patterns in this sector of the NW European Atlantic margin. In addition, it is necessary to consider the time correlation between denudation and deposition, so as to refine the spatial links between the source areas and sediment delivery into specific basins. Finally, it will be necessary to undertake detailed provenance studies on the sediments themselves and the potential sediment source regions to fingerprint and correlate their geochemical or geochronological signatures.

Conclusions

Although very little is known about the post-Variscan history of Ireland based on its fragmentary stratigraphic record, we are able to make some progress in assessing its thermal and denudational history through apatite fission-track analysis of samples collected from surface exposures. Reconstructed thermal histories are used to create maps of denudation for time slices since the beginning of Triassic time.

The denudation maps in general show a patchwork of relatively small-scale spatial variations in denudation. In Triassic time, most of Ireland underwent denudation at rates of 0–30 m Ma^{-1}, with some areas, such as NE Ireland, undergoing subsidence during this time. This pattern is supported by the preservation of Permo-Triassic stratigraphy in the Ulster Basin. Denudation continued to be relatively low in Jurassic time, with zones of subsidence, such as southern Ireland adjacent to known Jurassic depocentres in the North Celtic Sea, and local regions of crustal uplift, such as the Leinster Massif. During Early Cretaceous time two large zones of subsidence were established in the central–SE and north of Ireland, whereas basement massifs such as those of Connemara and Leinster underwent denudation at <40 m Ma^{-1}. In Late Cretaceous time most of Ireland subsided with the exception of a region centred on Dundalk Bay in NE Leinster and SE Ulster, which experienced denudation rates as high as 80 m Ma^{-1}. This wide distribution of heating or subsidence supports the idea that most of Ireland was covered by a veneer of Upper Cretaceous chalk. The denudation map for the Paleocene–Eocene interval is spatially variable, suggesting that fault blocks or geological compartments reacted like piano keys to a regional tectonic event. There is no strong indication of a regional increase in denudation across Ireland towards a putative Irish Sea dome, but the relatively local high denudation rates in central–east Ireland are compatible with the existence of an Early Tertiary focus of denudation in the Irish Sea that impinged on the region now occupied by the Irish landmass. Almost all of Ireland underwent moderate rates of denudation in the Oligocene–mid-Miocene time, with the notable exception of the Mourne region of County Down, which suffered *c.* 3 km of denudation in this time interval (120 m Ma^{-1}), consistent with the required rapid unroofing of the 56 Ma Mourne and Slieve Gullion granites.

The total discharge of sediment from Ireland as a function of time, and the average denudation rate, shows moderate values in Triassic time, falling through Jurassic time to very low values in Cretaceous time. Discharges increase significantly into Tertiary time. However, the maximum volumetric discharge of sediment in Cenozoic time is considerably smaller than that implied by the preserved solid volume of Cenozoic sediment in the Porcupine–Irish Rockall basin system, by a factor of about 5–10. Consequently, in searching for a source for the Cenozoic sediment of the offshore basins, one must look far beyond the present-day landmass of Ireland.

We are grateful to the Ireland Frontier Exploration Licence 02/97 (ARCO, BG, Anadarko) for funding during 1998–1999, and S. Bergman (ARCO) in particular for assistance and advice. We received assistance in sampling from G. Sevastopulo and J. Graham (Trinity College Dublin), and I. Meighan (Queen's University Belfast), and thank A. Gleadow, P. O'Sullivan and J. Murphy for providing us with their unpublished fission-track data from Donegal and the Mournes. We also gratefully acknowledge the collaboration and reviews of N. White and S. Jones at Cambridge University.

References

ALLEN, P.A., CORCORAN, D. & CLAYTON, G. 1998. Forward models of heat flow and inversion of vitrinite reflectance data in investigating Tertiary inversion events. *In*: SIMMS, M. (ed.) *Ireland Since the Carboniferous: a 300 Million Year Enigma, Handbook, Conference at Ulster Museum, Belfast, 9–10 September 1998*. Ulster Museum, Belfast.

BLUNDELL, D.J. 1979. The geology and structure of the Celtic Sea. *In*: BANNER, F.T., COLLINS, M.B. & MASSIE, K.S. (eds) *The Northwest European Shelf Seas: the Sea Bed and the Sea in Motion 1, Geology and Sedimentology*. Elsevier, Amsterdam, 43–60.

BRAY, R.J., GREEN, P.F. & DUDDY, I.R. 1992. Thermal history reconstruction using apatite fission track analysis and vitrinite reflectance: a case study from the UK East Midlands and Southern North Sea. *In*: HARDMAN, R.P.F. (ed.) *Exploration Britain: Geological Insights for the Next Decade*. Geological Society, London, Special Publications, **67**, 3–25.

BROCK, A. 1989. Heat flow measurements in Ireland. *Tectonophysics*, **164**, 231–236.

BROCK, A., BRÜCK, P. & ALDWELL, R. 1991. National Report of Ireland. *In*: HURTIG, E., CERMÁK, V., HAENEL, R. & ZUI, V. (eds) *Geothermal Atlas of Europe*. Hermann Haack, Gotha, 52–53.

BRODIE, J. & WHITE, N. 1994. Sedimentary basin inversion caused by igneous underplating: Northwest European continental shelf. *Geology*, **22**, 147–150.

BROUGHAN, F.M., NAYLOR, D. & SANSTEY, N.A. 1989. Jurassic rocks in the Kish Bank Basin. *Irish Journal of Earth Sciences*, **10**, 99–106.

BURNHAM, A.K. & SWEENEY, J.J. 1989. A chemical kinetic model of vitrinite reflectance maturation. *Geochimica et Cosmochimica Acta*, **53**, 2649–2656.

CERMÁK, V. 1979. Heat flow map of Europe. *In*: CERMÁK, V. & RYBACH, L. (eds) *Terrestrial Heat Flow in Europe*. Springer, Berlin, 1–40.

CHAPMAN, T.J., BROKS, T.M., CORCORAN, D.V., DUNCAN, L.A. & DANCER, P.N. 2000. The structural evolution of the Erris Trough, offshore northwest Ireland, and implications for hydrocarbon generation. *In*: FLEET, A.J. & BOLDY, S.A.R. (eds) *Petroleum Geology of Northwest Europe: Proceedings of the 5th Conference*. Geological Society, London, 455–469.

CLAYTON, G. 1989. Vitrinite reflectance from the Kinsale Harbour–Old Head of Kinsale area, southern Ireland and its bearing on the interpretation of the Munster basin. *Journal of the Geological Society, London*, **146**, 611–616.

CLAYTON, G., SEVASTOPULO, G.D. & BURNETT, R.D. 1986. Carboniferous (Dinantian and Silesian) and Permo-Triassic rocks in south County Wexford, Ireland. *Geological Journal*, **21**, 366–374.

CLAYTON, G., HAUGHEY, N., SEVASTOPULO, G.D. & BURNETT, R.D. 1989. *Thermal Maturation Levels in the Devonian and Carboniferous of Ireland*. Geological Survey of Ireland, Dublin.

COPE, J.C.W. 1994. A latest Cretaceous hotspot and the southeasterly tilt of Britain. *Journal of the Geological Society, London*, **151**, 905–908.

COPE, J.C.W. 1998. The Mesozoic and Tertiary history of the Irish Sea. *In*: MEADOWS, N.S., TRUEBLOOD, S.P., HARDIMAN, M. & COWAN, G. (eds) *Petroleum Geology of the Irish Sea and Adjacent Areas*. Geological Society, London, Special Publications, **124**, 47–59.

CORCORAN, D. & CLAYTON, G. 1999. Interpretation of vitrinite reflectance profiles in the Central Irish Sea area: implications for the timing of organic maturation. *Journal of Petroleum Geology*, **22**, 261–286.

DANCER, P.N., ALGAR, S.T. & WILSON, I.R. 1999. Structural evolution of the Slyne Trough. *In*: FLEET, A.J. & BOLDY, S.A.R. (eds) *Petroleum Geology of Northwest Europe: Proceedings of the 5th Conference*. Geological Society, London, 445–453.

DAVIES, G.L. 1970. The enigma of the Irish Tertiary. *In*: STEPHENS, N. & GLASSOCK, R.E. (eds) *Irish Geographical Studies*. The Queen's University, Belfast, 1–16.

DORÉ, A.G. 1992. Synoptic palaeogeography of the Northeast Atlantic Seaway: Late Permian to Cretaceous. *In*: PARNELL, J. (ed.) *Basins on the Atlantic Seaboard: Petroleum Geology, Sedimentology and Basin Evolution*. Geological Society, London, Special Publications, **62**, 421–446.

DORÉ, A.G., LUNDIN, E.R., JENSEN, N., BIRKELAND, Ø., ELIASSEN, P.E. & FICHLER, C. 1999. Principal tectonic events in the evolution of the northwest European Atlantic margin. *In*: FLEET, A.J. & BOLDY, S.A.R. (eds) *Petroleum Geology of Northwest Europe: Proceedings of the 5th Conference*. Geological Society, London, 41–61.

DUNCAN, W.I., GREEN, P.F. & DUDDY, I.R. 1998. Source rock burial history and seal effectiveness: key facets to understanding hydrocarbon exploration potential in the East and Central Irish Sea Basins. *AAPG Bulletin*, **82**, 1401–1415.

EVANS, A. & CLAYTON, G. 1998. The geological history of the Ballydeenlea Chalk Breccia, County Kerry, Ireland. *Marine and Petroleum Geology*, **15**, 299–307.

EVANS, D., KENOLTY, N., DOBSON, M.R. & WHITTINGTON, R.J. (1980) *The Geology of the Malin Sea*. Report, Institute of Geological Sciences, London, **79/15**.

GALBRAITH, R.F. & LASLETT, G.M. 1993. Statistical models for mixed fission track ages. *Nuclear Tracks and Radiation Measurement*, **21**, 459–470.

GALLAGHER, K. 1995. Evolving thermal histories from fission track data. *Earth and Planetary Science Letters*, **136**, 421–435.

GALLAGHER, K. & BROWN, R. 1999. Denudation and uplift at passive margins: the record of the Atlantic margin of southern Africa. *Philosophical Transactions of the Royal Society of London*, **357**, 835–859.

GREEN, P.F. 1986. On the thermo-tectonic evolution of Northern England: evidence from fission track analysis. *Geological Magazine*, **123**, 493–506.

GREEN, P.F. 1989. Thermal and tectonic history of the East Midlands Shelf (onshore UK) and surrounding regions assessed by apatite fission track analysis. *Journal of the Geological Society, London*, **146**, 755–773.

GREEN, P.F., DUDDY, I.R. & BRAY, R.J. 1997. Variation in thermal history styles around the Irish Sea and adjacent areas: implications for hydrocarbon occurrence and tectonic evolution. *In*: MEADOWS, N.S., TRUEBLOOD, S.P., HARDIMAN, M. & COWAN, G. (eds) *Petroleum Geology of the Irish Sea and Adjacent Areas*. Geological Society, London, Special Publications, **124**, 73–93.

GREEN, P.F., DUDDY, I.R., BRAY, R.J. & LEWIS, C.L.E. 1993. Elevated palaeotemperatures prior to Early Tertiary cooling throughout the UK region: implications for hydrocarbon generation. *In*: PARKER, J.R. (ed.) *Petroleum Geology of Northwest Europe: Proceedings of the 4th Conference*. Geological Society, London, 1067–1074.

GREEN, P.F., DUDDY, I.R., HEGARTY, K.A. & BRAY, R.J. 1999. Early Tertiary heat flow along the UK Atlantic margin and adjacent areas. *In*: FLEET, A.J. & BOLDY, S.A.R. (eds) *Petroleum Geology of Northwest Europe: Proceedings of the 5th Conference*. Geological Society, London, 349–357.

GREEN, P.F., DUDDY, I.R., HEGARTY, K.A., BRAY, R.J., SEVASTOPULO, G.S., CLAYTON, G. & JOHNSTON, D. 2000. The post-Carboniferous evolution of Ireland: evidence from thermal history reconstruction. *Proceedings of the Geologists' Association*, **111**, 307–320.

GUNNELL, Y., GALLAGHER, K., CARTER, A., WIDDOWSON, M. & HURFORD, A.J. in prep.. Denudation history of the continental margin of western peninsular India during the Mesozoic and Cenozoic.

HIGGS, K. & BEESE, A.P. 1986. A Jurassic microflora from the Colbond Clay of Cloyne, County Cork. *Irish Journal of Earth Sciences*, **7**, 99–110.

HIGGS, K.T. & JONES, G.Ll. 1998. Palynological evidence for Mesozoic karst at Piltown, County Kilkenny. *In*: SIMMS, M. (ed.) *Ireland Since the Carboniferous: A 300 Million Year Enigma, Handbook, Conference at Ulster Museum, Belfast, 9–10 September 1998*. Ulster Museum, Belfast.

HOLLIDAY, D.W. 1993. Mesozoic cover over northern England: interpretation of apatite fission track data. *Journal of the Geological Society, London*, **150**, 657–660.

HURFORD, A.J. 1990. Standardization of fission track dating calibration: recommendation by the Fission Track Working Group of the IUGS subcommision on geochronology. *Chemical Geology*, **80**, 177–178.

JONES, S. M. (2000). *Influence of the Iceland plume on Cenozoic sedimentation patterns*. PhD thesis, University of Cambridge.

KEELEY, M.L., LEWIS, C.L.E., SEVASTOPULO, G.D., CLAYTON, G. & BLACKMORE, R. 1993. Apatite fission track data from southeast Ireland: implications for post-Variscan burial history. *Geological Magazine*, **130**, 171–176.

LASLETT, G.M. & GALBRAITH, R. 1996. Statistical modelling of thermal annealing of fission tracks in apatite. *Geochimica et Cosmochimica Acta*, **60**, 5117–5131.

LASLETT, G.M., GREEN, P.F., DUDDY, I.R. & GLEADOW, A.J.W. 1987. Thermal annealing of fission tracks in apatite, 2: A quantitative analysis. *Chemical Geology*, **65**, 1–13.

LEWIS, C.L.E., GREEN, P.F., CARTER, A. & HURFORD, A.J. 1992. Elevated K/T palaeotemperatures throughout Northwest England: three kilometres of Tertiary erosion? *Earth and Planetary Science Letters*, **112**, 131–145.

MALONE, J.T. (2001). *Provenance and fluxes of Late Cenozoic sediments, onshore and offshore SE Ireland*. PhD thesis, Trinity College Dublin.

MCCAFFREY, R.J. & MCCANN, N. 1992. Post-Permian basin history of northeast Ireland. *In*: PARNELL, J. (ed.) *Basins on the Atlantic Seaboard: Petroleum Geology, Sedimentology and Basin Evolution*. Geological Society, London, Special Publications, **62**, 277–290.

MCCANN, N. 1988. An assessment of the subsurface geology between Magilligan Point and Fair Head, Northern Ireland. *Irish Journal of Earth Sciences*, **9**, 71–78.

MCCANN, N. 1990. The subsurface geology between Belfast and Larne, Northern Ireland. *Irish Journal of Earth Sciences*, **10**, 157–173.

MCCULLOCH, A.A. 1993. Apatite fission track results from Ireland and the Porcupine basin and their significance for the evolution of the North Atlantic. *Marine and Petroleum Geology*, **10**, 572–591.

MCCULLOCH, A.A. 1994. Low temperature thermal history of eastern Ireland: effects of fluid flow. *Marine and Petroleum Geology*, **11**, 389–399.

MCKENZIE, D. 1978. Some remarks on the development of sedimentary basins. *Earth and Planetary Science Letters*, **40**, 25–32.

MEIGHAN, I.G., GAMBLE, J.A. & WYSOCZANSKI, R.J. 1999. Constraints on the age of the British Tertiary Volcanic Province from ion microprobe U–Pb (SHRIMP) ages for acid igneous rocks from NE Ireland. *Journal of the Geological Society, London*, **156**, 291–299.

MENPES, R.J. & HILLIS, R.R. 1995. Quantification of Tertiary exhumation from sonic velocity data, Celtic Sea/South-Western Approaches. *In*: BUCHANAN, J.G. & BUCHANAN, P.G. (eds) *Basin Inversion*. Geological Society, London, Special Publications, **88**, 191–207.

MITCHELL, G.F. 1980. The search for the Tertiary in Ireland. *Journal of Earth Sciences, Royal Dublin Society*, **3**, 13–33.

MURDOCH, L.M., MUSGROVE, F.W. & PERRY, J.S. 1995. Tertiary uplift and inversion history in the North Celtic Sea Basin and its influence on source rock maturity. *In*: CROKER, P.F. & SHANNON, P.M. (eds) *The Petroleum Geology of Ireland's Offshore Basins*. Geological Society, London, Special Publications, **93**, 297–319.

MUSGROVE, F.W. & MITCHENER, B. 1996. Analysis of the pre-Tertiary rifting history of the Rockall Trough. *Petroleum Geoscience*, **2**, 353–360.

NAYLOR, D. 1992. The post-Variscan history of Ireland. *In*: PARNELL, J. (ed.) *Basins on the Atlantic Seaboard: Petroleum Geology, Sedimentology and Basin Evolution*. Geological Society, London, Special Publications, **62**, 255–275.

NAYLOR, D. (1998) *Irish Shorelines through Geological Time*. John Jackson Lecture. Royal Dublin Society, Occasional Papers in Irish Science and Technology **17**.

NAYLOR, D. & ANSTEY, N.A. 1987. A reflection seismic study of the Porcupine Basin, offshore West Ireland. *Irish Journal of Earth Sciences*, **8**, 187–210.

NAYLOR, D. & SHANNON, P.M. 1982. *The Geology of Offshore Ireland and West Britain*. Graham & Trotman, London.

PARNELL, J., SHUKLA, B. & MEIGHAN, I.G. 1989. The lignite and associated sediments of the Lough Neagh Basin. *Irish Journal of Earth Sciences*, **10**, 67–88.

PENNEY, S.R. 1980. A new look at the Old Red Sandstone succession of the Comeragh Mountains, County Waterford. *Irish Journal of Earth Sciences (Royal Dublin Society)*, **3**, 155–178.

PETRIE, S.H., BROWN, J.R., GRANGER, P.J. & LOVELL, J.P.B. 1989. Mesozoic history of the Celtic Sea basins. *In*: TANKARD, A.L. & BALKWILL, H.R. (eds) *Extensional Tectonics and Stratigraphy of the North Atlantic Margins*. Memoirs, American Association of Petroleum Geologists, **46**, 433–444.

ROWLEY, E. & WHITE, N. 1998. Inverse modelling of extension and denudation in the East Irish Sea and surrounding areas. *Earth and Planetary Science Letters*, **161**, 57–71.

SCOTCHMAN, I.C. & THOMAS, J.R.W. 1995. Maturity and hydrocarbon generation in the Slyne Trough, northwest Ireland. *In*: CROKER, P.F. & SHANNON, P.M. (eds) *The Petroleum Geology of Ireland's Offshore Basins*. Geological Society, London, Special Publications, **93**, 385–411.

SHANNON, P.M. 1991. Tectonic framework and petroleum potential of the Celtic Sea, Ireland. *First Break*, **9**, 107–122.

SHANNON, P.M., JACOB, A.W.B., MAKRIS, J., O'REILLY, B.M., HAUSER, F. & VOGT, U. 1995. Basin development and petroleum prospectivity of the Rockall and Hatton regions. *In*: CROKER, P.F. & SHANNON, P.M. (eds) *The Petroleum Geology of Ireland's Offshore Basins*. Geological Society, London, Special Publications, **93**, 435–457.

SHANNON, P.M., O'REILLY, B.M., HAUSER, F., READMAN, P.W. & MAKRIS, J. 1999. Structural setting, geological development and basin modelling in the Rockall Trough. *In*: FLEET, A.J. & BOLDY, S.A.R. (eds) *Petroleum Geology of Northwest Europe: Proceedings of the 5th Conference*. Geological Society, London, 421–431.

SIMMS, M.J. 1998. Sub-basalt palaeokarst and its implications for the Tertiary history of northeast Ireland. *In*: SIMMS, M. (ed.) *Ireland Since the Carboniferous: a 300 Million Year Enigma, Handbook, Conference at Ulster Museum, Belfast, 9–10 September 1998*. Ulster Museum, Belfast.

SPENCER, A.M., BIRKELAND, Ø., KNAG, B.Ø. & FREDSTED, R. 1999. Petroleum systems of the Atlantic margin of northwest Europe. *In*: FLEET, A.J. & BOLDY, S.A.R. (eds) *Petroleum Geology of Northwest Europe: Proceedings of the 5th Conference*. Geological Society, London, 231–246.

STOKER, M.S., VAN WEERING, T.C.E. & SVAERDBORG, T. 2001. A Mid- to Late-Cenozoic tectonostratigraphic framework for the Rockall Trough. *In*: SHANNON, P.M., HAUGHTON, P.D.W. & CORCORAN, D.V. (eds) *The Petroleum Exploration of Ireland's Offshore Basins*. Geological Society, London, Special Publications, **188**, 411–438.

TATE, M.P. 1993. Structural framework and tectonostratigraphic evolution of the Porcupine Seabight Basin, offshore western Ireland. *Marine and Petroleum Geology*, **10**, 95–123.

TRUEBLOOD, S. 1992. Petroleum geology of the Slyne Trough and adjacent basins. *In*: PARNELL, J. (ed.) *Basins on the Atlantic Seaboard: Petroleum Geology, Sedimentology and Basin Evolution*. Geological Society, London, Special Publications, **62**, 315–326.

TUCKER, P.M. & ARTER, G. 1987. The tectonic evolution of the North Celtic Sea and Cardigan Bay basins. *Tectonophysics*, **137**, 191–307.

VISSCHER, H. (1971). *The Permian and Triassic of the Kingscourt Outlier*. Geological Survey of Ireland, Special Paper, **1**.

WALSH, P.T. 1966. Cretaceous outliers in south west Ireland and their implications for Cretaceous palaeogeography. *Quarterly Journal of the Geological Society, London*, **122**, 63–84.

WATSON, D.F. & PHILIP, G.M. 1985. A refinement of inverse distance weighted interpolation. *Geo-Processing*, **2**, 315–327.

WHITE, N. & LOVELL, B. 1997. Measuring the pulse of a plume with the sedimentary record. *Nature*, **387**, 888–891.

WILKINSON, G.C., BAZLEY, R.A.B. & BOULTER, M.C. 1980. The geology and palynology of the Lough Neagh Clays, Northern Ireland. *Journal of the Geological Society, London*, **137**, 65–75.

WOODLAND, A.W. (1971). *The Llanbedr (Mochras) Borehole*. Report of the Institute of Geological Sciences, **71/18**.

YOUNG, E., MYERS, A.T., MUNSON, E.L. & CONKLIN, N.M. 1969. Mineralogy and geochemistry of fluorapatite from Cerro de Mercado. *In* US Geological Survey, Professional Paper, **650-D**, 84–93.

Prediction of the hydrocarbon system in exhumed basins, and application to the NW European margin

A. G. DORÉ[1], D. V. CORCORAN[2] & I. C. SCOTCHMAN[1]

[1]*Statoil (U.K.) Ltd, 11a Regent Street, London SW1Y 4ST, UK (e-mail: agdo@statoil.com)*

[2]*Statoil Exploration (Ireland) Ltd, Statoil House, 6, St George's Dock, IFSC, Dublin 1, Ireland*

Abstract: Uplift, erosion and removal of overburden have profound effects on sedimentary basins and the hydrocarbon systems they contain. These effects are predictable from theory and from observation of explored exhumed basins. Exhumed basins are frequently evaluated in the same way as 'normal' subsiding basins, leading to errors and unrealistic expectations. In this paper we discuss the consequences of exhumation in terms of prospect risk analysis, resource estimation, and overall basin characteristics.

Exhumation should be taken into account when assigning risk factors used to estimate the probability of discovery for a prospect. In general, exhumation reduces the probability of trapping or sealing hydrocarbons, except where highly ductile seals such as evaporites are present. Exhumation modifies the probability of reservoir in extreme cases; for example, where a unit may have been buried so deeply before uplift that it is no longer an effective reservoir, or where fracturing on uplift may have created an entirely new reservoir. The probability of sourcing or charging is affected by multiple factors, but primarily by the magnitude of the post-exhumation hydrocarbon budget and the efficiency of remigration. Generally gas will predominate as a result of methane liberation from oil, formation water and coal, and because of expansion of gas trapped before uplift. These factors in combination tend to result in gas flushing of exhumed hydrocarbon basins.

Compared with a similar prospect in a non-exhumed basin, resource levels of a prospect in an exhumed basin are generally lower. Higher levels of reservoir diagenesis influence the standard parameters used to calculate prospect resources. Porosity, water saturation and net-to-gross ratio are adversely affected, and (as a consequence of all three) lower recovery factors are likely. Hydrostatic or near-hydrostatic fluid pressure gradients (as observed in exhumed NE Atlantic margin basins) will also reduce the recovery factor and, in the case of gas, will adversely affect the formation volume factor.

Hydrocarbon systems in exhumed settings show a common set of characteristics. They can include: (1) large, basin-centred gas fields; (2) smaller, peripheral, remigrated oil accumulations; (3) two-phase accumulations; (4) residual oil columns; (5) biodegraded oils; (6) underfilled traps. Many basins on the NE Atlantic seaboard underwent kilometre-scale uplift during Cenozoic time and contain hydrocarbon systems showing the effects of exhumation. This knowledge can constrain risk and resource expectation in further evaluation of these basins, and in unexplored exhumed basins.

Global oil and gas resources are finite and depleting rapidly. Estimates as to when world oil production will begin its terminal decline vary, but all authorities agree that this must take place within a few decades (e.g. Campbell 1996). Natural gas, the logical short-term replacement for oil, is more abundant and may provide global supply for about a century at projected rates of consumption (Lerche 1996). Therefore, the search for oil may be said to be entering its 'end game', characterized by increasing difficulty in locating major new reserves. In the case of gas, economic attractiveness is tied to market availability and thus to location. In both cases, the result is a drive to explore the more difficult, higher risk basins.

Although no hydrocarbon basin is without exploration problems, an 'ideal' basin is perhaps one containing abundant reservoirs and rich source rocks, which is continuously subsiding and where hydrocarbons are being generated at the present day, replenishing those that leak to the surface. Examples include the Northern North Sea, the Gulf of Mexico and the South Caspian Basin. Most such basins are now known and under production. Exhumed basins, on the

From: DORÉ, A.G., CARTWRIGHT, J.A., STOKER, M.S., TURNER, J.P. & WHITE, N. 2002. *Exhumation of the North Atlantic Margin: Timing, Mechanisms and Implications for Petroleum Exploration*. Geological Society, London, Special Publications, **196**, 401–429. 0305-8719/02/$15.00

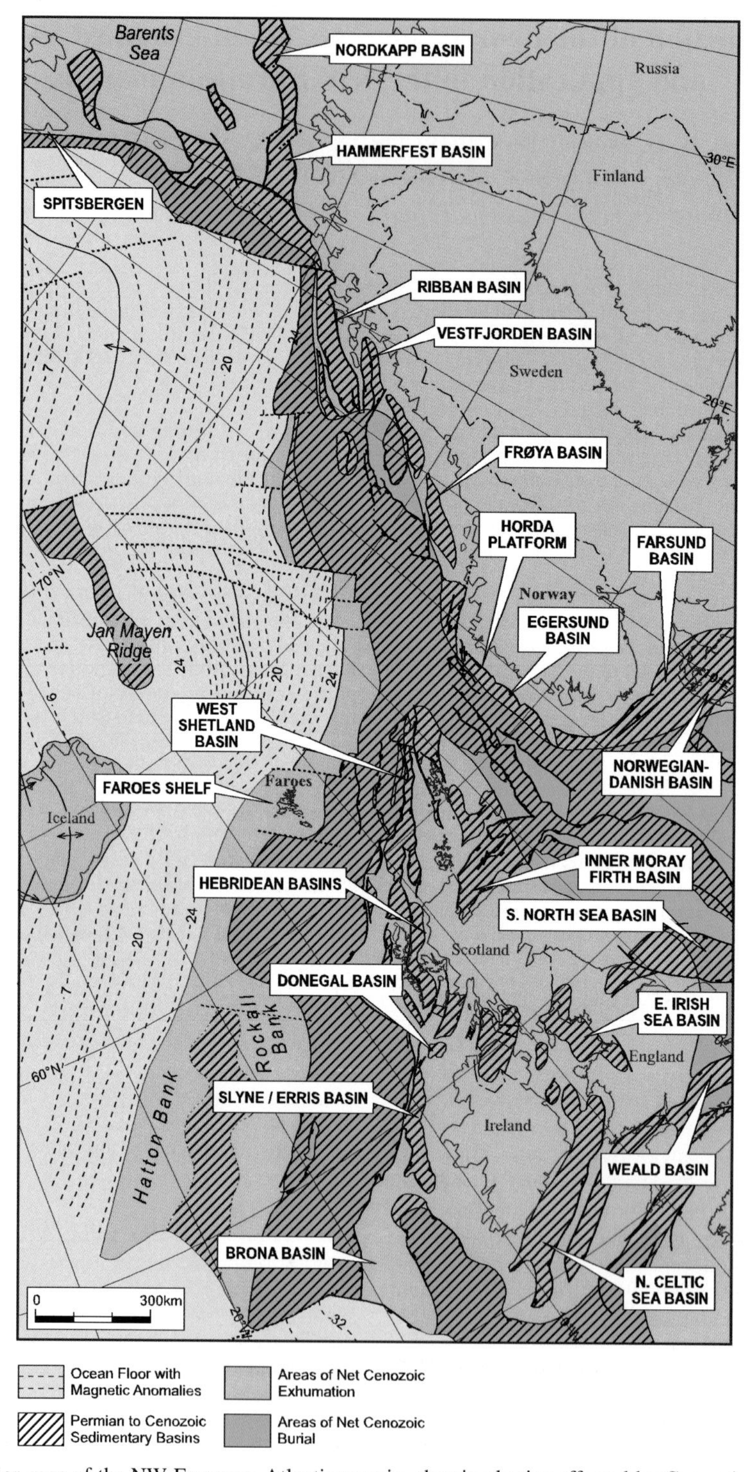

Fig. 1. Location map of the NW European Atlantic margin, showing basins affected by Cenozoic exhumation.

other hand, belong in the higher risk category, and will become increasingly important as global resources diminish.

For the purposes of this paper, an exhumed basin is loosely defined as one that has undergone uplift and erosion, such that the sedimentary rocks that constitute the petroleum system (source, reservoir and seal) are significantly shallower now than in the past (see more formal definitions given by Riis & Jensen (1992) and Doré & Jensen (1996) and Doré *et al.* (2002)). In such basins, the rock properties and hydrocarbon systems will be radically different from those at a similar depth in a continuously subsiding basin. These properties can be studied by observation of drilled exhumed basins, or predicted by modelling and experimentation (e.g. Nyland *et al.* 1992; Doré & Jensen 1996). Although most effects are individually understood, they are rarely studied systematically in the initial exploration of an exhumed basin. Together, however, they constitute a powerful predictive tool.

In the past, inappropriate comparison of the exploration potential of exhumed basins with 'classic' subsiding basins has resulted in unfulfilled expectations. Realization early in the exploration process that a basin has been exhumed gives rise to a different approach to hydrocarbon exploration, and can help to constrain resource prediction. In this paper we systematically describe the effects of exhumation with reference to two of the standard procedures of petroleum exploration: (1) estimation of the probability of finding hydrocarbons in a prospect; (2) calculation of its volumetric resource potential. Then, based on these discussions, we derive a generalized set of key characteristics for exhumed petroleum basins.

It is now generally understood that major uplift and erosion took place in the circum-North Atlantic borderlands during Cenozoic time, transforming a region dominated by low relief and shelf seas (Late Cretaceous) to one bordered by highlands such as Norway, Scotland and East Greenland (e.g. Riis & Fjeldskaar 1992; Riis 1996; Doré *et al.* 1999; Japsen & Chalmers 2000). It is also apparent that many of the offshore basins marginal to the landmasses underwent Cenozoic uplift and erosion. These include basins where hydrocarbon systems are proven (the western Barents Sea and Horda Platform (Norway), Inner Moray Firth, West Shetland Basin and East Irish Sea Basin (UK) and the North Celtic Sea and Slyne–Erris Basin (Ireland)) in addition to many unexplored basins (Fig. 1). Exhumation in these areas has been quantified using numerous methods, including seismic velocities, shale velocities, vitrinite reflectance, apatite fission track, mass balance and basin restoration (e.g. Riis & Jensen 1992). These measurements show that Cenozoic uplift around the North Atlantic was geographically variable and took place in several phases. Three of these phases have particularly widespread significance: an uplift of Paleocene age, generally thought to be associated with effects of the Iceland plume and incipient opening of the North Atlantic (e.g. White 1988; White & McKenzie 1989; Milton *et al.* 1990), an Oligo-Miocene episode usually associated with inversion (e.g. Underhill 1991; Murdoch *et al.* 1995; Parnell *et al.* 1999) and a Neogene (primarily Plio-Pleistocene) event of more enigmatic origin (e.g. Solheim *et al.* 1996; Doré *et al.* 1999; Japsen & Chalmers 2000). A discussion of alternative uplift mechanisms is not within the scope of this paper, except to note that explanations essentially fall into three groupings: (1) isostatic (response to erosional unloading); (2) thermal (associated with a mantle plume); (3) compressional (intraplate stress and inversion). Categories (1) and (2) are broad regional effects, whereas category (3) may be very local in nature.

In most cases, exploration of these basins has occurred without a full understanding of their exhumed nature. A particularly instructive example is the western Barents Sea (offshore Norway), where licencing in the early 1980s carried the hope of a major North Sea rift-type hydrocarbon province, but where expectations were radically revised as the effects of uplift became apparent during early exploration (Nyland *et al.* 1992; Doré 1995). This experience alerted researchers, particularly in Norway, to the widespread nature of the exhumation and, importantly, to its commercial consequences. We therefore believe that a template for prospect evaluation of the type presented here, although largely qualitative, can be beneficial in future exploration and resource evaluation of such provinces. Although specific reference is made to the NW European basins, this appproach is applicable to any exhumed basin.

Prospect risk analysis in exhumed basins

Introduction

Explorationists address prospect risk at two levels: (1) risk with respect to the validity of the prospect, i.e. the chance of success; (2) the range of possible reserves. In this account, we simply address the chance before drilling of discovering any hydrocarbon accumulation within a mapped prospect. Methods used to calculate this figure

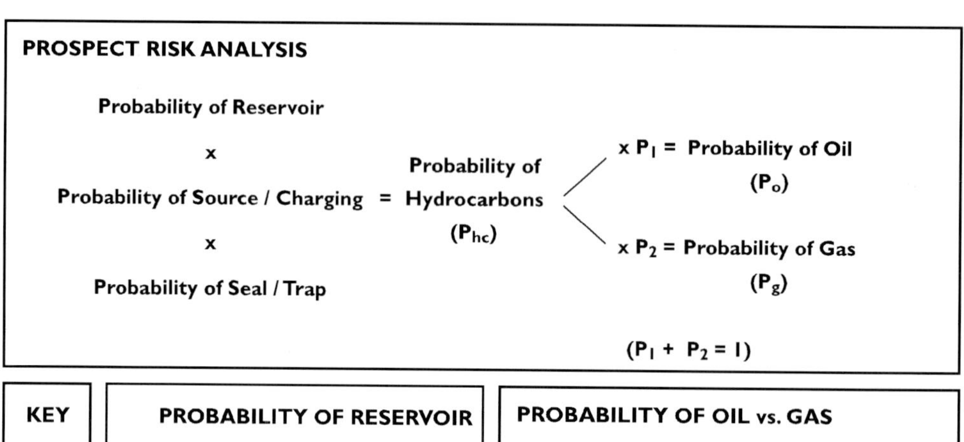

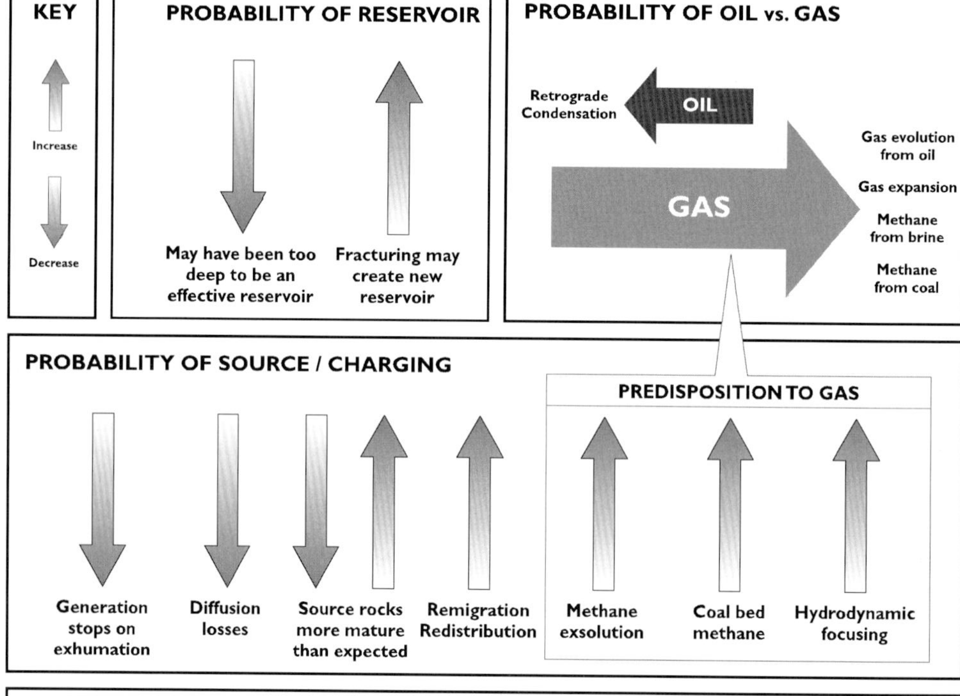

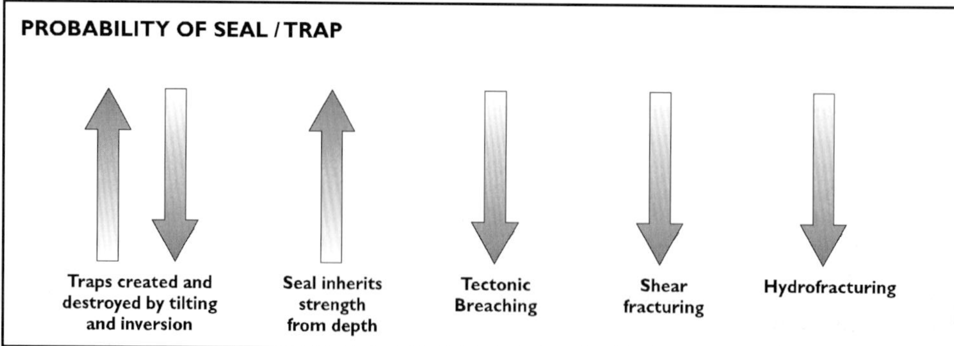

Fig. 2. Factors to be considered when carrying out risk analysis of a prospect in an exhumed basin. Arrows indicate increased or decreased probability in exhumed settings.

vary widely within the oil industry, but can be generalized in an equation of the type

$$P_{hc} = P_r \times P_s \times P_t \quad (1)$$

where P_{hc} is the probability of finding hydrocarbons, oil or gas (this value is sometimes regarded as the chance of finding any hydrocarbons at all, some specified minimum quantity, or hydrocarbons capable of flowing to the surface (see, e.g. Snow *et al.* 1996)), P_r is the probability that a reservoir rock is present, capable of holding oil or gas, P_s is the probability that there is a source rock that has charged the prospect and P_t is the probability that a sealed trap exists capable of holding oil or gas. The chance of a particular hydrocarbon phase being present (assuming a single-phase accumulation) is then given by

$$P_{hc} = P_o + P_g \quad (2)$$

where P_o and P_g are the probabilities of discovering oil and gas, respectively, as the main phase.

All of these factors are subjectively estimated by the petroleum geoscientist before drilling. Where basins are mature in terms of exploration they can be calibrated against known success rates. The influence of exhumation on these factors is summarized in Fig. 2 and discussed below.

Probability of reservoir

Reservoir rocks can be both enhanced or degraded in an exhumed terrane, depending on the type of reservoir and the nature of the process. Degradation compared with a reservoir at a similar depth in a subsiding basin can occur as a result of inheritance of a compactional and diagenetic state reflecting a previously greater burial (e.g. Walderhaug 1992). Improvement may occur because of fracture enhancement of porosity and permeability during uplift (e.g. Aguilera 1980). However, risk analysis does not take into account the quality of a reservoir, only whether a suitable reservoir exists or not. Thus, exhumation is probably neutral for the probability of reservoir except in extreme cases; for example, where a unit has been buried so deeply before uplift that it is no longer an effective reservoir, or where fracturing on extreme uplift has created an entirely new reservoir. The much more important effect on reservoir volumetric parameters (porosity, net/gross ratio and hydrocarbon saturation) and recovery factor is examined in the section on prospect resource estimation.

Probability of source and charge

The probability that a source rock is present is independent of whether or not the basin has been exhumed. However, the probability that such a source rock has been effective in charging a reservoir is strongly influenced by exhumation. A complex interplay of positive and negative factors must be considered.

In exhumed basins, any source rock will be more mature than expected for its present depth. Therefore, there will be an increased probability that source rocks now lying shallower than the oil generation window will have generated oil in the past. Whether such oil has survived uplift is a separate question. This reasoning has been applied to evaluate risk in marginal basins around Norway, where the Upper Jurassic source rock is shallow (Ghazi 1992: Jensen & Schmidt 1993). Conversely, a more deeply buried source rock may previously have been below oil window depths, increasing the chance of gas or that the source rock is 'burnt out' (postmature). This risk has been evaluated for uplifted Upper Palaeozoic source rocks in the eastern Norwegian Barents Sea (Theis *et al.* 1993). In most cases some indication of the degree of exhumation can be obtained even before drilling (e.g. from regional setting, seismic velocities and structural modelling) and can therefore be used to derive a first-order estimate of source rock maturity attained before uplift.

A critical observation is that, whatever the maturity state, generation is curtailed once uplift commences. Using the widely accepted kinetic model for generation of hydrocarbons from kerogen (Tissot & Espitalié 1975), which stresses the importance of temperature rather than time, it follows that any source rock that is significantly uplifted through a thermal frame of reference will cease hydrocarbon generation. No generation can occur until the basin subsides again and the previous maximum temperature is reached. In exhumed basins, no new hydrocarbons will be available to charge traps vacated during uplift by spillage and seal failure, or newly created traps (for example, folds formed during inversion-related uplift). Thus, any remaining original charge must have survived both the uplift and any subsequent leakage (e.g. by diffusion in the case of gas; Krooss *et al.* 1992), thereby significantly increasing risk.

An exception occurs where the uplifted basin is on the migration route for oil or gas generated in an adjacent, continuously subsiding basin. On the NW European margin such areas include parts of the Horda Platform, Inner Moray Firth Basin and West Shetland Basin. In such areas hydrocarbons lost during the uplift process can be replenished, and the probability of sourcing must therefore take into account migration timing, route and efficiency from the adjacent kitchen (e.g. Skjervøy & Sylta 1993; Goodchild *et al.* 1999; Parnell *et al.* 1999).

Hydrocarbon charging following an episode of uplift, in the absence of newly generated hydrocarbons, can take place only with hydrocarbons already present in the basin. The simplest of these processes is by remigration. In uplifted basins numerous processes can displace and redistribute hydrocarbons from pre-existing accumulations. Vertical remigration is a well-known phenomenon in subsiding basins, where reservoir overpressure builds until it exceeds the sealing capacity of the caprock, causing hydrocarbons to escape to shallower levels (e.g. in the Gulf of Mexico (Lopez 1990), in the North Sea Central Graben (Taylor *et al.* 1999) and in the Faeroe–Shetland Basin (Illiffe *et al.* 1999)). As shown later (see probability of trap and seal) and by Corcoran & Doré (2002), seal failure can also be anticipated during uplift. This means that hydrocarbons may subsequently accumulate in previously uncharged shallower levels. A notable example of such charging is in the Zagros fold belt in Iran and Iraq, where seal failure in Cretaceous reservoirs during the Late Miocene–Recent Zagros uplift expelled oil upwards into the highly prolific Oligocene–Miocene Asmari reservoir (Ala 1982; Bordenave & Burwood 1989). However, in many cases hydrocarbons must be lost to the surface during uplift. Assessment of this risk must take into account the position and geometry of the shallower reservoirs, and the amount of erosion of shallower levels that has taken place during uplift.

Assuming that uplift is not perfectly uniform, lateral remigration will occur through tilting and spilling, resulting in loss of some hydrocarbons from pre-existing traps and migration updip. Although some hydrocarbons will accumulate in updip traps, there will be a net decrease in hydrocarbon budget as a result of migration losses and escape to the surface. In pre-existing two-phase accumulations the oil leg will preferentially spill. In the Hammerfest Basin area of the Barents Sea, residual oil legs in fields such as Snøhvit and Askeladden show that significant oil spillage took place during Plio-Pleistocene uplift (Kjemperud & Fjeldskaar 1992; Nyland *et al.* 1992; Doré & Jensen 1996). It has been assumed by some workers that most of the oil from the Middle Jurassic reservoir has been lost to the surface, but recent careful modelling of the direction of remigration has led to the discovery of a new oilfield (Goliath) on the periphery of the basin (B. Wandaas, pers. comm.). All of the basins immediately adjacent to the Norwegian landmass (e.g. Horda Platform, Egersund Basin) will have been tilted westwards during the Cenozoic uplift of Scandinavia, and hydrocarbon redistribution by spillage can confidently be inferred there (Doré & Jensen 1996).

Risk associated with charging in an exhumed basin can be mitigated in the following ways: (1) by assessing the residual hydrocarbon budget after 'switching off' of source rock maturation; (2) by recognizing the hydrocarbon displacement drivers in the basin; (3) by identifying the post-exhumation regional spill direction; (4) by determining remigration pathways and bypassed areas ('shadows'). Timing of trap formation (pre-, syn- or post-uplift) is also critical with respect to charge adequacy.

Probability of oil versus gas

Perhaps the most radical effect of uplift on hydrocarbon basins is the shift towards gas-dominated systems. Several phenomena conspire to produce this effect, as follows.

(1) Gas exsolution from oil. Assuming a reservoired fluid is at or below its bubble-point pressure at maximum burial, a gaseous phase will be exsolved on uplift and consequent pressure–temperature reduction. Unless a trap is initially underfilled, oil will therefore be driven from the trap (Nyland *et al.* 1992; Doré & Jensen 1996).

(2) Gas expansion as a result of pressure–temperature reduction. This will result in net loss from gas-filled traps and, again, preferential displacement of the oil leg in two-phase accumulations. Nyland *et al.* (1992) estimated that over two billion barrels of oil were lost from the Snøhvit Field in the Barents Sea because of the combined effect of gas exsolution and expansion during regional uplift.

(3) Methane liberation from formation brine. At constant pressure, the solubility of methane in water attains a minimum between about 60 and 90 °C, then increases with temperature. At constant temperature, methane solubility increases with pressure. Therefore, in general, methane solubility in formation brines will increase with burial and, conversely, free methane will be liberated during uplift as pressure and temperature decrease (e.g. Culberson & McKetta 1951; Price 1979; Cramer *et al.* 1999; Cramer & Poelchau 2002). This mechanism is perhaps the most potent source of gas during exhumation as gas will be liberated

over the whole basin, tapping methane originally generated by dispersed organic matter as well as from rich source rocks. The potential for gas liberation is vast. All of the giant dry gas fields in the West Siberian Basin probably derive from this process (Cramer *et al.* 1999), as do the major basin-centred gas fields in uplifted basins in the western USA such as the Alberta, Denver and San Juan Basins (Doré & Jensen 1996: Price 2002). Gas fields in uplifted basins in Central Europe (Pannonian, Vienna and Po Basins) have been tied directly to a groundwater origin via noble gas markers (Ballentine *et al.* 1991). Exsolution from water probably also accounts for a significant part of the hydrocarbon budget in gas-dominated uplifted basins on the NW European margin such as the Barents Sea, East Irish Sea Basin, Slyne–Erris Basins and North Celtic Sea Basin. To date, however, the only quantitative work in such areas known to the authors is that reported for the Barents Sea by Doré & Jensen (1996).

(4) Methane expulsion from coal. Coal beds are widespread in many petroleum basins and expel gas as the coal is progressively buried through maturation thresholds. However, coal-bed methane studies show that significant quantities of gas are retained in the coal on internal surfaces (adsorption) or within the molecular framework of the organic matter (absorption). This methane will migrate out of the coal by desorption and diffusion during reduction in pressure (Littke & Leythaeuser 1993, Fig. 4; Rice 1993). Expulsion during uplift is also aided by increase of macroporosity and permeability in the coals compared with deeply buried coals of similar rank, presumably partly as a result of fracture dilation (Littke & Leythaeuser 1993, Fig. 5).

(5) Hydrodynamic flow. During subsidence, compaction-driven water flow outwards to the flanks of the basin is normal, whereas the outcrop of aquifers and development of topography during uplift and erosion may reverse this situation and create gravity-driven water flow towards the basin centre (see further discussion by Corcoran & Doré (2002)). As shown by Cramer *et al.* (1999), this flow provides a recharge mechanism whereby methane may be brought in from outside the normal drainage area of a gas field and liberated as a result of drop in reservoir pressure. Cramer *et al.* estimated that about 12% of the gas reserves of the giant Urengoy Field (Western Siberia) were emplaced in this way. Additional biogenic methane may be introduced during or after uplift by groundwater flow through coal beds, a process that can stimulate bacterial acivity and gas production from coal of any rank (Rice 1993). Finally, introduction of fresh groundwater into the system is also likely to promote bacterial biodegradation of any shallow oils within the aquifer, leading to the formation of heavy oil residues and again changing the oil–gas balance.

A changed oil–gas balance can also occur via retrograde condensation. In this case oil is favoured as a result of the dropping out of liquids from a wet gas during a reduction in pressure and temperature induced by uplift (e.g. Piggott & Lines 1991; Duncan *et al.* 1998). Otherwise the overwhelming tendency is for an increase in gas exsolution from oils, brines and coals. It may be predicted that significant exhumation of a petroliferous basin will produce a massive gas bloom in the basin centre, driving oil to peripheral locations, to more shallow depths where it will be biodegraded, or to the surface.

Assessment of the risk of gas flushing during uplift and pressure–temperature decrease can be carried out by geochemical modelling of the original oil–gas balance in a prospect. Input of gas by exsolution from formation brine can be assessed from volumetric calculations on the aquifer draining into the prospect (Doré & Jensen 1996; Cramer *et al.* 1999). Knowledge of formation water salinity will improve such calculations, because more saline brines can dissolve less methane (Maximov *et al.* 1984). Contribution of exsolution gas from hydrodynamic flow can be estimated by mapping hydraulic gradients, as shown by Cramer *et al.* (1999). In all cases control points such as nearby wells will, of course, increase the accuracy of such estimates.

Probability of trap and seal

Traps can be eliminated, or their volume decreased, by the effects of regional tilting during exhumation. Similarly, new traps can be created by tilting of three-way dip closures ('noses'). Extreme exhumation may, of course, breach pre-existing accumulations at the surface. In areas where uplift is associated with faulting, tectonic breaching of traps can occur through fault displacement of the seal (caprock), fault juxtaposition of hydrocarbon reservoirs against aquifers or thief zones, or the formation of crestal extension fractures over domes and anticlines. Where inversion (i.e. compressive reactivation of an extensional basin) is involved, traps such as horsts or tilted fault blocks may be destroyed, whereas new traps can be created by (for example) reverse rejuvenation of half-graben or bulge of the basin centre. MacGregor (1995) has shown from a global database that exploration success rates in strongly inverted rifts are lower than those in locally inverted rifts and much lower than those in uninverted rifts. In all cases, a key consideration is the timing of generation compared with the timing of uplift and restructuring. As already demonstrated, during uplift no new hydrocarbons can be generated from source rocks to compensate for redistribution losses. Newly created structures must therefore be filled by remigration or by hydrocarbons (principally gas) from exsolution.

A second and equally important consideration is the performance of sealing lithologies during uplift. In general, as a claystone seal is buried it becomes more compacted and stronger. When exhumed it should retain the tensile strength of its maximum burial depth, and thus will be stronger than a claystone at the same depth in a continuously subsiding basin. This observation is supported by LOTs (leak-off tests) and FITs (formation integrity tests) on seals in uplifted Atlantic margin basins (Corcoran & Doré 2002). Additionally, greater compaction than 'normal' (and corresponding decrease in pore throat size) should increase the capillary retention capacity of an exhumed claystone. Therefore, under certain conditions, the seal capacity of a prospect in an exhumed basin may be superior to that in a subsiding basin, at the equivalent depth. However, several important factors combine to diminish this capacity, as follows.

(1) Brittleness of the seal. A ductile rock can accommodate more strain (up to 10%) before fracturing than a brittle rock (<3%). Changes in ductility with increasing burial depth are complex and depend on composition, temperature, confining pressure and fluid pressure (Davis & Reynolds 1996). In general, however, brittleness in claystones can be said to increase with density, with the transition from ductile to brittle behaviour taking place over the range 2.2–2.5 g cm^{-3} (Hoshino *et al.* 1972). Consequently, exhumed claystone seals may be more brittle than normal seals at the same depth. Brittle seals will be more likely to rupture and leak in response to stress (e.g. from tectonic bending or hydrofracturing) than ductile ones, which will deform elastically and plastically before fracturing. A crucial question, therefore, is whether embrittlement of a potential claystone seal has taken place before uplift. Evaporites or mudstones containing evaporitic minerals, which deform plastically under a very wide range of pressure–temperature conditions, form the most efficient seals in uplifted basins (see, for example, work on the East Irish Sea Basin by Seedhouse & Racey (1997) and Cowan *et al.* (1999)).

(2) Hydraulic fracturing. Hydraulic leakage may occur when rapid exhumation, under conditions of low differential stress and disequilibrium fluid pressures, results in failure of brittle seals. This failure may be manifested as extensional shear fractures, dilation of fault planes, or hydrofracturing. Shear fractures will be formed in conditions of high differential stress in the caprock and will be promoted by disequilibrium pore pressures during rapid uplift. Pre-existing fractures and faults may also be induced to fail in these circumstances. The orientation of the new fractures that form, and of the pre-existing fractures that reactivate, will depend on the direction of the principal compressive stress (σ_1). Under conditions of low differential stress and similarly high retained disequilibrium fluid pressures, hydrofractures are likely to form by tensile failure of the caprock (Corcoran & Doré 2002; see also Sibson 1995).

(3) Diffusion. Leakage of hydrocarbons by means of molecular transport through caprocks is thought to be usual in petroleum basins (e.g. Krooss *et al.* 1992). Whereas diffusion rates for oil are probably negligible because of the large size of the oil molecules, gas will diffuse more readily through water-saturated claystone caprocks. There is considerable debate in the literature on diffusion rates, but there seems little doubt that over a moderate geological time scale (say, the length of the Cenozoic period, 65 Ma) diffusion losses from a shale-sealed gas field can be considerable (e.g. Leythaeuser *et al.* 1982; Krooss *et al.* 1992). In a subsiding basin the fill of a gas field will be determined by the ratio of diffusion losses through the seal to newly generated gas entering the trap. After exhumation, however, the supply of new hydrocarbons will be arrested, allowing the trap to be gradually depleted via diffusion. The diffusion rate of methane through evaporites is so low as to be negligible, again showing that evaporites are highly efficient seals that can preserve hydrocarbons over significant geological time in exhumed basin settings (see, for example, Kontorovitch *et al.* (1990), on Proterozoic gas reservoirs in the Lena–Tunguska Basin, Russia).

In summary, in exhumed basins the risk associated with trap and seal is significantly increased. Underfilled traps and near-hydrostatic reservoir pressures are commonly encountered in uplifted Atlantic margin basins (Corcoran & Doré 2002), presumably reflecting pressure depletion through the seal during exhumation, lack of new hydrocarbons from source rocks once uplift commenced, lack of new exsolution products once uplift stopped, subsequent escape of gas by diffusion and contraction of gas during post-exhumation reburial. Uplift-related depressuring in low-permeability rocks can also result in transient underpressuring, i.e. pressure gradients below hydrostatic (Luo & Vasseur 1995). Underpressuring is also a characteristic of the basin-centred gas fields in uplifted basins in the western USA. Some of these fields, such as Elmworth in the Alberta Basin, are actually synclines (Masters 1984) in which the gas accumulation probably represents a disequilibrium condition and where the underpressuring may be attributable to thermal contraction of formation fluids (Price 2002). Apart from a single well in the Barents Sea (Doré & Jensen 1996) underpressuring has not yet been reported in the Atlantic margin basins.

The probability of trapping in an exhumed basin setting is best assessed by structural modelling, whereby the timing of trap formation and modification is compared with the timing of charging. Knowledge of maximum burial depths (and hence maximum pressures and temperatures) can indicate whether a shale seal is likely to have become embrittled before exhumation. Evidence of fracture trends and present-day stress systems, combined with modelling of pressure evolution, can help in assessing whether hydraulic failure of seals is likely to have occurred. Published data on methane diffusion rates, in combination with estimation of the time elapsed since uplift of a trap, can quantify likely diffusion losses through a seal. Evidence of evaporites in an uplifted basin, even at a very preliminary stage of evaluation, significantly enhances the probability that some trapping capability will have been retained during exhumation.

Prospect resource estimation in exhumed basins

Introduction

Potential hydrocarbon reserves in a prospect are estimated from an equation of the type

$$RR = GRV \times N/G \times \phi \times S_{hc} \times FVF \times RF \quad (3)$$

where RR are the recoverable hydrocarbon reserves, expressed as a volume; GRV is the gross rock volume, i.e. the volume of the reservoir within the potential trap; N/G is the net-to-gross ratio, i.e. the fraction of the reservoir that is capable of containing movable hydrocarbons; ϕ is the interconnected porosity in the reservoir, expressed as a fraction; S_{hc} is the hydrocarbon saturation, i.e. the fraction of pore space taken up by hydrocarbons; FVF is the formation volume factor, a multiplier taking into account the expansion of gas or the contraction of oil (owing to liberation of dissolved gas) as the hydrocarbons are brought to surface pressure and temperature conditions during production; RF is the recovery factor, i.e. the proportion of the hydrocarbons in the prospect that can be recovered to surface given an assumed production method. These factors are estimated based on the predicted depth of the prospect, local and regional data, and experience.

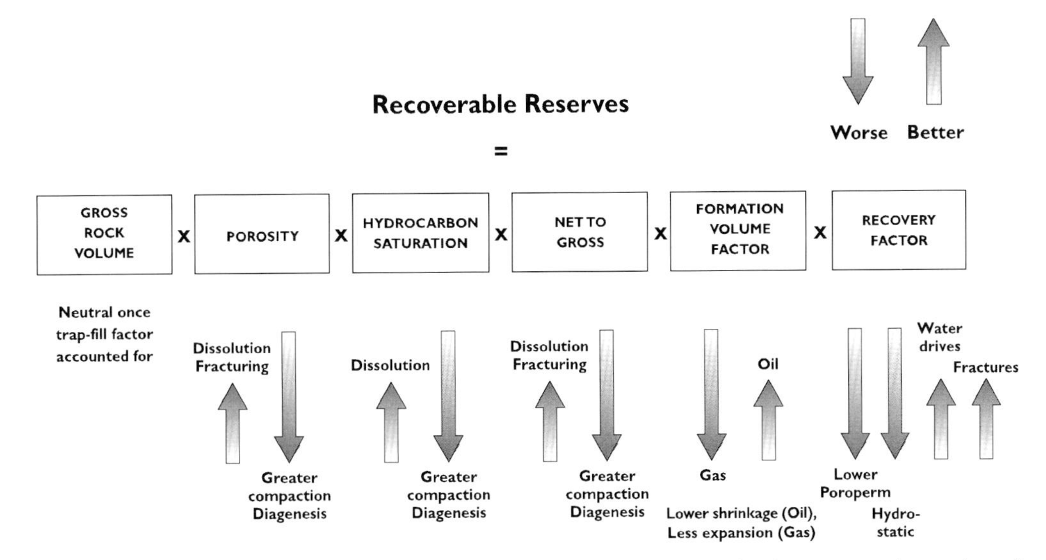

Fig. 3. Factors to be considered when carrying out recoverable reserves calculation for a prospect in an exhumed basin. Arrows indicate improvement or deterioration in reservoir parameters.

Uncertainty in reserves estimation is traditionally addressed by assuming a range of values for each prospect parameter (GRV, N/G, porosity, S_{hc}, FVF, RF) and by applying stochastic simulation procedures to generate a range of reserves. Predictive capability is, of course, mainly a function of local data density and quality. Where evaluation is mainly dependent on regional data, it is especially critical to know whether a basin has been exhumed. The effects of exhumation on the volumetric parameters are shown in Fig. 3 and discussed below.

Gross rock volume

Because the measurement of the total potential volume within a trap is made on a prospect-specific basis, this value is independent of whether the trap has been exhumed. However, the actual volume also depends on the degree-of-fill factor, i.e. the proportion of the available vertical closure taken up by the hydrocarbon column. For continuously subsiding basins, Sales (1993) and others have shown that a range of trap-fill values is possible, based on the relationship between the vertical closure (and hence the potential upward buoyancy pressure of any hydrocarbon fill) and the sealing capacity of the caprock. Because there is often a continuous supply of newly generated hydrocarbons in such basins, the sealing capacity becomes the main limiting factor. If the seal is efficient, the structure has a good chance of being completely hydrocarbon-filled: for example, Sales (1993) asserted that most gas fields in the Norwegian North Sea are full to spill. In contrast, in an exhumed basin, trap-fill is limited not only by sealing capacity (which, as shown in the section on trap and seal, can be catastrophically reduced during exhumation) but also by the hydrocarbon budget generated or liberated at the time of the last exhumation. Processes subsequent to exhumation (e.g. diffusion and reburial contraction of gas) will serve to diminish the trap-fill. This observation is strongly supported by observation of underfilled gas fields in NW European uplifted basins such as the Barents Sea (Spencer *et al.* 1987), West Shetland Basin (e.g. Goodchild *et al.* 1999), East Irish Sea (e.g. Stuart & Cowan 1991), Southern Gas Basin (e.g. Hillier & Williams 1991) and Slyne–Erris Basin (in-house data on the Corrib Field).

Net-to-gross ratio, porosity and hydrocarbon saturation

A sedimentary rock that has been uplifted with resulting removal of overburden will preserve the compactional and diagenetic state associated with its maximum burial depth. In most cases these higher levels of compaction and diagenesis (stylolitization, quartz precipitation and authigenic clay mineral formation) will involve porosity loss. Consequently, there will be a decrease in hydrocarbon saturation as a result of the increased proportion of a given pore occupied by the wetting water phase. Net-to-gross ratio will also be impaired as a result decrease in porosity or permeability of some rock below the threshold considered to define an effective reservoir. Thus, overall reservoir quality will usually be impaired compared with that at a similar depth in a subsiding basin. This general principle is illustrated in the Barents Sea, where Middle Jurassic sandstones in the Hammerfest Basin are petrographically similar to those that form major reservoirs off Mid-Norway and in the North Sea. The Barents Sea reservoirs, however, consistently show higher levels of stylolitization (Walderhaug 1992) and quartz precipitation (Berglund *et al.* 1986), with consequently lower porosities (Olaussen *et al.* 1984), because of a maximum burial depth some 1500 m greater than at present.

As indicated by Price (2002), cooling of formation waters during uplift will also result in the precipitation of solutes (e.g. silica) and hence the occluding of porosity. However, as Parnell (2002) points out, this effect may be minor and is not widely observed; furthermore, some mineral species (notably carbonates) become increasingly soluble at lower temperatures, thereby introducing the possibility of secondary porosity development.

As discussed in the section on prospect risk analysis, exhumation creates an increased probability of hydrodynamic flow and the introduction of meteoric water into the basin aquifers. Such groundwater will usually contain dissolved oxygen and will be acidic (principally as a result of dissolved carbon dioxide), leading to the possibility of oxidation and acid dissolution in reservoirs. The effect on reservoir quality will be complex and depend on the chemistry of the reservoir, the formation water and the introduced water. Oxidation of ferromagnesian minerals may form pore-clogging iron oxides, whereas acid dissolution of feldspars and carbonates can create substantial secondary porosity (see the much fuller discussion by Parnell (2002)). These effects are likely to be most prevalent close to the surface where meteoric water flow is strongest. Nevertheless, they can form an important modifier to the overall negative implications of exhumation on reservoir quality; in basins that have undergone repeated exhumation and reburial episodes, improved reservoir quality as a result of dissolution may be forecast below unconformity surfaces (e.g. Shanmugam 1988).

Knowledge that a basin has been exhumed allows the interpreter to place constraints on predicted reservoir quality. The most useful data for this process are, of course, local well descriptions, which will give direct evidence as to the detrital and authigenic mineralogy of the reservoir. However, even without such data, reconstruction of the maximum burial depth (maximum temperature exposure) of the reservoir is important. It allows a first-pass prediction as to whether the reservoir has exceeded temperature thresholds for kinetically controlled poroperm-reducing mineral transformations; for example, quartz cementation and the development of authigenic illite (e.g. Nadeau *et al.* 1985; Bjørkum *et al.* 1993).

Formation volume factor

Uplift of a hydrocarbon accumulation will result in lower temperatures and pressures, with exsolution of dissolved gas from oil and expansion of reservoired gas. Therefore, for a given hydrocarbon pore volume the expectation will be for more oil (because of lower shrinkage on production) and less gas (because of less expansion on production). It can be argued that these factors are independent of whether the reservoir has been uplifted, and are simply a function of pressure–temperature–volume relationships at a given depth. Although this is undoubtedly true, the occurrence of near-hydrostatic pressure gradients resulting from pressure dissipation on uplift (as observed on the NE Atlantic margin) will create a tendency towards lower shrinkage oils and lower expansion gases compared with a continuously subsiding basin.

Recovery factor

Recovery factor can be influenced positively and negatively by uplift and exhumation. Generally, the lower porosity and permeability of an uplifted reservoir for a given depth should impair recovery factor for both oil and gas. Additionally, the dissipation of overpressure during uplift will limit the amount of hydrocarbon that can be produced by simple pressure depletion, whereas a reservoir in a subsiding basin at the same depth may retain overpressure and hence have a higher initial reservoir pressure. Running counter to this argument, the development of open pressure systems and hydrodynamic flow as a result of exhumation may provide pressure maintenance (water drives) during commercial depletion of a field.

Fractures are an extremely important component of productive reservoirs worldwide, and can be attributed to diastrophism (for example, over fold axes, e.g. Aguilera 1980) or to removal of overburden stress by exhumation (e.g. Aguilera 1980: Sibson 1995: Corcoran & Doré 2002). As shown in the section on probability of trap and seal, shales that have been embrittled during burial may fracture during uplift. Reservoir lithologies are generally more brittle than sealing lithologies such as shales and evaporites. Tight sandstones, quartzites and dolomites are the most fracture-prone and limestones are the most ductile of the potential reservoir rocks (Handin *et al.* 1963: Stearns & Freidman 1972: Doré & Jensen 1996). Therefore, reservoirs may fracture without corresponding rupture of the caprock, a situation that creates the basis for globally important hydrocarbon resources such as the Asmari fractured carbonate fields of the Zagros fold belt (e.g. Daniel 1954).

Fracturing during uplift can create reservoirs from non-reservoir lithologies (e.g. basement or siliceous shales), contribute to both porosity and permeability in low-poroperm reservoirs, and enhance pore connectivity (and hence permeability) in higher poroperm reservoirs. Fractures therefore have the potential to significantly boost recovery in uplifted terranes where poroperms would otherwise be unacceptably low. A critical issue to well location, well completion and recovery in fractured reservoirs is identification of the fracture sets that are dilatational (and hence contribute the most to fluid flow). Dilation will occur most readily in fractures orthogonal to the least compressive stress direction (σ_3), which will be approximately horizontal in an extensional regime and approximately vertical in a compressional regime. For a simplified case of subvertical fractures, natural fractures are more likely to be open and support fluid flow if they strike close to the maximum horizontal stress (S_{hmax}) direction, an observation supported by global waterflood studies on producing fields (Heffer & Dowokpor 1990). The NE Atlantic margin at present is under a mild NW–SE compressive regime, probably attributable to ridge-push from the Atlantic spreading centre (Doré & Lundin 1996). Where evidence exists, it appears that open fractures have a strike close to the NW–SE S_{hmax} direction. The Clair Field in the uplifted West Shetland Basin has a reservoir consisting of fractured Upper Palaeozoic sandstone. Detailed studies show that specific fracture sets aligned close to S_{hmax} are dilatational, allowing a recovery program to be devised based on exploiting these fractures with directional wells (Coney *et al.* 1993).

Borehole data, regional geology, seismic reconstruction and basin modelling can help to assess whether fracture enhancement of an uplifted reservoir is likely to have occurred. Key criteria are: (1) prognosed reservoir lithology; (2) maximum burial depth and degree of uplift of the reservoir; (3) probability of overpressure development and dissipation; (4) direction of seismically mappable faults; (5) direction of fractures from boreholes or regional outcrop data; (6) present-day stress-field orientation.

Synopsis: key characteristics of hydrocarbon systems in exhumed basins

On the basis of the foregoing discussions, it is possible to describe a suite of phenomena prevalent in exhumed hydrocarbon systems (Fig. 4). Although these characteristics can also occur in subsiding basins, the occurrence of many or all factors together will be a strong signature of exhumation. Conversely, prior knowledge that a basin has been exhumed (from, for example, its truncated stratigraphic record or characteristic structural style) allows such factors to be anticipated. They are as follows:

(1) near-hydrostatically pressured or under-pressured reservoirs as a result of catastrophic pressure release during exhumation. Under-pressure may derive from depressuring and/or thermal contraction in low-permeability aquifers.

(2) Underfilled traps resulting from reduction in sealing capacity, spillage losses, and remigration inefficiency during exhumation, followed by cessation of the hydrocarbon supply, diffusion of gas and reburial shrinkage of gas after exhumation.

(3) Large, basin-centred gas deposits liberated during uplift from oils, formation brines and coals, and further displacing pre-existing oil by gas expansion. These accumulations often overlie the deepest part of the basin because of the thicker sedimentary succession available to generate exsolved gas, inversion of the basin centre to create new structural traps, slow dissipation of the gas bloom in low-permeability lithologies, and hydrodynamic focusing.

(4) Two-phase accumulations as a result of gas exsolution from oil and retrograde condensation of liquids from wet gas during uplift.

(5) Residual oil columns left behind by seal failure or in tectonically breached traps, by spillage, and by rising of the gas–oil and oil–water contacts during post-exhumation diffusion of the gas cap.

(6) Small, remigrated peripheral oil deposits: oil that is not driven completely from the system by seal failure, tilting and gas flooding is likely to accumulate in traps on the basin margin, for example, hanging-wall traps.

(7) Heavy oil deposits formed by the remigration of oils to shallow levels of the basin, where washing by meteoric water inflow and bacterial biodegradation can occur.

The occurrence of these characteristics on the NW European margin is discussed below with reference to selected basins (Figs 5–10) and summarized in Fig. 11.

Examples: exhumed provinces on the NW European margin

In the case histories and in Figs 5–10 the following terms of reference are used: (1) exhumation is uplift of key reference horizons above maximum burial depth; (2) two-phase accumulations are counted as both oil and gas; (3) success rate indicates the number of discovered pools of testable hydrocarbons divided by the number of exploration wells; it does not represent the rate of commercial success.

Western Barents Sea (Fig. 5)

The Barents Sea consists of a complex mosaic of basins and platforms, which in the western (Norwegian) sector become younger towards the North Atlantic Ocean. In the east, the Nordkapp Basin is a NE–SW-trending graben, initiated in Late Palaeozoic time and dominated today by near-surface salt domes and walls formed from Upper Carboniferous–Lower Permian halite. Farther west, the Hammerfest Basin, is an en echelon continuation of this trend, but in contrast the last significant rift episode was later (in Late Jurassic–Early Cretaceous time). The Hammerfest Basin is cross-cut to the west by a north–south line of deep Cretaceous depocentres (Bjørnoya and Tromsø Basins), which are in turn superceded to the west by Tertiary depocentres (e.g. Sørvestnaget Basin) close to the continent–ocean boundary (Gabrielsen *et al.* 1990).

Major exhumations took place during Cenozoic time, roughly synchronous with uplift of the Fennoscandian mainland. These included an episode of Paleogene uplift probably associated with incipient opening of the NE Atlantic, and a particularly severe Plio-Pleistocene episode emphasized by repeated glacial erosion and isostatic re-equilibration. The Nordkapp Basin is deeply exhumed, with a thin layer of Quaternary sediments overlying truncated Cretaceous rocks. Some Tertiary sediments are preserved in the Hammerfest Basin, but there is a major unconformity between the Paleocene and the Pliocene sequences (e.g. Westre 1983), and well data indicate removal of about 1500 m of overburden (e.g. Nyland *et al.* 1992; Walderhaug 1992).

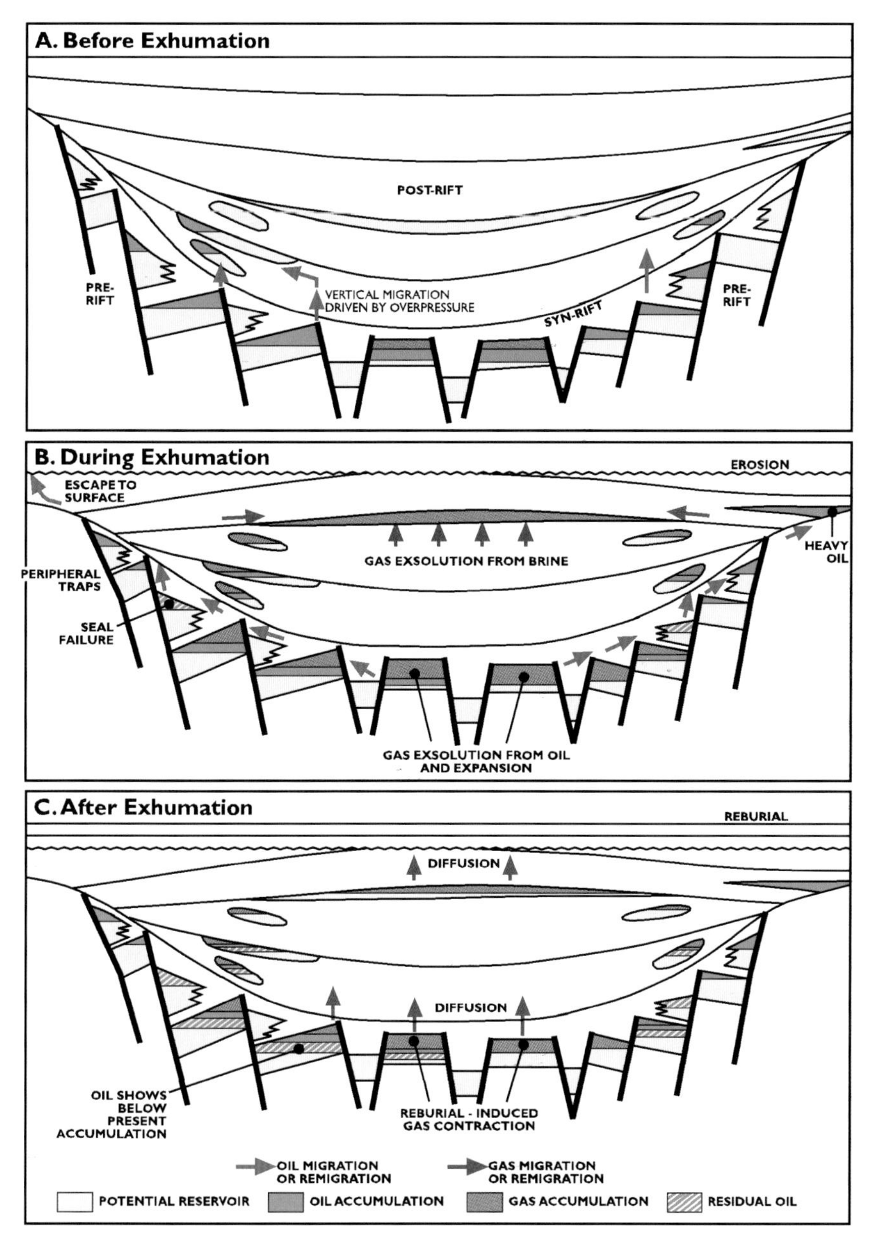

Fig. 4. Highly schematic and vertically exaggerated basin cross-section illustrating some effects of exhumation on the hydrocarbon system. (**a**) A simple rift geometry containing an oil-dominated hydrocarbon system, used as the starting point. (**b**) Effects taking place during exhumation. Regional exhumation, in this example, is accompanied by inversion of the basin centre. (**c**) Processes after exhumation has ceased and minor reburial has taken place.

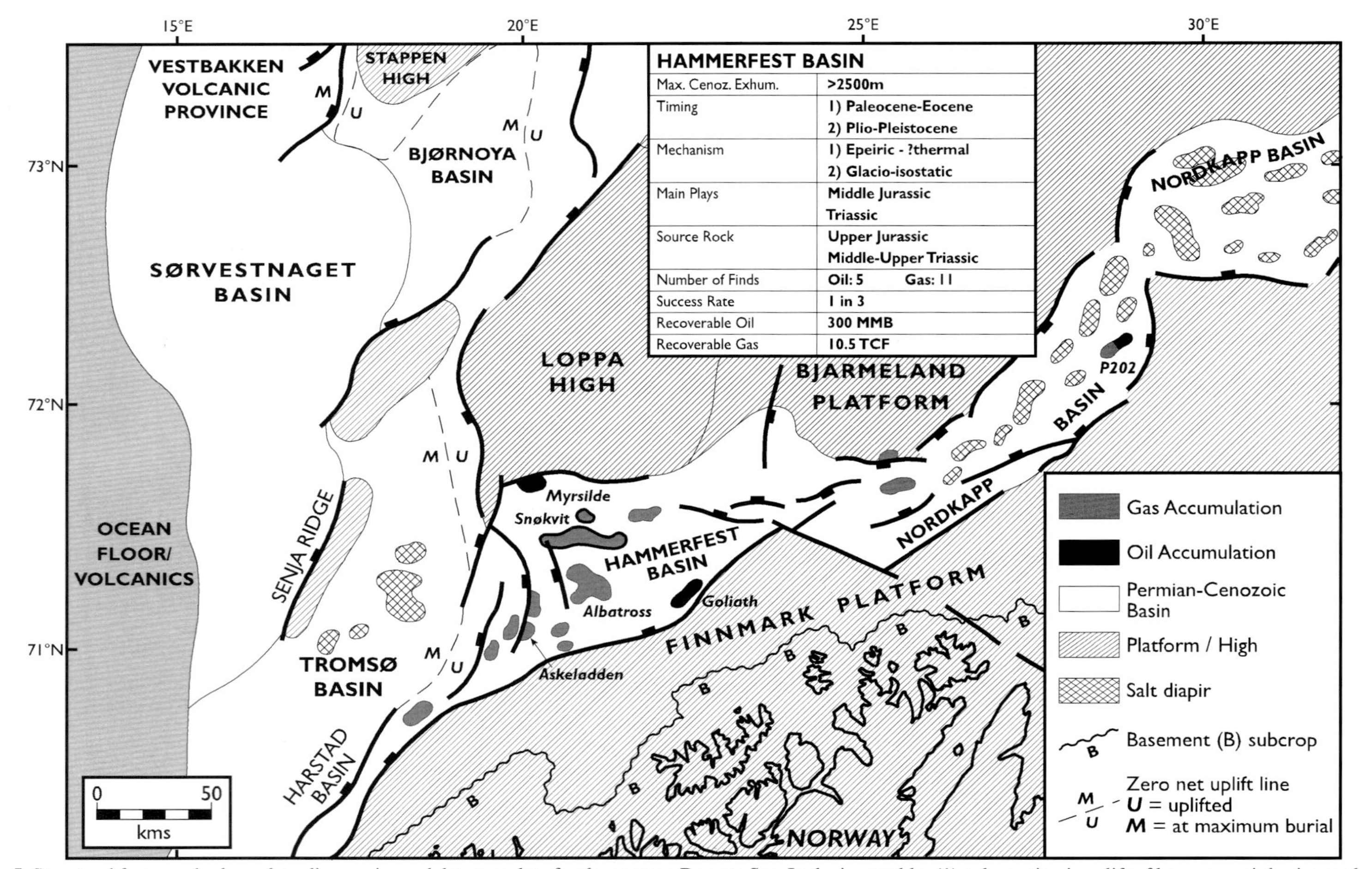

HAMMERFEST BASIN	
Max. Cenoz. Exhum.	>2500m
Timing	1) Paleocene-Eocene 2) Plio-Pleistocene
Mechanism	1) Epeiric - ?thermal 2) Glacio-isostatic
Main Plays	Middle Jurassic Triassic
Source Rock	Upper Jurassic Middle-Upper Triassic
Number of Finds	Oil: 5 Gas: 11
Success Rate	1 in 3
Recoverable Oil	300 MMB
Recoverable Gas	10.5 TCF

Fig. 5. Structural features, hydrocarbon discoveries and data template for the western Barents Sea. In the inset table: (1) exhumation is uplift of key reservoir horizons above maximum burial depth; (2) two-phase accumulations are counted as both oil and gas finds; (3) success rate does not represent commercial success, it indicates the number of discovered pools of testable hydrocarbons divided by the number of exploration wells.

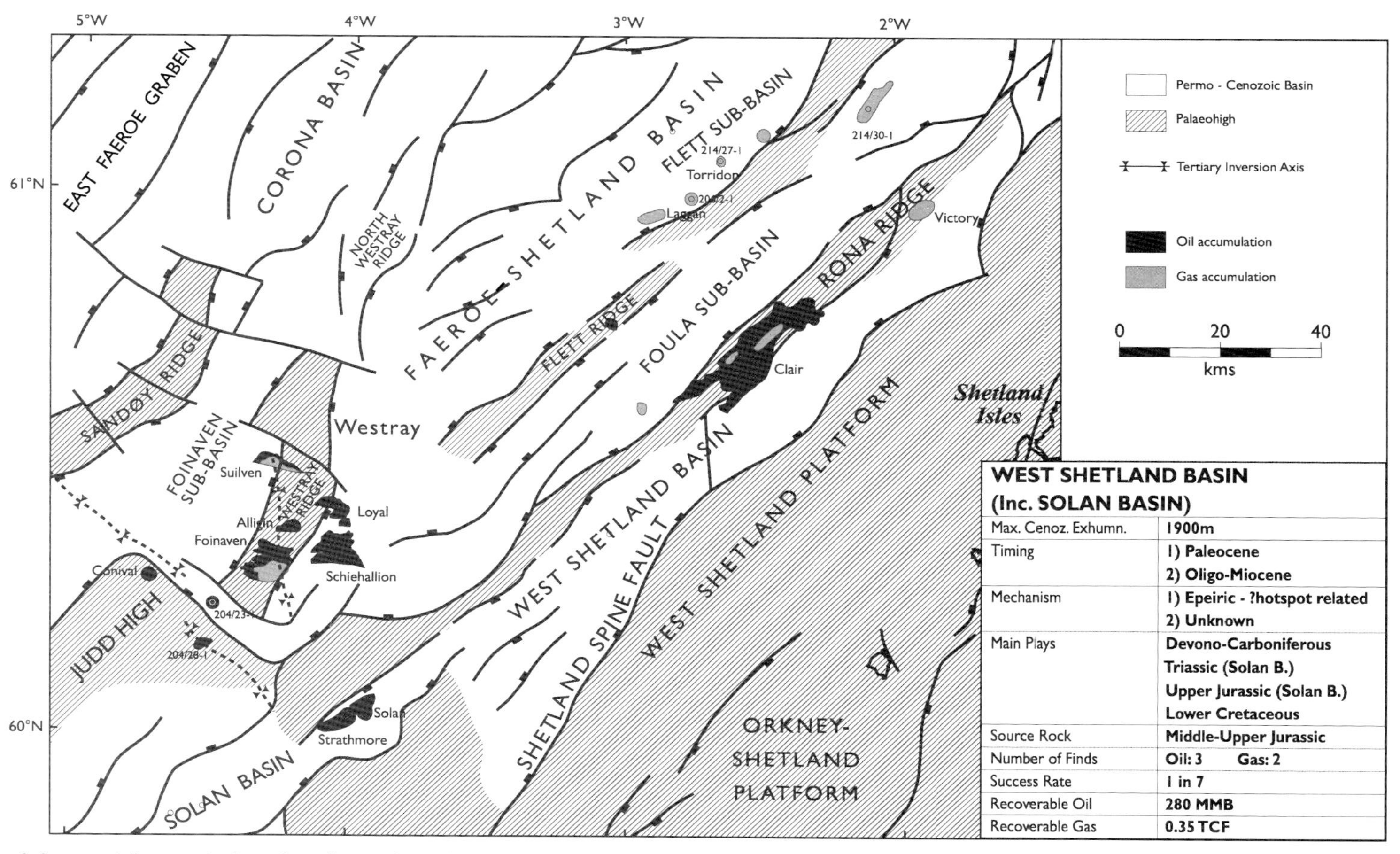

WEST SHETLAND BASIN (Inc. SOLAN BASIN)	
Max. Cenoz. Exhumn.	1900m
Timing	1) Paleocene 2) Oligo-Miocene
Mechanism	1) Epeiric - ?hotspot related 2) Unknown
Main Plays	Devono-Carboniferous Triassic (Solan B.) Upper Jurassic (Solan B.) Lower Cretaceous
Source Rock	Middle-Upper Jurassic
Number of Finds	Oil: 3 Gas: 2
Success Rate	1 in 7
Recoverable Oil	280 MMB
Recoverable Gas	0.35 TCF

Fig. 6. Structural features, hydrocarbon discoveries and data template for the West Shetland Basin. (For qualifiers to inset table, see Fig. 3.)

Exploration drilling of the Norwegian Barents Sea commenced in 1980 and at time of writing about 60 wells have been drilled. Although success rates have been fairly high (about one in three) the results have been commercially disappointing, largely because of the dominance of gas and the remoteness of the area from potential gas markets. Discovered resources are currently about 10.5 TCF (trillion cubic feet) gas with about 300 MMB (million barrels) oil. Most of the gas discovered to date is concentrated in three fields (Snøhvit, Askeladden and Albatross) in the axial part of the Hammerfest Basin, comprising fault blocks and horsts with a Middle Jurassic reservoir (Stø Formation) sealed by Upper Jurassic shales. Sourcing is from the Upper Jurassic Hekkingen Formation, which attained maturity before uplift in northern and western parts of the Hammerfest Basin.

The hydrocarbon system shows many classic attributes of exhumation. The Middle Jurassic reservoirs have anomalously high levels of diagenesis (e.g. Walderhaug 1992). The central gas accumulations are underlain by thin oil discs or residual oil legs, a result of spilling of pre-existing oil deposits. Modelling of the largest field, Snøhvit, suggests that the oil was evacuated by gas expansion and that current underfilling of the trap is consistent with diffusion losses and gas contraction during reburial (Nyland *et al.* 1992). Methane exsolution from oil and brine has probably also contributed to the dominance of gas (Doré & Jensen 1996). Discrete oil accumulations are small, and include the Myrsilde discovery, in Lower Cretaceous sands in a hanging-wall trap against the northern margin of the Hammerfest Basin, and the recent Goliath discovery of 70–100 MMB remigrated oil in a Middle Jurassic reservoir. Current exploration efforts focus on: (1) areas that have not been recently exhumed (e.g. the western margin of the Barents Sea, which received erosion products from the Cenozoic denudations); (2) areas with potential evaporite seals such as the Nordkapp Basin (where well 7228/7-1 in Licence P202 recently found oil and gas in Triassic rocks, partly sealed by a salt overhang); (3) prospects favourably located to receive oil spilled from the main Hammerfest Basin traps.

West Shetland Basin (Fig. 6)

The West Shetland Basin (WSB) and adjacent Faeroe–Shetland Basin (FSB) have a NE–SW grain, cross-cut by NW-trending transfer zones (e.g. Dean *et al.* 1999). The structural history of the area is complex, with multiple rifting events in Permo-Triassic, late Jurassic–early Cretaceous, mid–late Cretaceous and Paleocene times (Doré *et al.* 1999). In the WSB the dominant rifting event was in Permo-Triassic time, after which the basin underwent repeated exhumations while the FSB continued to subside. Major Cretaceous unconformities and/or non-sequences in the WSB presumably represent basin-flank uplift associated with rifting in the FSB. At the end of Cretaceous time the WSB and its southern continuation, the Solan Basin, were uplifted as part of a major emergence of the Scottish massif. Further uplift of the WSB in Oligo-Miocene time was probably connected to an episode of inversion that caused broad domal structuring in the FSB (e.g. Turner & Scrutton 1993: Herries *et al.* 1999; Parnell *et al.* 1999). As a result of repeated uplift and erosion, Cenozoic deposits are thin or absent over the WSB and Solan Basin.

About 60 wells have been drilled in the WSB and Solan Basin with a success rate of one in seven. Proven recoverable resources are in the order of 280 MMB oil and 0.35 TCF gas. Only one accumulation, the Clair oilfield, is currently considered commercial (Coney *et al.* 1993). Hydrocarbons in the WSB were sourced by Upper Jurassic marine shales with a probable Middle Jurassic lacustrine component (Bailey *et al.* 1987; Scotchman *et al.* 1998). Reservoirs range in age from Lewisian basement to Early Cretaceous. Charging occurred during latest Cretaceous and Cenozoic time as pulsed episodes of migration from the adjacent FSB (Parnell *et al.* 1999). Early oil charges have frequently been lost as a result of biodegradation or breaching of traps during uplift, as shown by fluid inclusions and residual shows (e.g. Goodchild *et al.* 1999). Replenishment from a continuously subsiding kitchen in the oil window probably explains the dominance of oil over gas in the basin, despite its exhumed nature.

In the Clair Field 3–5 billion barrels of heavy oil are held in fractured basement and Devono-Carboniferous red beds, although only about 200 MMB are thought to be recoverable because of the low porosity and permeability of the reservoir. Open fractures are important for optimizing deliverability in the reservoir (Coney *et al.* 1993: see also section on recovery factor). The oil is a mixture of biodegraded and fresh oil, a function of the multiple charging. Two small non-associated gas caps on Clair represent a late gas charge, or possibly exsolution gas. In the nearby Victory gas field the early oil charge to the Lower Cretaceous reservoir has been lost as a result of breaching of the trap during early to mid-Cenozoic uplift. This has left residual biodegraded oil within and below the current gas column, which occupies less than half the vertical closure of the structure (e.g. Goodchild *et al.* 1999). Farther south in the Solan Basin, small oil accumulations (Solan and Strathmore) totalling 60 MMB occur in Upper Jurassic and Triassic truncation traps. The non-biodegraded oil was sourced from a limited kichen area, the East Solan Basin, where the Upper Jurassic source rocks are at early oil maturity (Herries *et al.* 1999). Herries *et al.* considered that the fields were charged in two Cenozoic pulses, separated by the Oligo-Miocene inversion episode. We note, however, that because of the current very thin post-Paleocene cover, such generation must imply episodes of reburial and re-exhumation during Cenozoic time.

Inner Moray Firth Basin (Fig. 7)

The Inner Moray Firth Basin (IMFB) lies off the NE coast of Scotland, between the Grampian and Northern Highlands. The basin forms a westerly extension of the trilete Mesozoic graben system and is separated from the eastern part, the Outer Moray Firth Basin, by the Halibut Horst. The structural history of the IMFB is dominated by the effects of Permo-Triassic and Jurassic rifting, subsidence in Late Jurassic to Late Cretaceous–earliest Tertiary time (Andrews *et al.* 1990) and subsequent uplift.

Exhumation of the IMFB has been assigned to Paleocene time by Hillis *et al.* (1994), based on the assumption of synchronicity with the onset of denudation of the Scottish Highlands. Younger uplift episodes are not, however, precluded by the data. For example, Underhill (1991) identified an Oligo-Miocene inversion phase of possible far-field ‘Alpine’ origin, which reactivated Mesozoic faults and gave rise to differential relief within the basin. Uplift followed by subsidence towards the North Sea graben system has imparted an eastwards tilt to the IMFB, such that Jurassic and Lower Cretaceous rocks subcrop at the sea bed in the westernmost part of the basin and are succeeded eastward by Upper Cretaceous and Tertiary subcrops (Fig. 7). Hillis *et al.* (1994) used sonic velocity data to suggest that about 1 km of erosion took place over most of the basin during early Cenozoic time. However, because of later Cenozoic sedimentation the apparent erosion (net uplift *sensu* Riis & Jensen 1992) decreases eastwards to zero at about 1°W.

Approximately 80 exploration wells have been drilled in the IMFB with a success rate of one in eight, much lower than in the adjacent North Sea. Underhill (1991) attributed poor success rates in this area to breaching of traps by reactivated faults, some of which extend to the surface. Total proven resources are of the order of 680 MMB oil and 0.51 TCF gas. In the western, most uplifted part of the basin a single commercial oil discovery has been made, the Beatrice Field (155 MMB recoverable) along with some much smaller uncommercial oil and gas pools. The Beatrice hydrocarbon system consists of a Middle Jurassic reservoir in a tilted fault-block trap sealed by Oxfordian–Kimmeridgian shales. The oil was co-sourced by Devonian and Middle Jurassic mudrocks (Peters *et al.* 1989). Charging occurred during Late Cretaceous time, after which generation must have ceased as a result of uplift. Remarkably, the oil accumulation appears to have remained intact for the duration of Cenozoic time, preserving a 335 m oil column that fills the structure to spill (Stevens 1991). Retention may be partly due to the efficiency of the shale seal (which may have retained ductility before uplift: see Corcoran & Doré 2002), and partly due to the waxy, viscous nature of the crude. The very low energy of the oil (gas–oil ratio of 126 SCF per barrel, bubble point pressure 635 psig) may testify to gas depletion by diffusion or lack of an original gas charge.

A cluster of fields of mixed phase in the east of the IMF (Captain, Blake, Ross, Cromarty, Phoenix) lie within the uplifted area, although some of these fields may be receiving charge from currently generating kitchens. In the Captain Field, overlying the western end of the Halibut Horst, shallowly buried Lower Cretaceous sandstones contain recoverable oil reserves of 350 MMB sourced from Upper Jurassic rocks. Most of the Cenozoic succession is missing above the field, and the hydrocarbon accumulation carries a strong signature of exhumation. The oil is heavily biodegraded and includes a residual oil column in the east of the field attributed by Pinnock & Clitheroe (1997) to easterly tilting during Cenozoic time. The field has a small cap of thermogenic gas introduced as a late charge (Pinnock & Clitheroe 1997) and probably representing exsolution gas. Notably, however, the field is full to its spill point.

East Irish Sea Basin (Fig. 8)

The East Irish Sea Basin (EISB) is a preserved remnant of a late Palaeozoic to early Mesozoic extensional basin system (Knipe *et al.* 1993). Subsequent uplift and denudation has removed most of the post-Triassic cover from the basin. Post-Triassic burial–uplift history is therefore difficult to reconstruct, and relies on techniques

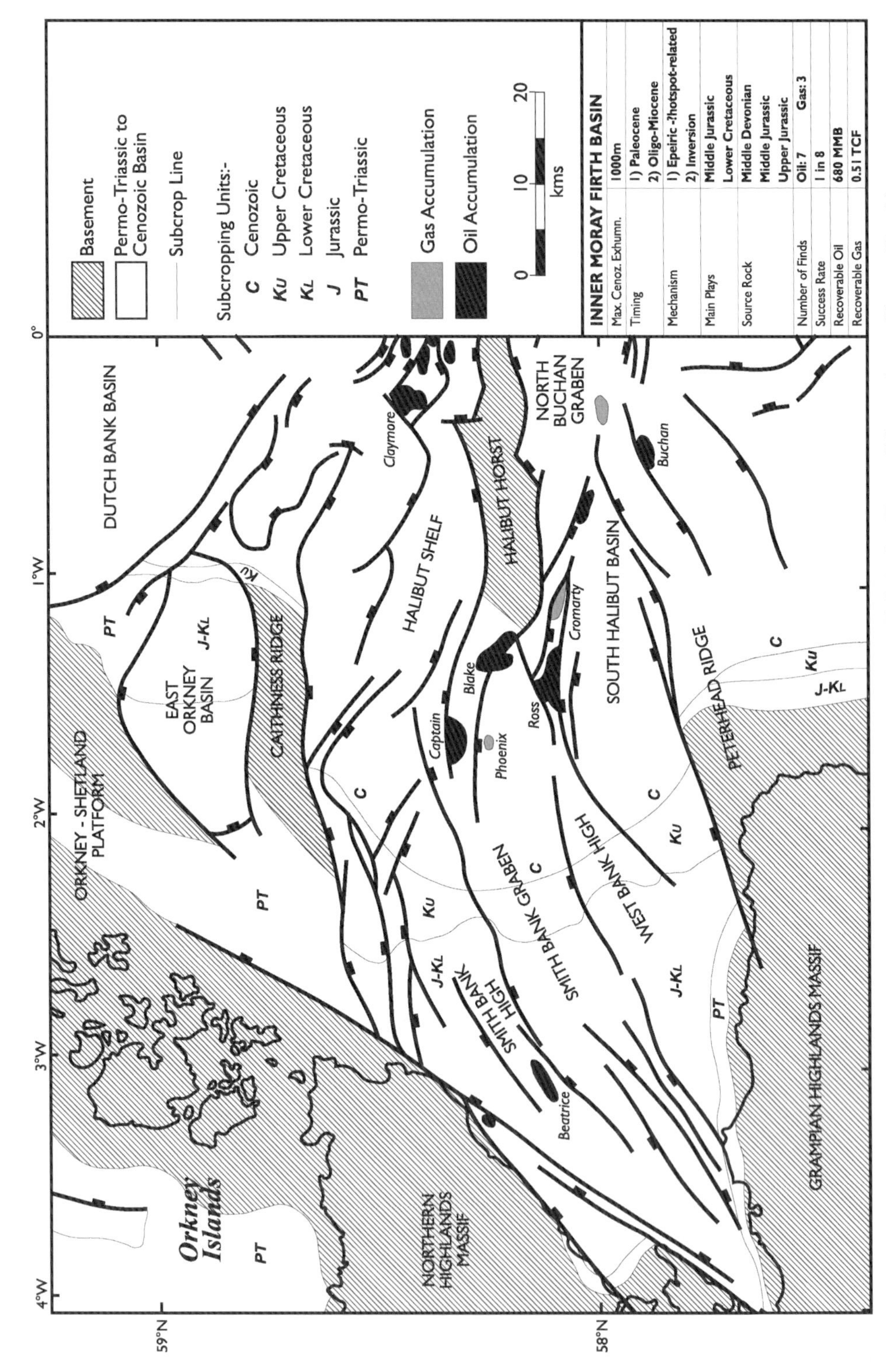

Fig. 7. Structural features, hydrocarbon discoveries and data template for the Inner Moray Firth Basin. (For qualifiers to inset table, see Fig. 5.)

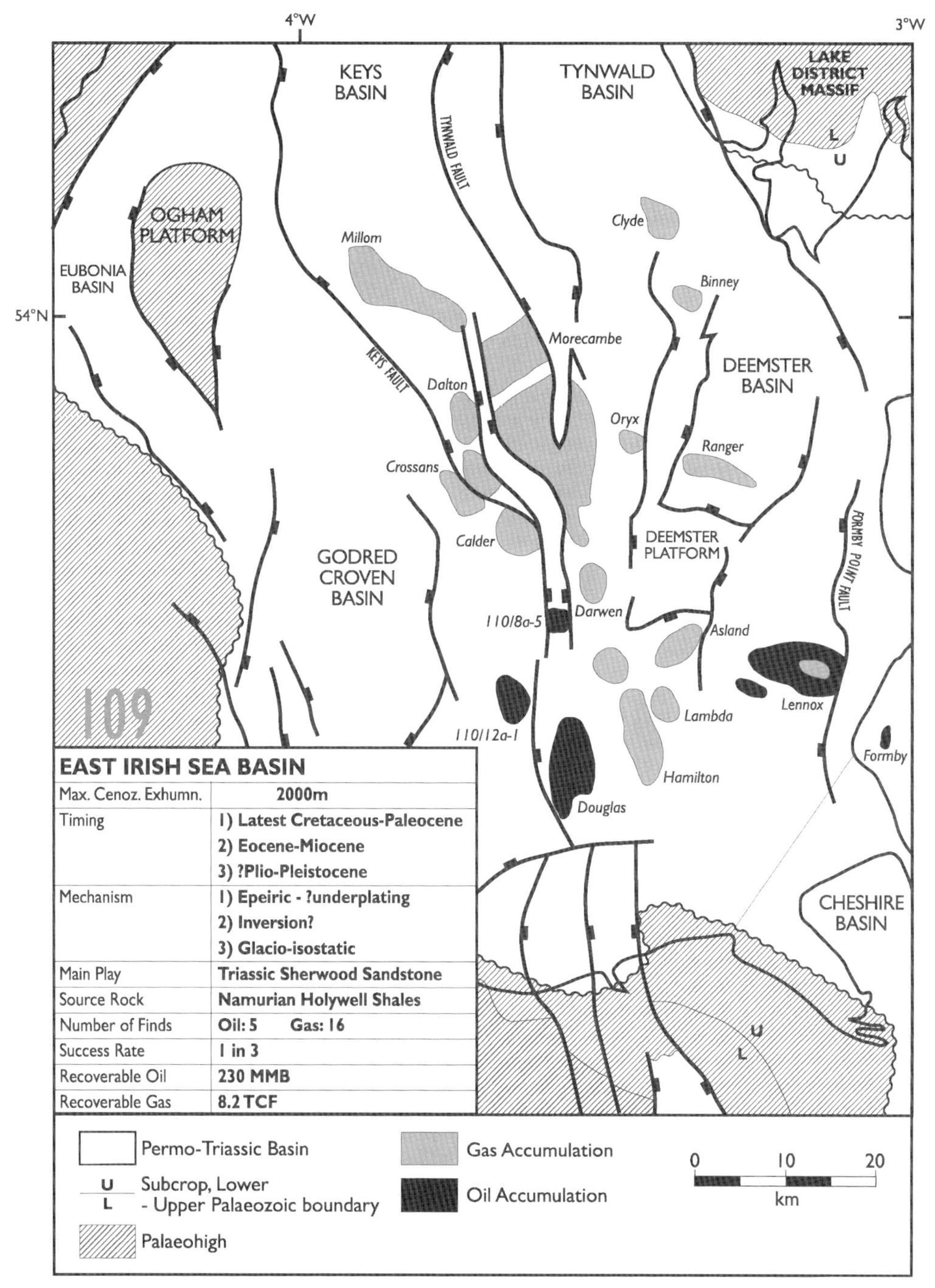

EAST IRISH SEA BASIN	
Max. Cenoz. Exhumn.	2000m
Timing	1) Latest Cretaceous-Paleocene 2) Eocene-Miocene 3) ?Plio-Pleistocene
Mechanism	1) Epeiric - ?underplating 2) Inversion? 3) Glacio-isostatic
Main Play	Triassic Sherwood Sandstone
Source Rock	Namurian Holywell Shales
Number of Finds	Oil: 5 Gas: 16
Success Rate	1 in 3
Recoverable Oil	230 MMB
Recoverable Gas	8.2 TCF

Fig. 8. Structural features, hydrocarbon discoveries and data template for the East Irish Sea Basin. (For qualifiers to inset table, see Fig. 5.)

such as apatite fission track, shale velocity and vitrinite reflectance. Some workers have modelled a major uplift phase in Early Cretaceous time (e.g. Duncan *et al.* 1998), whereas others have assigned earliest uplift to latest Cretaceous–Paleocene time, concident with North Atlantic opening and facilitated by thermal uplift or underplating (Cope 1994: Cowan *et al.*

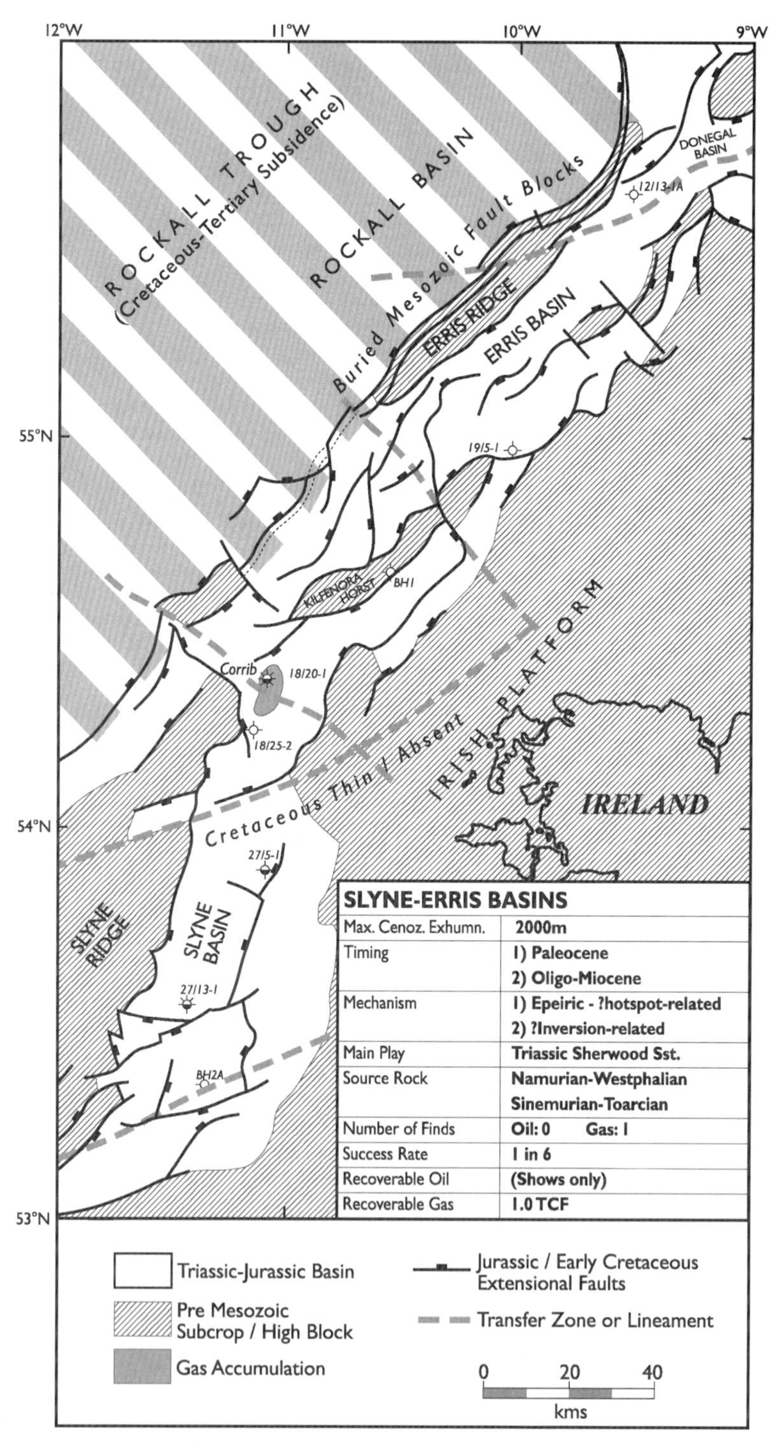

SLYNE-ERRIS BASINS	
Max. Cenoz. Exhumn.	**2000m**
Timing	**1) Paleocene** **2) Oligo-Miocene**
Mechanism	**1) Epeiric - ?hotspot-related** **2) ?Inversion-related**
Main Play	**Triassic Sherwood Sst.**
Source Rock	**Namurian-Westphalian** **Sinemurian-Toarcian**
Number of Finds	**Oil: 0 Gas: 1**
Success Rate	**1 in 6**
Recoverable Oil	**(Shows only)**
Recoverable Gas	**1.0 TCF**

Fig. 9. Structural features, hydrocarbon discoveries and data template for the Slyne and Erris Basins. (For qualifiers to inset table, see Fig. 5.)

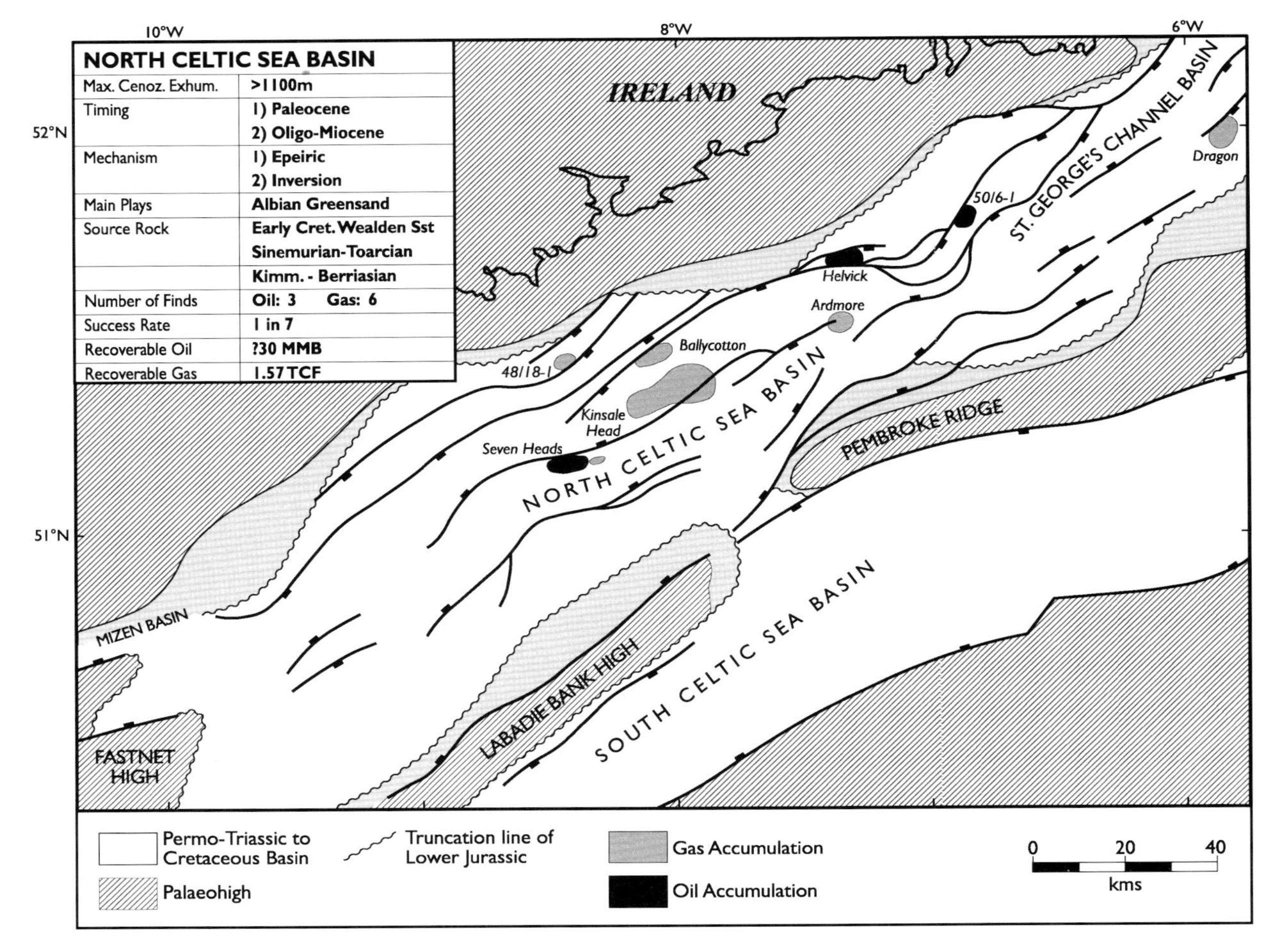

Fig. 10. Structural features, hydrocarbon discoveries and data template for the North Celtic Sea Basin. (For qualifiers to inset table, see Fig. 5.)

1999: Ware & Turner 2002). Later exhumation is even more difficult to constrain, but Ware & Turner (2002) have proposed a short-wavelength contribution from Eocene–Miocene inversion. The EISB was ice-covered during Pleistocene time, and a glacio-isostatic component cannot therefore be ruled out. Estimates of maximum exhumation vary between 1 and 3 km, with recent work suggesting 2 km as an upper limit (Cowan *et al.* 1999; Ware & Turner 2002).

The EISB is a prolific hydrocarbon province containing 10 gas fields, two oilfields and nine undeveloped hydrocarbon discoveries. About 60 wells have been drilled, with a one in three success record. Gas reserves of 8.2 TCF and oil reserves of about 230 MMBO have been identified in an area of 3500 km^2 (Quirk *et al.* 1999). The hydrocarbon system in the EISB consists of a Triassic aeolian–fluvial reservoir in structural traps, charged from a Namurian source rock and sealed by Upper Triassic evaporites and shales.

The hydrocarbon accumulations have a complex evolution intimately related to the exhumation history. The basin is characterized by a distinct northern gas province and a southern oil and gas province, with approximately 70% of proven reserves reservoired in the two Morecambe gas fields. Earliest gas and oil emplacement is believed to have occurred during Early Jurassic time. Breaching of seals during exhumation, for example in South Morecambe, resulted in loss of the initial oil-rich charge followed by later stage (?Early Tertiary) recharging with thermogenic gas and present-day underfilling (Stuart & Cowan 1991; Stuart 1993). Breaching or spillage of traps left behind palaeo-oil–water contacts, indicated on seismic data (Francis *et al.* 1997) and by illite cementation in Morecambe South. Residual columns of biodegraded oil also testify to the original charge (Bushell 1986; Woodward & Curtis 1987). Similarly, the Douglas and Lennox oilfields show evidence of multiple charging, with the earliest oil charge being totally degraded to bitumen, a subsequent higher maturity charge being partially biodegraded, followed by a final condensate charge (Haig *et al.* 1997; Yaliz 1997). The Formby oilfield, a pool of biodegraded oil trapped in the Sherwood Sandstone by Pleistocene till, is evidence of very recent remigration across the Formby Point Fault (Fig. 8) from a breached trap (Francis *et al.* 1997).

Recent modelling of the EISB has stressed the importance of remigration, driven by gas exsolution from oil and gas expansion during Cenozoic uplift (Cowan *et al.* 1999). Gas exsolution from formation water has not been incorporated into these models, but we suggest that it provides a powerful additional mechanism for the late gas charge. Oil was driven updip to the periphery of the basin by the late gas flux (Duncan *et al.* 1998). The complexity of the remigration process is, however, indicated by the juxtaposition of undersaturated oils (e.g. Douglas) with dry gas accumulations (e.g. Hamilton). Seedhouse & Racey (1997) and Cowan *et al.* (1999) have shown that the presence of halite beds in the basal part of the seal is a critical success factor for hydrocarbon entrapment and oil–gas balance. Gas will escape through the seal in shallow structures except where the basal evaporite is present. Conversely, this discharge of the gas leg allows oil to be preferentially trapped in the southern part of the basin where the basal evaporite is absent (Quirk *et al.* 1999).

Slyne–Erris Basin (Fig. 9)

The Slyne–Erris Basin (SEB) is a narrow, elongate, NE–SW-trending basin system 60 km off northwestern Ireland. It consists of a series of asymmetric half-grabens separated by cross-cutting transfer zones. It experienced a multiphase rifting and inversion history, although the preserved basin morphology is primarily the result of Mid–Late Jurassic rifting (Chapman *et al.* 1999; Dancer *et al.* 1999). A striking characteristic of the southerly Slyne Trough is the truncated stratigraphic record with an almost complete absence of post-rift sediments. A thin cover of Miocene sediments rests unconformably on synrift sediments of Late Bajocian to Bathonian age. However, in the northerly Erris Trough, more than 1 km of Cretaceous strata are locally preserved. Multiple phases of regional exhumation and local inversion affected the area. These included rift-related footwall uplift events in Late Jurassic–Early Cretaceous and Aptian time, regional uplift in Paleocene time probably associated with Atlantic opening, and inversion-related uplift in Oligo-Miocene time. Maximum exhumation is of the order of 2000 m, although it is difficult to establish what proportion took

Fig. 11. Summary map showing oil–gas balance and exhumation-related phenomena in exhumed basins with proven hydrocarbon systems on the NW European margin. Examples of fields or wells are given for exhumation-related characteristics in each area. Oil and gas quantities are related using oil industry standards, which are based on calorific value: 1 barrel of oil approximately equals 6000 standard cubic feet of gas.

place in Cenozoic time (Scotchman & Thomas 1995).

Sporadic exploration in this area over the past 25 years has resulted in the drilling of six exploration wells, which have yielded a single gas discovery (the Corrib Field) in the Slyne Trough. Total discovered resources to date are approximately 1 TCF gas, all in the Corrib Field, an underfilled faulted anticlinal structure (Corcoran & Doré 2002). The main gas exploration play consists of a Lower Triassic sandstone reservoir in structural traps, charged from the underlying Namurian–Westphalian claystones and coals and sealed by Upper Triassic evaporites and shales (Scotchman & Thomas 1995; Dancer *et al.* 1999). A Jurassic petroleum system is considered proven in the Slyne Trough by Spencer *et al.* (1999), based on the presence of palaeo-oil accumulations. Biodegraded, residual oil shows from a Lower Jurassic source have been encountered in Middle Jurassic reservoirs in wells 27/13-1, 27/5-1 and 18/20-1. These residual columns are consistent with breaching of traps and/or freshwater flushing during uplift to shallow levels. It has yet to be demonstrated that any producible accumulations from the Jurassic hydrocarbon system have survived the Cenozoic exhumation of the basin.

North Celtic Sea Basin (Fig. 10)

The North Celtic Sea Basin (NCSB) is a NE–SW trending Mesozoic extensional basin located to the south of Ireland. It is bounded by a series of Palaeozoic ridges and platforms and contains a thick Triassic to Cretaceous sedimentary fill. Major rifting episodes occurred during Late Jurassic and Early Cretaceous time (Rowell 1995), but post-rift subsidence was terminated by regional uplift and inversion during Cenozoic time. The exhumation resulted in subcrop of Cretaceous Chalk at the sea floor in the centre of the basin, and the complete removal of Cretaceous sediments in the NE of the basin. Two Cenozoic erosional events have been documented: regional uplift during Paleocene time and inversion characterized by basin doming and fault reversal during Oligo-Miocene time. Net exhumation in excess of 1100 m is interpreted in the NE of the basin (Murdoch *et al.* 1995).

The NCSB has proved to be a somewhat enigmatic petroleum province. About 70 exploration and appraisal wells have been drilled to date, with a success rate of one in six. Only two accumulations are producing, Kinsale Head and Ballycotton, containing proven reserves of 1.6 TCF (Taber *et al.* 1995). A further seven subcommercial oil and gas discoveries (e.g. Seven Heads, Helvick, Ardmore) have been identified. The main reservoir in the producing gas fields is the shallow marine Albian Greensand. Secondary production occurs from the fluvial Wealden reservoirs. Elsewhere in the basin oil has been tested from these stratigraphic levels and from Oxfordian fluvial sandstones and Middle Jurassic shelf limestones (Caston 1995). Seals are provided by the Albian–Cenomanian Gault Clay and intraformational claystones within the Bathonian to Aptian succession.

Typically for an exhumed basin, the NCSB is dominated by central gas deposits and in this case by a single accumulation, Kinsale Head. This structure, a basin-centre anticline, may have had some pre-Cenozoic expression but was greatly emphasized during Tertiary inversion (Taber *et al.* 1995). Atypically for fields in NE Atlantic exhumed basins, the trap is full to spill (Taber *et al.* 1995). This suggests that the maximum depth of burial of the Gault Clay caprock (1700–1800 m) may not have been enough for the claystone to achieve embrittlement, allowing it to deform plastically during exhumation and compressive overprint (Corcoran & Doré 2002). Published models for maturation and expulsion from the Lower Jurassic source rocks indicate that peak gas generation would have occurred towards the end of Cretaceous time (Murphy *et al.* 1995). Because the structure probably developed during Tertiary time, after generation from the source rocks would have been curtailed because of exhumation, it seems unlikely that the present gas represents the original thermogenic charge. Even given a ductile claystone caprock, gas losses as a result of diffusion and underfilling of the trap would be expected. Therefore, active charging of the Kinsale Head Field may have continued until recent geological time. A possible mechanism for additional gas charging is exsolution of methane from groundwater in the Greensand, Wealden and older aquifers, possibly focused by groundwater flow during uplift.

Other signatures of an exhumed hydrocarbon system in the NCSB include a residual oil column in Kinsale Head, probably indicative of an earlier oil charge later displaced by gas (Taber *et al.* 1995), minor peripheral hanging-wall oil accumulations such as Helvick (Caston 1995) and two-phase accumulations with biodegraded oil (Seven Heads).

Discussion

As summarized in Fig.11, many of the North Atlantic basins contain hydrocarbon systems

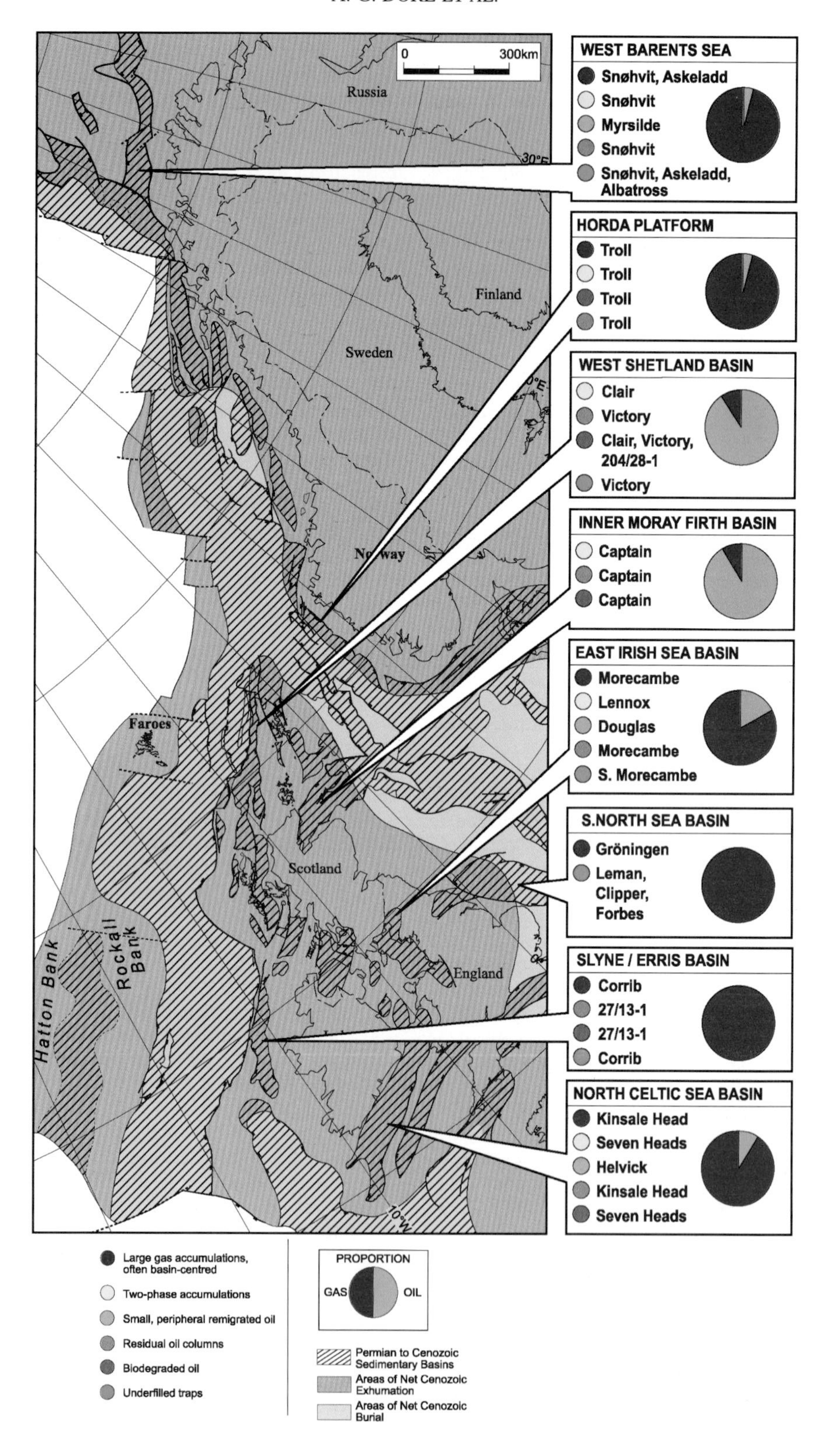

0 300km
Russia
Finland
Sweden
Norway
Faroes
Scotland
England
Rockall Bank
Hatton Bank
30°E
10°W
WEST BARENTS SEA
Snøhvit, Askeladd
Snøhvit
Myrsilde
Snøhvit
Snøhvit, Askeladd, Albatross
HORDA PLATFORM
Troll
Troll
Troll
Troll
WEST SHETLAND BASIN
Clair
Victory
Clair, Victory, 204/28-1
Victory
INNER MORAY FIRTH BASIN
Captain
Captain
Captain
EAST IRISH SEA BASIN
Morecambe
Lennox
Douglas
Morecambe
S. Morecambe
S.NORTH SEA BASIN
Gröningen
Leman, Clipper, Forbes
SLYNE / ERRIS BASIN
Corrib
27/13-1
27/13-1
Corrib
NORTH CELTIC SEA BASIN
Kinsale Head
Seven Heads
Helvick
Kinsale Head
Seven Heads
Large gas accumulations, often basin-centred
Two-phase accumulations
Small, peripheral remigrated oil
Residual oil columns
Biodegraded oil
Underfilled traps
PROPORTION
GAS
OIL
Permian to Cenozoic Sedimentary Basins
Areas of Net Cenozoic Exhumation
Areas of Net Cenozoic Burial

showing indicators of exhumation. Acknowledgement of the importance of exhumation can constrain future exploration strategy in these basins, and a similar approach can be applied to any exhumed basin.

Exploration risk analysis in exhumed terranes should take into account a decreased probability for seal or trap, and should address a complex interplay of positive and negative factors when assessing probability of source or charge. There is an increased chance that the dominant hydrocarbon phase will be gas, but oil can still be predicted by taking into account factors such as seal integrity, displacement from traps and remigration pathways. Although we have given a qualitative guide to such a risk analysis, it is not possible to provide numerical values. These will vary according to the unique geological characteristics of an area; for example, the quality of regional seals within the basin. It is frequently said that exploration risk analysis is a subjective procedure, of use as a comparative rather than an absolute measure. However, a more objective view of risk (in any basin, exhumed or otherwise) can be gained by carrying out an audit after a period of drilling, whereby the actual discovery rates are compared with the predicted ones. Thus, a risk analysis constrained by knowledge of exhumation can be checked and modified based on exploration history.

The strategy for targeting resources should also be contrained by knowledge of exhumation levels; for example, in the identification of areas of porosity preservation and potential fracture-prone lithologies. The recovery strategy for oil and gas in uplifted fields should be influenced by knowledge of present-day stress and fracture directions. As in the case of risk analysis, we provide no numerical values for resource assessment in this paper, but again point out that an audit of drilling results can provide an objective comparison of predicted and actual volumes at an intermediate stage of exploration of a basin.

Timing of exhumation is a key element in prediction of the hydrocarbon system. As shown in the case studies on the North Atlantic margin, most such areas have undergone multiple exhumations during Cenozoic time and in some cases exhumation began even earlier. It is difficult, and requires patient analysis, to disentangle the effects of the various events and to assign relative importance to them. In general, however, it may be predicted that the more extreme effects on the hydrocarbon system (for example, the flushing effect of a central gas bloom) are more likely to be observed where exhumation has been very recent; as is seen, for example, in the Barents Sea, where the Plio-Pleistocene regional uplift episode was particularly important (Nyland *et al.* 1992). The effects of older exhumations may be muted by the slow dissipation of gas by diffusion, and overprinted by reburial and the introduction of new hydrocarbons. The variability in degree of exhumation within a particular province is also important, and relates to the uplift mechanism. In an area of broad regional ('epeirogenic') uplift the effects on the hydrocarbon system should be similar over a wide area, whereas in basins inverted by compression these effects may be more local in nature as a result of selective uplift of intrabasinal structures. The fact that both forms of exhumation are superimposed in several of the North Atlantic basins (e.g. Inner Moray Firth, Eastern Irish Sea) provides an additional challenge.

The authors thank J. Parnell and M. Tate for thorough and constructive reviews of the manuscript, M. Stoker for editorial handling and J. Kipps for graphics. This paper was published by permission of Statoil.

References

AGUILERA, R. 1980. *Natural Fractured Reservoirs*. PennWell, Tulsa, OK.

ALA, M.A. 1982. Chronology of trap formation and migration of hydrocarbons in Zagros sector of southwest Iran. *AAPG Bulletin*, **66** (10), 1535–1541.

ANDREWS, I.J., LONG, D., RICHARDS, P.C., THOMSON, A.R., BROWN, S., CHESHER, J.A. & MCCORMAC, M. 1990. *The Geology of the Moray Firth. BGS UK Offshore Regional Report*. HMSO, London.

BAILEY, N.J.L., WALKO, P. & SAUER, M.J. 1987. Geochemistry and source rock potential west of Shetlands. *In*: BROOKS, J. & GLENNIE, K.W. (eds) *Petroleum Geology of North West Europe*. Graham and Trotman, London, 711–721.

BALLENTINE, C.J., O'NIONS, R.K., OXBURGH, E.R., HORVATH, F. & DEAK, J. 1991. Rare gas constraints on hydrocarbon accumulation, crustal degassing and groundwater flow in the Pannonian Basin. *Earth and Planetary Science Letters*, **105**, 229–246.

BERGLUND, L.T., AUGUSTSOM, J., FÆRSETH, R., GJELBERG, I., MOE-RAMBERG, H. *et al.* 1986. The evolution of the Hammerfest Basin. *In*: SPENCER, A.M. (ed.) *Habitat of Hydrocarbons on the Norwegian Continental Shelf*. Graham and Trotman, London, 319–338.

BJØRKUM, P.A., WALDERHAUG, O. & AASE, N.E. 1993. A model for the effect of illitization on porosity and quartz cementation of sandstones. *Journal of Sedimentary Petrology*, **63**, 1089–1091.

BORDENAVE, M.L. & BURWOOD, M.L. 1989. Source rock distribution and maturation in the Zagros orogenic belt: provenance of the Asmari and

Bangestan reservoir oil accumulations. *Organic Geochemistry*, **16** (1–3), 369–387.

BUSHELL, T.P. 1986. Reservoir geology of the Morecambe Field. *In*: BROOKS, J., GOFF, J.C. amp; VAN HORN, B. (eds) *Habitat of Palaeozoic Gas in NW Europe*. Geological Society, London, Special Publications, **23**, 189–203.

CAMPBELL, C.J. 1996. World oil: reserves, production, politics and prices. *In*: DORÉ, A.G. & SINGLING-LARSEN, R. (eds) *Quantification and Prediction of Hydrocarbon Resources*. Norwegian Petroleum Society Special Publication, **6**, 1–20.

CASTON, V.N.D. 1995. The Helvick oil accumulation, Block 49/9, North Celtic Sea Basin. *In*: CROKER, P.F. & SHANNON, P.M. (eds) *The Petroleum Geology of Ireland's Offshore Basins*. Geological Society, London, Special Publications, **93**, 209–225.

CHAPMAN, T.J., BROKS, T.M., CORCORAN, D.V., DUNCAN, L.A. & DANCER, P.N. 1999. The structural evolution of the Erris Trough, offshore northwest Ireland, and implications for hydrocarbon generation. *In*: FLEET, A.J. & BOLDY, S.A.R. (eds) *Petroleum Geology of Northwest Europe: Proceedings of the 5th Conference*. Geological Society, London, 455–469.

CONEY, D., FYFE, T.B., RETAIL, P. & SMITH, P.J. 1993. Clair appraisal: the benefits of a cooperative approach. *In*: PARKER, J.R. (ed.) *Petroleum Geology of Northwest Europe: Proceedings of the 4th Conference*. Geological Society, London, 1409–1420.

COPE, J.C.W. 1994. A latest Cretaceous hotspot and the southeasterly tilt of Britain. *Journal of the Geological Society, London*, **151**, 905–908.

CORCORAN, D.V. & DORÉ, A.G. 2002. Depressurization of hydrocarbon-bearing reservoirs in exhumed basin settings: evidence from Atlantic margin and borderland basins. *In*: DORÉ, A.G., CARTWRIGHT, J.A., STOKER, M.S., TURNER, J.P. & WHITE, N. (eds) *Exhumation of the North Atlantic Margin: Timing, Mechanisms and Implications for Petroleum Exploration*. Geological Society, London, Special Publications, **196**, 457–483.

COWAN, G., BURLEY, S.D., HOEY, A.N. & 5 OTHERS 1999. Oil and gas migration in the Sherwood Sandstone of the East Irish Sea Basin. *In*: Fleet, A.J. & Boldy, S.A.R. (eds) *Petroleum Geology of Northwest Europe: Proceedings of the 5th Conference*. Geological Society, London, 1383–1398.

CRAMER, B., POELCHAU, H.S., GERLING, P., LOPATIN, N.V. & LITTKE, R. 1999. Methane released from groundwater: the source of natural gas accumulations in northern West Siberia. *Marine and Petroleum Geology*, **16**, 225–244.

CRAMER, B.S., SCHÜMER, S. & POELCHAU, H.S. 2002. Uplift-related hydrocarbon accumulations: the release of natural gas from groundwater. *In*: DORÉ, A.G., CARTWRIGHT, J.A., STOKER, M.S., TURNER, J.P. & WHITE, N. (eds) *Exhumation of the North Atlantic Margin: Timing, Mechanisms and Implications for Petroleum Exploration*. Geological Society, London, Special Publications, **196**, 447–455.

CULBERSON, O.L. & MCKETTA, J.J. JR 1951. Phase equilibria in hydrocarbon-water systems, III—The solubility of methane in water at pressures to 10,000 PSIA. *Petroleum Transactions, American Institute of Mining and Metallurgical Engineers*, **192**, 223–226.

DANCER, P.N., ALGAR, S.T. & WILSON, I.R. 1999. Structural evolution of the Erris Trough. *In*: FLEET, A.J. & BOLDY, S.A.R. (eds) *Petroleum Geology of Northwest Europe: Proceedings of the 5th Conference*. Geological Society, London, 445–454.

DANIEL, E.J. 1954. Fractured reservoirs of the Middle East. *AAPG Bulletin*, **3**, 774–815.

DAVIS, G.H. & REYNOLDS, S.J. 1996. *Structural Geology of Rocks and Regions*. 2nd; Wiley, New York.

DEAN, K., MCLACHLAN, K. & CHAMBERS, A. 1999. Rifting and development of the Faroe–Shetland Basin. *In*: FLEET, A.J. & BOLDY, S.A.R. (eds) *Petroleum Geology of Northwest Europe: Proceedings of the 5th Conference*. Geological Society, London, 533–544.

DORÉ, A.G. 1995. Barents Sea geology, petroleum resources and commercial potential. *Arctic*, **48** (3), 207–221.

DORÉ, A.G. & JENSEN, L.N. 1996. The impact of late Cenozoic uplift and erosion on hydrocarbon exploration: offshore Norway and some other uplifted basins. *Global and Planetary Change*, **12**, 415–436.

DORÉ, A.G. & LUNDIN, E.R. 1996. Cenozoic compressional structures on the NE Atlantic margin: nature, origin and potential significance for hydrocarbon exploration. *Petroleum Geoscience*, **2**, 299–311.

DORÉ, A.G., CARTWRIGHT, J.A., STOKER, M.S., TURNER, J.P. & WHITE, N.J. 2002. Exhumation of the North Atlantic margin: introduction and background. *In*: DORÉ, A.J., CARTWRIGHT, J.A., STOKER, M.S., TURNER, J.P. & WHITE, N. (eds) *Exhumation of the North Atlantic Margin: Timing, Mechanisms and Implications for Petroleum Exploration*. Geological Society, London, Special Publications, **196**, 1–12.

DORÉ, A.G., LUNDIN, E.R., JENSEN, L.N., BIRKELAND, Ø., ELIASSEN, P.E. & FICHLER, C. 1999. Principal tectonic events in the evolution of the Northwest European Atlantic margin. *In*: FLEET, A.J. & BOLDY, S.A.R. (eds) *Petroleum Geology of Northwest Europe: Proceedings of the 5th Conference*. Geological Society, London, 41–62.

DUNCAN, W.I., GREEN, P.F. & DUDDY, I.R. 1998. Source rock burial history and seal effectiveness: key facets to understanding hydrocarbon exploration potential in the East and Central Irish Sea Basins. *AAPG Bulletin*, **82** (7), 1401–1415.

FRANCIS, A., MILWOOD HARGRAVE, M., MULHOLLAND, P. & WILLIAMS, D. 1997. Real and relict hydrocarbon indicators in the East Irish Sea Basin. *In*: MEADOWS, N.S., TRUEBLOOD, S.P., HARDMAN, M. & COWAN, G. (eds) *Petroleum Geology of the Irish Sea and Adjacent Areas*. Geological Society, London, Special Publications, **124**, 185–194.

GABRIELSEN, R.H., FÆRSETH, R.B., JENSEN, L.N., KALHEIM, J.E. & RIIS, F. 1990. *Structural Elements of the Norwegian Continental Shelf. Part 1, the Barents Sea Region*. Norwegian Petroleum Directorate Bulletin, **6**.

GHAZI, S.A. 1992. Cenozoic uplift in the Stord Basin area and its consequences for exploration. *Norsk Geologisk Tidsskrift*, **72**, 285–290.

GOODCHILD, M.W., HENRY, K.L., HINKLEY, R.J. & IMBUS, S.W. 1999. The Victory gas field, West of Shetland. *In*: FLEET, A.J. & BOLDY, S.A.R. (eds) *Petroleum Geology of Northwest Europe: Proceedings of the 5th Conference*. Geological Society, London, 713–724.

HAIG, D.B., PICKERING, S.C. & PROBERT, R. 1997. The Lennox oil and gas field. *In*: MEADOWS, N.S., TRUEBLOOD, S.P., HARDMAN, M. & COWAN, G. (eds) *Petroleum Geology of the Irish Sea and Adjacent Areas*. Geological Society, London, Special Publications, **124**, 417–436.

HANDIN, J., HAGER, R.V., FREIDMAN, M. & FEATHER, J.N. 1963. Experimental deformation of sedimentary rocks under confining pressure: pore pressure tests. *AAPG Bulletin*, **47**, 717–755.

HEFFER, K.J. & DOWOKPOR, A.B. 1990. Relationship between azimuths of flood anisotropy and local earth stresses in oil reservoirs. *In*: BULLER, A.T. (eds) *North Sea Oil and Gas Reservoirs—II*. Graham and Trotman, London, 251–260.

HERRIES, R., PODDUBIUK, R. & WILCOCKSON, P. 1999. Solan, Strathmore and the back basin play, West of Shetland. *In*: FLEET, A.J. & BOLDY, S.A.R. (eds) *Petroleum Geology of Northwest Europe: Proceedings of the 5th Conference*. Geological Society, London, 693–712.

HILLIER, A.P. & WILLIAMS, B.P. 1991. The Leman Field, Blocks 49/26, 49/27, 49/28, 53/1, 53/2, UK North Sea. *In*: ABBOTS, I.L. (ed.) *United Kingdom Oil and Gas Fields, 25 Years Commemorative Volume*. Geological Society, London, Memoir, **14**, 451–458.

HILLIS, R.R., THOMSON, K. & UNDERHILL, J.R. 1994. Quantification of Tertiary erosion in the Inner Moray Firth using sonic velocity data from the Chalk and the Kimmeridge Clay. *Marine and Petroleum Geology*, **11**, 283–293.

HOSHINO, K., KOIDE, H., INAMI, K., IWAMURA, S. & MITSUI, S. 1972. *Mechanical properties of Japanese Tertiary sedimentary rocks under high confining pressures*. Geological Survey of Japan, Report **244**.

ILLIFFE, J.E., ROBERTSON, A.G., WARD, G.H.F., WYNN, C., PEAD, S.D.M. & CAMERON, N. 1999. The importance of fluid pressures and migration to the hydrocarbon prospectivity of the Faeroe–Shetland White Zone. *In*: FLEET, A.J. & BOLDY, S.A.R. (eds) *Petroleum Geology of Northwest Europe: Proceedings of the 5th Conference*. Geological Society, London, 601–612.

JAPSEN, P. & CHALMERS, J.A. 2000. Neogene uplift and tectonics around the North Atlantic: overview. *Global and Planetary Change*, **24** (3–4), 165–174.

JENSEN, L.N. & SCHMIDT, B.J. 1993. Neogene uplift and erosion offshore south Norway: magnitude and consequences for hydrocarbon exploration in the Farsund Basin. *In*: SPENCER, A.M. (ed.) *Generation, Accumulation and Production of Europe's Hydrocarbons, III*. Special Publication of the European Association of Petroleum Geoscientists, **3**, 79–88.

KJEMPERUD, A. & FJELDSKAAR, W. 1992. Pleistocene glacial isostasy—implications for petroleum geology. *In*: LARSEN, R.M., BREKKE, H., LARSEN, B.T. & TALLERAAS, E. (eds) *Structural and Tectonic Modelling and its Application to Petroleum Geology*. Norwegian Petroleum Society Special Publication, **1**, 187–195.

KNIPE, R., COWAN, G. & BALENDRAN, B.S. 1993. The tectonic history of the East Irish Sea Basin with reference to the Morecambe Fields. *In*: PARKER, J.R. (ed.) *Petroleum Geology of Northwest Europe: Proceedings of the 4th Conference*. Geological Society, London, 857–866.

KONTOROVITCH, A.E., MANDELBAUM, V.S., SURKOV, V.S., TROFIMUK, A.A. & ZOLOTOV, A.N. 1990. Lena–Tuguska Upper Proterozoic–Palaeozoic petroleum superprovince. *In*: BROOKS, J. (ed.) *Classic Petroleum Provinces*. Geological Society, London, Special Publications, **50**, 473–489.

KROOSS, B.M., LEYTHAEUSER, D. & SCHAEFER, R.G. 1992. The quantification of diffusive hydrocarbon losses through cap rocks of natural gas reservoirs—a reevaluation. *AAPG Bulletin*, **76** (3), 403–406.

LERCHE, I. 1996. Gas in the 21st century: a world-wide perspective. *In*: DORÉ, A.G. & SINDING-LARSEN, R. (eds) *Quantification and Prediction of Hydrocarbon Resources*. Norwegian Petroleum Society Special Publication, **6**, 21–42.

LEYTHAEUSER, D., SCHAEFER, R.G. & YUKLER, A. 1982. Role of diffusion in primary migration of hydrocarbons. *AAPG Bulletin*, **66**, 408–429.

LITTKE, R. & LEYTHAEUSER, D. 1993. Migration of oil and gas in coals. *In*: LAW, B.E. & RICE, D.D. (eds) *Hydrocarbons from Coal*. American Association of Petroleum Geologists, Studies in Geology, **38**, 219–236.

LOPEZ, J.A. 1990. Structural styles of growth faults in the U.S. Gulf Coast Basin. *In*: BROOKS, J. (ed.) *Classic Petroleum Provinces*. Geological Society, London, Special Publications, **50**, 203–219.

LUO, X. & VASSEUR, G. 1995. Modelling of pore pressure evolution associated with sedimentation and uplift in sedimentary basins. *Basin Research*, **7**, 35–52.

MACGREGOR, D.S. 1995. Hydrocarbon habitat and classification of inverted rift basins. *In*: BUCHANAN, J.G. & BUCHANAN, P.G. (eds) *Basin Inversion*. Geological Society, London, Special Publications, **88**, 83–93.

MASTERS, J.A. 1984. Lower Cretaceous oil and gas in Western Canada. *In*: MASTERS, J.A. (ed.) *Elmworth: Case Study of a Deep Basin Gas Field*. American Association of Petroleum Geologists, Memoir, **38**, 1–33.

MAXIMOV, S.P., ZOLOTOV, A.N. & LODZHEVSKAYA, M.I. 1984. Tectonic conditions for oil and gas generation and distribution on ancient platforms. *Journal of Petroleum Geology*, **7** (3), 329–340.

MILTON, N.J., BERTRAM, G.T. & VANN, I.R. 1990. Early Palaeogene tectonics and sedimentation in the Central North Sea. *In*: HARDMAN, R.P.F., BROOKS, J. (ed.) *Tectonic Events Responsible for Britain's Oil and Gas Reserves*. Geological Society, London, Special Publication, **55**, 339–351.

MURDOCH, L.M., MUSGROVE, F.W. & PERRY, J.S. 1995. Tertiary uplift and inversion history in the North Celtic Sea Basin and its influence on source rock maturity. *In*: CROKER, P.F. & SHANNON, P.M. (eds) *The Petroleum Geology of Ireland's Offshore Basins*. Geological Society, London, Special Publications, **93**, 297–319.

MURPHY, N.J., SAUER, M.J. & ARMSTRONG, J.P. 1995. Toarcian source rock potential in the North Celtic Sea Basin, offshore Ireland. *In*: CROKER, P.F. & SHANNON, P.M. (eds) *The Petroleum Geology of Ireland's Offshore Basins*. Geological Society, London, Special Publications, **93**, 193–207.

NADEAU, P.H., WILSON, M.J., MCHARDY, W.J. & TAIT, J.M. 1985. The conversion of smectite to illite during diagenesis: evidence from some illitic clays from bentonites and sandstones. *Mineralogical Magazine*, **49**, 393–400.

NYLAND, B., JENSEN, L.N., SKAGEN, J., SKARPNES, O. & VORREN, T. 1992. Tertiary uplift and erosion in the Barents Sea; magnitude, timing and consequences. *In*: LARSEN, R.M., BREKKE, H., LARSEN, B.T. & TALLERAAS, E. (eds) *Structural and Tectonic Modelling and its Application to Petroleum Geology*. Norwegian Petroleum Society Special Publication, **1**, 153–162.

OLAUSSEN, S., GLOPPEN, T.G., JOHANNESEN, E. & DALLAND, A. 1984. Depositional environment and diagenesis of Jurassic reservoir sandstone in the eastern part of the Troms I area. *In*: SPENCER, A.M. (ed.) *Petroleum Geology of the North European Margin*. Graham and Trotman, London, 61–80.

PARNELL, J. 2002. Diagenesis and fluid flow in response to uplift and exhumation. *In*: DORÉ, A.G., CARTWRIGHT, J., STOKER, M.S., TURNER, J.P. & WHITE, N. (eds) *Exhumation of the North Atlantic Margin: Timing, Mechanisms and Implications for Petroleum Exploration*. Geological Society, London, Special Publications, **196**, 433–446.

PARNELL, J., CAREY, P.F., GREEN, P.F. & DUNCAN, W. 1999. Hydrocarbon migration history, West of Shetland: integrated fluid inclusion and fission track studies. *In*: FLEET, A.J. & BOLDY, S.A.R. (eds) *Petroleum Geology of Northwest Europe: Proceedings of the 5th Conference*. Geological Society, London, 613–626.

PETERS, K.E., MOLDOWAN, J.M., DRISCOLE, A.R. & DEMAISON, G.J. 1989. Origin of Beatrice oil by co-sourcing from Devonian and Middle Jurassic source rocks Inner Moray Firth, UK. *AAPG Bulletin*, **73**, 454–471.

PIGGOTT, N. & LINES, M.D. 1991. A case study of migration from the West Canada Basin. *In*: ENGLAND, W.A. & FLEET, A.J. (eds) *Petroleum Migration*. Geological Society, London, Special Publications, **59**, 207–225.

PINNOCK, S.J. & CLITHEROE, A.R.J. 1997. The Captain Field U.K. North Sea: appraisal and development of a viscous oil accumulation. *Petroleum Geoscience*, **3**, 305–312.

PRICE, L.C. 1979. Aqueous solubility of methane at elevated temperatures and pressures. *AAPG Bulletin*, **63**, 1527–1533.

PRICE, L.C. 2002. Geological and geochemical consequences of basin exhumation, and commercial implications (abstract). *In*: DORÉ, A.G., CARTWRIGHT, J.A., STOKER, M.S., TURNER, J.P. & WHITE, N. (eds) *Exhumation of the North Atlantic Margin: Timing, Mechanisms and Implications for Petroleum Exploration*. Geological Society, London, Special Publications, **196**, 431.

QUIRK, D.G., ROY, S., KNOTT, I., REDFERN, J. & HILL, L. 1999. Petroleum geology and future hydrocarbon potential of the Irish Sea. *Journal of Petroleum Geology*, **22**, 243–260.

RICE, D.D. 1993. Compositions and origins of coalbed gas. *In*: LAW, B.E. & RICE, D.D. (eds) *Hydrocarbons from Coal*. American Association of Petroleum Geologists, Studies in Geology, **38**, 159–184.

RIIS, F. 1996. Quantification of Cenozoic vertical movements of Scandinavia by correlation of morphological surfaces with offshore data. *Global and Planetary Change*, **12**, 331–357.

RIIS, F. & FJELDSKAAR, W. 1992. On the magnitude of the Late Tertiary and Quaternary erosion and its significance for the uplift of Scandinavia and the Barents Sea. *In*: LARSEN, R.M., BREKKE, H., LARSEN, B.T. & TALLERAAS, E. (eds) *Structural and Tectonic Modelling and its Application to Petroleum Geology*. Norwegian Petroleum Society Special Publication, **1**, 163–188.

RIIS, F. & JENSEN, L.N. 1992. Introduction: Measuring uplift and erosion—proposal for a terminology. *Norsk Geologisk Tidsskrift*, **72**, 223–228.

ROWELL, P. 1995. Tectono-stratigraphy of the North Celtic Sea Basin. *In*: CROKER, P.F. & SHANNON, P.M. (eds) *The Petroleum Geology of Ireland's Offshore Basins*. Geological Society, London, Special Publications, **93**, 101–138.

SALES, J.K. 1993. Closure vs. seal capacity—a fundamental control on the distribution of oil and gas. *In*: DORÉ, A.G. (ed.) *Basin Modelling: Advances and Applications*. Norwegian Petroleum Society Special Publication, **3**, 399–414.

SCOTCHMAN, I.C. & THOMAS, J.R.W. 1995. Maturity and hydrocarbon generation in the Slyne Trough, northwest Ireland. *In*: CROKER, P.F. & SHANNON, P.M. (eds) *The Petroleum Geology of Ireland's Offshore Basins*. Geological Society, London, Special Publications, **93**, 385–411.

SCOTCHMAN, I.C., GRIFFITH, C.E. & HOLMES, A.J. 1998. The Jurassic petroleum system north and west of Britain: a geochemical oil-source correlation study. *Organic Geochemistry*, **29**, 671–700.

SEEDHOUSE, J.K. & RACEY, A. 1997. Sealing capacity of the Mercia Mudstone Group in the East Irish Sea Basin: implications for petroleum exploration. *Journal of Petroleum Geology*, **20**, 261–286.

SHANMUGAM, G. 1988. Origin, recognition and importance of erosional unconformities in sedimentary basins. *In*: KLEINSPEHN, K.L. & PAOLA, C. (eds) *New Perspectives in Basin Analysis*. Springer, Berlin, 83–108.

SIBSON, R.H. 1995. Selective fault reactivation during basin inversion: potential for fluid redistribution through fault-valve action. *In*: BUCHANAN, J.G. & BUCHANAN, P.G. (eds) *Basin Inversion*. Geological Society, London, Special Publications, **88**, 3–19.

SKJERVØY, A. & SYLTA, Ø. 1993. Modelling of expulsion and secondary migration along the southwestern margin of the Horda Platform. *In*: DORÉ, A.G. (ed.) *Basin Modelling: Advances and Applications*. Norwegian Petroleum Society Special Publication, **3**, 499–538.

SNOW, J.H., DORÉ, A.G. & DORN-LOPEZ, D.W. 1996. Risk analysis and full-cycle probabilistic modelling of prospects: a prototype model developed for the Norwegian shelf. *In*: DORÉ, A.G. & SINDING-LARSEN, R. (eds) *Quantification and Prediction of Hydrocarbon Resources*. Norwegian Petroleum Society Special Publication, **6**, 153–166.

SOLHEIM, A., RIIS, F., ELVERHOI, A., FALEIDE, J.I., JENSEN, L.N. & CLOETINGH, S. 1996. Impact of glaciations on basin evolution: data and models from the Norwegian margin and adjacent areas—introduction and summary. *Global and Planetary Change*, **12**, 1–9.

SPENCER, A.M. *et al.* (eds) *Geology of Norwegian Oil and Gas Fields*. Graham and Trotman, London.

SPENCER, A.M., BIRKELAND, Ø., KNAG, G.Ø. & FREDSTAD, R. 1999. Petroleum systems of the Atlantic margin of northwest Europe. *In*: FLEET, A.J. & BOLDY, S.A.R. (eds) *Petroleum Geology of Northwest Europe: Proceedings of the 5th Conference*. Geological Society, London, 231–246.

STEARNS, D.W. & FREIDMAN, M. 1972. Reservoirs in fractured rock. *In*: KING, R.E. (ed.) *Stratigraphic Oil and Gas Fields: Classification, Exploration Methods and Case Histories*. American Association of Petroleum Geologists, Memoir, **16**, 82–106.

STEVENS, V. 1991. The Beatrice Field, Block 11/30a, UK North Sea. *In*: ABBOTS, J.L. (ed.) *United Kingdom Oil and Gas Fields, 25 Years Commemorative Volume*. Geological Society, London, Memoir, **14**, 242–252.

STUART, I.A. 1993. The geology of the North Morecambe Gas Field, East Irish Sea Basin. *In*: PARKER, J.R. (ed.) *Petroleum Geology of Northwest Europe: Proceedings of the 4th Conference*. Geological Society, London, 883–895.

STUART, I.A. & COWAN, G. 1991. The South Morecambe Field, Blocks 110/2a, 110/3a, 110/8a, UK East Irish Sea. *In*: ABBOTS, I.L. (ed.) *United Kingdom Oil and Gas Fields, 25 Years Commemorative Volume*. Geological Society, London, Memoir, **14**, 527–541.

TABER, D.R., VICKERS, M.K. & WINN, R.D. JR 1995. The definition of the Albian 'A' Sand reservoir fairway and aspects of associated gas accumulations in the North Celtic Sea Basin. *In*: CROKER, P.F. & SHANNON, P.M. (eds) *The Petroleum Geology of Ireland's Offshore Basins*. Geological Society, London, Special Publications, **93**, 227–244.

TAYLOR, M.S.G., LEROY, A. & FØRLAND, M. 1999. Hydrocarbon systems modelling of the Norwegian Central Graben Fairway trend. *In*: FLEET, A.J. & BOLDY, S.A.R. (eds) *Petroleum Geology of Northwest Europe: Proceedings of the 5th Conference*. Geological Society, London, 1325–1338.

THEIS, N.J., NIELSEN, H.H., SALES, J.K. & GALES, G.J. 1993. Impact of data integration on basin modelling in the Barents Sea. *In*: DORÉ, A.G. (ed.) *Basin Modelling: Advances and Applications*. Norwegian Petroleum Society Special Publication, **3**, 433–444.

TISSOT, B. & ESPITALIÉ, J. 1975. L'évolution thermique de la matiére organique des sédiments. Applications d'une simulation mathématique. *Revue de l'Institut Français de Pétrole*, **30**, 743–777.

TURNER, J.D. & SCRUTTON, R.A. 1993. Subsidence patterns in Western Margin basins: evidence from the Faeroe–Shetland Basin. *In*: PARKER, J.R. (ed.) *Petroleum Geology of Northwest Europe: Proceedings of the 4th Conference*. Geological Society, London, 975–984.

UNDERHILL, J.R. 1991. Implications of Mesozoic–Recent basin development in the western Inner Moray Firth, UK. *Marine and Petroleum Geology*, **8**, 359–369.

WALDERHAUG, O. 1992. Magnitude of uplift of the Stø and Nordmela Formations in the Hammerfest Basin—a diagenetic approach. *Norsk Geologisk Tidsskrift*, **72**, 321–323.

WARE, P.D. & TURNER, J.P. 2002. Sonic velocity analysis of the Tertiary denudation of the Irish Sea basin. *In*: DORÉ, A.G., CARTWRIGHT, J.A., STOKER, M.S., TURNER, J.P. & WHITE, N. (eds) *Exhumation of the North Atlantic Margin: Timing, Mechanisms and Implications for Petroleum Exploration*. Geological Society, London, Special Publications, **196**, 355–370.

WESTRE, S. 1983. The Askeladd gas field—Troms I. *In*: SPENCER, A.M. (ed.) *Petroleum Geology of the North European Margin*. Graham and Trotman, London, 33–39.

WHITE, R.S. 1988. A hot-spot model for Early Tertiary volcanism in the North Atlantic. *In*: MORTON, A.C. & PARSON, L.M. (eds) *Early Tertiary Volcanism in the North Atlantic*. Geological Society, London, Special Publications, **39**, 241–252.

WHITE, R.S. & MCKENZIE, D.P. 1989. Magmatism at rift zones: the generation of volcanic continental margins and flood basalts. *Journal of Geophysical Research*, **94b**, 7685–7729.

WOODWARD, K. & CURTIS, C.D. 1987. Predictive modelling for the distribution of production constraining illites—Morecambe Gas Field, Irish Sea, Offshore UK. *In*: BROOKS, J. & GLENNIE, K. (eds) *Petroleum Geology of North West Europe*. Graham and Trotman, London, 205–215.

YALIZ, A.M. 1997. The Douglas Oil Field. *In*: MEADOWS, N.S., TRUEBLOOD, S.P., HARDMAN, M. & COWAN, G. (eds) *Petroleum Geology of the Irish Sea and Adjacent Areas*. Geological Society, London, Special Publications, **124**, 399–416.

Geological and geochemical consequences of basin exhumation, and commercial implications

LEIGH C. PRICE

In the past, petroleum basins were traditionally largely viewed as either static or uniformly evolving entities. It is now more widely recognized that petroleum basins are subject to various intense geological processes, and hence substantial changes, during their evolutionary histories. One such geological process is exhumation, which is usually accompanied by significant erosion, decrease in both burial temperatures and fluid pressures, and often decrease in geothermal gradients. These changes can have profound consequences regarding hydrocarbon (HC) deposits. Besides all the obvious consequences (halting source-rock generation, damaging or destroying seals, expansion of gas with trap flushing, etc.), less obvious, but none the less equally meaningful, consequences also result.

For example, in basins with high geothermal gradients (e.g. Central Sumatra, Los Angeles), migration–accumulation processes result in most of the basin's reservoired oil being emplaced within the first 1–2 km of the surface. Strong erosion can largely destroy almost the entire oil resource of the basin, leaving a gas-only province (e.g. San Juan Basin, USA). Pore waters at high temperatures and pressures carry large amounts of both dissolved HC gases and inorganic mineral species (ions). Significant falls in burial temperatures from exhumation thus cause two results in the deeper regions of petroleum basins: (1) basin-centred gas deposits; (2) widespread destruction of deep-basin porosity, but especially permeability, from wholesale precipitation of dissolved mineral species as diagenetic minerals. In the case of basin-centred gas deposits, in going updip from the basin depocentre, eventually a location is reached where insufficient HC gas was dissolved in the pore water to allow gas to exsolve and form a free-gas phase that can exceed its critical fluid saturation level. Thus, the free-gas bubbles remain immobile and cause a two-phase permeability block (the Jamin effect), which becomes the leading updip seal for a downdip basin-centred gas deposit. Type examples of the results of this process are present in the Denver and San Juan Basins (USA) and the Alberta Basin (Canada). This process also results in tight-gas deposits in the basin deeps (e.g. Green River and Piceance Basins, USA), which are a variant of basin-centred gas deposits. With all basin-centred gas deposits, commercial production is dependent on finding 'sweet spots' where reservoir destruction from authigenic mineralization was impeded by a gas phase predating the exhumation.

Rejuvenation of basin formation with consequent basin downwarp and sedimentation can result in the most prospective of all oil-exploration targets: buried fault zones. With resurgent sedimentation, thick sequences of unfaulted sediments can be deposited over the previously exhumed sedimentary section, with two important results: (1) marine source sections that either were not buried deeply enough to generate and expel HCs, or had been generating but were interrupted by the exhumation, will now be buried deeply enough to achieve the high ranks necessary to commence HC generation in hydrogen-rich marine organic matter; (2) the thick section of overlying unfaulted shales capping the buried fault zones (a) focuses vertically migrating oil into the first reservoir of the trap, (b) prevents the HCs from substantial migration (except limited updip migration), and (c) serves as an excellent seal, allowing microseepage but not macroseepage. The key to such petroleum systems is to identify deeper reservoirs, which are connected to deeply buried known source rocks, while placing less exploration emphasis on the overlying unfaulted sediments.

Lee Price of the US Geological Survey, Denver, died in August 2000, shortly after the conference from which these papers are taken. We include Leigh's abstract to reflect his lively participation in the conference, and his significant contribution to the understanding of uplifted basins of the USA.

A. G. Doré for the editors

From: DORÉ, A.G., CARTWRIGHT, J.A., STOKER, M.S., TURNER, J.P. & WHITE, N. 2002. *Exhumation of the North Atlantic Margin: Timing, Mechanisms and Implications for Petroleum Exploration*. Geological Society, London, Special Publications, **196**, 431. 0305-8719/02/$15.00

Diagenesis and fluid flow in response to uplift and exhumation

JOHN PARNELL

Department of Geology and Petroleum Geology, King's College, University of Aberdeen, Aberdeen AB24 3UE, UK (e-mail: j.parnell@abdn.ac.uk)

Abstract: Uplift of sedimentary rocks is accompanied by a wide range of physical and chemical changes that contribute to diagenesis and modify fluid flow regimes. Topography becomes a major driving force behind fluid flow patterns, and meteoric water may penetrate to several kilometres below the surface. Typical diagenetic processes include alteration and leaching of feldspars and other unstable minerals, precipitation of iron oxides and kaolin, and leaching of carbonate and sulphate cements. Reservoired oil may be degraded by near-surface waters, but reservoir rocks may become more oil-wet.

Brittle fracturing is enhanced near the surface, and fluid flow may become predominantly fracture-bound as fractures dilate. Uplift also causes tilting of fluid contacts and remigration of hydrocarbons. Exsolution and expansion of gas similarly causes remigration of oil to peripheral traps.

Although basin uplift is generally regarded as being detrimental to hydrocarbon prospectivity, especially as a result of breaching of traps, there is also an enhanced potential for hydrocarbon plays based on reserves of exsolved gas, condensate dropout, peripheral traps and fractured reservoirs.

Although much research work on exhumed basins focuses upon the origins of uplift and exhumation, the resultant changes in geothermal patterns and the consequences of exhumation for regional sedimentation patterns, one of the most important economic aspects of exhumation is the effect upon basin diagenesis and fluid flow. This is particularly significant to hydrocarbon prospectivity where it influences reservoir quality. Although treatments of diagenesis rarely address exhumation in a specific manner, there are numerous diagenetic processes that are commonly associated with uplift and may therefore be predictable consequences of exhumation. As many case studies of diagenesis, particularly those undertaken before ready availability of offshore well cores, have been carried out in onshore basins, there are many data for rocks that have experienced exhumation. All rocks exposed at outcrop have been uplifted since maximum burial and therefore can be said to be exhumed. Exhumation is taken as the upward displacement of rocks with respect to the surface (England & Molnar 1990), the driving force for which in many cases is basin inversion. This review outlines the range of processes involved in diagenetic modification of exhumed basins and the consequences for reservoir quality, and the controls on changing fluid flow patterns. The review includes aspects of diagenesis and fluid flow relevant to uplift in general, including cases where rocks have subsequently been reburied. Given the importance of exhumation in NW Europe (e.g. west of Shetland, north Celtic Sea, east Irish Sea; Hillis 1995), it is valuable to be aware of these processes and their potential consequences.

Processes during exhumation

The effects of exhumation include a range of physical, chemical and biological processes. The physical processes include rheological effects, particularly fracturing, but also geometrical effects as a result of uplift. Chemical processes involve changes in pore fluid chemistry, and hence mineral precipitation or dissolution, owing to changes in pressure–temperature and fluid flow regimes. Biological processes involve degradational modification of hydrocarbon fluids as a result of the ingress of surficial waters to uplifted reservoirs. These various types of process are clearly interrelated. For example, new fracture systems play an important role in allowing circulation of fluids that cause chemical and biological changes.

From: Doré, A.G., Cartwright, J.A., Stoker, M.S., Turner, J.P. & White, N. 2002. *Exhumation of the North Atlantic Margin: Timing, Mechanisms and Implications for Petroleum Exploration*. Geological Society, London, Special Publications, **196**, 433–446. 0305-8719/02/$15.00

Physical processes

Changes in pressure and temperature, which both generally decrease during exhumation, have diverse consequences for the properties of rocks, minerals and entrained fluids, as follows.

(1) Rheological behaviour. At shallow depths, rocks that may have deformed in a ductile manner before inversion may behave instead in a brittle manner (Downey 1994; Ingram & Urai 1999). The important aspects of fracturing are discussed below.

(2) Mineral solubility. Changes in temperature have a strong influence on the solubility of the main minerals that cement sedimentary rocks, i.e. silica and carbonates. Silica solubility decreases markedly with falling temperature (Fournier 1983). Carbonate solubility, including calcite, by contrast increases with falling temperature, but is subject to a greater influence from PCO_2, which falls with pressure drop during uplift, and so can allow calcite precipitation (Wood 1986).

(3) Mineral transformation. The conversion of one mineral to another of a different chemistry necessitates interaction with ions in the pore fluids, but may ultimately be temperature driven, such as the conversion from smectite to illite with rising temperature. However, transformations of this type may not be readily reversible, hence the widespread occurrence of authigenic illite in exhumed rocks.

(4) Mineral phase changes. Temperature may also control transformation of one mineral polytype to another (e.g. kaolinite to dickite; chemically identical) or from one mineral phase to another within a single chemical system (e.g. gypsum to anhydrite plus water; effectively dehydration). These changes may also be less readily reversed, as shown by the metastability of dickite and anhydrite at the surface. Thus, exhumation generally has little consequence for changes in mineralogy. Nevertheless, prolonged interaction of anhydrite with water should convert it back to gypsum, with an accompanying volume increase.

(5) Fluid phase changes. Unlike mineral phase changes, fluid phase changes occur readily with falling temperature and pressure. In terms of hydrocarbons in the shallow crust, this involves exsolution of gas from oil and drop-out of some liquids from gas and gas condensates (e.g. Piggott & Lines 1991). Gas is also released from dissolution in brines as the pressure falls, although gas solubility can increase in solution with falling temperature (Tiab & Donaldson 1996). Existing gases experience expansion with falling pressure. Clearly, this release of gas causes an increase in the volume occupied by hydrocarbon fluids, and consequent shifting of fluid contacts within reservoirs, and may include displacement of hydrocarbons beyond the spill point into new reservoir compartments. This may result in successive trap structures updip holding more oil at shallower levels (e.g. Gussow 1954; Fig. 1), or loss from the trap structure (Nyland *et al.* 1992). Once a trap is filled with gas, no further oil will enter it, i.e. a trap filled with oil is still a potential gas trap, but a trap filled with gas is not an effective oil trap (Gussow 1954). The exsolution of gas from oil or water can also help to develop low permeabilities that function as seals. Where pores contain two fluid phases (e.g. oil and gas, water and gas) that have not become separated, the effective permeability to the individual phases is much lower than to a single pore-filling fluid phase (Chapman 1983; Osborne & Swarbrick 1997).

(6) Fluid viscosity. The viscosity of oil decreases with falling temperature, hence migration of oil is slower at shallower burial depths and may be effectively stopped in low-permeability pathways.

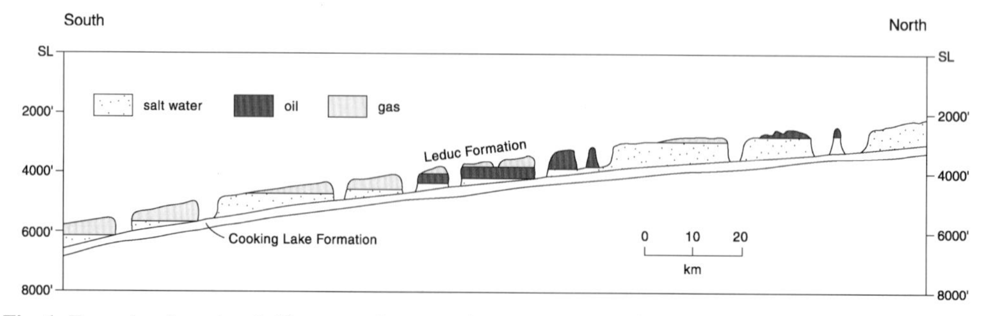

Fig. 1. Example of varying fluid contacts in successive trap structures in a tilted system, showing more oil at shallower levels, Leduc Reefs, Alberta (modified from Gussow 1954).

Chemical processes

Starting from first principles, the two media that are introduced from the surface upon exhumation are air and surface-derived water. In terms of reactive species, this involves oxygen, carbon dioxide and water. The processes that may be induced by these species are oxidation, acid dissolution (solution of carbon dioxide in water is acidic; dissolution in neutral water is negligible) and hydration. Protons for acidity are also contributed by the oxidation of organic matter and carbonate buffering reactions. As meteoric water penetrates to several kilometres depth (see below), these processes may affect significant proportions of a sedimentary basin.

An additional component introduced from the surface is bacteria, which mediate many chemical reactions, including sulphate reduction, and cause the degradation of oil. It is worth emphasizing the importance of the water, as the carrier of dissolved gases, salts and bacteria. Although rocks that are being exhumed will in most cases already be saturated with water, new water enhances the likelihood of diagenetic alteration.

Diagenesis involving near-surface fluids

The processes of oxidation, acid dissolution and hydration in uplifted rocks have important effects, particularly in sandstones, as follows.

(1) Oxidation causes alteration of any unstable detrital minerals (ferromagnesians, etc.) that had survived previous diagenesis, and alteration of authigenic minerals that had grown during burial diagenesis. This is most obvious in the formation of secondarily reddened sandstones, distinct from sands with red grain-coatings formed during deposition. The distinction is important, as iron oxides formed late in diagenesis can be sufficiently abundant to be pore-clogging, as opposed to the very fine grain-coatings that develop at the surface in arid to semi-arid environments (Walker *et al.* 1978). Secondary reddening can be widespread, as evinced by the distribution of reddened Carboniferous sandstones below the sub-Permian unconformity in the northern British Isles (Wang 1992); in that case, the reddened sandstones are porous because of associated carbonate cement dissolution. It is not simply a near-surface effect: the reddening is recorded at over 0.5 km below the palaeosurface, for example, in the eastern Irish Sea and North Sea (Jackson *et al.* 1987; Cowan 1989).

(2) One of the most widely quoted examples of alteration or dissolution of feldspars by meteoric waters is in the Jurassic Brent Group sandstones of the North Sea, where dissolution to create secondary porosity has been partly related to Cimmerian uplift (e.g. Sommer 1978; Glasmann *et al.* 1989; Bjørlykke *et al.* 1992). The degree to which this occurred during uplift as opposed to burial has been debated (Burley *et al.* 1985; Nedkvitne & Bjørlykke 1992), but the abundance of kaolin in the secondary porosity does suggest a meteoric influence (see below).

(3) Dissolution of carbonate and sulphate cements by surface-derived waters may give rise to substantial porosity. This is widely observed in the Permo-Triassic sediments of western Britain, which contain calcite, dolomite, gypsum and anhydrite cements, but which have been leached both by meteoric waters following Tertiary inversion and by modern groundwaters (e.g. Walton 1981; Burley 1984; Strong & Milodowski 1987). Of course, carbonate and evaporite rocks may be dissolved on a larger scale, to leave large cavities or collapse features.

(4) Another product of the acidity of near-surface waters is the precipitation of kaolin. The widespread occurrence of kaolin at unconformity surfaces (e.g. Esteoule-Choux 1983) is well known, but it is also common to find kaolin associated with secondary porosity after carbonate cement dissolution, as both are associated with low-pH fluids. Examples have been given by Curtis (1983) and Parnell (1987).

(5) As natural fluids tend to be silica saturated and reservoir fluids are often oversaturated (e.g. Bazin *et al.* 1997), this suggests that uplift should be a major cause of silica cementation as the solubility falls with temperature. However, except in cases where hot fluids are cooled very quickly (i.e. a more rapid process than uplift; see, e.g. Rossi *et al.* 2000), diagenetic studies do not record this as a process. The fact is that the volume of pore water is inadequate to cause much silica precipitation and at the low temperatures that accompany shallow depths the precipitation rate for silica is extremely low (Bjørlykke & Egeberg 1993). This exemplifies the general principle that chemical reaction rates increase with rise in temperature, i.e. low-temperature processes are not kinetically favoured. Small quartz outgrowths lining pores could reflect silica deposition as a result of uplift, but this process could never have a significant effect on reservoir quality.

(6) Given the widespread occurrence of basalts on the European Atlantic margin, particularly of Tertiary age, it is worth noting that basalt mineralogy is especially susceptible to alteration as it did not form in equilibrium with surface conditions. Alteration may also occur during

burial and hydrothermal processes, which may be difficult to distinguish from surficial alteration. An example of basalt alteration (to clays, zeolites and silica) that can be attributed to surficial processes has been described by Benson & Teague (1982).

The distinction between ingress by meteoric water and by seawater is important because meteoric water contains much lower concentrations of dissolved ions. The low sulphate content of meteoric water enhances the likelihood of siderite and dolomite precipitation (Kastner 1984). Where meteoric waters mix with other waters of different composition (e.g. connate waters), there is potential for mineral precipitation, although density contrasts between the fluids may limit the degree to which mixing occurs (Bjørlykke 1994).

Porosity trends

Table 1 summarizes controls on porosity decline or enhancement during burial and uplift. Porosities may remain higher than normal during burial for various reasons, particularly if hydrocarbon emplacement (Worden *et al.* 1998) or clay coating (Ehrenberg 1993; Sullivan *et al.* 1999) inhibits cementation, overpressuring inhibits compaction (Osborne & Swarbrick 1997) or secondary porosity is generated (Schmidt & McDonald 1979a). Thus pre-exhumation porosities can be high. On the other hand, deep burial may have largely eliminated porosity, such that 'over-compaction' is a typical sign that a basin has been uplifted since maximum burial. Over-compaction is widely observed in basins exhumed during Tertiary time in NW Europe, including those of the UK (Hillis 1995). Inverted basins inevitably exhibit lower porosities than expected for their current burial depth, as a result of the effects of additional compaction or cementation. Dronkers & Mrozek (1991) showed a marked difference in Triassic sandstones in the Broad Fourteens Basin, between tight strata in the inverted basin area and porous strata in a non-inverted platform site at the same current burial depth.

Upon uplift, different processes have very different consequences for increasing or decreasing porosity. Fracturing, and dissolution of carbonate cements or unstable grains, are both widespread processes that increase porosity, whereas any type of cementation (silica, kaolinite, iron oxide) will decrease it. As explained above, kaolinite precipitation may closely follow the creation of secondary porosity.

The importance of these trends for hydrocarbon systems depends on the relative timing of hydrocarbon migration. Exhumation inevitably turns off the hydrocarbon generation process, because of a drop in the temperature, as heat is needed to drive the chemical reactions involved. Thus post-exhumation porosity is relevant to remigrated existing accumulations or to hydrocarbons generated later after renewed burial and a suitable thermal history. Hillier & Marshall (1992) described an example in which hydrocarbon generation in the Devonian Orcadian Basin was arrested during Variscan inversion, then recommenced during subsequent Mesozoic burial. An example of gas generation interrupted by inversion before regeneration was recorded by van Wijhe *et al.* (1980) in the Sole Pit Basin.

The erosion of rock, associated with exhumation, has an effect on pore fluid pressure, which has been investigated by Luo & Vasseur (1995). The precise manner of fluid pressure decline depends upon permeability and rebound behaviour (a limited amount of compaction is elastic and reversible rather than plastic), but in general terms there is a time lag between commencement of erosion and reduction in effective stress (Fig. 2). This is because effective stress is not reduced until the pore pressure falls below the hydrostatic pressure, which does not occur instantaneously, especially if the system is initially overpressured (Luo & Vasseur 1995). The erosion rate controls the degree to which

Table 1. *Causes of change in porosity during burial and uplift*

Porosity loss	Porosity preservation or gain
Burial	
Compaction by sediment loading	Grain or cement dissolution at depth
Cementation (especially carbonates, quartz)	Preservation by oil or gas emplacement
	Preservation by overpressuring
Uplift	
Cementation (especially kaolinite, iron oxides)	Limited elastic rebound
	Dilation of fractures
	Grain or cement dissolution near surface

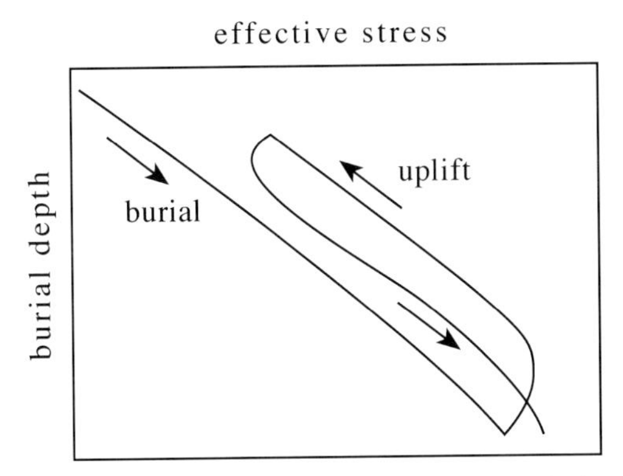

Fig. 2. Evolution pathway for effective stress during burial, erosion and reburial (i.e. development of unconformity), showing time lag between erosion and decline in effective stress (after Luo & Vasseur 1995).

porosity continues to decline after commencement of erosion (Fig. 3), which is another consequence of the delay in reduction of effective stress. At slower erosion rates, porosity is more likely to show an eventual reversal in the porosity–depth trend as a result of elastic rebound, although of very minor magnitude. These plots are based upon mechanical compaction, without the complication of cementation. In reality, cementation (chemical compaction) dominates over mechanical compaction above *c.* 90 °C (Bjørkum *et al.* 1998). Not only is overpressure usually dissipated during uplift, but there may be a slight development of underpressure related to decline in pore fluid pressure during sediment unloading (Fig. 4).

Secondary porosity and sandstone framework stability

The identification of secondary porosity rather than primary porosity is important because it has very different geometry and other properties that are of interest to petroleum engineers. It is inhomogeneously distributed, and the associated permeability and the pore surface area are different from those of primary porosity (Schmidt & McDonald 1979b).

Secondary porosity can of course be generated at considerable depths within a basin, particularly following decarboxylation of organic matter in source rocks (Schmidt & McDonald 1979a). Near-surface secondary porosity generation is also particularly a consequence of carbon dioxide availability, but from meteoric sources.

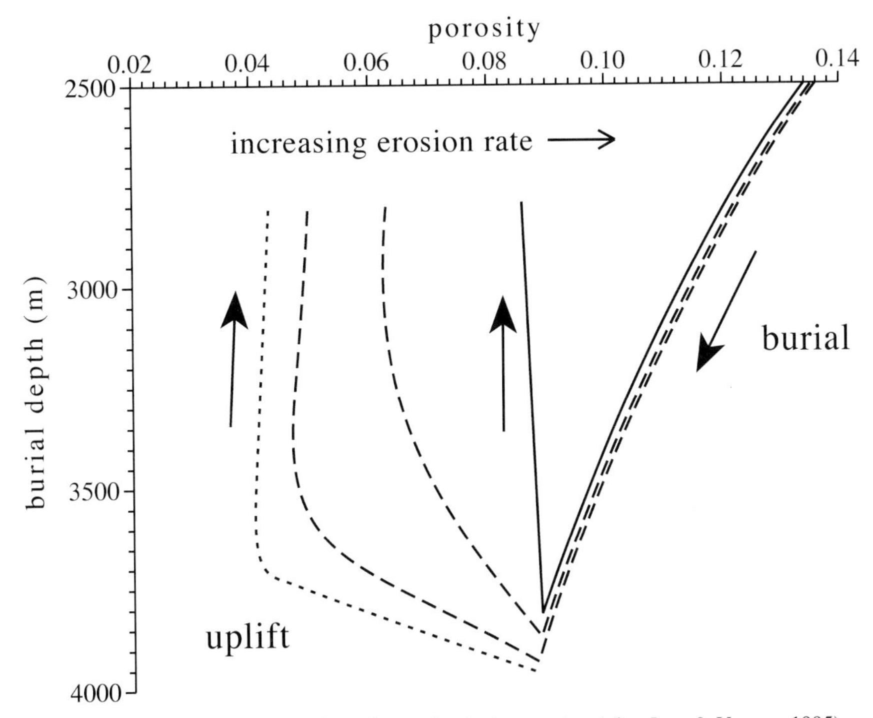

Fig. 3. Influence of erosion rate on evolution of porosity during erosion (after Luo & Vasseur 1995).

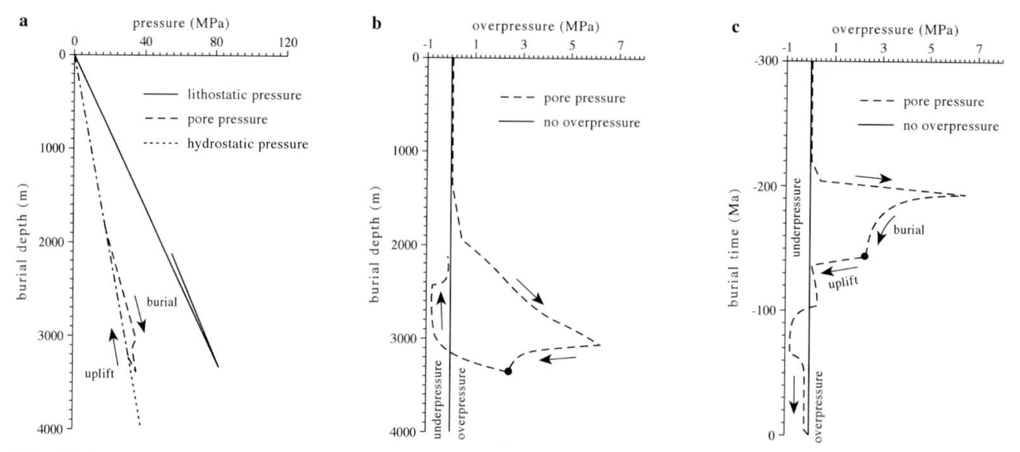

Fig. 4. Modelled effect of erosion on pore pressure evolution for a particular burial history (details given by Luo & Vasseur 1995): (**a**) pore pressure evolution with depth; (**b**) overpressure evolution with depth; (**c**) overpressure evolution with time. Arrows indicate direction of pathway. Continued porosity loss after erosion reflects continued increase in effective stress.

The significant difference in the response of the rock is that sandstone framework stability (Nagtegaal 1978) is critical at depth, but less so in the shallow subsurface, i.e. the surrounding fabric is more likely to collapse after cement dissolution under the high pressure of deep burial. However, if a sandstone has been deeply buried then exhumed it is likely to have developed a very rigid framework, conferred by tight grain packing, welded grain boundaries and quartz overgrowths. If secondary porosity can be generated in exhumed rocks it should be mechanically very stable unless it occupies so much volume that a 3D grain framework does not remain.

Secondary porosity may also be generated from the dissolution of unstable grains as exemplified in the Brent Group sandstones. Grain dissolution is more likely to happen during deep burial, for example, by late diagenetic illite formation from K-feldspar and kaolinite (e.g. Ehrenberg & Nadeau 1989), as by the time sandstones are exhumed most unstable grains are already lost. This is worth emphasizing: exhumed sandstones are, in general, mineralogically more mature than those at the same burial depth that have not been deeply buried. However, feldspars can survive deep burial (e.g. sandstones lacking kaolinite), and if abundant enough can lead to framework collapse upon dissolution after uplift. A dataset of uplifted Cambrian sandstones in Shropshire (Parnell 1987) shows low porosity where there are few feldspars to be leached, high porosities where more feldspars could be leached, but low porosities where so many feldspars were present that the sandstone framework collapsed after dissolution (Fig. 5).

Meteoric water flow can feasibly penetrate to several kilometres depth (Bethke *et al.* 1988), although the greatest flow rates will occur closest to the surface. Flow rates at shallow depth are up to several orders of magnitude greater than compaction-driven flow rates (Giles 1987; Harrison & Summa 1991), and so have the potential to account for relatively rapid mineral dissolution and precipitation. This is evident beneath present-day land surfaces, where, for example, Longstaffe (1984) reported kaolinite growth down to several hundred metres depth in the Alberta Basin, and Bath *et al.* (1987) recorded feldspar dissolution at similar depths in British Triassic reservoirs, both caused by penetration by meteoric fluids. Similarly, leaching by meteoric waters is evident beneath palaeo-exposure surfaces, i.e. below unconformities (Shanmugam 1988). Figure 6 shows how kaolinite dominates the clay mineralogy of the upper parts of Lower Carboniferous sections in Northern Ireland, whereas illite occurs in the lower parts. Where they occur together, illite occurs earlier in the diagenetic sequence than kaolinite, so this distribution is not simply a consequence of kaolinite reacting with feldspars to produce illite (see Ehrenberg & Nadeau 1989). Rather, the kaolinite, and associated high porosities, reflect diagenesis below the sub-Permian unconformity (Parnell 1991).

Sequences with complex burial histories may include multiple episodes of uplift-related

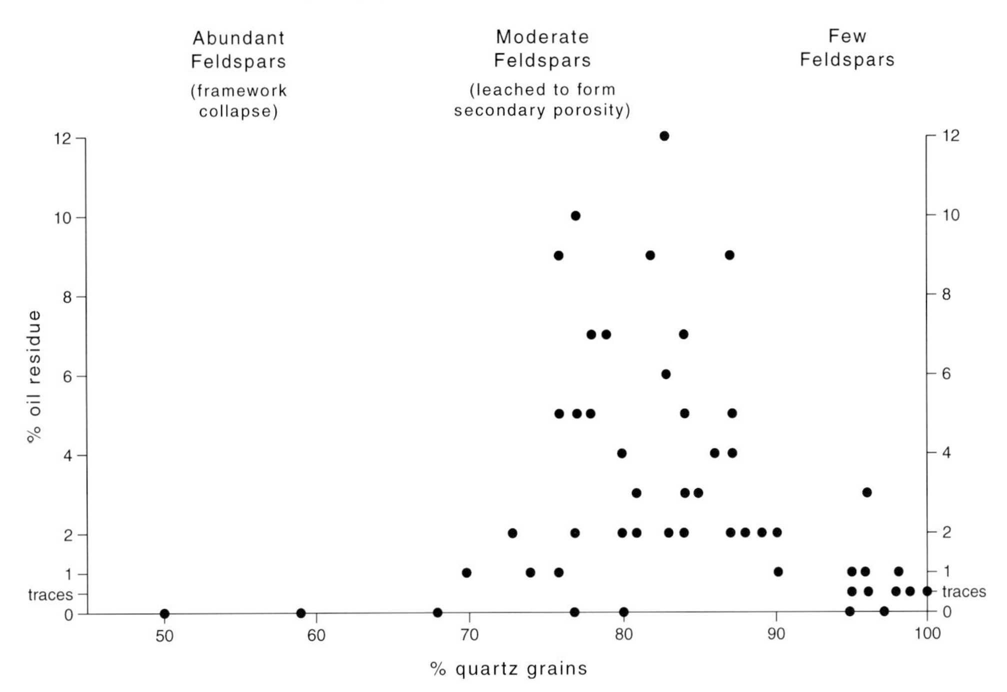

Fig. 5. Cross-plot of percentage oil residue (representing porosity) against quartz grain content in Cambrian sandstone, Shropshire (after Parnell 1987), showing importance of feldspar leaching to form secondary porosity, but framework collapse where feldspar content is too high.

diagenesis, including Atlantic margin basins where Cretaceous–early Tertiary uplift is widespread in addition to more recent exhumation. An example in Fig. 7 from NE Brazil (after Garcia *et al.* 1998) shows the predominance of kaolin and iron oxide cementation and cement dissolution during both uplift phases, and also degradation of oil during the palaeo-uplift event.

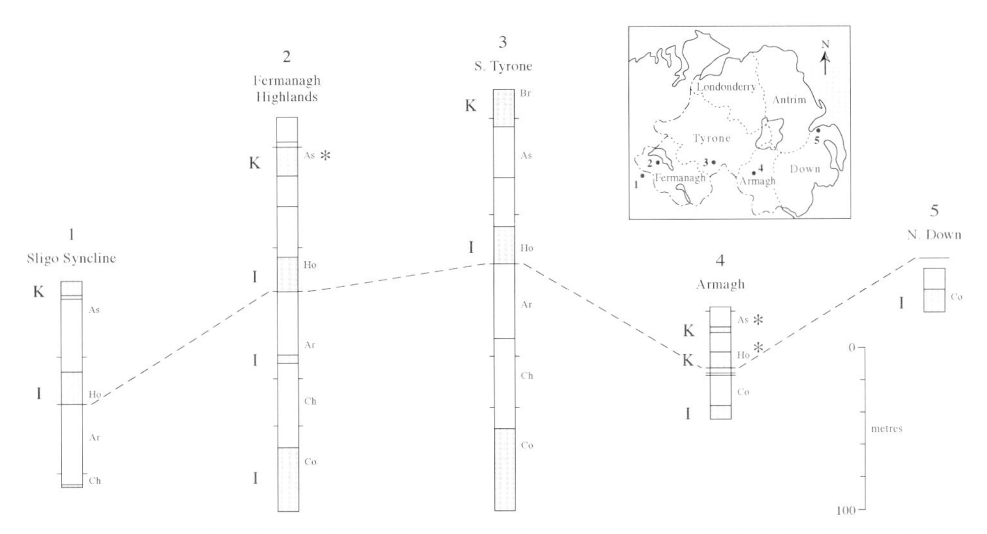

Fig. 6. Sections through the Lower Carboniferous succession across Northern Ireland, showing distribution of predominant clay mineral cement (K, kaolinite; I, illite), and most porous zones (starred). Kaolinite affects upper parts of sections as a result of ingress of oxidizing fluids during Permian uplift (see Parnell 1991). Co, Courceyan; Ch, Chadian; Ar, Arundian; Ho, Holkerian; As, Asbian; Br, Brigantian.

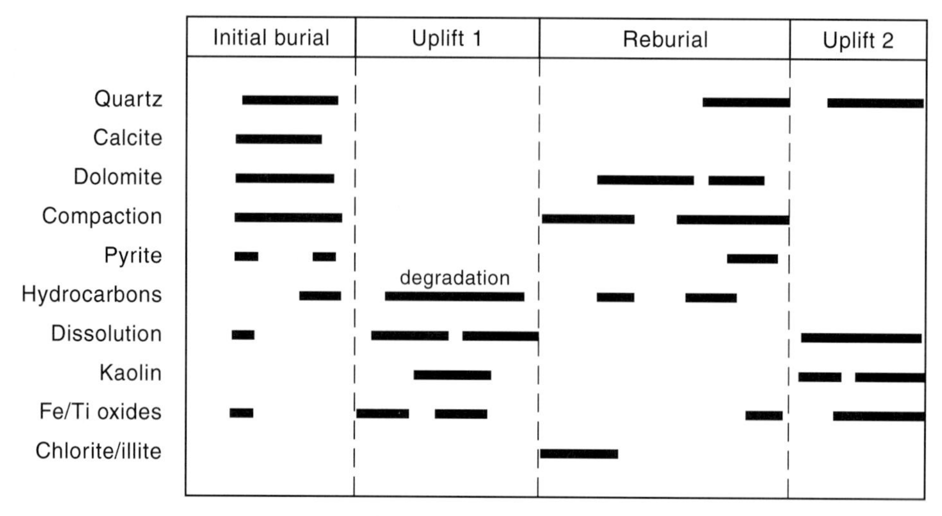

Fig. 7. Diagenetic sequence for Lower Cretaceous Serraria sandstones, central Sergipe–Alagoas Basin, NE Brazil, showing consequences of both palaeo-uplift and recent exhumation events (modified from Garcia *et al.* 1998).

Fracturing

Fracturing is widely developed in inverted basins, as tectonic activity is fundamental to the inversion process. On the Atlantic margin, fracture sets include those associated with extensional reactivation of syn-inversion compressional structures (Doré *et al.* 1999). Fracturing may also be a result of the reduction in effective stress owing to uplift, regardless of the tectonic setting (Sibson 1995).

Several aspects of the fracturing related to inversion are important, as follows.

(1) The adjustment from ductile to brittle deformation affects rocks that function as seals to trapped hydrocarbons, particularly in mudrocks, which may consequently allow leakage through new fractures. Ductility decreases with density, so mudrocks that have been buried then exhumed are particularly likely to exhibit fracturing (Ingram & Urai 1999; Corcoran & Doré 2002). Potentially, brittle deformation could also develop in evaporites (Downey 1994), although this is less likely than in mudrocks, as shown by hydrocarbon distribution in the east Irish Sea (Seedhouse & Racey 1997). The coherency of ductile seals is therefore critical to the prospectivity of exhumed basins (Gabrielsen & Kløvjan 1997; Spain & Conrad 1997; Corcoran & Doré 2002), although this also depends upon other factors, particularly the style of the inversion process.

(2) Fracturing, and increasing dilatancy of existing fractures, may cause fluid flow to become predominantly fracture-bound rather than through matrix permeability. Where permeabilities were already very low as a result of cementation or compaction during burial, fractures may account for almost all effective permeability. Figure 8 (after Dronkers & Mrozek 1991) shows an example from the margin of the inverted Broad Fourteens Basin, in which the effective permeability is in fractures related to reverse faulting and tight folding. Other planes of weakness, such as bedding planes or stylolites, may also show increased dilatancy. Bolton *et al.* (2000) suggested that horizontal effective stress decreases at a slower rate than vertical effective stress during unloading of clay-rich sediments, so that horizontal fractures develop, leading to a substantial increase in anisotropic permeability.

(3) Compressional deformation during inversion is likely to produce fracture networks with a marked anisotropy, i.e. fractures show a preferred alignment, or at least the degree of fracture dilatancy is greater in a preferred orientation. The result is fluid flow patterns that are focused along preferred fracture orientations. This is in contrast to the patterns of flow through matrix permeability, which tend to be controlled by the dip of the aquifers. An example of directional permeability in fractures related to inversion in the Bristol Channel basin has been recorded by Nemčok *et al.* (1995).Two other structural aspects of uplift have a significant influence on fluid flow, as follows.

(4) Uplift almost certainly will cause relative tilting of both lithological boundaries and fluid

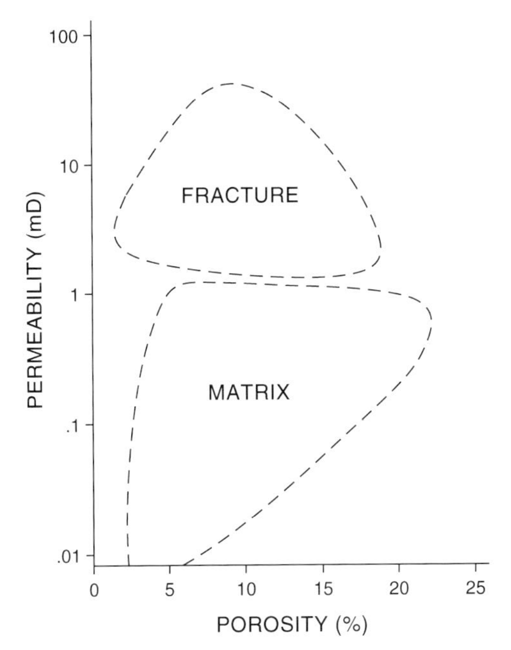

Fig. 8. Cross-plot of porosity and permeability for an oil well at the margin of the Broad Fourteens Basin, showing enhanced permeability in fractured sandstones compared with rocks in which permeability is through matrix (after Dronkers & Mrozek 1991).

contacts. Both may cause remigration of hydrocarbons (Gussow 1954). In some cases this will involve only minor movements within an aquifer, but in others it will push hydrocarbons beyond their spill point, which could involve migration into previously unused reservoir compartments (e.g. Cornford 1990). In reservoirs where cementation had been arrested or inhibited by hydrocarbon emplacement, but continued in the water-filled zone, tilting might move hydrocarbons into rocks of poorer reservoir quality, with the geometrical consequence that the hydrocarbons take up a greater volume of rock.

(5) Uplift increases the likelihood that topography is a driving force behind fluid flow (Bredehoeft *et al.* 1992; Deming 1994). This is partly because the aquifer is probably closer to the surface and more susceptible to downward-flowing meteoric waters, but more importantly if the exhumed basin is topographically higher than surrounding areas there may be a significant hydraulic head developed. In the Broad Fourteens Basin, offshore Netherlands, which has been inverted by at least 2 km, then reburied (Hooper *et al.* 1995), modelling (Verweij 1999; Verweij *et al.* 2000) emphasizes the importance of topography-driven flow syn- or post-inversion, where it has contributed to alteration of oils where they have been exposed to surface-derived waters (De Jager *et al.* 1996).In addition to these structural aspects of fluid flow, one of the most fundamental consequences of exhumation is of course that seals or aquitards are breached by erosion and allow fluid leakage (and also ingress of surficial fluids; see above). Leakage includes drainage of hydrocarbons, which is believed to have occurred in the Bristol Channel Basin via inversion-related fractures (Nemčok *et al.* 1995).

Fluid flow pathways and diagenetic processes are closely linked, as diagenesis requires import and export of ions. Once fracturing becomes the major flow pathway, this is where much mineral deposition occurs, i.e. fractures become sinks for ion precipitation, and can be infilled to the point of closure. The fracturing or faulting associated with inversion allows cross-formational flow of fluids that contributes to diagenesis. This includes accessing of waters from aquifers which may be either above and/or below the horizon of interest. For example, studies of Lower Permian Rotliegend sandstones in both the North Sea (Sullivan *et al.* 1994) and Germany (Platt 1994) show that cementation involved a contribution of fluids from the overlying Zechstein units, which could only enter the Rotliegend sandstones after fracturing related to Early Cretaceous inversion. In this example, rapid movement of large volumes of fluid occurred because overpressured Zechstein pore fluids were suddenly able to drain into the lower pressure Rotliegend sandstone.

Consequences for oil-bearing reservoirs

The influence of near-surface waters on the chemistry of oil has been widely described elsewhere (e.g. Tissot & Welte 1984), and will not be reviewed here except to emphasize the general effect of degrading oil quality (i.e. increasing its viscosity), particularly through water washing and biodegradation. In Europe, for example, the distribution of biodegraded oils offshore Netherlands reflects the effects of surface-derived waters following inversion (De Jager *et al.* 1996). There is debate over the degree to which biodegradation in reservoirs occurs before and after entrapment; for example, in the case of the huge deposits in the Lower Cretaceous sandstones of Alberta (Creany & Allan 1990). In inverted basins, biodegradation specifically linked to exposure to aquifers must be more likely, and a wide spectrum of oil preservation is found in this setting (Macgregor

1993, 1996). An interesting aspect of uplifted basins is that prior deep burial may 'sterilize' the oil so that biodegradation is prevented following uplift (Wilhelms *et al.* 2001).

The mineralogical changes that occur following uplift affect the wettability of sandstone reservoirs. The precipitation of grain-coatings of iron oxide (Wang & Guidry 1994), clays including kaolinite (Robin *et al.* 1995) and biodegraded oil residues (Gonzalez & Middea 1987) all tend to make the sandstones oil-wet rather than water-wet. The chemistry of the pore waters also influences wetting characteristics, in a complex way (Anderson 1986). Although oxidizing waters generally enhance oil-wetting, low-salinity waters enhance water-wetting. As near-surface waters may be both oxidizing and low salinity, their effect on wetting properties is difficult to predict. However, the grain-coating phases are likely to be predominant, and the oil-wet sandstones that they engender may both influence hydrocarbon flow behaviour and restrict subsequent cementation.

The influx of relatively fresh waters may reduce the resistivity contrast between oil and water, making difficult the detection of hydrocarbons based on resistivity measurements. This is comparable with the problems arising from the use of freshwater-based drilling muds (Rider 1996).

As explained above, gas expansion, reservoir tilting and leakage along fractures may all cause displacement of liquid hydrocarbons from reservoir compartments, so that underfilled traps, palaeo-fluid contacts, residual oil columns and peripheral oil accumulations are all typical features of exhumed basins, including those on the Atlantic margin (Doré & Jensen 1996; Doré *et al.* 2000). Basin-centred gas accumulations are also a potential result (Cramer *et al.* 1999). Gas

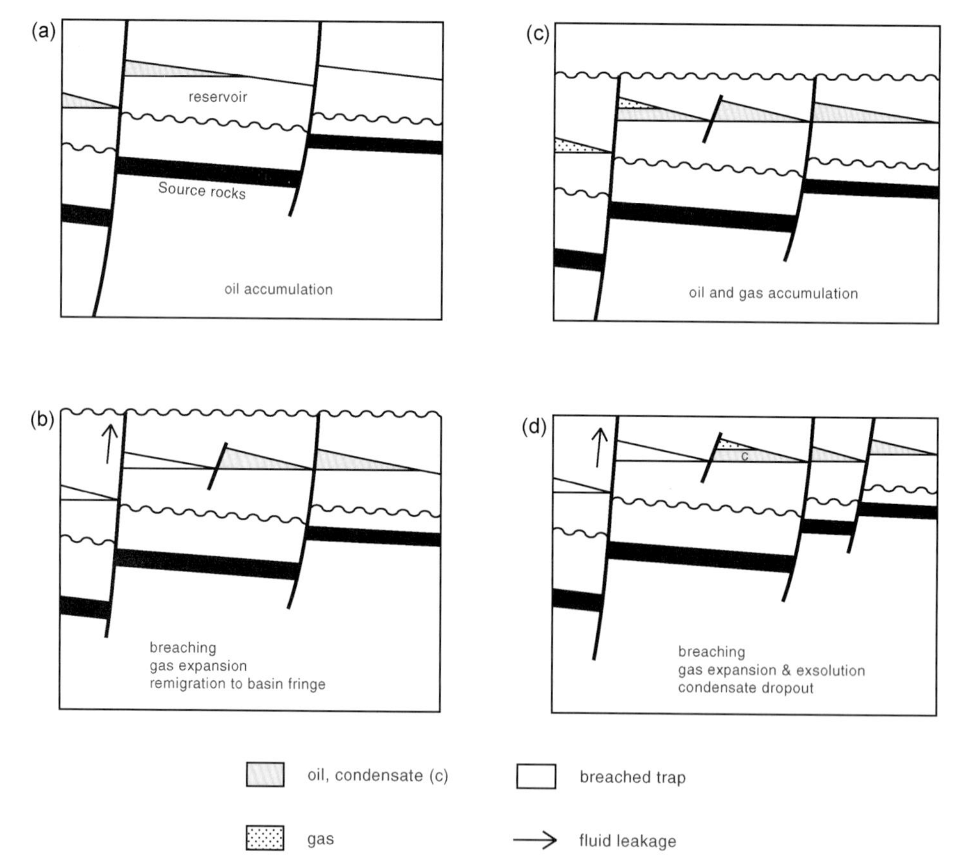

Fig. 9. Hydrocarbon fluid-fill history of Permo-Triassic reservoir compartments, eastern Irish Sea (modified from Duncan *et al.* 1998). Key episodes: (**a**) late Jurassic–early Cretaceous oil accumulation; (**b**) early Cretaceous inversion, breaching of traps, lateral migration to peripheral traps; (**c**) early Tertiary renewed burial, new oil and gas charge; (**d**) mid-Tertiary uplift, further breaching, gas expansion and condensate dropout.

migration may also be directly manifest as gas chimneys and sea-floor seepages. An example of the displacement processes in the eastern Irish Sea has been described by Duncan *et al.* (1998), and is summarized in Fig. 9. Permo-Triassic reservoirs, filled from Carboniferous source rocks, reveal a complex fluid-fill story related to their burial history. Oil generated before early Cretaceous (Cimmerian) inversion was lost or displaced by inversion, leaving breached traps of good reservoir quality, and new basin-fringe oil accumulations. Renewed burial during early Tertiary time caused refilling of traps with oil, and some charging by gas. Tertiary uplift caused renewed breaching, gas expansion and condensate dropout from gas. In this example, future exploration will require identification of areas to which remigration has occurred (Duncan *et al.* 1998).

Conclusions

In summary, many aspects of uplift and exhumation are detrimental to hydrocarbon prospectivity, including: (1) structural and erosional breaching of traps; (2) degrading of oil quality by influence of surface-derived water; (3) mineral deposition (except carbonates) as a result of decreasing solubility at lower temperatures. However, given the constraints of seal integrity, there is potential for distinct styles of hydrocarbon play: (1) large reserves of exsolved gas, especially at basin centres; (2) reserves of condensate dropout from gas; (3) remigrated oil at shallow field-peripheral sites; (4) fractured reservoirs, with enhanced porosity and permeability.

I am most grateful to A. G. Doré and an anonymous reviewer for critical comment on an earlier version of the manuscript, and to J. B. Fulton for essential cartographic assistance.

References

ANDERSON, W.G. 1986. Wettability literature survey. Part 1: rock/oil/brine interactions and the effects of core handling on wettability. *Journal of Petroleum Technology*, **38**, 1125–1149.

BATH, A.H., MILODOWSKI, I.E. & SPIRO, A.E. 1987. Diagenesis of carbonate cements in Permo-Triassic sandstones in the Wessex and East Yorkshire–Lincolnshire Basins, U.K.: a stable isotope study. *In*: GOFF, J.C. & WILLIAMS, B.P.J. (eds) *Fluid Flow in Sedimentary Basins and Aquifers*. Geological Society, London, Special Publications, **34**, 127–140.

BAZIN, B., BROSSE, E. & SOMMER, F. 1997. Chemistry of oil-field brines in relation to diagenesis of reservoirs. 2. Reconstruction of palaeo-water composition for modelling illite diagenesis in the Greater Alwyn area (North Sea). *Marine and Petroleum Geology*, **14**, 497–511.

BENSON, L.V. & TEAGUE, L.S. 1982. Diagenesis of basalts from the Pasco Basch, Washington. I. Distribution and composition of secondary mineral phases. *Journal of Sedimentary Petrology*, **52**, 595–613.

BETHKE, C.M., HARRISON, W.J., UPSON, C. & ALTANER, S.P. 1988. Supercomputer analysis of sedimentary basins. *Nature*, **239**, 261–267.

BJØRKUM, P.A., OELKERS, E.H., NADEAU, P.H., WALDERHAUG, O. & MURPHY, W.M. 1998. Porosity prediction in quartzose sandstones as a function of time, temperature, depth, stylolite frequency, and hydrocarbon saturation. *AAPG Bulletin*, **82**, 637–648.

BJØRLYKKE, K. 1994. Fluid-flow processes and diagenesis in sedimentary basins. *In*: PARNELL, J. (ed.) *Geofluids: Origin, Migration and Evolution of Fluids in Sedimentary Basins*. Geological Society, London, Special Publications, **78**, 127–140.

BJØRLYKKE, K. & EGEBERG, P.K. 1993. Quartz cementation in sedimentary basins. *AAPG Bulletin*, **77**, 1538–1548.

BJØRLYKKE, K., NEDKVITNE, T., RAMM, M. & SAIGAL, G. 1992. Diagenetic processes in the Brent Group (Middle Jurassic) reservoirs of the North Sea—an overview. *In*: MORTON, A.C., HASZELDINE, R.S., GILES, M.R. & BROWN, S. (eds) *Geology of the Brent Group*. Geological Society, London, Special Publications, **61**, 263–287.

BOLTON, A.J., MALTMAN, A.J. & FISHER, Q. 2000. Anisotropic permeability and bimodal pore-size distributions of fine-grained marine sediments. *Marine and Petroleum Geology*, **17**, 657–672.

BREDEHOEFT, J.D., BELITZ, K. & SHARP-HANSEN, S. 1992. The hydrodynamics of the Big Horn Basin: a study of the role of faults. *AAPG Bulletin*, **76**, 530–546.

BURLEY, S.D. 1984. Patterns of diagenesis in the Sherwood Sandstone Group (Triassic), United Kingdom. *Clay Minerals*, **19**, 403–440.

BURLEY, S.D., KANTOROWICZ, J.D. & WAUGH, B. 1985. Clastic diagenesis. *In*: BRENCHLEY, P.J. & WILLIAMS, B.P.J. (eds) *Sedimentology: Recent Developments and Applied Aspects*. Geological Society, London, Special Publications, **18**, 189–226.

CHAPMAN, R.E. 1983. *Petroleum Geology. Developments in Petroleum Geology, 16*. Elsevier, New York.

CORCORAN, D.V. & DORÉ, A.G. 2002. Depressurization of hydrocarbon-bearing reservoirs in exhumed basin settings: evidence from Atlantic margin and borderland basins. *In*: DORÉ, A.G., CARTWRIGHT, J.A., STOKER, M.S., TURNER, J.P. & WHITE, N. (eds) *Exhumation of the North Atlantic Margin: Timing, Mechanisms and Implications for Petroleum Exploration*. Geological Society, London, Special Publications, **196**, 457–483.

CORNFORD, C. 1990. Source rocks and hydrocarbons of the North Sea. *In*: GLENNIE, K.W. (eds)

Introduction to the Petroleum Geology of the North Sea, 3rd edn. Blackwell Science, Oxford, 294–361.

COWAN, G. 1989. Diagenesis of Upper Carboniferous sandstones: Southern North Sea Basin. *In*: WHATELEY, M.K.G. & PICKERING, K.T. (eds) *Deltas: Sites and Traps for Fossil Fuels*. Geological Society, London, Special Publications, **41**, 57–73.

CRAMER, B., POELCHAU, H.S., GERLING, P., LOPATIN, N.V. & LITTKE, R. 1999. Methane released from groundwater: the source of natural gas accumulations in West Siberia. *Marine and Petroleum Geology*, **16**, 225–244.

CREANY, S. & ALLAN, J. 1990. Hydrocarbon generation and migration in the Western Canada sedimentary basin. *In*: BROOKS, J. (ed.) *Classic Petroleum Provinces*. Geological Society, London, Special Publications, **50**, 189–202.

CURTIS, C.D. 1983. The link between aluminium mobility and destruction of secondary porosity. *AAPG Bulletin*, **67**, 380–384.

DE JAGER, J., DOYLE, M.A., GRANTHAM, P.J. & MABILLARD, J.E. 1996. Hydrocarbon habitat of the West Netherlands Basin. *In*: RONDEEL, H.E., BATJES, D.A.J. & NIEUWENHUIS, W.H. (eds) *Geology of Gas and Oil under the Netherlands*. Kluwer, Dordrecht, 191–209.

DEMING, D. 1994. Fluid flow and heat transport in the upper continental crust. *In*: PARNELL, J. (ed.) *Geofluids: Origin, Migration and Evolution of Fluids in Sedimentary Basins*. Geological Society, London, Special Publications, **78**, 27–42.

DORÉ, A.G. & JENSEN, L.N. 1996. The impact of late Cenozoic uplift and erosion on hydrocarbon exploration: offshore Norway and some other uplifted basins. *Global and Planetary Change*, **12**, 415–436.

DORÉ, A.G., LUNDIN, E.R., JENSEN, L.N., BIRKELAND, Ø., ELIASSEN, P.E. & FICHLER, C. 1999. Principal tectonic events in the evolution of the northwest European Atlantic margin. *In*: FLEET, A.J. & BOLDY, S.A.R. (eds) *Petroleum Geology of Northwest Europe: Proceedings of the 5th Conference*. Geological Society, London, 41–61.

DORÉ, A.G., SCOTCHMAN, I.C. & CORCORAN, D. 2000. Cenozoic exhumation and prediction of the hydrocarbon system on the NW European margin. *Journal of Geochemical Exploration*, **69–70**, 615–618.

DOWNEY, M.W. 1994. Hydrocarbon seal rocks. *In*: MAGOON, L.B. & DOW, W.G. (eds) *The Petroleum System—From Source to Trap*. Memoirs, American Association of Petroleum Geologists, **60**, 159–164.

DRONKERS, A.J. & MROZEK, F.J. 1991. Inverted basins of the Netherlands. *First Break*, **9**, 409–425.

DUNCAN, W.I., GREEN, P.F. & DUDDY, I.R. 1998. Source rock burial history and seal effectiveness: key facets to understanding hydrocarbon exploration potential in the East and Central Irish Sea Basins. *AAPG Bulletin*, **82**, 1401–1415.

EHRENBERG, S.N. 1993. Preservation of anomalously high porosity in deeply buried sandstones by grain-coating chlorite: examples from the Norwegian Continental Shelf. *AAPG Bulletin*, **77**, 1260–1286.

EHRENBERG, S.N. & NADEAU, P.H. 1989. Formation of diagenetic illite in sandstones of the Garn Formation, Haltenbanken area, mid-Norwegian Continental Shelf. *Clay Minerals*, **24**, 233–253.

ENGLAND, P. & MOLNAR, P. 1990. Surface uplift, uplift of rocks, and exhumation of rocks. *Geology*, **18**, 1173–1177.

ESTEOULE-CHOUX, J. 1983. Kaolinitic weathering profiles in Brittany: genesis and economic importance. *In*: WILSON, R.C.L. (ed.) *Residual Deposits: Surface Related Weathering Processes and Materials*. Geological Society, London, Special Publications, **11**, 33–38.

FOURNIER, R.O. 1983. A method of calculating quartz solubilities in aqueous sodium chloride solutions. *Geochimica et Cosmochimica Acta*, **47**, 579–586.

GABRIELSEN, R.H. & KLØVJAN, O.S. 1997. Late Jurassic–early Cretaceous caprocks of the southwest Barents Sea: fracture systems and rock mechanical properties. *In*: MOLLER-PEDERSEN, P. & KOESTLER, A.G. (eds) *Hydrocarbon Seals: Importance for Exploration and Production*. Norwegian Petroleum Society Special Publication, **7**, 73–89.

GARCIA, A.J.V., MORAD, S., DE ROS, L.F. & AL-AASM, I.S. 1998. Palaeogeographical, palaeoclimatic and burial history controls on the diagenetic evolution of reservoir sandstones: evidence from the Lower Cretaceous Serraria sandstones in the Sergipe–Alagoas Basin, NE Brazil. *In*: MORAD, S. (ed.) *Carbonate Cementation in Sandstones*. International Association of Sedimentologists, Special Publications, **26**, 107–140.

GILES, M.R. 1987. Mass transfer and problems of secondary porosity creation in deeply buried hydrocarbon reservoirs. *Marine and Petroleum Geology*, **4**, 188–201.

GLASMANN, J.R., LUNDEGARD, P.D., CLARK, R.A., PENNY, B.K. & COLLINS, I.D. 1989. Geochemical evidence for the history of diagenesis and fluid migration: Brent Sandstone, Heather Field, North Sea. *Clay Minerals*, **24**, 255–284.

GONZALEZ, G. & MIDDEA, A. 1987. Asphaltene adsorption by quartz and feldspar. *Journal of Dispersion Science and Technology*, **8**, 525–548.

GUSSOW, W.C. 1954. Differential entrapment of oil and gas: a fundamental principle. *AAPG Bulletin*, **38**, 816–853.

HARRISON, W.J. & SUMMA, L.L. 1991. Paleohydrology of the Gulf of Mexico Basin. *American Journal of Science*, **291**, 109–176.

HILLIER, S. & MARSHALL, J.E.A. 1992. Organic maturation, thermal history and hydrocarbon generation in the Orcadian Basin, Scotland. *Journal of the Geological Society, London*, **149**, 491–502.

HILLIS, R.R. 1995. Regional Tertiary exhumation in and around the United Kingdom. *In*: BUCHANAN, J.G. & BUCHANAN, P.G. (eds) *Basin Inversion*. Geological Society, London, Special Publications, **88**, 167–190.

HOOPER, R.J., GOH, L.S. & DEWEY, F. 1995. The inversion history of the northeastern margin of the Broad Fourteens Basin. *In*: BUCHANAN, J.G. &

BUCHANAN, P.G. (eds) *Basin Inversion*. Geological Society, London, Special Publications, **88**, 307–317.

INGRAM, G.M. & URAI, J.L. 1999. Top-seal leakage through faults and fractures: the role of mudrock properties. *In*: APLIN, A.C., FLEET, A.J. & MACQUAKER, J.H.S. (eds) *Muds and Mudstones: Physical and Fluid Flow Properties*. Geological Society, London, Special Publications, **158**, 125–135.

JACKSON, D.I., MULHOLLAND, P., JONES, S.M. & WARRINGTON, G. 1987. The geological framework of the East Irish Sea Basin. *In*: BROOKS, J. & GLENNIE, K. (eds) *Petroleum Geology of North West Europe*. Graham & Trotman, London, 191–203.

KASTNER, M. 1984. Control of dolomite formation. *Nature*, **311**, 410–411.

LONGSTAFFE, F.J. 1984. The role of meteoric water in diagenesis of shallow sandstones: stable isotope studies of the Milk River Aquifer and Gas Pool, southeastern Alberta. *In*: MCDONALD, D.A. & SURDAM, R.C. (eds) *Clastic Diagenesis*. Memoirs, American Association of Petroleum Geologists, **37**, 81–98.

LUO, X. & VASSEUR, G. 1995. Modelling of pore pressure evolution associated with sedimentation and uplift in sedimentary basins. *Basin Research*, **7**, 35–52.

MACGREGOR, D.S. 1993. Relationships between seepage, tectonics and subsurface petroleum reserves. *Marine and Petroleum Geology*, **10**, 606–619.

MACGREGOR, D.S. 1996. Factors controlling the destruction or preservation of giant light oilfields. *Petroleum Geoscience*, **2**, 197–217.

NAGTEGAAL, P.J.C. 1978. Sandstone-framework instability as a function of burial diagenesis. *Journal of the Geological Society, London*, **135**, 101–105.

NEDKVITNE, T. & BJØRLYKKE, K. 1992. Secondary porosity in the Brent Group (Middle Jurassic), Huldra Field, North Sea: implications for predicting lateral continuity of sandstones? *Journal of Sedimentary Petrology*, **62**, 23–34.

NEMČOK, M., GAYER, R. & MILIORIZOS, M. 1995. Structural analysis of the inverted Bristol Channel Basin: implications for the geometry and timing of fracture porosity. *In*: BUCHANAN, J.G. & BUCHANAN, P.G. (eds) *Basin Inversion*. Geological Society, London, Special Publications, **88**, 355–392.

NYLAND, B., JENSEN, L.N., SKARPNES, J., SKAGEN, O. & VORREN, T. 1992. Tertiary uplift and erosion in the Barents Sea: magnitude, timing and consequences. *In*: LARSEN, R.M. (ed.) *Structural and Tectonic Modelling and its Application to Petroleum Geology*. Norwegian Petroleum Society Special Publication, **1**, 153–162.

OSBORNE, M.J. & SWARBRICK, R.E. 1997. Mechanisms for generating overpressure in sedimentary basins: a reevaluation. *AAPG Bulletin*, **81**, 1023–1041.

PARNELL, J. 1987. Secondary porosity in hydrocarbon-bearing transgressive sandstones on an unstable Lower Palaeozoic continental shelf, Welsh Borderland. *In*: MARSHALL, J.D. (ed.) *Diagenesis of Sedimentary Sequences*. Geological Society, London, Special Publications, **36**, 297–312.

PARNELL, J. 1991. Hydrocarbon potential of Northern Ireland: Part II. Reservoir potential of the Carboniferous. *Journal of Petroleum Geology*, **14**, 143–160.

PIGGOTT, N. & LINES, M.D. 1991. A case study of migration from the West Canada Basin. *In*: ENGLAND, W.A. & FLEET, A.J. (eds) *Petroleum Migration*. Geological Society, London, Special Publications, **59**, 207–225.

PLATT, J.D. 1994. Geochemical evolution of pore waters in the Rotliegend (Early Permian) of northern Germany. *Marine and Petroleum Geology*, **11**, 66–78.

RIDER, M.H. 1996. *The Geological Interpretation of Well Logs*. Whittles Publishing, Latheronwheel.

ROBIN, M., ROSENBURG, E. & FASSI-FIHRI, O. 1995. Wettability studies at the pore level: a new approach by use of cryo-SEM. *Society of Petroleum Engineers Formation Evaluation*, **10**, 11–19.

ROSSI, C., GOLDSTEIN, R.H. & MARFIL, R. 2000. Pore fluid evolution and quartz diagenesis in the Khatatba Formation, Western Desert, Egypt. *Journal of Geochemical Exploration*, **69–70**, 91–96.

SCHMIDT, V. & MCDONALD, D.A. 1979*a*. The role of secondary porosity in the course of sandstone diagenesis. *In*: SCHOLLE, P.A. & SCHLUGER, P.R. (eds) *Aspects of Diagenesis*. Society of Economic Paleontologist and Mineralogists, Special Publications, **26**, 175–207.

SCHMIDT, V. & MCDONALD, D.A. 1979*b*. Texture and recognition of secondary porosity in sandstones. *In*: SCHOLLE, P.A. & SCHLUGER, P.R. (eds) *Aspects of Diagenesis*. Society of Economic Paleontologists and Mineralogists, Special Publications, **26**, 209–225.

SEEDHOUSE, J.K. & RACEY, A. 1997. Sealing capacity of the Mercia Mudstone Group in the East Irish Sea Basin: implications for petroleum exploration. *Journal of Petroleum Geology*, **20**, 261–286.

SHANMUGAM, G. 1988. Origin, recognition and importance of erosional unconformities in sedimentary basins. *In*: KLEINSPEHN, K.L. & PAOLA, C. (eds) *New Perspectives in Basin Analysis*. Springer, Berlin, 83–108.

SIBSON, R.H. 1995. Selective fault reactivation during basin inversion: potential for fluid redistribution through fault-valve action. *In*: BUCHANAN, J.G. & BUCHANAN, P.G. (eds) *Basin Inversion*. Geological Society, London, Special Publications, **88**, 3–19.

SOMMER, F. 1978. Diagenesis of Jurassic sandstones in the Viking graben. *Journal of the Geological Society, London*, **135**, 63–67.

SPAIN, D.R. & CONRAD, P.C. 1997. Quantitative analysis of top-seal capacity: offshore Netherlands, southern North Sea. *Geologie en Mijnbouw*, **76**, 217–226.

STRONG, G.E. & MILODOWSKI, A.E. 1987. Aspects of the diagenesis of the Sherwood Sandstones of the Wessex Basin and their influence on reservoir characteristics. *In*: MARSHALL, J.D. (ed.) *Diagenesis of Sedimentary Sequences*. Geological Society, London, Special Publications, **36**, 325–337.

SULLIVAN, M., COOMBES, T., IMBERT, P. & AHAMDACH-DEMARS, C. 1999. Reservoir quality and petrophysical evaluation of Paleocene sandstones in the West of Shetland area. *In*: FLEET, A.J. & BOLDY, S.A.R. (eds) *Petroleum Geology of Northwest Europe: Proceedings of the 5th Conference*. Geological Society, London, 627–633.

SULLIVAN, M.D., HASZELDINE, R.S., BOYCE, A.J., ROGERS, G. & FALLICK, A.E. 1994. Late anhydrite cements mark basin inversion: isotopic and formation water evidence, Rotliegend Sandstone, North Sea. *Marine and Petroleum Geology*, **11**, 46–54.

TIAB, D. & DONALDSON, E.C. 1996. *Petrophysics*. Gulf, Houston, TX.

TISSOT, B.P. & WELTE, D.H. 1984. *Petroleum Formation and Occurrence*. Springer, Berlin.

VAN WIJHE, D.H., LUTZ, M. & KAASSCHIETER, J.P.H. 1980. The Rotliegend in the Netherlands and its gas accumulations. *Geologie en Mijnbouw*, **59**, 3–24.

VERWEIJ, J.M. 1999. Application of fluid flow systems analysis to reconstruct the post-Carboniferous hydrogeohistory of the onshore and offshore Netherlands. *Marine and Petroleum Geology*, **16**, 561–579.

VERWEIJ, J.M., SIMMELINK, H.J., DAVID, P., VAN BALEN, R.T., VAN BERGEN, F. & VAN WEES, J.D.A.M. 2000. Geodynamic and hydrodynamic evolution of the Broad Fourteens Basin and the development of its petroleum systems: an integrated 2D basin modelling approach. *Journal of Geochemical Exploration*, **69–70**, 635–639.

WALKER, T.R., WAUGH, B. & CRONE, A.J. 1978. Diagenesis in first-cycle desert alluvium of Cenozoic age, southwestern United States and northwestern Mexico. *Geological Society of America Bulletin*, **89**, 19–32.

WALTON, N.R.G. 1981. *A Detailed Hydrogeochemical Study of Groundwaters from the Triassic Sandstone Aquifer of South West England*. Report of the Institute of Geological Sciences **81/5**.

WANG, F.H.L. & GUIDRY, L.J. 1994. Effect of oxidation–reduction conditions on wettability alteration. *Society of Petroleum Engineers Formation Evaluation*, **9**, 140–148.

WANG, W.H. 1992. Origin of reddening and secondary porosity in Carboniferous sandstones, Northern Ireland. *In*: PARNELL, J. (ed.) *Basins on the Atlantic Seaboard: Petroleum Geology, Sedimentology and Basin Evolution*. Geological Society, London, Special Publications, **62**, 243–254.

WILHELMS, A., LARTER, S.R., HEAD, I., FARRIMOND, P., DIPRIMIO, R. & ZWACH, C. 2001. Biodegradation of oil in uplifted basins prevented by deep-burial sterilization. *Nature*, **411**, 1034–1037.

WOOD, J.R. 1986. Thermal mass transfer in systems containing quartz and calcite. *In*: GAUTIER, D.L. (ed.) *Roles of Organic Matter in Sediment Diagenesis*. Society of Economic Paleontologists and Mineralogists, Special Publications, **38**, 169–180.

WORDEN, R.H., SMALLEY, P.C. & OXTOBY, N.H. 1998. Can oil emplacement stop quartz cementation in sandstones? *Petroleum Geoscience*, **4**, 129–138.

Uplift-related hydrocarbon accumulations: the release of natural gas from groundwater

BERNHARD CRAMER[1], STEFAN SCHLÖMER[2] & HARALD S. POELCHAU[3,4]

[1]*Federal Institute for Geosciences and Natural Resources (BGR), Stilleweg 2, 30655 Hannover, Germany (e-mail: bernhard.cramer@bgr.de)*

[2]*EniTecnologie SpA, Via F. Maritano 26, 20097 San Donato Milanese, Italy*

[3]*Institute of Petroleum and Organic Geochemistry, Forschungszentrum Jülich GmbH, D-52425 Jülich, Germany*

[4]*Present address: Kansas Geological Survey, University of Kansas, 1930 Constant Avenue, Lawrence, KS 66047, USA*

Abstract: Vertical tectonic movements often change the structural style and physico-chemical habitat of sedimentary basins. Changes in pressure, temperature and salinity of the groundwater caused by tectonic uplift may result in the release of previously dissolved gas. This process of gas exsolution from groundwater is shown to be an important mechanism in the formation of gas accumulations in uplifted basins. Two principal types of gas release are discussed. A hydrodynamic type is active when groundwater flows into areas of lower pressure or mixes with water of different temperature or salinity. It is anticipated that this effect is more of local importance, but over long periods of groundwater flow large volumes of gas may be exsolved. The hydrostatic type of gas release can occur in any sequence of sedimentary rocks where uplift causes a drop in pressure and temperature. This phenomenon may act basin-wide. Mass balance calculations show that the largest gas accumulations on Earth, such as the Urengoy field in West Siberia, could have been formed by this process.

During uplift of sedimentary rocks, maturation of sedimentary organic matter and associated hydrocarbon generation cease as a result of the drop in temperature. In addition, various uplift-related processes are known to be responsible for the destruction of petroleum reservoirs. The understanding of these mechanisms led to the evaluation of many inverted sedimentary basins as non-prospective for commercial hydrocarbon accumulations. The influence of uplift on hydrocarbon systems of sedimentary basins is much more complex and may either cause destruction of hydrocarbon accumulations or induce redistribution of hydrocarbons into new, uplift-related types of accumulations (Doré & Jensen 1996). Processes that influence the distribution of hydrocarbons in exhumed basin settings include: (1) the dismigration of hydrocarbons as a result of structural tilting and fracturing of cap rocks; (2) the diffusional losses of light hydrocarbons from the reservoir, which are not replenished as hydrocarbon generation ceases during uplift; (3) anomalous rock properties such as mature or cemented rocks at shallow depth; (4) the presence of fluids in disequilibrium leading to gas exsolution from pore water or liquid hydrocarbons, expansion of fluids, especially gas, and retrograde condensation. Of these phenomena we address here the exsolution of gas from formation water as an uplift-related process resulting in new accumulations of natural gas. After a short review of the solubility and occurrence of gas in the deep hydrosphere, the mechanisms of uplift-related gas release from groundwater are discussed.

Solubility of gas in water

The solubility of gas components in water depends on a variety of factors. Pressure (P), temperature (T), concentration and composition of inorganic components in the water, as well as the contribution of other gas components in solution are the most important known factors. A large number of measurements of gas solubility in water over wide ranges of pressure, temperature and salinity are available, with most published experimental data focused on the solubility of methane in pure water or water with a single electrolyte such as sodium chloride or

From: Doré, A.G., Cartwright, J.A., Stoker, M.S., Turner, J.P. & White, N. 2002. *Exhumation of the North Atlantic Margin: Timing, Mechanisms and Implications for Petroleum Exploration.* Geological Society, London, Special Publications, **196**, 447–455. 0305-8719/02/$15.00

calcium chloride (Culberson & McKetta 1951; O'Sullivan & Smith 1970; Sultanov *et al.* 1972; Bonham 1978; Price 1979; Cramer 1980; Price *et al.* 1981; Rettich *et al.* 1981). Some experimental data are available also for ethane (Culberson & McKetta 1950; Rettich *et al.* 1981; Crovetto *et al.* 1984). Investigations of the solubility of binary or ternary hydrocarbon gas mixtures in water are limited to moderate *P–T* conditions (Amirijafari & Campbell 1972). Only one series of measurements has been published on the solubility of a real natural gas in oilfield brines under a limited range of pressure and temperature conditions (Dodson & Standing 1945).

Calculated with a model based on experimental data (Haas 1978), Fig. 1 summarizes the solubility of methane for temperatures between 20 and 200 °C and pressures ranging from 1 to 100 MPa in brines with three NaCl concentrations. Some general conclusions can be drawn regarding the solubility of methane under pressure, temperature and salinity conditions typical for groundwater within the uppermost 6–8 km of the Earth's crust: (1) At constant temperature, the solubility of methane increases with increasing pressure. (2) Between about 60 and 90 °C the solubility of methane in water has a minimum for a constant pressure. At temperatures above this minimum, the solubility increases with rise in temperature. The influence of temperature exceeds the effect of pressure on methane solubility. (3) Increasing salinity of the brine suppresses the solubility of methane (salting out). Differences between electrolytes in the salting-out effect appear to be small compared with the overall effect. (4) Solubilities of hydrocarbon mixtures are greater than the solubilities of the pure components at the same pressure and temperature. (5) For the normal covariant rise in pressure and temperature with depth, methane solubility increases steadily, although not at constant rates.

To predict methane solubility under geologically relevant conditions a variety of mathematical models is available from the literature. The concepts for these models vary from semi-empirical equations mainly based on curve-fitting procedures (Haas 1978; Coco & Johnson 1981; Price *et al.* 1981; Battino 1984) to more theoretically based models applying the Pitzer phenomenology for the liquid phase (Barta & Bradley 1985) and an equation of state for the vapour phase (Duan *et al.* 1992). Although differences are obvious in the precision of these models and in the physicochemical conditions they cover, it is believed that they all are suited to predict methane solubility in water under *P–T* conditions relevant to sedimentary basins.

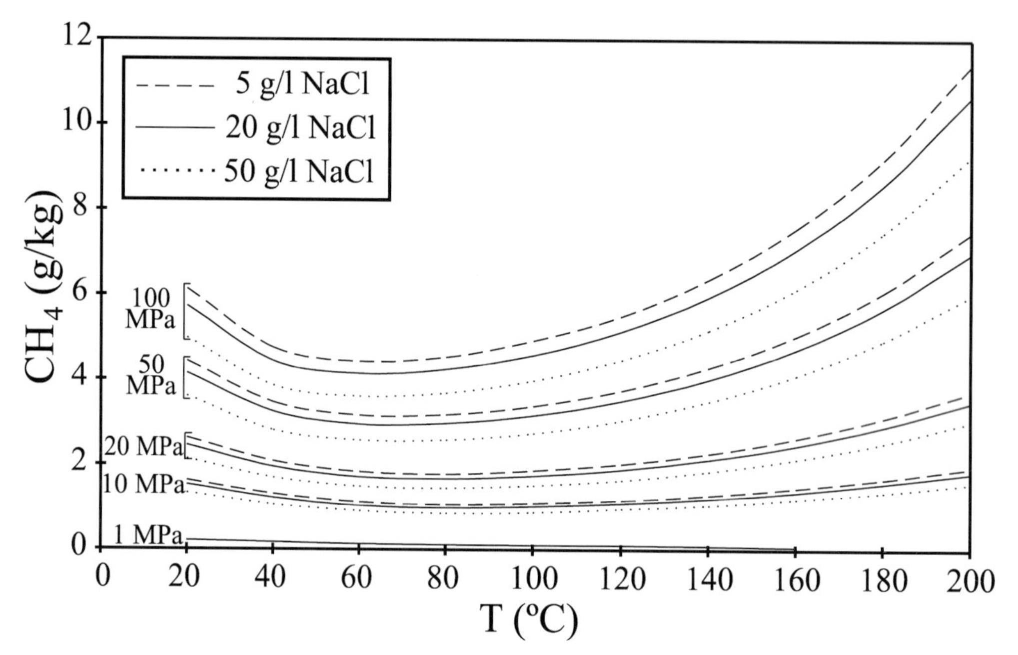

Fig. 1. Solubility of methane in water as a function of pressure, temperature and NaCl concentration, calculated with the model of Haas (1978).

Dissolved gas in the Earth's crust

The principal gas components dissolved in the crust down to the Mohorovičić discontinuity are methane, carbon dioxide, hydrogen and water vapour (Kortsenshtejn 1979). The composition of gas dissolved in near-surface groundwater is governed by the main atmospheric gas components nitrogen and oxygen. With increasing depth, the influence of the atmosphere decreases and gas from bacterial and thermal degradation of sedimentary organic matter (e.g. CH_4, CO_2), gas from recrystallization processes of minerals (e.g. N_2, Ar), as well as gas from mantle degassing (e.g. He, Ar) contribute to the dissolved gas phase of the groundwater. Secondary processes such as the decomposition or generation of individual gas components, as a result of bacterial activity or the thermal degradation of hydrocarbons, may significantly change the composition of the gas dissolved in deep groundwater. At least within the depth range of hydrocarbon generation from sedimentary organic matter, methane is by far the most important gas component dissolved in groundwater (Barkan & Yakutseni 1981).

The role of groundwater as a vast storage medium for gas in the subsurface was emphasized by Kortsenshtejn (1979). He estimated that at least 1×10^{19} m^3 gas (all volumes of gas are given in m^3 STP (standard temperature and pressure); 15.6 °C, 1.013 kPa) are dissolved in water of the subsurface hydrosphere. This equals about twice the volume of the Earth's atmospheric gas. The volume of gas dissolved in free groundwater of the upper 5 km of sedimentary

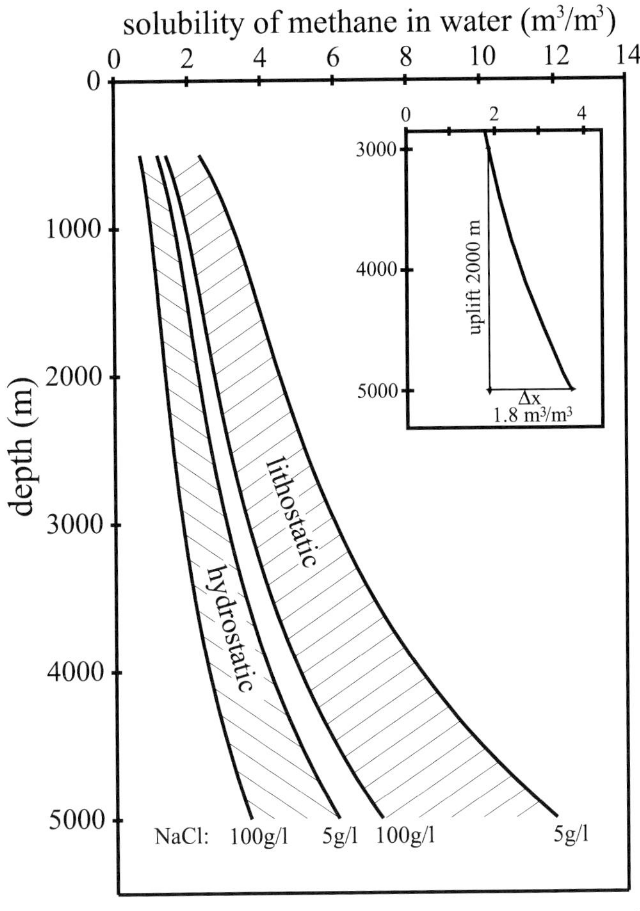

Fig. 2. Solubility regimes of methane in subsurface groundwater for hydrostatic and lithostatic pressure conditions and salinity range of 5–100 g l^{-1} NaCl. A geothermal gradient of 30 °C km^{-1} was assumed. In the example, 2 km uplift results in a drop in solubility of methane from 3.7 to 1.9 m^3 m^{-3}, with 1.8 m^3 m^{-3} released from the groundwater.

rocks is estimated to be at least $1.5 \times 10^{15}\,m^3$, with methane as the main constituent (Kortsenshtejn 1979). This is 10 times the estimated volume of the global conventional gas reserves (*c.* $0.15 \times 10^{15}\,m^3$; Barthel *et al.* 1999).

In the geological environment the factors influencing gas solubility in water described above lead to a general increase of gas solubility with increasing depth (Fig. 2). The temperature minimum below 90 °C shown in Fig. 1 is compensated in the subsurface by the effect of increasing pressure. Nevertheless, because of the temperature minimum in solubility, the temperature range up to 90 °C, corresponding to a depth down to 2500–3500 m, is characterized by a lower rate of increase in methane solubility than at greater depth (Fig. 2). Under hydrostatic pressure within the uppermost 5 km of the Earth's crust, the solubility of methane can exceed $5\,m^3\,m^{-3}$ (Fig. 2). Under lithostatic pressure $>10\,m^3\,m^{-3}$ of methane can be dissolved in the pore water at about 5 km depth.

In the past, the interest in dissolved gas in the subsurface was mainly focused on the economic potential of gas dissolved in brines (Kuuskraa & Meyers 1983; Marsden 1993) as well as on the role of dissolved gas for the deep gas potential of sedimentary basins. Because water is able to store gas effectively over a long time, dissolved gas in pore water is believed to support hydrocarbon potential even below the depth of main hydrocarbon generation (Barkan *et al.* 1984; Price 1997).

Uplift-related gas release from groundwater

Until recently, the processes responsible for gas release from groundwater in the deep hydrosphere have not been investigated in much detail. However, it is clear now that the principal reasons for gas exsolution are changes in the physicochemical habitat of the groundwater; uplift-related changes in pore pressure, temperature and salinity may cause the release of previously dissolved gas. Processes of gas release from groundwater can be classified as: (1) hydrostatic effect, not related to hydrodynamic activity; (2) hydrodynamic effect, related to the flow of groundwater. Hydrostatic gas release may occur in all sedimentary sequences where regional uplift causes a drop in subsurface pressure and temperature. Depending on the initial gas content of the pore water and on the amount of uplift, a critical point will be reached at which gas solubility drops sufficiently to initiate gas release from the water. In contrast, gas release related to hydrodynamic activity may occur if a change in pressure, temperature or salinity of the water is induced by groundwater flow, either by water flowing to a region of lower pressure or by mixing with cooler or more saline groundwater. During the exhumation of a sedimentary basin with an active hydrodynamic system, both effects, hydrostatic as well as hydrodynamic, will simultaneously contribute to the release of gas.

Hydrostatic effect

To illustrate the potential of this process to release gas, methane-saturated groundwater is assumed to have a salinity equivalent to $100\,g\,l^{-1}$ NaCl in a sedimentary formation at 5 km depth with a geothermal gradient of $30\,°C\,km^{-1}$ and hydrostatic pressure (Fig. 2). The basin is lifted 2000 m and the solubility of methane drops from *c.* 3.7 to $1.9\,m^3\,m^{-3}$. About $1.8\,m^3\,m^{-3}$ of methane could potentially be released from the water. In a water-saturated sedimentary layer of 10 m thickness with an average porosity of 8% and a lateral extent of $1\,km^2$ this uplift would release about $1.4 \times 10^6\,m^3$ of methane from the water. The duration of the uplift and the rate of upward movement are not considered in this calculation, because it is not believed that time plays a significant role in this type of gas exsolution. However, time becomes crucial when looking at the processes of gas migration to a trap and diffusional losses from an accumulation.

Hydrostatic gas release often occurs at a regional to basin-wide scale. The amount of gas released can be enormous. To evaluate the economic potential of this process, it is important to consider the geological factors promoting effective gas release from groundwater and accumulation of the gas in accessible hydrocarbon traps. Favourable conditions are: (1) thick aquifer systems with high porosity; (2) high contents of methane dissolved in the water (fully saturated) at maximum burial depth; (3) effective buoyancy-driven migration of the released gas into (4) hydrocarbon traps with effective cap rocks under prevailing conditions.

Factors (1) and (2), which determine the volumetrics of the gas release, seem to be mutually exclusive. In general, good aquifers are found at comparatively shallow depth, where the solubility of methane in water is still low. In contrast, high gas contents in the pore water are expected to be found at greater depth (Fig. 2), where the porosity of sediments is reduced. Exsolution of large volumes of gas requires that

Table 1. *Published mass balance calculations on the hydrostatic gas release during uplift*

Region	Gas field	Initial depth of aquifer (m)	Assumed amount of uplift (m)	Salinity of groundwater ($g\,l^{-1}$)	Change in solubility ($m^3\,m^{-3}$)	Methane release below drainage area (m^3)
General model (1)	–	6000	3000	350	4.1	–
General model (2)	–	6000	3000	100	8.3	–
Barents Sea (3)	Snøhvit	3400	1500	100–165	2.0	$(50–200)\times10^9$
West Siberia (4)	Urengoy	1800–4500	600–1000	10–25	≤1.7	$\leq9\times10^{12}$

1, Barkan & Yakutseni (1981); 2, Maximov *et al.* (1984); 3, Oygard and Eliassen, cited by Doré & Jensen (1996); 4, Cramer *et al.* (1999).

at least one of the two factors is favourable. Therefore, effective gas release from pore water can be expected at all depth ranges, but in the case of shallow aquifers down to *c.* 2500 m depth, sediments with high porosity are a prerequisite.

Although the formation of very large gas and gas-condensate accumulations as a result of gas release from groundwater is believed to be of global importance (Maximov *et al.* 1984), reports on actual cases are sparse. Table 1 summarizes published mass balance calculations of the effect of uplift-related gas exsolution. Whereas the first two estimates are more general in nature (Table 1), the calculations for the Snøhvit field in the Barents Sea and the Urengoy field in West Siberia are actual case studies. Doré & Jensen (1996) postulated that over the entire Barents Sea area vast amounts of gas may have been released during Plio-Pleistocene uplift and that the formation of major gas accumulations such as the Shtokmanovskoe field can be attributed to this process. From the overall model of the West Siberian Basin as a system of huge aquifers discharging to the north, Cramer *et al.* (1999) deduced that all dry gas fields in the north of West Siberia were sourced by this process, and that the region of a gas release from water extends into the Kara Sea. In summary, the entire region from the Middle Ob in West Siberia over the Kara Sea into the Barents Sea experienced uplift during Cenozoic time and was potentially subjected to release of gas from groundwater and charging of the giant gas accumulations identified in this area.

Basin-centred gas fields (such as the Alberta Deep Basin or the San Juan Basin) are also likely to have been sourced by an exsolution mechanism (Doré & Jensen 1996). These gas accumulations lie in deep parts of inverted basins, downdip from water with no apparent intervening permeability barrier. According to Price (cited by Doré & Jensen 1996) the underlying mechanism is, most probably, that gas is exsolved from water and forms static gas bubbles that block the pore throats and prevent further migration. The blocking of pore space by gas exsolved from groundwater has also been emphasized by Kuo (1997).

Hydrodynamic effect

In all cases where Darcy flow of water through a permeable rock as a result of lateral differences in pressure is active, the water may pass a point where the drop in hydrostatic pressure is sufficient to initiate a release of dissolved gas. From this point, the water will continuously release gas during its passage through the rock. This hydrodynamic gas release is not restricted to uplifted basin settings. However, the tectonic tilting of a sedimentary basin can activate hydrodynamic systems, because of an increase in pressure in the continuously subsiding region and a drop in pressure in the uplifted parts. This mechanism was shown to be active in the West Siberian Basin (Cramer *et al.* 1999), where a recent hydraulic gradient of $6\times10^{-5}\,m\;m^{-1}$ within the artesian Cretaceous aquifer causes a groundwater flow with an average linear velocity of about 20 km Ma^{-1}. The difference in pressure between the southern and the northern edge of the Urengoy anticline, which is in the flow direction of the groundwater, causes a drop in the solubility of methane of up to 0.017 $m^3\,m^{-3}$ over this distance. This amount of gas was potentially released by each cubic metre of groundwater passing the Urengoy recharge area, probably over a long time span. Cramer *et al.* (1999) calculated that up to $1 \times 10^{12}\,m^3$ of methane were released during Cenozoic uplift, as a result of the hydrodynamic effect within the recharge area of the Urengoy field. Therefore, this mechanism accounts for a considerable portion, up to 12%, of the gas in place within the Urengoy field.

Other possible mechanisms for activating groundwater flow are upward-directed water flow along faults that were opened as a result of the uplift, or an uplift-related depressurization of a sealed compartment, caused by cap rock or fault seal failure, in which large volumes of gas were dissolved in the pore water. In both cases, the subsequent mixing of groundwaters with different salinities may accelerate gas release. Methane exsolution from water may also be enhanced when water flows past salt domes and become more saline during migration (Kuo 1997).

Hydrodynamic gas release can be a regional or local phenomenon. In comparison with hydrostatic gas release, this process has a much smaller potential to generate economic gas accumulations. However, the time factor plays an important role; gas exsolution related to a long-lasting hydrodynamic system can potentially release large volumes of gas.

Efficiency of gas release from groundwater

Accumulations of natural gas are dynamic systems with continuous diffusional losses of gas through the cap rock. A process such as the release of gas from groundwater can generate significant accumulations only if the rate of charging exceeds the rate of loss. To illustrate the efficiency of gas release from groundwater, Fig. 3 displays a comparison of rates of thermal methane generation, gas release from groundwater and diffusional losses, calculated for the Urengoy gas field of West Siberia. For the calculations, the underlying model of the Urengoy field (including the thickness of the Pokur Formation and pressure and temperature data) was taken from Cramer *et al.* (1999).

It was assumed that methane thermally generated from the terrestrial organic matter of the Pokur Formation contributed to the West Siberian gas fields (Galimov 1988). To compare this gas generation with gas release from groundwater, the rate of thermal methane generation (Fig. 3) was calculated applying a specific set of reaction kinetic data for the Pokur Formation (Cramer *et al.* 1998). The continuous line in Fig. 3 indicates the generation rates (up to $3.4\,m^3\,m^{-2}\,Ma^{-1}$) for the thermal history of the Pokur Formation beneath the Urengoy field given

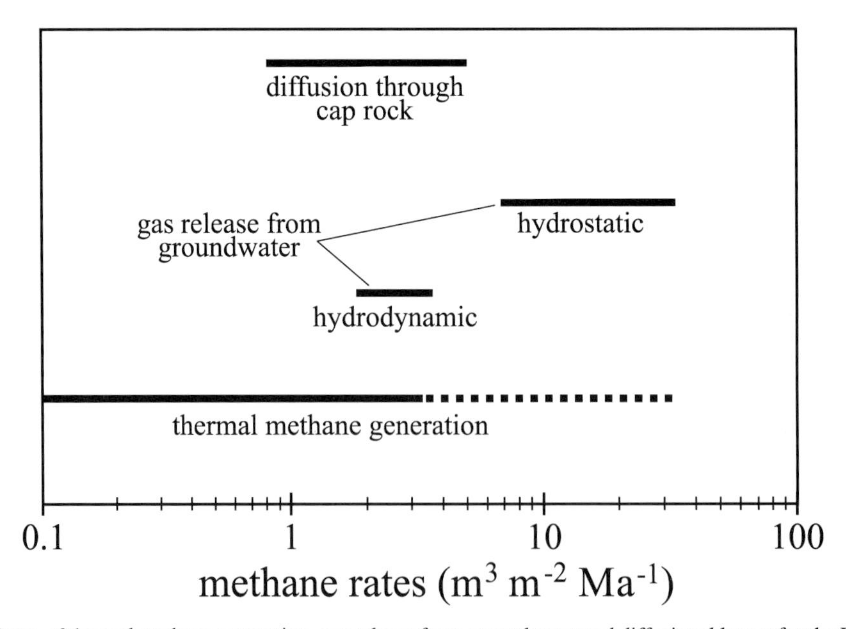

Fig. 3. Rates of thermal methane generation, gas release from groundwater and diffusional losses for the Urengoy gas field, West Siberia. Methane rates are normalised to cubic metre of methane (STP) per square metre and million years. (For further explanation see text.)

by Littke *et al.* (1999). The dotted line displays the maximum generation rates (up to $33.5\,m^3\,m^{-2}\,Ma^{-1}$) for a $2\,K\,Ma^{-1}$ heating rate. The rates for gas release from groundwater are taken from the detailed calculations of Cramer *et al.* (1999), with up to $3.6\,m^3\,m^{-2}\,Ma^{-1}$ methane released as a result of the hydrodynamic effect and up to $33.3\,m^3\,m^{-2}\,Ma^{-1}$ as a result of the hydrostatic effect.

Diffusional losses of gas from the Urengoy field (Fig. 3) were calculated based on experimental measurements of methane diffusion under *in situ* conditions through a sample of the cap rock, the Kuznetsov Formation. The methane diffusion through a water-saturated rock plug of about 1 cm thickness and about 2.8 cm diameter was measured in a triaxial flow cell at 35 °C and 10 MPa pore fluid pressure. Experimental details have been given by Schlömer (1998). The effective diffusion coefficient for the sample under investigation was determined to be $2.5 \times 10^{-10}\,m^2\,s^{-1}$. The cumulative amount of methane loss from the reservoir and the steady-state diffusion rates were calculated according to the relationship for diffusive transport through a plane sheet (Krooss *et al.* 1992a, 1992b). Considering the effective diffusion coefficient and the bulk-rock methane concentration under the relevant subsurface conditions (0.036 kg CH_4 m^{-3} rock) the highest steady-state diffusive loss rate through the 80 m rock sequence amounts to *c.* $5.0\,m^3\,m^{-2}\,Ma^{-1}$. Taking into account a larger thickness of the overlying cap rock (including the overlying 400 m thick, fine-grained rocks of the Berezov Formation) the computed rates of diffusive losses from the gas reservoir decrease to $0.8\,m^3\,m^{-2}\,Ma^{-1}$.

It should be mentioned that all methane rates presented in Fig. 3 are related to $1\,m^2$ of the reservoir area. Although diffusional losses from the accumulation are restricted to the reservoir area, the processes of charging, i.e. gas release and thermal generation, have to be related to the effective recharge area. In the case of the Urengoy anticline, this recharge area is up to 4.5 times larger than the recent area of the gas reservoir (Cramer *et al.* 1999). This enlarges the effective rates of the charging processes also by a factor of 4.5.

From the comparison of calculated methane rates it is obvious that gas release from groundwater was rapid enough to charge the Urengoy field. The rate of diffusional losses is at least one order of magnitude smaller than the gas release caused by the hydrostatic effect of uplift. In contrast, the hydrodynamic effect of gas release at the Urengoy field alone has similar rates to the diffusion loss. Also, methane generation rates from the Pokur Formation (Fig. 3, continuous line) in this case were not sufficient to keep up with diffusion loss and to charge the Urengoy field. Even if diffusional losses are neglected, the effective methane generation rate related to the recharge area cannot account for the huge gas accumulations (Schaefer *et al.* 1999). These findings also argue for gas release as the effective process of gas accumulation in the Urengoy field.

Conclusions

During subsidence of sedimentary basins gas solubility in water increases with increasing burial and thermally generated methane is continuously dissolved in the pore water. The estimated amount of methane dissolved in the Earth's groundwater by far exceeds the amount of conventional gas reserves. Subsurface water is a vast trap for long-term storage of gas in solution, because dissolved gas is excluded from rapid, buoyancy-driven migration processes. Uplift movements of basin settings can initiate the release of dissolved gas as a result of the associated drop in pressure and temperature (hydrostatic effect) and changes in groundwater flow (hydrodynamic effect). Both processes are shown to be appropriate to release sufficient gas to charge gas fields. For the huge Urengoy field of West Siberia, it is shown that hydrostatic gas release is the dominant gas charging process, overwhelming diffusional loss through the cap rock by at least one order of magnitude.

In general, the process of gas release from groundwater in exhumed basins is believed to be of global importance. Gas fields in basins that experienced uplift in the recent geological past should be re-evaluated with regard to the effect of gas release from groundwater.

References

AMIRIJAFARI, B. & CAMPBELL, J.M. 1972. Solubility of gaseous hydrocarbon mixtures in water. *Society of Petroleum Engineers Journal*, (Feb.), 21–27.

BARKAN, E.S. & YAKUTSENI, V.P. 1981. Perspective of the gas potential in great depth (in Russian). *Sovjetskaya Geologiya*, **4**, 6–15.

BARKAN, E.S., TIKHOMIROV, V.V., LEBEDEV, B.A. & ASTAF'EV, V.P. 1984. New data on the prospectivity of natural gas dissolved in brines at great depth (in Russian). *Sovetskaya Geologiya*, **2**, 11–20.

BARTA, L. & BRADLEY, D.J. 1985. Extension of the specific interaction model to include gas solubilities in high temperature brines. *Geochimica et Cosmochimica Acta*, **49**, 195–203.

BARTHEL, F., REMPEL, H., HILLER, K. & 12 OTHERS 1999. *Reserven, Ressourcen und*

Verfügbarkeit von Energierohstoffen. Bundesministerium für Wirtschaft und Technologie, BMWi-Dokumentation **465**, 62.

BATTINO, R. 1984. The solubility of methane in water between 298 and 627 K at a total pressure between 0.5 and 200 MPa. *In*: CLEVER, H.L. & YOUNG, C.L. (eds) *Methane*. Pergamon, Oxford, 24–44.

BONHAM, L.C. 1978. Solubility of methane in water at elevated temperatures and pressures. *AAPG Bulletin*, **62** (12), 2478–2488.

COCO, L.T. & JOHNSON, A.E. 1981. A correlation of published data on the solubility of methane in H_2O–NaCl solution. *In*: BEBOUT, D.G. & BACHMAN, A.L. (eds) *Geopressured–Geothermal Energy Conference. Proceedings of the 5th US Gulf Coast Conference, Baton Rouge*. Energy Programs Office, Baton Rouge, LA, 215–220.

CRAMER, B., KROOSS, B.M. & LITTKE, R. 1998. Modelling isotope fractionation during primary cracking of natural gas: a reaction kinetic approach. *Chemical Geology*, **149**, 235–250.

CRAMER, B., POELCHAU, H.S., GERLING, P., LOPATIN, N.V. & LITTKE, R. 1999. Methane release from groundwater—the source of natural gas accumulations in northern West Siberia. *Marine and Petroleum Geology*, **16**, 225–244.

CRAMER, S.D. 1980. *The solubility of methane, carbon dioxide, and oxygen in brines from 0° to 300°C*. US Bureau of Mines Report of Investigation **8706**.

CROVETTO, R., FERNANDEZ-PRIMI, R. & JAPAS, M.L. 1984. The solubility of ethane in water up to 473 K. *Berichte der Bunsengesellschaft für Physikalische Chemie*, **88**, 484–488.

CULBERSON, O.L. & MCKETTA, J.J.J. 1950. Phase equilibria in hydrocarbon–water systems—II—the solubility of ethane in water at pressures to 100.000 psi. *Petroleum Transactions, American Institute of Mining, Metallurgical and Petroleum Engineers*, **189**, 319–322.

CULBERSON, O.L. & MCKETTA, J.J.J. 1951. Phase equilibria in hydrocarbon–water systems—III—the solubility of methane in water at pressures to 10,000 PSIA. *Petroleum Transactions, American Institute of Mining, Metallurgical and Petroleum Engineers*, **192**, 223–226.

DODSON, C.R. & STANDING, M.B. 1945. Pressure–volume–temperature and solubility relations for natural-gas–water mixtures. *Drilling and Production Practice*, **1944**, 173–179.

DORÉ, A.G. & JENSEN, L.N. 1996. The impact of late Cenozoic uplift and erosion on hydrocarbon exploration: offshore Norway and some other uplifted basins. *Global and Planetary Change*, **12**, 415–436.

DUAN, Z., MOLLER, N., GREENBERG, J. & WEARE, J.H. 1992. The prediction of methane solubility in natural waters to high ionic strength from 0 to 250 °C and from 0 to 1600 bar. *Geochimica et Cosmochimica Acta*, **56**, 1451–1460.

GALIMOV, E.M. 1988. Sources and mechanisms of formation of gaseous hydrocarbons in sedimentary rocks. *Chemical Geology*, **71**, 77–95.

HAAS, J.L., JR 1978. *An empirical equation with tables of smoothed solubilities of methane in water and aqueous sodium chloride solutions up to 25 weight percent, 360 °C, and 138 MPa*. US Geological Survey Open File Report **78-1004**.

KORTSENSHTEJN, V.N. 1979. An estimate of global reserves of gas in the subsurface hydrosphere. *Doklady Akademii Nauk SSSR*, **235**, 223–224.

KROOSS, B.M., LEYTHAEUSER, D. & SCHÄFER, R.G. 1992*a*. The quantification of diffusive hydrocarbon losses through cap rocks of natural gas reservoirs—a re-evaluation. *AAPG Bulletin*, **76**, 403–406.

KROOSS, B.M., LEYTHAEUSER, D. & SCHÄFER, R.G. 1992*b*. The quantification of diffusive hydrocarbon losses through cap rocks of natural gas reservoirs—a re-evaluation: reply. *AAPG Bulletin*, **76**, 1842–1846.

KUO, L.-C. 1997. Gas exsolution during fluid migration and its relation to overpressure and petroleum accumulation. *Marine and Petroleum Geology*, **14** (3), 221–229.

KUUSKRAA, V.A. & MEYERS, R.F. 1983. Review of world resources of unconventional gas. *The Fifth IIASA Conference on Energy Resources*. International Institute for Applied System Analysis, Laxenberg, Austria, 409–458.

LITTKE, R., CRAMER, B., GERLING, P., LOPATIN, N.V., POELCHAU, H.S., SCHAEFER, R.G. & WELTE, D.H. 1999. Gas generation and accumulation in the West Siberian Basin. *AAPG Bulletin*, **83** (10), 1642–1665.

MARSDEN, S. 1993. A survey of natural gas dissolved in brine. *In*: HOWELL, D.G. (ed.) *The Future of Energy Gases*. US Geological Survey, Professional Papers, **1570**, 471–492.

MAXIMOV, S.P., ZOLOTOV, A.N. & LODZHEVSKAYA, M.I. 1984. Tectonic conditions for oil and gas generation and distribution on ancient platforms. *Journal of Petroleum Geology*, **7** (3), 329–340.

O'SULLIVAN, T.D. & SMITH, N.O. 1970. The solubility and partial molar volume of nitrogen and methane in water and aqueous sodium chloride from 50 to 125° and 100 to 600 Atm. *Journal of Physical Chemistry*, **74** (7), 1460–1466.

PRICE, L.C. 1979. Aqueous solubility of methane at elevated pressures and temperatures. *AAPG Bulletin*, **63**, 1527–1533.

PRICE, L.C. 1997. Origins, characteristics, evidence for, and economic viabilities of conventional and unconventional gas resource bases. *Geologic Controls of Deep Natural Gas Resources in the United States*. US Geological Survey Bulletin, **2146**, 181–207.

PRICE, L.C., BLOUNT, C.W., MACGOWAN, D. & WENGER, L. 1981. Methane solubility in brines with application to the geopressure resource. *In*: BEBOUT, D.G. & BACHMAN, A.L. (eds) *Geopressured–Geothermal Energy Conference. Proceedings of the 5th US Gulf Coast Conference, Baton Rouge*. Energy Programs Office, Baton Rouge, LA, 205–214.

RETTICH, T.R., HANDA, Y.P., BATTINO, R. & EMMERICH, W. 1981. Solubility of gases in liquids. 13. High-precision determination of Henry's constant for methane and ethane in liquid water at 275 to 328 K. *Journal of Physical Chemistry*, **85**, 3230–3237.

SCHAEFER, R.G., GALUSHKIN, Y., KOLLOFF, A. & LITTKE, R. 1999. Reaction kinetics of gas generation in selected source rocks of the West Siberian Basin: implications for the mass balance of early-thermogenic methane. *Chemical Geology*, **156**, 41–65.

SCHLÖMER, S. 1998. *Sealing Efficiency of Pelitic Rocks—Experimental Characterisation and Geological Relevance* (in German). Berichte des Forschungszentrums Jülich GmbH **3596**, 204.

SULTANOV, R.G., SKRIPKA, V.G. & NAMIOT, A.Y. 1972. Solubility of methane in water at elevated temperatures and pressures (in Russian). *Gazovaya Promyshlennost'*, **17**, 6–7.

Depressurization of hydrocarbon-bearing reservoirs in exhumed basin settings: evidence from Atlantic margin and borderland basins

D. V. CORCORAN[1] & A. G. DORÉ[2]

[1]*Statoil Exploration (Ireland) Ltd, Statoil House, 6, George's Dock, IFSC, Dublin, Ireland (e-mail: DVC@statoil.com)*

[2]*Statoil (UK) Ltd, 11a Regent Street, London SW1Y 4ST, UK*

Abstract: Depressurization of reservoirs in petroliferous basins commonly occurs through cap-rocks at structural crests where pore pressures are locally elevated because of either the presence of a hydrocarbon column or the redistribution of overpressures by water flow along laterally extensive inclined permeable aquifers. In exhumed petroliferous basins this deflation of excess pore pressures is enhanced by the denudation process, which results in the large-scale removal of overburden during regional uplift. Evidence from the exhumed basins of the Atlantic margin indicates that hydrocarbon accumulations in these basins are commonly characterized by underfilled traps and hydrostatically pressured or modestly overpressured reservoirs. These observations are reviewed in the context of the generic mechanisms by which top-seals leak, the properties of cap-rocks and the physical processes that occur during exhumation.

Water-wet shaly cap-rocks can form a capillary seal to a hydrocarbon column while simultaneously accommodating brine flow and equilibration of pressures between the reservoir and the top-seal. In contrast, thick, low-permeability shale or evaporite sequences may form pressure seals that restrict vertical brine and hydrocarbon flow and prevent the equilibration of aquifer pressures above and below the seal. In any sedimentary basin, the presence of regional pressure seals can result in a layered hydrogeological regime with hydrostatically pressured strata decoupled from over- or underpressured cells. Recently exhumed basins typically show limited overpressuring and in a number of these basins underpressured reservoirs have been described. Post-exhumation overpressure generation is primarily driven by tectonic compression, aquathermal pressuring and hydraulic head.

The fluid retention capacity of any cap-rock lithology during exhumation is dependent upon the physical and mechanical characteristics of the cap-rock at the time of exhumation and the timing and conditions of the associated deformation relative to the timing of hydrocarbon emplacement. The permeability and deformational characteristics of halite render it an excellent cap-rock with a high retention capacity, even under conditions of exhumation. However, mudrocks may also form effective cap-rocks in exhumed basins when the deformation associated with exhumation occurs before embrittlement and the shale cap-rock exhibits ductile behaviour.

Shale and evaporite cap-rocks form the main regional seals to hydrocarbon accumulations in exhumed basins of the Atlantic margin and borderlands. *Syn-exhumation* top-seal efficiency (fluid retention capacity) is a major exploration risk in these basins, although *post-exhumation* top-seal integrity in these basins may be relatively high under certain conditions. Consequently, a major exploration risk factor in exhumed basin settings pertains to the limited hydrocarbon budget available post-regional uplift and the efficiency of the remigration process.

Abnormal pore pressures are commonly observed in sedimentary basins. These abnormal pore pressures are denoted as *overpressures* where fluid pressures are in excess of the hydrostatic gradient at a specific depth and are denoted as *underpressures* where the fluid pressure is less than hydrostatic (Fig. 1) (Martinsen 1994). Present-day pore pressure distributions are a function of the mechanism of generation of the abnormal pore pressure conditions and the hydraulic evolution of the basin during and after the creation of these abnormal pore pressure conditions. In petroliferous basins, pore pressure evolution has an important bearing on the generation, migration and retention of petroleum in hydrocarbon traps. For example, differential pressures between the hydrocarbon source rock and carrier bed are

From: DORÉ, A.G., CARTWRIGHT, J.A., STOKER, M.S., TURNER, J.P. & WHITE, N. 2002. *Exhumation of the North Atlantic Margin: Timing, Mechanisms and Implications for Petroleum Exploration*. Geological Society, London, Special Publications, **196**, 457–483. 0305-8719/02/$15.00

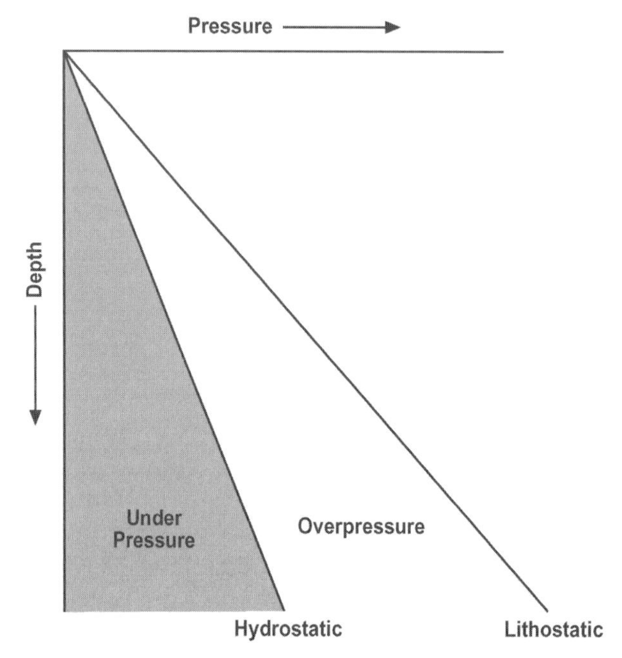

Fig. 1. Abnormal formation pressures; with pore pressure plotted v. depth. For any depth, formation pressures that are less than the hydrostatic pressure are termed underpressures and formation pressures in excess of hydrostatic are termed overpressures.

recognized as the driving force of primary migration (Magara 1968; Palciauskas & Domenico 1980). The role of overpressures in the secondary migration and remigration of hydrocarbons has been highlighted by Giles *et al.* (1999), Iliffe *et al.* (1999) and others. Temporal changes in hydraulic gradients, carrier bed interconnectivity, cap-rock and fault seal properties, and basin structure influence migration of oil and gas and the rates of ingress and egress of petroleum from the hydrocarbon trap.

Pore pressure increases above the hydrostatic level can be maintained only when the pore fluids, both hydrocarbons and brine, are retained in the reservoir by cap-rocks with sufficiently low permeability. Bradley & Powley (1994), in their review of pressure compartments in sedimentary basins, have differentiated *open hydraulic systems*, which are characterized by fluid (brine, oil and gas) continuity throughout, hydrostatic pressures in the aquifer and possible brine flow across a capillary seal, from *closed hydraulic systems*, which manifest no fluid continuity across the bounding seals, may be under-, over- or normally pressured and have no fluid flow across the pressure seal. Changes in the pore volume and fluid volume of any hydraulic system result from changes in temperature, effective stress and pore pressure, which are induced by uplift and denudation or subsidence and deposition. A closed hydraulic system will respond to these 'external' forces by a change in fluid pressure whereas an open hydraulic system will respond by fluid flow in or out of the system.

In the context of petroliferous basins, top-seals can be defined as rocks that prevent the vertical migration of hydrocarbons out of traps. Under conditions of a closed hydraulic system top-seals may also act as a barrier to brine flow (a pressure seal) and prevent the dissipation of elevated pore pressures in the reservoir. During the evolution of a petroliferous basin any lithology can serve as a top-seal for a hydrocarbon accumulation provided that its capillary entry pressure exceeds the upwards buoyancy pressure exerted by the hydrocarbon column in the underlying accumulation. In practice, however, the vast majority of effective seal rocks are evaporites, fine-grained clastic rocks and organic-rich mudrocks (Downey 1994).

The basic physical principles governing the effectiveness of petroleum cap-rocks and pressure seals are well established (Hubbert 1953; Berg 1975; Schowalter 1979; Watts 1987; Bradley & Powley 1994). Lithology, uniformity of stratigraphy and thickness are factors that influence seal capacity (Downey 1984). However, the fundamental rock properties that control seal performance are the capillary entry pressure of the seal (dominantly controlled by pore-throat diameter) and the ductility of the seal rock, which is a function of pressure, temperature and lithology. Although top-seal integrity is recognized as a major exploration risk factor in exhumed basins, few systematic studies of top-seal performance in these settings have been published to date (Gabrielsen & Kløvjan 1997; Seedhouse & Racey 1997; Spain & Conrad 1997; Cowan *et al.* 1999).

In all petroliferous basins the adequacy of the hydrocarbon charge together with the timing and rate of fill, spill and vertical or lateral leakage are key determinants of the present-day in-place oil and gas volumes preserved in hydrocarbon traps. However, in exhumed basins, the interplay between top-seal performance and hydrocarbon fill, spill and leakage is more critical as the 'switching off' of hydrocarbon generation during regional uplift may result in a lower probability of trap replenishment post-exhumation (Doré & Jensen 1996). Physical processes that may affect cap-rocks during exhumation include erosion, tectonic deformation, shear failure, hydrofracturing as a result of disequilibrium pore pressure conditions and a changing hydrodynamic regime. Furthermore, there is an increased risk of net hydrocarbon losses because of diffusion where gas accumulations are dependent upon porous and permeable shale seals.

The primary focus of this paper is to offer some insights with respect to the physical properties and processes that affect top-seal performance in exhumed basin settings. The implications of regional uplift for abnormal pore pressures (overpressures and underpressures) are also reviewed. Empirical evidence from exhumed basins of the Atlantic margin and borderland basins is then discussed in the context of these observations.

Top-seal leakage mechanisms: a summary

There are four generic mechanisms by which top-seals leak: tectonic breaching, capillary leakage, hydraulic leakage and molecular transport (diffusion). Figure 2 is a summary of the physical principles governing these mechanisms, which have been articulated by numerous workers (e.g. Schowalter 1979; Gretener 1981; Krooss *et al.* 1992; Davis & Reynolds 1996). These four generic leakage mechanisms are here reviewed in the context of exhumed basin settings of the Atlantic margin.

Tectonic breaching

Where deformation of a cap-rock occurs post-emplacement of hydrocarbons there is an increased risk of tectonic breaching and cap-rock leakage. Top-seal failure via tectonic breaching is the most readily recognized form of seal failure at the scale of seismically defined hydrocarbon traps. In addition, tectonic breaching may be facilitated by sub-seismic resolution faulting where the top-seal consists of mudrock layers interbedded with permeable siltstones and sandstones. Tectonic framework, basin evolution and fault reactivation are also important controls on fluid flow and distribution of overpressures in sedimentary basins. For example, in the North Sea Central Graben, pre-Cretaceous structure controls the flow of pore fluids below the regional Jurassic–Cretaceous pressure seal, with vertical fluid escape, through this pressure seal, focused on axial fault block crests (Darby *et al.* 1998).

The style and magnitude of tectonic deformation in any sedimentary basin is influenced by a number of factors, including plate tectonic setting, pre-existing structural grain and the presence or absence of detachment layers. During basin inversion, compressional, transpressional or reactivated extensional deformation may result in leakage through cross-fault juxtaposition of reservoirs from different stratigraphic levels (Fig. 2a(i)) or through the development of a connected network of juxtaposed leaky beds within the cap-rock interval (Fig. 2a(ii)). In addition, radial extension fractures will develop above the neutral surface of inversion folds and may result in leakage into the overlying sediments (Fig. 2a(iii)). Hall *et al.* (1997) have shown from case studies in the deep Central Graben of the North Sea that reservoir objectives lying above or close to the neutral surface of an inversion fold have a higher probability of being breached than reservoirs lying below.

Tectonic breaching is an important leakage mechanism in the exhumed basin settings of the Atlantic margin where syn-exhumation extensional fault reactivations are probable in addition to the overprint of compressional deformation resulting from the far-field signature of Alpine orogenesis and ridge-push phenomena (Underhill 1991; Murdoch *et al.* 1995; Doré *et al.* 1999).

Capillary leakage

The driving force for petroleum movement in the subsurface is buoyancy influenced by overpressure and hydrodynamics. The force opposing the movement of petroleum is the capillary

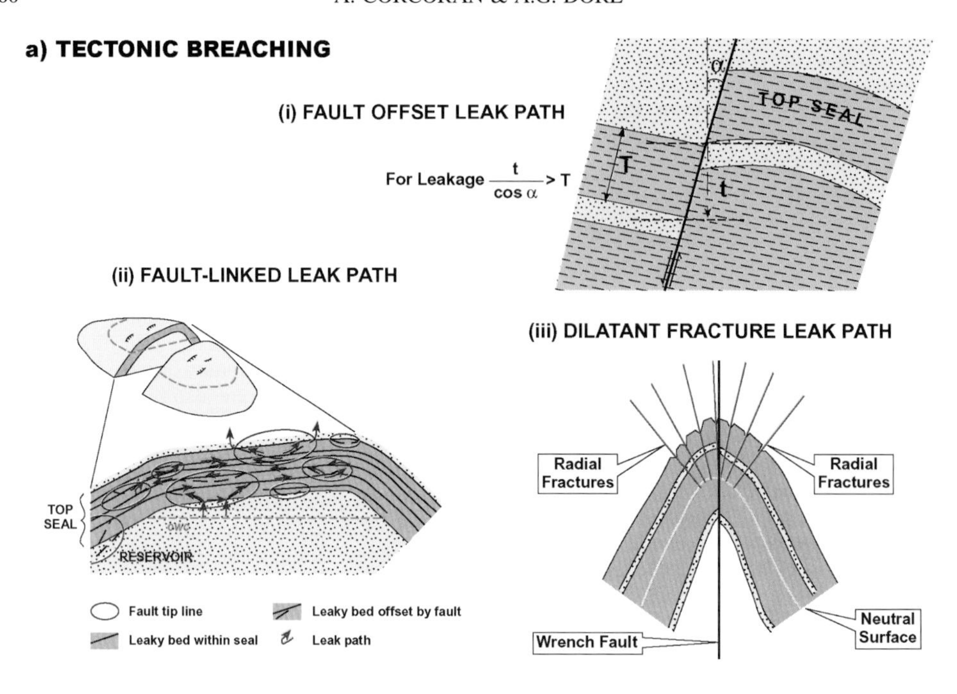

b) CAPILLARY LEAKAGE

P_{buoy} = buoyancy pressure HC column
ρ_w = density of formation water
ρ_{HC} = density of hydrocarbons
h = height of HC column
g = gravitational constant

P_{cap} = capillary entry pressure
Υ = HC-water interfacial tension
θ = contact angle against solid
R = pore throat radius

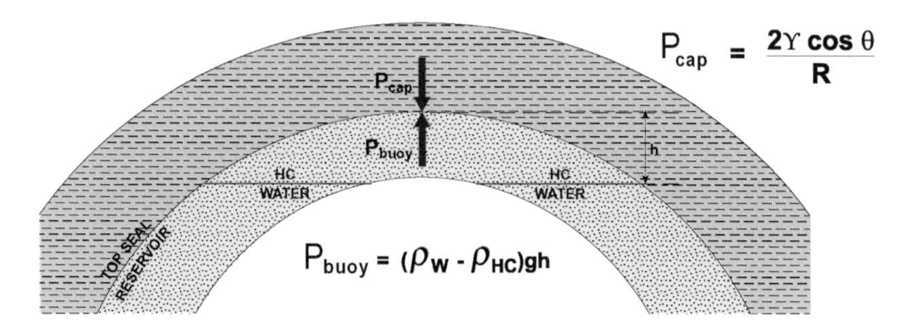

$$\text{MAX. HC COL. HEIGHT} \quad h = \frac{2\Upsilon \cos\theta}{R(\rho_W - \rho_{HC})g} - \frac{\Delta U}{(\rho_W - \rho_{HC})g}$$

ΔU = excess overpressure in reservoir relative to seal

Fig. 2. The four generic mechanisms by which top-seals leak. (**a**) Tectonic breaching; there are three common modes: (i) fault offset leak path; (ii) fault-linked leak path; (iii) dilatant fracture leak path. (**b**) Capillary leakage (for capillary leakage to occur the buoyancy force generated by the hydrocarbon column (P_{buoy}), plus any excess overpressure (ΔU) in the reservoir relative to the seal, must exceed the capillary resistance of the porous, water-wet, cap-rocks (P_{cap})). (**c**) Hydraulic leakage; this can result from the development of tension fractures (hydrofractures), which arise from changing effective stress conditions. Tension fractures occur under conditions

c) HYDRAULIC LEAKAGE

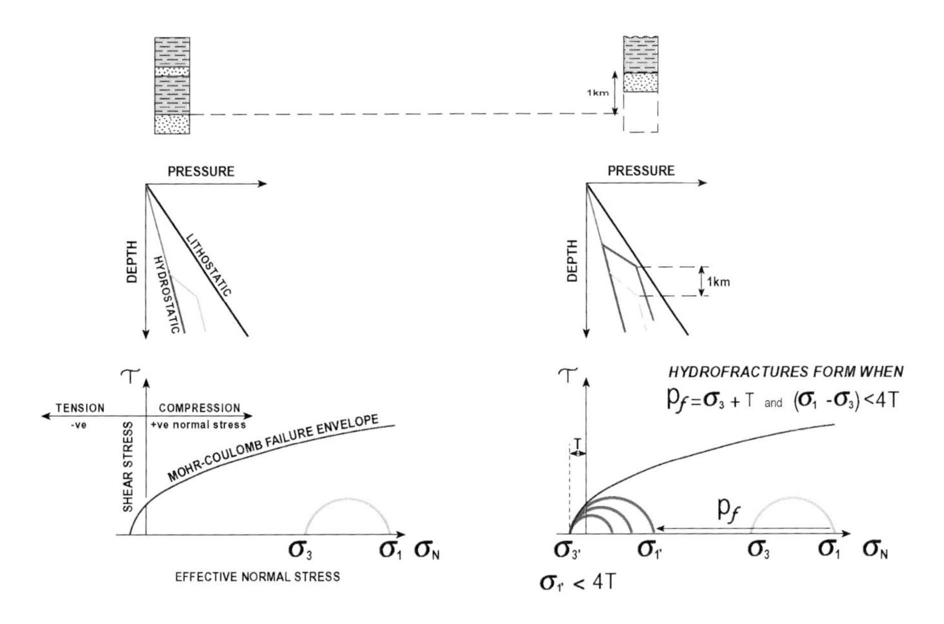

d) MOLECULAR TRANSPORT

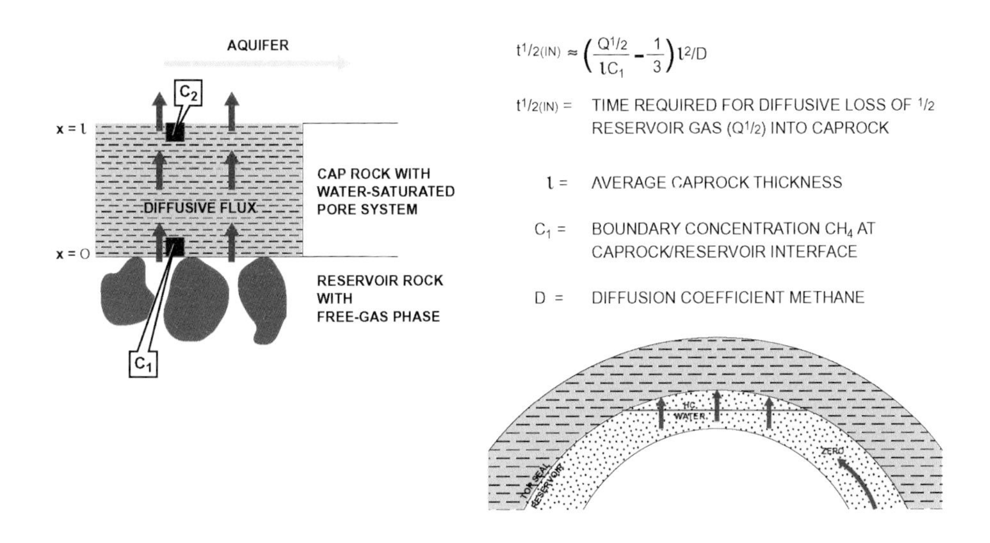

Fig 2. – *continued*

of low differential stress (small $\sigma_1 - \sigma_3$), when pore-fluid pressure in the cap-rock reduces the minimum effective horizontal stress below zero to the tensile strength of the rock. (**d**) Molecular transport; this is primarily the diffusion of methane through water-saturated shaly cap-rocks. (Adapted from Krooss *et al.* 1992; Davis & Reynolds 1996; Hall *et al.* 1997; Ingram & Urai 1999; capillary leakage equations from Schowalter 1979; Clayton & Hay 1994.)

resistance of porous rocks (Fig. 2b). Standard equations have been developed to describe these opposing forces P_{buoy} and P_{cap} at the interface of a hydrocarbon reservoir and cap-rock (Hubbert 1953; Berg 1975; Schowalter 1979; Watts 1987; Clayton & Hay 1994). Theoretically, for capillary leakage to occur the upwards buoyancy pressure of a hydrocarbon column plus any excess overpressure or hydraulic head must exceed the P_{cap} of the top-seal. However, recent discussions in the literature suggest that, in the case of a continuous water phase between a water-wet petroleum reservoir and a water-wet top-seal, overpressure in the aquifer will not contribute to capillary leakage when there is capillary pressure–gravity equilibrium throughout the hydrocarbon column (Bjørkum *et al.* 1998). Where a finite permeability to water is present at the junction of an oil column with a top-seal then there will be a minute but finite water flow from the reservoir into the cap-rock, which induces a dynamic pressure drop in the water phase of the reservoir, thereby increasing capillary pressure (Rodgers 1999). In contrast, Clayton (1999) argued that there is no continuity in the water phase, which is immobile, even in a water-wet reservoir, and that overpressure in the aquifer will indeed contribute to pushing hydrocarbons through a water-wet cap-rock.

Clayton & Hay (1994) have modelled the capillary seal capacity of a mudstone seal in a continuously subsiding basin, based on appropriate figures for interfacial tension, contact angle, largest interconnected pore-throat radii and subsurface density difference of gas and water (Fig. 3). The computed seal capacity curve for methane indicates that the modelled mudstone would retain a gas column of 500–1000 m depending upon depth.

In an exhumed basin the predicted capillary retention capacity of an average mudstone is likely to be higher at any present-day depth, as the higher compaction state of the exhumed mudstone will result in smaller interconnected pore throats. With respect to the exhumed basins of the Atlantic margin, the magnitude of all known gas columns discovered to date is less than 500 m, considerably less than the capillary seal retention capacity modelled for an average mudstone (Fig. 3).

Hydraulic leakage

Where the capillary entry pressures to a cap-rock (evaporite or super-tight shale) are so high that capillary failure is implausible, hydraulic leakage may occur through brittle top-seals as a result of the generation of new tension fractures (hydrofractures), shear fractures or the dilation of pre-existing fault planes (Fig. 2c). Hydraulic fracturing can occur independent of tectonic breaching and results from changes in effective stress conditions in the cap-rock. These changes may be induced by the development of disequilibrium pore pressure conditions or by changes in the tectonic load (Fig. 4). For example, a reduction in the minimum effective compressive stress (σ_3), induced by extension during regional uplift, may *per se* result in the formation of dilatant shear fractures of certain orientation within the cap-rock (Fig. 4a). Shear fractures will also be formed when the prevailing stress field results in conditions of high differential stress in the cap-rock (large $\sigma_1 - \sigma_3$). In this scenario the gradual elevation of pore-fluid pressures before exhumation, or the removal of overburden without the re-equilibration of elevated pore-fluid pressures, will result in Coulomb failure along planes in the rock that make appropriate angles with σ_1 (Fig. 4b). Pre-existing fractures, joints or faults (in fact, any planes of reduced cohesion) have important implications for the mechanical behaviour of cap-rocks during exhumation. When fractures are present their physical characteristics and their orientation must be known, to evaluate their structural significance (Gretener 1981).

Hydrofractures occur under conditions of low differential stress when pore-fluid pressure at the cap-rock–reservoir interface reduces the minimum effective horizontal stress below zero to the tensile strength of the rock (Fig. 2c). In extensional basins, where the minimum compressive stress (σ_3) is significantly less than the maximum compressive stress (σ_1), these hydrofractures are invariably vertical to semi-vertical in orientation and are perpendicular to the minimum horizontal stress (σ_3). For hydrofractures to develop in preference to shear fractures the conditions $P_f = \sigma_3 + T$ and $\sigma_1 - \sigma_3 < 4T$ must be satisfied (P_f is the pore-fluid pressure, σ_3 is the minimum horizontal stress and T is the tensile strength of the cap-rock) (Hubbert & Rubey 1959; Secor 1965; Sibson 1995). These conditions can occur in highly overpressured systems undergoing continuous subsidence or during exhumation, when rapid denudation, without re-equilibration of overpressures, results in tensile failure. In either case, when this state prevails, pervasive tension fractures may develop in the cap-rock, which will result in a catastrophic loss of any pre-existing hydrocarbon fill and the potential re-equilibration or reduction of overpressures. It has been argued by Bjørkum *et al.* (1998) that, in water-wet reservoir and cap-rock systems, the buoyancy force (overpressure)

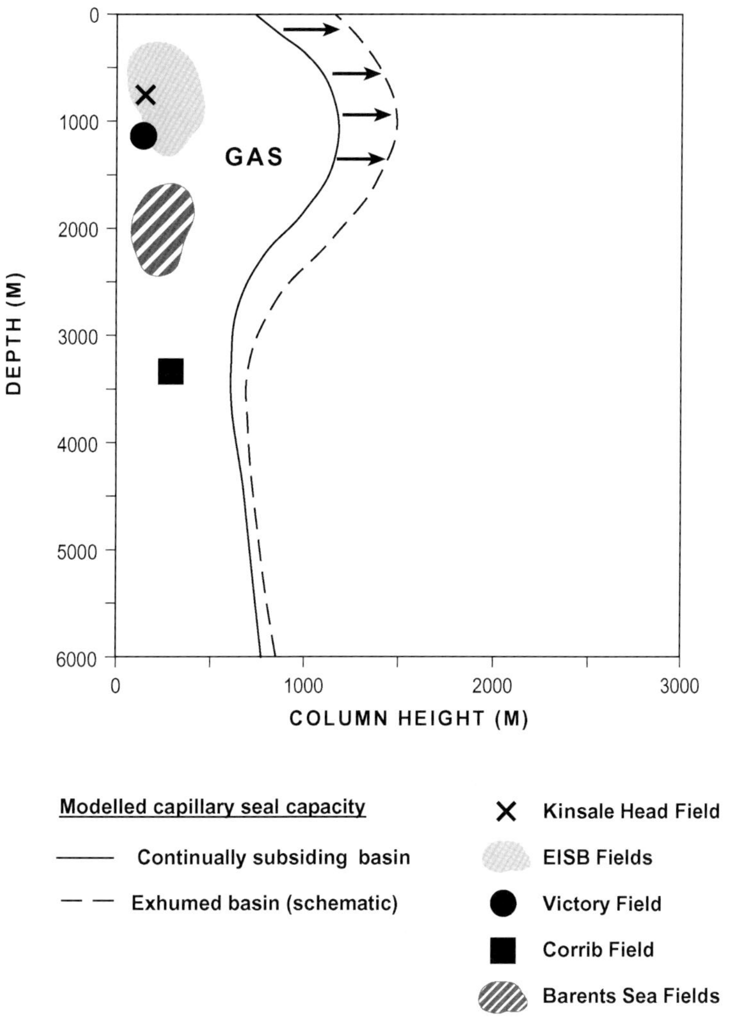

Fig. 3. Modelled capillary seal capacity with depth for a mudstone in a continuously subsiding basin (from Clayton & Hay 1994). The inferred equivalent curve for an exhumed basin setting is schematically shown (dashed line) together with the magnitude of the maximum hydrocarbon columns in gas accumulations from some exhumed Atlantic margin and borderland basins.

exerted by the hydrocarbon column relative to the water phase is balanced by the downward-directed interfacial tension force at the fluid interface and that the presence of a hydrocarbon column will not increase the risk of hydrofracturing the top-seal to the overpressured reservoir.

Hubbert & Rubey (1959) demonstrated that when the pore-fluid pressure in a sedimentary basin approaches the lithostatic pressure the fluid pressure is released by rock failure. Palciauskas & Domenico (1980) supported these observations by showing theoretically that microfractures can develop in overpressured sedimentary beds while undergoing burial. Capuano (1993) provided direct petrographic evidence of the occurrence of microfractures *in situ* at depths of 3–5 km, in geopressured Oligocene shales of the Gulf Coast Basin. Furthermore, the computed fracture permeabilities (of the order of 10^{-13} m^2) in these shales combined with paragenetic relationships indicate that fluid flow occurred preferentially through these microfractures rather than through the matrix of these shales. The development of these hydrofractures during burial is facilitated by a mechanism of episodic tensile failure in a low differential stress environment (Sibson 1995). Under these

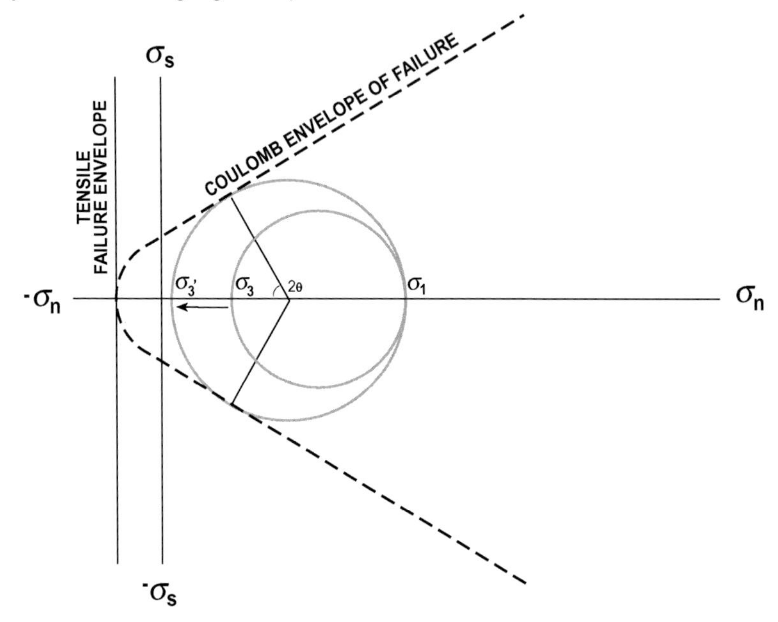

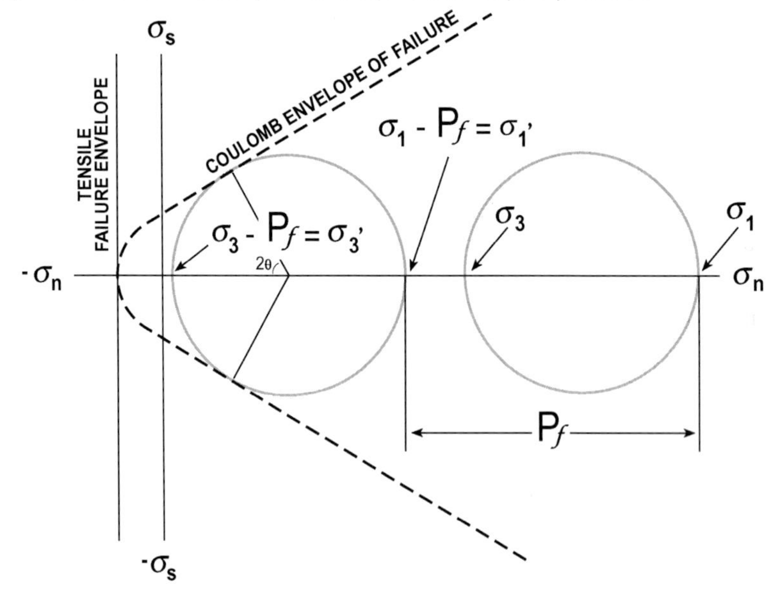

Fig. 4. Mohr circle representation of the development of dilatant shear fractures during exhumation: (**a**) induced by extension (reduction of minimum compressive stress, σ_3) during regional uplift, Coulomb failure in shear will occur along planes that make an angle θ with the orientation of σ_1; (**b**) under conditions of high differential stress (large $\sigma_1 - \sigma_3$) the removal of overburden during exhumation, without the re-equilibration of pore-fluid pressures, will result in the residual overburden load being disproportionally carried by the pore-fluid pressure (P_f), a lowering of the effective stress levels to σ_1' and σ_3', and Coulomb failure in shear along planes that make an angle θ with the σ_1 axis.

conditions, the raising of the pore-fluid pressure will result in the cracking of the rock in tension and the release of the fluid pressure, followed by a raising of fluid pressure and a repetition of this cycle (Fig. 2c).

In contrast to this pressure cyclicity, which is manifest in a continuously subsiding basin, depressurization of a reservoir during exhumation could potentially occur as a 'singular' catastrophic event. This may arise because of the reduced potential for build-up of overpressures, as some of the processes that produce overpressure (Osborne & Swarbrick 1997), such as disequilibrium compaction, dehydration reactions and kerogen transformation, will have been arrested once exhumation begins. Once catastrophic failure via fracturing has occurred during exhumation, the cap-rock unit can only regain 'seal status' when the high-permeability open fractures are healed or annealed. Fracture closure, during or post-exhumation, can occur through a range of mechanisms, including cementation and increased horizontal compressive stress. Cementation in the fracture may be caused by the cooling of upward-flowing fluids with the resulting redistribution of silica and other mineral phases or, in the absence of fluid flow, the chemical diffusion of solids into the fracture driven by local thermodynamic potentials (Pedersen & Bjørlykke 1994).

Molecular transport (diffusion)

Diffusion is a continuous and ubiquitous process in sedimentary basins and its role in hydrocarbon migration has been analysed by several workers (Fig. 2d) (Krooss *et al.* 1992; Montel *et al.* 1993; Schlomer & Krooss 1997). Migration of hydrocarbons via diffusion transport can be important under geological conditions that prohibit separate phase flow of hydrocarbons (Mann *et al.* 1997). Additional prerequisites include the presence of hydrocarbon components that have a high water solubility and a high concentration of these components in a specific area. Gas exsolution from groundwater is potentially an important mechanism for the formation of gas accumulations in uplifted basins (Cramer & Poelchau 2002).

The diffusive transport mechanism primarily pertains to the dismigration of natural gas accumulations in certain circumstances and has little relevance for oil dismigration because of the increased size of oil molecules relative to shale pore-throat dimensions. Molecular diffusion can occur independent of the pressure conditions in the reservoir and cap-rock. However, pressure is an important control on the solubility of methane in formation waters (Cramer & Poelchau 2002). Although the modelling studies of Kettel (1997) indicated that methane diffusion constants for rock salt are non-zero over geological time scales, diffusion losses from gas accumulations capped by thick evaporitic seals are considered minimal. Empirical observation of long-lived gas accumulations in Upper Proterozoic reservoirs sealed beneath Lower Cambrian salt in the extensively exhumed Lena–Tunguska province of the former Soviet Union supports this conclusion (Kontorovitch *et al.* 1990).

In contrast, Letythaeuser *et al.* (1982) have demonstrated that gas may diffuse through water-saturated cap-rocks over geological time scales. This diffusion model suggests that the evolution and preservation of natural gas accumulations is dependent upon the ratio of gas supply to the trap and gas losses through the cap-rock. For example, a case study of the Harlingen gas field, offshore Netherlands, indicated that half the 68 BCF (billion cubic feet) contained in the Lower Cretaceous reservoir would be lost by diffusion through the shales and marls of the Hauterivian cap-rock (390 m thick) in 4.5 Ma (Letythaeuser *et al.* 1982). Subsequent re-evaluation of these estimates by Krooss *et al.* (1992) suggested that the rate of diffusive hydrocarbon losses through the cap-rock at Harlingen are an order of magnitude lower (*c.* 70 Ma to dissipate half the in-place gas via diffusion through the cap-rock).

In a petroliferous basin that is characterized by continuous subsidence, hydrocarbon escape by diffusion and other processes can be wholly or partly offset by an active generation and migration system. However, in an exhumed basin setting, where the hydrocarbon generation and migration system is 'switched off' during regional uplift, diffusive losses through water-saturated shaly cap-rocks will increase with time since uplift and may be significant.

Cap-rocks: some physical properties

The examination of cap-rocks to hydrocarbon accumulations is primarily concerned with the properties of the weakest point of the reservoir–top-seal interface. As highlighted by Downey (1994), measured properties of a random core sample may not be relevant to the physical properties of the cap-rock at the leak point. Furthermore, geohistory is an important control on the sealing properties of top-seals (Knipe *et al.* 2000) and the location of the potential leak point throughout the evolution of the hydrocarbon trap. Water-wet shaly cap-rocks can form a capillary seal to a hydrocarbon column while simultaneously accommodating brine flow and equilibration of pressures between the reservoir and the top-seal. In contrast, thick, low-permeability shale or evaporite sequences may form pressure seals that restrict vertical brine and hydrocarbon flow and prevent the equilibration of aquifer pressures above and below the seal. Some of the petrophysical and mechanical properties that most influence top-seal performance are summarized below.

Lithology, porosity and permeability

Evaporites and mudrocks are commonly found as effective top-seals to hydrocarbon accumulations because they typically possess very low porosity and permeability, high capillary entry pressures, are relatively ductile and are often laterally continuous at the basin scale. However, other lithologies, such as siltstones and sandstones, have been identified as having capillary retention capacity and can form the top-seal to a hydrocarbon column (Spain & Conrad 1997).

In most sedimentary basins, mudstone porosities range from 5 to 80%, depending upon compaction state (Sclater & Christie 1980). Mudstone permeabilities vary by ten orders of magnitude (10^{-4}–10^{-15} mD) and by three orders of magnitude at a single porosity, primarily as a result of grain-size variations (Dewhurst *et al.* 1999). However, the largest interconnected pore-throat diameter is the critical factor with respect to the capillary entry pressure of the mudstone. In tight mudrocks (permeability 10^{-9} D range) the risk of capillary failure and Darcy flow through the matrix is low, as the capillary entry pressure commonly exceeds the buoyancy force of any potential hydrocarbon column (Fig. 3). In this case, the top-seal retention capacity of the mudrock is a function of the ductility of the mudstone and the potential for the formation of dilatant fractures under tectonic deformation or changing pore-fluid pressure conditions.

Halite forms an excellent top-seal (capillary and pressure seal) as a result of two characteristics: it has a practically infinite capillary entry pressure and it flows plastically under deformation. When it forms a continuous layer over the potential hydrocarbon trap and is immediately juxtaposed above the hydrocarbon-bearing reservoir, the seal risk for that trap is considerably reduced, even where the trap has experienced post-emplacement tectonic deformation and exhumation. Mudrocks that are exceptionally rich in organic matter also can form excellent capillary and pressure seals. For example, in the exhumed Williston Basin the thin but prolific Upper Devonian–Lower Mississippian Bakken Shale is an extremely rich source rock (with source potential Rock Eval pyrolysis parameters of total organic carbon (TOC) 10%, S2 30–70 mg g^{-1} rock and hydrogen index (HI) >600 mg g^{-1} TOC), has very low porosity (<3%) and vertical permeability (10^{-2}–10^{-3} nD) and has retained residual overpressures (>12 MPa present day), which were developed via oil generation at peak maturity between 75 and 50 Ma, before exhumation (Burrus 1998).

At both the hydrocarbon trap and basin scale, cap-rock units can manifest vertical and lateral heterogeneities. Internal lithostratigraphy of the unit can vary, with mud rocks often interbedded with large amounts of leaky strata such as siltstones or sandstones. These lithologies will be more prone to leakage than the mudrocks and can result in multimodal pore-throat diameter distributions within the cap-rock interval and the development of 'waste zones', if the siltstones are located immediately above the reservoir unit. In a study of hydrocarbon seals in the exhumed East Irish Sea Basin, Seedhouse & Racey (1997) utilized mercury injection porosimetry to identify pore-throat distributions in the Mercia Mudstone Group (MMG) seal and to describe some of the heterogeneities observed in this cap-rock interval. These workers identified the presence of halite immediately above the reservoir as a key factor in the retention of hydrocarbon columns in the exhumed Triassic Sherwood Sandstone Group reservoir. In addition, their study found that Sherwood Sandstone accumulations with significant hydrocarbon columns, but that are directly capped by mudrocks, invariably manifested hydrocarbon shows within the cap-rock interval up to the level of the first halite bed encountered above the reservoir. This suggested that the buoyancy forces exerted by the individual hydrocarbon columns were sufficient to overcome the capillary entry pressure of the heterogeneous MMG at these locations, but not the capillary resistance of the halite units.

Strength, ductility and brittleness

The mechanical response of rocks to an applied stress varies under different conditions so a valid comparison of the strength and ductility of rocks can be made only if the conditions of deformation are also known. Ductility is a rock property that pertains to the amount of strain that a material can withstand before brittle failure if it undergoes brittle failure at all. Ductile rocks respond to an applied stress by an initial, although limited, elastic deformation, followed by sustained plastic deformation before failure. Brittle rocks respond to an applied stress by first shortening elastically and then failing by the formation of discrete fractures and faults. A rock is considered ductile when it can accommodate strains of 8–10% without fracturing and brittle when strain is <3% before fracturing (Fig. 5). Rock ductility is a function of lithology, confining pressure, pore-fluid pressure, temperature, differential stress and strain rate (Davis & Reynolds 1996).

Because the matrix permeability of buried and compacted mudrocks is extremely low, it is the fracture permeability that primarily controls seal capacity of these rocks. More relevant definitions have been offered by Ingram & Urai (1999), who described a ductile mudrock as one that can deform without dilatancy and the creation of fracture permeability, and a brittle mudrock as one that dilates and develops fracture permeability.

Experimental studies have shown that, for most lithologies, both rock strength and ductility increase with rising confining pressure (Handin *et al.* 1963; Gretener 1981). This suggests that sedimentary rocks, including mudstones, increase in ductility during burial because confining pressure (*lithostatic pressure/total overburden stress*) increases with depth. However, this inference results from treating the

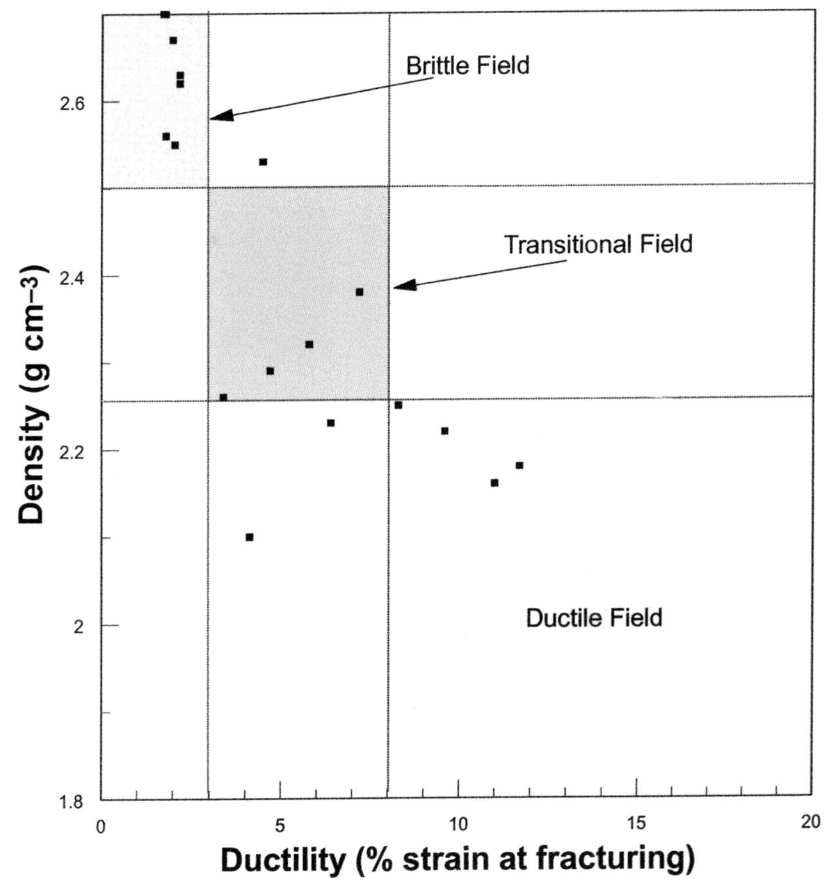

Fig. 5. Mechanical properties of mudrocks. Relationship between ductility (% strain at fracturing) and bulk density (g cm^{-3}), at a constant confining pressure of 1000 kg cm^{-2}, for a population of Neogene mudrocks from Japan. Mudrocks with densities in excess of *c.* 2.5 g cm^{-3} exhibit brittle behaviour and fracture at strains of <3% (data from Hoshino *et al.* 1972).

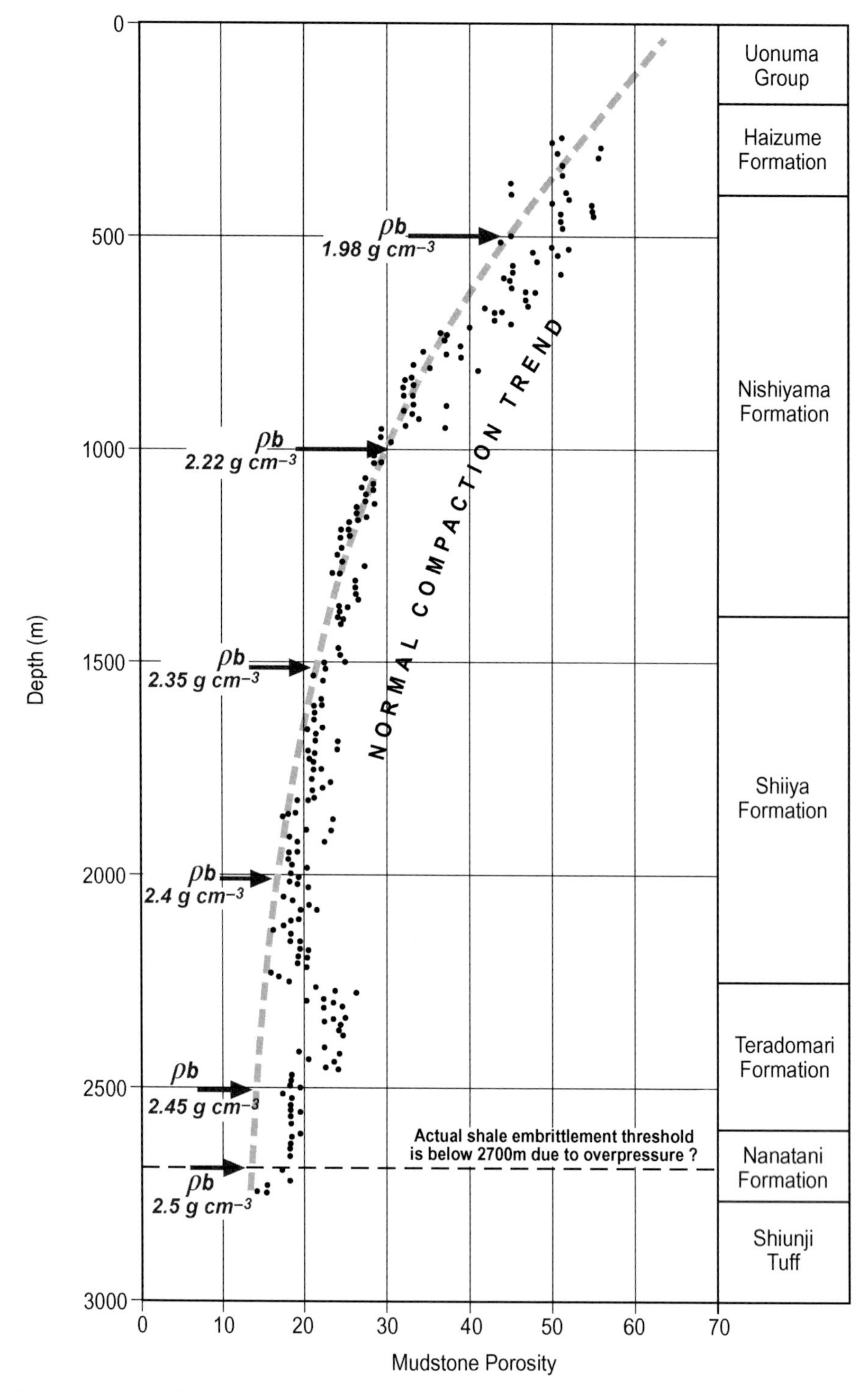

Fig. 6. Porosity–depth behaviour for a suite of Neogene mudrocks from Japan (after Magara 1968). The organically rich, overpressured, mudrocks of the Teradomari and Nanatani formations form the cap-rock to hydrocarbon accumulations in the underlying fractured volcanic reservoirs of the Miocene Shiunji Tuff Formation and equivalents. A general increase in shale density with depth (normal compaction trend) is observed down to

compaction of mudrocks as only a linear mechanical process and ignores the effects of overpressures and of chemical compaction in the deeper part of sedimentary basins (>2–3 km, 70–100 °C) (Bjørlykke 1999).

Bulk density (ρ_b) is one measured parameter that can be used to indicate the compaction state of mudrocks. However, mudrock density is also a function of matrix mineralogy, porosity, applied load, temperature and pore-fluid pressure. Experimental results (using a population of Neogene mudstones from Japan) of Hoshino *et al.* (1972) have indicated that, at constant confining pressures, ductility actually decreases with increasing density. This supports the view that chemical compaction and diagenetic changes can alter the picture, with respect to ductility, at increased burial depths in the sedimentary column (Fig. 5). Shales with densities less than *c.* 2.2 g cm^{-3} exhibit ductile behaviour; shales with densities of *c.* 2.2–2.5 g cm^{-3} are transitional and probably exhibit a wide range of mechanical behaviour; shales with densities >2.5 g cm^{-3} manifest brittle behaviour by fracturing at strains of <3%. If mudrock density can be used as a proxy for ductility then the onset of shale embrittlement can be estimated when the density–depth or porosity–depth behaviour of the claystone is known (Fig. 6). The collection of Neogene mudrocks presented by Magara (1968) also indicated the potential role of overpressure in defining the shale embrittlement threshold for any given basin. The presence of overpressure retards compaction by supporting more of the lithostatic load (confining pressure) and reducing effective stress within the mudrock. As a result, the shale embrittlement threshold (ρ_b *c.* 2.5 g cm^{-3}) will be reached at a deeper burial depth than in the case of a hydrostatically pressured sedimentary column. Geothermal gradient and palaeotemperature history also exert a critical control on diagenesis and hence the rheological evolution of mudrocks.

With respect to cap-rock integrity during basin inversion the key factors are the timing and magnitude of the deformation and the mechanical behaviour (brittle v. ductile) of the cap-rock at the time of deformation. Brittle shales are more likely to rupture and leak than ductile shales and evaporites, which may exhibit plastic flow under the applied deformation. However, even evaporitic rocks, which serve as extremely efficient ductile seals when overburden burial exceeds 1000 m, can manifest brittle behaviour at shallow depths (Downey 1994).

Bolton *et al.* (1998) have shown experimentally that although elevated pore-fluid pressures reduce effective stress and enhance shear deformation, it is the consolidation state at the onset of shear that is the crucial factor with respect to the deformation style and resulting permeability. Timing of overpressure is particularly relevant in this regard, as it can change the consolidation state of a mudstone cap-rock with respect to effective stress. Ingram & Urai (1999) have indicated that claystone cap-rocks that have undergone substantial uplift are prone to the formation of dilatant fractures, as they are likely to be over-consolidated and have anomalous strength. However, in the context of exhumed basins, the mechanical behaviour of a mudstone cap-rock will be dependent upon whether or not embrittlement has been achieved before exhumation.

Physical processes that occur during exhumation

A number of physical processes may affect cap-rocks during exhumation, including erosion, tectonic deformation, shear failure, hydrofracturing as a result of disequilibrium pore pressure conditions and a changing hydrodynamic regime. Each of these processes must be examined in the context of the evolutionary changes that may be occurring in the cap-rock, in the petroleum system and at the basin scale. For example, both the shear and tensile strength of mudrocks increase through burial and compaction (Fig. 7), hydrocarbons may migrate into the trap thereby locally increasing overpressure in the trap, and hydraulic head may be developed as a result of the uplift of the basin margins.

Changing hydrodynamic regime

Fluids in the subsurface may manifest static behaviour (hydrostatic condition) or dynamic

---

2200 m. Below this depth mudstone porosity deviates from the 'normal compaction' trend, as a result of overpressure. (The relative enrichment of the Teradomari Formation in low-density organic matter may also contribute to the deviation from the 'normal' trend.) Under conditions of 'normal compaction', embrittlement ($\rho_b = 2.5$ g cm^{-3}) of these Neogene claystones would occur at 2700 m; however, actual shale embrittlement probably occurs below a burial depth of 3000 m, because of the presence of overpressures.

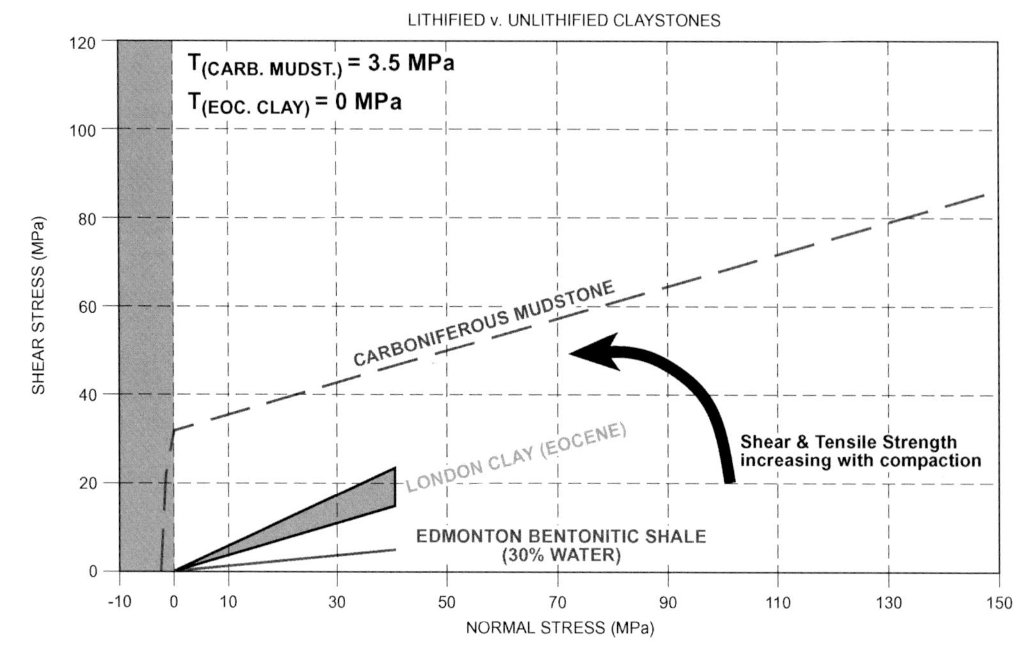

Fig. 7. Mohr–Coulomb failure envelopes for claystones at different levels of compaction. Both shear strength and tensile strength increase with mechanical and chemical compaction. (After Cartwright & Lonergan 1996.)

behaviour (hydrodynamic condition). Hubbert (1953) demonstrated that under hydrodynamic conditions accumulations of oil or gas will invariably exhibit inclined oil– or gas–water interfaces and in such cases the computation of hydrocarbon columns, based on the assumption of hydrostatic conditions, will be spurious. Water flow in sedimentary basins is driven by geographical variations in water potential, which can change considerably in nature and distribution during the evolution of a basin (Wells 1988). The early compaction history of an extensional basin, characterized by continuous subsidence, will result in up-dip water flow to the basin margins. The driving force of this system is the relatively high water potential in the basin centre generated by water released through the processes of mechanical compaction and clay mineral transformations. This water-potential system changes if significant topographic relief forms adjacent to the basin, owing to isostatic effects or some other mechanism. The earlier hydraulic system of the basin is then reversed. Elevated water tables along the basin margin create a hydraulic head, which drives water flow inward towards the basin centre, provided upward discharge is possible there.

A significant consequence of basin inversion is a change in hydrodynamic conditions within the basin. The pattern of exhumation convolved with the pre-existing basin morphology may result in the outcrop of key aquifers, a redistribution of recharge and discharge areas and a change in the direction of gravity-driven fluid flow within the basin. The existence of regional, topographically driven groundwater flow-systems has been documented for several exhumed sedimentary basins (Wells 1988; Deming *et al.* 1992; Bredehoeft *et al.* 1994; Deming 1994; Cramer *et al.* 1999).

Hydrodynamic effects on seal capacity may, for all practical purposes, be ignored except in those basins that manifest clear evidence of hydraulic gradients. In these basins hydrodynamic flow may modify seal retention capacity by either increasing or decreasing the driving pressure against the seal (Allen & Allen 1990). When the hydrodynamic force has an upward vector, it adds to the buoyancy force, thus reducing the hydrocarbon column heights the seal can support. In the case of a downward vector it reduces the buoyancy force on the seal and permits the retention of an increased hydrocarbon column.

Abnormal pore pressures in exhumed basins

Abnormal formation pressures, either above or below the hydrostatic condition, occur in a wide

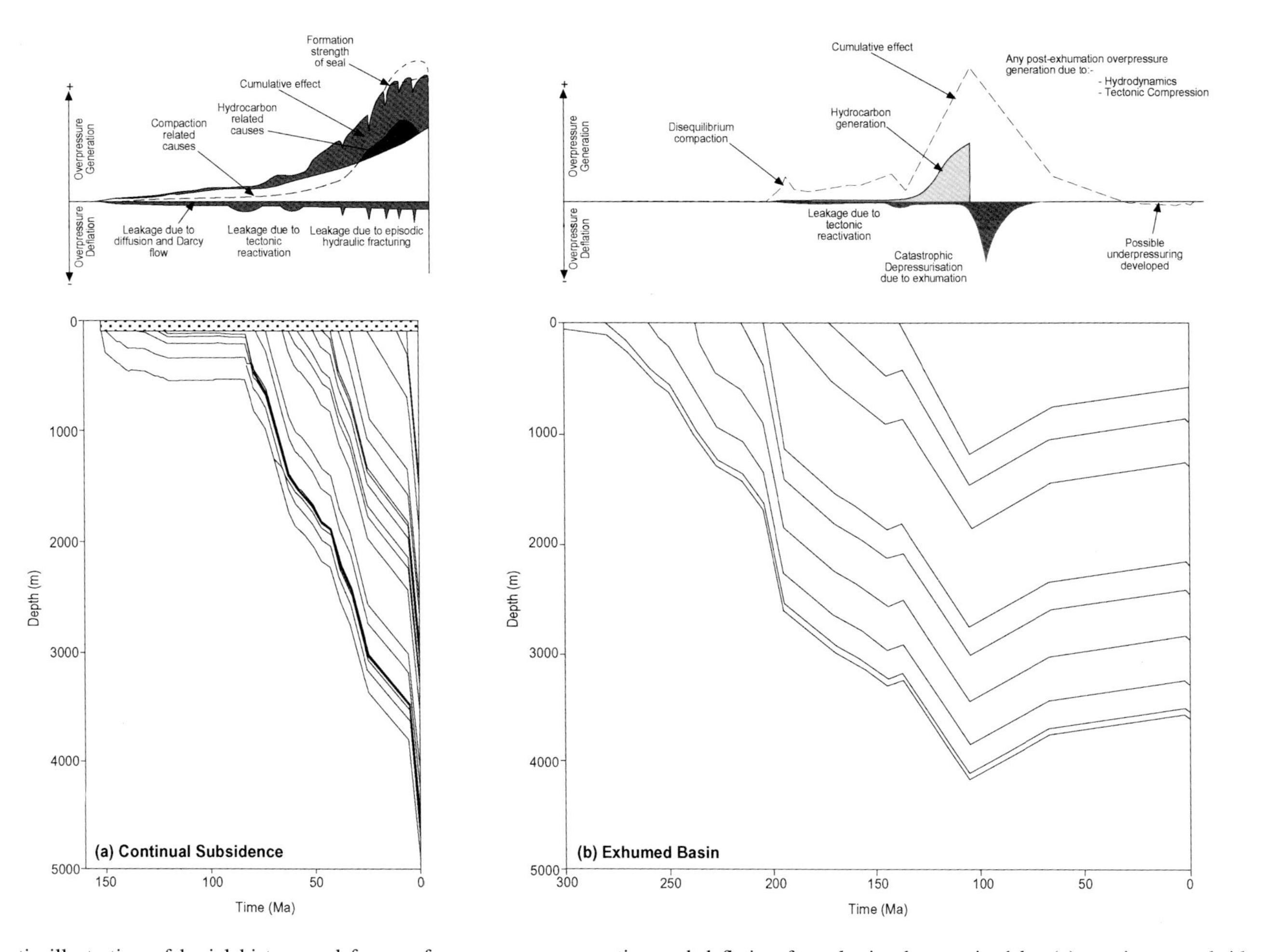

Fig. 8. Schematic illustration of burial history and forces of overpressure generation and deflation for a basin characterized by **(a)** continuous subsidence and **(b)** exhumation. (Adapted from Gaarenstroom *et al.* 1993; Luo & Vasseur 1995.)

range of geographical and geological settings (Swarbrick & Osborne 1998). The global compilation of Law & Spencer (1998) suggests that abnormal formation pressures are observed in some 150 basins with overpressure recorded in 148 of these regions and underpressures manifest in only about 12 areas. This apparent bias may in part be due to the fact that underpressure is more difficult to identify during conventional drilling operations than overpressure, as most wells are drilled with mud weight slightly 'over-balance'.

At any time during the evolution of a sedimentary basin, the distribution of abnormal pore pressures is a function of the balance between forces that generate overpressure and forces that dissipate overpressure (Luo & Vasseur 1995). For example, in a basin characterized by continuous subsidence, such as the North Sea Central Graben, the current 'snapshot' of pore pressure evolution suggests that forces of overpressure generation exceed the forces of dissipation (Fig. 8a) (Gaarenstroom *et al.* 1993). In contrast, in the exhumed basin, such as the Ordos Basin (China), a close to hydrostatic equilibrium has been achieved by pervasive depressurization during uplift and denudation between 105–65 Ma (Fig. 8b) (Luo & Vasseur 1995).

Mechanisms for the generation of abnormal pore pressures have been extensively reviewed in the literature (Gretener 1981; Neuzil & Pollock 1983; Martinsen 1994; Osborne & Swarbrick 1997; Law & Spencer 1998; Swarbrick & Osborne 1998). In summary, three generic mechanisms have been identified by these workers as primary potential contributors to the generation of overpressures in subsiding petroliferous basins: stress-related mechanisms (vertical loading, horizontal tectonic compression); pore-fluid volume expansion (increased temperature, dehydration reactions, hydrocarbon generation and oil to gas cracking); fluid movement (osmosis, hydrocarbon buoyancy and hydrodynamic flow).

Disequilibrium compaction is the most commonly cited mechanism for overpressure generation in young, rapidly subsiding basins with overpressures primarily controlled by the vertical permeability of the shaly facies containing the overpressures (Burrus 1998). The ephemeral nature of these overpressures has been emphasized by Deming (1994), who offered the view that rocks are not capable of sustaining zero effective permeability to water over extended periods of geological time. In older (pre-Cenozoic) rocks kerogen transformation is the most commonly cited mechanism for overpressure generation in subsiding basins (Law & Spencer 1998). When a subsiding basin is exhumed, mechanical compaction, mineral transformations involving dehydration reactions, and hydrocarbon generation and maturation processes will be arrested, and these forces will be unable to contribute to overpressure generation in an exhumed basin setting (Fig. 8).

Underpressured regimes are commonly characterized by shallowly buried (0.6–3.0 km) permeable rocks, which are, at least temporarily, hydraulically isolated within low-permeability mudrocks (Swarbrick & Osborne 1998). Underpressured rocks typically occur in exhumed basin settings, which suggests some causal linkage with the change in matrix and pore-fluid volumes resulting from decreases in pressure and temperature, and hydraulic realignment, induced by the exhumation process. Proposed mechanisms for underpressuring cited by Swarbrick & Osborne (1998) include: differential recharge and discharge rates in a topographically driven flow system; rapid migration of exsolved gas, from a low-permeability reservoir during uplift, relative to the rate of ingress of gas to the reservoir; dilation of pores in shallowly buried mudrocks as vertical load is removed via denudation; aquathermal contraction. Underpressured reservoirs are common in the exhumed Laramide basins of the USA and Canada, such as the Denver, Piceance, San Juan and Alberta basins (Belitz & Bredehoeft 1988; Bachu & Underschultz 1995). However, with the exception of one well from the Barents Sea Basin, reported by Doré & Jensen (1996), underpressuring has not been documented in exhumed Atlantic margin basins, although it is anticipated.

In any sedimentary basin the present-day pore pressure distributions are a function of the mechanism of generation of the abnormal pore pressure conditions and the hydraulic evolution of the basin during and after the creation of these abnormal pore pressure conditions. Both overpressuring and underpressuring are a disequilibrium condition, which will change with the evolution of the basin. For example, in a basin where thick low-permeability shaly successions are present as a pressure seal, the hydraulic evolution may involve pulses of hydrofracturing, which generates a transient permeability in the pressure seal and permits the escape of fluids. The timing of this hydraulic fracturing is controlled primarily by the prevailing stress conditions in the basin and the properties of the pressure seal, so that it can occur both during the burial and subsequent exhumation phase of the basin (Cosgrove 2001). An assessment of when and where fracturing can occur in the evolution

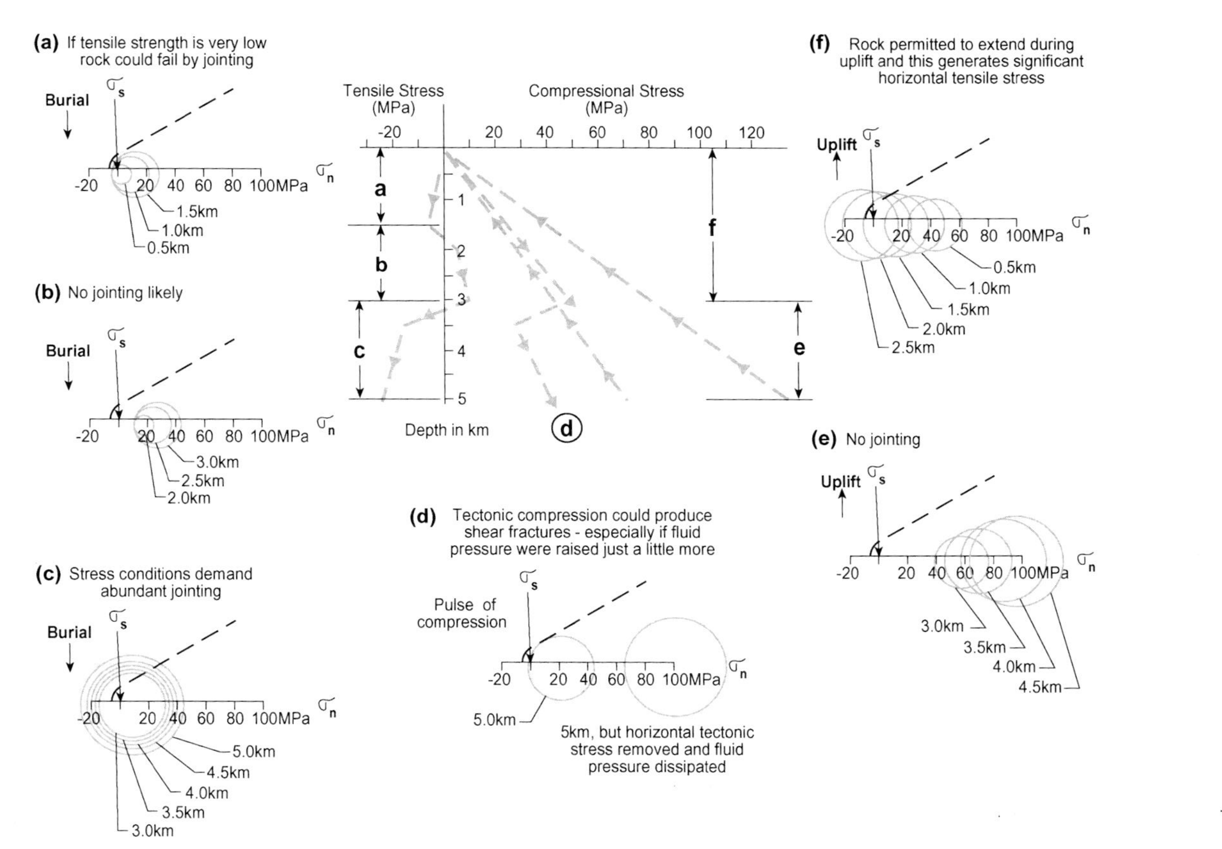

Fig. 9. Mohr diagrams with failure envelopes illustrating computed stress conditions in a sandstone, under assumed conditions, at 0.5 km intervals throughout burial and exhumation. Downward pointing arrows indicate the evolution of effective vertical stress (greatest effective principal stress σ_1) and effective horizontal stress (least effective principal stress σ_3) during burial. For example, a dramatic increase in pore-fluid pressure between 3 and 3.5 km results in a decrease in the magnitude of σ_1 and σ_3 through this interval, from 48 and 9 MPa to 22 and -17 MPa, respectively. In this scenario, the failure envelope is breached and fracturing will occur to relieve the pore-fluid pressure build-up. Upward pointing arrows indicate the evolution of vertical and horizontal compressive stresses during uplift. If the rock is permitted to extend during regional uplift significant fracturing will develop. Conditions favourable to hydraulic fracturing occur during burial at 0–1.5 km and 3–5 km, and during uplift at 3–0 km (from Davis & Reynolds 1996).

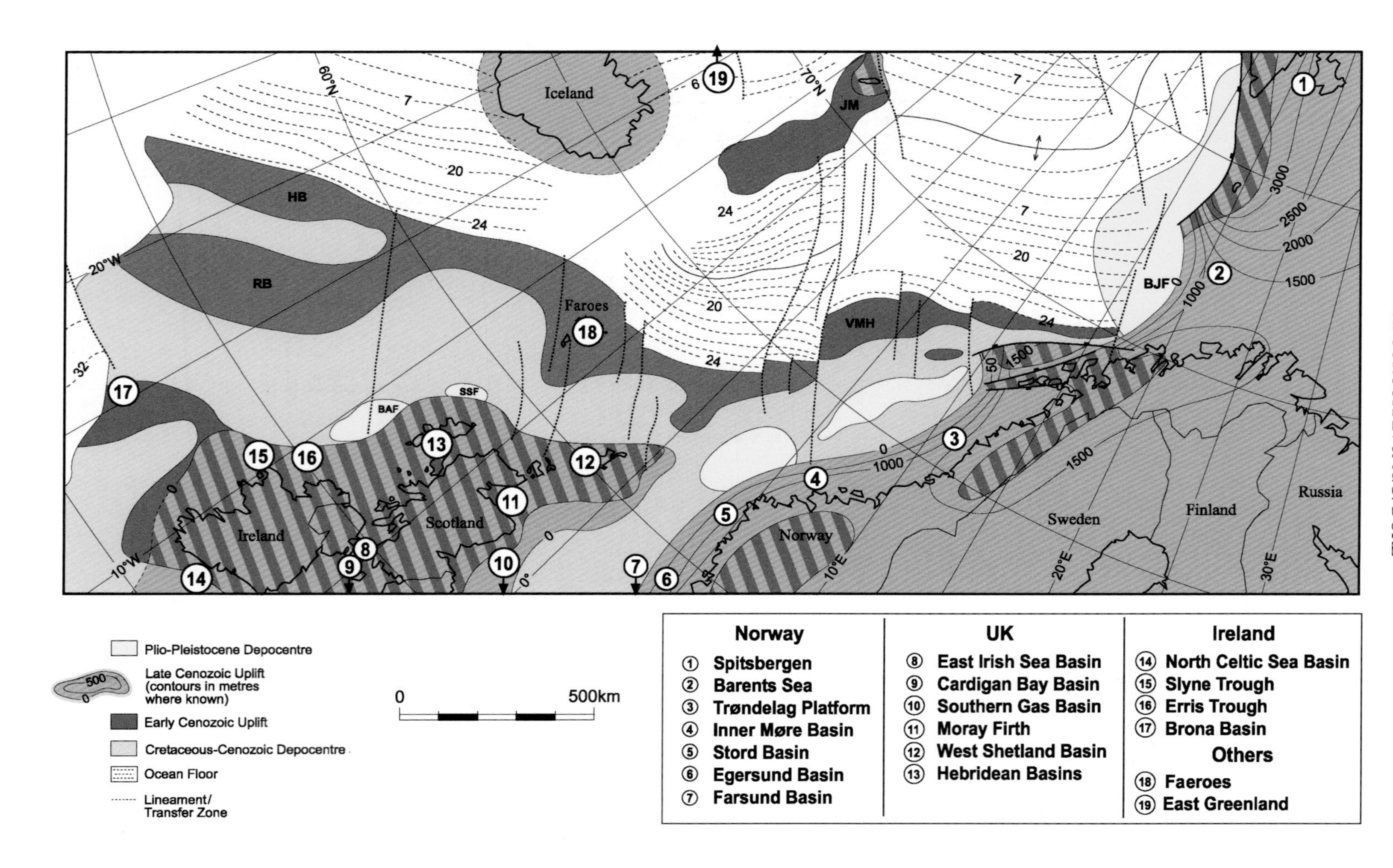

Norway
1 Spitsbergen
2 Barents Sea
3 Trøndelag Platform
4 Inner Møre Basin
5 Stord Basin
6 Egersund Basin
7 Farsund Basin
UK
8 East Irish Sea Basin
9 Cardigan Bay Basin
10 Southern Gas Basin
11 Moray Firth
12 West Shetland Basin
13 Hebridean Basins
Ireland
14 North Celtic Sea Basin
15 Slyne Trough
16 Erris Trough
17 Brona Basin
Others
18 Faeroes
19 East Greenland
Plio-Pleistocene Depocentre
Late Cenozoic Uplift (contours in metres where known)
Early Cenozoic Uplift
Cretaceous-Cenozoic Depocentre
Ocean Floor
Lineament/ Transfer Zone
0
500km
Iceland
Faroes
Scotland
Ireland
Norway
Sweden
Finland
Russia
JM
VMH
BJF
HB
RB
SSF
BAF
60°N
70°N
20°W
10°W
0°
10°E
20°E
30°E

of an exhumed basin can be made by analysing the stress conditions of a rock, under assumed conditions, at intervals of 0.5 km through burial and uplift (Fig. 9) (Davis & Reynolds 1996). Although the formation of hydraulic fractures during the early stages of burial and diagenesis (< 1.5 km burial) is an important mechanism for fluid escape from low-permeability, semilithified sediments, these fractures are rarely preserved in the rock record. However, Cosgrove (2001) has recently demonstrated the serendipitious preservation of these early hydraulic fractures in the low-permeability Mercia Mudstones of the Bristol Channel Basin, as a result of the injection of intra-formational sand bodies into the hydraulic fractures, thereby preserving them as sedimentary dykes and sills. Furthermore, a later set of sub-horizontal, bedding-parallel, hydraulic fractures, which pertain to the exhumation phase, have been preserved in outcrop, as a network of satin spar veins, within evaporite-rich horizons. The orientation and spatial organization of the satin spar veins is consistent with the tectonic compression of the Bristol Channel Basin during the exhumation that was initiated in Mid-Cretaceous time.

Empirical observations and discussion: exhumed Atlantic margin and borderland basins

Many Atlantic margin and borderland basins are characterized by exhumation during Cenozoic time (Fig. 10) Although prior exhumation events may have occurred during the evolution of these basins, in terms of top-seal assessment, Cenozoic exhumation is most critical, as it generally occurs in these basins after the initial migration of hydrocarbons into traps.

Shale and evaporite cap-rocks form the main regional seals to hydrocarbon accumulations in exhumed basins of the Atlantic margin and borderlands (Fig. 11). Shale cap-rocks of Jurassic–Cretaceous age are prevalent in the Celtic Sea, Inner Moray Firth, West of Shetlands and Barents Sea basins, and mixed evaporite and shale seals of Triassic age are encountered in three basins (Slyne Trough, East Irish Sea and Southern North Sea). The prodigious Zechstein evaporite seal is cap-rock to an estimated ultimate recovery (EUR) of > 150 TCF (trillion cubic feet) of gas in the southern Permian Basin, including the giant Groningen accumulation (97 TCF) and *c.* 35 TCF contained in 35 accumulations in the UK sector of the Southern North Sea Basin (Glennie 1997) (Fig. 11).

There are a number of hydrocarbon trapping and top-seal configurations observed in these exhumed basins. In the Celtic Sea Basin, a maximum gas column of 91 m in the Kinsale Head Gas Field is retained by 46 m of Gault claystone in a compressional anticlinal flexure (Fig. 12). This basin-centred accumulation experienced *c.* 900 m of exhumation during early Cenozoic time and has been overprinted by a compressional deformation that is poorly constrained but probably post-Paleocene in age (Murdoch *et al.* 1995). Local evidence suggests that the maximum burial depth of the Gault claystone was in the range of 1700–1800 m, which may not have been sufficient to achieve shale embrittlement before the applied deformation associated with exhumation and compressional inversion. In this scenario, it is postulated that fracture development was inhibited as the Gault claystone responded by plastic flow to Cenozoic deformation. This hypothesis is consistent with the experimental results of Bolton *et al.* (1998), which suggests that underconsolidated clayey sediments, undergoing shear, deform by bulk volume loss, which reduces permeability and results in weakly developed deformation fabrics that have little impact on the hydrological properties of the claystone.

All four generic leakage mechanisms (tectonic breaching, capillary leakage, hydraulic leakage and diffusion) operate in both continuously subsiding basins and exhumed basin settings. However, the critical aspect of trap leakage in exhumed basin settings is a lower probability of trap replenishment, as a result of the 'switching off' of hydrocarbon generation during regional uplift. In such cases, top-seal failure (induced by tectonic breaching or hydraulic leakage) during exhumation may result in the catastrophic loss of a pre-existing hydrocarbon fill whereas post-exhumation these traps can only be replenished from a curtailed hydrocarbon budget, which consists primarily of remigrating oil and gas. This is consistent with the observation that a number of hydrocarbon accumulations in

Fig. 10. Uplifted basins of the northeastern Atlantic margin and borderlands. Many of these basins are characterized by Cenozoic exhumation events, which have occurred after the initial migration of hydrocarbons into traps. HB, Hatton Bank; RB, Rockall Bank; BAF, Barra Fan; SSF, Sula Sgeir Fan; JM, Jan Mayen; VMH, Vøring Marginal High; BJF, Bjørnøya Fan.

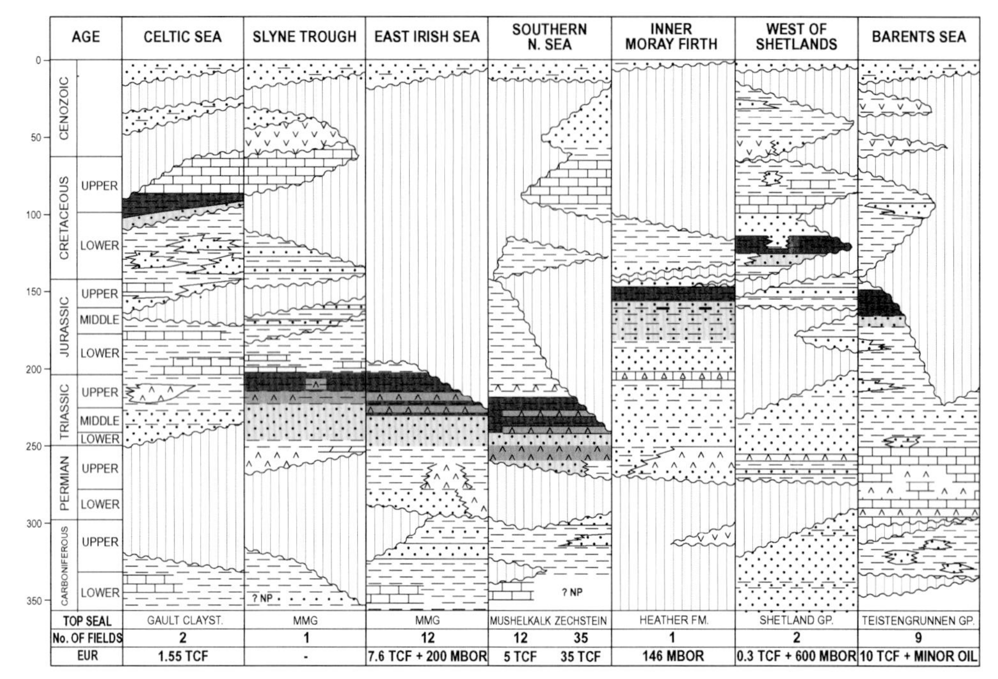

Fig. 11. Regional seals to hydrocarbon accumulations in exhumed basins of the Atlantic margin and borderlands. Shales and evaporites are the most common cap-rock lithologies in this setting. MMG, Mercia Mudstone Group; MBOR, million barrels of oil recoverable.

exhumed basin settings along the Atlantic margin are characterized by underfilled traps (Fig. 13). Empirical observation also indicates that hydrocarbon accumulations in exhumed basins are characterized by hydrostatically pressured or modestly overpressured reservoirs, whereas significantly overpressured reservoirs are common in basins that have experienced relatively continuous subsidence (Fig. 14). This suggests a close causal relationship between regional uplift, hydrocarbon remigration and dissipation of overpressures.

Exhumation may also have positive implications for the capillary and hydraulic retention capacity of mudrock seals. Increased mechanical compaction as a result of burial results in reduced interconnected pore-throat sizes and increasing shear strength and tensile strength for a claystone rock. For example, when a claystone is exhumed it retains the tensile strength of its maximum burial depth and, consequently, a higher pore-fluid pressure will be required to induce hydrofracturing than for a claystone in a continuously subsiding basin at the same depth. Leak-off tests (LOTs) and formation integrity tests (FITs) from a subset of the Atlantic margin basins support these observations (Fig. 15). These data indicate that, for any given burial depth, the minimum horizontal stress or fracture pressure (defined by the lower envelope of LOT pressures) is higher in exhumed basins than in those basins characterized by continuous subsidence. Critically, there are a number of FITs performed on exhumed claystones of Carboniferous, Triassic and Jurassic age, which indicate that these seal rocks have very high tensile strengths, appropriate to their maximum burial depth before exhumation. This suggests that post-exhumation top-seal integrity is relatively high in many of the Atlantic margin and borderland basins under low differential stresses. However, the anomalously high shear strength of exhumed mudstones may result in the development of dilatant shear fractures (under low confining pressures) if shale embrittlement has been achieved before exhumation. The absence of seismic chimneys across major gas accumulations in many of these basins (e.g. Celtic Sea, East Irish Sea Basin, Southern North Sea) suggests that dynamic leakage through the top-seal is not occurring at the present day and that pore-fluid pressures are below the top-seal capillary and hydraulic leakage thresholds for these accumulations.

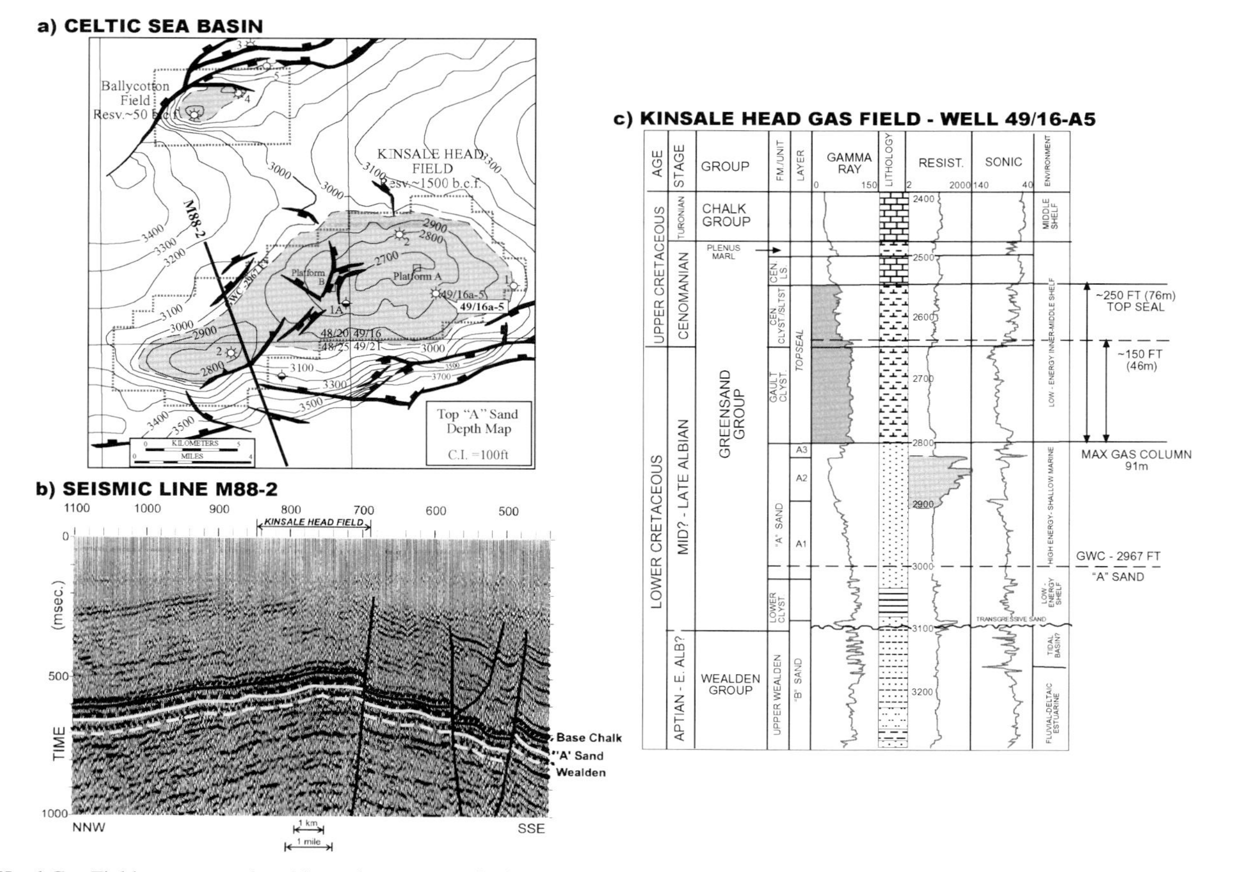

Fig. 12. Kinsale Head Gas Field, a compressional inversion structure in the exhumed Celtic Sea Basin. (**a**) Depth structure map on top main reservoir, indicating that the gas–water contact (GWC at − 2967 feet) is coincident with the spill point to the north of the structure. C.I., contour interval. (**b**) NNW–SSE seismic line, illustrating reverse faulting on the southern limb of the anticline. (**c**) Type log for the 'A' Sand–Gault Claystone reservoir–top-seal couplet at the Kinsale Head Field. A maximum gas column of 91 m, in the Greensand reservoir, is retained *in situ*, without apparent leakage, by 46 m of Gault Claystone. Maximum burial depth of the Gault claystone, in the Kinsale Head area, is estimated to have been <2 km. As a result, it is postulated that Cenozoic exhumation of the Gault Claystone occurred before shale embrittlement, thereby inhibiting the formation of dilatant fractures (see Fig. 6). (Data from Taber *et al.* 1995.)

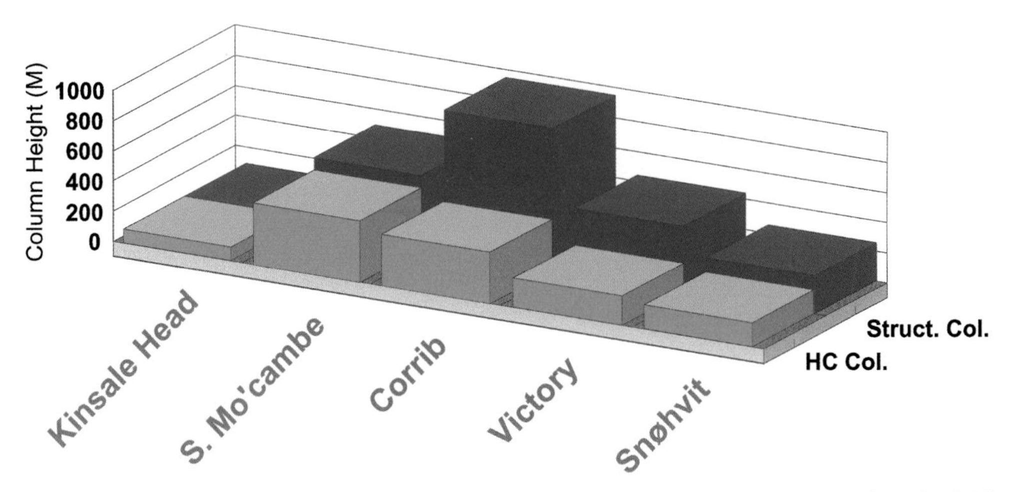

Fig. 13. Underfilled traps in exhumed Atlantic margin basins. Of the five major gas accumulations from the Celtic Sea to the Barents Sea (Kinsale Head, South Morecambe, Corrib, Victory, Snøhvit) only the Kinsale Head Field is full to structural spill point, at the present day.

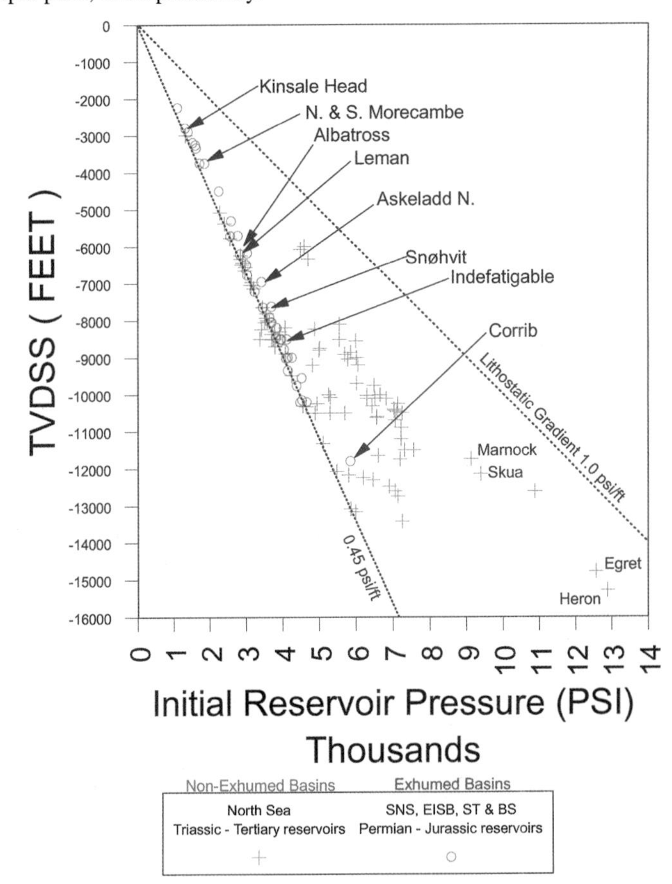

Fig. 14. Comparison of initial reservoir pressures v. depth for hydrocarbon accumulations in exhumed basins v. non-exhumed basins. Accumulations in exhumed basins are characterized by hydrostatically pressured or modestly overpressured reservoirs, whereas significantly overpressured reservoirs are common in basins that have experienced relatively continuous subsidence. SNS, Southern North Sea Basin; EISB, East Irish Sea Basin; ST, Slyne Trough; BS, Barents Sea Basin; TVDSS, true vertical depth, sub-sea. (Data compiled essentially from Spencer *et al.* 1986; Abbotts 1991; Pooler & Amory 1999.)

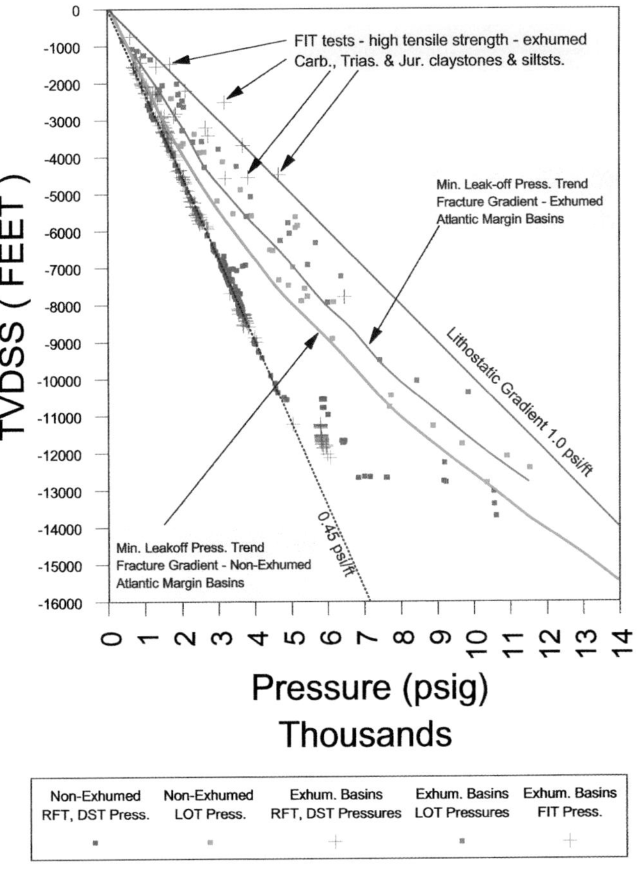

Fig. 15. Formation pressures (repeat formation tests (RFTs) and drill stem tests (DSTs)), formation integrity tests (FITs) and leak-off tests (LOTs) for basins offshore Ireland. Data indicate that, for any given burial depth, the minimum horizontal stress or fracture pressure (defined by the lower envelope of LOT pressures) is higher in exhumed basins than in those characterized by continuous subsidence.

Conclusions

A number of effective regional cap-rocks are recognized in the petroleum systems of exhumed Atlantic margin basins. However, the observation of underfilled traps, close to hydrostatically pressured reservoirs and breached traps with residual oil and gas shows, suggests that top-seal behaviour exercises a critical control on pore pressure evolution and hydrocarbon distribution and redistribution in exhumed basin settings. The following are the principal conclusions of this review.

(1) Depressurization of reservoirs during exhumation is a consequence of fault reactivation, fracturing of the cap-rock and the large-scale removal of overburden during regional uplift. As a result, exhumed basins typically show limited overpressuring except where residual overpressures have been retained by extremely low-permeability aquitards or exceptionally rich source rocks (with sub-nanodarcy permeabilities) or where overpressures have been generated after regional uplift.

(2) Rock pore dilatancy and fluid volume reduction, caused by a lowering of temperature during exhumation, may result in underpressured reservoirs, which are common in the exhumed Laramide basins of the USA and Canada. However, with the exception of one reported occurrence in the Barents Sea Basin, underpressuring has not been reported in exhumed Atlantic margin basins, although it is anticipated.

(3) Post-exhumation overpressure generation is primarily driven by tectonic compression, aquathermal pressuring and the evolved

hydrodynamic regime as disequilibrium compaction, mineral transformation and hydrocarbon generation mechanisms can no longer contribute to overpressure generation in the exhumed basin setting.

(4) Lithofacies is a major control on top-seal efficiency in exhumed basin settings. The juxtaposition of a halite directly above the hydrocarbon-bearing reservoir offers the most favourable condition for the retention of hydrocarbons in exhumed traps. However, the empirical evidence from the Atlantic margin suggests that mudrocks can form efficient top-seals in exhumed basins under certain conditions.

(5) The behaviour of any cap-rock lithology during exhumation is dependent upon the physical and mechanical characteristics of the cap-rock at the time of exhumation and the timing and conditions of the associated deformation relative to the timing of hydrocarbon emplacement.

(6) Mudrocks may form effective cap-rocks in exhumed basins when the deformation associated with exhumation occurs before embrittlement and the cap-rock exhibits ductile behaviour. Where exhumation occurs post-embrittlement the shale cap-rock will facilitate hydrocarbon leakage through the development of extensive fracture networks.

(7) *Syn-exhumation* top-seal efficiency (fluid retention capacity) is a major exploration risk in exhumed basin settings, although *post-exhumation* top-seal integrity in these basins may be relatively high. This suggests that a major exploration risk factor in exhumed basin settings pertains to the limited hydrocarbon budget available post-regional uplift and the efficiency of the remigration process.

The authors would like to thank A. Carr and J. Gluyas for their helpful reviews of this manuscript. We would also like to thank J. Kipps (Statoil (UK) Ltd), who draughted most of the figures in this paper. Finally, the authors would like to thank Statoil Exploration (Ireland) Ltd for their sponsorship of the color printing costs.

References

Abbotts, I.L. (ed.) 1991. *United Kingdom Oil and Gas Fields, 25 Years Commemorative Volume*. Geological Society, London, Memoirs, **14**.

ALLEN, P.A. & ALLEN, J.R. 1990. *Basin Analysis: Principles and Applications*. Blackwell Scientific, Oxford.

BACHU, S. & UNDERSCHULTZ, J.R. 1995. Large-scale underpressuring in the Mississipian–Cretaceous succession, southwestern Alberta Basin. *AAPG Bulletin*, **79**, 989–1004.

BELITZ, K. & BREDEHOEFT, J.D. 1988. Hydrodynamics of the Denver basin: an explanation of subnormal pressures. *AAPG Bulletin*, **72**, 1334–1359.

BERG, R.R. 1975. Capillary pressures in stratigraphic traps. *AAPG Bulletin*, **59** (6), 939–956.

BJØRKUM, P.A., WALDERHAUG, O. & NADEAU, P.H. 1998. Physical constraints on hydrocarbon leakage and trapping revisited. *Petroleum Geoscience*, **4**, 237–239.

BJØRLYKKE, K. 1999. Principal aspects of compaction and fluid flow in mudstones. *In*: ALPIN, A.C., FLEET, A.J. & MACQUAKER, J.H.S. (eds) *Muds and Mudstones: Physical and Fluid Flow Properties*. Geological Society, London, Special Publications, **158**, 73–78.

BOLTON, A.J., MALTMAN, A.J. & CLENNELL, M.B. 1998. The importance of overpressure timing and permeability evolution in fine-grained sediments undergoing shear. *Journal of Structural Geology*, **20** (8), 1013–1022.

BRADLEY, J.S. & POWLEY, D.E. 1994. Pressure compartments in sedimentary basins: a review. *In*: ORTOLEVA, P.J. (ed.) *Basin Compartments and Seals*. AAPG Memoirs, **61**, 3–26.

BREDEHOEFT, J.D., WESLEY, J.B. & FOUCH, T.D. 1994. Simulations of the origin of fluid pressure, fracture generation and the movement of fluids in the Uinta Basin, Utah. *AAPG Bulletin*, **78**, 1729–1747.

BURRUS, J. 1998. Overpressure models for clastic rocks, their relation to hydrocarbon expulsion: a critical reevaluation. *In*: LAW, B.E., ULMISHEK, G.F. & SLAVIN, V.I. (eds) *Abnormal Pressures in Hydrocarbon Environments*. AAPG Memoirs, **70**, 35–63.

CAPUANO, R.M. 1993. Evidence of fluid flow in microfractures in geopressured shales. *AAPG Bulletin*, **77**, 1303–1314.

CARTWRIGHT, J.A. & LONERGAN, L. 1996. Volumetric contraction during the compaction of mudrocks: a mechanism for the development of regional-scale polygonal fault systems. *Basin Research*, **8**, 183–193.

CLAYTON, C.J. 1999. Discussion: 'Physical constraints on hydrocarbon leakage and trapping revisited' by P.A. Bjørkum *et al. Petroleum Geoscience*, **5**, 99–101.

CLAYTON, C.J. & HAY, S.J. 1994. Gas migration mechanisms from accumulation to surface. *Bulletin of the Geological Society of Denmark*, **41**, 12–23.

COSGROVE, J.W. 2001. Hydraulic fracturing during the formation and deformation of a basin: a factor in the dewatering of low-permeability sediments. *AAPG Bulletin*, **85**, 737–748.

COWAN, G., BURLEY, S.D., HOEY, N., & 5 OTHERS 1999. Oil and gas migration in the Sherwood Sandstone of the East Irish Sea Basin. *In*: Fleet, A.J. & Boldy, S.A.R. (eds) *Petroleum Geology of Northwest Europe: Proceedings of the 5th Conference*. Geological Society, London, 41–61.

CRAMER, B. & POELCHAU, H.S. 2002. Gas exsolution from groundwater in exhumed basin settings. *In*:

DORÉ, A.G., CARTWRIGHT, J., STOKER, M.S., TURNER, J.P. & WHITE, N. (eds) *Exhumation of the North Atlantic Margin: Timing, Mechanisms and Implications for Petroleum Exploration*. Geological Society, London, Special Publications, **xx**, 111–222.

CRAMER, B., POELCHAU, H.S., GERLING, P., LOPATIN, N.V. & LITTKE, R. 1999. Methane released from groundwater: the source of natural gas accumulations in northern West Siberia. *Marine and Petroleum Geology*, **16**, 225–244.

DARBY, D., HASZELDINE, R.S. & COUPLES, G.D. 1998. Central North Sea overpressures: insights into fluid flow from one- and two-dimensional basin modelling. *In*: DÜPPENBECKER, S.J. & ILIFFE, J.E. (eds) *Basin Modelling: Practice and Progress*. Geological Society, London, Special Publications, **141**, 95–107.

DAVIS, G.H. & REYNOLDS, S.J. 1996. *Structural Geology of Rocks and Regions*. 2nd; John Wiley, New York.

DEMING, D. 1994. Factors necessary to define a pressure seal. *AAPG Bulletin*, **78**, 1005–1009.

DEMING, D., SASS, J.H., LACHENBRUCH, A.H. & DERITO, R.F. 1992. Heat flow and subsurface temperature as evidence for basin-scale groundwater flow. *Geological Society of America Bulletin*, **104**, 528–542.

DEWHURST, D.N., YANG, Y. & ALPIN, A.C. 1999. Permeability and fluid flow in natural mudstones. *In*: ALPIN, A.C., FLEET, A.J. & MACQUAKER, J.H.S. (eds) *Muds and Mudstones: Physical and Fluid Flow Properties*. Geological Society, London, Special Publications, **158**, 23–43.

DORÉ, A.G. & JENSEN, L.N. 1996. The impact of late Cenozoic uplift and erosion on hydrocarbon exploration: offshore Norway and some other uplifted basins. *Global and Planetary Change*, **12**, 415–436.

DORÉ, A.G., LUNDIN, E.R., JENSEN, L.N., BIRKELAND, Ø. & ELIASSEN, P.E. 1999. Principal tectonic events in the evolution of the northwest European Atlantic margin. *In*: FLEET, A.J. & BOLDY, S.A.R. (eds) *Petroleum Geology of Northwest Europe: Proceedings of the 5th Conference*. Geological Society, London, 41–61.

DOWNEY, M.W. 1984. Evaluating seals for hydrocarbon accumulations. *AAPG Bulletin*, **68**, 1752–1763.

DOWNEY, M.W. 1994. Hydrocarbon seal rocks. *In*: MAGOON, L.B. & DOW, W.G. (eds) *The Petroleum System—From Source to Trap*. AAPG Memoirs, **60**, 159–164.

GAARENSTROOM, L., TROMP, R.A.J., de Jong, M.C. & BRADENBERG, A.M. 1993. Overpressures in the Central North Sea: implications for trap integrity and drilling safety. *In*: PARKER, J.R. (ed.) *Petroleum Geology of Northwest Europe: Proceedings of the 4th Conference*. Geological Society, London, 1305–1313.

GABRIELSEN, R.H. & KLØVJAN, O.S. 1997. Late Jurassic–early Cretaceous caprocks of the southwestern Barents Sea: fracture systems and rock mechanical properties. *In*: MOLLER-PEDERSEN, P. & KOESTLER, A.G. (eds) *Hydrocarbon Seals: Importance for Exploration and Production*. Norwegian Petroleum Society Special Publication, **7**, 73–89.

GILES, M.R., INDRELID, S.L., KUSZNIR, N.J. & 13 OTHERS 1999. Charge and overpressure modelling in the North Sea: multi-dimensional modelling and uncertainty analysis. *In*: Fleet, A.J. & Boldy, S.A.R. (eds) *Petroleum Geology of Northwest Europe: Proceedings of the 5th Conference*. Geological Society, London, 1313–1324.

GLENNIE, K.W. 1997. History of exploration in the southern North Sea. *In*: ZIEGLER, K., TURNER, P. & DAINES, S.R. (eds) *Petroleum Geology of the Southern North Sea: Future Potential*. Geological Society, London, Special Publications, **123**, 5–16.

GRETENER, P.E. 1981. *Pore Pressure: Fundamentals, General Ramifications and Implications for Structural Geology, Revised Edition*. AAPG, Education Course Note Series, **4**, 15–33.

HALL, D.M., DUFF, B.A., ELIAS, M. & GYTRI, S.R. 1997. Pre-Cretaceous top-seal integrity in the greater Ekofisk area. *In*: MOLLER-PEDERSEN, P. & KOESTLER, A.G. (eds) *Hydrocarbon Seals: Importance for Exploration and Production*. Norwegian Petroleum Society Special Publication, **7**, 231–242.

HANDIN, J.H., HAGER JR., R.V., FRIEDMAN, M. & FEATHER, J.N. 1963. Experimental deformation of sedimentary rocks under confining pressure: pore pressure tests. *AAPG Bulletin*, **47**, 717–755.

HOSHINO, K., KOIDE, H., INAMI, K., IWAMURA, S. & MITSUI, S. 1972. *Mechanical Properties of Japanese Tertiary Sedimentary Rocks under High Confining Pressures*. Geological Survey of Japan, Report **244**.

HUBBERT, M.K. 1953. Entrapment of petroleum under hydrodynamic conditions. *AAPG Bulletin*, **37** (8), 1954–2026.

HUBBERT, M.K. & RUBEY, W.W. 1959. Role of pore fluid pressures in the mechanics of overthrust faulting. *Geological Society of America Bulletin*, **70**, 115–205.

ILIFFE, J.E., ROBERTSON, A.G., WARD, G.H.F., WYNN, C., PEAD, S.D.M. & CAMERON, N. 1999. The importance of fluid pressures and migration to the hydrocarbon prospectivity of the Faeroe–Shetland White Zone. *In*: FLEET, A.J. & BOLDY, S.A.R. (eds) *Petroleum Geology of Northwest Europe: Proceedings of the 5th Conference*. Geological Society, London, 601–611.

INGRAM, G.M. & URAI, J.L. 1999. Top-seal leakage through faults and fractures: the role of mudrock properties. *In*: ALPIN, A.C., FLEET, A.J. & MACQUAKER, J.H.S. (eds) *Muds and Mudstones: Physical and Fluid Flow Properties*. Geological Society, London, Special Publications, **158**, 125–135.

KETTEL, D. 1997. The dynamics of gas flow through rock salt in the scope of time. *In*: MOLLER-PEDERSEN, P. & KOESTLER, A.G. (eds) *Hydrocarbon Seals: Importance for Exploration and Production*. Norwegian Petroleum Society Special Publication, **7**, 175–186.

KNIPE, R.J., FISHER, Q.J., JONES, G. & 13 OTHERS 2000. Quantification and prediction of fault seal parameters: the importance of the geohistory. In: *Extended Abstracts, Hydrocarbon Seal Quantification*, Norwegian Petroleum Society (NPF) Conference, Stavanger, 16–18 October 2000.

KONTOROVITCH, A.E., MANDEL'BAUM, V.S., SURKOV, V.S., TROFIMUK, A.A. & ZOLOTOV, A.N. 1990. Lena–Tuguska Upper Proterozoic–Palaeozoic petroleum superprovince. *In*: BROOKS, J. (ed.) *Classic Petroleum Provinces*. Geological Society, London, Special Publications, **50**, 203–219.

KROOSS, B.M., LEYTHAEUSER, D. & SCHAEFER, R.G. 1992. The quantification of diffusive hydrocarbon losses through cap rocks of natural gas reservoirs—a reevaluation. *AAPG Bulletin*, **76** (3), 403–406.

LAW, B.E. & SPENCER, C.W. 1998. Abnormal pressures in hydrocarbon environments. *In*: LAW, B.E., ULMISHEK, G.F. & SLAVIN, V.I. (eds) *Abnormal Pressures in Hydrocarbon Environments*. AAPG Memoirs, **70**, 1–11.

LETYTHAEUSER, D., SCHAEFER, R.G. & YUKLER, A. 1982. Role of diffusion in primary migration of hydrocarbons. *AAPG Bulletin*, **66**, 408–429.

LUO, X. & VASSEUR, G. 1995. Modelling of pore pressure evolution associated with sedimentation and uplift in sedimentary basins. *Basin Research*, **7**, 35–52.

MAGARA, K. 1968. Compaction and migration of fluids in Miocene mudstones, Nagaoka Plain, Japan. *AAPG Bulletin*, **52**, 2466–2501.

MANN, U., HANTSCHEL, T., SCHAEFER, R.G., KROOSS, B., LEYTHAEUSER, D., LITTKE, R. & SACHSENHOFER, R.F. 1997. *In*: WELTE, D.H., HORSFIELD, B. & BAKER, D.R. (eds) Petroleum migration: mechanisms, pathways, efficiencies and numerical simulations. *Petroleum and Basin Evolution*. Springer, Berlin, 405–520.

MARTINSEN, R.E. 1994. Summary of published literature on anomalous pressures: implications for the study of pressure compartments. *In*: ORTOLEVA, P.J. (ed.) *Basin Compartments and Seals*. AAPG Memoirs, **61**, 27–38.

MONTEL, F., CAILLET, G., PUCHEU, A. & CALTAGIRONE, J.P. 1993. Diffusion model for predicting reservoir gas losses. *Marine and Petroleum Geology*, **10**, 51–57.

MURDOCH, L.M., MUSGROVE, F.W. & PERRY, J.S. 1995. Tertiary uplift and inversion history in the North Celtic Sea Basin and its influence on source rock maturity. *In*: CROKER, P.F. & SHANNON, P.M. (eds) *The Petroleum Geology of Ireland's Offshore Basins*. Geological Society, London, Special Publications, **93**, 297–319.

NEUZIL, C.E. & POLLOCK, D.W. 1983. Erosional unloading and fluid pressures in hydraulically tight rocks. *Journal of Geology*, **91**, 179–193.

OSBORNE, M.J. & SWARBRICK, R.E. 1997. Mechanisms for generating overpressure in sedimentary basins: a reevaluation. *AAPG Bulletin*, **81** (6), 1023–1041.

PALCIAUSKAS, V.V. & DOMENICO, P.A. 1980. Microfracture development in compacting sediments: relation to hydrocarbon maturation kinetics. *AAPG Bulletin*, **64**, 927–937.

PEDERSEN, T. & BJØRLYKKE, K. 1994. Fluid flow in sedimentary basins: model of pore water flow in a vertical fracture. *Basin Research*, **6**, 1–16.

POOLER, J. & AMORY, M. 1999. A subsurface perspective on ETAP—an integrated development of seven Central North Sea fields. *In*: FLEET, A.J. & BOLDY, S.A.R. (eds) *Petroleum Geology of Northwest Europe: Proceedings of the 5th Conference*. Geological Society, London, 993–1006.

RODGERS, S. 1999. Discussion: 'Physical constraints on hydrocarbon leakage and trapping revisited' by P.A. Bjørkum *et al.*—further aspects. *Petroleum Geoscience*, **5**, 421–423.

SCHLOMER, S. & KROOSS, B.M. 1997. Experimental characterisation of the hydrocarbon sealing efficiency of cap rocks. *Marine and Petroleum Geology*, **14** (5), 565–580.

SCHOWALTER, T.T. 1979. Mechanics of secondary hydrocarbon migration trapping. *AAPG Bulletin*, **63**, 723–760.

SCLATER, J.G. & CHRISTIE, P.A.B. 1980. Continental stretching: an explanation of the post-mid-Cretaceous subsidence of the Central North Sea basin. *Journal of Geophysical Research*, **85**, 3711–3739.

SECOR, D. JR 1965. Role of fluid pressure in jointing. *American Journal of Science*, **263**, 633–646.

SEEDHOUSE, J.K. & RACEY, A. 1997. Sealing capacity of the Mercia Mudstone Group in the East Irish Sea Basin: implications for petroleum exploration. *Journal of Petroleum Geology*, **20**, 261–286.

SIBSON, R.H. 1995. Selective fault reactivation during basin inversion: potential for fluid redistribution through fault-valve action. *In*: BUCHANAN, J.G. & BUCHANAN, P.G. (eds) *Basin Inversion*. Geological Society, London, Special Publications, **88**, 3–19.

SPAIN, D.R. & CONRAD, P.C. 1997. Quantitative analysis of top-seal capacity: offshore Netherlands, Southern North Sea. *Geologie en Mijnbouw*, **76**, 217–226.

SPENCER, A.M., CAMPBELL, C.J., HANSLIEN, S.H., HOLTER, E., NELSON, P.H.H., NYSAETHER, E. & ORMAASEN, E.G. (eds) 1986. *Habitat of Hydrocarbons on the Norwegian Continental Shelf*. Graham and Trotman, London.

SWARBRICK, R.E. & OSBORNE, M.J. 1998. Mechanisms that generate abnormal pressures: an overview. *In*: LAW, B.E., ULMISHEK, G.F. & SLAVIN, V.I. (eds) *Abnormal Pressures in Hydrocarbon Environments*. AAPG Memoirs, **70**, 13–34.

TABER, D.R., VICKERS, M.K. & WINN, R.D. JR 1995. The definition of the Albian 'A' Sand reservoir fairway and aspects of associated gas accumulations in the North Celtic Sea Basin. *In*: CROKER, P.F. & SHANNON, P.M. (eds) *The Petroleum Geology of Ireland's Offshore Basins*. Geological Society, London, Special Publications, **93**, 227–244.

UNDERHILL, J.R. 1991. Implications of Mesozoic–Recent basin development in the western Inner Moray Firth, UK. *Marine and Petroleum Geology*, **8**, 359–369.

WATTS, N.L. 1987. Theoretical aspects of cap-rock and fault seals for single- and two-phase hydrocarbon columns. *Marine and Petroleum Geology*, **4** (11), 274–307.

WELLS, P.R.A. 1988. Hydrodynamic trapping in the Cretaceous Nahr Umr Lower Sand of the North Area, Offshore Qatar. *Journal of Petroleum Technology*, March 1998, 357–361.

Index

Page numbers in italic, e.g. *70*, refer to figures. Page numbers in bold, e.g. **6**, signify entries in tables.